MySQL 实战

陈臣 ◎ 著

人民邮电出版社

北京

图书在版编目（CIP）数据

 MySQL实战 / 陈臣著. -- 北京 ：人民邮电出版社，2023.3
 （图灵原创）
 ISBN 978-7-115-61008-9

 Ⅰ．①M… Ⅱ．①陈… Ⅲ．①SQL语言－数据库管理系统 Ⅳ．①TP311.132.3

中国国家版本馆CIP数据核字(2023)第003580号

内 容 提 要

本书以 MySQL 8.0 为主，全面系统地阐述了 MySQL 日常使用及管理过程中的一些常用知识点：安装、复制、binlog、备份、监控、DDL、线程池、中间件、常用工具、组复制、InnoDB Cluster、JSON、MySQL 8.0 的新特性。

本书定位于实战，目的是让读者拿来即用，快速上手 MySQL。除了实战，本书还花费了大量的篇幅来讲解 MySQL 中一些常见操作、常用工具的实现原理。

组复制是 MySQL 官方推荐的高可用方案，本书会从源码角度分析组复制的一些核心模块的实现细节，包括分布式恢复、冲突检测、事务一致性以及流量控制机制等。

本书面向的读者对象包括 MySQL 初学者、系统运维人员、想要进阶的 DBA、架构师、MySQL 应用开发者，以及对 MySQL 感兴趣的所有人。

◆ 著　　　陈　臣
　责任编辑　王军花
　责任印制　胡　南

◆ 人民邮电出版社出版发行　　北京市丰台区成寿寺路11号
　邮编　100164　电子邮件　315@ptpress.com.cn
　网址　https://www.ptpress.com.cn
　北京联兴盛业印刷股份有限公司印刷

◆ 开本：800×1000　1/16
　印张：44.75　　　　　　　　2023年3月第1版
　字数：1170千字　　　　　　2023年3月北京第1次印刷

定价：149.80元

读者服务热线：(010)84084456-6009　印装质量热线：(010)81055316
反盗版热线：(010)81055315
广告经营许可证：京东市监广登字 20170147 号

序　一

MySQL 是最为流行的开源数据库。陈臣所著的《MySQL 实战》是基于他自己的大规模数据库管理实践，结合源码、官方文档和原理形成的一本书，系统地介绍了在 MySQL 的管理、运维和开发过程中使用范围较广但参考资料匮乏的工具。

在技术迭代和市场应用的双重驱动下，数据库近些年来的发展可以说纷繁复杂。数据库（database）的发展回归其本源，是沿着数据（data）和基座（base）两个方向演进。进入数字化社会，一切皆为数据，数据早已不再是原来的单一类型，而是极大地向多模化扩展；数据管理的基座也经由存储检索到数据仓库，再到数据湖，进一步发展到最近的湖仓一体化。

在 data 和 base 都被重新定义，其内涵和外延不断做加法的同时，另一个演进方向则是做减法，即"简化"（simplify）。运维的自动化和混合负载（HTAP）是简化的新方向，而 MySQL HeatWave 是基于 MySQL 单一数据库实例实现混合负载的新的分析引擎。

陈臣作为一线技术工程师，对于工作中的实操经验，勤于思考和研究，追求知其意、悟其理、守其则、践其行，难能可贵。故欣然为其《MySQL 实战》作序，也借此鼓励他基于 MySQL 以及数据库的新发展，笔耕不辍，今后再出新著。

谢鹏

甲骨文（Oracle）公司全球副总裁，中国区技术总经理

序 二

终于等到了。

早在三年前就听说陈臣打算写书，定位是"实战+原理"，而且当时就已经完成了大部分初稿，我还以为马上就要出版了。我个人深感写书不易，即使在平时整理公众号文章时，想要把一个知识点讲得透彻，也要花费大量的精力去查阅资料并付诸实践，甚至像陈臣这样去翻阅源码或者请教内核研发的专家。

看完这本书的目录及样章，深感陈臣满满的诚意。全书紧紧围绕 MySQL 实战的方方面面展开，有详细的实战细节，更是对几个关键知识点给了原理解读，让读者能更好地理解，知其然且知其所以然。以"组复制"这一章为例，前半部分详细演示了组复制的部署及管理操作，后半部分详细分析了组复制的一些关键实现原理，其中原理解读部分占了该章 60%以上的篇幅，详细程度可见一斑。

不多说了，强烈推荐这本书，绝对值得一看。

叶金荣

Oracle MySQL ACE Director，腾讯云 TVP 成员

序 三

关于——我与 MySQL

如果说，过去二十年间，MySQL 是中国乃至世界范围内最流行的数据库，应该没有人反对。我大学毕业后进入百度，做百万用户同时在线的即时通信系统，使用的是 MySQL；在 58 同城做商家平台、支付平台、推荐系统，面对千万用户量、十亿数据存储，使用的也是 MySQL；在到家集团做架构组件、中台业务，以及在快狗打车做讲究快速迭代的车货匹配，使用的仍然是 MySQL。可以说，我与 MySQL 结下了不解之缘，MySQL 也是我职业发展的见证者。

关于——MySQL 要学些什么

很多朋友问我，MySQL 这么流行，要学习哪些方面的内容。我认为，要学透 MySQL，至少三个层面的内容是需要了解与掌握的。

其一，应用层面：各行各业是如何使用 MySQL 来解决各类业务问题的。
其二，内核层面：MySQL 的内核细节是什么样的。
其三，上下游工具层面：MySQL 相关的工具有哪些，原理是怎样的。

关于——如何学习 MySQL

还有不少朋友问我，应该怎么学习 MySQL。我认为，看英文官网乃是首选。中文的话，可以看看这本《MySQL 实战》。这本书最大的特点就是，学透 MySQL 需要掌握的三个层面在其中均有覆盖：实战应用，讲解如何使用；内核原理，讲解实现细节；上下游工具，拿来就用。

关于——《MySQL 实战》的内容

读完样章，内容让人惊艳：GTID 的原理与并行复制，主从原理与 binlog 细节，XtraBackup 的巧妙原理，连接池与线程池，中间件的运用实践，包括最新的组复制实践……沉下心，嚼碎这些内容，我相信你一定能超越九成的 MySQL 从业人员。

感谢陈臣的诚意之作，让大家学习 MySQL 又多了一个选择。行业内能有这样的技术匠人，幸甚至哉！

沈剑

公众号"架构师之路"作者

序　四

很多年前，我在数据库社区发起过"ACMUG 特约撰稿人"技术分享的征文活动，希望通过这样的活动能鼓励技术人员通过总结、写作与分享，一起提升个人能力和数据库社区的技术氛围。那时候，记得有个少年提交过好几篇文章，每篇都写得洋洋洒洒的，虽然略显稚嫩，但绝对看得出来，他是用了心思的。此人正是这本书的作者，陈臣先生。

我非常赞同陈臣在前言中提到的，在写作过程中对"书读百遍，其义自见"的感想和解读。动手和动嘴是完全不一样的感觉，这的确是在写书的过程中才更容易感悟到的。古人讲："吟安一个字，捻断数茎须。"每当想象自己写出的字会像印章那样被一个一个刻出来，后续以实物的形式呈现在千千万万人的眼前时，作为一个作者，你生怕出一丝纰漏。

几年来，我虽因创办极数云舟而忙得焦头烂额，参与社区的活动少了，但对 MySQL 领域的关注一直不减。拿到这本书的书稿后，既觉得是意料之中，又确实出乎了意料。这本书算得上近年来国内 MySQL 领域出版物中内容最丰富、细节最清晰、篇幅也是最长的，其运维实践与源码解析相结合的思路正与《MySQL 运维内参》不谋而合，不愧为实战之作。

周彦伟

极数云舟创始人，

《MySQL 运维内参》作者，中国计算机行业协会数据库专委会会长

序 五

非常感谢陈臣老师邀请我为这本书写推荐序。我有幸和陈老师一起共事过几年,也见证了这本著作诞生的过程,陈老师结合 MySQL 源码反复进行实验论证的过程给我留下了非常深刻的印象。

回想自己初入数据库行业的时候,我的学习材料是一份 MySQL 官方文档。官方文档写得固然很不错,但我当初懵懵懂懂地看过之后只记得一些基本的操作命令,至于数据库的基本原理,我虽然很好奇,但始终不得要领。后来陆陆续续地阅读了一些数据库的相关书,对数据库的基本原理也有了基本的掌握,但始终感觉缺少了什么。后来我才意识到,MySQL 既是数据库也是用代码实现的软件产品,而之前学习的数据库原理和实践没有融合在一起,因为我在实际的工作中很少去验证和应用这些理论,也没有进一步去探究这些精妙绝伦的设计是如何用代码实现的。但理论与实践结合才能真正地掌握这些知识,之后,我也做过一些这方面的实验,在数据库的理解和认知方面收获巨大,而这需要耗费的精力和时间也不容小觑,所以我深知这本书的写作难度,非常钦佩陈老师的毅力,也非常感谢陈老师给 MySQL 从业者和爱好者总结和沉淀了这样一本宝贵的著作。

这本书不仅包含了 MySQL 的实战操作和对应的理论知识,还包含了 MySQL 的一些非常实用的周边工具和中间件,实属难得。不论对于数据库初学者还是有经验者,这本书都值得反复研读,实践是检验真理的唯一标准。

肖博

vivo 研发总监

序 六

当我收到书稿的时候,第一感觉是这本书真的厚。现在市面上的 MySQL 图书已经非常多了,再写一本会有什么不一样的东西呢?但当我阅读完样章之后,确实看到了这本书的价值。

陈臣是奋战在运维一线的 DBA 专家,见证过大型数据库平台从零开始建设的过程。他从一个数据库运维人员的角度帮大家把 MySQL 的整个运维体系梳理了出来,书中不仅介绍了 MySQL 的原理、使用方法,更重要的是贡献出了他自己的经验,包括对 MySQL 生态的周边工具、中间件、连接池等的介绍和使用经验。这些经验十分宝贵。

MySQL 能有今天的江湖地位,极其强大的生态是核心因素,因此我们学习 MySQL、使用 MySQL,不仅要看 MySQL 本身,对其周边生态软件的理解和使用也是非常重要的。这也是我觉得这本书非常大的价值所在。

非常推荐 MySQL DBA 们阅读一下这本书,无论对于新手入门还是老手进阶,这本书都是不错的参考。

彭立勋

ACMUG(中国 MySQL 用户组)主席,Oracle MySQL ACE Director

前　　言

写作感想与建议

反反复复，兜兜转转，本书写了好多年。

拖沓的原因，一是我的"野心至极"，试图覆盖 MySQL 所有常用的知识点；二是我确实想呈现一本"诚意之作"，所以在大大小小知识点的验证上花费了不少时间。

2021 年，当我将初稿交给王军花编辑的时候，她大吃一惊，发现竟然共有 1076 页。要知道我们当初约定的可是 350 页，即一般技术书的篇幅。收到她的反馈后，我也哭笑不得：怪不得总感觉写不完呢，原来是写得太多了。后来，尽管我反复删减，并且重写了部分章节，但交付的第二稿也有 878 页。这距离初稿交付，已经过去了 1 年之久。现在大家手上拿着的是第三稿。俗话说"三易其稿"，古人诚不欺我也。

首先聊聊我在写作和工作过程中的 4 个感想。

❑ 书读百遍，其义自见

这个我以前读书时没明白的道理，却在写书的过程中弄懂了。原来，"书读百遍"里的"读"，不是简单地知晓字面意思即可，而是要在不懂的地方反复推敲。例如我在写组复制章节的时候，操作和管理部分写得极快，毕竟我有实际的使用经验，而且官方文档也有相关内容；原理部分却很难下手，虽然官方文档中的相关资料也不少，但很多细节并没有讲透。无奈，我只能去翻看源码。当然，看源码是一个比较痛苦的过程，因为要反复调试。结果就是，写操作和管理部分只用了 1 个月，写原理部分却花了 3 个月。所幸，本书最后能从源码的角度展现组复制的一些核心模块的实现细节，这般付出还是值得的。

所以大家在碰到 MySQL 的相关问题时，一定要深挖，不要轻易放弃。从表面上看放弃的只是一个问题，但本质上放弃的是一次提升技术的机会。不仅如此，放弃还有"路径依赖"：一次放弃，后续大概率也会放弃。不如狠下心来深挖一次，虽然挑战自己的知识盲区很痛苦，但反复几次，自然就能提炼出分析问题、解决问题的方法论。

❑ 一万小时定律

这个定律的意思是，要成为某个领域的专家，至少需要在这个领域"刻意练习"10 000 小时。

按每天 8 小时、每周 5 天来计算的话，10 000 小时就是 5 年左右。由此可见，要成为一个领域的专家，至少需要 5 年的时间。故而，在学习这条道路上，不要急躁，要持续精进，静待花开。

❑ 艺不压身

我经常会收到一些后辈的咨询：某数据库是否可以学？我给的建议通常是：学。艺不压身，为什么不学呢？在年轻的时候多学点儿，总没有坏处。尽管如此，也不要盲目地学，毕竟人的精力是有限的。现在流行 T 型人才，强调的就是一专多能。当然，"专"和"能"里面，起核心作用的还是"专"。企业在招聘的时候，考查的首先就是应聘者的专业能力，只有专业能力满足要求了，才会考查其他方面的能力。所以我们在学习的时候，应该努力拓展自己在某个专业领域中的深度。在达到一定深度之前，不建议盲目学习与该领域无关的新技能。

❑ 分而治之

这其实是数据库领域的一个核心优化思想。任何瓶颈，都可以通过"拆"来解决。例如，我们可以通过读写分离来解决读瓶颈问题，可以通过分库分表来解决写瓶颈问题。在进行 SQL 优化时，这个思想同样适用。虽然有的 SQL 语句看上去很复杂，感觉无从下手，但我们可以利用分而治之的思想将一条"大 SQL 语句"拆分为多条"小 SQL 语句"，然后分别优化。

接下来看看本书的主要内容。

本书内容

本书一共 12 章，各章主要内容分别如下。

第 1 章介绍了 MySQL 的两种常用安装方法：基于二进制包的安装和基于源码包的安装。还在此基础上介绍了两种常用的 MySQL 服务管理方式：/etc/init.d/mysqld 和 systemd。

第 2 章从复制的基本原理出发，系统地介绍了 GTID 复制、半同步复制、并行复制、多源复制和延迟复制。此外，还介绍了事务的两阶段提交协议，并行复制方案的演进历史，以及如何利用延迟复制恢复误删的表。

第 3 章首先分析了 binlog 的 3 种格式及其优缺点；接着通过具体的示例演示了如何阅读 binlog 和 relay log 中的内容，包括一个事务由哪些事件组成，这些事件各个字段的具体含义；然后介绍了 binlog 中常见的事件类型；最后基于 python-mysql-replication 打造了一个 binlog 解析器。

第 4 章主要介绍了常用的复制管理操作，复制的监控，如何分析主从延迟，主从延迟的常见原因及解决方法，Seconds_Behind_Master 的计算逻辑，如何监控主从延迟，复制中的常见问题及解决方法。

第 5 章首先介绍了 MySQL 常见备份工具的具体用法及实现原理，包括 mysqldump、mydumper、XtraBackup、克隆插件、MySQL Shell Dump & Load；然后介绍了与备份相关的两个高频操作，即搭建从库和指定时间点（位置点）的恢复；最后介绍了如何搭建 binlog server 以及如何检测备份的有效性。

第 6 章介绍了业界流行的两个开源监控方案：Zabbix 和基于 Prometheus 开发的 PMM。还介绍了 MySQL 中常用的监控指标。

第 7 章首先介绍了 3 种常用的表结构变更方式：Online DDL、pt-online-schema-change 和 gh-ost，包括它们的使用方法、实现原理及各自的优缺点；接着介绍了元数据锁的基本概念和引入背景；最后分别分析了如何在 MySQL 5.6 和 MySQL 5.7（8.0）中定位 DDL 被阻塞的问题。

第 8 章围绕连接池和线程池展开。在连接池方面，介绍了连接池的运行原理和两个较有代表性的 Java 连接池（c3p0 和 DBCP）的参数配置。在线程池方面，介绍了 MySQL 线程池的实现原理和适用场景，并对比了 MySQL 企业版线程池和 Percona Server 线程池在使用方面及实现细节上的差异。最后，这一章还介绍了 MySQL server has gone away 错误的常见原因以及 interactive_timeout 和 wait_timeout 这两个参数的区别。

第 9 章介绍了 MySQL 中一些常用工具的具体用法及实现原理，包括 sysbench（数据库性能测试）、pt-archiver（数据归档）、pt-kill（杀掉连接）、pt-query-digest（汇总慢日志）、pt-show-grants（打印账号的授权信息）、pt-table-checksum（检查主从数据一致性）、pt-table-sync（修复从库与主库不一致的数据）、pt-upgrade（数据库升级）。

第 10 章是关于中间件的。这一章主要介绍 ProxySQL，包括如何使用 ProxySQL 实现读写分离、审计、SQL 黑名单、查询重写等。虽然开源的 MySQL 中间件不少，但 ProxySQL 是为数不多仍在持续迭代的中间件产品。第 12 章将介绍另一个中间件——MySQL Router。

第 11 章系统介绍了组复制的引入背景、部署、监控和常见的管理操作，并且在此基础上，基于源码分析了组复制一些核心模块的实现细节，包括分布式恢复、冲突检测、事务一致性和流量控制机制等。

第 12 章系统介绍了 InnoDB Cluster 的两大核心组件：MySQL Shell 和 MySQL Router。在此基础上，还介绍了如何使用 MySQL Shell 来管理和维护 InnoDB Cluster。

除此之外，本书还包括两个附录，主要内容分别如下。

附录 A 系统地介绍了 JSON，包括 JSON 字段常用的增删查改操作，如何对 JSON 字段创建索引、如何将存储 JSON 字符串的字符字段升级为 JSON 字段、使用 JSON 时的注意事项。最后，介绍了 MySQL 8.0 针对 JSON 文档引入的 Partial Update 特性，该特性极大提升了 JSON 字段的处理性能。

附录 B 系统介绍了 MySQL 8.0 的新特性，一共 112 项。这些新特性分为管理、开发、优化器、InnoDB、数据字典、复制、组复制、安全和账号、密码管理、performance_schema 这 10 个类别来阐述。最后，整理了 MySQL 8.0 中新增和移除的函数，以及默认值相对于 MySQL 5.7 发生变化的参数列表。

如何阅读本书

除了第 2～4 章和第 11～12 章建议按顺序阅读外，其他各章都是单独的主题，读者可以按需要决定阅读顺序。不仅如此，大部分章节由多个小主题组成，这些小主题也是相互独立的。大家可以按照自己感兴趣的主题自行决定阅读顺序，不要有心理负担。

本书特色

实战 + 实现原理！！！

实战的目的很简单，就是拿来即用。

所谓"取势、明道、优术"，明道才能优术。所以，除了讲解怎么用以外，本书还花了大量的篇幅来阐释各个操作背后的实现原理。如此，才能举一反三，在分析和定位问题时做到抽丝剥茧。

除了"实战 + 实现原理"，每章结尾还会通过列出问题的方式梳理这一章的重点。这些问题一方面可以用来检验对该章内容的掌握程度，另一方面也可以作为很好的面试题。

读者对象

- 想系统学习 MySQL 日常管理操作的运维人员、DBA。
- 想了解 MySQL 常见操作实现原理的 DBA。
- 对组复制源码感兴趣的 DBA。
- 使用 MySQL 的开发人员。本书会告诉你为什么不做读写分离，也要关注主从延迟；DBA 都是怎么做 DDL 操作的，为什么不会锁表；当 MySQL 数据库出现问题时，该关注哪些指标；在配置连接池时，需注意哪些参数……
- 对 MySQL 感兴趣的所有人。

勘误和支持

虽然我对很多知识点进行了反复验证，但由于精力和水平有限，很难保证书中没有错误和不足之处，恳请大家批评指正。如果发现错误，请到"MySQL 实战"公众号与我联系。我会把已确认的勘误及时发布在公众号上。

当然，本书只是你我之间缘分的起点。事实上，受限于篇幅，很多主题和细节没办法在书中完整呈现。例如，SQL_MODE、数据类型的最佳实践、复制相关参数讲解、Percona Toolkit 中的其他工具、分区表等。这些内容后续也会发布在公众号上。

"三人行，必有我师焉"，欢迎大家通过下面的二维码关注"MySQL 实战"公众号与我交流。大家相互切磋，共同提高！

致谢

感谢父母的养育和教导之恩。

感谢我的爱人，感谢她这些年来对家庭的辛勤付出，没有她的全力支持，我很难抽出时间完成本书的写作。对我的儿子说声抱歉，原本应该陪他玩的时间却用来写书了，实在是愧疚之极。感谢我的岳母，这些年她不辞辛劳地帮我们照顾孩子。

感谢王博先生、吴勇先生、刘国杏先生、左兴宇先生、肖博先生，让我非常有幸地参与了 vivo 互联网数据库平台从 0 到 1 的建设工作。

感谢谢鹏博士、嵇小峰先生、李珈女士、王伟先生，让我非常有幸地加入了甲骨文，为不同行业、不同规模的客户服务。

感谢周彦伟老师、彭立勋老师、沈剑老师在百忙之中抽出宝贵的时间为我的书作序。

感谢王军花编辑这么多年的耐心等待。

感谢叶金荣老师、吴炳锡老师、姜承尧老师在 MySQL 学习之路上对我的指点。感谢温正湖老师、宋利兵老师关于组复制的精彩分享。感谢 MySQL 社区的各位小伙伴源源不断的高质量分享，让我有机会站在前人的肩膀上学习。

在工作和学习的过程中，我得到过很多高人的指点、贵人的帮助，这里一并谢过（排名不分先后）：

张东海、李一多、吕明阳、韦国颂、程昂、张友平、罗健、杜霆、韩朝兵、张照仑、刘未未、黄博勇、刘煌、龙启东、宁林浩、汤文坚、朱慧鹏、刘石林、李永景、赖侯奇、黄小刚、邓松、周振如、周亚、张家齐、李阳冰、汪翔、袁健威、储敏、孙于田、付玉婷、林锋英、汪明寅、黄华亮、彭灵继、商永星、倪晓辉、龚兵、刘谦、蒋文辉、彭许生、彭元文、唐萌、张万、陈俊聪、王志强、郑志江、李泽平、舒雄、李明、姚嵩、黎锦红、廖光明、董红禹、姚良、蔡鹏、罗伟文、陶卫、顾铁军、唐承波、高胜杰、李墨宣、彭纬、谭贵天、缪念橙、陈瑾谦。

我想感谢的人太多，受限于篇幅，这里未能完全呈现。

最后，感谢各位读者，希望本书能让你们有所收获！

封面主图

本书封面主图来自 freepik。

目　　录

第 1 章　MySQL 入门、安装与服务的
　　　　管理 ·· 1
　1.1　MySQL 的历史 ··· 1
　1.2　MySQL 的安装 ··· 2
　　　1.2.1　下载 MySQL ··· 3
　　　1.2.2　基于二进制包的安装 ···························· 8
　　　1.2.3　基于源码包的安装 ····························· 14
　　　1.2.4　配置文件的读取顺序 ·························· 19
　1.3　MySQL 服务的管理 ······································ 21
　　　1.3.1　使用/etc/init.d/mysqld 管理
　　　　　　MySQL 服务 ·· 22
　　　1.3.2　使用 systemd 管理 MySQL
　　　　　　服务 ·· 27
　1.4　本章总结 ·· 33

第 2 章　复制 ·· 34
　2.1　复制的原理及搭建 ·· 35
　　　2.1.1　复制的搭建 ··· 36
　　　2.1.2　参考资料 ··· 41
　2.2　GTID 复制 ··· 41
　　　2.2.1　GTID 出现的背景 ······························· 41
　　　2.2.2　GTID 的搭建 ······································· 42
　　　2.2.3　GTID 的原理 ······································· 42
　　　2.2.4　GTID 的相关参数 ······························· 44
　　　2.2.5　GTID 的相关函数 ······························· 50
　　　2.2.6　在线修改复制模式 ····························· 51
　　　2.2.7　设置@@GLOBAL.GTID_PURGED
　　　　　　时的注意事项 ······································ 54
　　　2.2.8　参考资料 ··· 55
　2.3　半同步复制 ·· 55
　　　2.3.1　事务的两阶段提交协议 ····················· 56

　　　2.3.2　半同步复制的原理 ····························· 57
　　　2.3.3　半同步复制的安装 ····························· 58
　　　2.3.4　半同步复制的注意事项 ····················· 60
　　　2.3.5　半同步复制的常用参数 ····················· 60
　2.4　并行复制 ·· 63
　　　2.4.1　并行复制方案 ····································· 63
　　　2.4.2　如何开启并行复制 ····························· 71
　　　2.4.3　参考资料 ··· 71
　2.5　多源复制 ·· 72
　　　2.5.1　多源复制的搭建 ································· 72
　　　2.5.2　多源复制搭建过程中的注意
　　　　　　事项 ·· 73
　　　2.5.3　多源复制的管理 ································· 74
　2.6　延迟复制 ·· 74
　　　2.6.1　如何开启延迟复制 ····························· 75
　　　2.6.2　如何使用延迟复制恢复误删
　　　　　　的表 ·· 75
　　　2.6.3　延迟复制的总结 ································· 77
　2.7　本章总结 ·· 77

第 3 章　深入解析 binlog ······································ 79
　3.1　binlog 的格式 ·· 79
　　　3.1.1　STATEMENT ······································· 79
　　　3.1.2　ROW ·· 80
　　　3.1.3　MIXED ··· 81
　3.2　如何解读 binlog 的内容 ································ 82
　　　3.2.1　解析 STATEMENT 格式的
　　　　　　二进制日志 ·· 82
　　　3.2.2　解析 ROW 格式的二进制
　　　　　　日志 ·· 88
　3.3　如何解读 relay log 的内容 ··························· 92

3.4 binlog 中的事件类型 ································ 94
3.5 基于 python-mysql-replication
    打造一个 binlog 解析器 ················· 101
3.6 本章总结 ··· 109

第 4 章 深入 MySQL 的复制管理 110

4.1 常见的管理操作 ······································ 110
   4.1.1 查看主库的状态 ··························· 110
   4.1.2 查看从库复制的状态 ···················· 111
   4.1.3 搭建复制 ····································· 113
   4.1.4 开启复制 ····································· 115
   4.1.5 停止复制 ····································· 117
   4.1.6 在主库上查看从库 IP 和
         端口信息 ······································ 117
   4.1.7 查看实例当前拥有的 binlog ········ 118
   4.1.8 删除 binlog ································ 118
   4.1.9 查看 binlog 的内容 ···················· 118
   4.1.10 RESET MASTER、RESET
          SLAVE 和 RESET SLAVE
          ALL 的区别 ································ 119
   4.1.11 跳过指定事务 ···························· 120
   4.1.12 操作不写入 binlog ···················· 121
   4.1.13 判断主库的某个操作是否
          已经在从库上执行 ······················· 121
   4.1.14 在线设置复制的过滤规则 ········· 122
4.2 复制的监控 ·· 123
   4.2.1 连接 ··· 123
   4.2.2 事务重放 ······································ 125
   4.2.3 多线程复制 ·································· 126
   4.2.4 过滤规则 ······································ 128
   4.2.5 组复制 ·· 129
4.3 主从延迟 ·· 129
   4.3.1 如何分析主从延迟 ······················· 129
   4.3.2 主从延迟的常见原因及解决
         方法 ·· 132
   4.3.3 如何解读 Seconds_Behind_
         Master ·· 134
   4.3.4 参考资料 ······································ 142
4.4 复制中的常见问题及解决方法 ············· 142
   4.4.1 I/O 线程连接不上主库 ················ 142
   4.4.2 server_id 重复 ······························ 143
   4.4.3 包的大小超过 slave_max_
         allowed_packet 的限制 ············· 143
   4.4.4 从库需要的 binlog 在主库
         上不存在 ······································ 144
   4.4.5 从库的 GTID 多于主库的 ········· 147
   4.4.6 在执行插入操作时，提示
         唯一键冲突 ·································· 149
   4.4.7 在执行删除或更新操作时，
         提示记录不存在 ··························· 150
   4.4.8 主从数据不一致 ·························· 150
4.5 本章总结 ·· 150

第 5 章 备份 152

5.1 mysqldump ·· 155
   5.1.1 mysqldump 的实现原理 ·············· 155
   5.1.2 mysqldump 的常用选项 ·············· 160
   5.1.3 mysqldump 的常见用法 ·············· 165
   5.1.4 总结 ··· 166
5.2 mydumper ·· 166
   5.2.1 mydumper 的安装 ······················· 167
   5.2.2 mydumper 的实现原理 ················ 167
   5.2.3 mydumper 的参数解析 ················ 170
   5.2.4 myloader 的参数解析 ·················· 177
   5.2.5 mydumper 和 myloader 的
         常见用法 ······································ 179
   5.2.6 总结 ··· 180
5.3 XtraBackup ··· 180
   5.3.1 XtraBackup 的安装 ····················· 181
   5.3.2 基于源码分析 XtraBackup
         的实现原理 ·································· 182
   5.3.3 XtraBackup 的常见用法 ············· 192
   5.3.4 Xtrabackup 的重要参数 ·············· 205
   5.3.5 XtraBackup 的注意事项 ············· 206
   5.3.6 备份用户需要的权限 ·················· 207
   5.3.7 参考资料 ····································· 208
5.4 克隆插件 ··· 208
   5.4.1 克隆插件的安装 ·························· 209
   5.4.2 克隆插件的使用 ·························· 209
   5.4.3 查看克隆操作的进度 ·················· 212

	5.4.4	基于克隆数据搭建从库 ·············· 214		6.1.1	安装 Zabbix Server ·············· 242
	5.4.5	克隆插件的实现细节 ·············· 214		6.1.2	安装 Zabbix Agent ··············· 250
	5.4.6	克隆插件的限制 ···················· 215	6.2	安装 MySQL 监控插件 PMP ············· 251
	5.4.7	克隆插件与 XtraBackup 的异同 ······· 216	6.3	深入理解 PMP ···························· 256
	5.4.8	克隆插件的参数解析 ·············· 216		6.3.1	ss_get_mysql_stats.php 源码分析 ······ 258
	5.4.9	参考资料 ······························ 217		6.3.2	基于 ss_get_mysql_stats.php 自定义监控项 ···· 263
5.5	MySQL Shell Dump & Load ············· 217		6.4	Zabbix 常见问题定位及性能优化 ····· 266
	5.5.1	MySQL Shell Dump & Load 的用法 ········ 218		6.4.1	定位监控项的状态 Not supported ······· 266
	5.5.2	MySQL Shell Dump & Load 的关键特性 ····· 221		6.4.2	分区表 ······························ 268
	5.5.3	util.dumpInstance 的实现原理 ············ 221		6.4.3	Zabbix Server 的参数优化 ······· 270
	5.5.4	util.dumpInstance 的参数解析 ··········· 222		6.4.4	Zabbix API ·························· 273
				6.4.5	参考资料 ······························ 278
	5.5.5	util.loadDump 的参数解析 ······ 224	6.5	PMM ·································· 278
	5.5.6	MySQL Shell Dump & Load 的注意事项 ······ 226		6.5.1	PMM 的体系架构 ·············· 279
				6.5.2	安装 PMM Server ·············· 280
	5.5.7	参考资料 ······························ 226		6.5.3	安装 PMM Client ·············· 283
5.6	使用 XtraBackup 搭建从库 ············· 226		6.5.4	添加 MySQL 服务 ·············· 288
	5.6.1	使用 XtraBackup 搭建从库的基本步骤 ·· 227		6.5.5	Query Analytics ·············· 290
	5.6.2	基于从库备份搭建从库的注意事项 ······ 230		6.5.6	深入理解 PMM Server ·············· 292
				6.5.7	设置告警 ······················ 297
	5.6.3	设置 GTID_PURGED 的注意事项 ··············· 230		6.5.8	PMM 的常见问题 ·············· 299
				6.5.9	参考资料 ······························ 305
	5.6.4	使用 XtraBackup 8.0 搭建从库的注意事项 ······ 231	6.6	MySQL 中常用的监控指标 ············· 306
	5.6.5	总结 ································ 233		6.6.1	连接相关 ······················ 306
5.7	指定时间点（位置点）的恢复 ····· 234		6.6.2	Com 相关 ······················ 307
5.8	搭建 binlog server ···················· 236		6.6.3	Handler 相关 ·················· 308
	5.8.1	基于 mysqlbinlog 搭建 binlog server ······· 236		6.6.4	临时表相关 ······················ 309
				6.6.5	Table Cache 相关 ············· 310
	5.8.2	参考资料 ······························ 238		6.6.6	文件相关 ······················ 311
5.9	检测备份的有效性 ···················· 238		6.6.7	主从复制相关 ·················· 312
5.10	本章总结 ······························· 239		6.6.8	缓冲池相关 ······················ 312
				6.6.9	redo log 相关 ······················ 313
第 6 章	监控 ······································ 241		6.6.10	锁相关 ······························ 314
				6.6.11	排序相关 ······················ 315
				6.6.12	查询相关 ······················ 316
				6.6.13	其他重要指标 ·················· 316
6.1	Zabbix ································· 242	6.7	本章总结 ································ 317

第 7 章 DDL 318
7.1 Online DDL 319
7.1.1 Online DDL 的分类 320
7.1.2 Online DDL 的实现原理 324
7.1.3 如何检查 DDL 的进度 325
7.1.4 MySQL 8.0.12 引入的秒级加列特性 327
7.1.5 Online DDL 的优缺点 329
7.1.6 Online DDL 的注意事项 329
7.1.7 参考资料 331
7.2 pt-online-schema-change 331
7.2.1 pt-online-schema-change 的实现原理 331
7.2.2 pt-online-schema-change 的参数解析 337
7.2.3 pt-online-schema-change 的优缺点 345
7.2.4 pt-online-schema-change 的注意事项 345
7.3 gh-ost 346
7.3.1 gh-ost 的实现原理 346
7.3.2 gh-ost 的参数解析 354
7.3.3 与 gh-ost 进行交互 358
7.3.4 gh-ost 的优缺点 359
7.4 元数据锁 360
7.4.1 元数据锁引入的背景 360
7.4.2 元数据锁的基本概念 362
7.4.3 在 MySQL 5.7 和 8.0 中如何定位 DDL 被阻塞的问题 363
7.4.4 在 MySQL 5.6 中如何定位 DDL 被阻塞的问题 365
7.5 本章总结 367

第 8 章 连接池和线程池 369
8.1 连接池 369
8.1.1 连接池的运行原理 369
8.1.2 常用的 JDBC 连接池 370
8.1.3 c3p0 连接池 371
8.1.4 DBCP 连接池 374
8.1.5 参考配置 377
8.1.6 总结 378
8.2 MySQL 线程池 378
8.2.1 线程池的实现原理 379
8.2.2 如何开启线程池功能 380
8.2.3 MySQL 企业版线程池参数解析 380
8.2.4 Percona Server 线程池参数解析 381
8.2.5 MySQL 企业版线程池和 Percona Server 线程池的对比 382
8.2.6 线程池的适用场景 384
8.2.7 线程池的压测结果 384
8.2.8 线程池的监控 385
8.2.9 参考资料 389
8.3 MySQL server has gone away 深度解析 389
8.3.1 出现 MySQL server has gone away 错误的常见原因 389
8.3.2 interactive_timeout 和 wait_timeout 的区别 390
8.3.3 wait_timeout 设置为多大比较合适 393
8.4 本章总结 395

第 9 章 MySQL 的常用工具 396
9.1 sysbench 398
9.1.1 安装 sysbench 398
9.1.2 sysbench 用法讲解 399
9.1.3 对 MySQL 进行基准测试的基本步骤 399
9.1.4 如何分析 MySQL 的基准测试结果 401
9.1.5 如何使用 sysbench 对服务器性能进行测试 402
9.1.6 MySQL 常见的测试场景及对应的 SQL 语句 404
9.1.7 如何自定义 sysbench 测试脚本 407
9.1.8 总结 411

9.2 pt-archiver ································ 411
   9.2.1 安装 ··································· 411
   9.2.2 实现原理 ···························· 411
   9.2.3 常见用法 ···························· 413
   9.2.4 常用参数 ···························· 416
   9.2.5 总结 ··································· 418
9.3 pt-config-diff ···························· 418
9.4 pt-ioprofile ······························· 419
9.5 pt-kill ·· 420
   9.5.1 实现原理 ···························· 420
   9.5.2 过滤逻辑 ···························· 422
   9.5.3 常见用法 ···························· 424
9.6 pt-pmp ······································ 425
9.7 pt-query-digest ························· 426
   9.7.1 常见用法 ···························· 426
   9.7.2 常用参数 ···························· 429
9.8 pt-show-grants ·························· 430
9.9 pt-slave-restart ························· 431
9.10 pt-stalk ···································· 433
9.11 pt-table-checksum ·················· 435
   9.11.1 实现原理 ·························· 435
   9.11.2 常见用法 ·························· 440
   9.11.3 常用参数 ·························· 440
9.12 pt-table-sync ··························· 443
   9.12.1 实现原理 ·························· 443
   9.12.2 常见用法 ·························· 446
   9.12.3 常用参数 ·························· 446
9.13 pt-upgrade ······························ 448
9.14 本章总结 ································· 453

第 10 章 中间件 ··························· 455

10.1 ProxySQL 的安装 ··················· 458
10.2 ProxySQL 入门 ······················· 458
10.3 多层配置系统 ························· 460
10.4 读写分离 ································ 463
10.5 深入理解 ProxySQL 表 ··········· 470
10.6 ProxySQL 的高级特性 ············ 475
   10.6.1 定时器 ····························· 475
   10.6.2 SQL 审计 ························· 476

10.6.3 查询重写 ························· 477
10.6.4 mirroring ·························· 479
10.6.5 SQL 黑名单 ····················· 481
10.7 ProxySQL 连接池 ··················· 482
10.8 ProxySQL Cluster ··················· 490
   10.8.1 搭建 ProxySQL Cluster ······ 491
   10.8.2 添加一个新的节点 ··········· 493
10.9 ProxySQL 的常见参数 ············ 494
   10.9.1 管理参数 ························· 495
   10.9.2 监控参数 ························· 496
   10.9.3 MySQL 参数 ···················· 498
   10.9.4 如何修改参数 ·················· 501
10.10 ProxySQL 中的常见问题 ······· 501
   10.10.1 如何自定义 ProxySQL
             的数据目录 ··············· 501
   10.10.2 通过 USE DBNAME 切换
             数据库 ······················· 502
   10.10.3 ProxySQL 的高可用性 ···· 502
10.11 本章总结 ······························· 502

第 11 章 组复制 ··························· 504

11.1 部署组复制 ···························· 506
   11.1.1 准备安装环境 ·················· 506
   11.1.2 初始化 MySQL 实例 ········ 507
   11.1.3 启动组复制 ····················· 510
   11.1.4 添加节点 ························· 511
11.2 单主模式和多主模式 ·············· 512
   11.2.1 单主模式和多主模式的
           区别 ··································· 512
   11.2.2 单主模式和多主模式的
           在线切换 ·························· 515
11.3 监控组复制 ···························· 516
   11.3.1 replication_group_
           members ··························· 516
   11.3.2 replication_group_
           member_stats ···················· 517
11.4 组复制的要求和限制 ·············· 518
11.5 组复制的常见管理操作 ·········· 521
   11.5.1 强制组成员的重新配置 ···· 521
   11.5.2 如何设置 IP 白名单 ········· 523

- 11.5.3 如何查找单主模式下的 Primary 节点 ················ 524
- 11.5.4 新主选举算法 ················ 524
- 11.5.5 如何查看 Secondary 节点的延迟情况 ················ 529
- 11.5.6 大事务 ················ 530
- 11.5.7 查看组复制的内存使用 ···· 531
- 11.6 组复制的实现原理 ················ 533
  - 11.6.1 数据库状态机 ················ 533
  - 11.6.2 事务在组复制中的处理流程 ················ 533
  - 11.6.3 参考资料 ················ 535
- 11.7 组复制的实现细节 ················ 536
- 11.8 组复制的分布式恢复 ················ 545
  - 11.8.1 分布式恢复的实现原理 ···· 545
  - 11.8.2 分布式恢复的相关参数 ···· 552
- 11.9 组复制的冲突检测 ················ 553
  - 11.9.1 write_set ················ 553
  - 11.9.2 冲突检测数据库 ················ 556
  - 11.9.3 冲突检测的实现细节 ········ 557
  - 11.9.4 冲突检测数据库的清理逻辑 ················ 560
- 11.10 组复制的故障检测 ················ 562
  - 11.10.1 模拟网络分区 ················ 562
  - 11.10.2 故障检测流程 ················ 566
  - 11.10.3 XCom Cache ················ 567
  - 11.10.4 注意事项 ················ 569
  - 11.10.5 参考资料 ················ 569
- 11.11 组复制的事务一致性 ················ 569
  - 11.11.1 group_replication_consistency ················ 570
  - 11.11.2 总结 ················ 573
  - 11.11.3 参考资料 ················ 573
- 11.12 组复制的流量控制机制 ········ 573
  - 11.12.1 触发流控的条件 ············ 574
  - 11.12.2 配额的计算逻辑 ············ 576
  - 11.12.3 配额的作用时机 ············ 581
  - 11.12.4 流控的相关参数 ············ 583
  - 11.12.5 总结 ················ 584
  - 11.12.6 参考资料 ················ 585
- 11.13 组复制的重点参数 ················ 585
- 11.14 本章总结 ················ 588

第 12 章 InnoDB Cluster ················ 590

- 12.1 MySQL Shell ················ 591
  - 12.1.1 MySQL Shell 的安装 ········ 592
  - 12.1.2 MySQL Shell 的使用 ········ 592
  - 12.1.3 X DevAPI 的关键特性 ········ 595
  - 12.1.4 MySQL Shell 工具集 ········ 598
  - 12.1.5 MySQL Shell 的使用技巧 ················ 605
- 12.2 MySQL Router ················ 605
  - 12.2.1 MySQL Router 的安装 ···· 606
  - 12.2.2 MySQL Router 的使用 ···· 606
  - 12.2.3 启动 MySQL Router ········ 612
  - 12.2.4 测试 MySQL Router ········ 612
  - 12.2.5 MySQL Router 的注意事项 ················ 613
- 12.3 InnoDB Cluster 的搭建 ········ 614
  - 12.3.1 准备安装环境 ················ 614
  - 12.3.2 初始化 MySQL 实例 ········ 614
  - 12.3.3 创建超级管理员账号 ········ 616
  - 12.3.4 配置实例 ················ 616
  - 12.3.5 创建 InnoDB Cluster ········ 618
  - 12.3.6 添加节点 ················ 619
  - 12.3.7 查看集群的状态 ············ 622
  - 12.3.8 部署 MySQL Router ········ 623
- 12.4 InnoDB Cluster 的管理操作 ···· 624
  - 12.4.1 dba 对象支持的操作 ········ 624
  - 12.4.2 cluster 对象支持的操作 ················ 630
- 12.5 本章总结 ················ 643

附录 A JSON ················ 645

附录 B MySQL 8.0 的新特性 ················ 672

第 1 章

MySQL 入门、安装与服务的管理

作为最流行的开源数据库，MySQL 自 1995 年发布 1.0 版以来，迄今已发展了近 30 年的时间。从诞生之初的"玩具型"数据库，到如今互联网行业的事实标配，MySQL 之所以流行，可归结为以下两点。

- 互联网的兴起。这里不得不提到 LAMP。LAMP 是 Linux、Apache、MySQL 和 PHP 的缩写，是互联网发展之初最为流行的开源 Web 解决方案，也是很多网站搭建 Web 服务时的首选。使用这套方案，可以方便快捷地提供 Web 服务。可以说，互联网的兴起，LAMP 功不可没。
- 开源、简单、易用。这也是大家在项目之初就选用 MySQL 的原因。随着用户量的增多，社区得到的反馈也越来越多。随着 bug 的修复和功能的完善，MySQL 也变得越来越强大。这也会让更多的人选择 MySQL。

当然，MySQL 在创造历史的同时，也在见证历史。从它被 Sun 公司收购，到 Sun 公司被 Oracle 公司收购，这本身就体现了"眼见他起高楼，眼见他宴宾客，眼见他楼塌了"的无常感。

种一棵树最好的时间是十年前，其次是现在，学习 MySQL 同样如此。就让我们从 MySQL 的安装开始，踏上愉快的 MySQL 学习之旅吧！

本章主要包括以下内容。

- MySQL 的历史。主要介绍 MySQL 发展过程中的里程碑事件及各大版本的新特性。
- MySQL 的安装。主要介绍两种主流的安装方式：基于二进制包的安装和基于源码包的安装。
- MySQL 服务的管理。主要介绍两种常见的服务管理方式：/etc/init.d/mysqld 和 systemd。

1.1 MySQL 的历史

MySQL 的历史，相信很多人早已耳熟能详，这里就不再赘述了。下面仅从产品特性的角度梳理其发展过程中的里程碑事件。

1995 年，MySQL 1.0 发布，仅供内部使用。

1996 年，MySQL 3.19 发布，直接跳过了 MySQL 2.x 版本。

2000 年，MySQL 3.23 发布，集成了 Berkeley DB 存储引擎。该引擎由 Sleepycat 公司开发，支持事务。在集成该引擎的过程中，对 MySQL 代码进行了重构，为后续的可插拔式存储引擎架构奠定了基础。同年，MySQL 基于 GNU 通用公共许可证（General Public License，GPL）协议开放了源码。

2002 年，MySQL 4.0 发布，集成了后来大名鼎鼎的 InnoDB 存储引擎。该引擎由 Innobase 公司开发，不仅支持事务，也支持行级锁，适用于 OLTP 等高并发场景。

2005 年，MySQL 5.0 发布，开始支持游标、存储过程、触发器、视图、XA 事务等特性。同年，Oracle 公司收购了 Innobase 公司。

2008 年，Sun 公司以 10 亿美元收购了 MySQL AB。同年，MySQL 5.1 发布，开始支持事件调度器（event scheduler）、分区、基于行的复制等特性。

2010 年，Oracle 公司以 74 亿美元收购了 Sun 公司。同年年底，MySQL 5.5 发布，引入了半同步复制、utf8mb4 字符集、元数据锁等特性。同时，InnoDB 代替 MyISAM 成为 MySQL 默认的存储引擎。

2013 年，MySQL 5.6 发布，引入了 GTID 复制、延迟复制、基于库级别的并行复制、Online DDL、全文索引等特性。可以说，MySQL 5.6 是 MySQL 历史上的一个里程碑式的版本，也是目前生产中应用得较为广泛的一个版本。

2015 年，MySQL 5.7 发布，引入了组复制、InnoDB Cluster、多源复制、增强半同步（AFTER_SYNC）、基于 WRITESET 的并行复制、虚拟列、JSON 类型等特性。

2018 年，MySQL 8.0 发布，引入了不可见索引（invisible index）、降序索引、直方图、公用表表达式（common table expression，CTE）、窗口函数（window function）、快速加列（ALGORITHM=INSTANT）、哈希连接（hash join）、直方图、克隆插件（clone plugin）、备份锁、InnoDB ReplicaSet、InnoDB ClusterSet 等特性。

需要注意的是，上面提到的发布，一般指的是 GA（general availability）版本的发布。

表 1-1 列出了最近几个大版本的发布时间，以及截至本章写完时最新的小版本及其发布时间。

表 1-1 MySQL 的发布历史

版本	首个 GA 版本的发布时间	最新的小版本	最新小版本的发布时间	产品支持的结束时间
5.1	2008-11-14	5.1.73	2013-12-03	2013-12
5.5	2010-12-03	5.5.62	2018-10-22	2018-12
5.6	2013-02-05	5.6.51	2021-01-20	2021-02
5.7	2015-10-21	5.7.33	2021-01-18	2023-10
8.0	2018-04-19	8.0.25	2021-05-11	2026-04

从表 1-1 中，我们可以看到：

- 每两三年会发布一个大版本；
- 产品的支持周期一般是 8 年。

也许你以为 MySQL 5.5 是老古董了，但直到 2018 年 10 月 22 日，官方仍然在不断更新。

1.2 MySQL 的安装

介绍完 MySQL 的历史，接着介绍一下其安装方法。我们先下载 MySQL，然后基于二进制包或者源码包来安装它。

1.2.1 下载 MySQL

在 MySQL 官网上，点击 DOWNLOADS，此时会打开下载页面（如图 1-1 所示），这个页面提供了以下 3 个产品的下载入口。

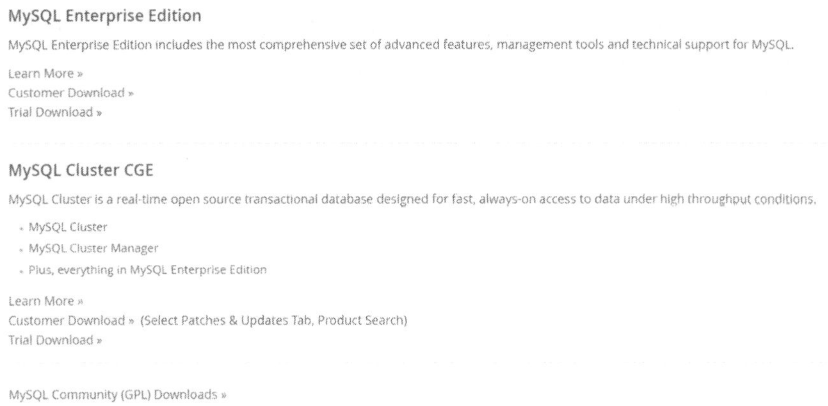

图 1-1 MySQL 下载页面

- MySQL 企业版（MySQL Enterprise Edition）：提供了企业版备份工具、线程池、防火墙、审计、监控等功能。
- MySQL Cluster 企业版（MySQL Cluster CGE）：MySQL Cluster 是一套基于内存、无共享的高可用方案，底层使用的是 NDB 存储引擎，不熟悉 MySQL 的人常常会将其与 InnoDB Cluster 混淆。
- MySQL 社区版：它是广受欢迎的免费下载版本，遵循 GPL 协议。

其中前两个是企业版，需要付费或者购买许可才能使用。

这里，我们点击 MySQL Community (GPL) Downloads 链接，之后会跳转到 MySQL 社区版的下载页面（如图 1-2 所示），这个页面提供了 MySQL 社区版产品的下载入口。

图 1-2 MySQL 社区版产品的下载入口

下面依次介绍图 1-2 中的部分产品。

MySQL Yum Repository、MySQL APT Repository、MySQL SUSE Repository 分别针对 RHEL（包括 CentOS 和 Fedora）、Debian、SUSE 提供的 MySQL 源，可分别使用 `yum`、`apt-get`、`zypper` 命令安装 MySQL Server。

MySQL Community Server 是 MySQL Server 的社区版。MySQL Router 是 MySQL 中间件，可实现读写分离。MySQL Router 与 MySQL Shell、Group Replication 一同组成 InnoDB Cluster。MySQL Shell 是 MySQL 客户端工具，实现了 X Protocol，它支持 JavaScript、Python 和 SQL 三种语言，提供 X DevAPI 和 AdminAPI 两种 API。X DevAPI 可将 MySQL 作为文档数据库进行操作，AdminAPI 用于管理 InnoDB Cluster。MySQL Workbench 是 MySQL 图形化管理工具，常用于 ER 建模。

Download Archives 是 MySQL 社区版产品历史版本的入口。

我们关注的重点是 MySQL Server，所以这里点击 MySQL Community Server，进入其下载页面（如图 1-3 所示）。

图 1-3　MySQL Community Server 的下载页面

在图 1-3 中，General Availability (GA) Releases 代表 MySQL Server 最新的 GA 版本，GA 版本是已经正式发布、可在线上使用的版本。另外还有 RC（release candidate）版本，指的是 GA 之前的候选版本。一般情况下，MySQL 在发布 GA 版本之前，会出两三个 RC 版本。General Availability (GA) Releases 右边的 Archives 是 MySQL Server 历史版本的入口，涵盖了 MySQL 5.0.15 以来的所有版本。Looking for previous GA versions? 是上一个大版本（MySQL 5.7）最新的 GA 版本的入口。

点击 Select Operating System 下拉框（如图 1-4 所示），可以看到 MySQL 几乎支持当前所有的主流操作系统。

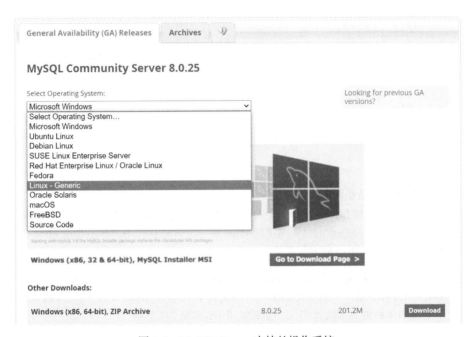

图 1-4　MySQL Server 支持的操作系统

比较常用的安装包有两个：Linux - Generic 和 Source Code，前者是 Linux 通用的二进制包（简称二进制包），后者是源码包。二进制包是经过源码编译，解压即可使用的软件包。rpm 包也是二进制包的一种，但由于安装路径较为固定，在单机多实例且实例版本不一致的场景中不太适用。所以如果要在线上使用二进制包，一般推荐使用 Linux - Generic。

下面看看二进制包的下载方式。

点击 Linux - Generic，进入二进制包的下载页面（如图 1-5 所示）。

从 MySQL 8.0.16 开始，Linux - Generic 提供了以下 3 个版本的下载：

- Linux - Generic (glibc 2.12) (x86, 32-bit)
- Linux - Generic (glibc 2.12) (x86, 64-bit)
- Linux - Generic (glibc 2.17) (x86, 64-bit)

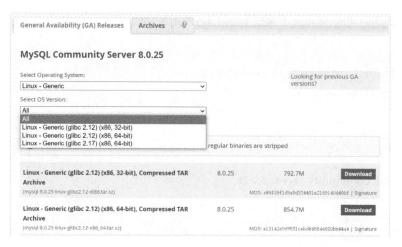

图 1-5　MySQL 二进制包的版本

在这里面，

- 32-bit、64-bit 指的是操作系统的位数。
- glibc 2.12、glibc 2.17 指的是编译 MySQL 的 glibc 版本。glibc 版本在 Linux 系统中可通过 ldd --version 查看。需要注意的是，这里的 glibc 版本没有限制作用，并不是说 glibc 2.12 版本的 MySQL 只能在 glibc 2.12 的操作系统上运行。

同样是 64 位的版本，Linux - Generic (glibc 2.12) 和 Linux - Generic (glibc 2.17) 最显著的区别并不是 glibc 版本不同，而是后者是一个最小化版本。最小化版本剔除了 debug 相关的二进制文件及 debug symbol，比常规版本体积更小（两者的大小分别如图 1-6 和图 1-7 所示）。

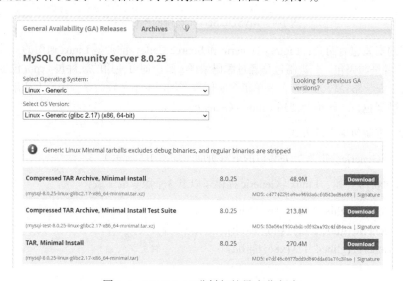

图 1-6　MySQL 二进制包的最小化版本

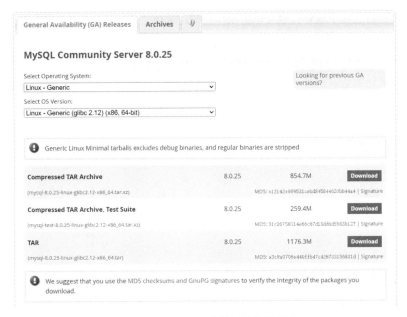

图 1-7　MySQL 二进制包的常规版本

除了上面介绍的 3 个版本，从 MySQL 8.0.31 开始，Linux - Generic 还提供了 Linux - Generic (glibc 2.17) (ARM, 64-bit) 版本的下载。

我们选择 MySQL 二进制包的常规版本，Compressed TAR Archive 和 TAR 均可。点击后面的 Download 按钮，会跳转到如图 1-8 所示的下载页面。这个页面乍一看需要登录或者注册，其实不然。点击页面左下角的 No thanks, just start my download.超链接，浏览器就会直接下载安装包。

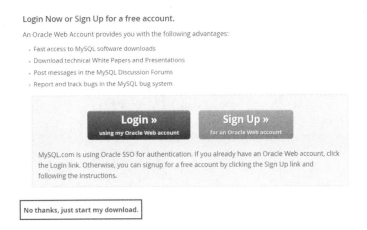

图 1-8　下载安装包

除了点击超链接直接下载外,也可单击鼠标右键,选择"复制链接地址"。这样就可以在 Linux 系统中直接通过 wget 来下载,比如:

```
# wget https://dev.mysql.com/get/Downloads/MySQL-8.0/mysql-8.0.25-linux-glibc2.12-x86_64.tar.xz
```

下载源码包的步骤基本相同,只不过要在图 1-6 中选择 Source Code。进入源码包的下载页面后,操作系统版本(OS Version)要选择 Generic Linux (Architecture Independent),如图 1-9 所示。注意,在新的版本中,Generic Linux (Architecture Independent) 已改名为 All Operating Systems (Generic) (Architecture Independent)。

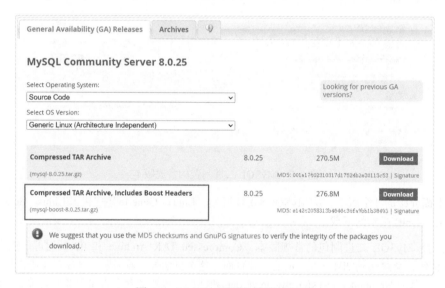

图 1-9 MySQL 源码包的下载页面

这里建议下载带有 Boost Headers 的版本,否则需要在 CMake 的过程中指定 -DDOWNLOAD_BOOST=1 -DWITH_BOOST=<directory>进行下载。

到这里,我们已经成功下载了安装包,接下来就可以安装 MySQL 了。常用的安装方式有两种:基于二进制包的安装和基于源码包的安装。

1.2.2 基于二进制包的安装

MySQL 8.0 二进制包的安装方式如下。

(1) 创建操作系统用户。

```
# groupadd mysql
# useradd -g mysql mysql
```

注意,这里也可以使用其他用户名,不一定是 mysql。

(2) 解压二进制包,建立软链接。

```
# cd /usr/local/
# tar xvf mysql-8.0.25-linux-glibc2.12-x86_64.tar.xz
# ln -s mysql-8.0.25-linux-glibc2.12-x86_64 mysql
```

这里是将二进制包解压到/usr/local/目录下。

(3) 编辑配置文件。

```
# vim /etc/my.cnf
```

为了方便，下面只给出几个主要参数：

```
[client]
socket      = /data/mysql/3306/data/mysql.sock

[mysqld]
basedir     = /usr/local/mysql
datadir     = /data/mysql/3306/data
user        = mysql
port        = 3306
socket      = /data/mysql/3306/data/mysql.sock
log_error   = /data/mysql/3306/data/mysqld.err
log_timestamps = system
```

各参数的作用如下。

- basedir：MySQL Server 的安装目录。对于二进制包，该参数的默认值为/usr/local/mysql。
- datadir：数据目录。
- user：MySQL Server 的运行用户。从安全角度出发，MySQL 禁止把 root 作为 MySQL Server 的运行用户，否则会提示以下错误：

  ```
  [ERROR] [MY-010123] [Server] Fatal error: Please read "Security" section of the manual to find out how to run mysqld as root!
  ```

  注意，MySQL 禁止的只是把 root 作为运行用户，并不影响我们以 root 用户的身份启动 MySQL。

- port：MySQL Server 的监听端口。
- socket：套接字文件。
- log_error：错误日志。
- log_timestamps：设置日志时间戳的时区，默认是 UTC（universal time coordinated，协调世界时）。这里设置的是 system，即使用系统的时区设置。

(4) 创建数据目录，并修改其属主、属组。

```
# mkdir -p /data/mysql/3306/data
# chown mysql.mysql /data/mysql/3306/data/
```

(5) 初始化实例。

```
# /usr/local/mysql/bin/mysqld --defaults-file=/etc/my.cnf --initialize
```

其中，--defaults-file 是参数文件的路径，--initialize 是初始化实例。

一般而言，如果终端没有错误输出且错误日志中没有错误信息，就意味着实例初始化成功。

上面的初始化命令使用了--initialize，其实也可使用--initialize-insecure，两者唯一的区别是前者会为 root@localhost 生成一个随机密码，而后者不会。线上一般建议使用--initialize。

在初始化实例的过程中，需注意以下两点。

- 如果指定了--defaults-file，那么它必须是命令行中的第一个参数，否则会没有效果。
- 数据目录必须为空，否则会报错：

```
[ERROR] --initialize specified but the data directory has files in it. Aborting.
```

(6) 启动实例。

```
# /usr/local/mysql/bin/mysqld_safe --defaults-file=/etc/my.cnf &
```

一般会通过以下两种方式来判断实例是否启动成功。

- 查看错误日志。如果启动成功，日志中会打印 ready for connections，具体如下：

```
[System] [MY-010931] [Server] /usr/local/mysql/bin/mysqld: ready for connections. Version: '8.0.25'  socket: '/data/mysql/3306/data/mysql.sock'  port: 3306  MySQL Community Server - GPL.
```

- 查看 mysqld 进程是否启动，具体命令如下：

```
# ps -ef | grep mysqld
root      27472 26119  0 20:55 pts/0    00:00:00 /bin/sh /usr/local/mysql/bin/mysqld_safe --defaults-file=/etc/my.cnf
mysql     27645 27472  0 20:55 pts/0    00:00:00 /usr/local/mysql/bin/mysqld --defaults-file=/etc/my.cnf --basedir=/usr/local/mysql --datadir=/data/mysql/3306/data --plugin-dir=/usr/local/mysql/lib/plugin --user=mysql --log-error=/data/mysql/3306/data/mysqld.err --pid-file=slowtech.pid --socket=/data/mysql/3306/data/mysql.sock --port=3306
```

如果启动失败，可通过以下步骤定位失败的原因。

(1) 查看错误日志。一般来说，启动过程中出现的问题都会打印到错误日志中。常见的错误有：端口被占用、参数名写错、参数在当前版本下不支持，等等。如果错误日志中没有错误信息，可尝试通过步骤(2)启动。

(2) 通过 mysqld 启动，具体命令如下：

```
# /usr/local/mysql/bin/mysqld --defaults-file=my.cnf &
```

如果依然无法定位，可尝试通过步骤(3)启动。

(3) 还是通过 mysqld 启动，但仅指定几个必要的参数，具体命令如下：

```
# /usr/local/mysql/bin/mysqld --no-defaults --basedir=/usr/local/mysql --datadir=/data/mysql/3306/data/ --user=mysql
```

为了避免读取默认位置的配置文件，这里使用了--no-defaults 参数。

通过这"三板斧"，基本上就能定位出 MySQL 启动失败的原因了。如果还是不行，可借助 strace 查看 MySQL 启动过程中的系统调用情况。

接着，我们来看看 MySQL 启动后，其数据目录中都有哪些文件：

```
# ll /data/mysql/3306/data/
total 188888
-rw-r----- 1 mysql mysql       56 Jun 25 20:55 auto.cnf
-rw-r----- 1 mysql mysql      156 Jun 25 20:55 binlog.000001
-rw-r----- 1 mysql mysql       16 Jun 25 20:55 binlog.index
-rw------- 1 mysql mysql     1676 Jun 25 20:55 ca-key.pem
-rw-r--r-- 1 mysql mysql     1112 Jun 25 20:55 ca.pem
-rw-r--r-- 1 mysql mysql     1112 Jun 25 20:55 client-cert.pem
-rw------- 1 mysql mysql     1680 Jun 25 20:55 client-key.pem
-rw-r----- 1 mysql mysql   196608 Jun 25 20:55 #ib_16384_0.dblwr
-rw-r----- 1 mysql mysql  8585216 Jun 25 20:55 #ib_16384_1.dblwr
-rw-r----- 1 mysql mysql     5472 Jun 25 20:55 ib_buffer_pool
-rw-r----- 1 mysql mysql 12582912 Jun 25 20:55 ibdata1
-rw-r----- 1 mysql mysql 50331648 Jun 25 20:55 ib_logfile0
-rw-r----- 1 mysql mysql 50331648 Jun 25 20:55 ib_logfile1
-rw-r----- 1 mysql mysql 12582912 Jun 25 20:55 ibtmp1
drwxr-x--- 2 mysql mysql     4096 Jun 25 20:55 #innodb_temp
drwxr-x--- 2 mysql mysql     4096 Jun 25 20:55 mysql
-rw-r----- 1 mysql mysql     1471 Jun 25 20:55 mysqld.err
-rw-r----- 1 mysql mysql 25165824 Jun 25 20:55 mysql.ibd
srwxrwxrwx 1 mysql mysql        0 Jun 25 20:55 mysql.sock
-rw------- 1 mysql mysql        6 Jun 25 20:55 mysql.sock.lock
drwxr-x--- 2 mysql mysql     4096 Jun 25 20:55 performance_schema
-rw------- 1 mysql mysql     1680 Jun 25 20:55 private_key.pem
-rw-r--r-- 1 mysql mysql      452 Jun 25 20:55 public_key.pem
-rw-r--r-- 1 mysql mysql     1112 Jun 25 20:55 server-cert.pem
-rw------- 1 mysql mysql     1676 Jun 25 20:55 server-key.pem
-rw-r----- 1 mysql mysql        6 Jun 25 20:55 slowtech.pid
drwxr-x--- 2 mysql mysql     4096 Jun 25 20:55 sys
-rw-r----- 1 mysql mysql 16777216 Jun 25 20:55 undo_001
-rw-r----- 1 mysql mysql 16777216 Jun 25 20:55 undo_002
```

各个文件（目录）的具体作用如下。

- auto.cnf：记录了实例的 server-uuid。
- ib_buffer_pool：文本文件，记录了缓冲池中数据页的地址（space_id 和 page_no）。这样，数据库在启动后，可直接将指定的数据页加载到缓冲池中，避免了较长的预热时间。
- ibdata1：系统表空间，主要包括数据字典信息、双写缓冲区（doublewrite buffer）和插入缓冲区（insert buffer）等。从 MySQL 8.0.20 开始，双写缓冲区会被存储在#ib_16384_0.dblwr 和 #ib_16384_1.dblwr 中。
- ib_logfile0 和 ib_logfile1：重做日志。
- ibtmp1：全局级别的临时表空间，主要用于存储对用户创建的临时表所做变更的回滚段。
- #innodb_temp：会话级别的临时表空间，主要用于存储用户创建的临时表和 SQL 在执行过程中产生的磁盘临时表。
- mysql：mysql 库，主要包括一些系统表，如授权相关的表（user、db、tables_priv、columns_priv、procs_priv 和 proxies_priv）、日志相关的表（general_log 和 slow_log）、复制相关的表（slave_master_info 和 slave_relay_log_info）。在 MySQL 8.0 之前，这些表都放在 mysql 目录下。在 MySQL 8.0 中，与日志相关的表依旧放在 mysql 目录下，其他表则放在一个单独的共享表空间中（mysql.ibd）。
- mysqld.err：错误日志。

- performance_schema：采集了 MySQL 运行过程中的性能数据。
- sys：MySQL 5.7 引入，库中只有一张基表（sys_config），其他都是基于 information_schema 和 performance_schema 的视图。这些视图简单直接，很好地弥补了 performance_schema 可读性差的缺点。
- undo_001 和 undo_002：回滚表空间，由回滚段组成。在 MySQL 5.6 之前，回滚段是放在系统表空间（ibdata1）中的。MySQL 5.6 则支持设置独立的回滚表空间，但可惜的是，该空间不能回收，一旦增大，就不会被释放，只能重复使用。MySQL 5.7 解决了这个问题，支持空间的自动回收，但在默认配置下，回滚段还是放在系统表空间中的，没有独立出来。直到 MySQL 8.0，回滚段才默认独立，这就是我们现在看到的 undo_001 和 undo_002。
- *.pem：与 RSA 相关的证书文件及私钥文件，用于开启 SSL 加密连接。

(7) 登录实例，这时需要使用初始化过程中生成的随机密码。

```
# grep password /data/mysql/3306/data/mysqld.err
2021-06-25T20:55:35.752122+08:00 6 [Note] [MY-010454] [Server] A temporary password is generated for root@localhost: VwmQ=H*4bkyv

# /usr/local/mysql/bin/mysql -uroot -S /data/mysql/3306/data/mysql.sock -p'VwmQ=H*4bkyv'
```

登录成功后，无论执行什么命令，系统都会提示重置密码：

```
mysql> select user();
ERROR 1820 (HY000): You must reset your password using ALTER USER statement before executing this statement.
mysql> select 1;
ERROR 1820 (HY000): You must reset your password using ALTER USER statement before executing this statement.
```

(8) 修改密码。

```
mysql> alter user user() identified by 'slowtech';
Query OK, 0 rows affected (0.00 sec)
```

这里的 user() 指的是 root@localhost。如果实例是通过 rpm 包安装的，同样的命令会报错：

```
mysql> alter user user() identified by 'slowtech';
ERROR 1819 (HY000): Your password does not satisfy the current policy requirements
```

究其原因，rpm 版本的 MySQL Server 会自动加载密码复杂度验证插件（validate_password）：

```
mysql> select * from mysql.component;
+--------------+--------------------+---------------------------------+
| component_id | component_group_id | component_urn                   |
+--------------+--------------------+---------------------------------+
|            1 |                  1 | file://component_validate_password |
+--------------+--------------------+---------------------------------+
1 row in set (0.00 sec)
```

validate_password 在 MySQL 8.0 中是通过 Component（组件）实现的，但在 MySQL 5.7 中还是通过 Plugin（插件）实现的，所以在 MySQL 5.7 中需通过 show plugins 查看：

```
mysql> show plugins;
+--------------------+----------+------+---------+---------+
| Name               | Status   | Type | Library | License |
```

```
+---------------------+----------+----------------------+--------------------+---------+
...
| validate_password   | ACTIVE   | VALIDATE PASSWORD    | validate_password.so | GPL   |
+---------------------+----------+----------------------+--------------------+---------+
```

除此之外，validate_password 在 MySQL 5.7 和 MySQL 8.0 中的加载方式也不同。

MySQL 5.7：

```
mysql> install plugin validate_password soname 'validate_password.so';
```

MySQL 8.0：

```
mysql> install component 'file://component_validate_password';
```

该组件可设置不同的密码验证策略，具体策略由参数 validate_password.policy 决定。可设置的策略及具体要求如表 1-2 所示，策略的默认值为 MEDIUM。

表 1-2　密码验证策略

策　　略	要　　求
0 或 LOW	满足长度要求即可，长度由参数 validate_password.length 决定，默认为 8
1 或 MEDIUM	相对于 0，密码中还需要有数字、大小写字母和特殊字符
2 或 STRONG	相对于 1，还需要字典文件

当密码验证策略为 MEDIUM 和 STRONG 时，还对数字、大小写字母和特殊字符有最小个数的要求。最小个数分别由参数 validate_password.number_count、validate_password.mixed_case_count、validate_password.special_char_count 决定，这 3 个参数默认都为 1。

不仅如此，这 3 个参数还决定了 validate_password.length 可设置的最小值。具体来说，validate_password.length 在设置时，不得小于以下值：

```
validate_password.number_count + validate_password.special_char_count + (2 *
validate_password.mixed_case_count)
```

由此来看，对于默认的密码验证策略（MEDIUM），密码中至少应包含 1 个数字、1 个特殊字符、1 个大写字母和 1 个小写字母，且总长度不小于 8。

看看下面这个示例：

```
mysql> alter user user() identified by 'PY@^&*1(';
ERROR 1819 (HY000): Your password does not satisfy the current policy requirements
mysql> alter user user() identified by 'Py@12345';
Query OK, 0 rows affected (0.09 sec)
```

直观上看，'PY@^&*1(' 的复杂度远大于 'Py@12345'，但因其缺少小写字母，还是不符合密码策略的要求。

MySQL 5.7 同样存在这个问题，即二进制包默认不会加载密码复杂度验证插件，而 rpm 包会。

关于 validate_password 的相关参数，有以下两点需要注意。

❑ 只有加载了插件，参数才会出现。

- MySQL 8.0 中的参数名与 MySQL 5.7 中的略有不同。具体来说，MySQL 8.0 中参数名的点号在 MySQL 5.7 中是下划线，比如，validate_password.policy 在 MySQL 5.7 中对应的参数是 validate_password_policy。

以上是 MySQL 8.0 二进制包的安装方式，这种方式也适用于 MySQL 5.7。对于 MySQL 5.6，上述安装方式同样适用，唯一不同的是实例的初始化方式：它使用的是 mysql_install_db，一个 Perl 脚本。相关命令如下：

```
# cd mysql-5.6.51-linux-glibc2.12-x86_64/
# scripts/mysql_install_db --defaults-file=/etc/my.cnf
Installing MySQL system tables...2021-06-25 21:03:09 0 [Warning] TIMESTAMP with implicit DEFAULT value is deprecated. Please use --explicit_defaults_for_timestamp server option (see documentation for more details).
2021-06-25 21:03:09 0 [Note] Ignoring --secure-file-priv value as server is running with --bootstrap.
2021-06-25 21:03:09 0 [Note] ./bin/mysqld (mysqld 5.6.51) starting as process 30475 ...
...
2021-06-25 21:03:11 30475 [Note] InnoDB: Starting shutdown...
2021-06-25 21:03:12 30475 [Note] InnoDB: Shutdown completed; log sequence number 1625977
OK

Filling help tables...2021-06-25 21:03:12 0 [Warning] TIMESTAMP with implicit DEFAULT value is deprecated. Please use --explicit_defaults_for_timestamp server option (see documentation for more details).
2021-06-25 21:03:12 0 [Note] Ignoring --secure-file-priv value as server is running with --bootstrap.
2021-06-25 21:03:12 0 [Note] ./bin/mysqld (mysqld 5.6.51) starting as process 30502 ...
...
2021-06-25 21:03:12 30502 [Note] InnoDB: Starting shutdown...
2021-06-25 21:03:14 30502 [Note] InnoDB: Shutdown completed; log sequence number 1625987
OK
...
```

出现两个 OK，一般就意味着初始化成功。

如果在指定配置文件的情况下初始化失败，很可能是因为参数文件中的某个参数有问题。这个时候，可先查看错误日志，里面一般会有提示。如果没有提示，可在命令行中显式指定 --no-defaults、--basedir、--datadir、--user 这 4 个参数进行初始化，初始化成功后再根据配置文件启动实例。如果某个参数存在问题，也会在启动阶段暴露出来。

1.2.3 基于源码包的安装

基于源码包的安装步骤和基于二进制包的安装步骤基本一致，唯一的区别是源码包不能直接使用。在使用之前，必须将它编译为二进制包。

下面我们看看 MySQL 源码包的编译步骤。

(1) 安装依赖包，解压源码包。

```
# yum install cmake3 gcc gcc-c++ glibc ncurses-devel openssl-devel libaio-devel
# cd /usr/src
# tar xvf mysql-boost-8.0.25.tar.gz
# cd mysql-8.0.25
```

(2) 创建一个临时的编译目录。

```
# mkdir build
# cd build
```

之所以要新建一个目录进行编译，主要是为了保证源码树的整洁。如果在当前目录下编译，会提示以下错误：

```
CMake Error at CMakeLists.txt:382 (MESSAGE):
  Please do not build in-source.  Out-of source builds are highly
  recommended: you can have multiple builds for the same source, and there is
  an easy way to do cleanup, simply remove the build directory (note that
  'make clean' or 'make distclean' does *not* work)

  You *can* force in-source build by invoking cmake with
  -DFORCE_INSOURCE_BUILD=1
```

(3) 配置编译文件。

```
# cmake3 /usr/src/mysql-8.0.25/ -DWITH_BOOST=/usr/src/mysql-8.0.25/boost/boost_1_73_0
-DENABLE_DOWNLOADS=1 -DBUILD_CONFIG=mysql_release
...
-- CMAKE_CXX_FLAGS_MINSIZEREL: -ffunction-sections -fdata-sections -Os -DNDEBUG
-- CMAKE_C_LINK_FLAGS:    -fuse-ld=gold -Wl,--gc-sections
-- CMAKE_CXX_LINK_FLAGS:   -fuse-ld=gold -Wl,--gc-sections
-- CMAKE_EXE_LINKER_FLAGS
-- CMAKE_MODULE_LINKER_FLAGS
-- CMAKE_SHARED_LINKER_FLAGS
-- Configuring done
-- Generating done
-- Build files have been written to: /usr/src/mysql-8.0.25/build
```

cmake 命令会基于源码包中的 CMakeLists.txt 生成 Makefile。这一步会检查编译器和依赖库是否已经安装，版本是否满足编译需求，等等。

> **注意**：如果在 CMake 过程中出现问题，那么当解决完问题重新 CMake 时，需要先删除当前目录下的 CMakeCache.txt 文件。

cmake 命令行中指定了 3 个选项，其具体含义如下。

- -DWITH_BOOST：指定 Boost 库的位置。
- -DENABLE_DOWNLOADS：自动下载某些扩展文件，如 Googletest。如果在内网环境中，则可不指定。
- -DBUILD_CONFIG=mysql_release：使用官方二进制包的编译选项。

当然，也可指定其他选项。常用的编译选项如下。

- -DCMAKE_BUILD_TYPE：build 的类型，如果要编译成 debug 版本，需将该选项设置为 Debug。
- -DWITH_DEBUG：是否编译成 debug 版本。若设置为 1，效果同 -DCMAKE_BUILD_TYPE=Debug 一样。
- -DCMAKE_INSTALL_PREFIX：指定安装目录的位置，默认为 /usr/local/mysql。
- -DMYSQL_DATADIR：指定数据目录的默认位置。
- -DDEFAULT_CHARSET：指定默认字符集。
- -DDEFAULT_COLLATION：指定字符集的默认校验规则。

- -DMYSQL_TCP_PORT：指定 MySQL Server 的监听端口，默认为 3306。
- -DMYSQL_UNIX_ADDR：指定套接字文件的位置，默认为 /tmp/mysql.sock。
- -WITHOUT_*xxx*_STORAGE_ENGINE：排除某些存储引擎的编译。比如不编译 BLACKHOLE 存储引擎，可指定以下选项：

    -DWITHOUT_BLACKHOLE_STORAGE_ENGINE=1

- -WITH_*xxx*_STORAGE_ENGINE：指定某些存储引擎的编译。在二进制包中，默认不支持 FEDERATED 引擎。如果要开启，只能编译源码，且需指定以下选项：

    -DWITH_FEDERATED_STORAGE_ENGINE=1

- -DWITHOUT_SERVER：是否编译 MySQL Server。如果设置为 1，则不会编译。

至于其他编译选项，可参考官方文档。

(4) 编译。

```
# make -j 16
...
[100%] Building CXX object unittest/gunit/CMakeFiles/merge_large_tests-t.dir/decoy_user-t.cc.o
[100%] Building CXX object unittest/gunit/CMakeFiles/merge_large_tests-t.dir/__/__/storage/example/ha_example.cc.o
[100%] Linking CXX executable ../../runtime_output_directory/merge_large_tests-t
[100%] Built target merge_large_tests-t
```

`make` 会基于 Makefile 中的规则将源码编译成可执行文件。命令行中的 `-j` 指的是可同时运行的作业数，相当于并发数。在一定范围内，并发数越高，编译时间越短。

(5) 安装。

```
# make install
...
-- Installing: /usr/local/mysql/man/man1/mysqld_safe.1
-- Installing: /usr/local/mysql/man/man8/mysqld.8
-- Installing: /usr/local/mysql/man/man1/mysqlrouter.1
-- Installing: /usr/local/mysql/man/man1/mysqlrouter_passwd.1
-- Installing: /usr/local/mysql/man/man1/mysqlrouter_plugin_info.1
```

`make install` 命令会将可执行文件和第三方依赖库拷贝到指定位置。编译后的文件默认安装在 /usr/local/mysql 目录下：

```
# ls /usr/local/mysql/
bin      lib              LICENSE-test           mysql-test      README-test    support-files
docs     LICENSE          man                    README          run            var
include  LICENSE.router   mysqlrouter-log-rotate README.router   share
```

源码编译过程中的常见问题

以下是源码编译过程中的常见问题及解决方法。

- GCC 版本过低，错误信息如下：

    ```
    CMake Error at cmake/os/Linux.cmake:80 (MESSAGE):
      GCC 5.3 or newer is required (-dumpversion says 4.8.5)
    Call Stack (most recent call first):
    ```

```
CMakeLists.txt:491 (INCLUDE)
```

解决方法：升级 GCC。但线上升级 GCC 毕竟是个高危操作，有可能导致程序出现兼容性问题。

这里推荐一种更为优雅的方法：Red Hat Software Collections（SCL）。SCL 出现的初衷是允许在同一个操作系统上安装和运行一个软件的多个版本，而不影响系统的其他软件。目前支持多个软件包，如 Ruby、Redis、PHP、Nginx 等，完整列表可参考其官网。在这些软件包里，有一个名为 Developer Toolset 的开发者工具包，它集成了一些版本较新的常见开发工具（如 GCC），类似于 yum 中的 Development tools。

SCL 的使用也比较简单。首先安装 SCL，具体命令如下：

```
# yum install centos-release-scl
```

接着，安装 devtoolset-7 中的 GCC，具体命令如下：

```
# yum install devtoolset-7-gcc devtoolset-7-gcc-c++
```

最后，创建一个开启 SCL 的 bash 会话，这个会话只在当前终端有效：

```
# gcc --version
gcc (GCC) 4.8.5 20150623 (Red Hat 4.8.5-44)
Copyright (C) 2015 Free Software Foundation, Inc.
This is free software; see the source for copying conditions.  There is NO
warranty; not even for MERCHANTABILITY or FITNESS FOR A PARTICULAR PURPOSE.

# scl enable devtoolset-7 bash

# gcc --version
gcc (GCC) 7.3.1 20180303 (Red Hat 7.3.1-5)
Copyright (C) 2017 Free Software Foundation, Inc.
This is free software; see the source for copying conditions.  There is NO warranty; not even for
MERCHANTABILITY or FITNESS FOR A PARTICULAR PURPOSE.
```

可以看到，在开启 SCL 后，GCC 的版本发生了变化，达到了升级的目的。

- 没有显式设置 CMAKE_C_COMPILER 和 CMAKE_CXX_COMPILER，错误信息如下：

```
-- This is .el7. as found from 'rpm -qf /'
-- Looking for a devtoolset compiler
CMake Warning at CMakeLists.txt:280 (MESSAGE):
  Could not find devtoolset compiler in /opt/rh/devtoolset-10

CMake Error at CMakeLists.txt:281 (MESSAGE):
  Please set CMAKE_C_COMPILER and CMAKE_CXX_COMPILER explicitly.

-- Configuring incomplete, errors occurred!
```

解决方法：按照错误信息中的提示，安装 devtoolset 编译器。相关代码如下：

```
# yum install centos-release-scl
# yum install devtoolset-10-gcc devtoolset-10-gcc-c++
```

- 在编译过程中，cc1plus 进程被杀掉，错误信息如下：

```
c++: internal compiler error: Killed (program cc1plus)
Please submit a full bug report,
```

```
with preprocessed source if appropriate.
See <http://bugzilla.redhat.com/bugzilla> for instructions.
```

问题原因：系统内存不足，cc1plus 因 OOM（out of memory，内存溢出）而被杀掉。

下面是 /var/log/messages 中的错误信息：

```
slowtech kernel: Out of memory: Kill process 18476 (cc1plus) score 591 or sacrifice child
slowtech kernel: Killed process 18476 (cc1plus) total-vm:1341708kB, anon-rss:1146228kB,
file-rss:40kB, shmem-rss:0kB
```

解决方法：创建一个 swap 分区。相关命令如下：

```
# dd if=/dev/zero of=/data/swapfile bs=1M count=4096
# mkswap /data/swapfile
# swapon /data/swapfile
```

❑ GCC 或 GLIBC 版本不兼容。

常见的错误有如下两个。

■ 'SYS_gettid' was not declared in this scope

具体错误信息如下：

```
[ 49%] Building CXX object storage/innobase/CMakeFiles/innobase.dir/buf/buf0buf.cc.o
/usr/src/mysql-8.0.25/storage/innobase/buf/buf0buf.cc: In function 'void
buf_pool_create(buf_pool_t*, ulint, ulint, std::mutex*, dberr_t&)':
/usr/src/mysql-8.0.25/storage/innobase/buf/buf0buf.cc:1228:44: error: 'SYS_gettid' was not
declared in this scope
 1228 |     setpriority(PRIO_PROCESS, (pid_t)syscall(SYS_gettid), -20);
      |                                            ^~~~~~~~~~
```

报错的常见原因是 GCC 版本不兼容。

解决方法：

(1) 通过 devtoolset 安装不同版本的 GCC；
(2) 修改源码，在 buf0buf.cc 文件头部导入头文件 #include <sys/syscall.h>。

这里优先使用方法(1)。

■ 'FALLOC_FL_ZERO_RANGE' was not declared in this scope

具体错误信息如下：

```
[ 51%] Building CXX object storage/innobase/CMakeFiles/innobase.dir/os/os0file.cc.o
/usr/src/mysql-8.0.25/storage/innobase/os/os0file.cc: In function 'bool
os_file_set_size_fast(const char*, pfs_os_file_t, os_offset_t, os_offset_t, bool, bool)':
/usr/src/mysql-8.0.25/storage/innobase/os/os0file.cc:5532:34: error: 'FALLOC_FL_ZERO_RANGE'
was not declared in this scope
         fallocate(pfs_file.m_file, FALLOC_FL_ZERO_RANGE, offset, size - offset);
                                   ^~~~~~~~~~~~~~~~~~~~
/usr/src/mysql-8.0.25/storage/innobase/os/os0file.cc:5532:34: note: suggested alternative:
'HAVE_FALLOC_FL_ZERO_RANGE'
         fallocate(pfs_file.m_file, FALLOC_FL_ZERO_RANGE, offset, size - offset);
                                   ^~~~~~~~~~~~~~~~~~~~
                                   HAVE_FALLOC_FL_ZERO_RANGE
```

报错的常见原因是 GLIBC 版本不兼容。

解决方法：

(1) 更新 GLIBC，具体命令如下。

```
# yum install glibc
```

(2) 按照提示修改源码，具体来说，是将 FALLOC_FL_ZERO_RANGE 修改为 HAVE_FALLOC_FL_ZERO_RANGE。

这里优先使用方法(1)。

最后谈谈，对于 MySQL，二进制包安装和源码包安装孰优孰劣的问题。很多人喜欢用源码包安装，原因不外乎以下 3 个。

- 可自定义路径。
- 可自行指定编译选项。
- 源码编译的性能更好。

关于第一个原因，对于 MySQL 来说，意义其实不是很大，毕竟 basedir 和 datadir 都可自定义。唯一有区别的是配置文件的默认加载位置，这一点后面会提到。

至于第二和第三个原因，我们不妨看看官方文档的说法：

在使用源码包进行安装之前，请检查 Oracle 是否为您的平台提供了预编译的二进制包，以及它是否适合您。为了实现最佳性能，我们投入了大量的精力来确保我们提供的二进制包是以最佳选项来构建的。

尤其对于第二个原因，官方文档甚至给出了"警告"：

使用非标准选项编译 MySQL 可能会导致功能性、性能或安全性下降。

不难看出，相对于源码包，官方更推荐使用二进制包来安装部署 MySQL。

1.2.4 配置文件的读取顺序

在启动 MySQL 时，建议使用 --defaults-file 选项显式指定配置文件，否则 mysqld 会依次读取下面 4 个配置文件：

- /etc/my.cnf
- /etc/mysql/my.cnf
- /usr/local/mysql/etc/my.cnf
- ~/.my.cnf

除了第 3 个配置文件，其他文件的路径基本固定。第 3 个配置文件的路径由源码编译时的 -DCMAKE_INSTALL_PREFIX 选项确定，在官方提供的二进制包中，该选项的默认值为 /usr/local/mysql，而在 rpm 包中，该选项的默认值又为 /usr。所以，这两个安装包的第 3 个配置文件分别是 /usr/local/mysql/etc/my.cnf 和 /usr/etc/my.cnf。如果不确定安装包的类型，可通过以下命令确认配置文件的读取顺序：

```
# /usr/local/mysql/bin/mysqld --verbose --help | grep -A1 "Default options"
Default options are read from the following files in the given order:
/etc/my.cnf /etc/mysql/my.cnf /usr/local/mysql/etc/my.cnf ~/.my.cnf
```

注意：配置文件的读取顺序不仅适用于 mysqld，同样也适用于 mysql、mysqladmin、mysqldump 等工具。

有时，我们通过 `ps -ef |grep mysqld` 看不到配置文件，这时不妨看看上面 4 个路径上是否存在配置文件。

在 MySQL 8.0 中，新增了一张 performance_schema.variables_info 表。通过这张表，我们可以很容易地看出一个参数的来源、修改时间、修改用户等。比如：

```
mysql> select * from performance_schema.variables_info where variable_name='max_connections'\G
*************************** 1. row ***************************
  VARIABLE_NAME: max_connections
VARIABLE_SOURCE: PERSISTED
  VARIABLE_PATH: /data/mysql/3306/data/mysqld-auto.cnf
      MIN_VALUE: 1
      MAX_VALUE: 100000
       SET_TIME: 2021-06-26 12:30:36.951278
       SET_USER: root
       SET_HOST: localhost
1 row in set (0.00 sec)
```

这里的 VARIABLE_SOURCE 代表参数的来源，它有如下取值。

- COMMAND_LINE：命令行。
- COMPILED：编译的默认值。
- DYNAMIC：在线动态修改，如 `set global max_connections=300`。
- EXPLICIT：来自 --defaults-file 的配置。
- EXTRA：来自 --defaults-extra-file 的配置。
- GLOBAL：全局配置文件，如 /etc/my.cnf、/etc/mysql/my.cnf 等。
- LOGIN：来自 ~/.mylogin.cnf 的配置。
- PERSISTED：持久化参数文件 mysqld-auto.cnf。
- SERVER：来自 $MYSQL_HOME/my.cnf 的配置。
- USER：来自 ~/.my.cnf 的配置。

最后，我们总结一下本节的内容。

- 无论是 MySQL Server 的初始化还是启动，必需的 3 个参数是 basedir、datadir 和 user。在初始化或启动的过程中，如果没有指定 basedir，二进制包也没有解压到默认位置（/usr/local/mysql/），通常会出现如下错误：

  ```
  [ERROR] Can't find error-message file '/usr/local/mysql/share/errmsg.sys'. Check error-message file location and 'lc-messages-dir' configuration directive.
  ```

- 如果使用 --defaults-file 来指定配置文件，一定要将它作为 mysqld（mysqld_safe）后的第一个参数。

- 使用 MySQL 8.0 之前版本的客户端连接 MySQL 8.0 的实例，通常会出现如下错误：

```
ERROR 2059 (HY000): Authentication plugin 'caching_sha2_password' cannot be loaded:
/usr/lib64/mysql/plugin/caching_sha2_password.so: cannot open shared object file: No such file or
directory
```

这个错误与创建用户时设置的密码认证插件的类型有关。我们在创建用户时，一般不会指定密码认证插件，如：

```
create user victor@'%' identified by 'Py@12345';
```

这条语句看上去直接指定了密码，但实际上，其完整语法如下：

```
identified with auth_plugin by 'auth_string'
```

这里的 `auth_plugin` 即密码认证插件的类型，如果没有显式指定，则使用默认的密码认证插件。默认的密码认证插件由参数 `default_authentication_plugin` 决定。在 MySQL 8.0 之前，该参数的默认值为 `mysql_native_password`，该插件基于 SHA-1 算法。在 MySQL 8.0 中，默认值改为了 `caching_sha2_password`，该插件基于 SHA-256 算法。

解决这种错误的方法有 3 种。

- 升级 MySQL 客户端或驱动。
- 将参数 `default_authentication_plugin` 设置为 `mysql_native_password`。
- 在创建用户时显式指定 `auth_plugin` 为 `mysql_native_password`，比如：

```
create user victor@'%' identified with mysql_native_password by 'Py@12345';
```

1.3 MySQL 服务的管理

安装完 MySQL Server 后，通常需要将它设置为一个服务，这主要基于以下两点考虑。

- 简化 MySQL 的管理。将 MySQL 设置为服务，就可通过 `service mysqld start|stop` 等通用的服务管理命令来开启或关闭 MySQL 服务。当线上出现故障，需要重启 MySQL 服务时，即便新手也能胜任。如果直接使用 mysqld_safe 或 mysqld 来管理，有两个弊端：启动时，对于不了解环境的人来说，参数文件位置的确定可能是一个难题；关闭时，对于不熟悉 MySQL 的人来说，有使用 kill -9 的风险。
- 可设置为开机自启动。避免服务器意外重启后，没有自动拉起 MySQL 服务。

当将 MySQL 作为服务来管理时，常用的管理方式有以下两种：

- 服务管理脚本（/etc/init.d/mysqld）；
- systemd。

无论是 /etc/init.d/mysqld 还是 systemd，都可使用 service 来管理服务：

```
service mysqld {start|stop|restart|status}
```

下面依次来看看如何使用 /etc/init.d/mysqld 和 systemd 来管理 MySQL 服务。

1.3.1 使用/etc/init.d/mysqld 管理 MySQL 服务

MySQL 安装包通常会提供 MySQL 服务脚本。以二进制包为例，服务脚本位于二进制包的 support-files 目录下：

```
# cd /usr/local/mysql
# ls support-files/
mysqld_multi.server  mysql-log-rotate  mysql.server
```

这里的 mysql.server 和 mysqld_multi.server 都是 MySQL 服务管理脚本，只不过前者用来管理单实例，而后者用来管理多实例。下面会重点讲解 mysql.server 及为什么不推荐使用 mysqld_multi.server 来管理多实例。

使用 mysql.server 管理单实例

使用 mysql.server 管理单实例的方法很简单，只需复制 mysql.server 到/etc/init.d/目录下：

```
# cp support-files/mysql.server /etc/init.d/mysqld
```

然后启动和关闭实例：

```
# /etc/init.d/mysqld start
Starting MySQL.. SUCCESS!

# /etc/init.d/mysqld stop
Shutting down MySQL. SUCCESS!
```

看上去一切运行正常。假如此时需要在这台主机上部署多个实例，为了让配置文件更有区分度，我们对它重命名。接着，再次启动实例：

```
# mv /etc/my.cnf /etc/my_3306.cnf
# /etc/init.d/mysqld start
Starting MySQL.Logging to '/usr/local/mysql/data/slowtech.err'.
 ERROR! The server quit without updating PID file (/usr/local/mysql/data/slowtech.pid).
```

结果就出现了大名鼎鼎的 The server quit without updating PID file 错误。

下面从代码的角度分析出现这个错误的根本原因：

```
wait_for_pid () {
  verb="$1"           # 动作名，created 或 removed
  pid="$2"            # 如果是启动 MySQL，指的是 mysqld_safe 的 pid；如果是关闭 MySQL，
                      # 则指的是 mysqld 的 pid
  pid_file_path="$3"  # pid 文件的路径

  i=0
  avoid_race_condition="by checking again"

  while test $i -ne $service_startup_timeout ; do

    case "$verb" in
      'created')
        # 如果 pid 文件存在，则意味着实例启动成功
        test -s "$pid_file_path" && i='' && break
        ;;
      'removed')
```

```
              # 如果pid文件不存在，则意味着实例关闭成功
              test ! -s "$pid_file_path" && i='' && break
              ;;
          *)
              echo "wait_for_pid () usage: wait_for_pid created|removed pid pid_file_path"
              exit 1
              ;;
      esac

      # 判断$pid是否不为空
if test -n "$pid"; then
# kill -0 不会发送任何信号，常用于检查一个进程是否存在，或当前用户是否有权限对该进程执行kill操作
      if kill -0 "$pid" 2>/dev/null; then
          :  # 继续循环判断
      else
# 在启动实例的时候，如果发现mysqld_safe进程不存在，则会通过avoid_race_condition进行二次判断
# 在关闭实例的时候，mysqld进程有可能是在检查pid文件之后、kill -0之前退出的，引入
# avoid_race_condition也是为了进行二次判断
          if test -n "$avoid_race_condition"; then
            avoid_race_condition=""
            continue
          fi
          log_failure_msg "The server quit without updating PID file ($pid_file_path)."
          return 1
      fi
  fi

  echo $echo_n ".$echo_c"
  i=`expr $i + 1`
  sleep 1

done

if test -z "$i" ; then
  log_success_msg
  return 0
else
  log_failure_msg
  return 1
fi
}
```

这个函数较为关键，无论是启动还是关闭MySQL服务，都会调用这个函数。

以启动为例，在执行完mysqld_safe命令后，就会调用该函数，具体代码如下：

```
$bindir/mysqld_safe --datadir="$datadir" --pid-file="$mysqld_pid_file_path" $other_args >/dev/null &
wait_for_pid created "$!" "$mysqld_pid_file_path"; return_value=$?
```

在调用wait_for_pid时，传入了3个参数。

- 参数1：动作名。因为是启动MySQL服务，所以传入的是created。如果是关闭MySQL服务，则会相应地传入removed。
- 参数2："$!"。在shell中用于获取最后运行的后台进程的pid，本例中是mysqld_safe进程的pid。
- 参数3："$mysqld_pid_file_path"。mysqld的pid文件。

下面看看 wait_for_pid 具体的处理逻辑：因为传入的是 created，所以首先会判断传入的 pid 文件是否存在。

- 如果存在，则意味着实例启动成功。此时会将变量 i 设置为空，并退出 while 循环。
- 如果不存在，则会继续判断 mysqld_safe 进程的 pid 是否不为空。如果不为空，则意味着在执行完 mysqld_safe 后，已经捕捉到了该进程的 pid。在这种情况下，会进一步通过 kill -0 判断 mysqld_safe 进程是否存在。
  - 如果存在，会将变量 i 加 1，并执行 sleep 1s 命令，继续 while 循环，从头开始判断。之所以要循环，是考虑到 mysqld 进程的启动需要一定的时间。
  - 如果不存在，则意味着 mysqld_safe 虽然启动了，但没有启动成功。这时会将变量 avoid_race_condition 设置为空，进行二次判断。如果判断的结果依旧，则会输出 The server quit without updating PID file 错误。

从代码的逻辑来看，出现 The server quit without updating PID file 错误有两个先决条件。

- mysqld 的 pid 文件不存在，即 mysqld 没有启动成功。
- mysqld_safe 虽然启动了，但没有启动成功。

这个时候就只能通过 MySQL 的错误日志来定位和分析了。

我们分析 mysql.server 的代码，发现如果不做任何修改，它只适用于以下这种场景。

- basedir 为/usr/local/mysql。
- 默认路径存在配置文件，具体来说，配置文件在以下 4 个位置上：
  - /etc/my.cnf
  - /etc/mysql/my.cnf
  - /usr/local/mysql/etc/my.cnf
  - ~/.my.cnf

如果要适配其他场景，必须修改代码。此时需要修改的地方有 4 处，具体如下。

- 显式设置 basedir 和指定配置文件。调整前的代码：

```
basedir=
datadir=
```

因为 basedir 和 datadir 没有设置，所以默认是/usr/local/mysql 和/usr/local/mysql/data。

调整后的代码：

```
basedir='/usr/local/mysql'
datadir=
config_file='/etc/my_3306.cnf'
```

这里新增了一个配置 config_file，它是 mysqld 的配置文件。之所以没有设置 datadir，是因为它可以从配置文件中获取。

- 调整 `parse_server_arguments` 函数。调整前的代码：

```
--pid-file=*) mysqld_pid_file_path=`echo "$arg" | sed -e 's/^[^=]*=//'` ;;
```

这段代码只能识别配置文件中的 pid-file。

调整后的代码：

```
--pid-file=*|--pid_file=*) mysqld_pid_file_path=`echo "$arg" | sed -e 's/^[^=]*=//'` ;;
```

除了 pid-file，这段代码还能识别配置文件中的 pid_file。

为什么 pid 文件的正确识别如此重要呢？因为 pid 文件记录了 mysqld 进程的 pid。在启动 MySQL 服务时，会将 pid 文件的存在作为 MySQL 服务启动成功的标志。在关闭 MySQL 服务时，会首先读取 pid 文件中的 pid，然后通过 kill $mysqld_pid 命令关闭 MySQL 服务。

既然 pid 文件如此重要，建议在配置文件中显式设置 pid_file，否则服务管理脚本默认会将 $datadir/`hostname`.pid 作为 mysqld 的 pid 文件。

- 为 my_print_defaults 指定配置文件。调整前的代码：

```
extra_args=""
if test -r "$basedir/my.cnf"
then
  extra_args="-e $basedir/my.cnf"
fi

parse_server_arguments `$print_defaults $extra_args mysqld server mysql_server mysql.server`
```

这里的 $print_defaults 其实是程序。my_print_defaults 会打印配置文件中指定区域的配置。看看下面这个示例：

```
# my_print_defaults --defaults-file=/etc/my_3306.cnf mysqld
--basedir=/usr/local/mysql
--datadir=/data/mysql/3306/data
--user=mysql
--port=3306
--socket=/data/mysql/3306/data/mysql.sock
--log_error=/data/mysql/3306/data/mysqld.err
--pid_file=/data/mysql/3306/data/mysqld.pid
--log_timestamps=system
```

输出的都是 /etc/my_3306.cnf 文件中 mysqld 部分的配置。

回到调整前的代码，因为没有指定 --defaults-file，所以 my_print_defaults 会默认读取以下 4 个文件的配置：

- /etc/my.cnf
- /etc/mysql/my.cnf
- /usr/local/mysql/etc/my.cnf
- ~/.my.cnf

除此之外，因为指定了 extra_args，所以还会读取 "$basedir/my.cnf" 中的配置，难免会造成干扰。

调整后的代码:

```
parse_server_arguments `$print_defaults --defaults-file="$config_file" mysqld server mysql_server mysql.server`
```

设置--defaults-file，让 my_print_defaults 只能读取 config_file 中的配置。

- 在启动 mysqld_safe 时，指定配置文件。调整前的代码:

```
$bindir/mysqld_safe --datadir="$datadir" --pid-file="$mysqld_pid_file_path" $other_args >/dev/null &
```

同 my_print_defaults 一样，会默认读取/etc/my.cnf 等 4 个文件的配置，同样会造成干扰。

调整后的代码:

```
$bindir/mysqld_safe --defaults-file="$config_file" --pid-file="$mysqld_pid_file_path" $other_args >/dev/null &
```

如此修改后，该脚本基本能覆盖各种个性化场景。修改后的脚本（mysqld_for_single_instance）已上传到了 GitHub 上（见仓库 slowtech/mysql）。如果未做特殊说明，后续其他脚本都会上传到这个仓库下。

mysql.server 主要用来管理单实例。对于多实例的管理，虽然官方提供了 mysqld_multi.server，但不推荐使用，因为它有以下 3 点不足。

- mysqld_multi 只是一个服务管理脚本，不能设置开机自启动。
- 不能针对单个实例指定--defaults-file，不仅如此，配置文件中的所有参数都会显示在命令行中。通过 ps -ef | grep mysqld 查看进程信息，可读性非常差，如下所示:

```
# ps -ef | grep mysqld
root       4283     1  0 21:22 pts/0    00:00:00 /bin/sh /usr/local/mysql/bin/mysqld_safe
--basedir=/usr/local/mysql --datadir=/data/mysql/3306/data --user=mysql --port=3306
--socket=/data/mysql/3306/data/mysql.sock --log_error=/data/mysql/3306/data/mysqld.err
--pid_file=/data/mysql/3306/data/mysqld.pid --mysqlx_port=33060
--mysqlx_socket=/data/mysql/3306/data/mysqlx.sock
mysql      4480  4283  0 21:22 pts/0    00:00:00 /usr/local/mysql/bin/mysqld --basedir=/usr/local/mysql
--datadir=/data/mysql/3306/data --plugin-dir=/usr/local/mysql/lib/plugin --user=mysql
--mysqlx-port=33060 --mysqlx-socket=/data/mysql/3306/data/mysqlx.sock
--log-error=/data/mysql/3306/data/mysqld.err --pid-file=/data/mysql/3306/data/mysqld.pid
--socket=/data/mysql/3306/data/mysql.sock --port=3306
```

- 将多个实例的配置放到一个配置文件中，风险还是比较高的。如果删除了这个文件，会影响所有实例。相对而言，我更推荐一个实例一个配置文件的管理方式。

对于第一点不足，就只能一个实例配置一个服务管理脚本。对于后两点不足，如果实例的部署遵循一定的规则（如表 1-3 所示），只需对 mysql.server 进行简单的调整就可用它来管理多实例。这也是我推荐的多实例管理方式。修改后的脚本（mysqld_for_multi_instance）已上传至 GitHub。限于篇幅，这里没有展现 mysqld_multi.server 的使用方式，相关内容放到了我的公众号上，感兴趣的读者可自行阅读。

表 1-3 实例的部署规则

目 录	作 用
/data/mysql/3306/base	basedir
/data/mysql/3306/conf	配置文件目录
/data/mysql/3306/data	数据目录
/data/mysql/3306/log	日志目录

1.3.2 使用 systemd 管理 MySQL 服务

systemd 是 Linux 系统新推出的初始化（Init）系统，用于代替之前的 SysV init。目前，绝大多数的 Linux 发行版已将 systemd 作为默认的初始化系统，包括如下几个发行版。

- Fedora 15 及后续版本。
- OpenSUSE 12.1 及后续版本。
- Red Hat Enterprise Linux 7 及后续版本，包括其衍生版本 CentOS、Oracle Linux 等。
- Ubuntu 15.04 及后续版本。
- Debian 8 及后续版本。

上面提到的 SysV init 指的是 System V 风格的初始化系统。在初始化过程中，它会首先读取 /etc/inittab 文件，获取默认的运行级别。常见的运行级别如下。

- 0：关机。
- 1：单用户模式。
- 3：字符界面的多用户模式。
- 5：图形界面的多用户模式。
- 6：重启。

在确定完运行级别后，开始执行对应级别下的脚本，脚本位于/etc/rc.d/rcX.d/目录下，目录中的 X 代表运行级别。脚本以 S 或者 K 开头。S 即 START，指的是开机时需要启动的服务；K 即 KILL，指的是关机时需要关闭的服务。S 和 K 后面的数字代表服务启动或关闭的顺序。比如：

```
# ll /etc/rc.d/rc3.d/
total 0
lrwxrwxrwx. 1 root root 20 Oct 14  2018 K50netconsole -> ../init.d/netconsole
lrwxrwxrwx  1 root root 17 Feb 22  2019 S10network -> ../init.d/network
```

SysV init 的主要优点是简单，只需编写对应的启动和停止脚本即可对服务进行管理，但其缺点也较为明显，主要有以下两个。

- 脚本串行执行导致服务器的启动时间长。
- 服务管理脚本冗长、复杂。

在这种背景下，systemd 应运而生，其设计初衷是尽可能地让多个进程并行启动，以缩短服务器的启动时间。

需要注意的是，虽然 systemd 已成为大多数 Linux 发行版默认的初始化系统，但这些系统依然向后兼容 SysV init，所以在上述版本中，依然可以通过/etc/init.d/xxxx 来管理服务。

从以下版本开始，通过 rpm 包或者 Debian 包安装的 MySQL，会默认使用 systemd 来管理 MySQL 服务，不再安装 mysqld_safe、/etc/init.d/mysqld（服务管理脚本）、mysqld_multi（多实例管理工具）和 mysqld_multi.server（多实例管理脚本）。

- Red Hat Enterprise Linux 7、Oracle Linux 7 和 CentOS 7。
- SUSE Linux Enterprise Server 12。
- Fedora 27 和 Fedora 28。
- Debian 8。
- Ubuntu 16。

但是二进制安装包还是会带有上述工具及脚本。

下面我们来看看如何使用 systemd 来管理 MySQL 实例。

1. 使用 systemd 管理单实例

使用 system 管理单实例的具体步骤如下。

(1) 编辑配置文件。

```
# vim /etc/my.cnf
[mysqld]
basedir    = /usr/local/mysql
datadir    = /data/mysql/3306/data
user       = mysql
port       = 3306
socket     = /data/mysql/3306/data/mysql.sock
log_error  = /data/mysql/3306/data/mysqld.err
log_timestamps = system
```

(2) 初始化实例。

```
# mkdir -p /data/mysql/3306/data
# /usr/local/mysql/bin/mysqld --defaults-file=/etc/my.cnf --initialize
```

(3) 创建 systemd 服务配置文件。

```
# vim /usr/lib/systemd/system/mysqld.service
[Unit]
Description=MySQL Server
Documentation=man:mysqld(8)
Documentation=http://dev.mysql.com/doc/refman/en/using-systemd.html
After=network.target
After=syslog.target

[Install]
WantedBy=multi-user.target

[Service]
User=mysql
Group=mysql

Type=notify
```

```
TimeoutSec=0

ExecStart=/usr/local/mysql/bin/mysqld --defaults-file=/etc/my.cnf $MYSQLD_OPTS

EnvironmentFile=-/etc/sysconfig/mysql

LimitNOFILE = 65536

Restart=on-failure

RestartPreventExitStatus=1

Environment=MYSQLD_PARENT_PID=1

PrivateTmp=false
```

这里的配置项主要包括 3 部分，其具体含义如下。

- [Unit]
  - Description：服务描述。
  - Documentation：服务文档。
  - After：依赖的服务，只有在依赖的服务启动后才能启动当前的服务。
- [Install]
  - WantedBy：设置依赖关系，multi-user.target 相当于 System V 中的 init 3，即多用户模式。在通过 systemctl enable mysqld 命令激活 mysqld 服务后，会在/etc/systemd/system/multi-user.target.wants/ 目录下创建一个 mysqld 服务的软链接。系统在以 multi-user 模式启动时，会自动启动该目录下的所有服务，相当于设置开机自启动。
- [Service]
  - User：指定服务的运行用户。
  - Group：指定用户的属组。
  - Type：指定服务的启动类型。可设置为 simple、exec、forking、oneshot、dbus、notify 和 idle，默认为 simple。这里重点说说 forking 和 notify 这两种类型。
    - forking：服务进程在启动过程中会调用 fork()创建子进程，当服务启动完毕后，父进程会退出，而子进程将作为主服务进程运行。这也是传统 Unix 服务的经典做法。在这种模式下，强烈建议设置 PIDFile，以便 systemd 识别服务的主进程。在 MySQL 5.7 中，只能将 Type 设置为 forking。
    - notify：服务启动成功后，会调用 sd_notify()通知 systemd。systemd 在收到通知后，会继续启动后续服务。相对于 forking，notify 这种方式更精确，但需要服务代码层的支持。所幸，MySQL 8.0 原生支持 notify 类型。
  - TimeoutSec：超时时长，包括 TimeoutStartSec 和 TimeoutStopSec。
    - TimeoutStartSec：服务启动的超时时长。如果服务在指定的时间内没有启动成功，systemd 会认为服务启动失败，此时会执行关闭操作。

- **TimeoutStopSec**：服务关闭的超时时长。当执行 STOP 操作时，systemd 会立刻发出 SIGTERM 信号来关闭服务进程。如果在指定时间内没有关闭成功，则会发出 SIGKILL 信号强行杀掉服务进程。如果将 TimeoutSec 设置为 0，则会关闭超时检测。
- **ExecStart**：服务的启动命令。注意，mysqld 必须使用绝对路径。
- **EnvironmentFile**：环境变量文件。可在该文件中配置服务启动时的环境变量。在 MySQL 中，该文件可用于设置其他内存分配器，如 tcmalloc 或 jemalloc。
- **Restart**：指定在什么情况下需要重启服务进程。可设置为 no、always、on-success、on-failure、on-abnormal、on-abort 和 on-watchdog，默认为 no，即不会自动重启服务。如果设置为 on-failure，则会在以下场景下重启服务：进程意外退出（返回码为非 0），接收到终止信号（包括 core dump，但不包括 SIGHUP、SIGINT、SIGTERM 和 SIGPIPE 这 4 种信号），操作超时（如服务的 reload 操作），配置的 Watchdog 被触发。
- **RestartPreventExitStatus**：如果程序的返回码或接收的信号在 RestartPreventExitStatus 中定义了，则即便满足了 Restart 中定义的条件，也不会触发程序重启。
- **Environment**：设置环境变量。这里的 MYSQLD_PARENT_PID 指的是 mysqld 父进程的 PID。在 MySQL 8.0 中，支持通过 restart 命令重启 MySQL 实例。在执行 restart 命令时，该进程会用来启动 mysqld。PID 为 1 的进程是 init 进程，它是 Linux 内核启动的第一个进程，是其他用户进程的父进程或祖先进程。在 MySQL 5.7 中，因为不支持 restart 命令，所以可不用设置 Environment。
- **PrivateTmp**：是否使用私有的 tmp 目录。若设置为 true，则在启动 mysqld 服务后，会在/tmp 目录下创建一个 systemd-private-aac808292059474b93e85832f0f1897a-mysqld.service-lMAPsF 之类的目录，并将该目录下的 tmp 目录作为临时目录。比如：

```
# ll /tmp/systemd-private-aac808292059474b93e85832f0f1897a-mysqld.service-lMAPsF/tmp/
total 4
srwxrwxrwx 1 mysql mysql 0 Jun 27 10:43 mysqlx.sock
-rw------- 1 mysql mysql 7 Jun 27 10:43 mysqlx.sock.lock
```

注意，这个服务配置文件只适用于 MySQL 8.0。对于 MySQL 5.7，因为它不支持 notify 这种服务启动类型，所以需调整以下几项配置，其他配置不变（该配置同样适用于 MySQL 8.0）：

```
Type=forking

PIDFile=/data/mysql/3306/data/mysqld.pid

ExecStart=/usr/local/mysql/bin/mysqld --defaults-file=/etc/my.cnf
--pid-file=/data/mysql/3306/data/mysqld.pid --daemonize $MYSQLD_OPTS
```

编辑完配置文件，执行以下命令让配置生效：

```
# systemctl daemon-reload
```

(4) 使用 systemd 管理服务。

启动 mysqld 服务：

```
# systemctl start mysqld
```

关闭 mysqld 服务：

```
# systemctl stop mysqld
```

查看 mysqld 服务的状态：

```
# systemctl status mysqld
```

将 mysqld 服务设置为开机自启动：

```
# systemctl enable mysqld
Created symlink from /etc/systemd/system/multi-user.target.wants/mysqld.service to /usr/lib/systemd/system/mysqld.service.
```

2. 使用 systemd 管理多实例

使用 systemd 管理多实例的具体步骤如下。

(1) 编辑配置文件。

```
# vim /etc/my.cnf
[mysqld@3306]
basedir         = /usr/local/mysql
datadir         = /data/mysql/3306/data
user            = mysql
port            = 3306
socket          = /data/mysql/3306/data/mysql.sock
log_error       = /data/mysql/3306/data/mysqld.err
pid_file        = /data/mysql/3306/data/mysqld.pid
mysqlx_port     = 33060
mysqlx_socket   = /data/mysql/3306/data/mysqlx.sock
log_timestamps  = system

[mysqld@3307]
basedir         = /usr/local/mysql
datadir         = /data/mysql/3307/data
user            = mysql
port            = 3307
socket          = /data/mysql/3307/data/mysql.sock
log_error       = /data/mysql/3307/data/mysqld.err
pid_file        = /data/mysql/3307/data/mysqld.pid
mysqlx_port     = 33070
mysqlx_socket   = /data/mysql/3307/data/mysqlx.sock
log_timestamps  = system
```

注意，组名中的@是 systemd 唯一支持的分隔符。@后面是实例别名，不一定是端口，也可使用其他更有意义的描述，如业务简称。

(2) 初始化实例。

```
# mkdir -p /data/mysql/3306/data/
# mkdir -p /data/mysql/3307/data/
# /usr/local/mysql/bin/mysqld --defaults-file=/etc/my.cnf --defaults-group-suffix=@3306 --initialize
# /usr/local/mysql/bin/mysqld --defaults-file=/etc/my.cnf --defaults-group-suffix=@3307 --initialize
```

(3) 创建 systemd 服务配置文件。

```
# vim /usr/lib/systemd/system/mysqld@.service
[Unit]
```

```
Description=MySQL Server
Documentation=man:mysqld(8)
Documentation=http://dev.mysql.com/doc/refman/en/using-systemd.html
After=network.target
After=syslog.target

[Install]
WantedBy=multi-user.target

[Service]
User=mysql
Group=mysql

Type=notify

TimeoutSec=0

ExecStart=/usr/local/mysql/bin/mysqld --defaults-file=/etc/my.cnf --defaults-group-suffix=@%I
$MYSQLD_OPTS

EnvironmentFile=-/etc/sysconfig/mysql

LimitNOFILE = 65536

Restart=on-failure

RestartPreventExitStatus=1

Environment=MYSQLD_PARENT_PID=1

PrivateTmp=false
```

注意,对于多实例,配置文件名使用的是 mysqld@,不是 mysqld。

编辑完配置文件,执行以下命令让配置生效:

```
# systemctl daemon-reload
```

(4) 使用 systemd 管理服务。

启动 mysqld 服务:

```
# systemctl start mysqld@3306
# systemctl start mysqld@3307
```

关闭 mysqld 服务:

```
# systemctl stop mysqld@3306
# systemctl stop mysqld@3307
```

查看 mysqld 服务的状态:

```
# systemctl status mysqld@3306
# systemctl status mysqld@3307
```

将 mysqld 服务设置为开机自启动:

```
# systemctl enable mysqld@3306
# systemctl enable mysqld@3307
```

最后,我们总结一下本节的内容。

- 通过服务管理脚本管理单实例，可使用官方自带的 mysql.server，不过脚本的默认设置很容易导致 The server quit without updating PID file 错误。如果要适配各种场景，需在 mysql.server 的基础上进行二次开发。
- 通过服务管理脚本管理多实例，可使用官方自带的 mysqld_multi，不过该工具不能为单个实例指定 --defaults-file，而且本身是用 Perl 开发的，所以如果要进行二次开发，还是有一定的门槛和难度的。相对而言，如果实例的部署遵循一定的规则，更推荐使用 mysql.server 来管理多实例。
- 在编写 systemd 服务配置文件时，注意 MySQL 5.7 和 MySQL 8.0 可设置的 Type 是不同的。在 MySQL 5.7 中，Type 只能设置为 forking，而在 MySQL 8.0 中，除了 forking，Type 还能设置为 notify。
- 无论是服务管理脚本还是 systemd，都可使用 service 命令来管理服务。

1.4 本章总结

本章首先简要介绍了 MySQL 的发展历史，并在此基础上重点介绍了 MySQL 5.5、MySQL 5.6、MySQL 5.7 和 MySQL 8.0 的新特性及重要更新。接着介绍了 MySQL 常见的两种安装方式：基于二进制包的安装和基于源码包的安装。很多人喜欢使用源码包来安装，但实际上官方更推荐使用二进制包。

在 MySQL 8.0 中，默认的密码认证插件由之前的 mysql_native_password 升级为 caching_sha2_password，但是很多客户端驱动没有及时更新，这样就会导致在密码认证时，出现 ERROR 2059 (HY000): Authentication plugin 'caching_sha2_password' cannot be loaded 错误。

无论是 mysqld 还是 mysql，在使用的时候，如果没有指定 --defaults-file，都要注意默认路径 4 个配置文件的影响。当然，强烈建议指定 --defaults-file。如果指定了 --defaults-file，它必须是命令行中的第一个参数。

本章最后介绍了 MySQL 服务的两种管理方式：服务管理脚本和 systemd。

重点问题回顾

- mysqld_safe 启动失败，如何定位？
- 如果指定了 --defaults-file，它必须是命令行中的第一个参数，否则会没有效果。
- 二进制包默认不会加载密码复杂度验证插件（validate_password），而 rpm 包会。
- 当 validate_password.policy 为 MEDIUM 时的密码复杂度要求。
- 配置文件的读取顺序。
- ERROR 2059 (HY000): Authentication plugin 'caching_sha2_password' cannot be loaded 错误的原因及解决方法。
- 出现 The server quit without updating PID file 错误的根本原因。
- 为什么不推荐使用 mysqld_multi 管理多实例？
- 在通过 systemd 管理 MySQL 服务时，注意 MySQL 5.7 和 MySQL 8.0 支持的服务启动类型（Type）的区别。

第 2 章

复　　制

MySQL 作为最流行的开源数据库，常年稳居 DB-Engines Ranking（数据库流行度排行榜）第二位（第一位是 Oracle）。个人觉得，支撑它如此出彩的因素有两个：InnoDB 存储引擎和复制。InnoDB 存储引擎支持事务、行级别锁、MVCC。基于此，MySQL 能实现更高程度的并发。而复制可将主库数据同步到另外一个实例，基于此，我们可以实现读写分离和数据灾备。不仅如此，MySQL 的复制协议也比较简单，只要实现了这个协议，就可将数据实时同步到其他组件中，从而在无形中拓展了 MySQL 的生态圈。此外，业界有很多基于 MySQL 复制协议的开源解决方案（如阿里巴巴的 Canal），基本上可以拿来即用，无须再造轮子。

具体来说，复制具有如下优点。

❑ 可横向扩展读能力。

对于读多写少的业务，如果主库负载很高，影响了写操作，那么可考虑读写分离，让从库承担部分读流量。从库越多，可承担的读流量也就越大。而且，新增一个从库，对主库来说，除了同步 binlog 需要部分网络带宽，基本上没有其他开销。从库分担业务的读操作，其实也是在间接提升主库的写性能。当然，如果业务做了读写分离，就要考虑主从延迟的影响。

❑ 很多关键的日常操作可放到从库上执行，如备份、数据分析等。

为什么不建议在主库上执行这些操作呢？首先看看数据分析等 OLAP 请求。因为会扫描大量数据，所以这些请求一方面会对磁盘的 I/O 造成较大的压力，另一方面会将业务的热点数据驱逐出缓冲池，进而影响主库性能的稳定性。再来说说备份。备份一方面是个 I/O 密集型操作，另一方面会加全局读锁，而在加锁期间，整个实例会处于只读状态，会阻塞主库的所有写操作，进而影响服务的可用性。

❑ 便于数据灾备。

通过部署从库，当主库发生故障时，利用高可用组件，业务流量可以很快地切换到从库上，进而保证数据库服务的可用性。这里的从库，既可在本地部署，也可跨机房、跨地区部署，没有距离限制，视业务的特性及对可用性的要求而定。MySQL 默认的复制方式是异步复制，对可用性要求高的业务可以考虑采用半同步复制或组复制。只不过半同步复制和组复制较适用于低时延的网络环境。

❑ 有丰富的生态圈。

很多数据库产品中的数据只能在本系统中流转，很难实时地（通常通过批处理）同步到其他产品中。反观 MySQL，其复制协议简单，而且只要实现了复制协议，就能将 MySQL 中的数据实时同步到其他产品（如 ES、ClickHouse、Hadoop 等）中，这样就可以充分利用这些产品的核心能力为业务赋能。从这个角度来看，MySQL 其实是一个数据分发商，而不是数据孤岛。

本章主要包括以下内容。

❑ 复制的原理及搭建。
❑ GTID 复制。
❑ 半同步复制。
❑ 并行复制。
❑ 多源复制。
❑ 延迟复制。

2.1 复制的原理及搭建

复制是基于 binlog 实现的。熟悉 Oracle 的读者可能会问：MySQL 不是有 redo log 吗？为什么不直接基于 redo log 来实现复制呢？

实际上，redo log 是 InnoDB 存储引擎所独有的。作为一个支持可插拔存储引擎的数据库，MySQL 不仅支持 InnoDB，还支持 MyISAM、CSV、Memory 等非事务型存储引擎，对这些存储引擎的操作同样需要持久化到 binlog 中。换言之，redo log 是在存储引擎层实现的，而 binlog 则是在 Server 层实现的。

复制的原理如图 2-1 所示。

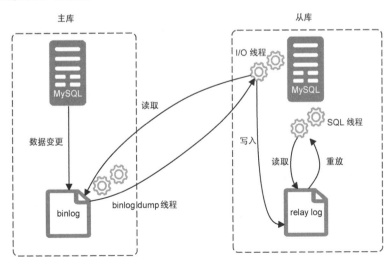

图 2-1　MySQL 复制原理图

复制涉及以下 3 个线程。

- 主库 binlog dump 线程。
- 从库 I/O 线程。
- 从库 SQL 线程。

复制的大致流程如下。

(1) 从库执行完 START SLAVE 命令后，会创建两个线程：I/O 线程和 SQL 线程。
(2) I/O 线程会建立一个到主库的连接，相应地，主库会创建一个 binlog dump 线程来响应这个连接的请求。此时，对于主库来说，从库的 I/O 线程就是一个普通的客户端。
(3) I/O 线程首先告诉主库应该从何处开始发送二进制日志事件。
(4) 主库的 binlog dump 线程开始从指定位置点读取二进制日志事件，并发送给 I/O 线程。
(5) I/O 线程接收到二进制日志事件后，会将其写入 relay log。
(6) SQL 线程读取 relay log 中的二进制日志事件，然后进行重放。

2.1.1　复制的搭建

演示版本为 MySQL 8.0.25，搭建环境如表 2-1 所示。

表 2-1　搭建环境

角　色	IP
主库	192.168.244.10
从库	192.168.244.20

下面介绍复制搭建的步骤。

(1) 编辑配置文件。

主库：

```
[mysqld]
log-bin  = mysql-bin
server-id = 1
```

从库：

```
[mysqld]
server-id = 2
```

这里给出的只是搭建复制的最简参数：

- `log-bin` 开启 binlog，这里的 `mysql-bin` 是 binlog 文件的前缀；
- `server-id` 是服务端 ID，在一个复制组内必须全局唯一，有效值为 $1 \sim 2^{32}-1$。

在设置时，注意以下两点。

- `server-id` 支持在线调整，但开启 binlog 需重启实例。

- 在 MySQL 8.0 之前，binlog 是默认关闭的，不显式设置 log-bin 则不会开启 binlog。而在 MySQL 8.0 中，binlog 是默认开启的，如果要关闭 binlog，必须显式设置 skip_log_bin 或 disable_log_bin。

(2) 在主库上创建复制用户。

```
mysql> create user 'repl'@'192.168.244.20' identified by 'repl123';
mysql> grant replication slave on *.* to 'repl'@'192.168.244.20';
```

在 MySQL 8.0 之前，直接执行第二条命令会隐式创建用户，但官方不推荐这种方式。

为了验证创建的复制用户是否有效及主从库之间的网络是否畅通，可在从库上进行登录测试。

```
# mysql -h 192.168.244.10 -urepl -prepl123
```

(3) 获取主库的备份。

为了方便起见，这里直接使用 mysqldump 进行备份。备份集通过 scp 命令远程拷贝到从库上。

```
# mysqldump -S /data/mysql/3306/data/mysql.sock --single-transaction --master-data=2 -E -R --triggers -A > full_backup.sql
# scp full_backup.sql 192.168.244.20:/backup
```

备份命令中各参数的具体含义可参考第 5 章。从 MySQL 8.0.26 开始，建议使用 --source-data，而不是 --master-data，否则备份文件中会写入以下信息，影响后续的导入。

```
WARNING: --master-data is deprecated and will be removed in a future version. Use --source-data instead.
```

(4) 基于主库的备份恢复从库。

```
# mysql -S /data/mysql/3306/data/mysql.sock < /backup/full_backup.sql
```

(5) 建立主从复制。

在这里，要执行 CHANGE MASTER TO 命令，具体用法如下：

```
change master to
    master_host='master_host_name',
    master_port=master_port,
    master_user='replication_user_name',
    master_password='replication_password',
    master_log_file='recorded_log_file_name',
    master_log_pos=recorded_log_position;
```

下面看看各配置项的具体含义。

- master_host：主库的主机信息，可指定为主机名或 IP。
- master_port：主库端口，若不指定，则默认为 3306。
- master_user：复制用户。
- master_password：密码。
- master_log_file 和 master_log_pos：从库 I/O 线程启动时，应该从主库的哪个 binlog（由 master_log_file 确定）的哪个位置（由 master_log_pos 确定）开始读取二进制日志事件。

前 4 个配置项很好确定，后 2 个配置项又该如何填写呢？

事实上，在用 mysqldump 进行备份时，因为指定了 --master-data=2，所以它会将备份时的 binlog

位置点信息记录到备份文件中。这个位置点信息可通过 grep 命令获得，如下所示：

```
# grep -m 1 "CHANGE MASTER TO" full_backup.sql
-- CHANGE MASTER TO MASTER_LOG_FILE='mysql-bin.000002', MASTER_LOG_POS=38185149;
```

命令行中的 -m 1 指的是只输出第一个匹配项。

基于上面的位置点信息，在从库中执行如下命令：

```
change master to
    master_host='192.168.244.10',
    master_user='repl',
    master_password='repl123',
    master_log_file='mysql-bin.000002',
    master_log_pos=38185149;
```

在 MySQL 8.0 中，还需加上 get_master_public_key=1，否则在开启复制后，会提示以下错误：

```
Last_IO_Error: error connecting to master 'repl@192.168.244.10:3306' - retry-time: 60 retries: 1
message: Authentication plugin 'caching_sha2_password' reported error: Authentication requires secure
connection.
```

究其原因，是创建的复制用户（repl）使用了 caching_sha2_password（MySQL 8.0 默认的密码认证插件）。在主从复制中，如果复制用户使用了这个插件，则要么开启 SSL 通信，要么支持使用 RSA 密钥对进行密码交换。

这里的 get_master_public_key=1 允许从库直接从主库获取公钥。

当从库执行完 CHANGE MASTER TO 命令后，复制的相关信息会保存在两张表中：mysql.slave_master_info 和 mysql.slave_relay_log_info。mysql.slave_master_info 主要记录了两类信息：(1) 主库的连接信息，包括主库的 IP、端口、复制用户和密码；(2) I/O 线程读取的主库 binlog 的位置点信息。注意，这个位置点不是实时更新的，它的更新频率与 sync_master_info 参数有关。该参数默认为 10000，即 I/O 线程每写入 10 000 个事务会更新 mysql.slave_master_info 一次。mysql.slave_relay_log_info 主要记录了 SQL 线程重放 relay log 的位置点信息。事务每次提交时都会更新 mysql.slave_relay_log_info。这两张表是 MySQL 5.6 引入的。在此之前，复制的相关信息保存在 master.info（对应 mysql.slave_master_info 表）和 relay-log.info（对应 mysql.slave_relay_log_info 表）文件中，这两个文件默认位于从库的数据目录下。复制的相关信息是保存在文件还是系统表中由 master_info_repository 参数决定，其中，FILE 代表文件，TABLE 代表系统表。从 MySQL 8.0.2 开始，该参数的默认值由 FILE 调整为 TABLE。

(6) 开启主从复制。

在从库上执行如下命令：

```
mysql> start slave;
```

(7) 查看主从复制的状态。

```
mysql> show slave status\G
*************************** 1. row ***************************
               Slave_IO_State: Waiting for master to send event
                  Master_Host: 192.168.244.10
                  Master_User: repl
                  Master_Port: 3306
```

```
              Connect_Retry: 60
            Master_Log_File: mysql-bin.000003
        Read_Master_Log_Pos: 156
             Relay_Log_File: node2-relay-bin.000004
              Relay_Log_Pos: 371
      Relay_Master_Log_File: mysql-bin.000003
           Slave_IO_Running: Yes
          Slave_SQL_Running: Yes
            Replicate_Do_DB:
        Replicate_Ignore_DB:
         Replicate_Do_Table:
     Replicate_Ignore_Table:
    Replicate_Wild_Do_Table:
Replicate_Wild_Ignore_Table:
                 Last_Errno: 0
                 Last_Error:
               Skip_Counter: 0
        Exec_Master_Log_Pos: 156
            Relay_Log_Space: 627
            Until_Condition: None
             Until_Log_File:
              Until_Log_Pos: 0
         Master_SSL_Allowed: No
         Master_SSL_CA_File:
         Master_SSL_CA_Path:
            Master_SSL_Cert:
          Master_SSL_Cipher:
             Master_SSL_Key:
      Seconds_Behind_Master: 0
Master_SSL_Verify_Server_Cert: No
              Last_IO_Errno: 0
              Last_IO_Error:
             Last_SQL_Errno: 0
             Last_SQL_Error:
  Replicate_Ignore_Server_Ids:
           Master_Server_Id: 1
                Master_UUID: a9227af1-edb5-11eb-8f68-fa163ecbac93
           Master_Info_File: mysql.slave_master_info
                  SQL_Delay: 0
        SQL_Remaining_Delay: NULL
     Slave_SQL_Running_State: Slave has read all relay log; waiting for more updates
          Master_Retry_Count: 86400
                 Master_Bind:
      Last_IO_Error_Timestamp:
     Last_SQL_Error_Timestamp:
              Master_SSL_Crl:
          Master_SSL_Crlpath:
          Retrieved_Gtid_Set:
           Executed_Gtid_Set:
               Auto_Position: 0
        Replicate_Rewrite_DB:
                Channel_Name:
          Master_TLS_Version:
     Master_public_key_path:
       Get_master_public_key: 1
           Network_Namespace:
1 row in set, 1 warning (0.01 sec)
```

重点关注两列的输出：Slave_IO_Running 和 Slave_SQL_Running，两个均为 Yes 代表主从复制搭建成功。

细心的读者可能会发现，在执行 SHOW SLAVE STATUS 命令时提示了 warning。我们不妨先看看 warning 的内容。

```
mysql> show warnings;
+---------+------+--------------------------------------------------------------------------------------------------+
| Level   | Code | Message                                                                                          |
+---------+------+--------------------------------------------------------------------------------------------------+
| Warning | 1287 | 'SHOW SLAVE STATUS' is deprecated and will be removed in a future release. Please use SHOW REPLICA STATUS instead |
+---------+------+--------------------------------------------------------------------------------------------------+
1 row in set (0.06 sec)
```

提示很明显，SHOW SLAVE STATUS 被弃用（deprecated），取而代之的是 SHOW REPLICA STATUS。

实际上，从 MySQL 8.0.22 开始，START SLAVE、STOP SLAVE、SHOW SLAVE STATUS、SHOW SLAVE HOSTS 和 RESET SLAVE 命令都被弃用，取而代之的是 START REPLICA、STOP REPLICA、SHOW REPLICA STATUS、SHOW REPLICAS 和 RESET REPLICA。

从 MySQL 8.0.23 开始，CHANGE MASTER TO 命令被弃用，取而代之的是 CHANGE REPLICATION SOURCE TO。

从 MySQL 8.0.26 开始，标识符（如系统参数、状态变量）中的 MASTER、SLAVE 和 MTS（multithreaded slave 的缩写）将分别被 SOURCE、REPLICA 和 MTA（multithreaded applier 的缩写）替换，这一点需要注意。

在主库上执行 SHOW PROCESSLIST 命令：

```
mysql> show processlist\G
...
*************************** 2. row ***************************
     Id: 14
   User: repl
   Host: 192.168.244.20:41814
     db: NULL
Command: Binlog Dump
   Time: 154
  State: Master has sent all binlog to slave; waiting for more updates
   Info: NULL
...
```

ID 为 14 的连接对应的即 binlog dump 线程。

在从库上执行 SHOW PROCESSLIST 命令：

```
mysql> show processlist\G
...
*************************** 3. row ***************************
     Id: 11
   User: system user
   Host: connecting host
     db: NULL
Command: Connect
   Time: 250
  State: Waiting for master to send event
   Info: NULL
*************************** 4. row ***************************
     Id: 12
   User: system user
```

```
      Host:
        db: NULL
   Command: Query
      Time: 250
     State: Slave has read all relay log; waiting for more updates
      Info: NULL
4 rows in set (0.00 sec)
```

ID 为 11 和 12 的连接分别对应从库的 I/O 线程和 SQL 线程。

至此，一个简单的异步复制环境搭建完毕。

2.1.2 参考资料

- 官方文档"Setting Up Binary Log File Position Based Replication"。
- 《MariaDB 原理与实现》中有关复制的章节。

2.2 GTID 复制

GTID（global transaction identifier，全局事务 ID）是 MySQL 5.6 引入的新特性，它会为每个事务分配一个唯一的事务 ID。事务 ID 的格式如下：

`source_id:transaction_id`

它由两部分组成，其中

- `source_id` 表示事务是在哪个实例上产生的，通常用实例的 `server_uuid` 来表示。
- `transaction_id` 是事务的序列号，按照事务提交的顺序从 1 开始顺序分配。

这样，可保证 GTID 在整个复制拓扑中都是全局唯一的。

2.2.1 GTID 出现的背景

传统的复制是基于 binlog 位置点的。在搭建的时候，需指定 `master_log_file` 和 `master_log_pos`。这个位置的准确性十分关键，如果弄错了，很容易导致主从数据不一致，甚至主从复制中断。

在一主多从的架构中，在主库出现故障，将一个从库提升为主库后，剩下的从库并不知道要基于哪个位置点与新的主库建立复制关系。毕竟，主库上的同一个操作，在不同从库上对应的 binlog 位置点基本上都不相同。这也是 MHA（MySQL 高可用管理工具）出现的背景，它能自动计算出搭建复制时要指定的位置点。但在更加复杂的复制拓扑（如级联复制）中，即便 MHA 也无能为力。

引入 GTID 的初衷其实就是为了简化复制的日常管理。有了 GTID，无论是搭建复制、拓扑变更，还是高可用切换，都无须再关心具体的 binlog 位置点信息。

另外一个流行的 MySQL 高可用管理工具 Orchestrator 支持复杂的、基于 binlog 位置点的复制拓扑的管理。在实现上，它会定期（默认每 5 秒）向主库注入一个 Pseudo-GTID 事务，该事务只有一个简单的删除视图（DROP VIEW）操作。比如：

```
drop view if exists `_pseudo_gtid_`.`_asc:5dcfce93:00000001:96b185b1e799c12e`
```

视图名其实就是一个很好的 GTID。

2.2.2 GTID 的搭建

GTID 复制与基于 binlog 位置点的传统复制，在搭建上的区别主要体现在以下两方面。

- **参数配置**

  如果要开启 GTID 复制，主从节点必须设置如下参数：

  ```
  gtid-mode=on
  enforce-gtid-consistency=1
  ```

  注意：gtid-mode 是枚举值，不是布尔值，只能设置为 ON 或 OFF，不能设置为 1 或 0。

  在 MySQL 5.6 中，除了上述两个参数，还需要设置以下参数：

  ```
  log-bin=mysql-bin
  log-slave-updates=1
  ```

  否则 MySQL 会启动失败，错误日志中会提示如下信息：

  ```
  [ERROR] --gtid-mode=ON or UPGRADE_STEP_1 or UPGRADE_STEP_2 requires --log-bin and
  --log-slave-updates
  ```

  这也就意味着，在 MySQL 5.6 中，如果要开启 GTID 复制，从库也必须开启 binlog。

- **CHANGE MASTER TO 命令**

  相对于基于位置点的复制，GTID 复制只需设置 master_auto_position = 1，无须指定 master_log_file 和 master_log_pos。

  ```
  change master to
      master_host='master_host_name',
      master_port=master_port,
      master_user='replication_user_name',
      master_password='replication_password',
      master_auto_position = 1;
  ```

2.2.3 GTID 的原理

在讲述原理之前，首先介绍一个概念——GTID 集（gtid_set）。它的语法如下：

```
gtid_set:
    uuid_set [, uuid_set] ...
    | ''

uuid_set:
    uuid:interval[:interval]...

uuid:
    hhhhhhhh-hhhh-hhhh-hhhh-hhhhhhhhhhhh

h:
    [0-9|A-F]
```

```
interval:
    n[-n]
    (n >= 1)
```

从语法结构上可以看到，gtid_set 的基本组成单位是 uuid_set，而 uuid_set 是以区间的形式来表示同一个 uuid 上的连续多个 GTID。例如，a9227af1-edb5-11eb-8f68-fa163ecbac93:1、a9227af1-edb5-11eb-8f68-fa163ecbac93:2、a9227af1-edb5-11eb-8f68-fa163ecbac93:3 这 3 个连续的 GTID，用 uuid_set 表示就是 a9227af1-edb5-11eb-8f68-fa163ecbac93:1-3。相对于简单地罗列多个 GTID，uuid_set 更直观，也更节省存储空间。

下面以一个事务为例来看看 GTID 的原理。

```
mysql> show master status;
+------------------+----------+--------------+------------------+-------------------------------------------+
| File             | Position | Binlog_Do_DB | Binlog_Ignore_DB | Executed_Gtid_Set                         |
+------------------+----------+--------------+------------------+-------------------------------------------+
| mysql-bin.000001 |      562 |              |                  | a9227af1-edb5-11eb-8f68-fa163ecbac93:1-2  |
+------------------+----------+--------------+------------------+-------------------------------------------+
1 row in set (0.00 sec)

mysql> insert into slowtech.t1 values(1);
Query OK, 1 row affected (0.01 sec)

mysql> show master status;
+------------------+----------+--------------+------------------+-------------------------------------------+
| File             | Position | Binlog_Do_DB | Binlog_Ignore_DB | Executed_Gtid_Set                         |
+------------------+----------+--------------+------------------+-------------------------------------------+
| mysql-bin.000001 |      835 |              |                  | a9227af1-edb5-11eb-8f68-fa163ecbac93:1-3  |
+------------------+----------+--------------+------------------+-------------------------------------------+
1 row in set (0.00 sec)
```

在事务提交之前，主库上的 GTID 集是 a620daa6-c54f-11e8-bc4f-000c29f66609:1-2。当事务提交时，MySQL 会为该事务分配一个没有使用的最小序列号作为事务 GTID 的序列号，在本例中是 3。

在将 GTID 事务持久化到 binlog 的过程中，会在事务之前原子性地写入一个 GTID 事件——Gtid_log_event。这一点可从 binlog 中直观地看出来。

```
mysql> show binlog events in 'mysql-bin.000001';
+------------------+-----+----------------+-----------+-------------+-------------------------------------------------------------------+
| Log_name         | Pos | Event_type     | Server_id | End_log_pos | Info                                                              |
+------------------+-----+----------------+-----------+-------------+-------------------------------------------------------------------+
| mysql-bin.000001 |   4 | Format_desc    |         1 |         125 | Server ver: 8.0.25, Binlog ver: 4                                 |
| mysql-bin.000001 | 125 | Previous_gtids |         1 |         156 |                                                                   |
| mysql-bin.000001 | 156 | Gtid           |         1 |         233 | SET @@SESSION.GTID_NEXT= 'a9227af1-edb5-11eb-8f68-fa163ecbac93:1' |
| mysql-bin.000001 | 233 | Query          |         1 |         353 | create database slowtech /* xid=38 */                             |
| mysql-bin.000001 | 353 | Gtid           |         1 |         430 | SET @@SESSION.GTID_NEXT= 'a9227af1-edb5-11eb-8f68-fa163ecbac93:2' |
| mysql-bin.000001 | 430 | Query          |         1 |         562 | create table slowtech.t1(id int primary key) /* xid=39 */         |
| mysql-bin.000001 | 562 | Gtid           |         1 |         641 | SET @@SESSION.GTID_NEXT= 'a9227af1-edb5-11eb-8f68-fa163ecbac93:3' |
| mysql-bin.000001 | 641 | Query          |         1 |         712 | BEGIN                                                             |
| mysql-bin.000001 | 712 | Table_map      |         1 |         764 | table_id: 134 (slowtech.t1)                                       |
| mysql-bin.000001 | 764 | Write_rows     |         1 |         804 | table_id: 134 flags: STMT_END_F                                   |
| mysql-bin.000001 | 804 | Xid            |         1 |         835 | COMMIT /* xid=41 */                                               |
+------------------+-----+----------------+-----------+-------------+-------------------------------------------------------------------+
11 rows in set (0.00 sec)
```

其中，Event_type 指的是二进制日志的事件类型。可以看到，在每个操作之前，都有一个 Event_type 为 Gtid 的事件，这个事件就是 Gtid_log_event。具体到上面的 INSERT 操作，其对应的 GTID 事件是 SET @@SESSION.GTID_NEXT= 'a9227af1-edb5-11eb-8f68-fa163ecbac93:3'。

当从库接收到该事务，准备应用时，它会首先应用该事务对应的 GTID 事件。具体来说，在会话级别设置 GTID_NEXT，并告诉从库不用为接下来的事务分配新的 GTID，直接复用刚设置的 GTID_NEXT 即可。理解这一点很关键，在 GTID 复制中，跳过指定事务利用的就是这个原理。

这就是一个 GTID 事务大致的处理流程。

2.2.4　GTID 的相关参数

GTID 的相关参数有 9 个，具体如下。

```
mysql> show variables like '%gtid%';
+----------------------------------+------------------------------------------+
| Variable_name                    | Value                                    |
+----------------------------------+------------------------------------------+
| binlog_gtid_simple_recovery      | ON                                       |
| enforce_gtid_consistency         | ON                                       |
| gtid_executed                    | a9227af1-edb5-11eb-8f68-fa163ecbac93:1-3 |
| gtid_executed_compression_period | 0                                        |
| gtid_mode                        | ON                                       |
| gtid_next                        | AUTOMATIC                                |
| gtid_owned                       |                                          |
| gtid_purged                      |                                          |
| session_track_gtids              | OFF                                      |
+----------------------------------+------------------------------------------+
9 rows in set (0.00 sec)
```

其中，gtid_owned 是只读参数，主要供内部使用。下面看看其他几个参数的具体含义。

- **gtid_mode**

该参数用来表示是否开启 GTID 复制。

在 MySQL 5.7.6 之前，所有的事务可分为两类：GTID 事务和非 GTID 事务。开启 GTID 的关键参数 gtid_mode 和 enforce_gtid_consistency 是只读参数，不能动态修改，这就意味着如果要开启或关闭 GTID 复制，必须重启实例。

从 MySQL 5.7.6 开始，支持在线开关 GTID 复制。相对于之前版本的 GTID 复制，有如下变化。

☐ gtid_mode 和 enforce_gtid_consistency 可在线动态修改。
☐ 所有的事务分为两类：GTID 事务和匿名事务（anonymous transaction）。匿名事务即之前的非 GTID 事务，与之前的版本相比，它会在每个事务之前新增一个 ANONYMOUS_GTID_LOG_EVENT。
☐ gtid_mode 新增了两个选项：OFF_PERMISSIVE 和 ON_PERMISSIVE。在此之前，只有 ON 和 OFF 两个选项。各选项的具体含义如下。

　■ OFF：所有新的事务和复制中的事务都必须是匿名事务。

- OFF_PERMISSIVE：新的事务是匿名事务，复制中的事务既可以是匿名事务，也可以是 GTID 事务。
- ON_PERMISSIVE：新的事务是 GTID 事务，复制中的事务既可以是匿名事务，也可以是 GTID 事务。
- ON：所有新的事务和复制中的事务都必须是 GTID 事务。

同时，对于该参数，只能逐步修改，不能直接修改。如将 OFF 调整为 ON，只能按照 OFF -> OFF_PERMISSIVE -> ON_PERMISSIVE -> ON 的顺序进行修改。如果直接修改，会提示以下错误：

```
mysql> show global variables like 'gtid_mode';
+---------------+-------+
| Variable_name | Value |
+---------------+-------+
| gtid_mode     | OFF   |
+---------------+-------+
1 row in set (0.31 sec)

mysql> set global gtid_mode = ON;
ERROR 1788 (HY000): The value of @@GLOBAL.GTID_MODE can only be changed one step at a time: OFF <-> OFF_PERMISSIVE <-> ON_PERMISSIVE <-> ON. Also note that this value must be stepped up or down simultaneously on all servers. See the Manual for instructions.
```

- **enforce_gtid_consistency**

在 GTID 复制中，为了保证主从数据的一致性，会存在以下限制。

☐ 不支持 CREATE TABLE ... SELECT ... 语句。

在基于 ROW 格式的复制中，对于此语句，会在 binlog 中记录两个事件：一是针对 CREATE TABLE 操作的 QUERY_EVENT，二是针对 INSERT 操作的 WRITE_ROWS_EVENT。这样就会导致两个 EVENT 对应同一个 GTID，在从库应用的时候，这会导致 INSERT 操作被忽略。

针对这个限制，可将该语句拆成两步来执行：先通过 CREATE TABLE 建表，再通过 INSERT INTO ... SELECT ... 插入数据。

从 MySQL 8.0.21 开始，该限制被取消，CREATE TABLE ... SELECT ... 会被当作一个原子 DDL 来处理。

☐ 不允许在事务、存储过程、函数、触发器中执行 CREATE TEMPORARY TABLE 或 DROP TEMPORARY TABLE 操作。

从 MySQL 8.0.13 开始，如果 binlog 格式为 ROW 或 MIXED，则无此限制。

☐ 不能将 InnoDB 表和非 InnoDB 表放到一个事务内操作。

参数 enforce_gtid_consistency 可用来检测操作是否违反了上述限制，其有如下取值。

☐ OFF：关闭检测。
☐ ON：开启检测。如果操作违反了上述限制，会直接报错。
☐ WARN：开启检测。如果操作违反了上述限制，不会报错，只是会在错误日志中打印 warning。

注意，如果要将 gtid_mode 设置为 ON，则 enforce_gtid_consistency 的取值必须为 ON。

- **gtid_executed**

该参数用来表示 MySQL 已经执行过的 GTID 集。它是一个内存值，并且实时更新。SHOW MASTER STATUS 和 SHOW SLAVE STATUS 中的 Executed_Gtid_Set 取的就是该值。因为是内存值，所以在实例关闭的时候，它会被清除。在实例启动的时候，它会被重新初始化，初始化的逻辑与 binlog_gtid_simple_recovery 参数有关，这一点后面会具体提到。注意，执行 RESET MASTER 命令会清空 gtid_executed。

- **gtid_purged**

该参数用来表示 MySQL 已经执行过但在 binlog 中不存在的 GTID 集。gtid_purged 是 gtid_executed 的子集，它通常来源于以下 3 个场景。

❏ 从库没有开启 binlog。此时 gtid_purged 等于 gtid_executed，当 gtid_executed 发生变化的时候，gtid_purged 也会随之改变。

❏ 被删除（purge）的 binlog 中的 GTID 集。这里的删除既包括手动删除，即通过 PURGE BINARY LOGS 命令删除 binlog，也包括自动删除，即 binlog 因过期（过期时间由 expire_logs_days 或 binlog_expire_logs_seconds 参数控制）被 MySQL 自动删除。

```
mysql> show global variables where variable_name in ('gtid_executed','gtid_purged');
+----------------+----------------------------------------+
| Variable_name  | Value                                  |
+----------------+----------------------------------------+
| gtid_executed  | a9227af1-edb5-11eb-8f68-fa163ecbac93:1-3 |
| gtid_purged    |                                        |
+----------------+----------------------------------------+
2 rows in set (0.00 sec)

mysql> show master status;
+------------------+----------+--------------+------------------+------------------------------------------+
| File             | Position | Binlog_Do_DB | Binlog_Ignore_DB | Executed_Gtid_Set                        |
+------------------+----------+--------------+------------------+------------------------------------------+
| mysql-bin.000001 |      835 |              |                  | a9227af1-edb5-11eb-8f68-fa163ecbac93:1-3 |
+------------------+----------+--------------+------------------+------------------------------------------+
1 row in set (0.00 sec)

mysql> flush logs;
Query OK, 0 rows affected (0.07 sec)

mysql> purge binary logs to 'mysql-bin.000002';
Query OK, 0 rows affected (0.01 sec)

mysql> show global variables where variable_name in ('gtid_executed','gtid_purged');
+----------------+------------------------------------------+
| Variable_name  | Value                                    |
+----------------+------------------------------------------+
| gtid_executed  | a9227af1-edb5-11eb-8f68-fa163ecbac93:1-3 |
| gtid_purged    | a9227af1-edb5-11eb-8f68-fa163ecbac93:1-3 |
+----------------+------------------------------------------+
2 rows in set (0.00 sec)
```

- 通过 SET @@GLOBAL.GTID_PURGED 命令显式设置。

在 MySQL 8.0 之前，如果要显式设置 gtid_purged 的值，gtid_executed 必须为空。在 MySQL 8.0 中，则无此限制。

同 gtid_executed 一样，gtid_purged 也是内存值，在实例启动的时候，同样需要初始化，初始化的逻辑与 binlog_gtid_simple_recovery 参数有关。

- **binlog_gtid_simple_recovery**

在解释这个参数之前，首先看看两个二进制日志事件：Previous_gtids_log_event 和 Gtid_log_event。Gtid_log_event 在前面介绍过，它是事务对应的 GTID 事件。Previous_gtids_log_event 代表之前已经分配的 GTID 集，这个 GTID 集等于上一个 binlog 的 Previous_gtids_log_event 中的 GTID 加上上一个 binlog 中所有事务的 GTID。对于 MySQL 中的第一个 binlog，其 Previous_gtids_log_event 会为空。看看下面这个示例：

```
mysql> show binlog events in 'mysql-bin.000001';
+------------------+-----+----------------+-----------+-------------+----------------------------------------------------------------------+
| Log_name         | Pos | Event_type     | Server_id | End_log_pos | Info                                                                 |
+------------------+-----+----------------+-----------+-------------+----------------------------------------------------------------------+
| mysql-bin.000001 |   4 | Format_desc    |         1 |         125 | Server ver: 8.0.25, Binlog ver: 4                                    |
| mysql-bin.000001 | 125 | Previous_gtids |         1 |         156 |                                                                      |
| mysql-bin.000001 | 156 | Gtid           |         1 |         233 | SET @@SESSION.GTID_NEXT= 'a9227af1-edb5-11eb-8f68-fa163ecbac93:1'     |
| mysql-bin.000001 | 233 | Query          |         1 |         353 | create database slowtech /* xid=82 */                                |
+------------------+-----+----------------+-----------+-------------+----------------------------------------------------------------------+
4 rows in set (0.00 sec)

mysql> flush logs;
Query OK, 0 rows affected (0.03 sec)

mysql> show binlog events in 'mysql-bin.000002';
+------------------+-----+----------------+-----------+-------------+----------------------------------------------------+
| Log_name         | Pos | Event_type     | Server_id | End_log_pos | Info                                               |
+------------------+-----+----------------+-----------+-------------+----------------------------------------------------+
| mysql-bin.000002 |   4 | Format_desc    |         1 |         125 | Server ver: 8.0.25, Binlog ver: 4                  |
| mysql-bin.000002 | 125 | Previous_gtids |         1 |         196 | a9227af1-edb5-11eb-8f68-fa163ecbac93:1             |
+------------------+-----+----------------+-----------+-------------+----------------------------------------------------+
2 rows in set (0.00 sec)
```

结合这个示例，我们看看 binlog_gtid_simple_recovery 的具体处理逻辑。

若 binlog_gtid_simple_recovery 为 FALSE，对于 gtid_executed 的初始化，MySQL 会从最新的一个 binlog 开始扫描，判断其是否包含 Previous_gtids_log_event。

- 如果包含，则退出扫描，并将该 binlog 中的 Previous_gtids_log_event 中的 GTID 加上该 binlog 中所有事务的 GTID 赋值给 MySQL 内部变量 gtids_in_binlog。
- 如果不包含，则继续扫描上一个 binlog。

gtid_executed 取 gtids_in_binlog 和 mysql.gtid_executed 表中 GTID 的并集。需要注意的是，mysql.gtid_executed 是 MySQL 5.7 引入的。所以，在 MySQL 5.6 中，gtid_executed 就等于 gtids_in_binlog。

gtid_purged 的初始化逻辑与 gtid_executed 的相反，它从最旧的一个 binlog 开始扫描，退出扫描的依据是 Previous_gtids_log_event 不为空，或者 Previous_gtids_log_event 为空但 binlog（实际上就是第一个 binlog）中存在 GTID 事务。在退出扫描后，读取该 binlog 中的 Previous_gtids_log_event。

gtid_purged 取自 gtid_executed -（gtids_in_binlog - Previous_gtids_log_event）。在开启 binlog 的情况下，gtid_executed 等于 gtids_in_binlog，所以在这种情况下，gtid_purged 就等于 Previous_gtids_log_event。如果是从库且没有开启 binlog，gtids_in_binlog 和 Previous_gtids_log_event 会为空，这个时候，gtid_purged 就等于 mysql.gtid_executed 表中的 GTID。

若 binlog_gtid_simple_recovery 为 TRUE，无论是 gtid_executed 还是 gtid_purged 的初始化，MySQL 都只会读取一个 binlog。具体来说，gtid_executed 会读取最新的 binlog，而 gtid_purged 则读取最旧的 binlog。从 MySQL 5.7.7 开始，该参数默认为 TRUE。

- **gtid_executed_compression_period**

这个参数与 mysql.gtid_executed 表有关。该表是 MySQL 5.7 引入的，引入的初衷是为了消除在 MySQL 5.6 中使用 GTID 必须开启 binlog 的限制。很多时候，对于不承担故障切换任务的从库，我们并不想开启 binlog，毕竟不开启 binlog 的好处显而易见：既能节省磁盘空间，又能减轻从库服务器的 I/O 压力。

下面看看开启和不开启 binlog 对 mysql.gtid_executed 的影响。

- 若开启，那么因为 binlog 自身就保存了 Previous_gtids_log_event 和 Gtid_log_event 信息，所以无须通过 mysql.gtid_executed 来持久化 GTID 信息。对 mysql.gtid_executed 表的写入，只发生在日志切换或实例关闭时。这一点既适用于主库，也适用于从库。不过从 MySQL 8.0.17 开始，因为克隆插件的引入，即使开启了 binlog，GTID 信息也还是会实时写入 mysql.gtid_executed 表。
- 若不开启，GTID 信息则会实时写入 mysql.gtid_executed 表。这一点只适用于从库。对于主库，如果没有开启 binlog（当然，没开启 binlog，其实不能称之为主库），那么即使开启了 GTID，也不会在 mysql.gtid_executed 中记录任何 GTID 信息。

既然是实时更新，最快的办法当然就是在 mysql.gtid_executed 中直接插入记录，即一个事务对应一条记录，如下所示。

```
mysql> select * from mysql.gtid_executed;
+--------------------------------------+----------------+--------------+
| source_uuid                          | interval_start | interval_end |
+--------------------------------------+----------------+--------------+
| a9227af1-edb5-11eb-8f68-fa163ecbac93 |              1 |            1 |
| a9227af1-edb5-11eb-8f68-fa163ecbac93 |              2 |            2 |
| a9227af1-edb5-11eb-8f68-fa163ecbac93 |              3 |            3 |
+--------------------------------------+----------------+--------------+
3 rows in set (0.00 sec)
```

但这样会导致 mysql.gtid_executed 表中的记录越来越多。为了节省空间，MySQL 会定期对 mysql.gtid_executed 表进行压缩，压缩的频率由 gtid_executed_compression_period 参数控制。在 MySQL 8.0.23 之前，该参数的默认值为 1000，即每插入 1000 个事务就会对表进行一次压缩。从

MySQL 8.0.23 开始，该参数的默认值为 0，这时会根据需要进行压缩。压缩后，连续的 GTID 会以区间的形式表示，如下所示。

```
mysql> select * from mysql.gtid_executed;
+--------------------------------------+----------------+--------------+
| source_uuid                          | interval_start | interval_end |
+--------------------------------------+----------------+--------------+
| a9227af1-edb5-11eb-8f68-fa163ecbac93 |              1 |         6702 |
| a9227af1-edb5-11eb-8f68-fa163ecbac93 |           6703 |         6726 |
| a9227af1-edb5-11eb-8f68-fa163ecbac93 |           6727 |         6743 |
+--------------------------------------+----------------+--------------+
3 rows in set (0.00 sec)
```

- **gtid_next**

该参数用来设置下一个 GTID 值，但只能在会话级别设置。它有以下取值。

- ❏ `AUTOMATIC`：自动获取下一个 GTID 值。此为默认值。
- ❏ `ANONYMOUS`：匿名事务。
- ❏ `UUID:NUMBER`：显式设置 GTID 值。一般用于主从复制中断后，从库跳过指定事务。

在实际测试过程中发现，如果设置的 `gtid_next` 大于当前的 `gtid_executed`，那么将其恢复为 `AUTOMATIC` 后，新分配的 GTID 会首先填补 gtid_executed 和 gtid_next 之间的缺口，如下所示：

```
mysql> create table slowtech.t1(id int);
Query OK, 0 rows affected (0.13 sec)

mysql> show global variables like 'gtid_executed';
+---------------+----------------------------------------+
| Variable_name | Value                                  |
+---------------+----------------------------------------+
| gtid_executed | a9227af1-edb5-11eb-8f68-fa163ecbac93:1 |
+---------------+----------------------------------------+
1 row in set (0.00 sec)

mysql> set session gtid_next='a9227af1-edb5-11eb-8f68-fa163ecbac93:4';
Query OK, 0 rows affected (0.00 sec)

mysql> insert into slowtech.t1 values(2);
Query OK, 1 row affected (0.01 sec)

mysql> show global variables like 'gtid_executed';
+---------------+------------------------------------------+
| Variable_name | Value                                    |
+---------------+------------------------------------------+
| gtid_executed | a9227af1-edb5-11eb-8f68-fa163ecbac93:1:4 |
+---------------+------------------------------------------+
1 row in set (0.00 sec)

mysql> set session gtid_next='automatic';
Query OK, 0 rows affected (0.00 sec)

mysql> insert into slowtech.t1 values(3);
Query OK, 1 row affected (0.01 sec)

mysql> show global variables like 'gtid_executed';
```

```
+---------------+---------------------------------------------+
| Variable_name | Value                                       |
+---------------+---------------------------------------------+
| gtid_executed | a9227af1-edb5-11eb-8f68-fa163ecbac93:1-2:4 |
+---------------+---------------------------------------------+
1 row in set (0.00 sec)

mysql> insert into slowtech.t1 values(4);
Query OK, 1 row affected (0.01 sec)

mysql> show global variables like 'gtid_executed';
+---------------+-------------------------------------------+
| Variable_name | Value                                     |
+---------------+-------------------------------------------+
| gtid_executed | a9227af1-edb5-11eb-8f68-fa163ecbac93:1-4 |
+---------------+-------------------------------------------+
1 row in set (0.00 sec)
```

- **session_track_gtids**

在事务执行完毕后，该参数决定了是否将 GTID 信息返回给客户端。它有以下取值。

❑ OFF：不返回。默认值。

❑ OWN_GTID：只返回事务对应的 GTID 值。

❑ ALL_GTIDS：返回 gtid_executed 的值。这个值是在事务成功提交后获取的。

有了 GTID 信息，结合 WAIT_FOR_EXECUTED_GTID_SET()函数，可以解决读写分离场景，从从库读到旧数据的问题。

2.2.5 GTID 的相关函数

- **GTID_SUBSET(set1,set2)**

给定两个 GTID 集：set1 和 set2，判断 set1 是否为 set2 的子集。若是，返回 1，否则返回 0。例如：

```
mysql> select GTID_SUBSET('a9227af1-edb5-11eb-8f68-fa163ecbac93:1-3',
    ->      'a9227af1-edb5-11eb-8f68-fa163ecbac93:1-5') as GTID_SUBSET;
+-------------+
| GTID_SUBSET |
+-------------+
|           1 |
+-------------+
1 row in set (0.00 sec)
```

- **GTID_SUBTRACT(set1,set2)**

返回属于 set1 但不属于 set2 的 GTID 集。

```
mysql> select GTID_SUBTRACT('a9227af1-edb5-11eb-8f68-fa163ecbac93:1-3:60-2000',
    ->      'a9227af1-edb5-11eb-8f68-fa163ecbac93:1-2') as GTID_SUBTRACT;
+----------------------------------------------+
| GTID_SUBTRACT                                |
+----------------------------------------------+
| a9227af1-edb5-11eb-8f68-fa163ecbac93:3:60-2000 |
+----------------------------------------------+
1 row in set (0.00 sec)
```

- **WAIT_FOR_EXECUTED_GTID_SET(gtid_set[, timeout])**

在从库上执行，判断给定的 GTID 集对应的操作是否已经执行完毕。如果没有，客户端会一直阻塞。如果不想客户端一直阻塞，也可通过 timeout 参数指定超时时长。

```
mysql> select WAIT_FOR_EXECUTED_GTID_SET('a9227af1-edb5-11eb-8f68-fa163ecbac93:1-3');
+-----------------------------------------------------------------------+
| WAIT_FOR_EXECUTED_GTID_SET('a9227af1-edb5-11eb-8f68-fa163ecbac93:1-3') |
+-----------------------------------------------------------------------+
|                                                                     0 |
+-----------------------------------------------------------------------+
1 row in set (18.71 sec)

mysql> select WAIT_FOR_EXECUTED_GTID_SET('a9227af1-edb5-11eb-8f68-fa163ecbac93:1-4',60);
+--------------------------------------------------------------------------+
| WAIT_FOR_EXECUTED_GTID_SET('a9227af1-edb5-11eb-8f68-fa163ecbac93:1-4',60) |
+--------------------------------------------------------------------------+
|                                                                        1 |
+--------------------------------------------------------------------------+
1 row in set (1 min 0.00 sec)
```

如果指定的 GTID 集对应的操作已执行完毕，则返回 0。如果在指定的时间内还没有执行完毕，则返回 1。

该函数可用来判断主库的某个操作是否已在从库上执行完毕。

2.2.6 在线修改复制模式

在介绍具体的操作之前，来看看表 2-2，该表列出了不同的 gtid_next 在不同的 gtid_mode 下的处理逻辑。了解这背后的原理，有助于我们更好地理解后面的操作。

表 2-2 不同的 gtid_next 在不同的 gtid_mode 下的处理逻辑

gtid_mode	gtid_next			
	AUTOMATIC（开启 binlog）	AUTOMATIC（关闭 binlog）	ANONYMOUS	UUID:NUMBER
OFF	ANONYMOUS	ANONYMOUS	ANONYMOUS	Error
OFF_PERMISSIVE	ANONYMOUS	ANONYMOUS	ANONYMOUS	UUID:NUMBER
ON_PERMISSIVE	New GTID	ANONYMOUS	ANONYMOUS	UUID:NUMBER
ON	New GTID	ANONYMOUS	Error	UUID:NUMBER

从表 2-2 中可以得出以下 3 个结论。

- 当 gtid_next 为 AUTOMATIC 时，
    - 如果关闭了 binlog，则无论 gtid_mode 设置为何值，产生的都是匿名事务。这也解释了为什么对于关闭了 binlog 但 gtid_mode 为 ON 的实例，新增的事务不会在 mysql.gtid_executed 表中插入对应的 GTID 值。
    - 如果开启了 binlog，则只有当 gtid_mode 设置为 ON_PERMISSIVE 或 ON 时，才会为新增的事务分配 GTID 值。
- 当 gtid_next 为 ANONYMOUS 时，无论是否关闭 binlog，新增的事务都是匿名事务。

- 只要 gtid_mode 不为 OFF，无论是否关闭 binlog，都能通过 gtid_next 为事务显式设置 GTID 值。

在线修改复制模式包括在线开启 GTID 复制和在线关闭 GTID 复制，下面我们看看具体的操作步骤。

在线开启 GTID 复制

- 分别在主从库上将 enforce_gtid_consistency 设置为 WARN。

    ```
    mysql> set global enforce_gtid_consistency = WARN;
    ```

    设置完毕，打开错误日志，观察一段时间，看看业务的正常操作中是否有违反 GTID 限制的 SQL。若有，则会在错误日志中提示警告。例如，将 InnoDB 表和 MyISAM 表放到一个事务内操作，错误日志中会出现以下提示：

    ```
    [Warning] [MY-013098] [Server] Statement violates GTID consistency: Updates to non-transactional tables can only be done in either autocommitted statements or single-statement transactions, and never in the same statement as updates to transactional tables.
    ```

    这个时候，就需要业务调整 SQL。

    如果没有问题，则继续下一步。

- 分别在主从库上将 enforce_gtid_consistency 设置为 ON。

    ```
    mysql> set global enforce_gtid_consistency = ON;
    ```

- 分别在主从库上将 gtid_mode 设置为 OFF_PERMISSIVE。

    ```
    mysql> set global gtid_mode = OFF_PERMISSIVE;
    ```

- 分别在主从库上将 gtid_mode 设置为 ON_PERMISSIVE。

    ```
    mysql> set global gtid_mode = ON_PERMISSIVE;
    ```

    设置完毕，则为会新产生的事务分配 GTID 值。

- 分别在主从库上检查状态变量 ONGOING_ANONYMOUS_TRANSACTION_COUNT 的值。

    ```
    mysql> show status like 'ONGOING_ANONYMOUS_TRANSACTION_COUNT';
    ```

    该参数用来统计正在执行的匿名事务的数量，若为 0，则继续下一步。

- 等待从库上匿名事务应用完毕。

    这个时候，可通过 MASTER_POS_WAIT 函数进行判断。具体步骤如下。

    首先，通过 SHOW MASTER STATUS 获取主库当前二进制日志的位置点。

    ```
    mysql> show master status;
    +------------------+----------+--------------+------------------+-------------------------------------------+
    | File             | Position | Binlog_Do_DB | Binlog_Ignore_DB | Executed_Gtid_Set                         |
    +------------------+----------+--------------+------------------+-------------------------------------------+
    | mysql-bin.000015 | 78996220 |              |                  | a9227af1-edb5-11eb-8f68-fa163ecbac93:1-67346 |
    +------------------+----------+--------------+------------------+-------------------------------------------+
    1 row in set (0.00 sec)
    ```

接着，在从库上执行如下操作：

```
mysql> select MASTER_POS_WAIT('mysql-bin.000015',78996220);
```

该命令会一直阻塞，直到指定位置点的事务在从库上执行完毕。

- 分别在主从库上将 gtid_mode 设置为 ON。

  ```
  mysql> set global gtid_mode = ON;
  ```

- 修改主从库的参数文件，加上 GTID 的相关参数。

  ```
  gtid_mode=ON
  enforce_gtid_consistency=ON
  ```

- 将基于 binlog 位置点的复制修改为 GTID 复制。

  ```
  mysql> stop slave;
  mysql> change master to master_auto_position = 1;
  mysql> start slave;
  ```

  如果是在多源复制中开启，还需指定具体的 channel 值。

在线关闭 GTID 复制

- 将从库上基于 GTID 的复制修改为基于 binlog 位置点的复制。

  首先，关闭从库。

  ```
  mysql> stop slave;
  ```

  接着，查看 SHOW SLAVE STATUS 的值。

  ```
  mysql> show slave status\G
  *************************** 1. row ***************************
  ...
         Relay_Master_Log_File: mysql-bin.000006
  ...
            Exec_Master_Log_Pos: 15688006
  ...
  ```

  获取 Relay_Master_Log_File 和 Exec_Master_Log_Pos 的值。

  然后，执行 CHANGE MASTER TO 命令。

  ```
  mysql> change master to master_auto_position = 0, master_log_file = 'mysql-bin.000006',
  master_log_pos = 15688006;
  ```

  最后，开启复制。

  ```
  mysql> start slave;
  ```

- 分别在主从库上将 gtid_mode 设置为 ON_PERMISSIVE。

  ```
  mysql> set global gtid_mode = ON_PERMISSIVE;
  ```

- 分别在主从库上将 gtid_mode 设置为 OFF_PERMISSIVE。

  ```
  mysql> set global gtid_mode = OFF_PERMISSIVE;
  ```

- 分别在主从库上检查 gtid_owned 变量的值。

  ```
  mysql> select @@global.gtid_owned;
  ```

  如果值为空，则继续下一步。

- 等待所有的 GTID 事务应用完毕。

  同样可通过 MASTER_POS_WAIT 函数进行判断，具体步骤可参考在线开启 GTID 复制部分。

- 分别在主从库上将 gtid_mode 设置为 OFF。

  ```
  mysql> set global gtid_mode = OFF;
  ```

- 修改主从库的配置文件，将 GTID 的相关参数显式设置为 OFF。

  ```
  gtid_mode=OFF
  enforce_gtid_consistency=OFF
  ```

2.2.7 设置@@GLOBAL.GTID_PURGED 时的注意事项

在 MySQL 8.0 之前，设置 gtid_purged 的前提是 gtid_executed 必须为空，而要让 gtid_executed 为空，只能执行 RESET MASTER 命令。看看下面这个示例。

```
mysql> show global variables where variable_name in ('gtid_executed','gtid_purged');
+----------------+------------------------------------------+
| Variable_name  | Value                                    |
+----------------+------------------------------------------+
| gtid_executed  | 7d1724b6-f27f-11eb-bed0-fa163ecbac93:1-7 |
| gtid_purged    |                                          |
+----------------+------------------------------------------+
2 rows in set (0.00 sec)

mysql> set global gtid_purged='7d1724b6-f27f-11eb-bed0-fa163ecbac93:1-2';
ERROR 1840 (HY000): @@GLOBAL.GTID_PURGED can only be set when @@GLOBAL.GTID_EXECUTED is empty.

mysql> reset master;
Query OK, 0 rows affected (0.05 sec)

mysql> set global gtid_purged='7d1724b6-f27f-11eb-bed0-fa163ecbac93:1-2';
Query OK, 0 rows affected (0.00 sec)

mysql> show global variables where variable_name in ('gtid_executed','gtid_purged');
+----------------+------------------------------------------+
| Variable_name  | Value                                    |
+----------------+------------------------------------------+
| gtid_executed  | 7d1724b6-f27f-11eb-bed0-fa163ecbac93:1-2 |
| gtid_purged    | 7d1724b6-f27f-11eb-bed0-fa163ecbac93:1-2 |
+----------------+------------------------------------------+
2 rows in set (0.00 sec)
```

MySQL 8.0 移除了这个限制，但需注意以下两点。

- 设置的 gtid_purged 不能包含 gtid_executed 中任何尚未清除的 GTID 集，换言之，设置的 gtid_purged 不能与 gtid_subtract(gtid_executed,gtid_purged)存在交集。看看下面这个示例。

```
mysql> show global variables where variable_name in ('gtid_executed','gtid_purged');
+----------------+------------------------------------------+
| Variable_name  | Value                                    |
+----------------+------------------------------------------+
| gtid_executed  | a9227af1-edb5-11eb-8f68-fa163ecbac93:1-8 |
| gtid_purged    | a9227af1-edb5-11eb-8f68-fa163ecbac93:1-4 |
+----------------+------------------------------------------+
2 rows in set (0.00 sec)

mysql> set global gtid_purged='a9227af1-edb5-11eb-8f68-fa163ecbac93:1-5';
ERROR 3546 (HY000): @@GLOBAL.GTID_PURGED cannot be changed: the added gtid set must not overlap with @@GLOBAL.GTID_EXECUTED
```

- 设置的 gtid_purged 必须是当前 gtid_purged 的超集。

```
mysql> show global variables where variable_name in ('gtid_executed','gtid_purged');
+----------------+------------------------------------------+
| Variable_name  | Value                                    |
+----------------+------------------------------------------+
| gtid_executed  | a9227af1-edb5-11eb-8f68-fa163ecbac93:1-8 |
| gtid_purged    | a9227af1-edb5-11eb-8f68-fa163ecbac93:1-4 |
+----------------+------------------------------------------+
2 rows in set (0.00 sec)

mysql> set global gtid_purged='a9227af1-edb5-11eb-8f68-fa163ecbac93:1-3';
ERROR 3546 (HY000): @@GLOBAL.GTID_PURGED cannot be changed: the new value must be a superset of the old value

mysql> set global gtid_purged='a9227af1-edb5-11eb-8f68-fa163ecbac93:1-4,d7664b79-e0ee-11e8-b1ad-000c2927cfea:1';
Query OK, 0 rows affected (0.00 sec)

mysql> show global variables where variable_name in ('gtid_executed','gtid_purged')\G
*************************** 1. row ***************************
Variable_name: gtid_executed
        Value: a9227af1-edb5-11eb-8f68-fa163ecbac93:1-8,
d7664b79-e0ee-11e8-b1ad-000c2927cfea:1
*************************** 2. row ***************************
Variable_name: gtid_purged
        Value: a9227af1-edb5-11eb-8f68-fa163ecbac93:1-4,
d7664b79-e0ee-11e8-b1ad-000c2927cfea:1
2 rows in set (0.01 sec)
```

可以看到，新增加的 GTID 集同样添加到了 gtid_executed 中。

除此之外，还可通过加号添加新的 GTID 集，如下所示：

```
mysql> set global gtid_purged='+d7664b79-e0ee-11e8-b1ad-000c2927cfea:1'
```

2.2.8 参考资料

- 官方文档 "Replication with Global Transaction Identifiers"。

2.3 半同步复制

从 MySQL 5.5 开始，MySQL 以插件的形式支持半同步复制。如何理解半同步复制呢？我们首先看看异步复制和全同步复制的定义。

- **异步复制（asynchronous replication）**

MySQL 默认的复制方式。主库在执行完客户端提交的事务后，会立刻给客户端反馈。主库并不关心从库是否已经接受或者应用完该事务，这就会带来一个问题：假如主库出现故障了，主库上已经提交的事务很可能没有完全传到从库上，此时如果强行将从库提升为主库，可能会导致新主库上的数据不完整。

- **全同步复制（fully synchronous replication）**

主库执行完一个事务，需要等待所有从库应用完事务才给客户端反馈。这种同步方式最安全，但因为需要等待所有从库应用完，所以对数据库的性能影响相对较大。

- **半同步复制（semisynchronous replication）**

介于异步复制和全同步复制之间。主库在执行完客户端提交的事务后，不会立刻给客户端反馈，而是需要等待至少一个从库接收到（注意，只是接收，不是应用）该事务对应的二进制日志事件。

相对于异步复制，半同步复制无疑提高了数据的安全性，同时相对于全同步复制，它对数据库的性能影响也没有那么大。

2.3.1 事务的两阶段提交协议

两阶段提交协议是为了实现分布式事务而提出的。首先看看维基百科中对于分布式事务的定义：

> 分布式事务是涉及两个或多个主机的数据库事务。通常，主机提供事务资源，而事务管理器则负责创建和管理针对此类资源所有操作的全局事务。与任何其他事务一样，分布式事务必须满足 ACID（原子性、一致性、隔离性、持久性）特性。

简而言之，分布式事务也是事务。所以，它必须满足 ACID 特性。与单机事务不同的是，它涉及的资源分散在多个主机上，而主机与主机之间必然需要网络通信，但是网络本身就是不可靠的，网络硬件损坏、网络质量较差、网络延迟较大等问题都会不可避免地带来网络分区问题，这也给分布式事务的实现带来了极大的挑战。

尽管如此，业内还是出现了一大批经典的一致性协议和算法，其中较为有名的有两阶段提交协议、三阶段提交协议、Paxos 算法。

在分布式事务中，每个节点虽然能够知道自己的处理结果，但其他节点不知道。在两阶段提交协议和三阶段提交协议中，会引入一个协调者的角色来统一调度所有分布式节点（参与者）的行为，并基于它们的行为结果来判断是否需要提交事务。

下面重点说说两阶段提交协议。顾名思义，两阶段提交协议会将事务的整个操作分成两个阶段进行处理。

- **准备阶段**

协调者向所有的参与者发出 Prepare 请求，要求所有参与者进行本地事务的处理，并反馈处理结果。

- 提交阶段

协调者基于准备阶段的反馈结果来决定事务是否需要提交。

之所以这样处理,主要是因为要尽可能地让事务在提交之前将所有操作都准备好。这样,最后的提交将是一个耗时较小的操作,从而大大降低整个分布式事务执行失败的概率。

具体在 MySQL 中,由于存储引擎层和 Server 层是分开的,为了保证事务在存储引擎和二进制日志之间的一致性,MySQL 5.0 引入了 XA(分布式事务)模型,同样将事务的处理分成两个阶段,具体如下。

- 第一阶段

在存储引擎内部进行事务的 Prepare 操作(InnoDB Prepare),写 redo log 并刷盘(write/fsync redo log)。此时,binlog 不做任何操作。

- 第二阶段

□ 写 binlog,并刷盘(write/fsync binlog)。
□ 在存储引擎内部进行事务的 Commit 操作(InnoDB Commit)。

MySQL 事务的两阶段提交协议保证了只要事务被写入 binlog,该事务就一定会在存储引擎内部进行提交。如果实例在事务 Commit 阶段崩溃,那么在实例恢复的过程中,对于已经存在于 binlog 中的事务,会在存储引擎内部进行提交,这也就保证了主从之间的数据一致性。

2.3.2 半同步复制的原理

半同步复制的原理如图 2-2 所示。

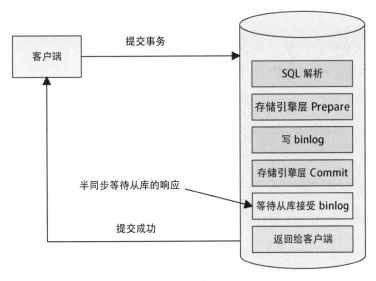

图 2-2 半同步复制原理图

在图 2-2 中，等待从库的反馈发生在存储引擎层提交之后，这会带来一个潜在的问题：在存储引擎层提交之后，从库反馈之前，所有其他的会话都可以看到这次提交的事务。如果此时主库出现故障，切换到从库上，客户端就会发现切换前后的数据不一致，也就是切换之前看到的数据丢失了。

针对上述问题，MySQL 5.7 引入了一种新的半同步方案：无损复制（lossless replication）。

相对于之前的半同步方案，它将图 2-2 中的等待从库反馈的阶段调整到了存储引擎层提交之前。

具体使用哪种方案，由 rpl_semi_sync_master_wait_point 参数控制。

- AFTER_SYNC：无损复制。默认值。
- AFTER_COMMIT：旧的半同步复制方案。

2.3.3 半同步复制的安装

使用半同步复制，有以下几个先决条件。

- MySQL 5.5 及以上版本。
- 允许插件的动态加载，即变量 have_dynamic_loading 必须为 YES。
- 异步复制已经存在。

接下来看看具体的安装步骤。

(1) 加载插件。

分别在主库和从库上执行以下命令。

主库：

```
mysql> install plugin rpl_semi_sync_master soname 'semisync_master.so';
```

从库：

```
mysql> install plugin rpl_semi_sync_slave soname 'semisync_slave.so';
```

如何判断插件是否加载成功呢？可采用以下两种方式。

- SHOW PLUGINS 命令。

```
mysql> show plugins;
+----------------------+--------+-------------+--------------------+---------+
| Name                 | Status | Type        | Library            | License |
+----------------------+--------+-------------+--------------------+---------+
...
| rpl_semi_sync_master | ACTIVE | REPLICATION | semisync_master.so | GPL     |
...
```

- 查询 information_schema.plugins 表。

```
mysql> select plugin_name, plugin_status from information_schema.plugins where plugin_name like '%semi%';
+----------------------+---------------+
| plugin_name          | plugin_status |
+----------------------+---------------+
```

```
| rpl_semi_sync_master   | ACTIVE          |
+------------------------+-----------------+
1 row in set (0.00 sec)
```

输出以上结果，即意味着插件加载成功。

(2) 启动半同步复制。

在安装完插件后，半同步复制默认是关闭的，此时需要设置参数来开启半同步复制。具体命令如下。

主库：

```
mysql> set global rpl_semi_sync_master_enabled = 1;
```

从库：

```
mysql> set global rpl_semi_sync_slave_enabled = 1;
```

以上启动方式是在 MySQL 客户端中操作的，也可写在配置文件中，这样在实例启动后，就会自动开启半同步复制。具体配置如下。

主库：

```
plugin-load = rpl_semi_sync_master=semisync_master.so
rpl_semi_sync_master_enabled = 1
```

从库：

```
plugin-load = rpl_semi_sync_slave=semisync_slave.so
rpl_semi_sync_slave_enabled = 1
```

不过一般建议如下配置。这样，即使发生了高可用切换，也能继续使用半同步复制。

```
plugin-load = "rpl_semi_sync_master=semisync_master.so;rpl_semi_sync_slave=semisync_slave.so"
rpl-semi-sync-master-enabled = 1
rpl-semi-sync-slave-enabled = 1
```

(3) 重启从库上的 I/O 线程。

```
mysql> stop slave io_thread;
mysql> start slave io_thread;
```

如果没有重启，则默认还是异步复制。

(4) 查看半同步复制是否运行。

分别在主库和从库上执行以下命令。

主库：

```
mysql> show status like 'Rpl_semi_sync_master_status';
+-----------------------------+-------+
| Variable_name               | Value |
+-----------------------------+-------+
| Rpl_semi_sync_master_status | ON    |
+-----------------------------+-------+
1 row in set (0.01 sec)
```

从库：

```
mysql> show status like 'Rpl_semi_sync_slave_status';
+----------------------------+-------+
| Variable_name              | Value |
+----------------------------+-------+
| Rpl_semi_sync_slave_status | ON    |
+----------------------------+-------+
1 row in set (0.00 sec)
```

只有这两个变量都为 ON，才意味着半同步复制正在运行。

至此，MySQL 半同步复制搭建完毕。

2.3.4 半同步复制的注意事项

- 半同步复制在主库和从库上必须同时开启。只有一方开启时，使用的还是异步复制。
- 从库响应的条件是事务对应的二进制日志事件写入 relay log。
- 如果在规定的时间内（由 rpl_semi_sync_master_timeout 参数决定，默认是 10 秒），主库没有收到从库的响应，半同步复制会切换为异步复制，同时，错误日志中会提示以下信息：

  ```
  [Warning] [MY-011153] [Repl] Timeout waiting for reply of binlog (file: mysql-bin.000004, pos: 9720138), semi-sync up to file mysql-bin.000004, position 9718392.
  ```

  后续如果收到了从库的响应，则异步复制又会恢复成半同步复制。

  如果不想让半同步复制退化为异步复制，可将 rpl_semi_sync_master_timeout 设置为一个较大的值。

- 在一主多从的架构中，如果要开启半同步复制，并不要求所有的从库都开启半同步复制。
- 在 MySQL 5.7 之前，binlog dump 线程既要发送 binlog 给从库，又要等待从库的反馈，而且这两个操作是串行执行的，这无疑会影响半同步复制的性能。MySQL 5.7 针对这个瓶颈，新增了一个 Ack Collector 线程，专门用于接收从库的反馈信息。
- 半同步复制会给事务的提交带来一定程度的延迟，这个延迟至少是一个 TCP/IP 往返的时间。所以，建议在低延时的网络中使用半同步复制。

2.3.5 半同步复制的常用参数

首先，看看与半同步复制相关的参数。

主库：

```
mysql> show global variables like '%semi%';
+-------------------------------------------+-------+
| Variable_name                             | Value |
+-------------------------------------------+-------+
| rpl_semi_sync_master_enabled              | ON    |
| rpl_semi_sync_master_timeout              | 10000 |
| rpl_semi_sync_master_trace_level          | 32    |
| rpl_semi_sync_master_wait_for_slave_count | 1     |
| rpl_semi_sync_master_wait_no_slave        | ON    |
```

```
| rpl_semi_sync_master_wait_point         | AFTER_SYNC |
+-----------------------------------------+------------+
6 rows in set (0.00 sec)
```

从库：

```
mysql> show global variables like '%semi%';
+--------------------------------+-------+
| Variable_name                  | Value |
+--------------------------------+-------+
| rpl_semi_sync_slave_enabled    | ON    |
| rpl_semi_sync_slave_trace_level| 32    |
+--------------------------------+-------+
2 rows in set (0.00 sec)
```

rpl_semi_sync_master_enabled、rpl_semi_sync_slave_enabled、rpl_semi_sync_master_timeout、rpl_semi_sync_master_wait_point 在前面提到过。

下面解释一下其他几个重要参数。

- **rpl_semi_sync_master_wait_for_slave_count**

这个参数是 MySQL 5.7.3 引入的，用于设置主库需要等待反馈的从库的个数，默认为 1。

- **rpl_semi_sync_master_wait_no_slave**

这个参数与上面的 rpl_semi_sync_master_wait_for_slave_count 参数有关。正如前面提到的，在执行一个事务时，如果在指定的时间内（由 rpl_semi_sync_master_timeout 参数决定，默认是 10 秒），主库没有收到从库的响应，则半同步复制会切换为异步复制。

如果在等待的过程中，发现连接到主库上的处于半同步复制状态的从库的个数（可由状态变量 Rpl_semi_sync_master_clients 确定）小于 rpl_semi_sync_master_wait_for_slave_count，那么主库有两种选择：继续等待；立即切换为异步复制。具体选择哪个由 rpl_semi_sync_master_wait_no_slave 参数决定：若值为 ON（默认值），则继续等待；若值为 OFF，则会立即切换为异步复制。

- **rpl_semi_sync_master_trace_level**

用于设置半同步复制相关的日志输出级别。

接下来，看看半同步复制相关的状态变量。

主库：

```
mysql> show global status like '%semi%';
+--------------------------------------------+---------+
| Variable_name                              | Value   |
+--------------------------------------------+---------+
| Rpl_semi_sync_master_clients               | 1       |
| Rpl_semi_sync_master_net_avg_wait_time     | 0       |
| Rpl_semi_sync_master_net_wait_time         | 0       |
| Rpl_semi_sync_master_net_waits             | 2114    |
| Rpl_semi_sync_master_no_times              | 0       |
| Rpl_semi_sync_master_no_tx                 | 0       |
| Rpl_semi_sync_master_status                | ON      |
| Rpl_semi_sync_master_timefunc_failures     | 0       |
| Rpl_semi_sync_master_tx_avg_wait_time      | 4015    |
```

```
| Rpl_semi_sync_master_tx_wait_time            | 8175737 |
| Rpl_semi_sync_master_tx_waits                | 2036    |
| Rpl_semi_sync_master_wait_pos_backtraverse   | 0       |
| Rpl_semi_sync_master_wait_sessions           | 0       |
| Rpl_semi_sync_master_yes_tx                  | 2078    |
+----------------------------------------------+---------+
14 rows in set (0.00 sec)
```

从库：

```
mysql> show global status like '%semi%';
+----------------------------+-------+
| Variable_name              | Value |
+----------------------------+-------+
| Rpl_semi_sync_slave_status | ON    |
+----------------------------+-------+
1 row in set (0.00 sec)
```

下面解释一下各个变量的具体含义。

- **Rpl_semi_sync_master_status 和 Rpl_semi_sync_slave_status**

分别用于判断主从库半同步复制的状态。

- **Rpl_semi_sync_master_clients**

连接到主库的处于半同步复制状态的从库的个数。

- **Rpl_semi_sync_master_net_avg_wait_time 和 Rpl_semi_sync_master_net_wait_time**

主库等待从库反馈的平均等待时间和总等待时间。已弃用，计划在未来版本中删除。这两个变量总是显示为 0。

- **Rpl_semi_sync_master_net_waits**

主库等待从库反馈的次数。

- **Rpl_semi_sync_master_no_times**

主库切换为异步复制的次数。

- **Rpl_semi_sync_master_no_tx**

主库切换为异步复制状态下执行的事务的个数。

- **Rpl_semi_sync_master_tx_avg_wait_time 和 Rpl_semi_sync_master_tx_wait_time**

在半同步复制状态下，主库上事务的平均等待时间和总等待时间，单位是微秒，不包括等待超时的事务。

- **Rpl_semi_sync_master_tx_waits**

在半同步复制状态下，主库上事务等待的次数，不包括等待超时的事务。

- **Rpl_semi_sync_master_yes_tx**

主库在半同步复制状态下执行的事务的个数。

- `Rpl_semi_sync_master_wait_sessions`

当前等待从库反馈的会话的个数。

可基于上面的状态变量来衡量半同步复制的性能、主从库网络的质量。

注意：半同步复制相关的参数及状态变量只有在加载了半同步复制插件后才能看到。

2.4 并行复制

有线上 MySQL 维护经验的读者都知道，主从延迟往往是一个让人头疼不已的问题。不仅是其造成的潜在问题比较严重，而且问题的定位尤其考验 DBA 的综合能力：既要熟悉复制的内部原理，又能解读主机层面的资源使用情况，甚至还要会分析 binlog。

关于主从延迟的原因，通过前面几节的介绍我们知道，对于写入 binlog 的二进制日志事件，从库上只有一个 SQL 线程进行重放，但是这些二进制日志事件在主库中是并发写入的。就好比几个人（多线程）挖坑一个人（单线程）填，在挖坑速度不快的情况下，填坑人尚能应付，一旦挖坑人稍微加速，就只能眼睁睁地看着坑越来越深。具体到 MySQL 中，这就意味着 `Seconds_Behind_Master` 的值会越来越大。

主从延迟带来的问题，主要体现在以下两个方面。

- 对于读写分离的业务，主从延迟意味着业务会读取到旧数据。
- 主从延迟过大，会影响数据库的高可用切换。这一点尤其需要注意。
  - 如果等待从库应用完差异的 binlog 才做高可用切换，无疑会影响数据库服务的可用性。
  - 如果不等待，直接切换，则意味着没应用完的这部分 binlog 的数据会丢失，业务不一定能接受这种情况。

2.4.1 并行复制方案

MySQL 官方先后提出了多个不同的并行复制方案，具体如下。

- MySQL 5.6 基于库级别的并行复制方案。
- MySQL 5.7 基于组提交的并行复制方案。
- MySQL 8.0 基于 WRITESET 的并行复制方案。

因为线上的大部分环境是单库多表的，所以基于库级别的并行复制实际上用得并不多。

下面重点看看后两个方案的实现原理。

1. 基于组提交的并行复制方案

MySQL 5.7 基于组提交的并行复制方案，先后经历了两个版本的迭代：Commit-Parent-Based 方案和 Lock-Based 方案。

- **Commit-Parent-Based 方案**

在 2.3 节中，我们提到过 MySQL 会将一个事务拆分为两个阶段进行处理：Prepare 阶段和 Commit 阶段。另外，InnoDB 使用的锁机制是悲观锁。在悲观锁中，事务是在操作之初执行加锁操作，如果锁资源被其他事务占用了，则该事务会被阻塞。基于这两点，我们不难推断，两个事务如果都进入了 Prepare 阶段，则意味着它们之间是没有锁冲突的，在从库重放时可并行执行。这就是 Commit-Parent-Based 方案的核心思想。

具体实现上，

- 主库有个全局计数器（global counter），每次在事务存储引擎层提交之前，都会增加这个计数器的计数。
- 在事务进入 Prepare 阶段之前，会将全局计数器的当前值记录在事务中，这个值称为事务的 commit-parent。
- 这个 commit-parent 会写入 binlog，记录在事务的头部。
- 从库重放时，如果发现两个事务的 commit-parent 相同，会并行执行这两个事务。

以下面这 7 个事务为例，看看它们在从库中的并行执行情况。

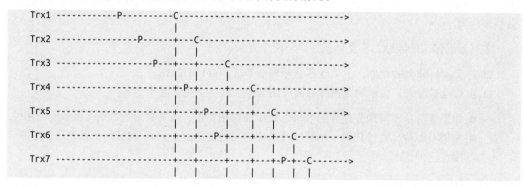

这里的 Trx 指的是事务，P 指的是事务在进入 Prepare 阶段之前，读取 commit-parent 的时间点。C 指的是事务在进入 Commit 阶段之前，增加全局计数器的时间点。

下面看看这 7 个事务的并行执行情况。

- Trx1、Trx2、Trx3 并行执行。
- Trx4 串行执行。
- Trx5、Trx6 并行执行。
- Trx7 串行执行。

这在很大程度上实现了并行，但还不够完美。实际上，Trx4、Trx5、Trx6 也可并行执行，因为它们同时进入了 Prepare 阶段。同理，Trx6、Trx7 也可并行执行。

基于此，官方迭代了并行复制方案，推出了新的 Lock-Based 方案。

- **Lock-Based 方案**

该方案引入了锁区间（locking interval）的概念，锁区间定义了一个事务持有锁的时间范围。具体来说：

- 将 Prepare 阶段最后一个 DML 语句获取锁的时间点定义为锁区间的开始点；
- 将存储引擎层提交之前锁释放的时间点定义为锁区间的结束点。

如果两个事务的锁区间存在交集，则意味着这两个事务没有锁冲突，可并行重放。例如：

```
Trx1 -----L---------C------------->
Trx2 ----------L---------C------->
```

反之，则不可并行重放。例如：

```
Trx1 -----L----C------------------>
Trx2 --------------L----C-------->
```

这里的 L 代表锁区间的开始点，C 代表锁区间的结束点。

在具体实现上，主库引入了以下 4 个变量。

- `global.transaction_counter`：事务计数器。
- `transaction.sequence_number`：事务序列号。在事务进入 Prepare 阶段之前，会将 `global.transaction_counter` 自增加 1 并赋值给 `transaction.sequence_number`。序列号不是一直递增的，每切换一个 binlog，都会将 `transaction.sequence_number` 重置为 1。
- `global.max_committed_transaction`：当前已提交事务的最大序列号。在事务进行存储引擎层提交之前，会取 `global.max_committed_transaction` 和当前事务的 `sequence_number` 的最大值，赋值给 `global.max_committed_transaction`。代码如下：

  ```
  global.max_committed_transaction = max(global.max_committed_transaction,
                                         transaction.sequence_number)
  ```

- `transaction.last_committed`：在事务进入 Prepare 阶段之前，已提交事务的最大序列号。代码如下：

  ```
  transaction.last_committed = global.max_committed_transaction
  ```

在这 4 个变量中，`transaction.sequence_number` 和 `transaction.last_committed` 会写入 binlog。具体来说，对于 GTID 复制，它们会写入 `GTID_LOG_EVENT`；对于非 GTID 复制，则写入 `ANONYMOUS_GTID_LOG_EVENT`。

对于示例中的 7 个事务，记录在 binlog 中的 `last_committed` 和 `sequence_number` 如下所示：

```
Trx1: last_committed=0 sequence_number=1
Trx2: last_committed=0 sequence_number=2
Trx3: last_committed=0 sequence_number=3
Trx4: last_committed=1 sequence_number=4
Trx5: last_committed=2 sequence_number=5
Trx6: last_committed=2 sequence_number=6
Trx7: last_committed=5 sequence_number=7
```

下面说说从库并行重放的逻辑。

从库引入了一个事务队列（transaction_sequence），包含了当前正在执行的事务。该队列是有序的，按照事务的 sequence_number 从小到大排列。这个队列中的事务可并行执行。

一个新的事务能否插入这个队列，唯一的判断标准是，事务的 last_committed 是否小于队列中第一个事务的 sequence_number。只有小于才能允许插入，代码如下：

```
transaction.last_committed < transaction_sequence[0].sequence_number
```

最后，回到示例中的 7 个事务。结合 binlog 中的 last_committed 和 sequence_number，我们看看它们的并行执行情况。

- Trx1、Trx2、Trx3 并行执行。
- Trx1 执行完毕后，Trx4 可加入队列。
- Trx2 执行完毕后，Trx5、Trx6 可加入队列。
- Trx5 执行完毕后，Trx7 可加入队列。

不难发现，相对于 Commit-Parent-Based 方案，Lock-Based 方案的并行度确实大大提高了。

无论是 Commit-Parent-Based 方案，还是 Lock-Based 方案，依赖的都是组提交（Group Commit）。组提交方案有以下两个特点。

- 适用于高并发场景。因为只有在高并发场景下，才会有更多的事务放到一个组中提交。
- 在级联复制中，层级越深，并行度越低。

针对低并发场景，如果要提升从库的并行效率，可调整以下两个参数。

- binlog_group_commit_sync_delay：binlog 刷盘（fsync）之前等待的时间。单位是微秒，默认值为 0，即不等待。该值越大，一个组内的事务就越多，从库的并行度也就越高。但与此同时，客户端的响应时间也会越长。
- binlog_group_commit_sync_no_delay_count：在 binlog_group_commit_sync_delay 时间内，允许等待的最大事务数。如果 binlog_group_commit_sync_delay 设置为 0，则此参数无效。

2. 基于 WRITESET 的并行复制方案

MySQL 8.0 推出了基于 WRITESET 的并行复制方案，简称为 WRITESET 方案。该方案推出的初衷实际上是为组复制服务，主要用于认证阶段（Certification）的冲突检测。WRITESET 方案的核心思想是，两个来自不同节点的并发事务，只要没修改同一行，就不存在冲突。对于没有冲突的并发事务，在写入 relay log 时，可以共享一个 last_committed。这里的冲突检测，实际上比较的是两个事务之间的写集合（writeset）。注意 writeset 和 WRITESET 两者的区别，前者指的是事务的写集合，后者则特指 WRITESET 方案。

下面来看看事务写集合的生成过程。具体步骤如下。

- 首先提取被修改行的主键、唯一索引、外键信息。一张表，如果有主键和一个唯一索引，则每修改一行，会提取两条约束信息：一条针对主键，另一条针对唯一索引。针对主键提取的信息包括主键名、库名、表名、主键值，这些信息会拼凑为一个字符串。

❏ 计算该字符串的哈希值，具体的哈希算法由 transaction_write_set_extraction 参数指定。
❏ 将计算后的哈希值插入当前事务的写集合。

接下来，结合源码看看 WRITESET 方案的实现原理。

```cpp
void Writeset_trx_dependency_tracker::get_dependency(THD *thd,
                                                     int64 &sequence_number,
                                                     int64 &commit_parent) {
  Rpl_transaction_write_set_ctx *write_set_ctx =
      thd->get_transaction()->get_transaction_write_set_ctx();
  std::vector<uint64> *writeset = write_set_ctx->get_write_set();

#ifndef NDEBUG
  /* 空事务的写集合必须为空 */
  if (is_empty_transaction_in_binlog_cache(thd))
    assert(writeset->size() == 0);
#endif

  /*
    判断一个事务能否使用写集合
  */
  bool can_use_writesets =
      // 事务写集合的大小不为0或者事务为空事务
      (writeset->size() != 0 || write_set_ctx->get_has_missing_keys() ||
       is_empty_transaction_in_binlog_cache(thd)) &&
      // 事务的 transaction_write_set_extraction 必须与全局设置一致
      (global_system_variables.transaction_write_set_extraction ==
       thd->variables.transaction_write_set_extraction) &&
      // 不能被其他表外键关联
      !write_set_ctx->get_has_related_foreign_keys() &&
      // 事务写集合的大小不能超过 binlog_transaction_dependency_history_size
      !write_set_ctx->was_write_set_limit_reached();
  bool exceeds_capacity = false;

  if (can_use_writesets) {
    /*
      检查 m_writeset_history 加上事务写集合的大小是否超过 m_writeset_history 的上限，
      m_writeset_history 的上限由参数 binlog_transaction_dependency_history_size 决定
    */
    exceeds_capacity =
        m_writeset_history.size() + writeset->size() > m_opt_max_history_size;

    /*
      计算所有冲突行中最大的 sequence_number，并将被修改行的哈希值插入 m_writeset_history
    */
    int64 last_parent = m_writeset_history_start;
    for (std::vector<uint64>::iterator it = writeset->begin();
         it != writeset->end(); ++it) {
      Writeset_history::iterator hst = m_writeset_history.find(*it);
      if (hst != m_writeset_history.end()) {
        if (hst->second > last_parent && hst->second < sequence_number)
          last_parent = hst->second;

        hst->second = sequence_number;
      } else {
        if (!exceeds_capacity)
          m_writeset_history.insert(
```

```
              std::pair<uint64, int64>(*it, sequence_number));
    }
  }
  // 如果表上都存在主键，则会取 last_parent 和 commit_parent 的较小值作为事务的 commit_parent
  if (!write_set_ctx->get_has_missing_keys()) {
    commit_parent = std::min(last_parent, commit_parent);
  }
}

if (exceeds_capacity || !can_use_writesets) {
  m_writeset_history_start = sequence_number;
  m_writeset_history.clear();
}
}
```

该函数的处理流程如下。

- 调用函数时，会传入事务的 sequence_number 和 commit_parent（last_committed），这两个值是基于 Lock-Based 方案生成的。
- 获取事务的写集合。可以看到，事务的写集合是数组类型。
- 判断一个事务能否使用 WRITESET 方案。

  以下场景不能使用 WRITESET 方案，只能使用 Lock-Based 方案生成的 last_committed。

  - 事务没有写集合。常见的原因是表上没有主键。
  - 当前事务 transaction_write_set_extraction 的设置与全局不一致。
  - 表被其他表外键关联。
  - 事务写集合的大小超过 binlog_transaction_dependency_history_size。

- 如果能使用 WRITESET 方案，

  - 首先判断 m_writeset_history 的容量是否超标。具体来说，m_writeset_history + writeset 的大小是否超过 binlog_transaction_dependency_history_size 的设置。
  - 将 m_writeset_history_start 赋值给变量 last_parent。

    m_writeset_history_start 代表不在 m_writeset_history 中最后一个事务的 sequence_number。它的初始值为 0，当参数 binlog_transaction_dependency_tracking 发生变化或清空 m_writeset_history 时，会更新 m_writeset_history_start。

  - 循环遍历事务的写集合，判断被修改行对应的哈希值是否在 m_writeset_history 中存在。

    若存在，则意味着 m_writeset_history 存在同一行的操作。既然是同一行的不同操作，自然就不能并行重放。这个时候，会将 m_writeset_history 中该行的 sequence_number 赋值给 last_parent。需要注意的是，这里会循环遍历完事务的写集合，毕竟这个事务中可能有多条记录在 m_writeset_history 中存在。在遍历的过程中，会判断 m_writeset_history 中冲突行的 sequence_number 是否大于 last_parent，只有大于才会赋值。换言之，这里会取所有冲突行中最大的 sequence_number，赋值给 last_parent。

若不存在，则判断 m_writeset_history 的容量是否超标，若不超标，则会将被修改行的哈希值插入 m_writeset_history。可以看到，m_writeset_history 是字典类型，其中键存储的是被修改行的哈希值，值存储的是事务的 sequence_number。
- 判断被操作的表上是否都存在主键。若存在，才会取 last_parent 和 commit_parent 的较小值作为事务的 commit_parent。否则，使用的还是 Lock-Based 方案生成的 commit_parent。
- 如果 m_writeset_history 容量超标或者事务不能使用 WRITESET 方案，则会将当前事务的 sequence_number 赋值给 m_writeset_history_start，同时清空 m_writeset_history。

WRITESET 方案的相关参数

下面看看 WRITESET 方案的 3 个参数。

- **binlog_transaction_dependency_tracking**

指定基于何种方案决定事务的依赖关系。对于同一个事务，不同的方案可生成不同的 last_committed。该参数有以下取值。

- COMMIT_ORDER：默认值，基于 Lock-Based 方案决定事务的依赖关系。
- WRITESET：基于 WRITESET 方案决定事务的依赖关系。
- WRITESET_SESSION：与 WRITESET 类似，只不过同一个会话中的事务不能并行执行。

- **transaction_write_set_extraction**

指定事务写集合的哈希算法，可设置的值有 OFF、MURMUR32、XXHASH64（默认值）。对于组复制，该参数必须设置为 XXHASH64。注意，若要将 binlog_transaction_dependency_tracking 设置为 WRITESET 或 WRITESET_SESSION，则该参数不能设置为 OFF。

- **binlog_transaction_dependency_history_size**

m_writeset_history 的上限，默认为 25000。一般来说，binlog_transaction_dependency_history_size 越大，m_writeset_history 能存储的行的信息就越多。在不出现行冲突的情况下，m_writeset_history_start 也会越小。相应地，新事务的 last_committed 也会越小，在从库重放的并发程度也会越高。

接下来，看看 MySQL 官方对 COMMIT_ORDER、WRITESET_SESSION、WRITESET 这 3 种方案的压测结果。

- 主库环境：16 核，SSD，1 个数据库，16 张表，共 800 万条数据。
- 压测场景：OLTP Read/Write、Update Indexed Column 和 Write-only。
- 压测方案：在关闭复制的情况下，在不同的线程数下，注入 100 万个事务。开启复制，观察不同线程数下，不同方案的从库重放速度。

3 个场景下的压测结果分别如图 2-3、图 2-4 和图 2-5 所示。

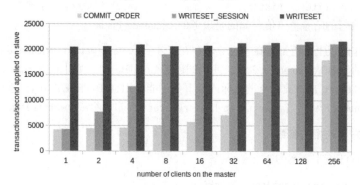

图 2-3　Read/Write 压测结果

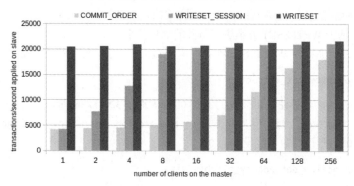

图 2-4　Update Indexed Column 压测结果

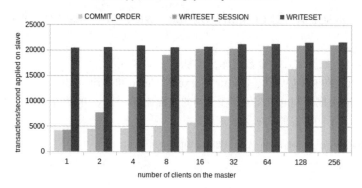

图 2-5　Write-only 压测结果

分析压测结果，我们可以得出以下结论。

- 对于 COMMIT_ORDER 方案，主库的并发度越高，从库的重放速度越快。
- 对于 WRITESET 方案，主库的并发线程数对其几乎没有影响。甚至单线程下 WRITESET 的重放速度都超过了 256 线程下的 COMMIT_ORDER。
- 与 COMMIT_ORDER 一样，WRITESET_SESSION 也依赖于主库并发。只不过，在主库并发线程数较低（4 线程、8 线程）的情况下，WRITESET_SESSION 也能实现较高的吞吐量。

2.4.2 如何开启并行复制

在从库上设置以下 3 个参数。

- slave_parallel_type = LOGICAL_CLOCK
- slave_parallel_workers = 16
- slave_preserve_commit_order = ON

下面看看这 3 个参数的的具体含义。

- slave_parallel_type：设置从库并行复制的类型。该参数有以下取值。
  - DATABASE：基于库级别的并行复制。
  - LOGICAL_CLOCK：基于组提交的并行复制。
- slave_parallel_workers：设置 Worker 线程的数量。开启多线程复制，原来的 SQL 线程将演变为一个 Coordinator 线程和多个 Worker 线程。
- slave_preserve_commit_order：事务在从库上的提交顺序是否与主库保持一致，建议将该参数设置为 ON。

需要注意的是，调整这 3 个参数后，需要重启复制才能生效。

从 MySQL 5.7.22 和 MySQL 8.0 开始，可使用 WRITESET 方案进一步提升并行复制的效率。此时，需在主库上设置以下参数。

```
binlog_transaction_dependency_tracking = WRITESET_SESSION
transaction_write_set_extraction = XXHASH64
binlog_transaction_dependency_history_size = 25000
binlog_format = ROW
```

注意，基于 WRITESET 的并行复制方案，只在 binlog 格式为 ROW 的情况下才生效。

2.4.3 参考资料

- 官方文档 "WL#6314: MTS: Prepared transactions slave parallel applier"。
- 官方文档 "WL#6813: MTS: ordered commits (sequential consistency)"。
- 官方文档 "WL#7165: MTS: Optimizing MTS scheduling by increasing the parallelization window on master"。

- 官方文档"WL#8440: Group Replication: Parallel applier support"。
- 官方文档"WL#9556: Writeset-based MTS dependency tracking on master"。
- 简书文章"WriteSet 并行复制",作者:真之棒 2016。
- 官网博客文章"Improving the Parallel Applier with Writeset-based Dependency Tracking"。

2.5 多源复制

多源复制是 MySQL 5.7 引入的一个新功能,允许将多个主库的数据复制到同一个从库上。常见的应用场景有如下 3 个。

- 数据灾备。
- 分库分表场景。将多个分片的数据汇总到一个实例上进行分析。
- 数据聚合。将多个实例的数据聚集到一个实例上。

图 2-6 是多源复制的原理图。

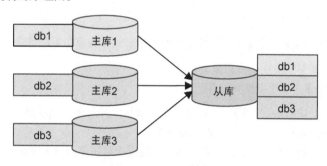

图 2-6　多源复制原理图

从上面的原理图可以看出,不同主库上的数据库名不能相同,不然复制到从库会引起冲突。

为了与之前的复制(姑且称之为"单源复制")进行区分,多源复制引入了 channel(通道)的概念,每个 channel 对应一个主库。

2.5.1 多源复制的搭建

多源复制的搭建相对比较简单,相比于单源复制,它有以下两处不同。

- 参数 master-info-repository 和 relay-log-info-repository 必须设置为 TABLE。

  在单源复制中,这两个参数可设置为 FILE,即将 I/O 线程和 SQL 线程的相关信息分别记录在 master.info 和 relay-log.info 文件中。但在多源复制中,这并不现实。如果为每个 channel 都生成一个 master.info 和 relay-log.info 文件,不仅会增加磁盘的 I/O 压力,更不便于管理。

- 执行 CHANGE MASTER TO 命令时,需显式指定 channel 的值。例如:

```
change master to
    master_host='192.168.244.10',
```

```
master_user='repl',
master_password='repl123',
master_auto_position = 1
for channel 'master-1';
```

2.5.2 多源复制搭建过程中的注意事项

- 在多源复制中，对主库的复制模式没有限制，可以其中一个主库是 GTID 复制，另一个主库是基于 binlog 位置点的复制。

  如果主库的复制模式不一样，则从库的 GTID_MODE 必须设置为 ON_PERMISSIVE 或 OFF_PERMISSIVE，不能设置为 ON 或 OFF。

- 如果是基于现有的主库搭建多源复制，有以下几点需要注意。

  - 备份还原工具的选择

    要么全部使用逻辑备份工具，要么就第一个主库使用物理备份工具。为什么不能全部使用物理备份工具呢？因为物理备份工具会拷贝所有文件，在恢复的时候，不同实例之间的系统表（ibdata）会存在冲突。

  - GTID_PURGED 的设置

    在 MySQL 8.0 之前，如果主库都开启了 GTID，则在导入第二个主库时会报如下错误：

    ```
    $ mysql < full_backup_master1.sql
    $ mysql < full_backup_master2.sql
    ERROR 1840 (HY000) at line 24: @@GLOBAL.GTID_PURGED can only be set when @@GLOBAL.GTID_EXECUTED is empty.
    ```

    提示很明显，设置 GTID_PURGED 的前提是 GTID_EXECUTED 必须为空。

    此时，可在从库上执行 RESET MASTER 命令，待第二个主库导入后，同样执行 RESET MASTER 命令，最后在从库上执行 SET @@GLOBAL.GTID_PURGED。命令中 GTID_PURGED 的值取 master1 和 master2 的 GTID_PURGED 的交集，如：

    ```
    $ grep "GTID_PURGED" full_backup_master1.sql
    ...
    SET @@GLOBAL.GTID_PURGED='0ed33867-d1b0-11e8-8f58-000c2914fb06:1-9';
    ...
    $ grep "GTID_PURGED" full_backup_master2.sql
    ...
    SET @@GLOBAL.GTID_PURGED='ff693ae0-d1b2-11e8-b732-000c2927cfea:1-4';
    ```

    最后 GTID_PURGED 的取值如下：

    ```
    SET @@GLOBAL.GTID_PURGED='0ed33867-d1b0-11e8-8f58-000c2914fb06:1-9,ff693ae0-d1b2-11e8-b732-000c2927cfea:1-4';
    ```

    如果是基于 MySQL 8.0 搭建，则不会出现这个问题。为什么呢？因为 MySQL 8.0 支持通过加号添加新的 GTID 集。

    ```
    mysql> set global gtid_purged='+d7664b79-e0ee-11e8-b1ad-000c2927cfea:1'
    ```

而 MySQL 8.0 中的 mysqldump 就是这么输出的，如：

```
SET @@GLOBAL.GTID_PURGED=/*!80000 '+'*/ '453a5124-020e-11ec-8719-000c29f66609:1-12';
```

命令中的/*!80000 '+'*/是注释，仅当 MySQL 服务端版本大于 8.0.00 时才会执行。

- 二进制日志是 ROW 格式

如果二进制日志是 ROW 格式，那么不建议复制 mysql 库，否则很容易导致主从复制中断。

2.5.3 多源复制的管理

相对于单源复制，多源复制中引入了 channel 的概念来对不同主库所对应的主从复制进行管理，所以在执行相关管理操作时需指定 channel 的值。例如：

```
show slave status for channel 'master-1'\G
start slave for channel 'master-1';
stop slave for channel 'master-1';
reset slave for channel 'master-1';
reset slave all for channel 'master-1';
```

如果不指定 channel，则默认对所有 channel 生效。

2.6 延迟复制

延迟复制是 MySQL 5.6 推出的，它的核心思想是，一个事务在主库执行后，会等待若干秒才在从库上执行。

延迟复制常见的使用场景有如下几个。

- 应对主库的误操作，尤其是 DROP TABLE 操作。

    影响较大且较为常见的主库误操作有两类：DELETE（或 UPDATE）操作没加条件和 DROP TABLE 操作。对于前者，如果二进制日志的格式为 ROW，可通过 binlog 闪回工具进行恢复。而后者是 DDL 操作，属于 QUERY_EVENT，在 binlog 中只会记录其原生 SQL 语句，不会保存相关记录的前镜像，所以该操作是不可逆的，无法通过工具来恢复。

    在这种情况下，常用的方法是基于备份搭建一个临时实例，然后应用 binlog，将实例恢复到表被删的前一个时间点，最后从临时实例中导出被删的表，导入到线上。显然，这种方式不仅耗时，还很烦琐，对 DBA 的综合能力要求也高。

- 查看数据库的历史状态。
- 人为模拟主从延迟。

    要模拟主从延迟，除了延迟复制这种方式，还可通过加锁的方式来实现，如全局读锁（FLUSH TABLES WITH READ LOCK）。

2.6.1 如何开启延迟复制

开启延迟复制较为简单,只需设置 CHANGE MASTER TO 命令中的 MASTER_DELAY 选项。

既可在搭建主从库时设置:

```
change master to
    master_host='192.168.244.10',
    master_user='repl',
    master_password='repl123',
    master_port=3306,
    master_log_file='mysql-bin.000001',
    master_log_pos=155,
    master_delay=28800;
```

也可针对一个已经运行的从库设置,具体命令如下:

```
stop slave;
change master to master_delay = 28800;
start slave;
```

延迟复制的相关信息可通过 SHOW SLAVE STATUS\G 查看。

```
            SQL_Delay: 28800
  SQL_Remaining_Delay: 28774
Slave_SQL_Running_State: Waiting until MASTER_DELAY seconds after master executed event
```

输出中,SQL_Delay 指的是期望延迟时间,即 CHANGE MASTER TO 命令中设置的 MASTER_DELAY。SQL_Remaining_Delay 指的是当前被暂停的事务需要等待多久,才能到达期望延迟时间。

2.6.2 如何使用延迟复制恢复误删的表

下面演示一个恢复案例。

(1) 在主库中创建测试数据。

```
master> create database slowtech;
master> create table slowtech.t1(id int primary key);
master> insert into slowtech.t1 values(1),(2);
master> select * from slowtech.t1;
+----+
| id |
+----+
|  1 |
|  2 |
+----+
2 rows in set (0.00 sec)
```

(2) 在从库中检查数据是否同步过来。

```
slave> select * from slowtech.t1;
+----+
| id |
+----+
|  1 |
```

```
|  2 |
+----+
2 rows in set (0.00 sec)
```

(3) 从库开启延迟复制。

```
slave> stop slave;
slave> change master to master_delay = 28800;
slave> start slave;
```

延迟 28 800 秒，即 8 小时。

(4) 在主库中新增一条记录，并删除 t1 表。

```
master> insert into slowtech.t1 values(3);
master> drop table slowtech.t1;
```

此时查询从库，会发现 t1 表依然存在，且只有两条记录。

(5) 在主库中，查看 DROP 操作在 binlog 中的位置点信息。

```
master> show master status;
+------------------+----------+--------------+------------------+-------------------+
| File             | Position | Binlog_Do_DB | Binlog_Ignore_DB | Executed_Gtid_Set |
+------------------+----------+--------------+------------------+-------------------+
| mysql-bin.000003 |     1260 |              |                  |                   |
+------------------+----------+--------------+------------------+-------------------+
1 row in set (0.00 sec)

master> pager grep -iB 5 drop
PAGER set to 'grep -iB 5 drop'
master> show binlog events in 'mysql-bin.000003';
| mysql-bin.000003 |  875 | Query          | 1 |  942 | BEGIN                                                          |
| mysql-bin.000003 |  942 | Table_map      | 1 |  990 | table_id: 113 (slowtech.t1)                                    |
| mysql-bin.000003 |  990 | Write_rows     | 1 | 1026 | table_id: 113 flags: STMT_END_F                                |
| mysql-bin.000003 | 1026 | Xid            | 1 | 1053 | COMMIT /* xid=84 */                                            |
| mysql-bin.000003 | 1053 | Anonymous_Gtid | 1 | 1126 | SET @@SESSION.GTID_NEXT= 'ANONYMOUS'                           |
| mysql-bin.000003 | 1126 | Query          | 1 | 1260 | DROP TABLE `slowtech`.`t1` /* generated by server */ /* xid=85 */ |
18 rows in set (0.00 sec)
```

由于 binlog 中的事务较多，直接查看不太方便，这里使用了 grep 进行过滤，命令行中的 -i 表示忽略大小写，-B 5 会打印匹配文本的前 5 行。基于 binlog 的输出可以看到，DROP 操作前一个事务的结束位置点为 1053。

(6) 将从库恢复到指定位置点。

```
slave> stop slave;
slave> change master to master_delay = 0;
slave> start slave until master_log_file='mysql-bin.000003',master_log_pos=1053;
```

检查 SHOW SLAVE STATUS 的输出，确保 SQL 线程已执行到该位置点。

确定依据，(Relay_Master_Log_File == Until_Log_File) && (Exec_Master_Log_Pos == Until_Log_Pos) && (Slave_SQL_Running == 'No')。例如：

```
slave> show slave status\G
*************************** 1. row ***************************
               Slave_IO_State: Waiting for master to send event
```

```
                    ...
      Relay_Master_Log_File: mysql-bin.000003
           Slave_IO_Running: Yes
          Slave_SQL_Running: No
                    ...
         Exec_Master_Log_Pos: 1053
                    ...
           Until_Condition: Master
            Until_Log_File: mysql-bin.000003
             Until_Log_Pos: 1053
1 row in set (0.00 sec)
```

(7) 再来看看 t1 表的内容。

```
slave> select * from slowtech.t1;
+----+
| id |
+----+
|  1 |
|  2 |
|  3 |
+----+
3 rows in set (0.00 sec)
```

id 为 3 的记录已复制过来，此时可导出 t1 表，并将其导入主库。

2.6.3 延迟复制的总结

- 延迟复制在本质上是暂停 SQL 线程的应用，并不影响 I/O 线程接受主库的 binlog。需要注意的是，在执行 `CHANGE MASTER TO MASTER_DELAY = N` 命令时，从库会清除已有的 relay log，并基于 `Relay_Master_Log_File` 和 `Exec_Master_Log_Pos` 的值来重新初始化 `Master_Log_File` 和 `Read_Master_Log_Pos`。
- 对于 MySQL 5.6 之前的版本，可利用 Percona Toolkit 中的 `pt-slave-delay` 来实现类似功能。

2.7 本章总结

本章从复制的基本原理出发，延伸到复制的高阶特性，如 GTID 复制、半同步复制、并行复制、多源复制、延迟复制。

从数据安全的角度出发，强烈建议线上开启半同步复制且使用 `AFTER_SYNC` 方案。需要注意的是，如果在指定的时间内没有收到从库的反馈，半同步复制会退化为异步复制。如果不想让半同步复制退化为异步复制，可将超时时间设置为一个较大的值。

强烈建议线上开启 GTID 复制，原因有二。其一，能简化复制的日常管理。其二，如果项目中使用了 Canal 之类的 binlog 同步工具，无论高可用如何切换，都不影响数据的正常抽取。

对于线上核心库，建议添加一个延迟复制的从库。如果依赖备份加上 binlog 来做指定时间点的恢复，则恢复周期都比较长。

重点问题回顾

- 复制的基本原理。
- GTID 复制的优点。
- 如何在线开启、关闭 GTID 复制。
- MySQL 事务的两阶段提交。
- 半同步复制的原理。`AFTER_SYNC` 和 `AFTER_COMMIT` 的区别。半同步复制在什么情况下会退化为异步复制？如何避免半同步复制退化为异步复制？在半同步复制中，还会出现主从延迟吗？
- 并行复制的实现原理。
- 使用延迟复制进行指定位置点的恢复，重点是能解析 binlog 找到待恢复的位置点（`master_log_file` 和 `master_log_pos`），这就要求大家对 binlog 的格式和事件类型有一定的了解。

第 3 章
深入解析 binlog

在第 2 章中，我们曾多次提到 binlog 和二进制日志事件（binlog event），实际上，binlog 的基本组成单位就是二进制日志事件。为什么要关注 binlog 和二进制日志事件呢？因为 binlog 是复制得以实现的一个重要载体，熟悉 binlog 有助于我们加深对 MySQL 复制的理解，也有助于我们处理与主从复制相关的问题，比如主从复制中断、主从延迟、大事务等。而熟悉二进制日志事件，有助于我们在基于 binlog 进行指定位置点（时间点）的恢复时，找到准确的位置点。

本章主要包括以下内容。

- binlog 的格式。
- 如何解读 binlog 的内容。
- 如何解读 relay log 的内容。
- binlog 中的事件类型。
- 如何基于 python-mysql-replication 打造一个 binlog 解析器。

3.1 binlog 的格式

binlog 的格式决定了 SQL 记录在 binlog 中的内容。对于同一条 SQL 语句，binlog 的格式不同，记录在 binlog 中的内容也会不同，相应地，这条 SQL 语句在从库重放时的执行方式和注意事项也不同。

binlog 的格式由参数 `binlog_format` 决定，它的值有 3 个选项：STATEMENT、ROW 和 MIXED，可在全局或会话级别修改。在 MySQL 5.7.7 之前，该参数的默认值为 STATEMENT；从 MySQL 5.7.7 开始，其默认值为 ROW。

下面具体来看看 binlog 这 3 种格式的区别及各自的优缺点。

3.1.1 STATEMENT

基于语句（statement）的复制是 MySQL 3.23 引入的。在这种复制格式下，主库上执行的 SQL 语句会原封不动地记录在 binlog 中。

基于语句的复制有如下优点。

- 节省 binlog 的空间，尤其是当 SQL 涉及的记录比较多的时候。

- 可用于审核，毕竟所有的 DML 语句都会记录在 binlog 中。

基于语句的复制有如下缺点。

- 可能会导致主从数据不一致。

  在执行 SQL 语句时，如果碰到以下提示，则意味着该语句是不安全的。这里的不安全指的是，同一条语句在主从库中执行的结果有可能不一致。

  ```
  mysql> insert into slowtech.t1 values(sleep(1));
  Query OK, 1 row affected, 1 warning (1.01 sec)

  mysql> show warnings\G
  *************************** 1. row ***************************
    Level: Note
     Code: 1592
  Message: Unsafe statement written to the binary log using statement format since BINLOG_FORMAT = STATEMENT. Statement is unsafe because it uses a system function that may return a different value on the slave.
  1 row in set (0.00 sec)
  ```

  不安全的语句通常分为以下 3 类。

  - 使用了以下函数：

    LOAD_FILE()、UUID()、UUID_SHORT()、USER()、FOUND_ROWS()、SYSDATE()、GET_LOCK()、IS_FREE_LOCK()、IS_USED_LOCK()、MASTER_POS_WAIT()、RAND()、RELEASE_LOCK()、SLEEP()和VERSION()。

  - DELETE 和 UPDATE 操作使用了 LIMIT 子句，却没有带 ORDER BY。
  - 使用了自定义函数或存储过程，但函数或存储过程的执行结果是不确定的。这里的不确定指的是相同的输入并不总是产生相同的输出。

- 在执行下述操作时，会锁定更多的行。

  - INSERT ... SELECT。
  - 涉及自增主键的 INSERT 操作。
  - UPDATE 或 DELETE 操作中的 WHERE 条件没有索引。

之所以不推荐使用 STATEMENT 格式，最主要还是因为它会导致主从数据不一致。

3.1.2　ROW

基于行（row）的复制是 MySQL 5.1 引入的。相对于基于语句的复制，它不会记录具体的 SQL，而是记录 DML 操作涉及的行。在从库重放时，直接由存储引擎层处理，而基于语句的复制，因为执行的是具体的 SQL，所以还会经历词法解析、语法解析、选择执行计划等阶段。

基于行的复制有如下优点。

- 安全，不会导致主从数据不一致。
- 某些操作的锁定范围比 STATEMENT 格式小。

基于行的复制有如下缺点。

- 会产生大量日志。

  假如一张表有 10 000 条记录，如果我不带任何条件地执行 DELETE 操作，那么在基于语句的复制中，binlog 只会记录 DELETE 这一条命令，但是在基于行的复制中，则会记录 10 000 条记录。产生大量日志有以下弊端：

  (1) 利用 binlog 进行恢复，所需时间相对较长。
  (2) 在把数据写入 binlog 时，数据量大会导致 binlog 的锁定时间较长，进而影响数据库的并发。
  (3) 日志越大，对磁盘 I/O 和网络 I/O 的压力也越大。

- 不能看到具体的 SQL 语句。

3.1.3 MIXED

在这种混合格式下，MySQL 会根据执行的语句自动在 STATEMENT 和 ROW 这两种格式间切换。默认情况下，使用的是 STATEMENT 格式，在遇到不安全的语句时，会切换为 ROW 格式。

需要注意的是，数据库的事务隔离级别还会影响 binlog 的格式。

- 在 RC（read committed，读已提交）隔离级别下，binlog 格式只能为 ROW 或 MIXED，不能为 STATEMENT。看看下面这个示例。

  ```
  mysql> set session transaction isolation level read committed;
  Query OK, 0 rows affected (0.00 sec)

  mysql> set session binlog_format='statement';
  Query OK, 0 rows affected (0.00 sec)

  mysql> insert into slowtech.t1 values(1);
  ERROR 1665 (HY000): Cannot execute statement: impossible to write to binary log since BINLOG_FORMAT = STATEMENT and at least one table uses a storage engine limited to row-based logging. InnoDB is limited to row-logging when transaction isolation level is READ COMMITTED or READ UNCOMMITTED.
  ```

  为什么 RC 隔离级别和 STATEMENT 日志格式不能共存呢？因为这会导致主从数据不一致。相关的 bug 信息可参考 MySQL bug #23051：READ COMMITTED breaks mixed and statement-based replication。

- 在 RC 隔离级别下，虽然能将 binlog_format 设置为 MIXED，但实际生效的还是 ROW 格式。看看下面这个示例。

  ```
  mysql> set session transaction isolation level read committed;
  Query OK, 0 rows affected (0.00 sec)

  mysql> set session binlog_format='mixed';
  Query OK, 0 rows affected (0.00 sec)

  mysql> insert into slowtech.t1 values(1);
  Query OK, 1 row affected (0.01 sec)

  mysql> show binlog events in 'mysql-bin.000001';
  +------------------+-----+-------------+-----------+-------------+------------------------------------------+
  | Log_name         | Pos | Event_type  | Server_id | End_log_pos | Info                                     |
  ```

```
+------------------+-----+----------------+-----------+-------------+---------------------------------------------+
| mysql-bin.000001 |   4 | Format_desc    |         1 |         125 | Server ver: 8.0.25, Binlog ver: 4           |
| mysql-bin.000001 | 125 | Previous_gtids |         1 |         156 |                                             |
| mysql-bin.000001 | 156 | Anonymous_Gtid |         1 |         235 | SET @@SESSION.GTID_NEXT= 'ANONYMOUS'        |
| mysql-bin.000001 | 235 | Query          |         1 |         306 | BEGIN                                       |
| mysql-bin.000001 | 306 | Table_map      |         1 |         358 | table_id: 257 (slowtech.t1)                 |
| mysql-bin.000001 | 358 | Write_rows     |         1 |         398 | table_id: 257 flags: STMT_END_F             |
| mysql-bin.000001 | 398 | Xid            |         1 |         429 | COMMIT /* xid=209 */                        |
+------------------+-----+----------------+-----------+-------------+---------------------------------------------+
7 rows in set (0.00 sec)
```

这里的 Write_rows 即 ROW 格式下针对 INSERT 操作的事件类型。对于类似的 INSERT 操作，再来看看 RR（repeatable read，可重复读）隔离级别下的效果。

```
mysql> set session transaction isolation level repeatable read;
Query OK, 0 rows affected (0.00 sec)

mysql> set session binlog_format='mixed';
Query OK, 0 rows affected (0.00 sec)

mysql> insert into slowtech.t1 values(1);
Query OK, 1 row affected (0.01 sec)

mysql> show binlog events in 'mysql-bin.000001';
+------------------+-----+----------------+-----------+-------------+---------------------------------------------+
| Log_name         | Pos | Event_type     | Server_id | End_log_pos | Info                                        |
+------------------+-----+----------------+-----------+-------------+---------------------------------------------+
| mysql-bin.000001 |   4 | Format_desc    |         1 |         125 | Server ver: 8.0.25, Binlog ver: 4           |
| mysql-bin.000001 | 125 | Previous_gtids |         1 |         156 |                                             |
| mysql-bin.000001 | 156 | Anonymous_Gtid |         1 |         235 | SET @@SESSION.GTID_NEXT= 'ANONYMOUS'        |
| mysql-bin.000001 | 235 | Query          |         1 |         317 | BEGIN                                       |
| mysql-bin.000001 | 317 | Query          |         1 |         427 | insert into slowtech.t1 values(1)           |
| mysql-bin.000001 | 427 | Xid            |         1 |         458 | COMMIT /* xid=227 */                        |
+------------------+-----+----------------+-----------+-------------+---------------------------------------------+
6 rows in set (0.00 sec)
```

可以看到具体的 SQL 语句，这就是 STATEMENT 格式下 binlog 的内容。

3.2 如何解读 binlog 的内容

下面我们通过具体的示例来看看 binlog 的内容。

3.2.1 解析 STATEMENT 格式的二进制日志

首先，构造以下测试数据。

```
mysql> set session binlog_format='statement';
Query OK, 0 rows affected (0.00 sec)

mysql> create database slowtech;
Query OK, 1 row affected (0.00 sec)

mysql> create table slowtech.t1(id int primary key, c1 varchar(10));
Query OK, 0 rows affected (0.02 sec)

mysql> insert into slowtech.t1 values(1,'a');
```

```
Query OK, 1 row affected (0.00 sec)

mysql> show master status;
+------------------+----------+--------------+------------------+------------------------------------------+
| File             | Position | Binlog_Do_DB | Binlog_Ignore_DB | Executed_Gtid_Set                        |
+------------------+----------+--------------+------------------+------------------------------------------+
| mysql-bin.000001 |      884 |              |                  | 12ad7127-0466-11ec-8c74-000c292c1f7b:1-3 |
+------------------+----------+--------------+------------------+------------------------------------------+
1 row in set (0.00 sec)
```

接下来看看这些操作对应的 binlog 的内容。查看 binlog 的内容，通常有以下两种方式。

- mysqlbinlog

  官方自带的工具，用于离线解析 binlog，支持基于时间点和位置点的解析，功能强大且输出信息详尽。

- SHOW BINLOG EVENTS [IN 'log_name'] [FROM pos] [LIMIT [offset,] row_count]

  在线解析 binlog，功能有限。

首先看看 SHOW BINLOG EVENTS 命令的解析结果。

```
mysql> show binlog events in 'mysql-bin.000001';
+------------------+-----+----------------+-----------+-------------+-----------------------------------------------------------------------------+
| Log_name         | Pos | Event_type     | Server_id | End_log_pos | Info                                                                        |
+------------------+-----+----------------+-----------+-------------+-----------------------------------------------------------------------------+
| mysql-bin.000001 |   4 | Format_desc    |         1 |         125 | Server ver: 8.0.25, Binlog ver: 4                                           |
| mysql-bin.000001 | 125 | Previous_gtids |         1 |         156 |                                                                             |
| mysql-bin.000001 | 156 | Gtid           |         1 |         233 | SET @@SESSION.GTID_NEXT= '12ad7127-0466-11ec-8c74-000c292c1f7b:1'           |
| mysql-bin.000001 | 233 | Query          |         1 |         353 | create database slowtech /* xid=58 */                                       |
| mysql-bin.000001 | 353 | Gtid           |         1 |         430 | SET @@SESSION.GTID_NEXT= '12ad7127-0466-11ec-8c74-000c292c1f7b:2'           |
| mysql-bin.000001 | 430 | Query          |         1 |         578 | create table slowtech.t1(id int primary key, c1 varchar(10)) /* xid=59 */   |
| mysql-bin.000001 | 578 | Gtid           |         1 |         657 | SET @@SESSION.GTID_NEXT= '12ad7127-0466-11ec-8c74-000c292c1f7b:3'           |
| mysql-bin.000001 | 657 | Query          |         1 |         739 | BEGIN                                                                       |
| mysql-bin.000001 | 739 | Query          |         1 |         853 | insert into slowtech.t1 values(1,'a')                                       |
| mysql-bin.000001 | 853 | Xid            |         1 |         884 | COMMIT /* xid=60 */                                                         |
+------------------+-----+----------------+-----------+-------------+-----------------------------------------------------------------------------+
10 rows in set (0.00 sec)
```

每一行都是一个事件。下面解释一下各个字段的含义。

- Log_name：当前正在解析的 binlog 文件。
- Pos：事件的起始位置点。
- Event_type：事件类型。
- Server_id：生成该事件的实例的 server_id。
- End_log_pos：事件的结束位置点。
- Info：事件的内容。

熟悉了 SHOW BINLOG EVENTS 的输出，接下来看看 mysqlbinlog 的解析结果。

```
# mysqlbinlog /data/mysql/3306/data/mysql-bin.000001
/*!50530 SET @@SESSION.PSEUDO_SLAVE_MODE=1*/;
/*!50003 SET @OLD_COMPLETION_TYPE=@@COMPLETION_TYPE,COMPLETION_TYPE=0*/;
DELIMITER /*!*/;
```

```
# at 4
#210824 12:46:35 server id 1  end_log_pos 125 CRC32 0x7c4165cd  Start: binlog v 4, server v 8.0.25 created
210824 12:46:35 at startup
# Warning: this binlog is either in use or was not closed properly.
ROLLBACK/*!*/;
BINLOG '
q3kkYQ8BAAAAeQAAAH0AAAABAAQAOC4wLjI1AAAAAAAAAAAAAAAAAAAAAAAAAAAAAAAA
AAAAAAAAAAAAAAAAAACreSRhEwANAAgAAAAABAAEAAAAYQAEGggAAAAICAgCAAAACgoKKioAEjQA
CigBzWVBfA==
'/*!*/;

# at 125
#210824 12:46:35 server id 1  end_log_pos 156 CRC32 0x9aa32f82  Previous-GTIDs
# [empty]

# at 156
#210824 16:18:07 server id 1  end_log_pos 233 CRC32 0xe4fd5d6b  GTID    last_committed=0
sequence_number=1       rbr_only=no
original_committed_timestamp=1629793087585337immediate_commit_timestamp=1629793087585337
transaction_length=197
# original_commit_timestamp=1629793085585337 (2021-08-24 16:18:07.585337 CST)
# immediate_commit_timestamp=1629793087585337 (2021-08-24 16:18:07.585337 CST)
/*!80001 SET @@session.original_commit_timestamp=1629793087585337*//*!*/;
/*!80014 SET @@session.original_server_version=80025*//*!*/;
/*!80014 SET @@session.immediate_server_version=80025*//*!*/;
SET @@SESSION.GTID_NEXT= '12ad7127-0466-11ec-8c74-000c292c1f7b:1'/*!*/;
# at 233
#210824 16:18:07 server id 1  end_log_pos 353 CRC32 0xe4aff954  Query   thread_id=16    exec_time=0
error_code=0    Xid = 58
SET TIMESTAMP=1629793087/*!*/;
SET @@session.pseudo_thread_id=16/*!*/;
SET @@session.foreign_key_checks=1, @@session.sql_auto_is_null=0, @@session.unique_checks=1,
@@session.autocommit=1/*!*/;
SET @@session.sql_mode=1168113696/*!*/;
SET @@session.auto_increment_increment=1, @@session.auto_increment_offset=1/*!*/;
/*!\C utf8mb4 *//*!*/;
SET
@@session.character_set_client=255,@@session.collation_connection=255,@@session.collation_server=
255/*!*/;
SET @@session.lc_time_names=0/*!*/;
SET @@session.collation_database=DEFAULT/*!*/;
/*!80011 SET @@session.default_collation_for_utf8mb4=255*//*!*/;
/*!80016 SET @@session.default_table_encryption=0*//*!*/;
create database slowtech
/*!*/;

# at 353
#210824 16:18:13 server id 1  end_log_pos 430 CRC32 0x9f37bceb  GTID    last_committed=1
sequence_number=2       rbr_only=no
original_committed_timestamp=1629793093490326immediate_commit_timestamp=1629793093490326
transaction_length=225
# original_commit_timestamp=1629793093490326 (2021-08-24 16:18:13.490326 CST)
# immediate_commit_timestamp=1629793093490326 (2021-08-24 16:18:13.490326 CST)
/*!80001 SET @@session.original_commit_timestamp=1629793093490326*//*!*/;
/*!80014 SET @@session.original_server_version=80025*//*!*/;
/*!80014 SET @@session.immediate_server_version=80025*//*!*/;
SET @@SESSION.GTID_NEXT= '12ad7127-0466-11ec-8c74-000c292c1f7b:2'/*!*/;
# at 430
#210824 16:18:13 server id 1  end_log_pos 578 CRC32 0xa984042a  Query   thread_id=16    exec_time=0
error_code=0    Xid = 59
```

```
SET TIMESTAMP=1629793093/*!*/;
/*!80013 SET @@session.sql_require_primary_key=0*//*!*/;
create table slowtech.t1(id int primary key, c1 varchar(10))
/*!*/;

# at 578
#210824 16:18:23 server id 1  end_log_pos 657 CRC32 0xb1028a94   GTID     last_committed=2       sequence_number=3       rbr_only=no
original_committed_timestamp=1629793103625235immediate_commit_timestamp=1629793103625235       transaction_length=306
# original_commit_timestamp=1629793103625235 (2021-08-24 16:18:23.625235 CST)
# immediate_commit_timestamp=1629793103625235 (2021-08-24 16:18:23.625235 CST)
/*!80001 SET @@session.original_commit_timestamp=1629793103625235*//*!*/;
/*!80014 SET @@session.original_server_version=80025*//*!*/;
/*!80014 SET @@session.immediate_server_version=80025*//*!*/;
SET @@SESSION.GTID_NEXT= '12ad7127-0466-11ec-8c74-000c292c1f7b:3'/*!*/;
# at 657
#210824 16:18:23 server id 1  end_log_pos 739 CRC32 0x2b7c870e  Query    thread_id=16   exec_time=0  error_code=0
SET TIMESTAMP=1629793103/*!*/;
BEGIN
/*!*/;
# at 739
#210824 16:18:23 server id 1  end_log_pos 853 CRC32 0xb4628b67  Query    thread_id=16   exec_time=0  error_code=0
SET TIMESTAMP=1629793103/*!*/;
insert into slowtech.t1 values(1,'a')
/*!*/;
# at 853
#210824 16:18:23 server id 1  end_log_pos 884 CRC32 0x92b1b8c9  Xid = 60
COMMIT/*!*/;
SET @@SESSION.GTID_NEXT= 'AUTOMATIC' /* added by mysqlbinlog */ /*!*/;
DELIMITER ;
# End of log file
/*!50003 SET COMPLETION_TYPE=@OLD_COMPLETION_TYPE*/;
/*!50530 SET @@SESSION.PSEUDO_SLAVE_MODE=0*/;
```

为了提升可读性，我们在一些事件前后加了空行。输出中包括 2 个公共事件和 3 个与事务相关的事件。

- at 4：FORMAT_DESCRIPTION_EVENT。这是 binlog 中的第一个事件，mysqlbinlog 会根据这个事件的内容来解析 binlog 中的其他事件。
- at 125：PREVIOUS_GTIDS_LOG_EVENT。这个事件会记录之前已经分配的 GTID 集，因为 mysql-bin.000001 是第一个 binlog，所以这里的内容为空。
- at 156、at 353、at 578 分别对应一个事务，每个事务由若干个事件组成。

下面以一个完整的事务为例，看看一个事务由哪些事件组成。

```
# at 578
#210824 16:18:23 server id 1  end_log_pos 657 CRC32 0xb1028a94   GTID     last_committed=2       sequence_number=3       rbr_only=no
original_committed_timestamp=1629793103625235immediate_commit_timestamp=1629793103625235       transaction_length=306
# original_commit_timestamp=1629793103625235 (2021-08-24 16:18:23.625235 CST)
# immediate_commit_timestamp=1629793103625235 (2021-08-24 16:18:23.625235 CST)
/*!80001 SET @@session.original_commit_timestamp=1629793103625235*//*!*/;
/*!80014 SET @@session.original_server_version=80025*//*!*/;
```

```
/*!80014 SET @@session.immediate_server_version=80025*//*!*/;
SET @@SESSION.GTID_NEXT= '12ad7127-0466-11ec-8c74-000c292c1f7b:3'/*!*/;

# at 657
#210824 16:18:23 server id 1  end_log_pos 739 CRC32 0x2b7c870e  Query    thread_id=16    exec_time=0
error_code=0
SET TIMESTAMP=1629793103/*!*/;
BEGIN
/*!*/;

# at 739
#210824 16:18:23 server id 1  end_log_pos 853 CRC32 0xb4628b67  Query    thread_id=16    exec_time=0
error_code=0
SET TIMESTAMP=1629793103/*!*/;
insert into slowtech.t1 values(1,'a')
/*!*/;

# at 853
#210824 16:18:23 server id 1  end_log_pos 884 CRC32 0x92b1b8c9  Xid = 60
COMMIT/*!*/;
```

整个事务由如下 4 个事件组成。

- at 578 ~ at 657：GTID_LOG_EVENT。在开启 GTID 的情况下，每个事务都会分配一个 GTID 值，这个值会以 GTID_LOG_EVENT 的形式写入 binlog。如果没有开启 GTID，此处会写入 ANONYMOUS_GTID_LOG_EVENT，例如：

  ```
  SET @@SESSION.GTID_NEXT= 'ANONYMOUS'
  ```

- at 657 ~ at 739：QUERY_EVENT。binlog 中最常用的事件，对于 DDL、STATEMENT 格式的 DML 操作、事务开始的 BEGIN 操作，都会用 QUERY_EVENT 表示。
- at 739 ~ at 853：QUERY_EVENT。记录 STATEMENT 格式的 DML 操作。
- at 853 ~ at 884：XID_EVENT。记录 InnoDB 表的 COMMIT 操作。

每个事件前面都有注释，注释以#号开头。例如：

```
# at 739
#210824 16:18:23 server id 1  end_log_pos 853 CRC32 0xb4628b67  Query    thread_id=16    exec_time=0
error_code=0
```

下面看看这个注释中各个字段的具体含义。

- at 739：事件的起始位置点。需要注意的是，at xxx 不仅仅是一个逻辑位置点，更重要的是，它也是这个文件的物理位置点。

  例如，对于示例中的日志，最后一个事件的 end_log_pos 是 884，实际上，这个 binlog 的大小也是 884 字节。

  ```
  # ll /data/mysql/3306/data/mysql-bin.000001
  -rw-r-----. 1 mysql mysql 884 Aug 24 16:18 /data/mysql/3306/data/mysql-bin.000001
  ```

  再来看另外一个示例，仅通过截取文件来获取 binlog 中指定事务的内容。

  ```
  #!/usr/bin/python
  # -*- coding: utf-8 -*-
  ```

```
with open("/data/mysql/3306/data/mysql-bin.000001", "r") as f1:
    with open("/tmp/mysql-bin.tmp", "w") as f2:
        format_description_event = f1.read(155)
        f2.write(format_description_event)
        f1.seek(577)
        insert_related_event = f1.read(884 - 577)
        f2.write(insert_related_event)
```

这段代码截取了 insert into slowtech.t1 values(1,'a') 这个操作的相关事件。除此之外，还截取了 FORMAT_DESCRIPTION_EVENT。之所以要截取这个事件，是因为 mysqlbinlog 需要基于这个事件来解析 binlog 中的其他事件。

下面来看看截取的效果。

```
# mysql -e 'truncate table slowtech.t1'
# mysql -e 'select * from slowtech.t1'
# mysqlbinlog /tmp/mysql-bin.tmp --skip-gtids | mysql
# mysql -e 'select * from slowtech.t1'
+----+------+
| id | c1   |
+----+------+
|  1 | a    |
+----+------+
```

INSERT 操作确实执行了。

- 210824 16:18:23：事件的开始执行时间。
- server id 1：生成该事件的实例的 server_id。
- end_log_pos 853：事件的结束位置点。
- CRC32 0xb4628b67：从 MySQL 5.6 开始，为了保证事件的完整性，引入了校验和机制。MySQL 在写入每个事件时，都会添加一个校验和。默认的校验和算法是 CRC32。
- Query：事件类型为 QUERY_EVENT。
- thread_id=16：执行这个事件的线程 ID。
- exec_time=0：事件的执行时长，单位是秒。
- error_code=0：事件的错误码。

在使用 mysqlbinlog 进行基于位置点的数据恢复时，很多人看到这些位置点后十分茫然，不知道哪个位置点代表着事务的结束。他们常常以为 COMMIT 操作对应的位置点是 853，于是在指定结束位置点时将 --stop-position 设置为 853，但实际上，853 只是事件的开始，一个事件的结束位置点应该是 COMMIT 操作对应的注释部分的 end_log_pos。所以，如果要包含这个事务，必须将 --stop-position 设置为 884。

需要注意的是，如果指定的位置点不对，没有包括 COMMIT 操作，则 mysqlbinlog 会自动在解析结果的末尾添加一个 ROLLBACK 操作。

下面还以上述事务为例，看看不包括 COMMIT 操作时的输出。

```
# mysqlbinlog /data/mysql/3306/data/mysql-bin.000001 --start-position=578 --stop-position=853
...
# at 739
```

```
#210824 16:18:23 server id 1  end_log_pos 853 CRC32 0xb4628b67  Query   thread_id=16    exec_time=0
error_code=0
SET TIMESTAMP=1629793103/*!*/;
insert into slowtech.t1 values(1,'a')
/*!*/;
ROLLBACK /* added by mysqlbinlog */ /*!*/;
SET @@SESSION.GTID_NEXT= 'AUTOMATIC' /* added by mysqlbinlog */ /*!*/;
DELIMITER ;
# End of log file
/*!50003 SET COMPLETION_TYPE=@OLD_COMPLETION_TYPE*/;
/*!50530 SET @@SESSION.PSEUDO_SLAVE_MODE=0*/;
```

3.2.2 解析 ROW 格式的二进制日志

同样的测试数据，如果在 ROW 格式下生成，对应 binlog 的内容会有何不同呢？

```
# mysqlbinlog /data/mysql/3306/data/mysql-bin.000001
/*!50530 SET @@SESSION.PSEUDO_SLAVE_MODE=1*/;
/*!50003 SET @OLD_COMPLETION_TYPE=@@COMPLETION_TYPE,COMPLETION_TYPE=0*/;
DELIMITER /*!*/;
# at 4
#210824 16:40:03 server id 1  end_log_pos 125 CRC32 0xd19cf4c1  Start: binlog v 4, server v 8.0.25 created 210824 16:40:03 at startup
# Warning: this binlog is either in use or was not closed properly.
ROLLBACK/*!*/;
BINLOG '
Y7AkYQ8BAAAAeQAAAH0AAAABAAQAOC4wLjI1AAAAAAAAAAAAAAAAAAAAAAAAAAAAAAAA
AAAAAAAAAAAAAAAABjsCRhEwANAAgAAAAABAAEAAAAYQAEGggAAAAICAgCAAAACgoKKioAEjQA
CigBwfSc0Q==
'/*!*/;

# at 125
#210824 16:40:03 server id 1  end_log_pos 156 CRC32 0x409f2679  Previous-GTIDs
# [empty]

# at 156
#210824 16:40:13 server id 1  end_log_pos 233 CRC32 0xabfc200c  GTID    last_committed=0
sequence_number=1       rbr_only=no
original_committed_timestamp=1629794413463441immediate_commit_timestamp=1629794413463441
transaction_length=197
# original_commit_timestamp=1629794413463441 (2021-08-24 16:40:13.463441 CST)
# immediate_commit_timestamp=1629794413463441 (2021-08-24 16:40:13.463441 CST)
/*!80001 SET @@session.original_commit_timestamp=1629794413463441*//*!*/;
/*!80014 SET @@session.original_server_version=80025*//*!*/;
/*!80014 SET @@session.immediate_server_version=80025*//*!*/;
SET @@SESSION.GTID_NEXT= '12ad7127-0466-11ec-8c74-000c292c1f7b:1'/*!*/;
# at 233
#210824 16:40:13 server id 1  end_log_pos 353 CRC32 0xe4bbb101  Query   thread_id=17    exec_time=0
error_code=0    Xid = 71
SET TIMESTAMP=1629794413/*!*/;
SET @@session.pseudo_thread_id=17/*!*/;
SET @@session.foreign_key_checks=1, @@session.sql_auto_is_null=0, @@session.unique_checks=1, @@session.autocommit=1/*!*/;
SET @@session.sql_mode=1168113696/*!*/;
SET @@session.auto_increment_increment=1, @@session.auto_increment_offset=1/*!*/;
/*!\C utf8mb4 *//*!*/;
SET @@session.character_set_client=255,@@session.collation_connection=255,@@session.collation_server=2
```

```
55/*!*/;
SET @@session.lc_time_names=0/*!*/;
SET @@session.collation_database=DEFAULT/*!*/;
/*!80011 SET @@session.default_collation_for_utf8mb4=255*//*!*/;
/*!80016 SET @@session.default_table_encryption=0*//*!*/;
create database slowtech
/*!*/;

# at 353
#210824 16:40:17 server id 1  end_log_pos 430 CRC32 0x944c4efb  GTID    last_committed=1
sequence_number=2       rbr_only=no
original_committed_timestamp=1629794417906773immediate_commit_timestamp=1629794417906773
transaction_length=225
# original_commit_timestamp=1629794417906773 (2021-08-24 16:40:17.906773 CST)
# immediate_commit_timestamp=1629794417906773 (2021-08-24 16:40:17.906773 CST)
/*!80001 SET @@session.original_commit_timestamp=1629794417906773*//*!*/;
/*!80014 SET @@session.original_server_version=80025*//*!*/;
/*!80014 SET @@session.immediate_server_version=80025*//*!*/;
SET @@SESSION.GTID_NEXT= '12ad7127-0466-11ec-8c74-000c292c1f7b:2'/*!*/;
# at 430
#210824 16:40:17 server id 1  end_log_pos 578 CRC32 0x9053ebcb  Query   thread_id=17    exec_time=0
error_code=0    Xid = 72
SET TIMESTAMP=1629794417/*!*/;
/*!80013 SET @@session.sql_require_primary_key=0*//*!*/;
create table slowtech.t1(id int primary key, c1 varchar(10))
/*!*/;

# at 578
#210824 16:40:22 server id 1  end_log_pos 657 CRC32 0xba5e1986  GTID    last_committed=2
sequence_number=3       rbr_only=yes
original_committed_timestamp=1629794422205691immediate_commit_timestamp=1629794422205691
transaction_length=283
/*!50718 SET TRANSACTION ISOLATION LEVEL READ COMMITTED*//*!*/;
# original_commit_timestamp=1629794422205691 (2021-08-24 16:40:22.205691 CST)
# immediate_commit_timestamp=1629794422205691 (2021-08-24 16:40:22.205691 CST)
/*!80001 SET @@session.original_commit_timestamp=1629794422205691*//*!*/;
/*!80014 SET @@session.original_server_version=80025*//*!*/;
/*!80014 SET @@session.immediate_server_version=80025*//*!*/;
SET @@SESSION.GTID_NEXT= '12ad7127-0466-11ec-8c74-000c292c1f7b:3'/*!*/;
# at 657
#210824 16:40:22 server id 1  end_log_pos 728 CRC32 0x4dda952a  Query   thread_id=17    exec_time=0
error_code=0
SET TIMESTAMP=1629794422/*!*/;
BEGIN
/*!*/;
# at 728
#210824 16:40:22 server id 1  end_log_pos 788 CRC32 0x0767e99e  Table_map: `slowtech`.`t1` mapped to number 209
# at 788
#210824 16:40:22 server id 1  end_log_pos 830 CRC32 0x21209717  Write_rows: table id 209 flags: STMT_END_F

BINLOG '
drAkYRMBAAAAPAAAABQDAAAAAANEAAAAAAAAEACHNsb3d0ZWNoAAJ0MQACAw8CKAACAQEAAgP8/wCe
6WcH
drAkYR4BAAAAKgAAAD4DAAAANEAAAAAAEAAgAC/wABAAAAAWEXlyAh
'/*!*/;
# at 830
#210824 16:40:22 server id 1  end_log_pos 861 CRC32 0x568c0bad  Xid = 73
COMMIT/*!*/;
SET @@SESSION.GTID_NEXT= 'AUTOMATIC' /* added by mysqlbinlog */ /*!*/;
```

```
DELIMITER ;
# End of log file
/*!50003 SET COMPLETION_TYPE=@OLD_COMPLETION_TYPE*/;
/*!50530 SET @@SESSION.PSEUDO_SLAVE_MODE=0*/;
```

竟然只能看到 CREATE 操作，没有 INSERT 操作！

实际上，这跟事件类型有关。对于 ROW 格式的 DML 操作，MySQL 会通过 Base64 来进行编码，所以我们看到的只是下面这样看似毫无意义的字符串。

```
drAkYRMBAAAAPAAAABQDAAAAANEAAAAAAAEACHNsb3d0ZWNoAAJ0MQACAw8CKAACAQEAAgP8/wCe
6WcH
drAkYR4BAAAAKgAAAD4DAAAAANEAAAAAAAEAAgAC/wABAAAAAWEXlyAh
```

除此之外，通过 SHOW BINLOG EVENTS 命令也无法看到具体的 SQL 语句，例如：

```
mysql> show binlog events in 'mysql-bin.000001';
+------------------+-----+----------------+-----------+-------------+----------------------------------------------------------------------------------+
| Log_name         | Pos | Event_type     | Server_id | End_log_pos | Info                                                                             |
+------------------+-----+----------------+-----------+-------------+----------------------------------------------------------------------------------+
| mysql-bin.000001 |   4 | Format_desc    |         1 |         125 | Server ver: 8.0.25, Binlog ver: 4                                                |
| mysql-bin.000001 | 125 | Previous_gtids |         1 |         156 |                                                                                  |
| mysql-bin.000001 | 156 | Gtid           |         1 |         233 | SET @@SESSION.GTID_NEXT= '12ad7127-0466-11ec-8c74-000c292c1f7b:1'                 |
| mysql-bin.000001 | 233 | Query          |         1 |         353 | create database slowtech /* xid=71 */                                            |
| mysql-bin.000001 | 353 | Gtid           |         1 |         430 | SET @@SESSION.GTID_NEXT= '12ad7127-0466-11ec-8c74-000c292c1f7b:2'                 |
| mysql-bin.000001 | 430 | Query          |         1 |         578 | create table slowtech.t1(id int primary key, c1 varchar(10)) /* xid=72 */        |
| mysql-bin.000001 | 578 | Gtid           |         1 |         657 | SET @@SESSION.GTID_NEXT= '12ad7127-0466-11ec-8c74-000c292c1f7b:3'                 |
| mysql-bin.000001 | 657 | Query          |         1 |         728 | BEGIN                                                                            |
| mysql-bin.000001 | 728 | Table_map      |         1 |         788 | table_id: 209 (slowtech.t1)                                                      |
| mysql-bin.000001 | 788 | Write_rows     |         1 |         830 | table_id: 209 flags: STMT_END_F                                                  |
| mysql-bin.000001 | 830 | Xid            |         1 |         861 | COMMIT /* xid=73 */                                                              |
+------------------+-----+----------------+-----------+-------------+----------------------------------------------------------------------------------+
11 rows in set (0.00 sec)
```

要想看到具体的 SQL 语句，需要添加 -v（--verbose 的缩写）参数。下面以一个事务为例。

```
# mysqlbinlog /data/mysql/3306/data/mysql-bin.000001 --start-position=578 --stop-position=861 -v
...
BEGIN
/*!*/;
# at 728
#210824 16:40:22 server id 1  end_log_pos 788 CRC32 0x0767e99e  Table_map: `slowtech`.`t1` mapped to number 209
# at 788
#210824 16:40:22 server id 1  end_log_pos 830 CRC32 0x21209717  Write_rows: table id 209 flags: STMT_END_F

BINLOG '
drAkYRMBAAAAPAAAABQDAAAAANEAAAAAAAEACHNsb3d0ZWNoAAJ0MQACAw8CKAACAQEAAgP8/wCe
6WcH
drAkYR4BAAAAKgAAAD4DAAAAANEAAAAAAAEAAgAC/wABAAAAAWEXlyAh
'/*!*/;
### INSERT INTO `slowtech`.`t1`
### SET
###   @1=1
###   @2='a'
# at 830
#210824 16:40:22 server id 1  end_log_pos 861 CRC32 0x568c0bad  Xid = 73
COMMIT/*!*/;
SET @@SESSION.GTID_NEXT= 'AUTOMATIC' /* added by mysqlbinlog */ /*!*/;
```

```
DELIMITER ;
# End of log file
/*!50003 SET COMPLETION_TYPE=@OLD_COMPLETION_TYPE*/;
/*!50530 SET @@SESSION.PSEUDO_SLAVE_MODE=0*/;
```

当然，严格来说，我们看到的这个语句也不能称为 SQL 语句，充其量是伪 SQL 语句。

如果将 -v 指定两次，还会输出列的基本信息，如列的类型、是否为空等。例如：

```
# mysqlbinlog /data/mysql/3306/data/mysql-bin.000001 --start-position=578 --stop-position=861 -vv
...
### INSERT INTO `slowtech`.`t1`
### SET
###   @1=1 /* INT meta=0 nullable=0 is_null=0 */
###   @2='a' /* VARSTRING(40) meta=40 nullable=1 is_null=0 */
...
```

我看过很多资料在解析 ROW 格式的 binlog 时会加上 --base64-output=decode-rows。起初我以为这个参数是导致 SQL 语句出现的决定性因素，但实际上并不是。正如上面所演示的，-v 参数才是导致 SQL 语句出现的关键因素。那么，添加 --base64-output=decode-rows 有什么作用呢？

首先看看加上这个参数时的输出。

```
# mysqlbinlog /data/mysql/3306/data/mysql-bin.000001 --start-position=578 --stop-position=861 -v
--base64-output=decode-rows
...
BEGIN
/*!*/;
# at 728
#210824 16:40:22 server id 1  end_log_pos 788 CRC32 0x0767e99e  Table_map: `slowtech`.`t1` mapped to number 209
# at 788
#210824 16:40:22 server id 1 end_log_pos 830 CRC32 0x21209717 Write_rows: table id 209 flags: STMT_END_F
### INSERT INTO `slowtech`.`t1`
### SET
###   @1=1
###   @2='a'
# at 830
#210824 16:40:22 server id 1  end_log_pos 861 CRC32 0x568c0bad  Xid = 73
COMMIT/*!*/;
SET @@SESSION.GTID_NEXT= 'AUTOMATIC' /* added by mysqlbinlog */ /*!*/;
DELIMITER ;
# End of log file
/*!50003 SET COMPLETION_TYPE=@OLD_COMPLETION_TYPE*/;
/*!50530 SET @@SESSION.PSEUDO_SLAVE_MODE=0*/;
```

相对于没有添加这个参数时的输出，少了 BINLOG 部分的内容，具体如下。

```
BINLOG '
drAkYRMBAAAAPAAAABQDAAAAANEAAAAAAEACHNsb3d0ZWNoAAJ0MQACAw8CKAACAQEAAgP8/wCe
6WcH
drAkYR4BAAAAKgAAAD4DAAAAANEAAAAAAEAAgAC/wABAAAAAWEXlyAh
'/*!*/;
```

实际上，这部分内容才是真正可被 MySQL 执行的数据，而下面这部分内容只是被注释掉的伪 SQL 语句，不能被 MySQL 执行。

```
### INSERT INTO `slowtech`.`t1`
### SET
###   @1=1
###   @2='a'
```

接下来对比验证一下。

```
# mysql -e 'truncate table slowtech.t1'

# mysqlbinlog mysql-bin.000001 --start-position=578 --stop-position=861 --skip-gtids
--base64-output=decode-rows | mysql

# mysql -e 'select * from slowtech.t1'

# mysqlbinlog mysql-bin.000001 --start-position=578 --stop-position=861 --skip-gtids | mysql

# mysql -e 'select * from slowtech.t1'
+----+------+
| id | c1   |
+----+------+
|  1 | a    |
+----+------+
```

不难发现，第二次执行时才有效果，而第一次没有。这是因为第一次执行时指定了 `--base64-output=decode-rows`，这个选项会过滤掉 INSERT 操作对应的二进制数据。因此，如果要基于 binlog 进行恢复，千万不要指定 `--base64-output=decode-rows`。

上面在使用 mysqlbinlog 时指定了 `--skip-gtids` 参数，这个参数会过滤掉 `GTID_LOG_EVENT`。如果不指定这个参数，因为 INSERT 操作对应的 GTID 在目标实例中存在，所以 INSERT 操作在实际应用的过程中会被直接忽略掉。这一点同样需要注意。

3.3 如何解读 relay log 的内容

还是以 STATEMENT 格式下的测试数据为例，我们通过 SHOW RELAYLOG EVENTS 命令看看从库 relay log 的内容。

```
mysql> show relaylog events in 'node2-relay-bin.000002';
+------------------------+------+----------------+-----------+-------------+----------------------------------------------------------------------------------+
| Log_name               | Pos  | Event_type     | Server_id | End_log_pos | Info                                                                             |
+------------------------+------+----------------+-----------+-------------+----------------------------------------------------------------------------------+
| node2-relay-bin.000002 |    4 | Format_desc    |         2 |         125 | Server ver: 8.0.25, Binlog ver: 4                                                |
| node2-relay-bin.000002 |  125 | Previous_gtids |         2 |         156 |                                                                                  |
| node2-relay-bin.000002 |  156 | Rotate         |         1 |           0 | mysql-bin.000001;pos=4                                                           |
| node2-relay-bin.000002 |  203 | Format_desc    |         1 |         125 | Server ver: 8.0.25, Binlog ver: 4                                                |
| node2-relay-bin.000002 |  324 | Rotate         |         0 |           0 | mysql-bin.000001;pos=156                                                         |
| node2-relay-bin.000002 |  371 | Gtid           |         1 |         233 | SET @@SESSION.GTID_NEXT= '12ad7127-0466-11ec-8c74-000c292c1f7b:1'                |
| node2-relay-bin.000002 |  448 | Query          |         1 |         353 | create database slowtech /* xid=58 */                                            |
| node2-relay-bin.000002 |  568 | Gtid           |         1 |         430 | SET @@SESSION.GTID_NEXT= '12ad7127-0466-11ec-8c74-000c292c1f7b:2'                |
| node2-relay-bin.000002 |  645 | Query          |         1 |         578 | create table slowtech.t1(id int primary key, c1 varchar(10)) /* xid=59 */        |
| node2-relay-bin.000002 |  793 | Gtid           |         1 |         657 | SET @@SESSION.GTID_NEXT= '12ad7127-0466-11ec-8c74-000c292c1f7b:3'                |
| node2-relay-bin.000002 |  872 | Query          |         1 |         739 | BEGIN                                                                            |
| node2-relay-bin.000002 |  954 | Query          |         1 |         853 | insert into slowtech.t1 values(1,'a')                                            |
| node2-relay-bin.000002 | 1068 | Xid            |         1 |         884 | COMMIT /* xid=60 */                                                              |
+------------------------+------+----------------+-----------+-------------+----------------------------------------------------------------------------------+
13 rows in set (0.00 sec)
```

对比看一下主库 binlog 的内容。

```
mysql> show binlog events in 'mysql-bin.000001';
+------------------+-----+----------------+-----------+-------------+-----------------------------------------------------------------------------+
| Log_name         | Pos | Event_type     | Server_id | End_log_pos | Info                                                                        |
+------------------+-----+----------------+-----------+-------------+-----------------------------------------------------------------------------+
| mysql-bin.000001 |   4 | Format_desc    |         1 |         125 | Server ver: 8.0.25, Binlog ver: 4                                           |
| mysql-bin.000001 | 125 | Previous_gtids |         1 |         156 |                                                                             |
| mysql-bin.000001 | 156 | Gtid           |         1 |         233 | SET @@SESSION.GTID_NEXT= '12ad7127-0466-11ec-8c74-000c292c1f7b:1'            |
| mysql-bin.000001 | 233 | Query          |         1 |         353 | create database slowtech /* xid=58 */                                       |
| mysql-bin.000001 | 353 | Gtid           |         1 |         430 | SET @@SESSION.GTID_NEXT= '12ad7127-0466-11ec-8c74-000c292c1f7b:2'            |
| mysql-bin.000001 | 430 | Query          |         1 |         578 | create table slowtech.t1(id int primary key, c1 varchar(10) /* xid=59 */    |
| mysql-bin.000001 | 578 | Gtid           |         1 |         657 | SET @@SESSION.GTID_NEXT= '12ad7127-0466-11ec-8c74-000c292c1f7b:3'            |
| mysql-bin.000001 | 657 | Query          |         1 |         739 | BEGIN                                                                       |
| mysql-bin.000001 | 739 | Query          |         1 |         853 | insert into slowtech.t1 values(1,'a')                                       |
| mysql-bin.000001 | 853 | Xid            |         1 |         884 | COMMIT /* xid=60 */                                                         |
+------------------+-----+----------------+-----------+-------------+-----------------------------------------------------------------------------+
10 rows in set (0.01 sec)
```

不难发现，同一个事件，除了起始位置点不一样，其他内容基本一致，包括结束位置点。为什么起始位置点不一样，但结束位置点一样呢？我们不妨使用 mysqlbinlog 解析一下 relay log 的内容。

```
# mysqlbinlog /data/mysql/3306/data/node2-relay-bin.000002
/*!50530 SET @@SESSION.PSEUDO_SLAVE_MODE=1*/;
/*!50003 SET @OLD_COMPLETION_TYPE=@@COMPLETION_TYPE,COMPLETION_TYPE=0*/;
DELIMITER /*!*/;
# at 4
#210824 16:10:34 server id 2  end_log_pos 125 CRC32 0x9c6bac93  Start: binlog v 4, server v 8.0.25 created
210824 16:10:34
# This Format_description_event appears in a relay log and was generated by the slave thread.

# at 125
#210824 16:10:34 server id 2  end_log_pos 156 CRC32 0x0d413a54  Previous-GTIDs
# [empty]

# at 156
#700101  8:00:00 server id 1  end_log_pos 0 CRC32 0x68ac0e1f    Rotate to mysql-bin.000001  pos: 4

# at 203
#210824 12:46:35 server id 1  end_log_pos 125 CRC32 0x7c4165cd  Start: binlog v 4, server v 8.0.25 created
210824 12:46:35 at startup
ROLLBACK/*!*/;
BINLOG '
q3kkYQ8BAAAAeQAAAH0AAAAAAAQAOC4wLjI1AAAAAAAAAAAAAAAAAAAAAAAAAAAAAAAA
AAAAAAAAAAAAAAAAACreSRhEwANAAgAAAAABAAEAAAAYQAEGggAAAAICAgCAAAACgoKKKioAEjQA
CigBzWVBfA==
'/*!*/;
# at 324
#210824 16:10:34 server id 0  end_log_pos 0 CRC32 0xfd0f480b    Rotate to mysql-bin.000001  pos: 156

# at 371
#210824 16:18:07 server id 1  end_log_pos 233 CRC32 0xe4fd5d6b  GTID    last_committed=0
sequence_number=1       rbr_only=no
original_committed_timestamp=1629793087585337immediate_commit_timestamp=1629793087585337
transaction_length=197
# original_commit_timestamp=1629793087585337 (2021-08-24 16:18:07.585337 CST)
# immediate_commit_timestamp=1629793087585337 (2021-08-24 16:18:07.585337 CST)
/*!80001 SET @@session.original_commit_timestamp=1629793087585337*///*!*/;
/*!80014 SET @@session.original_server_version=80025*///*!*/;
```

```
/*!80014 SET @@session.immediate_server_version=80025*//*!*/;
SET @@SESSION.GTID_NEXT= '12ad7127-0466-11ec-8c74-000c292c1f7b:1'/*!*/;
# at 448
#210824 16:18:07 server id 1  end_log_pos 353 CRC32 0xe4aff954   Query    thread_id=16    exec_time=0
error_code=0    Xid = 58
SET TIMESTAMP=1629793087/*!*/;
SET @@session.pseudo_thread_id=16/*!*/;
SET @@session.foreign_key_checks=1, @@session.sql_auto_is_null=0, @@session.unique_checks=1,
@@session.autocommit=1/*!*/;
SET @@session.sql_mode=1168113696/*!*/;
SET @@session.auto_increment_increment=1, @@session.auto_increment_offset=1/*!*/;
/*!\C utf8mb4 *//*!*/;
SET
@@session.character_set_client=255,@@session.collation_connection=255,@@session.collation_server=2
55/*!*/;
SET @@session.lc_time_names=0/*!*/;
SET @@session.collation_database=DEFAULT/*!*/;
/*!80011 SET @@session.default_collation_for_utf8mb4=255*//*!*/;
/*!80016 SET @@session.default_table_encryption=0*//*!*/;
create database slowtech
/*!*/;
...
```

事件的起始位置点不一样很容易理解。从上面的输出可以看到，relay log 不仅写入了 binlog 中的所有事件，还额外加入了多个事件，如 FORMAT_DESCRIPTION_EVENT、ROTATE_EVENT。除此之外，relay log 的切换时机也和 binlog 不一样。基于这两点，很难保证同一个事件在 binlog 和 relay log 中的起始位置点是一致的。既然如此，为什么事件的结束位置点又是一样的呢？因为事件的结束位置点（end_log_pos）来自事件头的注释，而注释作为事件的一部分，无论是在 binlog 中还是在 relay log 中，都是完全一样的。

既然 binlog 的二进制日志事件只是被简单地复制到 relay log 中，那么在基于位置点的复制中，要判断一个事务是否已被应用，在 MySQL 5.6 之前，其实没有很可靠的方法，只能简单地依赖 relay-log.info 文件中的 Relay_log_name 和 Relay_log_pos。如果 relay-log.info 文件更新不及时，当数据库服务器异常重启后，很容易导致同一个事务被重放两次。同理，如果 master.info 文件更新不及时，当数据库服务器异常重启后，也容易导致同一个事务被拉取两次。所以，从 MySQL 5.6 开始，在基于位置点的复制中，强烈建议设置 relay_log_recovery = ON 和 relay_log_info_repository=TABLE。

对于 GTID 复制，判断一个事务是否已被应用很简单，只需要看看该事务的 GTID 是否在 gtid_executed 中。如果在，则意味着该事务已被应用。在这种情况下，我们其实不用关心 relay-log.info 和 master.info 的更新是否及时。只需保证数据库在异常重启后，gtid_executed 还能准确无误地反映实例已经执行过的事务。为了保证这一点，需将参数 sync_binlog 和 innodb_flush_log_at_trx_commit 设置为 1。

3.4 binlog 中的事件类型

之前我们多次提到了事件。实际上，binlog 中记录的所有操作都有对应的事件类型。

首先，我们看看 binlog 中有哪些事件类型。

mysql-8.0.25/libbinlogevents/include/binlog_event.h

```
enum Log_event_type {
  /**
      每次添加新的事件类型时，都要注意以下两点：
      1. 为每个事件类型指定一个数，避免在移除之前被弃用的事件类型时出现问题
      2. 更新 Format_description_event::Format_description_event()
  */
  UNKNOWN_EVENT = 0,
  START_EVENT_V3 = 1,
  QUERY_EVENT = 2,
  STOP_EVENT = 3,
  ROTATE_EVENT = 4,
  INTVAR_EVENT = 5,

  SLAVE_EVENT = 7,

  APPEND_BLOCK_EVENT = 9,
  DELETE_FILE_EVENT = 11,

  RAND_EVENT = 13,
  USER_VAR_EVENT = 14,
  FORMAT_DESCRIPTION_EVENT = 15,
  XID_EVENT = 16,
  BEGIN_LOAD_QUERY_EVENT = 17,
  EXECUTE_LOAD_QUERY_EVENT = 18,

  TABLE_MAP_EVENT = 19,

  /** ROWS_EVENT 的 V1 版本，用于 MySQL 5.1.16 和 MySQL 5.6（不包括 MySQL 5.6）之间的版本*/
  WRITE_ROWS_EVENT_V1 = 23,
  UPDATE_ROWS_EVENT_V1 = 24,
  DELETE_ROWS_EVENT_V1 = 25,

  INCIDENT_EVENT = 26,

  /** 主库空闲时发送的心跳事件 */
  HEARTBEAT_LOG_EVENT = 27,

  IGNORABLE_LOG_EVENT = 28,
  ROWS_QUERY_LOG_EVENT = 29,

  /** ROWS_EVENT 的 V2 版本 */
  WRITE_ROWS_EVENT = 30,
  UPDATE_ROWS_EVENT = 31,
  DELETE_ROWS_EVENT = 32,

  GTID_LOG_EVENT = 33,
  ANONYMOUS_GTID_LOG_EVENT = 34,

  PREVIOUS_GTIDS_LOG_EVENT = 35,

  TRANSACTION_CONTEXT_EVENT = 36,

  VIEW_CHANGE_EVENT = 37,

  XA_PREPARE_LOG_EVENT = 38,

  /** UPDATE_ROWS_EVENT 的扩展，支持 JSON 文档的原地更新 */
  PARTIAL_UPDATE_ROWS_EVENT = 39,
```

```
    TRANSACTION_PAYLOAD_EVENT = 40,

    ENUM_END_EVENT
};
```

共 34 个事件类型。在这些事件中，比较常见的有以下 16 个。

- **QUERY_EVENT**

QUERY_EVENT 是 binlog 中最常见的一个事件，常用在以下场景中。

❑ DDL 操作。
❑ STATEMENT 格式的 DML 操作。
❑ 事务开始的 BEGIN 操作。

QUERY_EVENT 会记录具体的 SQL 语句，这也就是为什么我们在用 mysqlbinlog 解析 STATEMENT 格式的二进制日志时，不需要解码。

看看下面这个示例。

```
mysql> show binlog events in 'mysql-bin.000001';
+------------------+-----+----------------+-----------+-------------+----------------------------------------------------------------------------+
| Log_name         | Pos | Event_type     | Server_id | End_log_pos | Info                                                                       |
+------------------+-----+----------------+-----------+-------------+----------------------------------------------------------------------------+
| mysql-bin.000001 |   4 | Format_desc    |         1 |         125 | Server ver: 8.0.25, Binlog ver: 4                                          |
| mysql-bin.000001 | 125 | Previous_gtids |         1 |         156 |                                                                            |
| mysql-bin.000001 | 156 | Gtid           |         1 |         235 | SET @@SESSION.GTID_NEXT= 'bd6b3216-04d6-11ec-b76f-000c292c1f7b:1'           |
| mysql-bin.000001 | 235 | Query          |         1 |         317 | BEGIN                                                                      |
| mysql-bin.000001 | 317 | Query          |         1 |         431 | insert into slowtech.t1 values(1,'a')                                      |
| mysql-bin.000001 | 431 | Xid            |         1 |         462 | COMMIT /* xid=838 */                                                       |
+------------------+-----+----------------+-----------+-------------+----------------------------------------------------------------------------+
6 rows in set (0.00 sec)
```

- **ROWS_EVENT**

ROWS_EVENT 用于记录 ROW 格式的 DML 操作，具体包括以下 3 个事件。

❑ WRITE_ROWS_EVENT：记录 INSERT 操作。
❑ UPDATE_ROWS_EVENT：记录 UPDATE 操作。
❑ DELETE_ROWS_EVENT：记录 DELETE 操作。

看看下面这个示例。

```
mysql> set session binlog_format='row';
Query OK, 0 rows affected (0.00 sec)

mysql> insert into slowtech.t1 values(1,'a');
Query OK, 1 row affected (0.00 sec)

mysql> update slowtech.t1 set c1='b' where id=1;
Query OK, 1 row affected (0.01 sec)
Rows matched: 1  Changed: 1  Warnings: 0

mysql> delete from slowtech.t1 where id=1;
Query OK, 1 row affected (0.01 sec)
```

```
mysql> show binlog events in 'mysql-bin.000001';
+------------------+------+----------------+-----------+-------------+---------------------------------------------+
| Log_name         | Pos  | Event_type     | Server_id | End_log_pos | Info                                        |
+------------------+------+----------------+-----------+-------------+---------------------------------------------+
| mysql-bin.000001 |    4 | Format_desc    |         1 |         125 | Server ver: 8.0.25, Binlog ver: 4           |
| mysql-bin.000001 |  125 | Previous_gtids |         1 |         156 |                                             |
| mysql-bin.000001 |  156 | Anonymous_Gtid |         1 |         235 | SET @@SESSION.GTID_NEXT= 'ANONYMOUS'        |
| mysql-bin.000001 |  235 | Query          |         1 |         306 | BEGIN                                       |
| mysql-bin.000001 |  306 | Table_map      |         1 |         366 | table_id: 378 (slowtech.t1)                 |
| mysql-bin.000001 |  366 | Write_rows     |         1 |         408 | table_id: 378 flags: STMT_END_F             |
| mysql-bin.000001 |  408 | Xid            |         1 |         439 | COMMIT /* xid=248 */                        |
| mysql-bin.000001 |  439 | Anonymous_Gtid |         1 |         518 | SET @@SESSION.GTID_NEXT= 'ANONYMOUS'        |
| mysql-bin.000001 |  518 | Query          |         1 |         598 | BEGIN                                       |
| mysql-bin.000001 |  598 | Table_map      |         1 |         658 | table_id: 378 (slowtech.t1)                 |
| mysql-bin.000001 |  658 | Update_rows    |         1 |         708 | table_id: 378 flags: STMT_END_F             |
| mysql-bin.000001 |  708 | Xid            |         1 |         739 | COMMIT /* xid=249 */                        |
| mysql-bin.000001 |  739 | Anonymous_Gtid |         1 |         818 | SET @@SESSION.GTID_NEXT= 'ANONYMOUS'        |
| mysql-bin.000001 |  818 | Query          |         1 |         889 | BEGIN                                       |
| mysql-bin.000001 |  889 | Table_map      |         1 |         949 | table_id: 378 (slowtech.t1)                 |
| mysql-bin.000001 |  949 | Delete_rows    |         1 |         991 | table_id: 378 flags: STMT_END_F             |
| mysql-bin.000001 |  991 | Xid            |         1 |        1022 | COMMIT /* xid=250 */                        |
+------------------+------+----------------+-----------+-------------+---------------------------------------------+
17 rows in set (0.00 sec)
```

对于 ROWS_EVENT，写入 binlog 的内容是相关记录经过 Base64 编码后的结果，而不是具体的 SQL。即便使用 mysqlbinlog 加 -v 参数解析，看到的也只是伪 SQL 语句。由于无法像 QUERY_EVENT 那样看到具体的 SQL 语句，所以 ROWS_EVENT 在很多场景下还是存在不足的，如审计、大事务分析等场景。

所幸，MySQL 5.6 引入了 binlog_rows_query_log_events 参数，可将具体的 SQL 语句以 ROWS_QUERY_LOG_EVENT 的形式记录到 binlog 中。例如：

```
mysql> show binlog events in 'mysql-bin.000001';
+------------------+------+----------------+-----------+-------------+---------------------------------------------+
| Log_name         | Pos  | Event_type     | Server_id | End_log_pos | Info                                        |
+------------------+------+----------------+-----------+-------------+---------------------------------------------+
| mysql-bin.000001 |    4 | Format_desc    |         1 |         125 | Server ver: 8.0.25, Binlog ver: 4           |
| mysql-bin.000001 |  125 | Previous_gtids |         1 |         156 |                                             |
| mysql-bin.000001 |  156 | Anonymous_Gtid |         1 |         235 | SET @@SESSION.GTID_NEXT= 'ANONYMOUS'        |
| mysql-bin.000001 |  235 | Query          |         1 |         306 | BEGIN                                       |
| mysql-bin.000001 |  306 | Rows_query     |         1 |         367 | # insert into slowtech.t1 values(1,'a')     |
| mysql-bin.000001 |  367 | Table_map      |         1 |         427 | table_id: 398 (slowtech.t1)                 |
| mysql-bin.000001 |  427 | Write_rows     |         1 |         469 | table_id: 398 flags: STMT_END_F             |
| mysql-bin.000001 |  469 | Xid            |         1 |         500 | COMMIT /* xid=258 */                        |
| mysql-bin.000001 |  500 | Anonymous_Gtid |         1 |         579 | SET @@SESSION.GTID_NEXT= 'ANONYMOUS'        |
| mysql-bin.000001 |  579 | Query          |         1 |         659 | BEGIN                                       |
| mysql-bin.000001 |  659 | Rows_query     |         1 |         723 | # update slowtech.t1 set c1='b' where id=1  |
| mysql-bin.000001 |  723 | Table_map      |         1 |         783 | table_id: 398 (slowtech.t1)                 |
| mysql-bin.000001 |  783 | Update_rows    |         1 |         833 | table_id: 398 flags: STMT_END_F             |
| mysql-bin.000001 |  833 | Xid            |         1 |         864 | COMMIT /* xid=259 */                        |
| mysql-bin.000001 |  864 | Anonymous_Gtid |         1 |         943 | SET @@SESSION.GTID_NEXT= 'ANONYMOUS'        |
| mysql-bin.000001 |  943 | Query          |         1 |        1014 | BEGIN                                       |
| mysql-bin.000001 | 1014 | Rows_query     |         1 |        1072 | # delete from slowtech.t1 where id=1        |
| mysql-bin.000001 | 1072 | Table_map      |         1 |        1132 | table_id: 398 (slowtech.t1)                 |
| mysql-bin.000001 | 1132 | Delete_rows    |         1 |        1174 | table_id: 398 flags: STMT_END_F             |
| mysql-bin.000001 | 1174 | Xid            |         1 |        1205 | COMMIT /* xid=260 */                        |
+------------------+------+----------------+-----------+-------------+---------------------------------------------+
20 rows in set (0.00 sec)
```

在本例中，Event_type 为 Rows_query 的事件即为 ROWS_QUERY_LOG_EVENT，后面的 Info 是具体的 SQL 语句。

- **FORMAT_DESCRIPTION_EVENT**

FORMAT_DESCRIPTION_EVENT 是 binlog 中的第一个事件，它记录了 MySQL 的版本、binlog 的版本、binlog 的创建时间、其他事件的事件头的长度等。mysqlbinlog 会基于这些信息来解析 binlog 中的其他事件。

```
# mysqlbinlog /data/mysql/3306/data/mysql-bin.000001
/*!50530 SET @@SESSION.PSEUDO_SLAVE_MODE=1*/;
/*!50003 SET @OLD_COMPLETION_TYPE=@@COMPLETION_TYPE,COMPLETION_TYPE=0*/;
DELIMITER /*!*/;
# at 4
#210824 19:09:25 server id 1  end_log_pos 125 CRC32 0xa685f289  Start: binlog v 4, server v 8.0.25 created
210824 19:09:25 at startup
...
```

- **XID_EVENT**

XID 中的 X 指的是分布式事务。在 MySQL 中，存储引擎层和 Server 层是分开的，为了保证事务在存储引擎和 binlog 之间的一致性，MySQL 5.0 引入了 XA 模型。对于 XA 事务，对应的 COMMIT 操作就是 XID_EVENT 类型。在 MySQL 自带的存储引擎中，只有 InnoDB 支持事务，所以 InnoDB 表的 COMMIT 操作对应的事件类型就是 XID_EVENT。对于其他不支持事务的存储引擎，对应的 COMMIT 操作对应的事件类型是 QUERY_EVENT。下面以 MyISAM 表的操作为例。

```
mysql> create table slowtech.t2(id int) engine=myisam;
Query OK, 0 rows affected (0.01 sec)

mysql> insert into slowtech.t2 values(1);
Query OK, 1 row affected (0.00 sec)

mysql> show binlog events in 'mysql-bin.000001';
+------------------+-----+----------------+-----------+-------------+----------------------------------------------------+
| Log_name         | Pos | Event_type     | Server_id | End_log_pos | Info                                               |
+------------------+-----+----------------+-----------+-------------+----------------------------------------------------+
| mysql-bin.000001 |   4 | Format_desc    |         1 |         125 | Server ver: 8.0.25, Binlog ver: 4                  |
| mysql-bin.000001 | 125 | Previous_gtids |         1 |         156 |                                                    |
| mysql-bin.000001 | 156 | Anonymous_Gtid |         1 |         233 | SET @@SESSION.GTID_NEXT= 'ANONYMOUS'               |
| mysql-bin.000001 | 233 | Query          |         1 |         358 | create table slowtech.t2(id int) engine=myisam     |
| mysql-bin.000001 | 358 | Anonymous_Gtid |         1 |         437 | SET @@SESSION.GTID_NEXT= 'ANONYMOUS'               |
| mysql-bin.000001 | 437 | Query          |         1 |         508 | BEGIN                                              |
| mysql-bin.000001 | 508 | Table_map      |         1 |         560 | table_id: 489 (slowtech.t2)                        |
| mysql-bin.000001 | 560 | Write_rows     |         1 |         600 | table_id: 489 flags: STMT_END_F                    |
| mysql-bin.000001 | 600 | Query          |         1 |         672 | COMMIT                                             |
+------------------+-----+----------------+-----------+-------------+----------------------------------------------------+
9 rows in set (0.00 sec)
```

- **GTID_LOG_EVENT**

在开启 GTID 的情况下，MySQL 会为每个事务分配一个 GTID，这个 GTID 对应的事件类型即为 GTID_LOG_EVENT。例如：

3.4 binlog 中的事件类型

```
mysql> show binlog events in 'mysql-bin.000001';
+------------------+-----+----------------+-----------+-------------+--------------------------------------------------------------------------+
| Log_name         | Pos | Event_type     | Server_id | End_log_pos | Info                                                                     |
+------------------+-----+----------------+-----------+-------------+--------------------------------------------------------------------------+
| mysql-bin.000001 |   4 | Format_desc    |         1 |         125 | Server ver: 8.0.25, Binlog ver: 4                                        |
| mysql-bin.000001 | 125 | Previous_gtids |         1 |         156 |                                                                          |
| mysql-bin.000001 | 156 | Gtid           |         1 |         233 | SET @@SESSION.GTID_NEXT= '12ad7127-0466-11ec-8c74-000c292c1f7b:1'        |
| mysql-bin.000001 | 233 | Query          |         1 |         353 | create database slowtech /* xid=306 */                                   |
| mysql-bin.000001 | 353 | Gtid           |         1 |         430 | SET @@SESSION.GTID_NEXT= '12ad7127-0466-11ec-8c74-000c292c1f7b:2'        |
| mysql-bin.000001 | 430 | Query          |         1 |         565 | create table slowtech.t1(id int,c1 varchar(10)) /* xid=307 */            |
+------------------+-----+----------------+-----------+-------------+--------------------------------------------------------------------------+
6 rows in set (0.01 sec)
```

- **ANONYMOUS_GTID_LOG_EVENT**

在没开启 GTID 的情况下，每个事务在 MySQL 内部称为匿名事务。对于匿名事务，同样会设置 @@SESSION.GTID_NEXT，只不过它的值是 ANONYMOUS，而不是具体的 GTID。

- **PREVIOUS_GTIDS_LOG_EVENT**

PREVIOUS_GTIDS_LOG_EVENT 是 binlog 中的第二个事件，仅次于 FORMAT_DESCRIPTION_EVENT，代表之前已经分配的 GTID 集。例如：

```
mysql> show binlog events in 'mysql-bin.000002';
+------------------+-----+----------------+-----------+-------------+-----------------------------------------+
| Log_name         | Pos | Event_type     | Server_id | End_log_pos | Info                                    |
+------------------+-----+----------------+-----------+-------------+-----------------------------------------+
| mysql-bin.000002 |   4 | Format_desc    |         1 |         125 | Server ver: 8.0.25, Binlog ver: 4       |
| mysql-bin.000002 | 125 | Previous_gtids |         1 |         196 | 12ad7127-0466-11ec-8c74-000c292c1f7b:1-2|
+------------------+-----+----------------+-----------+-------------+-----------------------------------------+
2 rows in set (0.00 sec)
```

- **ROTATE_EVENT**

当 binlog 发生切换时，会在当前的 binlog 中添加一个 ROTATE_EVENT 事件，用于记录下一个 binlog 的文件名。例如：

```
mysql> show binlog events in 'mysql-bin.000002';
+------------------+-----+----------------+-----------+-------------+-----------------------------------------+
| Log_name         | Pos | Event_type     | Server_id | End_log_pos | Info                                    |
+------------------+-----+----------------+-----------+-------------+-----------------------------------------+
| mysql-bin.000002 |   4 | Format_desc    |         1 |         125 | Server ver: 8.0.25, Binlog ver: 4       |
| mysql-bin.000002 | 125 | Previous_gtids |         1 |         196 | 12ad7127-0466-11ec-8c74-000c292c1f7b:1-2|
| mysql-bin.000002 | 196 | Rotate         |         1 |         243 | mysql-bin.000003;pos=4                  |
+------------------+-----+----------------+-----------+-------------+-----------------------------------------+
3 rows in set (0.00 sec)
```

- **TABLE_MAP_EVENT**

TABLE_MAP_EVENT 仅用于 ROW 格式，是 ROWS_EVENT 的前一个事件，用于描述表的内部 ID 和列的数据类型等。例如：

```
#210824 19:17:54 server id 1  end_log_pos 406 CRC32 0xce7f135c  Table_map: `slowtech`.`t1` mapped to number 534
```

注意，这个表 ID（534）只用于 binlog 内部，与数据字典表中的 table_id（1078）和 space_id（19）

没有任何关系，后面的两个 ID 可通过以下 SQL 语句来获取。

```
mysql> select table_id,name from information_schema.innodb_tables where name='slowtech/t1';
+----------+-------------+
| table_id | name        |
+----------+-------------+
|     1078 | slowtech/t1 |
+----------+-------------+
1 row in set (0.01 sec)

mysql> select space,name from information_schema.innodb_tablespaces where name='slowtech/t1';
+-------+-------------+
| space | name        |
+-------+-------------+
|    19 | slowtech/t1 |
+-------+-------------+
1 row in set (0.01 sec)
```

- **BEGIN_LOAD_QUERY_EVENT、EXECUTE_LOAD_QUERY_EVENT 和 APPEND_BLOCK_EVENT**

这 3 个事件主要用来解决 STATEMENT 格式下 LOAD DATA INFILE 语句在主从之间如何复制的问题。这 3 个事件的具体作用如下。

- BEGIN_LOAD_QUERY_EVENT：代表 LOAD DATA INFILE 操作的开始，同时会记录文件的一部分数据，默认是 128KB。
- APPEND_BLOCK_EVENT：文件中没有记录完的数据将以一个或多个 APPEND_BLOCK_EVENT 来记录。APPEND_BLOCK_EVENT 中 BLOCK 的大小默认是 128KB。
- EXECUTE_LOAD_QUERY_EVENT：记录 LOAD DATA INFILE 语句。

看看下面这个示例。

```
mysql> load data infile '/tmp/t1.txt' into table slowtech.t1;
```

对应的 binlog 内容如下。

```
...
BEGIN
/*!*/;
# at 318
#210824 19:28:04 server id 1  end_log_pos 131417 CRC32 0x23bc94ab
#Begin_load_query: file_id: 3  block_len: 131072
# at 131417
#210824 19:28:04 server id 1  end_log_pos 262516 CRC32 0x0dc255ea
#Append_block: file_id: 3  block_len: 131072
# at 262516
#210824 19:28:04 server id 1  end_log_pos 363955 CRC32 0x6113a075
#Append_block: file_id: 3  block_len: 101412
# at 363955
#210824 19:28:04 server id 1 end_log_pos 364197 CRC32 0x3d3241aa    Execute_load_query    thread_id=11
exec_time=0     error_code=0
SET TIMESTAMP=1629804484/*!*/;
LOAD DATA LOCAL INFILE '/tmp/SQL_LOAD_MB-3-0' INTO TABLE `slowtech`.`t1` FIELDS TERMINATED BY '\t'
ENCLOSED BY '' ESCAPED BY '\\' LINES TERMINATED BY '\n' (`id`, `c1`)
/*!*/;
# file_id: 3
# at 364197
```

```
#210824 19:28:04 server id 1  end_log_pos 364228 CRC32 0x95e09031      Xid = 39
COMMIT/*!*/;
SET @@SESSION.GTID_NEXT= 'AUTOMATIC' /* added by mysqlbinlog */ /*!*/;
DELIMITER ;
# End of log file
/*!50003 SET COMPLETION_TYPE=@OLD_COMPLETION_TYPE*/;
/*!50530 SET @@SESSION.PSEUDO_SLAVE_MODE=0*/;
```

- STOP_EVENT

当实例正常关闭时，会向当前的 binlog 写入一个 STOP_EVENT 事件。例如：

```
mysql> show binlog events in 'mysql-bin.000004';
+------------------+-----+----------------+-----------+-------------+----------------------------------------------+
| Log_name         | Pos | Event_type     | Server_id | End_log_pos | Info                                         |
+------------------+-----+----------------+-----------+-------------+----------------------------------------------+
| mysql-bin.000004 |   4 | Format_desc    |         1 |         125 | Server ver: 8.0.25, Binlog ver: 4            |
| mysql-bin.000004 | 125 | Previous_gtids |         1 |         196 | 12ad7127-0466-11ec-8c74-000c292c1f7b:1-3     |
| mysql-bin.000004 | 196 | Rotate         |         1 |         243 | mysql-bin.000005;pos=4                       |
+------------------+-----+----------------+-----------+-------------+----------------------------------------------+
3 rows in set (0.00 sec)
```

3.5 基于 python-mysql-replication 打造一个 binlog 解析器

python-mysql-replication 是一个实现了 MySQL 复制协议的 Python 库。本质上，任何一个工具，只要实现了 MySQL 复制协议，就可以将自己伪装成一个从库，来实时获取主库的所有操作。不仅是 Python，其他语言也有相应的开源实现，如 Java（open-replicator）和 Go（go-mysql）。

基于这些库衍生了很多我们耳熟能详的开源组件，如 Canal（阿里巴巴开源的跨机房数据同步工具）、gh-ost（GitHub 开源的 MySQL 表结构变更工具）、binlog2sql（大众点评开源的 binlog 解析及 DML 回滚工具）。

前面介绍了 binlog 的事件类型，看起来可能比较抽象，下面我们通过 python-mysql-replication 库直观地展示这些常见的事件类型。

首先，准备测试环境，具体步骤如下。

(1) 安装软件。

软件的版本信息如表 3-1 所示。

表 3-1 软件的版本信息

软件	版本
MySQL	8.0.25
Python	3.6.14
PyMySQL	1.0.2
mysql-replication	0.25

(2) 设置参数。

```
log-bin = mysql-bin
server_id = 1
binlog_format = row
```

参数的主要作用是开启 binlog 并将 binlog_format 设置为 ROW。

(3) 创建复制用户。

```
create user 'repl_user'@'%' identified with mysql_native_password by 'repl_pass';
grant select, replication slave, replication client on *.* to 'repl_user'@'%';
```

这里授予了 3 个权限。理论上，复制用户只需要 REPLICATION SLAVE 权限。新增的 SELECT 权限用于查询 information_schema.columns 以获取表的元数据信息，如列名、主键等；而 REPLICATION CLIENT 权限则用来执行 SHOW MASTER STATUS 以获取主库当前的 binlog 文件名。

首先做一个简单的测试，测试脚本内容如下。

```
#!/usr/bin/python3
# -*- coding: utf-8 -*-

from pymysqlreplication import BinLogStreamReader

mysql_settings = {'host': '192.168.244.10', 'port': 3306, 'user': 'repl_user', 'passwd': 'repl_pass'}

stream = BinLogStreamReader(connection_settings = mysql_settings, server_id=100,blocking=True)

for binlogevent in stream:
    binlogevent.dump()

stream.close()
```

然后构造测试数据。

```
mysql> create database slowtech;
Query OK, 1 row affected (0.00 sec)

mysql> create table slowtech.t1 (id int primary key, c1 varchar(10));
Query OK, 0 rows affected (0.01 sec)
```

脚本对应的输出如下。

```
=== RotateEvent ===
Position: 4
Next binlog file: mysql-bin.000001

=== FormatDescriptionEvent ===
Date: 2021-08-25T07:35:54
Log position: 125
Event size: 98
Read bytes: 0

=== GtidEvent ===
Date: 2021-08-25T07:35:55
Log position: 233
Event size: 54
Read bytes: 25
```

```
Commit: True
GTID_NEXT: bd6b3216-04d6-11ec-b76f-000c292c1f7b:1

=== QueryEvent ===
Date: 2021-08-25T07:35:55
Log position: 353
Event size: 97
Read bytes: 97
Schema: b'slowtech'
Execution time: 0
Query: create database slowtech

=== GtidEvent ===
Date: 2021-08-25T07:36:03
Log position: 430
Event size: 54
Read bytes: 25
Commit: True
GTID_NEXT: bd6b3216-04d6-11ec-b76f-000c292c1f7b:2

=== QueryEvent ===
Date: 2021-08-25T07:36:03
Log position: 579
Event size: 126
Read bytes: 126
Schema: b''
Execution time: 0
Query: create table slowtech.t1 (id int primary key, c1 varchar(10))
```

一共打印了 6 个事件,依次是 RotateEvent、FormatDescriptionEvent、GtidEvent、QueryEvent、GtidEvent 和 QueryEvent。GtidEvent 和 QueryEvent 对应一个 DDL 操作。

下面重点看看 QueryEvent 中的内容。

- Date:事件的开始执行时间。
- Log position:事件的结束位置点。
- Schema:库名。
- Execution time:操作耗时。这个字段仅对于 QueryEvent 有意义。对于 RowsEvent,它永远为 0。
- Query:执行的具体 SQL。

接下来,分别执行 INSERT、UPDATE 和 DELETE 操作。

```
mysql> insert into slowtech.t1 values(1,'a');
Query OK, 1 row affected (0.00 sec)

mysql> update slowtech.t1 set c1='b' where id=1;
Query OK, 1 row affected (0.00 sec)
Rows matched: 1  Changed: 1  Warnings: 0

mysql> delete from slowtech.t1 where id=1;
Query OK, 1 row affected (0.00 sec)
```

首先看看 INSERT 操作的相关输出。

```
=== GtidEvent ===
Date: 2021-08-25T07:39:15
Log position: 658
```

```
Event size: 56
Read bytes: 25
Commit: False
GTID_NEXT: bd6b3216-04d6-11ec-b76f-000c292c1f7b:3

=== QueryEvent ===
Date: 2021-08-25T07:39:15
Log position: 729
Event size: 48
Read bytes: 48
Schema: b''
Execution time: 0
Query: BEGIN

=== TableMapEvent ===
Date: 2021-08-25T07:39:15
Log position: 789
Event size: 37
Read bytes: 28
Table id: 129
Schema: slowtech
Table: t1
Columns: 2

=== WriteRowsEvent ===
Date: 2021-08-25T07:39:15
Log position: 831
Event size: 19
Read bytes: 12
Table: slowtech.t1
Affected columns: 2
Changed rows: 1
Values:
--
* id : 1
* c1 : a

=== XidEvent ===
Date: 2021-08-25T07:39:15
Log position: 862
Event size: 8
Read bytes: 8
Transaction ID: 26
```

一共打印了 5 个事件，依次是 GtidEvent、QueryEvent、TableMapEvent、WriteRowsEvent 和 XidEvent。ROW 格式的一个事务需要这 5 个事件来表示。

下面重点看看 WriteRowsEvent 中的内容。

❑ Table：表名。

❑ Affected columns：列的数量。

❑ Changed rows：受影响的行的个数。

❑ Values：新增值。

可以看到，Values 中包括了列名，但实际上，在 binlog 中并不会记录具体的列名，看下面这个示例。

```
### INSERT INTO `slowtech`.`t1`
### SET
###   @1=1 /* INT meta=0 nullable=0 is_null=0 */
###   @2='a' /* VARSTRING(40) meta=40 nullable=1 is_null=0 */
```

所以 `Values` 中的列名实际上是 python-mysql-replication 基于 `information_schema.columns` 中的内容实现的。

对于 UPDATE 和 DELETE 操作，同样会有上面的 `QueryEvent`、`TableMapEvent` 和 `XidEvent`。这里重点看看 `RowsEvent` 的内容。

```
=== UpdateRowsEvent ===
Date: 2021-08-25T07:39:34
Log position: 1131
Event size: 27
Read bytes: 13
Table: slowtech.t1
Affected columns: 2
Changed rows: 1
Affected columns: 2
Values:
--
*id:1=>1
*c1:a=>b

=== DeleteRowsEvent ===
Date: 2021-08-25T07:39:41
Log position: 1414
Event size: 19
Read bytes: 12
Table: slowtech.t1
Affected columns: 2
Changed rows: 1
Values:
--
* id : 1
* c1 : b
```

输出内容与 INSERT 操作的内容相差不大，只不过，对于 UPDATE 操作，`Values` 记录了各列修改前后的值。

接下来看看 STATEMENT 格式的 DML 操作。

```
mysql> set session binlog_format=statement;
Query OK, 0 rows affected (0.01 sec)

mysql> insert into slowtech.t1 values (2,'b');
Query OK, 1 row affected (0.01 sec)
```

对应的输出如下。

```
=== GtidEvent ===
Date: 2021-08-25T07:44:26
Log position: 1524
Event size: 56
Read bytes: 25
```

```
Commit: True
GTID_NEXT: bd6b3216-04d6-11ec-b76f-000c292c1f7b:6

=== QueryEvent ===
Date: 2021-08-25T07:44:26
Log position: 1606
Event size: 59
Read bytes: 59
Schema: b''
Execution time: 0
Query: BEGIN

=== QueryEvent ===
Date: 2021-08-25T07:44:26
Log position: 1721
Event size: 92
Read bytes: 92
Schema: b''
Execution time: 0
Query: insert into slowtech.t1 values (2,'b')

=== XidEvent ===
Date: 2021-08-25T07:44:26
Log position: 1752
Event size: 8
Read bytes: 8
Transaction ID: 33
```

可以看到，对于 STATEMENT 格式的 DML 操作，对应的 binlog 事件类型是 QueryEvent。

注意，QueryEvent 中的 Schema 指的是 SQL 语句执行时所在的 schema，并不是表本身所属的 schema。看看下面这个示例。

```
mysql> set session binlog_format=statement;
Query OK, 0 rows affected (0.00 sec)

mysql> use mysql

mysql> insert into slowtech.t1 values (3,'c');
Query OK, 1 row affected (0.00 sec)

=== QueryEvent ===
Date: 2021-08-25T07:45:31
Log position: 2038
Event size: 97
Read bytes: 97
Schema: b'mysql'
Execution time: 0
Query: insert into slowtech.t1 values (3,'c')
```

因为这个 SQL 语句是在 mysql 库下执行的，所以这里显示的 Schema 是 mysql。

之所以提到这一点，是因为它与部分复制的处理逻辑有关。在进行部分库的复制时，我们通常会使用 --replicate-do-db 或 --replicate-ignore-db 这两个参数。但实际上，在 STATEMENT 格式中，这两个参数的判断依据是 USE 操作中的 schema，并不是 SQL 中的实际操作对象。在进行跨库操作时，很容易出现该复制的操作没有复制，该过滤的操作没有过滤的情况，进而导致主从数据不一致。

所以在 STATEMENT 格式下，不建议使用 --replicate-do-db 或 --replicate-ignore-db 进行库级别的复制。此时，可用 --replicate-wild-do-table=db_name.% 或 --replicate-wild-ignore-table=db_name.% 来代替。

对于 ROW 格式，我们可以看到，binlog 并没有记录 Schema 的信息，而是直接记录 Table 的信息。使用 --replicate-do-db 或 --replicate-ignore-db 进行库级别的复制不会导致主从数据不一致。

以上就是常见操作的事件类型，下面看看如何打造一个 binlog 解析器。

下面是一个简单的演示，可实时捕捉 ROW 格式的 DML 操作。

```python
#!/usr/bin/python3
# -*- coding: utf-8 -*-
import pymysql
from pymysqlreplication import BinLogStreamReader
from pymysqlreplication.row_event import (
    DeleteRowsEvent,
    UpdateRowsEvent,
    WriteRowsEvent
)
from pymysqlreplication.event import (QueryEvent, XidEvent)

host = "192.168.244.10"
port = 3306
user = "repl_user"
passwd = "repl_pass"

mysql_settings = {'host': host, 'port': port, 'user': user, 'passwd': passwd}

stream = BinLogStreamReader(connection_settings=mysql_settings, server_id=100, blocking=True)

conn = pymysql.connect(host=host, user=user, passwd=passwd, port=port, charset='utf8', autocommit=False)
cursor = conn.cursor()

for binlog_event in stream:
    if isinstance(binlog_event, QueryEvent) and binlog_event.query == "BEGIN":
        print("BEGIN;")
    elif isinstance(binlog_event, (DeleteRowsEvent, UpdateRowsEvent, WriteRowsEvent)):
        table_schema = binlog_event.schema
        table_name = binlog_event.table
        rows = binlog_event.rows
        if isinstance(binlog_event, DeleteRowsEvent):
            for each_row in rows:
                col_name = ' AND '.join(
                    ['`%s`=%%s' % (k) if v != None else '`%s` is %%s' % (k) for k, v in each_row['values'].items()])
                delete_sql = "DELETE FROM `{0}`.`{1}` WHERE {2};".format(table_schema, table_name, col_name)
                print(cursor.mogrify(delete_sql, list(each_row['values'].values())))
        elif isinstance(binlog_event, WriteRowsEvent):
            for each_row in rows:
                col_name = ','.join('`%s`' % (k) for k in each_row['values'].keys())
                format_str = ','.join('%s' for k in each_row['values'].keys())
                insert_sql = 'INSERT INTO `{0}`.`{1}` ({2}) VALUES ({3});'.format(table_schema, table_name, col_name,
                                                                                    format_str)
```

```
                print(cursor.mogrify(insert_sql, list(each_row['values'].values())))
        elif isinstance(binlog_event, UpdateRowsEvent):
            for each_row in rows:
                before_values = each_row["before_values"]
                after_values = each_row["after_values"]
                set_str = ','.join(
                    ['`%s`=%%s' % (k) for k, v in after_values.items()])
                where_str = ' AND '.join(
                    ['`%s`=%%s' % (k) if v != None else '`%s` is %%s' % (k) for k, v in before_values.items()])
                update_sql = "UPDATE `{0}`.`{1}` SET {2} WHERE {3};".format(table_schema, table_name, set_str,
                                                                                                where_str)
                values = list(after_values.values()) + list(before_values.values())
                print(cursor.mogrify(update_sql, values))
        elif isinstance(binlog_event, XidEvent):
            print("COMMIT;\n")
stream.close()
cursor.close()
conn.close()
```

该脚本（binlog_parser.py）已上传至 GitHub。

下面结合具体的测试语句看看脚本的实际执行效果。

```
mysql> insert into slowtech.t1 values(1,'a');
Query OK, 1 row affected (0.01 sec)

mysql> begin;
Query OK, 0 rows affected (0.00 sec)

mysql> insert into slowtech.t1 values(2,'a');
Query OK, 1 row affected (0.00 sec)

mysql> update slowtech.t1 set c1='b' where c1='a';
Query OK, 2 rows affected (0.00 sec)
Rows matched: 2  Changed: 2  Warnings: 0

mysql> delete from slowtech.t1;
Query OK, 2 rows affected (0.01 sec)

mysql> commit;
Query OK, 0 rows affected (0.00 sec)
```

脚本的输出结果如下。

```
BEGIN;
INSERT INTO `slowtech`.`t1` (`id`,`c1`) VALUES (1,'a');
COMMIT;

BEGIN;
INSERT INTO `slowtech`.`t1` (`id`,`c1`) VALUES (2,'a');
UPDATE `slowtech`.`t1` SET `id`=1,`c1`='b' WHERE `id`=1 AND `c1`='a';
UPDATE `slowtech`.`t1` SET `id`=2,`c1`='b' WHERE `id`=2 AND `c1`='a';
DELETE FROM `slowtech`.`t1` WHERE `id`=1 AND `c1`='b';
DELETE FROM `slowtech`.`t1` WHERE `id`=2 AND `c1`='b';
COMMIT;
```

可以看到，脚本打印的 SQL 语句虽然在形式上跟原生 SQL 语句有出入，但实际效果是一样的。

之所以能实现这样的效果，是因为在 ROW 格式下，binlog 会记录修改行的前后镜像。这里虽然生成的是正向操作，但实际上，基于前后镜像，我们同样可以生成回滚操作。

3.6 本章总结

本章主要围绕 binlog 展开，深入分析了 binlog 的 3 种格式及各自的优缺点。还在此基础上通过具体的示例展示了在不同的 binlog 格式下，同一条 SQL 语句记录到 binlog 的内容也会不同。这在本质上与二进制日志事件的类型有关。

通过对 relay log 的分析，我们发现同一个事务的起始位置点在从库 relay log 和主库 binlog 中不一样。同样，在一主多从的架构中，主库的一个操作在各个从库 binlog 中的位置点也不一样。

最后，我们基于 python-mysql-replication 打造了一个 binlog 解析器。通过解析 binlog，我们可以实现一种实时、对数据库无侵入的数据同步方式。

重点问题回顾

- binlog 的 3 种格式及其各自的优缺点。为什么不推荐线上使用基于 STATEMENT 的日志格式？
- 在 RC 事务隔离级别下，可否将 binlog 的格式设置为 STATEMENT？
- 常见的二进制日志事件类型。
- 在 binlog 中，一个事务结束的位置点。
- 在 mysqlbinlog 中，`--base64-output=decode-rows` 的作用。
- 如何在 ROW 格式下看到原生 SQL 语句？

第 4 章

深入 MySQL 的复制管理

前面两章从复制的基本原理出发，系统介绍了 GTID 复制、半同步复制、并行复制、多源复制和延迟复制，并在此基础上对 binlog 进行了深入的分析，包括 binlog 的 3 种格式及其各自的优缺点，如何解读 binlog 的内容，以及 binlog 中常见的事件类型等。

本章围绕与 MySQL 复制相关的一些常见管理操作展开，主要包括以下内容。

- 常见的管理操作。
- 复制的监控。
- 主从延迟。
- 复制中的常见问题及解决方法。

4.1 常见的管理操作

本节总结了与 MySQL 复制相关的一些常见管理操作。

4.1.1 查看主库的状态

查看主库状态的命令是 SHOW MASTER STATUS。

```
mysql> show master status;
+------------------+----------+--------------+------------------+-------------------+
| File             | Position | Binlog_Do_DB | Binlog_Ignore_DB | Executed_Gtid_Set |
+------------------+----------+--------------+------------------+-------------------+
| mysql-bin.000001 |      156 |              |                  |                   |
+------------------+----------+--------------+------------------+-------------------+
1 row in set (0.00 sec)
```

只有开启了 binlog，该命令才会有结果输出。下面看看各个字段的具体含义。

- `File`：当前正在写入的 binlog。
- `Position`：binlog 的具体位置点。
- `Binlog_Do_DB`：只将某些库的操作记录在 binlog 中，由 binlog-do-db 参数决定。
- `Binlog_Ignore_DB`：忽略某些库的操作，由 binlog-ignore-db 参数决定。
- `Executed_Gtid_Set`：在 GTID 复制中，已经执行过的 GTID 事务的集合。该列只有在开启 GTID 的情况下才有输出。

4.1.2 查看从库复制的状态

查看从库复制状态的命令是 SHOW SLAVE STATUS [FOR CHANNEL channel]。

```
mysql> show slave status\G
*************************** 1. row ***************************
               Slave_IO_State: Waiting for master to send event
                  Master_Host: 192.168.244.10
                  Master_User: repl
                  Master_Port: 3306
                Connect_Retry: 60
              Master_Log_File: mysql-bin.000001
          Read_Master_Log_Pos: 136054
               Relay_Log_File: node2-relay-bin.000003
                Relay_Log_Pos: 136269
        Relay_Master_Log_File: mysql-bin.000001
             Slave_IO_Running: Yes
            Slave_SQL_Running: Yes
              Replicate_Do_DB:
          Replicate_Ignore_DB:
           Replicate_Do_Table:
       Replicate_Ignore_Table:
      Replicate_Wild_Do_Table:
  Replicate_Wild_Ignore_Table:
                   Last_Errno: 0
                   Last_Error:
                 Skip_Counter: 0
          Exec_Master_Log_Pos: 136054
              Relay_Log_Space: 136478
              Until_Condition: None
               Until_Log_File:
                Until_Log_Pos: 0
           Master_SSL_Allowed: No
           Master_SSL_CA_File:
           Master_SSL_CA_Path:
              Master_SSL_Cert:
            Master_SSL_Cipher:
               Master_SSL_Key:
        Seconds_Behind_Master: 0
Master_SSL_Verify_Server_Cert: No
                Last_IO_Errno: 0
                Last_IO_Error:
               Last_SQL_Errno: 0
               Last_SQL_Error:
  Replicate_Ignore_Server_Ids:
             Master_Server_Id: 1
                  Master_UUID: bd6b3216-04d6-11ec-b76f-000c292c1f7b
             Master_Info_File: mysql.slave_master_info
                    SQL_Delay: 0
          SQL_Remaining_Delay: NULL
      Slave_SQL_Running_State: Slave has read all relay log; waiting for more updates
           Master_Retry_Count: 86400
                  Master_Bind:
      Last_IO_Error_Timestamp:
     Last_SQL_Error_Timestamp:
               Master_SSL_Crl:
           Master_SSL_Crlpath:
           Retrieved_Gtid_Set: bd6b3216-04d6-11ec-b76f-000c292c1f7b:1-16
            Executed_Gtid_Set: bd6b3216-04d6-11ec-b76f-000c292c1f7b:1-16
```

```
                  Auto_Position: 1
            Replicate_Rewrite_DB:
                    Channel_Name:
              Master_TLS_Version:
          Master_public_key_path:
           Get_master_public_key: 1
               Network_Namespace:
1 row in set, 1 warning (0.00 sec)
```

下面看看各个字段的具体含义。

- `Slave_IO_State`：I/O 线程的状态，常见的是 `Waiting for master to send event`，代表 I/O 线程在等待主库发送新的二进制日志事件。
- `Master_Host`、`Master_User` 和 `Master_Port`：分别代表主库的地址（主机名或 IP）、复制用户及端口。
- `Connect_Retry`：当 I/O 线程出现问题时，重试的时间间隔，对应 CHANGE MASTER TO 中的 `MASTER_CONNECT_RETRY`。
- `Master_Log_File` 和 `Read_Master_Log_Pos`：I/O 线程当前正在接收的二进制日志事件在主库 binlog 中的位置。
- `Relay_Log_File` 和 `Relay_Log_Pos`：SQL 线程当前正在重放的二进制日志事件在从库 relay log 中的位置，一般用得不多。
- `Relay_Master_Log_File` 和 `Exec_Master_Log_Pos`：SQL 线程当前正在重放的二进制日志事件在主库 binlog 中的位置。
- `Slave_IO_Running`：I/O 线程是否处于正常运行的状态。有 `Yes`、`No` 和 `Connecting` 这 3 种状态。
- `Slave_SQL_Running`：SQL 线程是否处于正常运行的状态。
- `Replicate_Do_xxx` 和 `Replicate_Ignore_xxx`：基于库表的复制。
- `Replicate_Wild_Do_Table` 和 `Replicate_Wild_Ignore_Table`：也是基于库表的复制，只不过支持通配符（%和_）。注意，如果要复制指定库，推荐使用 `Replicate_Wild_Do_Table`，而不是 `Replicate_Do_DB`。
- `Last_Errno` 和 `Last_Error`：分别是 `Last_SQL_Errno` 和 `Last_SQL_Error` 的别名。
- `Skip_Counter`：参数 `sql_slave_skip_counter` 的当前值。该参数可用来跳过 STATEMENT 格式的事务。
- `Relay_Log_Space`：当前所有 relay log 的总大小。relay log 的总大小受参数 `relay_log_space_limit` 的限制，默认为 0，即没有限制。
- `Until_Condition`、`Until_Log_File` 和 `Until_Log_Pos`：SQL 线程在应用到指定位置点时停止。
- `Master_SSL_xxx` 和 `Master_TLS_Version`：与 SSL 相关。
- `Seconds_Behind_Master`：从库相对于主库的延迟。
- `Last_IO_Errno`、`Last_IO_Error` 和 `Last_IO_Error_Timestamp`：当 I/O 线程出现问题时，对应的错误码、错误信息及发生时间。
- `Last_SQL_Errno`、`Last_SQL_Error` 和 `Last_SQL_Error_Timestamp`：当 SQL 线程出现问题时，对应的错误码、错误信息及发生时间。

- Replicate_Ignore_Server_Ids：忽略来自指定 server_id 的二进制日志事件。常用在环形复制中，实际生产中很少用到。
- Master_Server_Id 和 Master_UUID：主库的 server_id 和 server_uuid。
- Master_Info_File：复制相关信息保存的位置。若 master_info_repository 为 TABLE，则这里是 mysql.slave_master_info，否则是文件名。
- SQL_Delay 和 SQL_Remaining_Delay：常用在延迟复制中。SQL_Delay 指的是期望延迟时间，对应 CHANGE MASTER TO 中的 MASTER_DELAY。SQL_Remaining_Delay 指的是当前被暂停的事务需要等待多久，才能到达期望延迟时间。两者的单位都是秒。
- Slave_SQL_Running_State：SQL 线程的状态。
- Master_Retry_Count：I/O 线程出现问题时可以重试的次数，对应 CHANGE MASTER TO 中的 MASTER_RETRY_COUNT。
- Master_Bind：从库绑定的网卡地址。
- Retrieved_Gtid_Set：从库接受过的事务所对应的 GTID 集。
- Executed_Gtid_Set：从库执行过的事务所对应的 GTID 集。该值与 gtid_executed 和 SHOW MASTER STASUS 中的 Executed_Gtid_Set 的值相同。
- Auto_Position：对应 CHANGE MASTER TO 中的 MASTER_AUTO_POSITION，在开启 GTID 的情况下，我们一般将其设置为 1。这样，在建立复制时就无须指定 MASTER_LOG_FILE 和 MASTER_LOG_POS 了。
- Replicate_Rewrite_DB：将主库某个数据库的操作重写到从库另一个数据库中。
- Channel_Name：Channel 名，常用在多源复制中。
- Master_public_key_path：指定主库公钥的地址。如果复制用户使用了 caching_sha2_password，在搭建复制时，需设置 Master_public_key_path 或 Get_master_public_key。
- Get_master_public_key：是否允许从库直接从主库获取公钥。
- Network_Namespace：网络命名空间，常用在容器或虚拟环境中。

4.1.3 搭建复制

在第 2 章中执行的 CHANGE MASTER TO 命令只给出了必需且常用的几个选项，实际上，一个完整的 CHANGE MASTER TO 命令包括众多选项，不同的选项可用来实现不同的功能，如 GTID、多源复制、延迟复制等。

下面具体看看一个完整的 CHANGE MASTER TO 命令包括哪些选项。

```
CHANGE MASTER TO option [, option] ... [ channel_option ]

option: {
    MASTER_BIND = 'interface_name'
  | MASTER_HOST = 'host_name'
  | MASTER_USER = 'user_name'
  | MASTER_PASSWORD = 'password'
  | MASTER_PORT = port_num
  | PRIVILEGE_CHECKS_USER = {'account' | NULL}
  | REQUIRE_ROW_FORMAT = {0|1}
  | REQUIRE_TABLE_PRIMARY_KEY_CHECK = {STREAM | ON | OFF}
  | ASSIGN_GTIDS_TO_ANONYMOUS_TRANSACTIONS = {OFF | LOCAL | uuid}
```

```
    | MASTER_LOG_FILE = 'source_log_name'
    | MASTER_LOG_POS = source_log_pos
    | MASTER_AUTO_POSITION = {0|1}
    | RELAY_LOG_FILE = 'relay_log_name'
    | RELAY_LOG_POS = relay_log_pos
    | MASTER_HEARTBEAT_PERIOD = interval
    | MASTER_CONNECT_RETRY = interval
    | MASTER_RETRY_COUNT = count
    | SOURCE_CONNECTION_AUTO_FAILOVER = {0|1}
    | MASTER_DELAY = interval
    | MASTER_COMPRESSION_ALGORITHMS = 'value'
    | MASTER_ZSTD_COMPRESSION_LEVEL = level
    | MASTER_SSL = {0|1}
    | MASTER_SSL_CA = 'ca_file_name'
    | MASTER_SSL_CAPATH = 'ca_directory_name'
    | MASTER_SSL_CERT = 'cert_file_name'
    | MASTER_SSL_CRL = 'crl_file_name'
    | MASTER_SSL_CRLPATH = 'crl_directory_name'
    | MASTER_SSL_KEY = 'key_file_name'
    | MASTER_SSL_CIPHER = 'cipher_list'
    | MASTER_SSL_VERIFY_SERVER_CERT = {0|1}
    | MASTER_TLS_VERSION = 'protocol_list'
    | MASTER_TLS_CIPHERSUITES = 'ciphersuite_list'
    | MASTER_PUBLIC_KEY_PATH = 'key_file_name'
    | GET_MASTER_PUBLIC_KEY = {0|1}
    | NETWORK_NAMESPACE = 'namespace'
    | IGNORE_SERVER_IDS = (server_id_list)
    | GTID_ONLY = {0|1}
}

channel_option:
    FOR CHANNEL channel

server_id_list:
    [server_id [, server_id] ... ]
```

大部分选项在第 2 章介绍过，这里重点说说其余的几个选项。

- `RELAY_LOG_FILE` 和 `RELAY_LOG_POS`：当从库 SQL 线程启动时，它应该从哪个 relay log（由 `RELAY_LOG_FILE` 确定）的哪个位置（由 `RELAY_LOG_POS` 确定）开始重放操作。
- `MASTER_BIND`：当从库存在多个网络接口时，用来指定从库使用哪个网络接口来连接主库。
- `PRIVILEGE_CHECKS_USER`：指定重放用户。默认情况下，从库会重放主库的所有操作。在某些场景下，这可能会不安全。有了重放用户，就可以通过限制重放用户的权限，来达到限制重放操作的效果。

例如下面这个用户只能重放 slowtech 库的 CREATE、INSERT、DELETE 和 UPDATE 操作。对于 binlog 中的其他操作，SQL 线程会因权限不足而中断。

```
CREATE USER 'rpl_applier_user'@'localhost';

GRANT REPLICATION_APPLIER, SESSION_VARIABLES_ADMIN ON *.* TO 'rpl_applier_user'@'localhost';

GRANT CREATE, INSERT, DELETE, UPDATE ON slowtech.* TO 'rpl_applier_user'@'localhost';
```

- `REQUIRE_ROW_FORMAT`：检查接收的二进制日志事件是否为 ROW 格式。若不是，I/O 线程会中断，

- REQUIRE_TABLE_PRIMARY_KEY_CHECK：检查表上是否存在主键。严格来说，是检查两类操作：CTEATE TABLE 是否会定义主键；ALTER TABLE 是否会删除主键。不会检查无主键表的 DML 操作。如果违反了约束，会提示以下错误。

  ```
  Last_SQL_Error: Error 'Unable to create or change a table without a primary key, when the system variable 'sql_require_primary_key' is set. Add a primary key to the table or unset this variable to avoid this message. Note that tables without a primary key can cause performance problems in row-based replication, so please consult your DBA before changing this setting.' on query. Default database: ''. Query: 'create table slowtech.t1(id int)'
  ```

- ASSIGN_GTIDS_TO_ANONYMOUS_TRANSACTIONS：是否为匿名事务分配 GTID。使用这个选项，即使主库没有开启 GTID，从库也能使用 GTID。
- MASTER_COMPRESSION_ALGORITHMS：指定主从连接的压缩算法。
- MASTER_ZSTD_COMPRESSION_LEVEL：如果使用的压缩算法是 zstd，该选项用来指定 zstd 算法的压缩等级。
- MASTER_SSL_xxx 和 MASTER_TLS_VERSION：与 SSL 相关。
- MASTER_CONNECT_RETRY 和 MASTER_RETRY_COUNT：当从库的 I/O 线程无法与主库建立连接时，会不断重试，重试的时间间隔由 MASTER_CONNECT_RETRY 确定，默认为 60 秒。重试次数由 MASTER_RETRY_COUNT 决定，默认为 86 400。
- SOURCE_CONNECTION_AUTO_FAILOVER：当主库出现故障时，是否自动切换到已定义的其他节点上。
- IGNORE_SERVER_IDS：忽略来自指定 server_id 的事务。常用在基于位置点的环形复制中。
- GTID_ONLY：设置为 1，则只会使用 GTID 来作为事务是否已经重放的依据，不会再更新 mysql.slave_relay_log_info 和 mysql.slave_worker_info 表中的位置点信息。将 GTID_ONLY 设置为 1 的前提是 MASTER_AUTO_POSITION = 1 和 REQUIRE_ROW_FORMAT = 1。

在开启复制后，如果要修改某个选项的值，需在停止复制后重新执行 CHANGE MASTER TO 命令，但这次只需指定对应的选项，不用带上其他选项。例如，要修改复制用户的密码，就可通过以下命令实现。

```
change master to master_password = '123456';
```

另外，在某些特殊场合下，可能需要重放 relay log 中的内容，这时可执行以下操作。

```
change master to
  relay_log_file = 'node2-relay-bin.000002',
  relay_log_pos = 704;
```

4.1.4 开启复制

开启复制的命令是 START SLAVE。但其实，该命令支持众多选项，其完整语法如下。

```
START {SLAVE | REPLICA} [thread_types] [until_option] [connection_options] [channel_option]

thread_types:
    [thread_type [, thread_type] ... ]

thread_type:
```

```
    IO_THREAD | SQL_THREAD

until_option:
    UNTIL {   {SQL_BEFORE_GTIDS | SQL_AFTER_GTIDS} = gtid_set
          |   MASTER_LOG_FILE = 'log_name', MASTER_LOG_POS = log_pos
          |   SOURCE_LOG_FILE = 'log_name', SOURCE_LOG_POS = log_pos
          |   RELAY_LOG_FILE = 'log_name', RELAY_LOG_POS = log_pos
          |   SQL_AFTER_MTS_GAPS    }

connection_options:
    [USER='user_name'] [PASSWORD='user_pass'] [DEFAULT_AUTH='plugin_name'] [PLUGIN_DIR='plugin_dir']

channel_option:
    FOR CHANNEL channel
```

如果不指定任何选项，则默认开启 I/O 线程和 SQL 线程。也可通过以下命令单独启动某个线程。

```
start slave io_thread;
start slave sql_thread;
```

这里重点说说 until_option，这个选项会让 SQL 线程应用在指定位置点时停止。它常见的使用场景如下。

- 在一主多从且基于位置点的复制场景中，变更从库之间的拓扑。例如，一主两从（master->slave1,slave2）的场景，将 slave1 切换为 slave2 的从库。
- 在延迟复制中，恢复到指定位置点。

下面看看 until_option 的具体选项。

- SQL_BEFORE_GTIDS 和 SQL_AFTER_GTIDS：针对 GTID 复制。这里的 GTIDS 是个集合，以下面的 GTIDS 为例。

```
bd6b3216-04d6-11ec-b76f-000c292c1f7b:31-39
```

对于 SQL_BEFORE_GTIDS，SQL 线程在应用到 bd6b3216-04d6-11ec-b76f-000c292c1f7b:31 这个事务上时（不包括这个事务）停止。

对于 SQL_AFTER_GTIDS，SQL 线程在应用完 bd6b3216-04d6-11ec-b76f-000c292c1f7b:39 这个事务后停止。

- MASTER_LOG_FILE 和 MASTER_LOG_POS：主库 binlog 的位置点，对应 SHOW SLAVE STATUS 中的 Relay_Master_Log_File 和 Exec_Master_Log_Pos。
- RELAY_LOG_FILE 和 RELAY_LOG_POS：从库 relay log 的位置点，对应 SHOW SLAVE STATUS 中的 Relay_Log_File 和 Relay_Log_Pos。
- SQL_AFTER_MTS_GAPS：在多线程复制且 slave_preserve_commit_order = 0 时，如果(1) coordinator 线程在运行过程中被杀掉，(2) Applier 线程在运行过程中出现问题，或者(3) MySQL 意外关闭，有可能会出现 binlog 中某些位置靠后的事务先执行，而在它之前的事务没有执行的情况。这些没有执行的事务称为 GAP 事务。如果碰到这种情况，就需指定 SQL_AFTER_MTS_GAPS，让 SQL 线程先执行完 GAP 事务。

在使用 until_option 时，需要注意以下几点。

- until_option 只对 SQL 线程有效果，对 I/O 线程没有效果。
- 在 GTID 复制中，也可指定位置点。
- 执行 STOP SLAVE 会清空 until_option。

4.1.5 停止复制

停止复制的命令是 STOP SLAVE [IO_THREAD | SQL_THREAD][FOR CHANNEL channel]。

如果不指定任何选项，则默认停止 I/O 线程和 SQL 线程。也可通过以下命令单独停止某个线程。

```
stop slave io_thread;
stop slave sql_thread;
```

命令中的 channel 用来指定 channel 名，常用于多源复制。如果不指定，默认会对所有 channel 生效。

4.1.6 在主库上查看从库 IP 和端口信息

要在主库上查看从库 IP 和端口信息，可采用以下两种方式。

- SHOW PROCESSLIST

```
mysql> show processlist;
+-----+-----------------+---------------------+------+-------------+--------+-------------------------------------------------------------+-----------------+
| Id  | User            | Host                | db   | Command     | Time   | State                                                       | Info            |
+-----+-----------------+---------------------+------+-------------+--------+-------------------------------------------------------------+-----------------+
|   5 | event_scheduler | localhost           | NULL | Daemon      | 195841 | Waiting on empty queue                                      | NULL            |
| 243 | root            | localhost           | NULL | Query       |      0 | init                                                        | SHOW PROCESSLIST|
| 245 | repl            | 192.168.244.20:35674| NULL | Binlog Dump |     11 | Master has sent all binlog to slave; waiting for more updates| NULL            |
+-----+-----------------+---------------------+------+-------------+--------+-------------------------------------------------------------+-----------------+
3 rows in set (0.00 sec)
```

输出中，Command 为 Binlog Dump 的是 binlog dump 线程，通过 Host 这一列的输出，我们可以知道从库的 IP 信息。但遗憾的是，SHOW PROCESSLIST 并没有输出从库的端口信息。这一点可通过 SHOW SLAVE HOSTS 命令来弥补。

- SHOW SLAVE HOSTS | SHOW REPLICAS

```
mysql> show slave hosts;
+-----------+------+------+-----------+--------------------------------------+
| Server_id | Host | Port | Master_id | Slave_UUID                           |
+-----------+------+------+-----------+--------------------------------------+
|         2 |      | 3306 |         1 | 7e23f07e-04cc-11ec-9386-000c294fd13c |
+-----------+------+------+-----------+--------------------------------------+
1 row in set, 1 warning (0.00 sec)
```

SHOW SLAVE HOSTS 也有缺陷，它默认只显示端口信息。在一主一从的架构中，确实能基于 SHOW PROCESSLIST 和 SHOW SLAVE HOSTS 的输出定位到从库的 IP 和端口信息。但在一主多从的环境中，如果从库的端口不一致，就很难进行区分了。

此时，可在从库上设置 report_host 和 report_port，显式指定从库的 IP 和端口。设置后的结果会在 SHOW SLAVE HOSTS 中体现出来。例如：

```
mysql> show slave hosts;
+-----------+---------------+------+-----------+--------------------------------------+
| Server_id | Host          | Port | Master_id | Slave_UUID                           |
+-----------+---------------+------+-----------+--------------------------------------+
|         2 | 192.168.244.20| 3306 |         1 | 7e23f07e-04cc-11ec-9386-000c294fd13c |
+-----------+---------------+------+-----------+--------------------------------------+
1 row in set, 1 warning (0.03 sec)
```

4.1.7 查看实例当前拥有的 binlog

查看实例当前拥有的 binlog 可使用命令 SHOW {BINARY | MASTER} LOGS。

```
mysql> show binary logs;
+------------------+-----------+-----------+
| Log_name         | File_size | Encrypted |
+------------------+-----------+-----------+
| mysql-bin.000001 |    139239 | No        |
| mysql-bin.000002 |   1055792 | No        |
| mysql-bin.000003 |   1053529 | No        |
+------------------+-----------+-----------+
3 rows in set (0.00 sec)
```

SHOW BINARY LOGS 反映的实际上是 binlog 索引文件（mysql-bin.index）的内容。

从 MySQL 8.0.14 开始，支持 binlog 和 relay log 的加密存储。

4.1.8 删除 binlog

删除 binlog 的命令是 PURGE { BINARY | MASTER } LOGS { TO 'log_name' | BEFORE datetime_expr }。

既可基于文件名，也可基于时间点来删除。例如：

```
purge binary logs to 'mysql-bin.000004';
purge binary logs before '2017-06-13 17:28:29';
```

在执行 PURGE 操作时，需注意以下两点。

❑ 第一个 PURGE 操作只会删除 mysql-bin.000004 之前的 binlog。
❑ 在 GTID 复制中，执行 PURGE BINARY LOGS 操作还会同步修改 gtid_purged 的值。

除了 PURGE 操作，还可通过操作系统命令（如 Linux 系统下的 rm）删除 binlog，只不过要相应地修改 binlog 索引文件（mysql-bin.index）的内容。需要注意的是，手动修改 mysql-bin.index 文件，需执行 FLUSH BINARY LOGS 让其生效，否则在执行 SHOW BINARY LOGS 时还会看到被删除的 binlog。

4.1.9 查看 binlog 的内容

查看 binlog 内容的命令是 SHOW BINLOG EVENTS [IN 'log_name'] [FROM pos] [LIMIT [offset,] row_count]。

```
mysql> show binlog events in 'mysql-bin.000004';
+------------------+-----+----------------+-----------+-------------+---------------------------------------------------+
| Log_name         | Pos | Event_type     | Server_id | End_log_pos | Info                                              |
+------------------+-----+----------------+-----------+-------------+---------------------------------------------------+
| mysql-bin.000004 |   4 | Format_desc    |         1 |         125 | Server ver: 8.0.25, Binlog ver: 4                 |
| mysql-bin.000004 | 125 | Previous_gtids |         1 |         196 | bd6b3216-04d6-11ec-b76f-000c292c1f7b:1-38         |
+------------------+-----+----------------+-----------+-------------+---------------------------------------------------+
2 rows in set (0.00 sec)
```

如果不指定 log_name，则默认查看的是 SHOW BINARY LOGS 中的第一个 binlog。

4.1.10 RESET MASTER、RESET SLAVE 和 RESET SLAVE ALL 的区别

在管理 MySQL 的过程中，RESET MASTER、RESET SLAVE 和 RESET SLAVE ALL 是 3 个经常会用到的命令，很容易混淆。

下面具体来看看这 3 个命令之间的区别。

1. RESET MASTER

该命令会执行以下操作。

- 删除所有的 binlog，并从头开始生成一个新的 binlog。
- 清空 binlog 索引文件（mysql-bin.index）。
- 在 GTID 复制中，会同时清空变量 gtid_executed、gtid_purged 及表 mysql.gtid_executed 的内容。

例如：

```
mysql> show binary logs;
+------------------+-----------+-----------+
| Log_name         | File_size | Encrypted |
+------------------+-----------+-----------+
| mysql-bin.000001 |    139239 | No        |
| mysql-bin.000002 |   1055792 | No        |
| mysql-bin.000003 |   1053576 | No        |
| mysql-bin.000004 |       196 | No        |
+------------------+-----------+-----------+
4 rows in set (0.01 sec)

mysql> reset master;
Query OK, 0 rows affected (0.09 sec)

mysql> show binary logs;
+------------------+-----------+-----------+
| Log_name         | File_size | Encrypted |
+------------------+-----------+-----------+
| mysql-bin.000001 |       156 | No        |
+------------------+-----------+-----------+
1 row in set (0.00 sec)
```

该命令既可在主库上执行，也可在从库上执行。在执行的时候需注意以下两点。

- 在正常的主从复制环境中，切记不要在主库上执行 RESET MASTER 命令，否则很容易导致主从复制中断，甚至主从数据不一致。
- 执行完 RESET MASTER，默认情况下生成的第一个 binlog 的序号是 000001，也可通过 RESET MASTER TO *xxx* 命令显式指定第一个 binlog 的序号。

2. RESET SLAVE

该命令会执行以下操作。

- 删除所有的 relay log，并从头开始生成一个新的 relay log。
- 清空 relay log 索引文件。
- 清空 mysql.slave_relay_log_info 和 mysql.slave_worker_info 表的内容。
- 清除 mysql.slave_master_info 表中 binlog 的位置点（Master_log_name 和 Master_log_pos）信息，但是会保留连接信息。

既然连接信息依然存在，就可直接通过 START SLAVE 重启复制。在 GTID 复制中，因为有 gtid_executed，这样操作是没有问题的，复制可以继续。但如果是基于位置点的复制，切记不要这样操作，因为 I/O 线程会从主库现有的第一个 binlog 开始重新拉取数据，这样很容易导致主从数据不一致及主从复制中断。

3. RESET SLAVE ALL

与 RESET SLAVE 不同，RESET SLAVE ALL 会直接清空 mysql.slave_master_info 表的所有内容，包括连接信息。这个时候，如果要重启复制，只能执行 CHANGE MASTER TO 命令。

无论是 RESET SLAVE 还是 RESET SLAVE ALL，在执行的时候都需注意以下两点。

- 在执行之前，必须停止复制。
- 执行时不会清除与 GTID 相关的任何信息。

4.1.11 跳过指定事务

主从复制中断，绝大多数情况下是 SQL 线程应用错误。这个时候，SHOW SLAVE STATUS 中的 Last_SQL_Error 会显示具体的错误信息，例如：

```
Last_SQL_Error: Error 'Table 't1' already exists' on query. Default database: ''. Query: 'create table slowtech.t1(id int primary key)'
```

为了恢复主从复制，我们有时候会先跳过这个事务，事后再分析主从复制中断的原因。

在基于位置点的复制和 GTID 复制中，跳过指定事务需要执行的操作不一样。

- 在基于位置点的复制中，跳过复制中的事务需设置 sql_slave_skip_counter。sql_slave_skip_counter 的单位可简单地理解为事务。例如，下面的操作会跳过复制中的一个事务。

```
mysql> stop slave;
mysql> set global sql_slave_skip_counter = 1;
mysql> start slave;
```

- 在 GTID 复制中，跳过事务的本质是注入空事务。看看下面这个示例。

```
mysql> show slave status\G
*************************** 1. row ***************************
...
            Last_SQL_Error: Error 'Table 't1' already exists' on query. Default database: ''.
Query: 'create table slowtech.t1(id int primary key)'
...
        Retrieved_Gtid_Set: bd6b3216-04d6-11ec-b76f-000c292c1f7b:11664-11711
         Executed_Gtid_Set: bd6b3216-04d6-11ec-b76f-000c292c1f7b:1-11690
...

mysql> stop slave;
mysql> set session gtid_next='bd6b3216-04d6-11ec-b76f-000c292c1f7b:11691';
mysql> begin;
mysql> commit;
mysql> set session gtid_next='automatic';
mysql> start slave;
```

这里，第一个 gtid_next 的设置是关键，需取 Executed_Gtid_Set 的最大值加 1。

4.1.12 操作不写入 binlog

如果不想操作写入 binlog，可在会话级别设置 sql_log_bin。

```
mysql> set session sql_log_bin=0;
```

如果是 GTID 复制，也不会为该操作分配 GTID 值。

4.1.13 判断主库的某个操作是否已经在从库上执行

要判断主库的某个操作是否已经在从库上执行，常规做法如下。

- 主库执行完操作。
- 执行 SHOW MASTER STATUS，获取主库当前 binlog 的位置点信息，即 File 和 Position。
- 接着在从库上执行 SHOW SLAVE STATUS，获取当前重放的事务在主库 binlog 中的位置点信息，即 Relay_Master_Log_File 和 Exec_Master_Log_Pos。
- 如果 Relay_Master_Log_File >= File && Exec_Master_Log_Pos >= Position，则意味着这个操作已经在从库上执行了。

但这种方式相对比较烦琐，需要不断地执行 SHOW SLAVE STATUS 来进行比较。其实，更优雅的方法是使用 MySQL 自带的函数。

具体来说，对于位置点复制，推荐使用 MASTER_POS_WAIT 函数，其使用方式如下。

```
SELECT MASTER_POS_WAIT('master_log_file', master_log_pos [, timeout][, channel])
```

给出具体的 binlog 及位置点，如果从库 SQL 线程还没重放到这个位置，该命令会一直阻塞。也可指定 timeout，设置命令的超时时长，如果在超时时长内还没重放到指定位置，该命令会返回 -1。

对于 GTID 复制，推荐使用 WAIT_FOR_EXECUTED_GTID_SET 函数，其使用方式如下。

```
SELECT WAIT_FOR_EXECUTED_GTID_SET(gtid_set[, timeout])
```

4.1.14 在线设置复制的过滤规则

MySQL 5.7 支持在线设置复制的过滤规则，具体命令如下。在此之前，只能通过参数来设置，并且重启实例才能生效。

```
CHANGE REPLICATION FILTER filter[, filter]
       [, ...] [FOR CHANNEL channel]

filter: {
    REPLICATE_DO_DB = (db_list)
  | REPLICATE_IGNORE_DB = (db_list)
  | REPLICATE_DO_TABLE = (tbl_list)
  | REPLICATE_IGNORE_TABLE = (tbl_list)
  | REPLICATE_WILD_DO_TABLE = (wild_tbl_list)
  | REPLICATE_WILD_IGNORE_TABLE = (wild_tbl_list)
  | REPLICATE_REWRITE_DB = (db_pair_list)
}
db_list:
    db_name[, db_name][, ...]
tbl_list:
    db_name.table_name[, db_name.table_name][, ...]
wild_tbl_list:
    'db_pattern.table_pattern'[, 'db_pattern.table_pattern'][, ...]
db_pair_list:
    (db_pair)[, (db_pair)][, ...]
db_pair:
    from_db, to_db
```

下面看看各个规则的具体含义。

- REPLICATE_DO_DB：复制指定库的操作。
- REPLICATE_IGNORE_DB：忽略指定库的操作。
- REPLICATE_DO_TABLE：复制指定表的操作。
- REPLICATE_IGNORE_TABLE：忽略指定表的操作。
- REPLICATE_WILD_DO_TABLE：复制指定表的操作，支持模糊匹配。
- REPLICATE_WILD_IGNORE_TABLE：忽略指定表的操作，支持模糊匹配。
- REPLICATE_REWRITE_DB：将主库某个数据库的操作重写到从库另一个数据库中。

每个规则都对应有一个同名参数，如 REPLICATE_DO_DB 对应 replicate_do_db 参数，REPLICATE_IGNORE_DB 对应 replicate_ignore_db 参数。在执行完 CHANGE REPLICATION FILTER 命令后，建议将对应的规则写入参数文件，否则，在实例重启后，会丢失之前通过命令配置的规则。

以下是 CHANGE REPLICATION FILTER 命令的一些常见用法。

```
change replication filter replicate_do_db=(db1);
change replication filter replicate_ignore_table=(db1.t1,db2.t2);
change replication filter replicate_wild_do_table = ('db1.t1%');
change replication filter replicate_rewrite_db = ((master_db1, slave_db2), (master_db3, slave_db4));
```

在设置过滤规则时，需注意以下几点。

- 如果没有指定 channel，则默认创建的规则是全局规则，会应用于所有 channel。全局规则可通过 performance_schema.replication_applier_global_filters 查看。如果指定了 channel，则创建的规则是 channel 级别的规则，只应用于指定的 channel。channel 级别的规则可通过 performance_schema.replication_applier_filters 查看。
- 同样的过滤规则如果设置两次，第二次的会覆盖第一次的。所以，对于同一个规则，如果要新增内容，必须将当前的过滤规则加上。
- 要清除过滤规则，可使用以下命令。

```
change replication filter replicate_do_db = (), replicate_ignore_db = ();
```

4.2 复制的监控

在 MySQL 5.7 之前，对复制进行监控所依赖的数据源主要有以下两类。

- SHOW SLAVE STATUS 的输出。
- 复制相关的状态变量，具体如下。

```
Slave_heartbeat_period
Slave_last_heartbeat
Slave_received_heartbeats
Slave_retried_transactions
Slave_running
```

这些变量里面用得较多的是 Slave_running，该变量可用来判断从库是否正常运行，只有当 Slave_IO_Running 和 Slave_SQL_Running 都为 YES 时，该变量才为 ON。

在 MySQL 5.7 中，因为多源复制的引入，这些变量默认不再支持，取而代之的是在 performance_schema 中引入了多张与复制相关的表。但为了保持向后兼容，MySQL 5.7 同时引入了参数 show_compatibility_56 来开启这些变量。不过在 MySQL 8.0 中，已经完全移除了这些变量及 show_compatibility_56。

下面重点说说 performance_schema 中与复制相关的表，一共有 10 张，具体可分为如下几类。

4.2.1 连接

1. replication_connection_configuration

该表记录了复制的配置信息。例如：

```
mysql> select * from replication_connection_configuration\G
*************************** 1. row ***************************
               CHANNEL_NAME: 
                       HOST: 192.168.244.10
                       PORT: 3306
                       USER: repl
          NETWORK_INTERFACE: 
              AUTO_POSITION: 1
                SSL_ALLOWED: NO
                SSL_CA_FILE: 
```

```
                        SSL_CA_PATH: 
                    SSL_CERTIFICATE: 
                         SSL_CIPHER: 
                            SSL_KEY: 
       SSL_VERIFY_SERVER_CERTIFICATE: NO
                       SSL_CRL_FILE: 
                       SSL_CRL_PATH: 
          CONNECTION_RETRY_INTERVAL: 60
             CONNECTION_RETRY_COUNT: 86400
                 HEARTBEAT_INTERVAL: 30.000
                        TLS_VERSION: 
                    PUBLIC_KEY_PATH: 
                     GET_PUBLIC_KEY: YES
                  NETWORK_NAMESPACE: 
              COMPRESSION_ALGORITHM: uncompressed
             ZSTD_COMPRESSION_LEVEL: 3
                    TLS_CIPHERSUITES: NULL
      SOURCE_CONNECTION_AUTO_FAILOVER: 0
1 row in set (0.00 sec)
```

可以看到，这些配置基本上都是 CHANGE MASTER TO 的选项。

2. replication_connection_status

该表记录了 I/O 线程的状态信息。例如：

```
mysql> select * from replication_connection_status\G
*************************** 1. row ***************************
                                      CHANNEL_NAME: 
                                        GROUP_NAME: 
                                       SOURCE_UUID: bd6b3216-04d6-11ec-b76f-000c292c1f7b
                                         THREAD_ID: 60
                                     SERVICE_STATE: ON
                         COUNT_RECEIVED_HEARTBEATS: 1
                          LAST_HEARTBEAT_TIMESTAMP: 2021-09-05 09:12:34.496022
                          RECEIVED_TRANSACTION_SET: bd6b3216-04d6-11ec-b76f-000c292c1f7b:1-353
                                 LAST_ERROR_NUMBER: 0
                                LAST_ERROR_MESSAGE: 
                              LAST_ERROR_TIMESTAMP: 0000-00-00 00:00:00.000000
                           LAST_QUEUED_TRANSACTION: bd6b3216-04d6-11ec-b76f-000c292c1f7b:353
 LAST_QUEUED_TRANSACTION_ORIGINAL_COMMIT_TIMESTAMP: 2021-09-05 09:13:43.757521
LAST_QUEUED_TRANSACTION_IMMEDIATE_COMMIT_TIMESTAMP: 2021-09-05 09:13:43.757521
     LAST_QUEUED_TRANSACTION_START_QUEUE_TIMESTAMP: 2021-09-05 09:13:43.910597
       LAST_QUEUED_TRANSACTION_END_QUEUE_TIMESTAMP: 2021-09-05 09:13:43.910754
                              QUEUEING_TRANSACTION: 
      QUEUEING_TRANSACTION_ORIGINAL_COMMIT_TIMESTAMP: 0000-00-00 00:00:00.000000
     QUEUEING_TRANSACTION_IMMEDIATE_COMMIT_TIMESTAMP: 0000-00-00 00:00:00.000000
         QUEUEING_TRANSACTION_START_QUEUE_TIMESTAMP: 0000-00-00 00:00:00.000000
1 row in set (0.00 sec)
```

部分字段的具体含义如下。

- CHANNEL_NAME：channel 名。
- GROUP_NAME：组复制中的组名。
- SOURCE_UUID：主库的 server_uuid。
- THREAD_ID：I/O 线程的线程 ID。注意，它不是 PROCESSLIST ID。
- SERVICE_STATE：I/O 线程的状态，对应 SHOW SLAVE STATUS 中的 Slave_IO_Running。

- COUNT_RECEIVED_HEARTBEATS：复制启动后，从库接收到的心跳包的个数。对应 MySQL 5.6 中的 Slave_received_heartbeats。
- LAST_HEARTBEAT_TIMESTAMP：收到最近一个心跳包的时间戳。对应 MySQL 5.6 中的 Slave_last_heartbeat。
- RECEIVED_TRANSACTION_SET：从库接受过的事务所对应的 GTID 集。对应 SHOW SLAVE STATUS 中的 Retrieved_Gtid_Set。
- LAST_ERROR_NUMBER、LAST_ERROR_MESSAGE 和 LAST_ERROR_TIMESTAMP：当 I/O 线程出现问题时，对应的错误码、错误信息及时间戳，分别对应 SHOW SLAVE STATUS 中的 Last_IO_Errno、Last_IO_Error 和 Last_IO_Error_Timestamp。

除此之外，replication_connection_status 表还记录了以下两个事务的相关信息。

- 最近一个写入 relay log 的事务。

  与之相关的字段有，

  - LAST_QUEUED_TRANSACTION：事务的 GTID。
  - LAST_QUEUED_TRANSACTION_START_QUEUE_TIMESTAMP：该事务写入 relay log 队列的开始时间。
  - LAST_QUEUED_TRANSACTION_END_QUEUE_TIMESTAMP：该事务写入 relay log 队列的结束时间。
  - LAST_QUEUED_TRANSACTION_ORIGINAL_COMMIT_TIMESTAMP：该事务在主库提交的时间戳。
  - LAST_QUEUED_TRANSACTION_IMMEDIATE_COMMIT_TIMESTAMP：该事务在上一级节点提交的时间戳。

  这里重点说说 LAST_QUEUED_TRANSACTION_ORIGINAL_COMMIT_TIMESTAMP 和 LAST_QUEUED_TRANSACTION_IMMEDIATE_COMMIT_TIMESTAMP 这两个时间戳的具体含义。以一个级联复制为例：A -> B -> C -> D。对于同一个事务，无论是在哪个实例上查看 LAST_QUEUED_TRANSACTION_ORIGINAL_COMMIT_TIMESTAMP，看到的值都是一样的，都是该事务在 A 上提交的时间戳。但 IMMEDIATE_COMMIT_TIMESTAMP 不一样，它对应的是事务在上一级节点提交的时间戳。具体来说，在 C 上看到的 LAST_QUEUED_TRANSACTION_IMMEDIATE_COMMIT_TIMESTAMP 是事务在 B 上提交的时间戳，在 D 上看到的 LAST_QUEUED_TRANSACTION_IMMEDIATE_COMMIT_TIMESTAMP 是事务在 C 上提交的时间戳。

- 当前正在写入 relay log 的事务。

  相关字段以 QUEUEING_TRANSACTION 开头，字段的具体含义可参考 LAST_QUEUED_TRANSACTION。

4.2.2 事务重放

1. replication_applier_configuration

该表记录了复制中会影响从库重放事务的配置信息。

```
mysql> select * from replication_applier_configuration\G
*************************** 1. row ***************************
              CHANNEL_NAME:
             DESIRED_DELAY: 0
      PRIVILEGE_CHECKS_USER: NULL
         REQUIRE_ROW_FORMAT: NO
```

```
            REQUIRE_TABLE_PRIMARY_KEY_CHECK: STREAM
  ASSIGN_GTIDS_TO_ANONYMOUS_TRANSACTIONS_TYPE: OFF
 ASSIGN_GTIDS_TO_ANONYMOUS_TRANSACTIONS_VALUE: NULL
1 row in set (0.00 sec)
```

这些配置基本上都是 CHANGE MASTER TO 的选项。DESIRED_DELAY 对应 CHANGE MASTER TO 中的 MASTER_DELAY，用于延迟复制。ASSIGN_GTIDS_TO_ANONYMOUS_TRANSACTIONS_TYPE 和 ASSIGN_GTIDS_TO_ANONYMOUS_TRANSACTIONS_VALUE 对应 CHANGE MASTER TO 中的 ASSIGN_GTIDS_TO_ANONYMOUS_TRANSACTIONS。ASSIGN_GTIDS_TO_ANONYMOUS_TRANSACTIONS_TYPE 是类型，有 3 个取值：OFF、LOCAL、UUID。ASSIGN_GTIDS_TO_ANONYMOUS_TRANSACTIONS_VALUE 是具体的 UUID 值。

2. replication_applier_status

该表记录了从库应用线程的总体状态信息。注意，它记录的是总体状态信息，不针对任何线程。

```
mysql> select * from replication_applier_status;
+--------------+---------------+-----------------+----------------------------+
| CHANNEL_NAME | SERVICE_STATE | REMAINING_DELAY | COUNT_TRANSACTIONS_RETRIES |
+--------------+---------------+-----------------+----------------------------+
|              | ON            |            NULL |                          0 |
+--------------+---------------+-----------------+----------------------------+
1 row in set (0.00 sec)
```

部分字段的具体含义如下。

- SERVICE_STATE：应用线程的总体状态。对应 SHOW SLAVE STATUS 中的 Slave_SQL_Running。
- REMAINING_DELAY：与延迟复制有关，对应 SHOW SLAVE STATUS 中的 SQL_Remaining_Delay。
- COUNT_TRANSACTIONS_RETRIES：事务因重放失败而进行重试的次数。单个事务的最大重试次数由参数 slave_transaction_retries 决定，默认为 10。

4.2.3　多线程复制

1. replication_applier_status_by_coordinator

该表记录了 Coordinator 线程的状态信息。只有开启多线程复制，该表才会有内容。

```
mysql> select * from replication_applier_status_by_coordinator\G
*************************** 1. row ***************************
                                      CHANNEL_NAME:
                                         THREAD_ID: 63
                                     SERVICE_STATE: ON
                                 LAST_ERROR_NUMBER: 0
                                LAST_ERROR_MESSAGE:
                              LAST_ERROR_TIMESTAMP: 0000-00-00 00:00:00.000000
                          LAST_PROCESSED_TRANSACTION: bd6b3216-04d6-11ec-b76f-000c292c1f7b:1324
 LAST_PROCESSED_TRANSACTION_ORIGINAL_COMMIT_TIMESTAMP: 2021-09-05 09:15:40.046353
LAST_PROCESSED_TRANSACTION_IMMEDIATE_COMMIT_TIMESTAMP: 2021-09-05 09:15:40.046353
    LAST_PROCESSED_TRANSACTION_START_BUFFER_TIMESTAMP: 2021-09-05 09:15:40.205151
      LAST_PROCESSED_TRANSACTION_END_BUFFER_TIMESTAMP: 2021-09-05 09:15:40.205384
                             PROCESSING_TRANSACTION:
       PROCESSING_TRANSACTION_ORIGINAL_COMMIT_TIMESTAMP: 0000-00-00 00:00:00.000000
      PROCESSING_TRANSACTION_IMMEDIATE_COMMIT_TIMESTAMP: 0000-00-00 00:00:00.000000
          PROCESSING_TRANSACTION_START_BUFFER_TIMESTAMP: 0000-00-00 00:00:00.000000
1 row in set (0.00 sec)
```

部分字段的具体含义如下。

- SERVICE_STATE：Coordinator 线程的状态。
- LAST_ERROR_NUMBER、LAST_ERROR_MESSAGE 和 LAST_ERROR_TIMESTAMP：Coordinator 线程出现问题时，对应的错误码、错误信息及时间戳，分别对应 SHOW SLAVE STATUS 中的 Last_SQL_Errno、Last_SQL_Error 和 Last_SQL_Error_Timestarm。
- LAST_PROCESSED_TRANSACTION：Coordinator 线程处理的最近一个事务的 GTID。
- LAST_PROCESSED_TRANSACTION_START_BUFFER_TIMESTAMP：该事务写入 Worker 线程缓冲区的开始时间。
- LAST_PROCESSED_TRANSACTION_END_BUFFER_TIMESTAMP：该事务写入 Worker 线程缓冲区的结束时间。
- PROCESSING_TRANSACTION：当前正在处理的事务的 GTID。其他状态值与 LAST_PROCESSED_TRANSACTION 中的类似。

ORIGINAL_COMMIT_TIMESTAMP 和 IMMEDIATE_COMMIT_TIMESTAMP 相关字段的具体含义可参考 replication_connection_status 表。

2. replication_applier_status_by_worker

该表记录了 Worker 线程的状态信息。

```
mysql> select * from replication_applier_status_by_worker limit 1\G
*************************** 1. row ***************************
                                         CHANNEL_NAME:
                                            WORKER_ID: 1
                                            THREAD_ID: 64
                                        SERVICE_STATE: ON
                                    LAST_ERROR_NUMBER: 0
                                   LAST_ERROR_MESSAGE:
                                 LAST_ERROR_TIMESTAMP: 0000-00-00 00:00:00.000000
                             LAST_APPLIED_TRANSACTION: bd6b3216-04d6-11ec-b76f-000c292c1f7b:1919
   LAST_APPLIED_TRANSACTION_ORIGINAL_COMMIT_TIMESTAMP: 2021-09-05 09:16:49.472341
  LAST_APPLIED_TRANSACTION_IMMEDIATE_COMMIT_TIMESTAMP: 2021-09-05 09:16:49.472341
        LAST_APPLIED_TRANSACTION_START_APPLY_TIMESTAMP: 2021-09-05 09:16:49.635567
          LAST_APPLIED_TRANSACTION_END_APPLY_TIMESTAMP: 2021-09-05 09:16:49.638315
                                  APPLYING_TRANSACTION:
       APPLYING_TRANSACTION_ORIGINAL_COMMIT_TIMESTAMP: 0000-00-00 00:00:00.000000
      APPLYING_TRANSACTION_IMMEDIATE_COMMIT_TIMESTAMP: 0000-00-00 00:00:00.000000
            APPLYING_TRANSACTION_START_APPLY_TIMESTAMP: 0000-00-00 00:00:00.000000
                 LAST_APPLIED_TRANSACTION_RETRIES_COUNT: 0
  LAST_APPLIED_TRANSACTION_LAST_TRANSIENT_ERROR_NUMBER: 0
 LAST_APPLIED_TRANSACTION_LAST_TRANSIENT_ERROR_MESSAGE:
LAST_APPLIED_TRANSACTION_LAST_TRANSIENT_ERROR_TIMESTAMP: 0000-00-00 00:00:00.000000
                     APPLYING_TRANSACTION_RETRIES_COUNT: 0
      APPLYING_TRANSACTION_LAST_TRANSIENT_ERROR_NUMBER: 0
     APPLYING_TRANSACTION_LAST_TRANSIENT_ERROR_MESSAGE:
   APPLYING_TRANSACTION_LAST_TRANSIENT_ERROR_TIMESTAMP: 0000-00-00 00:00:00.000000
1 row in set (0.00 sec)
```

如果没有开启多线程复制，这里的 Worker 线程指的就是 SQL 线程。如果开启了多线程复制，这里的 Worker 线程指的就是工作线程，每个工作线程对应一条记录。

部分字段的具体含义如下。

- SERVICE_STATE：工作线程的状态。
- LAST_ERROR_NUMBER、LAST_ERROR_MESSAGE 和 LAST_ERROR_TIMESTAMP：当工作线程出现问题时，对应的错误码、错误信息及时间戳，分别对应 SHOW SLAVE STATUS 中的 Last_SQL_Errno、Last_SQL_Error 和 Last_SQL_Error_Timestamp。
- LAST_APPLIED_TRANSACTION：最近一个重放的事务的 GTID。
- LAST_APPLIED_TRANSACTION_START_APPLY_TIMESTAMP：该事务重放的开始时间。
- LAST_APPLIED_TRANSACTION_END_APPLY_TIMESTAMP：该事务重放的结束时间。
- LAST_APPLIED_TRANSACTION_RETRIES_COUNT：该事务重试的次数。
- LAST_APPLIED_TRANSACTION_LAST_TRANSIENT_ERROR_NUMBER、LAST_APPLIED_TRANSACTION_LAST_TRANSIENT_ERROR_MESSAGE 和 LAST_APPLIED_TRANSACTION_LAST_TRANSIENT_ERROR_TIMESTAMP：导致事务重试的最近一个错误的错误码、错误信息及时间戳。
- APPLYING_TRANSACTION：当前正在重放的事务的 GTID。其他状态值同 LAST_APPLIED_TRANSACTION 中的类似。

通过 APPLYING_TRANSACTION_ORIGINAL_COMMIT_TIMESTAMP，我们可以很精确地算出主从延迟时间。

4.2.4 过滤规则

1. replication_applier_filters

该表记录了复制中 channel 级别的过滤规则。

```
mysql> select * from replication_applier_filters;
+--------------+------------------------+--------------+--------------------------------------+----------------------------+---------+
| CHANNEL_NAME | FILTER_NAME            | FILTER_RULE  | CONFIGURED_BY                        | ACTIVE_SINCE               | COUNTER |
+--------------+------------------------+--------------+--------------------------------------+----------------------------+---------+
| channel_1    | REPLICATE_DO_DB        | db1          | CHANGE_REPLICATION_FILTER_FOR_CHANNEL | 2021-09-05 09:24:54.408321 |       1 |
| channel_1    | REPLICATE_IGNORE_TABLE | db1.t1,db2.t2 | CHANGE_REPLICATION_FILTER_FOR_CHANNEL | 2021-09-05 09:24:54.896588 |       0 |
+--------------+------------------------+--------------+--------------------------------------+----------------------------+---------+
2 rows in set (0.00 sec)
```

部分字段的具体含义如下。

- CONFIGURED_BY：规则配置的方式，有如下取值。
  - CHANGE_REPLICATION_FILTER：通过 CHANGE REPLICATION FILTER 进行的配置。
  - STARTUP_OPTIONS：通过配置文件或命令行进行的配置。
  - CHANGE_REPLICATION_FILTER_FOR_CHANNEL：通过 CHANGE REPLICATION FILTER xxx FOR CHANNEL channel 进行的配置。
  - STARTUP_OPTIONS_FOR_CHANNEL：同 STARTUP_OPTIONS 一样，只不过指定了具体的 CHANNEL，如 replicate-do-db=channel_1:db_name。
- ACTIVE_SINCE：规则配置的时间。
- COUNTER：规则被使用的次数。

2. replication_applier_global_filters

该表记录了复制中全局级别的过滤规则。这里的全局指的是对所有 channel 都生效的规则，对应 replication_applier_filters 表中 CONFIGURED_BY 为 CHANGE_REPLICATION_FILTER 或 STARTUP_OPTIONS 的规则。

```
mysql> select * from replication_applier_global_filters;
+---------------------+-------------+---------------------------+----------------------------+
| FILTER_NAME         | FILTER_RULE | CONFIGURED_BY             | ACTIVE_SINCE               |
+---------------------+-------------+---------------------------+----------------------------+
| REPLICATE_IGNORE_DB | db2         | CHANGE_REPLICATION_FILTER | 2021-09-05 09:27:51.663448 |
+---------------------+-------------+---------------------------+----------------------------+
1 row in set (0.00 sec)
```

4.2.5 组复制

replication_group_members 和 replication_group_member_stats 与组复制相关，暂且不谈，具体可见第 11 章。

需要注意的是，如果禁用了 performance_schema，则不会采集 START 和 END 之类的时间点，如 LAST_APPLIED_TRANSACTION_START_APPLY_TIMESTAMP 和 LAST_APPLIED_TRANSACTION_END_APPLY_TIMESTAMP。

4.3 主从延迟

主从延迟作为 MySQL 的痛点已经存在很多年了，以至于大家都有一种错觉：有 MySQL 复制的地方就有主从延迟。很多人将主从延迟的原因归结为从库的单线程重放。实际上，这个说法比较片面，因为有很多问题并不是并行复制能解决的，比如从库 SQL 线程被阻塞，从库磁盘 I/O 存在瓶颈等。

很多人在分析此类问题时缺乏一个系统的方法论，以致无法准确地定位主从延迟的根本原因。下面就如何分析主从延迟进行系统、全面的总结，本节主要包括以下 3 方面的内容。

- 如何分析主从延迟。
- 主从延迟的常见原因及解决方案。
- 如何解读 Seconds_Behind_Master。

4.3.1 如何分析主从延迟

分析主从延迟时一般会采集以下 3 类信息。

- 从库服务器的负载情况。

  为什么要首先查看服务器的负载情况呢？因为软件层面的所有操作都需要系统资源来支撑。常见的系统资源有四类：CPU、内存、I/O、网络。对于主从延迟，一般会重点关注 CPU 和 I/O。

  分析 CPU 是否达到瓶颈，常用的命令是 top，通过它可以直观地看到主机的 CPU 使用情况。以下是 top 中与 CPU 相关的输出。

  ```
  Cpu(s):  0.2%us,  0.2%sy,  0.0%ni, 99.5%id,  0.0%wa,  0.0%hi,  0.2%si,  0.0%st
  ```

下面看看各个指标的具体含义。

- `us`：处理用户态（user）任务的 CPU 时间占比。
- `sy`：处理内核态（system）任务的 CPU 时间占比。
- `ni`：处理低优先级进程用户态任务的 CPU 时间占比。进程的优先级由 `nice` 值决定，`nice` 的范围是 –20 ~ 19，值越大，优先级越低。1 ~ 19 称为低优先级。
- `id`：处于空闲（idle）状态的 CPU 时间占比。
- `wa`：等待 I/O 的 CPU 时间占比。
- `hi`：处理硬中断（irq）的 CPU 时间占比。
- `si`：处理软中断（softirq）的 CPU 时间占比。
- `st`：当系统运行在虚拟机中的时候，被其他虚拟机占用（steal）的 CPU 时间占比。

一般来说，当 CPU 使用率（1 减去处于空闲状态的 CPU 时间占比）超过 90% 时，需要格外关注。毕竟，对于数据库应用来说，CPU 很少是瓶颈，除非有大量的慢 SQL。

接下来看看 I/O。

查看磁盘 I/O 负载情况，常用的命令是 `iostat`。

```
# iostat -xm 1
avg-cpu:  %user   %nice %system %iowait  %steal   %idle
           4.21    0.00    1.77    0.35    0.00   93.67

Device:    rrqm/s   wrqm/s     r/s     w/s    rMB/s    wMB/s avgrq-sz avgqu-sz   await r_await w_await  svctm  %util
sda          0.00     0.00    0.00    0.00     0.00     0.00     0.00     0.00    0.00    0.00    0.00   0.00   0.00
sdb          0.00     0.00  841.00 3234.00    13.14    38.96    26.19     0.60    0.15    0.30    0.11   0.08  32.60
```

命令中指定了如下 3 个选项。

- `-x`：打印扩展信息。
- `-m`：指定吞吐量的单位是 MB/s，默认是 KB/s。
- `1`：每隔 1 秒打印一次。

下面看看输出中各指标的具体含义。

- `rrqm/s`：每秒被合并的读请求的数量。
- `wrqm/s`：每秒被合并的写请求的数量。
- `r/s`：每秒发送给磁盘的读请求的数量。
- `w/s`：每秒写入磁盘的写请求的数量。注意，这里的请求是合并后的请求。r/s 加 w/s 等于 IOPS。
- `rMB/s`：每秒从磁盘读取的数据量。
- `wMB/s`：每秒写入磁盘的数据量。rMB/s + wMB/s 等于吞吐量。
- `avgrq-sz`：I/O 请求的平均大小，单位是扇区，扇区的大小是 512 字节。一般而言，I/O 请求越大，耗时越长。
- `avgqu-sz`：队列里的平均 I/O 请求数量。
- `await`：I/O 请求的平均耗时，包括磁盘的实际处理时间及队列中的等待时间，单位是毫秒。r_await 是读请求的平均耗时，w_await 是写请求的平均耗时。

- svctm：I/O 请求的平均服务时间，单位是 ms。注意，这个指标已弃用，会在后续版本中移除。
- %util：磁盘饱和度，反映了一个采样周期内有多少时间在做 I/O 操作。对于只能串行处理 I/O 请求的设备来说，该值接近 100%就意味着设备饱和。但对于 RAID、SSD 等能并行处理 I/O 请求的设备来说，该值的参考意义不大，即使达到了 100%也不意味设备出现了饱和。至于是否达到了性能上限，需参考性能压测下的 IOPS 和吞吐量。

☐ 主从复制状态。

对于主库，执行 SHOW MASTER STATUS。

```
mysql> show master status;
+------------------+----------+--------------+------------------+------------------------------------------+
| File             | Position | Binlog_Do_DB | Binlog_Ignore_DB | Executed_Gtid_Set                        |
+------------------+----------+--------------+------------------+------------------------------------------+
| mysql-bin.000004 | 1631495  |              |                  | bd6b3216-04d6-11ec-b76f-000c292c1f7b:1-5588 |
+------------------+----------+--------------+------------------+------------------------------------------+
1 row in set (0.00 sec)
```

在 SHOW MASTER STATUS 的输出中，需重点关注 File 和 Position 这两个指标的值。

对于从库，执行 SHOW SLAVE STATUS。

```
mysql> show slave status\G
*************************** 1. row ***************************
                ...
      Master_Log_File: mysql-bin.000004
  Read_Master_Log_Pos: 1631495
                ...
   Relay_Master_Log_File: mysql-bin.000004
                ...
      Exec_Master_Log_Pos: 1631495
                ...
```

在 SHOW SLAVE STATUS 的输出中，需重点关注 Master_Log_File、Read_Master_Log_Pos、Relay_Master_Log_File、Exec_Master_Log_Pos 这 4 个指标的值。

接下来重点比较以下两对值。

第一对是(File, Position)和(Master_Log_File, Read_Master_Log_Pos)，其中(File, Position)记录了主库 binlog 的位置，(Master_Log_File, Read_Master_Log_Pos)记录了 I/O 线程当前正在接收的二进制日志事件在主库 binlog 中的位置。如果(File, Position) 小于 (Master_Log_File, Read_Master_Log_Pos)，则意味着 I/O 线程存在延迟。

第二对是(Master_Log_File, Read_Master_Log_Pos)和(Relay_Master_Log_File, Exec_Master_Log_Pos)，其中，(Relay_Master_Log_File, Exec_Master_Log_Pos)记录了 SQL 线程当前正在重放的二进制日志事件在主库 binlog 中的位置。如果(Relay_Master_Log_File, Exec_Master_Log_Pos) 小于(Master_Log_File, Read_Master_Log_Pos)，则意味着 SQL 线程存在延迟。

☐ 主库 binlog 的写入量。

主要是看主库 binlog 的生成速度。

4.3.2 主从延迟的常见原因及解决方法

下面分别从 I/O 线程和 SQL 线程这两个方面展开介绍。

1. I/O 线程存在延迟

下面看看 I/O 线程出现延迟的常见原因及解决方法。

- 网络延迟。

  判断是否为网络带宽限制。如果是，可开启 `slave_compressed_protocol` 参数，启用 binlog 的压缩传输。

- 磁盘 I/O 存在瓶颈。

  可调整从库的双一设置或关闭 binlog。注意，在 MySQL 5.6 中，如果开启了 GTID，则会强制要求开启 binlog。从 MySQL 5.7 开始无此限制。

- 网卡存在问题。

一般情况下，I/O 线程很少会出现延迟。

2. SQL 线程存在延迟

下面看看 SQL 线程出现延迟的常见原因及解决方法。

- 主库写入量过大，SQL 线程单线程重放。

  具体体现如下。

  - 从库磁盘 I/O 无明显瓶颈。
  - `Relay_Master_Log_File` 和 `Exec_Master_Log_Pos` 在不断变化。
  - 主库写入量过大。如果磁盘使用的是 SATA SSD，当 binlog 的生成速度快于 5 分钟一个时，从库重放就会有瓶颈。

  这是 MySQL 软件层面的硬伤。要解决该问题，可升级到 MySQL 5.7，开启并行复制。

- STATEMENT 格式下的慢 SQL。

  具体体现为 `Relay_Master_Log_File` 和 `Exec_Master_Log_Pos` 在一段时间内没有变化。

  看看下面这个示例，对一张有 1000 万条数据的表进行 DELETE 操作。表上没有任何索引，在主库上执行用了 7.52 秒，观察从库的 `Seconds_Behind_Master`，发现它最大达到了 7 秒。

```
mysql> show variables like 'binlog_format';
+---------------+-----------+
| Variable_name | Value     |
+---------------+-----------+
| binlog_format | STATEMENT |
+---------------+-----------+
1 row in set (0.00 sec)

mysql> select count(*) from sbtest.sbtest1;
```

```
+----------+
| count(*) |
+----------+
| 10000000 |
+----------+
1 row in set (1.41 sec)

mysql> show create table sbtest.sbtest1\G
*************************** 1. row ***************************
       Table: sbtest1
Create Table: CREATE TABLE `sbtest1` (
  `id` int NOT NULL,
  `k` int NOT NULL DEFAULT '0',
  `c` char(120) NOT NULL DEFAULT '',
  `pad` char(60) NOT NULL DEFAULT ''
) ENGINE=InnoDB DEFAULT CHARSET=utf8mb4 COLLATE=utf8mb4_0900_ai_ci
1 row in set (0.00 sec)

mysql> delete from sbtest.sbtest1 where id <= 100;
Query OK, 100 rows affected (7.52 sec)
```

对于这种执行较慢的 SQL 语句，并行复制实际上也无能为力，此时只能优化 SQL。

在 MySQL 5.6.11 中，引入了参数 log_slow_slave_statements，可将 SQL 重放过程中执行时长超过 long_query_time 的操作记录在慢日志中。

❑ 表上没有任何索引，且二进制日志格式为 ROW。

同样，在一段时间内，Relay_Master_Log_File 和 Exec_Master_Log_Pos 不会变化。

对于没有任何索引的表进行操作，在主库上只会进行一次全表扫描。但在从库重放时，在 ROW 格式下，对于每条记录的操作都会进行一次全表扫描。

还是针对上面的表执行同样的操作，只不过二进制日志格式为 ROW，在主库上执行用了 7.53 秒，但 Seconds_Behind_Master 最大达到了 723 秒，是 STATEMENT 格式下的 100 多倍。

```
mysql> show variables like 'binlog_format';
+---------------+-------+
| Variable_name | Value |
+---------------+-------+
| binlog_format | ROW   |
+---------------+-------+
1 row in set (0.00 sec)

mysql> delete from sbtest.sbtest1 where id <= 100;
Query OK, 100 rows affected (7.53 sec)
```

如果因为表上没有任何索引，导致主从延迟过大，常见的优化方案如下。

- 在从库上临时创建一个索引，加快记录的重放。注意，尽量选择一个区分度高的列添加索引。列的区分度越高，重放的速度就越快。
- 将参数 slave_rows_search_algorithms 设置为 INDEX_SCAN,HASH_SCAN。

   设置后，对于同样的操作，Seconds_Behind_Master 最大只有 53 秒。

- 大事务。

    这里的大事务，指的是二进制日志格式为 ROW 的情况下，操作涉及的记录数较多。

    还是使用上面的表，只不过这次的 id 列是自增主键，执行批量更新操作。更新操作如下，其中 N 是记录数，M 是一个随机字符，每次操作的字符均不一样。

    update sbtest.sbtest1 **set** c=repeat(M,120) **where** id<=N

    接下来我们看看表 4-1，该表列出了不同记录数对应的 Seconds_Behind_Master 的最大值。

    表 4-1　记录数与 Seconds_Behind_Master 的关系

记录数	主库执行时长（秒）	Seconds_Behind_Master 的最大值（秒）
50 000	0.76	1
200 000	3.10	8
500 000	17.32	39
1 000 000	63.47	122

    可以看到，随着记录数的增加，Seconds_Behind_Master 是不断增大的。

    所以对于大事务操作，建议分而治之，每次小批量执行。

- 从库上有查询操作。

    从库上有查询操作，通常会有两方面的影响：

    - 消耗系统资源；
    - 锁等待。

    常见的是从库的查询操作阻塞了主库的 DDL 操作。

- 从库上存在备份。

    常见的情况是备份的全局读锁阻塞了 SQL 线程的重放。

- 磁盘 I/O 存在瓶颈。

4.3.3　如何解读 Seconds_Behind_Master

在衡量主从延迟时，Seconds_Behind_Master 是一个我们常常会看也比较关注的指标。虽然关于它不准确的说法很多，但具体哪里不准确，很多人就语焉不详了，以至于到最后，不准确成了它的代名词，它本身的意义反而被忽略了。所以就出现了一种奇怪的现象：我们明明内心对它是抗拒的，却还总是基于它来定位问题。

但事实上，存在即合理。在绝大多数情况下，Seconds_Behind_Master 还是能很好地反映主从延迟情况的。

下面从源码的角度来看看它的实现逻辑及局限性。

mysql-8.0.25/sql/rpl_slave.cc

```
/*
  if (SQL thread is running)
  {
    if (SQL thread processed all the available relay log)
    {
      if (IO thread is running)
         print 0;
      else
         print NULL;
    }
     else
        compute Seconds_Behind_Master;
  }
    else
     print NULL;
*/
```

以上是计算 Seconds_Behind_Master 的伪代码。

通过上述代码，我们可以得出如下结论。

- Seconds_Behind_Master 为 0 的条件是 SQL 线程已经重放完所有的 relay log 且 Slave_IO_Running 为 Yes。
- 在以下两种情况中，Seconds_Behind_Master 会为 NULL。
    - Slave_SQL_Running 为 No。
    - Slave_SQL_Running 为 Yes，在重放完所有的 relay log 时，Slave_IO_Running 不为 Yes。

但是上面的伪代码并没有说明 Seconds_Behind_Master 具体是如何计算的，所以接下来我们看看对应的源码。

mysql-8.0.25/sql/rpl_slave.cc

```
if (mi->rli->slave_running) {
  if ((mi->get_master_log_pos() == mi->rli->get_group_master_log_pos()) &&
      (!strcmp(mi->get_master_log_name(),
               mi->rli->get_group_master_log_name()))) {
    if (mi->slave_running == MYSQL_SLAVE_RUN_CONNECT)
      protocol->store(0LL);
    else
      protocol->store_null();
  } else {
    long time_diff = ((long)(time(nullptr) - mi->rli->last_master_timestamp) -
                      mi->clock_diff_with_master);
    protocol->store(
        (longlong)(mi->rli->last_master_timestamp ? max(0L, time_diff) : 0));
  }
} else {
  protocol->store_null();
}
```

通过对比伪代码和源码，不难看出 Seconds_Behind_Master 的计算逻辑是由如下代码实现的。

```
long time_diff = ((long)(time(nullptr) - mi->rli->last_master_timestamp) -
                  mi->clock_diff_with_master);
```

```
protocol->store(
    (longlong)(mi->rli->last_master_timestamp ? max(0L, time_diff) : 0));
```

在这段代码里，

- time(0)表示从库当前的系统时间。
- mi->rli->last_master_timestamp 表示 SQL 线程当前重放的事务在主库上开始执行的时间戳，具体到 binlog 中，即事件注释部分的时间戳，如下面这个事件中的 210912 9:53:25。

```
#210912  9:53:25 server id 1  end_log_pos 26045 CRC32 0xcca7e8ab     GTID    last_committed=66
sequence_number=67       rbr_only=no
```

- mi->clock_diff_with_master 表示当 I/O 线程启动时主从之间的系统时间差。

综合起来，Seconds_Behind_Master = 从库当前的系统时间 – SQL 线程当前重放的事务在主库的开始时间戳 – 主从之间的系统时间差。

但在具体计算时，针对 STATEMENT 格式和 ROW 格式的计算方法又不相同，来看下面这个测试。

首先针对 STATEMENT 格式进行测试。

(1) 测试语句。

```
mysql> set session binlog_format='statement';
Query OK, 0 rows affected (0.00 sec)

mysql> delete from sbtest.sbtest1 where id=1;
Query OK, 1 row affected (8.95 sec)
```

(2) 每隔 1 秒采集一次从库 Seconds_Behind_Master 的值。

```
# while true; do mysql -e 'show slave status\G' |grep -E
'Seconds_Behind_Master|Relay_Log_File|Relay_Log_Pos';date;sleep 1;done
            Relay_Log_File: node2-relay-bin.000003
             Relay_Log_Pos: 26181
     Seconds_Behind_Master: 0
Sun Sep 12 09:53:34 CST 2021
            Relay_Log_File: node2-relay-bin.000003
             Relay_Log_Pos: 26181
     Seconds_Behind_Master: 1
...
            Relay_Log_File: node2-relay-bin.000003
             Relay_Log_Pos: 26181
     Seconds_Behind_Master: 8
Sun Sep 12 09:53:42 CST 2021
            Relay_Log_File: node2-relay-bin.000003
             Relay_Log_Pos: 26181
     Seconds_Behind_Master: 9
Sun Sep 12 09:53:43 CST 2021
            Relay_Log_File: node2-relay-bin.000003
             Relay_Log_Pos: 26495
     Seconds_Behind_Master: 0
Sun Sep 12 09:53:44 CST 2021
```

(3) 查看 DELETE 操作对应的 relay log 的内容。

```
# at 26181
#210912  9:53:25 server id 1  end_log_pos 26045 CRC32 0xcca7e8ab     GTID    last_committed=66
```

```
sequence_number=67        rbr_only=no       original_committed_timestamp=1631411614812820
immediate_commit_timestamp=1631411614812820        transaction_length=314
# original_commit_timestamp=1631411614812820 (2021-09-12 09:53:34.812820 CST)
# immediate_commit_timestamp=1631411614812820 (2021-09-12 09:53:34.812820 CST)
/*!80001 SET @@session.original_commit_timestamp=1631411614812820*//*!*/;
/*!80014 SET @@session.original_server_version=80025*//*!*/;
/*!80014 SET @@session.immediate_server_version=80025*//*!*/;
SET @@SESSION.GTID_NEXT= 'bd6b3216-04d6-11ec-b76f-000c292c1f7b:67'/*!*/;
# at 26260
#210912  9:53:25 server id 1  end_log_pos 26131 CRC32 0x6332aa0b        Query   thread_id=85    exec_time=9
error_code=0
SET TIMESTAMP=1631411605/*!*/;
SET @@session.pseudo_thread_id=85/*!*/;
SET @@session.foreign_key_checks=1, @@session.sql_auto_is_null=0, @@session.unique_checks=1,
@@session.autocommit=1/*!*/;
SET @@session.sql_mode=1168113696/*!*/;
SET @@session.auto_increment_increment=1, @@session.auto_increment_offset=1/*!*/;
/*!\C utf8mb4 *//*!*/;
SET
@@session.character_set_client=255,@@session.collation_connection=255,@@session.collation_server=2
55/*!*/;
SET @@session.lc_time_names=0/*!*/;
SET @@session.collation_database=DEFAULT/*!*/;
/*!80011 SET @@session.default_collation_for_utf8mb4=255*//*!*/;
BEGIN
/*!*/;
# at 26346
#210912  9:53:25 server id 1  end_log_pos 26249 CRC32 0x820b04fb        Query   thread_id=85    exec_time=9
error_code=0
use `sbtest`/*!*/;
SET TIMESTAMP=1631411605/*!*/;
delete from sbtest.sbtest1 where id=1
/*!*/;
# at 26464
#210912  9:53:25 server id 1  end_log_pos 26280 CRC32 0x3328f601        Xid = 45170
COMMIT/*!*/;
```

这个事务在主库上开始执行的时间戳是 210912 9:53:25。

当 Seconds_Behind_Master 达到最大值时，从库的系统时间是 09:53:43，且主从时间一致。09:53:43 减去 9:53:25 等于 18 秒，但为何此时 Seconds_Behind_Master 的值是 9 秒呢？

接下来针对 ROW 格式进行测试。

(1) 测试语句。

```
mysql> set session binlog_format='row';
Query OK, 0 rows affected (0.00 sec)

mysql> delete from sbtest.sbtest1 where id=1;
Query OK, 1 row affected (9.00 sec)
```

(2) 每隔 1 秒采集一次从库 Seconds_Behind_Master 的值。

```
# while true; do mysql -e 'show slave status\G' |grep -E
'Seconds_Behind_Master|Relay_Log_File|Relay_Log_Pos';date;sleep 1;done
            Relay_Log_File: node2-relay-bin.000003
             Relay_Log_Pos: 27796
     Seconds_Behind_Master: 0
```

```
Sun Sep 12 10:07:27 CST 2021
            Relay_Log_File: node2-relay-bin.000003
             Relay_Log_Pos: 27796
     Seconds_Behind_Master: 10
Sun Sep 12 10:07:28 CST 2021
            Relay_Log_File: node2-relay-bin.000003
             Relay_Log_Pos: 27796
     Seconds_Behind_Master: 11
Sun Sep 12 10:07:29 CST 2021
      ...
            Relay_Log_File: node2-relay-bin.000003
             Relay_Log_Pos: 27796
     Seconds_Behind_Master: 24
Sun Sep 12 10:07:42 CST 2021
            Relay_Log_File: node2-relay-bin.000003
             Relay_Log_Pos: 28275
     Seconds_Behind_Master: 0
Sun Sep 12 10:07:43 CST 2021
```

(3) 查看 DELETE 操作对应的 relay log 的内容。

```
# at 27796
#210912 10:07:18 server id 1  end_log_pos 27660 CRC32 0xfdbbc98a       GTID       last_committed=70
sequence_number=71      rbr_only=yes     original_committed_timestamp=1631412447730254
immediate_commit_timestamp=1631412447730254      transaction_length=479
/*!50718 SET TRANSACTION ISOLATION LEVEL READ COMMITTED*//*!*/;
# original_commit_timestamp=1631412447730254 (2021-09-12 10:07:27.730254 CST)
# immediate_commit_timestamp=1631412447730254 (2021-09-12 10:07:27.730254 CST)
/*!80001 SET @@session.original_commit_timestamp=1631412447730254*//*!*/;
/*!80014 SET @@session.original_server_version=80025*//*!*/;
/*!80014 SET @@session.immediate_server_version=80025*//*!*/;
SET @@SESSION.GTID_NEXT= 'bd6b3216-04d6-11ec-b76f-000c292c1f7b:71'/*!*/;
# at 27875
#210912 10:07:18 server id 1  end_log_pos 27737 CRC32 0xd6f1066e       Query   thread_id=85    exec_time=9
error_code=0
SET TIMESTAMP=1631412438/*!*/;
SET @@session.pseudo_thread_id=85/*!*/;
SET @@session.foreign_key_checks=1, @@session.sql_auto_is_null=0, @@session.unique_checks=1,
@@session.autocommit=1/*!*/;
SET @@session.sql_mode=1168113696/*!*/;
SET @@session.auto_increment_increment=1, @@session.auto_increment_offset=1/*!*/;
/*!\C utf8mb4 *//*!*/;
SET
@@session.character_set_client=255,@@session.collation_connection=255,@@session.collation_server=2
55/*!*/;
SET @@session.lc_time_names=0/*!*/;
SET @@session.collation_database=DEFAULT/*!*/;
/*!80011 SET @@session.default_collation_for_utf8mb4=255*//*!*/;
BEGIN
/*!*/;
# at 27952
#210912 10:07:18 server id 1  end_log_pos 27804 CRC32 0x366916c9       Table_map: `sbtest`.`sbtest1`
mapped to number 207
# at 28019
#210912 10:07:18 server id 1  end_log_pos 28029 CRC32 0x272cf311       Delete_rows: table id 207 flags:
STMT_END_F
### DELETE FROM `sbtest`.`sbtest1`
### WHERE
###   @1=1
###   @2=4992833
```

```
###
@3='83868641912-28773972837-60736120486-75162659906-27563526494-20381887404-41576422241-9342679396
4-56405065102-33518432330'
###    @4='67847967377-48000963322-62604785301-91415491898-96926520291'
# at 28244
#210912 10:07:18 server id 1  end_log_pos 28060 CRC32 0x196ee952        Xid = 45178
COMMIT/*!*/;
```

同样取 Seconds_Behind_Master 最大值的时间戳进行计算。

从库的主机时间是 10:07:42，事务开始执行的时间是 10:07:18，由于主从时间同步，10:07:42 减去 10:07:18 等于 24 秒，和 Seconds_Behind_Master 的最大值 24 秒吻合。

为何和 STATEMENT 格式下的输出不一样呢？

细心的读者可能会发现，在 ROW 格式下，Seconds_Behind_Master 第一次出现延迟的延迟时间是 10 秒，而在 STATEMENT 格式下，第一次出现延迟的延迟时间则是 1 秒。实际上，这是由二进制日志事件的类型决定的。

对于 QUERY_EVENT，包括 DDL 和 STATEMENT 格式的 DML 操作，在计算 mi->rli->last_master_timestamp 时，会加上 QUERY_EVENT 中的 exec_time，即 mi->rli->last_master_timestamp 等于 SQL 线程当前重放的事务在主库的开始时间戳加上 exec_time。所以，在 STATEMENT 格式下：Seconds_Behind_Master = 从库的系统时间 − SQL 线程当前重放的事务在主库的开始时间戳 − exec_time − 主从之间的系统时间差。这也就是为什么在 STATEMENT 格式下，Seconds_Behind_Master 会从 1 秒开始，以及当从库的系统时间减去 SQL 线程当前重放的事务在主库的开始时间戳等于 18 秒时，Seconds_Behind_Master 只有 9 秒。

而对于 ROWS_EVENT，它没有记录 exec_time。所在，在 ROW 格式下：Seconds_Behind_Master = 从库的系统时间 − SQL 线程当前重放的事务在主库的开始时间戳 − 主从之间的系统时间差。这也就是为什么在 ROW 格式下，Seconds_Behind_Master 第一次出现延迟的延迟时间就是 10 秒。

1. Seconds_Behind_Master 的局限性

通过上面的分析，我们对 Seconds_Behind_Master 已经有了比较深入的了解。接下来，我们说说它的局限性。

- Seconds_Behind_Master 只能衡量 SQL 线程的延迟情况，不能衡量 I/O 线程的延迟情况。

  考虑以下 3 种场景。
  - 出于网络原因，主库的 binlog 不能及时发送到从库上。
  - 从库磁盘 I/O 达到瓶颈，导致主库的二进制日志事件无法及时写入 relay log。
  - 参数 slave_net_timeout 设置得较大，在复制的过程中，如果主从网络断开，从库不会及时感知到。

  在上面 3 个场景中，因为从库的 relay log 已重放完，所以 Seconds_Behind_Master 会显示为 0。但实际上，从库已经延迟了。

- 对于 STATEMENT 格式和 ROW 格式的事务，Seconds_Behind_Master 的计算逻辑并不一样。如果不熟悉这两者之间的区别，就很容易感到困惑。
- 不适用于级联复制场景。如果二级从库出现了延迟，对于其他级别的从库，Seconds_Behind_Master 大概率会为 0。

2. 如何监控主从延迟

既然我们看到了 Seconds_Behind_Master 作为主从延迟监控指标的局限性，那么线上又该如何监控主从延迟呢？在 MySQL 8.0 之前，业界通用的解决方案是 pt-heartbeat。pt-heartbeat 是 Percona Toolkit 中的一个工具。

下面看看 pt-heartbeat 的具体用法。

(1) 首先针对主库执行，定期更新主库的心跳表。

```
pt-heartbeat -h 192.168.244.10 -u pt_user -p pt_pass --database percona --update --daemonize
```

其中的参数解释如下。

- --update：对于主库，必须指定 --update，它会以每秒一次的频率更新 heartbeat 表。
- --database：heartbeat 表所属的数据库。在使用前，必须手动创建。
- --daemonize：让脚本以后台进程的方式运行。

注意，如果是第一次运行，还需指定 --create-table，创建 heartbeat 表。

(2) 其次针对从库执行，检查从库的主从延迟情况。

此时，需指定 --monitor 或者 --check，其中 --monitor 是持续检测，每秒输出一次。例如：

```
# pt-heartbeat -h 192.168.244.20 -u pt_user -p pt_pass --database percona --monitor
113.00s [  1.88s,  0.38s,  0.13s ]
114.00s [  3.78s,  0.76s,  0.25s ]
115.00s [  5.70s,  1.14s,  0.38s ]
```

输出一共有 4 列，第 1 列是当前的延迟时间，第 2、第 3、第 4 列分别是最近 1 分钟、5 分钟和 15 分钟内的平均延迟情况。

也可指定 --check，只检测一次。例如：

```
# pt-heartbeat -h 192.168.244.20 -u pt_user -p pt_pass --database percona --check
303.00
```

接下来看看 pt-heartbeat 的实现原理。

pt-heartbeat 首先会在主库上创建一张心跳表，并插入一条心跳记录。

```
mysql> show create table percona.heartbeat\G
*************************** 1. row ***************************
       Table: heartbeat
Create Table: CREATE TABLE `heartbeat` (
  `ts` varchar(26) NOT NULL,
  `server_id` int unsigned NOT NULL,
  `file` varchar(255) DEFAULT NULL,
  `position` bigint unsigned DEFAULT NULL,
```

```
  `relay_master_log_file` varchar(255) DEFAULT NULL,
  `exec_master_log_pos` bigint unsigned DEFAULT NULL,
  PRIMARY KEY (`server_id`)
) ENGINE=InnoDB DEFAULT CHARSET=utf8mb4 COLLATE=utf8mb4_0900_ai_ci
1 row in set (0.00 sec)

mysql> select * from percona.heartbeat;
+----------------------------+-----------+----------------+----------+-----------------------+---------------------+
| ts                         | server_id | file           | position | relay_master_log_file | exec_master_log_pos |
+----------------------------+-----------+----------------+----------+-----------------------+---------------------+
| 2021-09-11T17:34:40.010430 |         1 | mysql-bin.000003 |   807269 | NULL                  |                NULL |
+----------------------------+-----------+----------------+----------+-----------------------+---------------------+
1 row in set (0.00 sec)
```

在心跳表中，重点关注两列：ts 和 server_id，前者代表记录插入或更新时的系统时间，后者是主库的 server_id。既然会存储实例的 server_id，就意味着这里的节点不一定是一级主库，也可以是级联复制中的中间节点。

随后，针对主库的 pt-heartbeat 会以每秒一次的频率更新这条记录，而针对从库的 pt-heartbeat 同样会以每秒一次的固定频率进行检查。两者的执行频率虽然一样（都由 --interval 参数控制），但执行时间点不同。更新操作通常是整秒执行的，如上面 ts 中的 40 秒，而 check 操作则会滞后 0.5 秒（由 --skew 参数控制）执行，主从延迟时间等于系统当前时间减去心跳表中的 ts 值和 skew。相关的源码如下。

```
my ($ts, $hostname, $server_id) = $sth->fetchrow_array();
my $now = time;
PTDEBUG && _d("Heartbeat from server", $server_id, "\n",
   "   now:", ts($now, $utc), "\n",
   "    ts:", $ts, "\n",
   "  skew:", $skew);
my $delay = $now - unix_timestamp($ts, $utc) - $skew;
PTDEBUG && _d('Delay', sprintf('%.6f', $delay), 'on', $hostname);
$delay = 0.00 if $delay < 0;
```

无论是更新还是 check 操作，pt-heartbeat 都极度依赖于系统时间。如果这两个操作是在不同的主机上执行的，务必确保这两台主机的系统时间是一致的。建议在同一台主机上执行这两个操作。

介绍完 pt-heartbeat，接下来看看 MySQL 8.0 原生的解决方案。它主要基于 performance_schema.replication_applier_status_by_worker 中的 APPLYING_TRANSACTION_ORIGINAL_COMMIT_TIMESTAMP，这个字段记录了当前正在重放的事务在主库的提交时间。

看看下面这个示例。

```
mysql> select * from performance_schema.replication_applier_status_by_worker\G
*************************** 1. row ***************************
                                         CHANNEL_NAME:
                                            WORKER_ID: 0
                                            THREAD_ID: 3112
                                        SERVICE_STATE: ON
                                    LAST_ERROR_NUMBER: 0
                                   LAST_ERROR_MESSAGE:
                                 LAST_ERROR_TIMESTAMP: 0000-00-00 00:00:00.000000
                              LAST_APPLIED_TRANSACTION: bd6b3216-04d6-11ec-b76f-000c292c1f7b:2232
LAST_APPLIED_TRANSACTION_ORIGINAL_COMMIT_TIMESTAMP: 2021-09-11 17:39:58.012221
LAST_APPLIED_TRANSACTION_IMMEDIATE_COMMIT_TIMESTAMP: 2021-09-11 17:39:58.012221
      LAST_APPLIED_TRANSACTION_START_APPLY_TIMESTAMP: 2021-09-11 17:39:58.026871
```

```
              LAST_APPLIED_TRANSACTION_END_APPLY_TIMESTAMP: 2021-09-11 17:39:58.029345
                                  APPLYING_TRANSACTION: bd6b3216-04d6-11ec-b76f-000c292c1f7b:2233
                 APPLYING_TRANSACTION_ORIGINAL_COMMIT_TIMESTAMP: 2021-09-11 17:39:59.006270
                APPLYING_TRANSACTION_IMMEDIATE_COMMIT_TIMESTAMP: 2021-09-11 17:39:59.006270
                    APPLYING_TRANSACTION_START_APPLY_TIMESTAMP: 2021-09-11 17:39:59.021139
                     LAST_APPLIED_TRANSACTION_RETRIES_COUNT: 0
         LAST_APPLIED_TRANSACTION_LAST_TRANSIENT_ERROR_NUMBER: 0
        LAST_APPLIED_TRANSACTION_LAST_TRANSIENT_ERROR_MESSAGE:
      LAST_APPLIED_TRANSACTION_LAST_TRANSIENT_ERROR_TIMESTAMP: 0000-00-00 00:00:00.000000
                          APPLYING_TRANSACTION_RETRIES_COUNT: 0
              APPLYING_TRANSACTION_LAST_TRANSIENT_ERROR_NUMBER: 0
             APPLYING_TRANSACTION_LAST_TRANSIENT_ERROR_MESSAGE:
           APPLYING_TRANSACTION_LAST_TRANSIENT_ERROR_TIMESTAMP: 0000-00-00 00:00:00.000000
1 row in set (0.00 sec)
```

所以，可通过以下 SQL 查看主从延迟情况。

```sql
select case
       when min_commit_timestamp is null then 0
       else unix_timestamp(now(6)) - unix_timestamp(min_commit_timestamp)
       end as seconds_behind_master
from (
   select min(applying_transaction_original_commit_timestamp) as min_commit_timestamp
   from performance_schema.replication_applier_status_by_worker
   where applying_transaction <> ''
) t;
```

这条 SQL 语句同样适用于多线程复制。

4.3.4 参考资料

- 博客文章"深入理解 iostat"，作者：Bean Li。
- 博客文章"容易被误读的 IOSTAT"，作者：vmunix。
- 博客文章"iostat(1) — Linux manual page"，作者：Michael Kerrisk。
- 官网介绍"pt-heartbeat"。

4.4 复制中的常见问题及解决方法

下面看看复制中的常见问题及解决方法。

4.4.1 I/O 线程连接不上主库

I/O 线程连接不上从库，常见的错误信息如下。

```
Last_IO_Error: error connecting to master 'repl@192.168.244.10:3306' - retry-time: 60 retries: 1 message: Can't connect to MySQL server on '192.168.244.10:3306' (110)
```

报错的常见原因有两个。

- 防火墙的限制。

    这个时候，可通过 `iptables -L` 查看主从服务器上的防火墙规则。

- 主机名或端口指定错误，对应的实例其实并不存在。

这个时候，可在主库服务器上通过 `netstat -ntlup | grep 3306` 命令查看实例是否存在。

除了上述错误，还有 Access denied 错误，例如：

```
Last_IO_Error: error connecting to master 'repl@192.168.244.10:3306' - retry-time: 60 retries: 1
message: Access denied for user 'repl'@'192.168.244.20' (using password: YES)
```

报错的常见原因是复制用户账号或密码错误，这时可通过以下 SQL 获取当前正在使用的复制用户的账号和密码，然后在主库上确认这个账号是否存在、密码是否匹配等。

```
mysql> select user_name, user_password from mysql.slave_master_info;
+-----------+---------------+
| user_name | user_password |
+-----------+---------------+
| repl      | 123           |
+-----------+---------------+
1 row in set (0.00 sec)
```

4.4.2　server_id 重复

如果从库和主库的 server_id 相同，会提示如下错误。

```
Last_IO_Errno: 13117
Last_IO_Error: Fatal error: The slave I/O thread stops because master and slave have equal MySQL server ids; these ids must be different for replication to work (or the --replicate-same-server-id option must be used on slave but this does not always make sense; please check the manual before using it).
```

解决方法是修改从库的 server_id，然后重启复制。所幸，server_id 支持在线修改，无须重启实例。

4.4.3　包的大小超过 slave_max_allowed_packet 的限制

以下是相关的错误信息。

```
Last_IO_Errno: 13125
Last_IO_Error: Got a packet bigger than 'slave_max_allowed_packet' bytes
```

slave_max_allowed_packet 是 MySQL 5.6 引入的，用于限制主从复制中数据包的大小，它的默认值也是最大值，为 1GB。在此之前，主从复制中包的大小受 max_allowed_packet 参数的限制。

在 ROW 格式下，如果表中存在 BLOB、TEXT 等大字段，可能会导致包的大小超过 slave_max_allowed_packet 的限制。

常见的解决方法如下。

- 如果开启了 binlog_rows_query_log_events，可分析 binlog，拿到对应的 SQL，直接在从库上执行。注意，如果是 GTID 复制，在执行之前，需设置 gtid_next。
- 先跳过这个事务，事后再通过 pt-table-sync 修复。
- 重建复制。

如无十足的把握，建议重建复制。

4.4.4 从库需要的 binlog 在主库上不存在

如果从库需要的 binlog 在主库上不存在，在基于位置点的复制和 GTID 复制中，错误信息分别如下。

```
Last_IO_Error: Got fatal error 1236 from master when reading data from binary log: 'Could not find first log file name in binary log index file'
```

```
Last_IO_Error: Got fatal error 1236 from master when reading data from binary log: 'Cannot replicate because the master purged required binary logs. Replicate the missing transactions from elsewhere, or provision a new slave from backup. Consider increasing the master's binary log expiration period. To find the missing transactions, see the master's error log or the manual for GTID_SUBTRACT.'
```

上述错误常出现在以下场景中。

- 复制中断太久，当重启复制时，发现所需的 binlog 在主库上不存在。
- 在 GTID 复制中，DBA 不小心在从库上执行了一些管理操作，导致从库产生了新的 GTID，而随着时间的推移，这些 GTID 对应的 binlog 会因为过期而被删除。删除之后，如果发生了高可用切换，且这个从库提升为了主库，则当其他节点指向这个新的主库时，会出现这个错误。

首先看看规避方法。

- 对于第一种场景，一方面，对复制进行监控，当复制中断后，立即告警；另一方面，在磁盘空间允许的情况下，适当增大 expire_logs_days 或 binlog_expire_logs_seconds 的值，让 binlog 保留更长的时间。
- 对于第二种场景，一方面，对 gtid_executed 进行监控，当从库的 gtid_executed 不是主库的子集时，立即告警；另一方面，在从库上设置 super_read_only，避免误操作。

下面看看解决方法。

首先检查是否有 binlog 备份，如果没有，就只能重新搭建从库了。

如果有 binlog 备份，需要区分两种场景：基于位置点的复制和 GTID 复制。不同的场景有不同的解决方法。

- 对于基于位置点的复制，可将所需的 binlog 重新拷贝回主库，接着修改主库的 mysql-bin.index，并执行 FLUSH BINARY LOGS 让修改生效，最后重启主从复制。
- 对于 GTID 复制，同样需要将所需的 binlog 拷贝回主库，但此时的难点是如何修改 gtid_executed，使其包括拷贝的这部分 binlog 所对应的 GTID。而 gtid_executed 是个只读参数，只能通过调整 gtid_purged 来间接调整。在 MySQL 8.0 之前，gtid_purged 可设置的前提是 gtid_executed 为空，而要让 gtid_executed 为空，只能执行 RESET MASTER 命令。在 MySQL 8.0 中，虽然不需要执行 RESET MASTER 命令，但新设置的 gtid_purged 必须包含原来的 gtid_purged。所以，在 GTID 复制中，通过常规方案是不能将 binlog 重新加回主库的。

当然，这个时候可重启主库，MySQL 会根据最旧一个 binlog 的 Previous_gtids_log_event 来初始化 gtid_purged。但重启主库会影响数据库服务的可用性，业务一般很难接受。

下面看看另外一种方案，这种方案无须重启主库。

4.4 复制中的常见问题及解决方法

首先创建测试数据,重现现场。

(1) 在主库中创建测试数据。

```
master> create database slowtech;
Query OK, 1 row affected (0.01 sec)

master> create table slowtech.t1(id int primary key,c1 varchar(10));
Query OK, 0 rows affected (0.03 sec)

master> insert into slowtech.t1 values (1,'a');
Query OK, 1 row affected (0.00 sec)
```

(2) 在从库中关闭复制。

```
slave> select * from slowtech.t1;
+----+------+
| id | c1   |
+----+------+
|  1 | a    |
+----+------+
1 row in set (0.00 sec)

slave> stop slave;
Query OK, 0 rows affected, 1 warning (0.01 sec)
```

(3) 在主库中手动删除从库还未应用的 binlog。

```
master> insert into slowtech.t1 values (2,'b');
Query OK, 1 row affected (0.01 sec)

master> flush logs;
Query OK, 0 rows affected (0.01 sec)

master> show master status;
+------------------+----------+--------------+------------------+-------------------------------------------+
| File             | Position | Binlog_Do_DB | Binlog_Ignore_DB | Executed_Gtid_Set                         |
+------------------+----------+--------------+------------------+-------------------------------------------+
| mysql-bin.000002 |      196 |              |                  | bd6b3216-04d6-11ec-b76f-000c292c1f7b:1-4  |
+------------------+----------+--------------+------------------+-------------------------------------------+
1 row in set (0.00 sec)

# cp /data/mysql/3306/data/mysql-bin.000001 /tmp/

master> purge binary logs to 'mysql-bin.000002';
Query OK, 0 rows affected (0.01 sec)
```

(4) 在从库中重启复制,提示预期错误。

```
slave> start slave;
Query OK, 0 rows affected, 1 warning (0.02 sec)

slave> show slave status\G
*************************** 1. row ***************************
               Slave_IO_State:
                  Master_Host: 192.168.244.10
                  ...
          Relay_Master_Log_File: mysql-bin.000001
             Slave_IO_Running: No
            Slave_SQL_Running: Yes
```

```
            ...
         Exec_Master_Log_Pos: 860
            ...
            Last_IO_Error: Got fatal error 1236 from master when reading data from binary log:
'Cannot replicate because the master purged required binary logs. Replicate the missing transactions
from elsewhere, or provision a new slave from backup. Consider increasing the master's binary log
expiration period. The GTID set sent by the slave is 'bd6b3216-04d6-11ec-b76f-000c292c1f7b:1-3', and
the missing transactions are 'bd6b3216-04d6-11ec-b76f-000c292c1f7b:4''
            ...
         Retrieved_Gtid_Set: bd6b3216-04d6-11ec-b76f-000c292c1f7b:1-3
         Executed_Gtid_Set: bd6b3216-04d6-11ec-b76f-000c292c1f7b:1-3
            Auto_Position: 1
            ...
1 row in set, 1 warning (0.01 sec)
```

下面看看具体的修复步骤。

(1) 将 GTID 复制切换为基于位置点的复制。

```
slave> stop slave;
Query OK, 0 rows affected, 1 warning (0.01 sec)

slave> change master to master_log_file = 'mysql-bin.000001', master_log_pos = 860,
master_auto_position = 0;
Query OK, 0 rows affected, 4 warnings (0.02 sec)
```

命令中的 master_log_file 和 master_log_pos 分别取自 SHOW SLAVE STATUS 中的 Relay_Master_Log_File 和 Exec_Master_Log_Pos。

(2) 将主库备份的 binlog 拷贝回数据目录,并修改 mysql-bin.index 文件。

修改完 mysql-bin.index,执行 FLUSH BINARY LOGS 操作,否则新添加的 binlog 无法被 MySQL 识别。

```
# cd /data/mysql/3306/data/
# cp /tmp/mysql-bin.000001 .
# chown mysql.mysql mysql-bin.000001
# vim mysql-bin.index
./mysql-bin.000001
./mysql-bin.000002

mysql> show binary logs;
+------------------+-----------+-----------+
| Log_name         | File_size | Encrypted |
+------------------+-----------+-----------+
| mysql-bin.000002 |       196 | No        |
+------------------+-----------+-----------+
1 row in set (0.00 sec)

mysql> flush binary logs;
Query OK, 0 rows affected (0.01 sec)

mysql> show binary logs;
+------------------+-----------+-----------+
| Log_name         | File_size | Encrypted |
+------------------+-----------+-----------+
| mysql-bin.000001 |      1190 | No        |
| mysql-bin.000002 |       243 | No        |
```

```
| mysql-bin.000003 |       196 | No        |
+------------------+-----------+-----------+
3 rows in set (0.00 sec)
```

(3) 重启从库复制，查看复制状态。

```
slave> start slave;
Query OK, 0 rows affected, 1 warning (0.01 sec)

slave> show slave status\G
*************************** 1. row ***************************
               Slave_IO_State: Waiting for master to send event
                  Master_Host: 192.168.244.10
                  ...
          Relay_Master_Log_File: mysql-bin.000003
             Slave_IO_Running: Yes
            Slave_SQL_Running: Yes
                  ...
          Exec_Master_Log_Pos: 196
                  ...
           Retrieved_Gtid_Set: bd6b3216-04d6-11ec-b76f-000c292c1f7b:4
            Executed_Gtid_Set: bd6b3216-04d6-11ec-b76f-000c292c1f7b:1-4
                Auto_Position: 0
                  ...
1 row in set, 1 warning (0.01 sec)

slave> select * from slowtech.t1;
+----+------+
| id | c1   |
+----+------+
|  1 | a    |
|  2 | b    |
+----+------+
2 rows in set (0.00 sec)
```

(4) 主从恢复同步后，将 MASTER_AUTO_POSITION 设置回 1。

```
slave> stop slave;
Query OK, 0 rows affected, 1 warning (0.01 sec)

slave> change master to master_auto_position = 1;
Query OK, 0 rows affected, 2 warnings (0.01 sec)

slave> start slave;
Query OK, 0 rows affected, 1 warning (0.01 sec)
```

至此，因为缺少 binlog，导致主从复制中断的问题解决完毕。

4.4.5 从库的 GTID 多于主库的

从库的 GTID 多于主库，多出的 GTID 部分对应的 server_uuid 有两种情况。

- server_uuid 属于从库，即多出的 GTID 是从库产生的。
- server_uuid 属于主库，即多出的 GTID 是主库产生的。

对于第一种情况，比较常见的是 DBA 不小心在从库上执行了一些管理操作。规避方法在 4.4.4 节中介绍过，主要是在从库上设置 super_read_only。下面我们看看常见的解决方法。

(1) 如果 GTID 对应的 binlog 存在，可通过分析 binlog 定位具体的操作。如果该操作是简单的管理操作，可在主库上重新执行一次或撤销从库上的操作。如果不想如此麻烦，或者 GTID 对应的 binlog 不存在，可使用方法(2)。

(2) 通过 pt-table-checksum 检查主从数据的一致性，不一致的地方通过 pt-table-sync 来修复。

(3) 重建复制。

注意，无论是方法(1)还是方法(2)，在修复完数据后，都需要在关闭复制的情况下重新设置 gtid_purged。

以上是 server_uuid 属于从库这种情况的规避方法和解决方法。

下面重点说说 server_uuid 属于主库这种情况，以下是相关的错误信息。

```
Last_IO_Error: Got fatal error 1236 from master when reading data from binary log: 'Slave has more GTIDs than the master has, using the master's SERVER_UUID. This may indicate that the end of the binary log was truncated or that the last binary log file was lost, e.g., after a power or disk failure when sync_binlog != 1. The master may or may not have rolled back transactions that were already replica'
```

这个错误很容易模拟，注入一个空事务即可，具体步骤如下。

```
mysql> show slave status\G
*************************** 1. row ***************************
               Slave_IO_State: Waiting for master to send event
                  Master_Host: 192.168.244.10
                  ...
           Retrieved_Gtid_Set: bd6b3216-04d6-11ec-b76f-000c292c1f7b:1-3
            Executed_Gtid_Set: bd6b3216-04d6-11ec-b76f-000c292c1f7b:1-3
                Auto_Position: 1
                  ...
1 row in set, 1 warning (0.01 sec)

mysql> stop slave;
Query OK, 0 rows affected, 1 warning (0.01 sec)

mysql> set session gtid_next='bd6b3216-04d6-11ec-b76f-000c292c1f7b:4';
Query OK, 0 rows affected (0.00 sec)

mysql> begin;
Query OK, 0 rows affected (0.00 sec)

mysql> commit;
Query OK, 0 rows affected (0.00 sec)

mysql> set session gtid_next='automatic';
Query OK, 0 rows affected (0.00 sec)

mysql> start slave;
Query OK, 0 rows affected, 1 warning (0.02 sec)

mysql> show slave status\G
*************************** 1. row ***************************
               Slave_IO_State:
                  Master_Host: 192.168.244.10
                  ...
               Last_IO_Error: Got fatal error 1236 from master when reading data from binary log: 'Slave has more GTIDs than the master has, using the master's SERVER_UUID. This may indicate that the end of
```

```
the binary log was truncated or that the last binary log file was lost, e.g., after a power or disk
failure when sync_binlog != 1. The master may or may not have rolled back transactions that were already
replicated to the slave. Suggest to replicate any transactions that master has rolled back from slave
to master, and/or commit empty transactions on master to account for transactions that have been'
          ...
          Retrieved_Gtid_Set: bd6b3216-04d6-11ec-b76f-000c292c1f7b:1-3
           Executed_Gtid_Set: bd6b3216-04d6-11ec-b76f-000c292c1f7b:1-4
               Auto_Position: 1
          ...
1 row in set, 1 warning (0.00 sec)
```

这个错误的常见原因及解决方法如下。

- 正如 Last_IO_Error 所提示的，未将 sync_binlog 设置为 1。这样，就有可能出现，主库服务器异常宕机时，binlog 中的 binlog event 已经发送给从库，但还未持久化到硬盘上。

  按照 Last_IO_Error 中的建议，解析从库的 binlog，找出 bd6b3216-04d6-11ec-b76f-000c292c1f7b:4 这个 GTID 对应的操作，然后判断它是否在主库上执行过。如果执行过，则在主库注入一个空事务。如果没有，则在主库上执行该操作，最后重启复制。

  当然，这只是一种理想方案，如果碰到以下两个场景可能就不适用了。

  - 主库恢复后，马上就有业务操作，导致 bd6b3216-04d6-11ec-b76f-000c292c1f7b:4 这个 GTID 被另外一个操作使用了。因为这个 GTID 在从库上存在，所以从库会跳过这个操作。这样，相同的 GTID 却对应主从两个不同的操作，主从数据也就不一致了。
  - 示例中只是相差一个事务，分析起来比较简单。如果相差的事务过多呢？

- 主库不小心执行了 RESET MASTER 操作，导致主库的 binlog 被清除，gtid_executed 为空。

  这种场景下，如果想快速恢复复制，需要在从库上执行 RESET MASTER 和 RESET SLAVE 操作，然后重启复制。

注意，上面介绍的只是一种临时性的解决方法，目的是快速恢复复制。不过即使复制恢复了，主从数据也有可能已经不一致了。最安全且彻底的解决方法还是重建复制。

针对第一个原因，如果要从根本上杜绝这个问题，还是建议设置双一。这里的双一，指的是将参数 sync_binlog 和 innodb_flush_log_at_trx_commit 设置为 1，这样可确保提交的事务一定会持久化到 binlog 和 redo log 中。

4.4.6 在执行插入操作时，提示唯一键冲突

以下是相关的错误信息。

```
Last_SQL_Errno: 1062
Last_SQL_Error: Could not execute Write_rows event on table slowtech.t1; Duplicate entry '6' for key
'PRIMARY', Error_code: 1062; handler error HA_ERR_FOUND_DUPP_KEY; the event's master log
mysql-bin.000003, end_log_pos 463
```

出现这个错误的直接原因是主从数据不一致。至于主从数据不一致是什么原因导致的，后面会具体提到。最安全的解决方法是重建复制。

但如果想快速恢复复制,可直接在从库上删除这条记录,然后重启复制。需要注意的是,如果是 GTID 复制,在从库执行删除操作时,务必将 `sql_log_bin` 设置为 OFF,否则会产生新的 GTID。

切忌直接跳过这个事务,有以下两个原因。

- 这条记录有可能只是主键相同,其他列的值并不完全相同。
- 这个事务可能包含多个操作。

4.4.7 在执行删除或更新操作时,提示记录不存在

以下是相关的错误信息。

```
Last_SQL_Errno: 1032
Last_SQL_Error: Could not execute Delete_rows event on table slowtech.t1; Can't find record in 't1',
Error_code: 1032; handler error HA_ERR_KEY_NOT_FOUND; the event's master log mysql-bin.000003,
end_log_pos 762
```

出现这个错误的直接原因是主从数据不一致,建议重建复制。

但如果想快速恢复复制,可基于主键找到主库对应的记录,然后插入从库,最后重启复制。

切忌直接跳过这个事务,因为这个事务可能包含多个操作。

4.4.8 主从数据不一致

主从数据不一致的常见原因如下。

- 从库被误写入。

  从库被误写入的常见场景有:开发人员误将从库配置为主库,以及 DBA 在从库上执行管理操作。

  针对这些场景,建议从库开启 `super_read_only`。

- 二进制日志格式是 STATEMENT。

  在基于 STATEMENT 的复制格式下,很多操作会被官方视为不安全,这些不安全的操作可能会导致主从数据不一致。建议将二进制日志格式设置为 ROW。

- 从库没有设置 `relay_log_recovery = ON` 和 `relay_log_info_repository = TABLE`。
- 主库没有设置双一。
- 从库没有设置双一。

4.5 本章总结

本章主要介绍了复制的一些常见管理操作。很多命令虽然常用,但里面还是蕴藏了很多高级"玩法"。例如,START SLAVE 命令中的 `until_option` 不仅可以用于延迟复制中基于位置点的恢复,还可用于一主多从且基于位置点的复制场景中,变更从库之间的拓扑。

对于复制的监控，MySQL 5.6 和 MySQL 5.7 有很大的不同。在 MySQL 5.6 中，这主要依赖 SHOW SLAVE STATUS 和复制相关的状态变量。但在 MySQL 5.7 中，随着多源复制的引入，很多监控项被集成到了 performance_schema 中。

在线上，为什么要关注主从延迟呢？一是因为读写分离场景会读到旧的数据；二是因为如果主库在主从延迟期间出现问题，会影响数据库的高可用切换。虽然这种情况发生的概率很小，但按照墨菲定律的说法，任何有可能发生的事情最后都一定会发生。对于数据库服务，还是不要心存侥幸。

最后，再来说说主从数据不一致的问题。严格来说，只要遵循以下几点，主从之间数据不一致的概率极小。

- binlog_format 为 ROW。
- 从库设置了 super_read_only 或 read_only。
- 主从设置了双一。
- 从库设置了 relay_log_recover = ON 和 relay_log_info_repository = TABLE。

如果主从数据不一致导致了复制中断，建议重建复制，不要轻易跳过。

重点问题回顾

- 对于基于位置点的复制和 GTID 复制，如何跳过指定事务？
- 在一主多从且基于位置点的复制中，如何将一个从库切换为另一个从库的从库？
- RESET MASTER、RESET SLAVE 和 RESET SLAVE ALL 之间的区别。
- 复制的监控。
- 如何定位主从延迟问题。
- 主从延迟的常见原因及解决方法。
- Seconds_Behind_Master 的计算方法及局限性。
- pt-heartbeat 的实现原理。
- 计算主从延迟，MySQL 8.0 原生的解决方案 performance_schema.replication_applier_status_by_worker。
- 如果从库需要的 binlog 被主库删除了，在有 binlog 备份的情况下，如何恢复？
- 主从数据不一致的常见原因。

第 5 章

备　　份

数据库备份的重要性毋庸置疑，可以说，这是数据安全的最后一道防线。鉴于此，对于备份，我们通常有以下要求。

- 多地部署

    对于核心数据库，我们通常有"两地三中心"的部署要求。对于备份来说，也是如此。一个备份应该有多个副本，每个副本存储在不同区域。

- 多介质部署

    一个备份的多个副本应存储在不同介质上，如磁盘和磁带上，以防单一介质失效。

- 定期检查备份的有效性

    备份只是在做正确的事情，至于有没有把事情做对，还得依靠备份的有效性检查。

下面看看备份的常见分类。

物理备份与逻辑备份

物理备份，顾名思义，就是备份物理文件。物理备份的优点如下。

- 备份、恢复速度快。尤其是恢复速度直接关系着数据库服务的 RTO。
- 无须实例在线。在实例关闭的情况下，可直接拷贝文件，而不用担心备份的一致性。关闭实例进行备份，也称为"冷备"。

物理备份的缺点如下。

- 备份文件大。
- 恢复时，对平台、操作系统、MySQL 版本有要求，必须一致或兼容。
- 只能在本地发起备份。
- 因为是拷贝物理文件，所以即使文件中存在很多（大量 DELETE 操作导致的）"空洞"，也无法通过恢复来收缩。
- 对表的存储引擎有要求，无法备份 MEMORY 表。

逻辑备份，即备份数据库对象的逻辑记录。逻辑备份的优点如下。

- 可移植性强。恢复时，对平台、操作系统、MySQL 版本无要求。

- 灵活。尤其是在恢复时，可只恢复一个库或一张表。
- 对表的存储引擎没有要求，任何类型的表都可备份。
- 备份文件较小。
- 可远程发起备份。
- 恢复后，能有效收缩空间。

逻辑备份的缺点如下。

- 备份、恢复速度慢。单论备份速度，多线程备份其实不慢。主要是恢复速度慢，即使多线程恢复，也比较慢。
- 备份会"污染"缓冲池。业务热点数据会被备份数据驱逐出缓冲池。

离线备份与在线备份

离线备份，又称为"冷备"，即在实例关闭的情况下进行的备份。此时，只能进行物理备份，即全量拷贝物理文件。

在线备份，又称为"热备"，即在实例运行过程中进行的备份。此时，既可进行物理备份，又可进行逻辑备份。

因对业务侵入较小，线上一般使用在线备份。

全量备份与增量备份

全量备份，即备份整个实例的全量数据。

增量备份，即只备份自上次备份以来发生了"变化"的数据。

通常来说，基于物理备份来实现增量备份较为简单。以 MySQL 为例，只需判断数据页的 LSN 是否发生了变化。

而基于逻辑备份来实现增量备份则很难，如常见的基于某个时间字段来进行增量备份，但其实很难保证某个时间段之前的数据不被修改或删除。

接下来，我们看看 MySQL 中的备份工具。

物理备份的相关工具如下。

- XtraBackup：Percona 公司开源的备份工具，适用于 MySQL 和 Percona Server。
- MySQL Enterprise Backup：MySQL 企业级备份工具，常用于 MySQL 企业版。
- 克隆插件：MySQL 8.0.17 引入的克隆插件，初衷是为了方便组复制添加新的节点。有了克隆插件，无须借助其他备份工具就能很方便地搭建一个从库。

这 3 个工具的实现原理基本相同，都是先在备份的过程中拷贝物理文件和 redo log，恢复时利用 InnoDB Crash Recovery（崩溃恢复）将物理文件恢复到备份结束时的一致性状态。

逻辑备份的相关工具如下。

- mysqldump：MySQL 安装包自带的备份工具，单线程备份。

- mydumper：开源工具，可实现单表 chunk 级别的并行备份。
- mysqlpump：MySQL 5.7 引入的备份工具，可实现表级别的并行备份。
- MySQL Shell Dump & Load：MySQL 官方新推出的多线程逻辑备份工具，可实现单表 chunk 级别的并行备份。
- SELECT ... INTO OUTFILE：SQL 命令，可将表记录直接导出到文件中。

下面说说这 5 个工具的异同点。

- 从实现原理来看，mysqldump、mydumper、mysqlpump、MySQL Shell Dump & Load 可归为一类。它们本质上都是通过 SELECT * FROM TABLE 的方式备份数据，只不过在此基础上，通过全局读锁与 RR 事务隔离级别，实现了数据库的一致性备份。
- SELECT ... INTO OUTFILE 只是一个命令，无法实现数据库的一致性备份。
- 从导出的内容来看，mysqldump、mydumper、mysqlpump 会以 INSERT 语句的形式保存备份结果，例如：

  ```
  INSERT INTO `t1` VALUES (1,'aaa'),(2,'bbb'),(3,'ccc');
  ```

  而 MySQL Shell Dump & Load 和 SELECT ... INTO OUTFILE 则以 TSV 格式保存备份结果，例如：

  ```
  1    aaa
  2    bbb
  3    ccc
  ```

- 各个工具对应的恢复工具也不一样。

  mysqldump、mysqlpump 对应的恢复工具是 mysql 客户端，所以是单线程恢复。

  mydumper 对应的恢复工具是 myloader，支持多线程恢复。

  MySQL Shell Dump & Load 中的恢复工具是 util.loadDump()，该工具实际调用的是 LOAD DATA LOCAL INFILE 命令，支持多线程恢复。

  SELECT ... INTO OUTFILE 对应的恢复命令是 LOAD DATA。

了解了备份的常见分类和 MySQL 中常用的备份工具，我们看看这些工具的常见用法及背后的实现原理，主要包括以下内容。

- mysqldump 的常见用法和实现原理。
- mydumper 的常见用法和实现原理。
- XtraBackup 的常见用法和实现原理。
- 克隆插件的常见用法和实现原理。
- MySQL Shell Dump & Load 的常见用法和实现原理。
- 使用 XtraBackup 搭建从库。
- 基于备份进行指定时间点的恢复。
- 搭建 binlog server。
- 检测备份的有效性。

5.1 mysqldump

mysqldump 是 MySQL 官方提供的，也是大家日常使用得比较多的逻辑备份工具。mysqldump 内置在 MySQL 安装包中，无须额外安装。

为了演示方便，这里首先构建测试数据，具体如下。

```
create database db1;
create table db1.t1(id int,name varchar(10));
insert into db1.t1 values(1,'db1.t1'),(2,'db1.t1');

create table db1.t2(id int,name varchar(10));
insert into db1.t2 values(1,'db1.t2'),(2,'db1.t2');

create database db2;
create table db2.t3(id int,name varchar(10));
insert into db2.t3 values(1,'db2.t3'),(2,'db2.t3');

create table db2.t4(id int,name varchar(10));
insert into db2.t4 values(1,'db2.t4'),(2,'db2.t4');
```

5.1.1 mysqldump 的实现原理

mysqldump 用多了，很多人就会好奇它背后的实现原理。当然，了解一个工具、一个软件的原理，最直接的方法就是去看源码，但是阅读 MySQL 源码对绝大部分人来说有一定的门槛。尽管如此，我们还是可以通过一个"利器"来间接获悉 mysqldump 的实现原理。这个"利器"就是 general log。

general log 会记录 MySQL 服务端接收到的所有（SQL）操作，无论其是否执行成功。开启 general log 对 MySQL 的性能会有一定的影响，所以在线上一般很少开启，除非出于定位问题的需要。

general log 可通过以下命令开启。

```
mysql> set global general_log=on;
```

general log 的位置可通过以下命令查看。

```
mysql> show variables like 'general_log_file';
```

执行全库备份，具体命令如下。

```
# mysqldump --all-databases --master-data=2 --single-transaction --triggers --routines --events > all_databases.sql
```

各个选项的具体含义后面会详细介绍。为了简化起见，这里是在本地备份，且 root 用户没有设置密码。在实际生产环境中，一般需设置主机 IP、端口、用户名和密码，例如：

```
# mysqldump -h 192.168.244.10 -P 3306 -u root -p'123456' --all-databases --master-data=2 --single-transaction --triggers --routines --events > all_databases.sql
```

下面看看 general log 中的内容。为了简化起见，这里过滤了 general log 中的时间戳信息。

```
540 Connect     root@localhost on  using Socket
540 Query       /*!40100 SET @@SQL_MODE='' */
540 Query       /*!40103 SET TIME_ZONE='+00:00' */
540 Query       /*!80000 SET SESSION information_schema_stats_expiry=0 */
540 Query       SET SESSION NET_READ_TIMEOUT= 86400, SESSION NET_WRITE_TIMEOUT= 86400
```

```
540 Query         FLUSH /*!40101 LOCAL */ TABLES
540 Query         FLUSH TABLES WITH READ LOCK
540 Query         SET SESSION TRANSACTION ISOLATION LEVEL REPEATABLE READ
540 Query         START TRANSACTION /*!40100 WITH CONSISTENT SNAPSHOT */
540 Query         SHOW VARIABLES LIKE 'gtid\_mode'
540 Query         SELECT @@GLOBAL.GTID_EXECUTED
540 Query         SHOW MASTER STATUS
540 Query         UNLOCK TABLES
540 Query         SELECT LOGFILE_GROUP_NAME, FILE_NAME, TOTAL_EXTENTS, INITIAL_SIZE, ENGINE, EXTRA FROM
INFORMATION_SCHEMA.FILES WHERE FILE_TYPE = 'UNDO LOG' AND FILE_NAME IS NOT NULL AND LOGFILE_GROUP_NAME
IS NOT NULL GROUP BY LOGFILE_GROUP_NAME, FILE_NAME, ENGINE, TOTAL_EXTENTS, INITIAL_SIZE ORDER BY
LOGFILE_GROUP_NAME
540 Query         SELECT DISTINCT TABLESPACE_NAME, FILE_NAME, LOGFILE_GROUP_NAME, EXTENT_SIZE,
INITIAL_SIZE, ENGINE FROM INFORMATION_SCHEMA.FILES WHERE FILE_TYPE = 'DATAFILE' ORDER BY
TABLESPACE_NAME, LOGFILE_GROUP_NAME
540 Query         SHOW DATABASES
540 Query         SHOW VARIABLES LIKE 'ndbinfo\_version'
```

这段日志里各个操作的具体含义如下。

❏ FLUSH /*!40101 LOCAL */ TABLES

FLUSH TABLES 会关闭所有打开的表，并刷新 PreparedStatement 缓存。

❏ FLUSH TABLES WITH READ LOCK（FTWRL，全局读锁）

该命令会关闭所有打开的表并加全局读锁。

很多人可能会好奇：为什么不直接执行 FTWRL，非要先执行 FLUSH TABLES 操作呢？这么做，主要是为了降低 FTWRL 被阻塞的概率。FTWRL 一旦被阻塞，在它后面执行的所有 DML 操作都会被阻塞。

下面通过表 5-1 和表 5-2 看看 FLUSH TABLES 和 FTWRL 的区别。

表 5-1 FLUSH TABLES 阻塞对 DML 的影响

会话 1	会话 2	会话 3
mysql> select sleep(100) from t1; 执行中……		
	mysql> flush tables; 阻塞中……	
		mysql> delete from t2 where id=1; Query OK, 1 row affected (0.01 sec)

表 5-2 FTWRL 阻塞对 DML 的影响

会话 1	会话 2	会话 3
mysql> select sleep(100) from t1; 执行中……		
	mysql> flush tables with read lock; 阻塞中……	
		mysql> delete from t2 where id=1; 阻塞中……

- **SET SESSION TRANSACTION ISOLATION LEVEL REPEATABLE READ**

    将当前会话的事务隔离等级设置为 REPEATABLE READ（RR），可避免不可重复读和幻读。

- **START TRANSACTION /*!40100 WITH CONSISTENT SNAPSHOT */**

    开启事务的一致性快照，在效果上等价于开启事务（START TRANSACTION）并对所有 InnoDB 表执行一次 SELECT 操作。这样就可以确保备份时，在任意时间点看到的数据和执行 START TRANSACTION WITH CONSISTENT SNAPSHOT 时的数据一致。WITH CONSISTENT SNAPSHOT 修饰符只在 RR 隔离级别下有效。

    下面我们通过表 5-3 和表 5-4 看看 START TRANSACTION WITH CONSISTENT SNAPSHOT 和 START TRANSACTION 的区别。注意，会话 2 是自动提交的。

表 5-3 START TRANSACTION WITH CONSISTENT SNAPSHOT 的效果

会话 1	会话 2
mysql> create table t1(id int primary key, c1 varchar(10)); Query OK, 0 rows affected (0.05 sec) mysql> insert into t1 values(1,'a'); Query OK, 1 row affected (0.02 sec) mysql> start transaction with consistent snapshot; Query OK, 0 rows affected (0.00 sec)	
	mysql> insert into t1 values(2,'b'); Query OK, 1 row affected (0.01 sec)
mysql> select * from t1; +----+------+ | id | c1 | +----+------+ | 1 | a | +----+------+ 1 row in set (0.00 sec)	

表 5-4 START TRANSACTION 的效果

会话 1	会话 2
mysql> create table t1(id int primary key, c1 varchar(10)); Query OK, 0 rows affected (0.13 sec) mysql> insert into t1 values(1,'a'); Query OK, 1 row affected (0.00 sec) mysql> start transaction; Query OK, 0 rows affected (0.00 sec)	
	mysql> insert into t1 values(2,'b'); Query OK, 1 row affected (0.01 sec)
mysql> select * from t1; +----+------+ | id | c1 | +----+------+ | 1 | a | | 2 | b | +----+------+ 2 rows in set (0.00 sec)	

- SHOW MASTER STATUS

  它是由 --master-data 选项决定的，记录了备份开始时 binlog 的位置点信息，包括 MASTER_LOG_FILE 和 MASTER_LOG_POS。

- UNLOCK TABLES

  释放锁。

下面看看针对具体库的备份操作。

```
540 Init DB     db1
540 Query       SHOW CREATE DATABASE IF NOT EXISTS `db1`
540 Query       SAVEPOINT sp
540 Query       show tables
540 Query       show table status like 't1'
540 Query       SET SQL_QUOTE_SHOW_CREATE=1
540 Query       SET SESSION character_set_results = 'binary'
540 Query       show create table `t1`
540 Query       SET SESSION character_set_results = 'utf8mb4'
540 Query       show fields from `t1`
540 Query       show fields from `t1`
540 Query       SELECT /*!40001 SQL_NO_CACHE */ * FROM `t1`
540 Query       SET SESSION character_set_results = 'binary'
540 Query       use `db1`
540 Query       select @@collation_database
540 Query       SHOW TRIGGERS LIKE 't1'
540 Query       SET SESSION character_set_results = 'utf8mb4'
540 Query       SET SESSION character_set_results = 'binary'
540 Query       SELECT COLUMN_NAME, JSON_EXTRACT(HISTOGRAM, '$."number-of-buckets-specified"')
FROM information_schema.COLUMN_STATISTICS WHERE SCHEMA_NAME = 'db1' AND TABLE_NAME = 't1'
540 Query       SET SESSION character_set_results = 'utf8mb4'
540 Query       ROLLBACK TO SAVEPOINT sp
540 Query       show table status like 't2'
...
540 Query       RELEASE SAVEPOINT sp
540 Query       show events
540 Query       use `db1`
540 Query       select @@collation_database
540 Query       SET SESSION character_set_results = 'binary'
540 Query       SHOW FUNCTION STATUS WHERE Db = 'db1'
540 Query       SHOW PROCEDURE STATUS WHERE Db = 'db1'
540 Query       SET SESSION character_set_results = 'utf8mb4'
```

从上面的输出中可以看到以下几点。

- 备份的核心是 SELECT /*!40001 SQL_NO_CACHE */ * FROM `t1`。

  设置 SQL_NO_CACHE 的作用有两个：不检查 Query Cache，不缓存查询结果。

- SHOW CREATE DATABASE IF NOT EXISTS `db1`，show create table `t1`

  生成建库语句和建表语句。

- SHOW TRIGGERS LIKE 't1'

  备份触发器。

- 设置 SAVEPOINT，每备份完一张表就执行一次 ROLLBACK TO SAVEPOINT 操作。

之所以这么做，是为了释放已经备份完的表的元数据锁，这样可以提高备份期间 DDL 操作的并发性。

我们分 3 个场景来测试一下。

场景 1：先查询，再执行 DDL。此时，DDL 操作被阻塞，如表 5-5 所示。

表 5-5　测试场景 1

会话 1	会话 2
mysql> start transaction with consistent snapshot; Query OK, 0 rows affected (0.00 sec) mysql> select * from t1; +----+------+ | id | c1 | +----+------+ | 1 | a | +----+------+ 1 row in set (0.00 sec)	
	mysql> alter table t1 add column c2 varchar(10); 阻塞中……

场景 2：先执行 DDL，再查询。此时，DDL 操作可以成功执行，如表 5-6 所示。

表 5-6　测试场景 2

会话 1	会话 2
mysql> start transaction with consistent snapshot; Query OK, 0 rows affected (0.00 sec)	
	mysql> alter table t1 add column c2 varchar(10); Query OK, 0 rows affected (0.05 sec) Records: 0 Duplicates: 0 Warnings: 0
mysql> select * from t1; +----+------+------+ | id | c1 | c2 | +----+------+------+ | 1 | a | NULL | +----+------+------+ 1 row in set (0.00 sec)	

如果在 MySQL 5.6 和 MySQL 5.7 中测试相同的场景，查询会报错。

```
mysql> select * from t1;
ERROR 1412 (HY000): Table definition has changed, please retry transaction
```

是否报错与 DDL 的类型有关。具体来说，对于需要 Rebuild Table 的 DDL 才会导致查询报错。

场景 3：模仿 mysqldump 的备份原理，设置断点，如表 5-7 所示。不难看出，在执行完 ROLLBACK TO SAVEPOINT sp 操作后，之前被阻塞的 DDL 马上就能执行了。

表 5-7 测试场景 3

会话 1	会话 2
`mysql> start transaction with consistent snapshot;` `Query OK, 0 rows affected (0.00 sec)` `mysql> savepoint sp;` `Query OK, 0 rows affected (0.00 sec)` `mysql> select * from t1;` `+----+------+` `| id	c1 |` `+----+------+` `| 1	a |` `+----+------+` `1 row in set (0.00 sec)`  `mysql> rollback to savepoint sp;` `Query OK, 0 rows affected (0.00 sec)`	`mysql> alter table t1 add column c2 varchar(10);` 阻塞中…… `mysql> alter table t1 add column c2 varchar(10);` `Query OK, 0 rows affected (20.35 sec)` `Records: 0  Duplicates: 0  Warnings: 0`

❑ show events

备份定时器。

❑ SHOW FUNCTION STATUS WHERE Db = 'db1', SHOW PROCEDURE STATUS WHERE Db = 'db1'

备份函数和存储过程。

上面重点分析了 db1 的备份实现，db2 和 mysql 的备份同样如此，唯一需要注意的是，对于 general_log 和 slow_log 这两张表，mysqldump 只会备份表结构，不会备份数据。

5.1.2 mysqldump 的常用选项

mysqldump 的选项众多，这里不妨参考官方文档，对选项进行分类阐述。

1. 过滤相关

- **-A, --all-databases**

备份所有数据库。默认不会备份 information_schema、performance_schema 和 sys 库。

- **-B, --databases**

备份指定库。如果要备份多个库，库之间需用空格隔开。

- **--tables**

备份指定表。

- **--ignore-table=db_name.tbl_name**

忽略指定表的备份。

- **-E, --events**

备份定时器。使用该选项，会通过 SHOW EVENTS 命令生成 CREATE EVENT 语句。

- **-R, --routines**

备份存储过程和自定义函数。

- **--triggers**

备份触发器，默认开启。可通过 --skip-triggers 将其关闭。

- **--no-data**

只备份表结构，不备份数据。

- **-w, --where=name**

只导出满足 WHERE 条件的数据。因为 mysqldump 的底层实现是 SELECT * FROM TABLE_NAME，所以可通过 WHERE 子句设置过滤条件。

2. DDL 相关

- **--add-drop-database**

在 CREATE DATABASE 操作之前，添加 DROP DATABASE 操作。默认不会添加。

- **--add-drop-table**

在 CREATE TABLE 操作之前，添加 DROP TABLE 操作。该选项是默认开启的。指定 --skip-add-drop-table 则不会添加。

- **-n, --no-create-db**

在执行全库备份（--all-databases）或部分库备份（--databases）时，默认会添加 CREATE DATABASE 操作，而使用该选项则不会添加。

- **-t, --no-create-info**

在导出数据之前，默认都会添加 CREATE TABLE 操作，而使用该选项则不会添加。

- **--replace**

使用 REPLACE INTO 代替 INSERT INTO，例如：

```
REPLACE INTO `t1` VALUES (1,'db1.t1'),(2,'db1.t1');
```

- **--insert-ignore**

使用 INSERT IGNORE 代替 INSERT INTO。

3. 复制相关

- `--master-data[=value]`

执行 SHOW MASTER STATUS，获取一致性备份时的 binlog 位置点信息。获取到的位置点信息会通过 CHANGE MASTER TO 语句写入备份文件。

```
CHANGE MASTER TO MASTER_LOG_FILE='mysql-bin.000002', MASTER_LOG_POS=156;
```

这个位置点可用来搭建从库，以上是将 `--master-data` 设置为 1 的效果。

若将 `--master-data` 设置为 2，则会在该语句前加上注释。例如：

```
-- CHANGE MASTER TO MASTER_LOG_FILE='mysql-bin.000002', MASTER_LOG_POS=156;
```

为了避免在导入备份文件时，CHANGE MASTER TO 语句对目标实例造成不必要的影响，一般建议将 `--master-data` 设置为 2。

注意，如果只指定 `--master-data`，而没有指定 `--single-transaction`，则会隐式开启 `--lock-all-tables`。

从 MySQL 8.0.26 开始，建议使用 `--source-data`，而不是 `--master-data`，否则备份文件中会有 warning，导入时会报错。

```
WARNING: --master-data is deprecated and will be removed in a future version. Use --source-data instead.
```

- `--dump-slave[=value]`

同 `--master-data` 的作用类似，只不过这里获取的位置点是 SHOW SLAVE STATUS 中的 Relay_Master_Log_File 和 Exec_Master_Log_Pos，在对从库进行备份时可指定该选项。该选项会在备份开始前关闭 SQL 线程，直到备份结束才开启。

从 MySQL 8.0.26 开始，`--dump-slave` 被弃用，取而代之的是 `--dump-replica`。

- `--include-master-host-port`

在使用 `--dump-slave` 时，生成的 CHANGE MASTER TO 语句中只有 MASTER_LOG_FILE 和 MASTER_LOG_POS，如果指定了 `--include-master-host-port`，则会同时加上对应主库的 IP 和端口。

```
CHANGE MASTER TO MASTER_HOST='192.168.244.10', MASTER_PORT=3306, MASTER_LOG_FILE='mysql-bin.000005', MASTER_LOG_POS=493;
```

从 MySQL 8.0.26 开始，`--include-master-host-port` 被弃用，取而代之的是 `--include-source-host-port`。

- `--set-gtid-purged=value`

指定是否在备份文件中添加 SET @@GLOBAL.GTID_PURGED 操作，可选值有 ON、COMMENTED、OFF 和 AUTO。默认为 AUTO，即目标实例如果开启了 GTID，则会添加，反之则不会添加。若设置为 COMMENTED，则该操作会以注释的形式添加。

注意，在 SET @@GLOBAL.GTID_PURGED 操作之前，会将 SQL_LOG_BIN 设置为 0。

```
SET @MYSQLDUMP_TEMP_LOG_BIN = @@SESSION.SQL_LOG_BIN;
SET @@SESSION.SQL_LOG_BIN= 0;

--
-- GTID state at the beginning of the backup
--

SET @@GLOBAL.GTID_PURGED=/*!80000 '+'*/ 'bd6b3216-04d6-11ec-b76f-000c292c1f7b:1-3';
```

4. 事务相关

- **--single-transaction**

在备份开始前，会执行以下操作，获取事务表的一致性快照，从而保证事务表在整个备份期间的一致性。

```
SET SESSION TRANSACTION ISOLATION LEVEL REPEATABLE READ
START TRANSACTION /*!40100 WITH CONSISTENT SNAPSHOT */
```

注意，该选项只适用于事务表。对于非事务表（如 MyISAM），则无法保证一致性。

- **-x, --lock-all-tables**

在备份开始前加全局读锁，直到备份结束才释放。如果数据库中的非事务表比较多，又要保证备份的一致性，就可使用该选项。但在加锁期间，只能读、不能写，这一点尤其需要注意。

- **-l, --lock-tables**

在备份每个数据库前，会通过以下操作锁定该数据库下的所有表，这样就能保证这个数据库备份的一致性。

```
LOCK TABLES `t1` READ /*!32311 LOCAL */,`t2` READ /*!32311 LOCAL */
```

在使用此选项时需注意以下两点。

- ❑ 使用 --lock-tables 备份数据库，只能保证每个数据库的一致性，不能保证整个实例的备份一致性。
- ❑ --lock-tables 和 --single-transaction 是互斥的，因为 LOCK TABLES 操作会导致事务隐式提交。同时指定两者不会报错，此时 --single-transaction 会覆盖 --lock-tables。

- **--add-locks**

在 INSERT 操作之前添加锁表操作。

```
LOCK TABLES `t1` WRITE;
/*!40000 ALTER TABLE `t1` DISABLE KEYS */;
INSERT INTO `t1` (`id`, `name`) VALUES (1,'db1.t1'),(2,'db1.t1');
/*!40000 ALTER TABLE `t1` ENABLE KEYS */;
UNLOCK TABLES;
```

5. 性能相关

- **--opt**

该选项等同于 --add-drop-table、--add-locks、--create-options、--quick、--extended-insert、

`--lock-tables`、`--set-charset`、`--disable-keys`。默认开启。

- **-q, --quick**

如果指定了 `--quick`，则会通过 `mysql_use_result()` 获取结果集，反之则通过 `mysql_store_result()` 获取。

```
if (quick)
  res = mysql_use_result(mysql);
else
  res = mysql_store_result(mysql);
```

两者的区别是，`mysql_store_result()` 会一次性读取结果集，然后缓存在客户端中，而 `mysql_use_result()` 返回的只是一个游标，后续需通过 `mysql_fetch_row()` 逐行读取数据。因为整个结果集需要在客户端中维护，所以相对来说，`mysql_store_result()` 比 `mysql_use_result()` 需要更多的内存和更大的开销。如果结果集很大，客户端还存在内存溢出的风险。`mysql_use_result()` 每次只需为一行数据分配内存，而且不用为整个结果集设置复杂的数据结构，所以相对来说速度更快。

- **-K, --disable-keys**

数据在导入前，先禁用索引，等导入后，再开启索引。

```
/*!40000 ALTER TABLE `t1` DISABLE KEYS */;
INSERT INTO `t1` VALUES (1,'db1.t1'),(2,'db1.t1');
/*!40000 ALTER TABLE `t1` ENABLE KEYS */;
```

这样，可提升数据的导入速度。只不过这两个命令只适用于 MyISAM 存储引擎。

- **-e, --extended-insert**

将多行数据放到一个 INSERT 操作中。

```
INSERT INTO `t1` VALUES (1,'db1.t1'),(2,'db1.t1');
```

如果禁用该选项，则一个 INSERT 操作只有一行数据。

6. 其他重要选项

- **--flush-privileges**

在备份完 mysql 库后，会添加 FLUSH PRIVILEGES 操作。

- **--set-charset**

将 SET NAMES character_set 操作添加到输出中。指定 –skip-set-charset 则不会添加。这里的字符集由 `--default-character-set` 指定。在 MySQL 8.0 中，default-character-set 默认为 utf8mb4。

- **-c, --complete-insert**

在 INSERT 操作中指定列名。

```
INSERT INTO `t1` (`id`, `name`) VALUES (1,'db1.t1'),(2,'db1.t1');
```

- **-r file_name, --result-file=file_name**

指定备份文件。

5.1.3 mysqldump 的常见用法

1. 全库备份

```
# mysqldump --all-databases --master-data=2 --single-transaction --triggers --routines --events > all_databases.sql
```

注意，`--master-data=2 --single-transaction` 只能保证 InnoDB 表的备份一致性，无法保证非事务表（如 MyISAM）的备份一致性。如果要保证后者的备份一致性，只能指定 `--lock-all-tables` 加全局读锁。但全局读锁的影响较大，一般不建议使用。这也从另一个角度说明了为什么要在线上禁用非事务表。

2. 备份指定库

```
# mysqldump --databases db1 db2 --result-file=db1_db2.sql
```

在备份指定库时，默认会开启 `--opt` 选项。这样，在备份每个数据库时，会对该数据库下的所有表执行锁表操作。

```
LOCK TABLES `t1` READ /*!32311 LOCAL */,`t2` READ /*!32311 LOCAL */
```

3. 压缩备份

```
# mysqldump --all-databases --master-data=2 --single-transaction | gzip > all_databases.gz
```

通过管道的方式使用 gzip 进行压缩，除了 gzip，也可指定其他压缩算法。

4. 备份指定表

备份指定表的写法比较灵活，如对 db1 中的 t1 和 t2 表进行备份，常用的备份方式有：

- mysqldump --databases db1 --tables t1 t2 > mysql_backup.sql
- mysqldump db1 t1 t2 > mysql_backup.sql
- mysqldump db1 --tables t1 t2 > mysql_backup.sql

在备份指定表时，要注意以下两点。

- 不支持 db_name.tbl_name 这种写法。例如：

  ```
  # mysqldump --tables db1.t1> mysql_backup.sql
  mysqldump: Got error: 1049: Unknown database 'db1.t1' when selecting the database
  ```

- 无法对不同数据库中的多张表同时进行备份，如 db1.t1 和 db2.t3。

5. 忽略指定表

```
# mysqldump --databases db1 db2 --ignore-table=db1.t1 --ignore-table=db2.t3 > mysql_backup.sql
```

注意，`--ignore-table` 后面只能接一张表，如果要忽略多张表，需指定多次。之前提到过，使用 `--tables` 无法对不同数据库中的多张表进行备份，此时可通过 `--ignore-table` 间接实现，但总的来说，还是比较麻烦。对于类似需求，更推荐使用 mydumper 或 MySQL Shell Dump 来实现。

6. 导出表的部分数据

```
# mysqldump db1 t1 --where='id=1' > db1_t1.sql
```

id=1 是过滤条件，对应的导出语句为

```
SELECT /*!40001 SQL_NO_CACHE */ * FROM `t1` WHERE id=1
```

7. 导入备份文件

```
# mysql < mysql_backup.sql
```

通过 mysqldump 生成的备份数据一般是通过 mysql 客户端来恢复的。除了上述命令，也可通过 source 命令来恢复。

```
mysql> source mysql_backup.sql
```

8. 导入压缩文件

```
# gunzip < all_databases.gz | mysql
```

导入之前，先通过 gunzip 进行解压。

9. 边备份，边导入

```
# mysqldump --databases db1 | mysql -h 192.168.244.128 -uroot -p123456
```

通过管道，可以实现边备份、边导入的效果。

5.1.4 总结

- mysqldump 是通过 SELECT * FROM TABLE_NAME 来实现数据备份的。
- mysqldump 是单线程备份。可以看到，备份过程中的线程 ID 自始至终都是 540。
- START TRANSACTION /*!40100 WITH CONSISTENT SNAPSHOT */ 的影响比较大，从上面与 START TRANSACTION 的对比中可以看出，即使一张表备份完了，也要维持它的一致性快照，直到整个备份结束。这就意味着，在备份过程中，对于表的任何 DML 操作都会将相应记录的前镜像保存到回滚段中。如果数据操作很频繁，会导致回滚段越来越大。好在，从 MySQL 5.7 开始，支持回滚表空间的回收。
- mysqldump 生成的备份文件一般是通过 mysql 客户端来导入的。
- 如果实例开启了 GTID，且备份时没有指定 --set-gtid-purged=OFF，则生成的备份文件中会将 SQL_LOG_BIN 设置为 0。

5.2 mydumper

mydumper 是一个多线程逻辑备份工具，可实现单表 chunk 级别的并行备份。相对于 mysqldump 的单线程备份，mydumper 的备份时间更短。不仅如此，mydumper 会将一张表的数据写入一个甚至多个备份文件。这样在恢复时，不仅能实现表与表之间的并行恢复，还能实现同一张表在不同备份文件之间的并行恢复。相对来说，恢复时间也更短。反观 mysqldump 生成的备份文件，一般是通过 mysql

客户端导入的，而 mysql 客户端是单线程处理的。

官方在 MySQL 5.7 中也发布了一个多线程备份工具 mysqlpump，不过它只能实现表级别的并行备份，生成的备份文件也只能通过 mysql 客户端导入。

5.2.1 mydumper 的安装

截至本章写作完成时，mydumper 的最新版本是 0.12.1。

下面看看 mydumper 的具体安装步骤。

```
# tar xvf v0.12.1.tar.gz -C /usr/src/
# cd /usr/src/mydumper-0.12.1/
# yum install glib2-devel mysql-devel zlib-devel pcre-devel openssl-devel gcc-c++ cmake
# cmake .
# make
# make install
```

安装完毕后，会在 /usr/local/bin 下生成两个可执行文件：mydumper 和 myloader。mydumper 是备份工具，myloader 是恢复工具。

如果在 `make` 的过程中出现了 `error: 'SSL_MODE_REQUIRED' undeclared (first use in this function)` 或 `error: 'MYSQL_TYPE_JSON' undeclared (first use in this function)` 之类的错误，则意味着 mysql-devel 版本过低。这个时候，可通过显式指定 MYSQL_CONFIG 来解决，如：

```
# cmake -DMYSQL_CONFIG=/usr/local/mysql/bin/mysql_config .
```

命令中的 /usr/local/mysql 是新版本 MySQL 的安装目录。重新编译时，需删除 CMakeCache.txt。

5.2.2 mydumper 的实现原理

我们还是结合 general log 来分析。

为了直观地展示 mydumper 的备份过程，这里新建了两个测试库：sbtest_innodb 和 sbtest_myisam。每个库下分别有 5 张表，其中，sbtest_innodb 库中是 InnoDB 表，sbtest_myisam 库中是 MyISAM 表。

首先，执行全库备份。备份命令如下。

```
# mydumper -u mydumper_user -p mydumper_pass -t 2 -o /data/backup/mydumper
```

其中，-u 是备份用户，-p 是密码，-t 是备份的并发线程数，-o 是备份目录。注意，这里备份的是本地实例，所以没有设置主机 IP（-h）和端口（-P）。

打开 general log。

```
112 Connect    mydumper_user@localhost on  using Socket
112 Query      SET SESSION wait_timeout = 2147483
112 Query      SET SESSION net_write_timeout = 2147483
112 Query      SHOW PROCESSLIST
112 Query      SELECT @@version_comment, @@version
112 Query      FLUSH TABLES WITH READ LOCK
112 Query      START TRANSACTION /*!40108 WITH CONSISTENT SNAPSHOT */
```

```
112 Query      /*!40101 SET NAMES binary*/
112 Query      SHOW MASTER STATUS
112 Query      SHOW SLAVE STATUS
```

这段日志里主要执行了以下操作。

- 执行 SHOW PROCESSLIST，判断是否有长查询（长查询的标准由 --long-query-guard 参数确定，默认为 60 秒）。如果有长查询，且没有指定 --kill-long-queries，则会终止备份。如果指定了 --kill-long-queries，则会杀掉长查询。

  如果不想杀掉长查询，可设置重试次数（由 --long-query-retries 参数决定，默认为 0，即不重试）及重试之间的时间间隔（由 --long-query-retry-interval 参数决定，默认为 60 秒）。

- 加全局读锁。
- 通过 START TRANSACTION /*!40108 WITH CONSISTENT SNAPSHOT */ 命令开启事务的一致性快照。
- 执行 SHOW MASTER STATUS，获取 binlog 的位置点信息，并写入备份目录下的元数据（metadata）文件。
- 执行 SHOW SLAVE STATUS。如果有输出，则会将 Relay_Master_Log_File、Exec_Master_Log_Pos、Master_Host、Executed_Gtid_Set 写入元数据文件。

```
113 Connect    mydumper_user@localhost on  using Socket
113 Query      SET SESSION wait_timeout = 2147483
113 Query      SET SESSION TRANSACTION ISOLATION LEVEL REPEATABLE READ
113 Query      START TRANSACTION /*!40108 WITH CONSISTENT SNAPSHOT */
113 Query      /*!40103 SET TIME_ZONE='+00:00' */
113 Query      /*!40101 SET NAMES binary*/

114 Connect    mydumper_user@localhost on  using Socket
114 Query      SET SESSION wait_timeout = 2147483
114 Query      SET SESSION TRANSACTION ISOLATION LEVEL REPEATABLE READ
114 Query      START TRANSACTION /*!40108 WITH CONSISTENT SNAPSHOT */
114 Query      /*!40103 SET TIME_ZONE='+00:00' */
114 Query      /*!40101 SET NAMES binary*/
```

这段日志里主要执行了以下操作。

- 创建两个子线程，这里的线程数由 -t 参数指定，默认为 4。
- 将会话的事务隔离级别设置为 REPEATABLE READ，并开启事务的一致性快照。因为这些操作是在全局读锁的背景下执行的，所以可以确保每个线程看到的 InnoDB 表的数据都是一致的。
- SET TIME_ZONE='+00:00'，设置时区。这个操作由 --tz-utc 参数决定，默认开启。如果要禁用，需设置 --skip-tz-utc。
- SET NAMES binary，设置客户端字符集。字符集由 --set-names 参数决定，默认是 binary。

```
112 Query      SHOW DATABASES
114 Init DB    mysql
113 Query      SHOW CREATE DATABASE IF NOT EXISTS `mysql`
114 Query      SHOW TABLE STATUS
113 Query      SHOW CREATE DATABASE IF NOT EXISTS `sbtest_innodb`
113 Init DB    sbtest_innodb
113 Query      SHOW TABLE STATUS
113 Query      SHOW CREATE DATABASE IF NOT EXISTS `sbtest_myisam`
113 Init DB    sbtest_myisam
```

```
113 Query      SHOW TABLE STATUS
114 Query      SHOW CREATE DATABASE IF NOT EXISTS `sys`
114 Init DB    sys
114 Query      SHOW TABLE STATUS
```

主线程执行 SHOW DATABASES，对于每个数据库（information_schema、performance_schema 除外），都会执行 SHOW TABLE STATUS 获取表的相关信息，包括表的存储引擎、是否是视图等。

相关的库表信息会插入 innodb_tables、non_innodb_table、table_schemas、view_schemas、schema_post 这 5 个链表。innodb_tables 对应 InnoDB 表，non_innodb_table 对应非 InnoDB 表，table_schemas 对应需要备份表结构的表，view_schemas 对应视图，schema_post 对应定时器和存储过程（自定义函数）。

接下来，会依次开始备份表结构、非 InnoDB 表、InnoDB 表、视图、定时器和存储过程（自定义函数）。虽然这几个备份项有先后顺序，但并不是前者备份完才会备份后者。因为是多线程并发处理，所以只要有线程空闲下来，就会继续处理下一个备份项，无论当前的备份项是否备完。

(1) 备份表结构。

```
114 Query      SHOW CREATE TABLE `sbtest_innodb`.`sbtest2`
112 Query      select COLUMN_NAME from information_schema.COLUMNS where TABLE_SCHEMA='sbtest_myisam'
 and TABLE_NAME='sbtest1' and extra like '%GENERATED%' and extra not like '%DEFAULT_GENERATED%'
114 Query      SHOW CREATE TABLE `sbtest_innodb`.`sbtest3`
112 Query      select COLUMN_NAME from information_schema.COLUMNS where TABLE_SCHEMA='sbtest_myisam'
 and TABLE_NAME='sbtest2' and extra like '%GENERATED%' and extra not like '%DEFAULT_GENERATED%'
114 Query      SHOW CREATE TABLE `sbtest_innodb`.`sbtest4`
112 Query      select COLUMN_NAME from information_schema.COLUMNS where TABLE_SCHEMA='sbtest_myisam'
 and TABLE_NAME='sbtest3' and extra like '%GENERATED%' and extra not like '%DEFAULT_GENERATED%'
...
```

(2) 备份非 InnoDB 表。

```
114 Query      SELECT /*!40001 SQL_NO_CACHE */ * FROM `sbtest_myisam`.`sbtest2`
113 Query      SELECT /*!40001 SQL_NO_CACHE */ * FROM `sbtest_myisam`.`sbtest3`
114 Query      SELECT /*!40001 SQL_NO_CACHE */ * FROM `sbtest_myisam`.`sbtest4`
113 Query      SELECT /*!40001 SQL_NO_CACHE */ * FROM `sbtest_myisam`.`sbtest5`
114 Query      SELECT /*!40001 SQL_NO_CACHE */ * FROM `sbtest_innodb`.`sbtest1`
114 Query      SELECT /*!40001 SQL_NO_CACHE */ * FROM `sbtest_innodb`.`sbtest2`
112 Query      UNLOCK TABLES /* FTWRL */
112 Quit
```

在备份完非 InnoDB 表后，主线程会释放全局读锁。可以看到，在释放全局读锁之前，就已经开始了 InnoDB 表的备份。

(3) 备份 InnoDB 表。

```
113 Query      SELECT /*!40001 SQL_NO_CACHE */ * FROM `sbtest_innodb`.`sbtest3`
114 Query      SELECT /*!40001 SQL_NO_CACHE */ * FROM `sbtest_innodb`.`sbtest4`
113 Query      SELECT /*!40001 SQL_NO_CACHE */ * FROM `sbtest_innodb`.`sbtest5`
114 Query      SELECT /*!40001 SQL_NO_CACHE */ * FROM `mysql`.`columns_priv`
114 Query      SELECT /*!40001 SQL_NO_CACHE */ * FROM `mysql`.`component`
114 Query      SELECT /*!40001 SQL_NO_CACHE */ * FROM `mysql`.`db`
114 Query      SELECT /*!40001 SQL_NO_CACHE */ * FROM `mysql`.`default_roles`
...
```

(4) 备份视图。

```
114 Init DB     sys
114 Query       SHOW FIELDS FROM `sys`.`host_summary`
114 Query       SHOW CREATE VIEW `sys`.`host_summary`
114 Init DB     sys
114 Query       SHOW FIELDS FROM `sys`.`host_summary_by_file_io`
114 Query       SHOW CREATE VIEW `sys`.`host_summary_by_file_io`
114 Init DB     sys
114 Query       SHOW FIELDS FROM `sys`.`host_summary_by_file_io_type`
114 Query       SHOW CREATE VIEW `sys`.`host_summary_by_file_io_type`
114 Init DB     sys
114 Query       SHOW FIELDS FROM `sys`.`host_summary_by_stages`
114 Query       SHOW CREATE VIEW `sys`.`host_summary_by_stages`
...
```

备份完毕，我们看看备份集的内容。

```
# ls /data/backup/mydumper/ | grep -Ev 'sys|mysql'
metadata
sbtest_innodb.sbtest1.00000.sql
sbtest_innodb.sbtest1-metadata
sbtest_innodb.sbtest1-schema.sql
sbtest_innodb.sbtest2.00000.sql
sbtest_innodb.sbtest2-metadata
sbtest_innodb.sbtest2-schema.sql
...
sbtest_innodb-schema-create.sql
...
```

备份集主要包括以下几类文件。

- metadata，即备份的元数据文件。记录了 binlog 的位置点信息。基于这些信息，可以搭建从库。

  ```
  Started dump at: 2022-04-06 10:31:47
  SHOW MASTER STATUS:
      Log: mysql-bin.000006
      Pos: 196
      GTID:1ad08e93-b425-11ec-bd69-525400d51a16:1-2775

  Finished dump at: 2022-04-06 10:31:48
  ```

- 数据库定义文件，如 sbtest_innodb-schema-create.sql。
- 表结构定义文件，如 sbtest_innodb.sbtest1-schema.sql。
- 数据文件，如 sbtest_innodb.sbtest1.00000.sql。
- 表的元数据信息文件，记录了表的大小，如 sbtest_innodb.sbtest1-metadata。

5.2.3　mydumper 的参数解析

mydumper 的参数可分为如下几类。

1. 过滤相关

- `-B, --database`

备份指定库。可指定多个库，中间需用逗号隔开。

- `-T, --tables-list`

备份指定表，表必须是 `database.table` 的形式。可指定多张表，中间需用逗号隔开。

- `-O, --omit-from-file`

忽略文件中指定表的备份。

文件中每个表占据一行，表必须是 `database.table` 的形式。

- `-x, --regex`

指定库表备份，支持正则表达式。

在执行全库备份时，如果不想备份 sys 库，可用 `--regex` 过滤。

- `-i, --ignore-engines`

忽略指定存储引擎表的备份。可指定多个存储引擎，中间需用逗号隔开。

- `--triggers`、`--events`、`--routines`

分别用于备份触发器、定时器、存储过程及函数。

- `-W, --no-views`

不备份视图。

- `-m, --no-schemas`

不备份表结构。

- `-d, --no-data`

不备份数据。

- `--where`

只导出满足 WHERE 条件的数据。

2. 分片选项

- `-r, --rows`

使用该选项，mydumper 会将表分成多个 chunk 进行并行备份，否则只进行表级别的并行备份。

当然，并不是所有的表都能进行 chunk 级别的并行备份。下面结合源码，看看怎样的表才支持并行备份。

```
GList *get_chunks_for_table(MYSQL *conn, char *database, char *table,
                            struct configuration *conf) {

  GList *chunks = NULL;
  MYSQL_RES *indexes = NULL, *minmax = NULL, *total = NULL;
  MYSQL_ROW row;
  // field是分片键
```

```c
char *field = NULL;
int showed_nulls = 0;

// 通过 SHOW INDEX 查看表的索引信息，例如：
// mysql> show index from sbtest_innodb.sbtest1;
// +----------+------------+----------+--------------+-------------+-----------+-------------+
// | Table    | Non_unique | Key_name | Seq_in_index | Column_name | Collation | Cardinality |
// +----------+------------+----------+--------------+-------------+-----------+-------------+
// | sbtest1  |          0 | PRIMARY  |            1 | id          | A         |      986400 |
// | sbtest1  |          1 | k_1      |            1 | k           | A         |      187075 |
// +----------+------------+----------+--------------+-------------+-----------+-------------+
gchar *query = g_strdup_printf("SHOW INDEX FROM `%s`.`%s`", database, table);
mysql_query(conn, query);
g_free(query);
indexes = mysql_store_result(conn);

// 如果 indexes 不为 NULL，代表表上存在索引，这个时候，会选择合适的索引作为分片键
if (indexes){
  while ((row = mysql_fetch_row(indexes))) {
    // 优先选择主键，如果主键由多个列组成，则取第一列作为分片键
    // 这里的 row[2] 是 Key_name 列，row[3] 是 Seq_in_index 列
    if (!strcmp(row[2], "PRIMARY") && (!strcmp(row[3], "1"))) {
      field = row[4];
      break;
    }
  }

  // 其次选择唯一索引，这里的 row[1] 是 Non_unique 列。为 0，代表是唯一索引
  if (!field) {
    mysql_data_seek(indexes, 0);
    while ((row = mysql_fetch_row(indexes))) {
      if (!strcmp(row[1], "0") && (!strcmp(row[3], "1"))) {
        field = row[4];
        break;
      }
    }
  }

  // 最后，选择 Cardinality 最大的索引。Cardinality 可理解为区分度，指的是列中不同值的个数
  if (!field && conf->use_any_index) {
    guint64 max_cardinality = 0;
    guint64 cardinality = 0;

    mysql_data_seek(indexes, 0);
    while ((row = mysql_fetch_row(indexes))) {
      if (!strcmp(row[3], "1")) {
        // 第 6 列是 Cardinality
        if (row[6])
          cardinality = strtoul(row[6], NULL, 10);
        if (cardinality > max_cardinality) {
          field = row[4];
          max_cardinality = cardinality;
        }
      }
    }
  }
}
```

```c
// 如果 field 依然为 NULL, 代表表上没有索引, 此时不能对表进行并行备份
if (!field)
  goto cleanup;

// 获取分片键的最大值和最小值
mysql_query(conn, query = g_strdup_printf(
                    "SELECT %s MIN(`%s`),MAX(`%s`) FROM `%s`.`%s`",
                    (detected_server == SERVER_TYPE_MYSQL)
                        ? "/*!40001 SQL_NO_CACHE */"
                        : "",
                    field, field, database, table));
g_free(query);
minmax = mysql_store_result(conn);

if (!minmax)
  goto cleanup;

row = mysql_fetch_row(minmax);
MYSQL_FIELD *fields = mysql_fetch_fields(minmax);

if (row[0] == NULL)
  goto cleanup;

char *min = row[0];
char *max = row[1];

guint64 estimated_chunks, estimated_step, nmin, nmax, cutoff, rows;

/*
   重点来了, 分片键只能是 MYSQL_TYPE_LONG、MYSQL_TYPE_LONGLONG、MYSQL_TYPE_INT24、MYSQL_TYPE_SHORT
 这 4 种数据类型中的一种。
   这 4 种类型分别对应 MySQL 中的 INT、BIGINT、MEDIUMINT 和 SMALLINT。
   鉴于 MySQL 多使用自增主键, 虽然 mydumper 只支持这 4 种数据类型, 倒也能涵盖大部分场景
 */
switch (fields[0].type) {
case MYSQL_TYPE_LONG:
case MYSQL_TYPE_LONGLONG:
case MYSQL_TYPE_INT24:
case MYSQL_TYPE_SHORT:
  // 通过 estimate_count 获取表的总行数
  // 总行数取的是 EXPLAIN SELECT `id` FROM `sbtest_innodb`.`sbtest1` WHERE `id` >= 1
  // AND `id` <= 1000000 中的 rows 值
  rows = estimate_count(conn, database, table, field, min, max);
  // 这里的 rows_per_file 由 --rows 选项决定, 可理解为分片的大小
  // 如果表太小, 总行数小于 rows_per_file, 则无须对表进行并行备份
  if (rows <= rows_per_file)
    goto cleanup;

  // 根据 rows / rows_per_file 获取 chunk 的数量
  estimated_chunks = rows / rows_per_file;
  nmin = strtoul(min, NULL, 10);
  nmax = strtoul(max, NULL, 10);
  estimated_step = (nmax - nmin) / estimated_chunks + 1;
  // max_rows 由 --max-rows 选项决定, 代表单个 chunk 的最大行数, 默认为 1000000
  if (estimated_step > max_rows)
    estimated_step = max_rows;
  cutoff = nmin;
  while (cutoff <= nmax) {
```

```c
        // chunks 是一个列表，列表元素是基于分片键的范围查询，如
        // `id` IS NULL OR (`id` >= 1 AND `id` < 2030), (`id` >= 2030 AND `id` < 4059)
        chunks = g_list_prepend(
            chunks,
            g_strdup_printf("%s%s%s%s(`%s` >= %llu AND `%s` < %llu)",
                            !showed_nulls ? "`" : "",
                            !showed_nulls ? field : "",
                            !showed_nulls ? "`" : "",
                            !showed_nulls ? " IS NULL OR " : "", field,
                            (unsigned long long)cutoff, field,
                            (unsigned long long)(cutoff + estimated_step)));
        cutoff += estimated_step;
        showed_nulls = 1;
      }
      chunks = g_list_reverse(chunks);
    default:
      goto cleanup;
  }
// 清理阶段，释放查询结果
cleanup:
  if (indexes)
    mysql_free_result(indexes);
  if (minmax)
    mysql_free_result(minmax);
  if (total)
    mysql_free_result(total);
  return chunks;
}
```

- `-F, --chunk-filesize`

对于不能进行并行备份的表，默认情况下，该表的数据只会写到一个文件中。在恢复时，也只能启动一个线程来恢复。对于这些表，我们可以使用 `--chunk-filesize`，将单表数据写到多个文件中，这样在恢复时，myloader 就可以启动多个线程进行并行恢复。

- `--max-rows`

限制单个 chunk 的最大行数，默认为 `1000000`。

3. 事务相关

- `-k, --no-locks`

备份过程中不加任何锁。

- `--no-backup-locks`

Percona Server 支持备份锁。使用 mydumper 备份 Percona Server 时，默认会使用备份锁来代替全局读锁。若指定该选项，则会禁用备份锁。

- `--less-locking`

同样会加全局读锁，只不过全局读锁会在子线程对非 InnoDB 表加完表锁后释放，而表锁是在备份完非 InnoDB 表后释放。

下面是单线程下（将 -t 设置为 1）general log 的输出。

```
...
388 Query     FLUSH TABLES WITH READ LOCK
...
389 Query     LOCK TABLES `sbtest_myisam`.`sbtest1` READ LOCAL, `sbtest_myisam`.`sbtest2` READ LOCAL,
`sbtest_myisam`.`sbtest3` READ LOCAL, `sbtest_myisam`.`sbtest4` READ LOCAL, `sbtest_myisam`.`sbtest5`
READ LOCAL
...
389 Query     SELECT /*!40001 SQL_NO_CACHE */ * FROM `sbtest_myisam`.`sbtest1`
...
388 Query     UNLOCK TABLES /* FTWRL */
388 Quit
389 Query     SELECT /*!40001 SQL_NO_CACHE */ * FROM `sbtest_myisam`.`sbtest2`
389 Query     SELECT /*!40001 SQL_NO_CACHE */ * FROM `sbtest_myisam`.`sbtest3`
389 Query     SELECT /*!40001 SQL_NO_CACHE */ * FROM `sbtest_myisam`.`sbtest4`
389 Query     SELECT /*!40001 SQL_NO_CACHE */ * FROM `sbtest_myisam`.`sbtest5`
389 Query     UNLOCK TABLES /* Non Innodb */
...
```

- `--use-savepoints`

同 mysqldump 一样，在每个表备份之前设置 SAVEPOINT，可提高 DDL 的并发性。只不过如果设置了 `--rows`，则会禁用 `--use-savepoints`。

```
if (rows_per_file && use_savepoints) {
  use_savepoints = FALSE;
  g_warning("--use-savepoints disabled by --rows");
}
```

- `--lock-all-tables`

使用 LOCK TABLE ... READ 命令加表锁来代替全局读锁，在备份完非 InnoDB 表后释放。

- `--trx-consistency-only`

同样会加全局读锁，只不过释放的时间点是在各个子线程初始化（设置事务隔离级别，开启事务的一致性快照）之后。这样只能保证 InnoDB 表的一致性。

4. 导出选项

默认情况下，备份结果以 INSERT INTO tbl_name VALUES (value_list), (value_list) ... 的形式记录在文件中。相关的参数如下。

- `-N, --insert-ignore`

使用 INSERT IGNORE 导出数据。

- `--complete-insert`

在 INSERT 操作中指定列名。

- `-s, --statement-size`

一个 INSERT 操作的最大长度。不指定该参数，则默认为 1000000，单位是字节。

除此之外，还可将备份集以 CSV 格式保存。这个时候，需指定 `--csv`，该参数会自动开启

--load-data，并将导出格式设置为 csv。如果要保存为其他格式，需设置 --fields-terminated-by、--fields-enclosed-by、--fields-escaped-by、--lines-starting-by、--lines-terminated-by。

指定的导出格式会以 LOAD DATA LOCAL INFILE 命令的形式记录在数据文件中。

```
# cat sbtest_innodb.sbtest1.00000.sql
/*!40101 SET NAMES binary*/;
/*!40014 SET FOREIGN_KEY_CHECKS=0*/;
/*!40103 SET TIME_ZONE='+00:00' */;
LOAD DATA LOCAL INFILE 'sbtest_innodb.sbtest1.00000.dat' REPLACE INTO TABLE `sbtest1` FIELDS TERMINATED
BY ',' ENCLOSED BY '"' ESCAPED BY '\\' LINES STARTING BY '' TERMINATED BY '
' (`id`,`k`,`c`,`pad`);
```

5. 其他选项

- **-a, --ask-password**

提示输入用户密码。这样，就不用通过 -p 显式设置密码。相对而言，通过 -a 输入密码更安全。

- **-c, --compress**

压缩备份。

- **--stream**

流式备份。

- **-t, --threads**

指定备份线程的数量，默认为 4。

- **-L, --logfile**

将日志输出到指定文件。默认是输出到终端。

- **-D, --daemon, -I, --snapshot-interval**

以后台进程的方式运行 mydumper。启动后，会在备份目录中生成两个子目录：0 和 1。备份集会依次放到这两个目录中，即第一次的备份集会放到 0 目录下，第二次的备份集会放到 1 目录下，第三次又重新放到 0 目录下，依此类推。备份的时间间隔由 --snapshot-interval 参数指定，默认为 60，单位是分钟。子目录的数量由 --snapshot-count 决定，默认为 2。

- **-U, --updated-since**

只备份最近几天修改过的表。

判断的依据是 information_schema.tables 表中的 update_time 字段，没有备份的表会写到备份目录的 not_updated_tables 文件中。

```
gchar *query =
    g_strdup_printf("SELECT CONCAT(TABLE_SCHEMA,'.',TABLE_NAME) FROM "
                    "information_schema.TABLES WHERE TABLE_TYPE = 'BASE "
                    "TABLE' AND UPDATE_TIME < NOW() - INTERVAL %d DAY",
                    updated_since);
```

5.2.4 myloader 的参数解析

myloader 的参数可分为如下几类。

1. 连接相关

- `-h, --host`

主机名或 IP。

- `-P, --port`

端口。

- `-u, --user`

用户名。

- `-p, --password`

密码。

- `-S, --socket`

套接字文件地址。

- `-a, --ask-password`

提示输入用户密码。

2. 过滤相关

- `-s, --source-db`

指定要导入的数据库，注意只能指定单个库。如果不指定，则默认导入备份目录下的所有库。

- `-B, --database`

将表导入 -B 指定的数据库。

- `--skip-triggers`

不导入触发器。默认导入。

- `--skip-post`

不导入定时器、存储过程和自定义函数。默认导入。

- `--no-data`

不导入数据。

- `-O, --omit-from-file`

忽略文件中指定表的导入。

- **-T, --tables-list**

只导入指定表，表必须是 database.table 的形式。可指定多张表，中间需用逗号隔开。

- **-x, --regex**

基于正则匹配导入指定库表。

- **--skip-definer**

移除 CREATE 语句中的 DEFINER 子句。

3. 导入速度相关

- **-q, --queries-per-transaction**

多久提交一次。默认是 1000 个 INSERT 操作。

- **--innodb-optimize-keys**

先导入数据，再添加索引。

- **--disable-redo-log**

通过 ALTER INSTANCE DISABLE INNODB REDO_LOG 命令禁用 redo log，这样可加快导入速度。

4. 其他选项

- **-d, --directory**

备份目录。

- **-o, --overwrite-tables**

导入时，对于已经存在的表，会首先执行 DROP TABLE 操作。

如果不指定该选项，则会直接导入备份数据并提示以下错误。

```
** (myloader:27919): CRITICAL **: 21:23:25.027: Thread 1 issue restoring //data/backup/mydumper/slowtech.t1-schema.sql: Table 't1' already exists
```

- **-e, --enable-binlog**

导入时，开启 binlog。默认情况下，会将 SQL_LOG_BIN 设置为 0。

- **--purge-mode**

导入时，如果对已经存在的表指定了 --overwrite-tables，默认会执行 DROP TABLE 操作。

如果要执行其他操作，可通过 --purge-mode 指定，可选值有 NONE、DROP、TRUNCATE、DELETE。

- **--serialized-table-creation**

顺序创建表。在导入的过程中，如果报错 Deadlock found when trying to get lock; try restarting transaction，可指定该选项。

- `-r, --rows`

指定一个 INSERT 操作的行数。

- `--max-threads-per-table`

针对单张表的最大线程数，默认为 4。

- `-t, --threads`

设置导入线程的数量，默认为 4。注意，导入的粒度是文件。

- `--stream`

通过流式备份的文件，在恢复时，需指定 `--stream`。

5.2.5 mydumper 和 myloader 的常见用法

1. 全库备份

```
# mydumper -u mydumper_user -p mydumper_pass -G -E -R -r 100000 -F 64 -t 10 -o /data/backup/mydumper
```

其中，`-G` `-E` `-R` 分别用来备份触发器、定时器和存储过程（函数），`-r` 指定了分片的大小，`-F` 指定了单个 chunk 的大小，`-t` 指定了备份线程的数量。

2. 备份单个库

```
# mydumper -B db1 -u mydumper_user -p mydumper_pass -o /data/backup/mydumper
```

这里会备份 db1 库。

3. 备份多个表

```
# mydumper -T db1.t1,db2.t2 -u mydumper_user -p mydumper_pass -o /data/backup/mydumper
```

这里会备份 db1.t1 和 db2.t2 两张表。

4. 备份多个库

```
# mydumper -B 'db1,db2' -u mydumper_user -p mydumper_pass -o /data/backup/mydumper
```

这里会备份 db1 和 db2 两个库。

5. 忽略指定库的备份

```
# mydumper -x '^(?!db1[.]|db2[.])' -u mydumper_user -p mydumper_pass -o /data/backup/mydumper
```

这里会忽略 db1 和 db2 库的备份，`[.]` 是对点（`.`）的转义。

6. 忽略指定表的备份

```
# mydumper -O /tmp/skip_tables -u mydumper_user -p mydumper_pass -o /data/backup/mydumper
```

其中，/tmp/skip_tables 的内容如下，

```
# cat /tmp/skip_tables
```

```
db1.t1
db2.t2
```

也可使用正则表达式来过滤，例如：

```
# mydumper -x '^(?!db1[.]t1|db2[.]t2)' -u mydumper_user -p mydumper_pass -o /data/backup/mydumper
```

7. 流式备份

```
# mydumper -u mydumper_user -p mydumper_pass --stream | lz4 > /data/backup/mydumper/full_backup.lz4
```

lz4 是压缩算法，这里是备份到本地。

也可直接备份到远程服务器，例如：

```
mydumper -u mydumper_user -p mydumper_pass --stream | lz4 | ssh root@10.0.20.4 "cat - > /data/backup/mydumper/full_backup.lz4"
```

除此之外，还能边备份边导入。

```
# mydumper -u mydumper_user -p mydumper_pass --stream | myloader -h 10.0.20.4 -P 3307 -u mydumper_user -p mydumper_pass -d /data/backup/3307 --stream
```

8. 全库还原

```
# myloader -u mydumper_user -p mydumper_pass -o -e -t 8 -d /data/backup/mydumper/
```

其中，-o 用于删除已经存在的表或视图，-e 用于开启 binlog，-t 指定导入的并发线程数。

9. 导入指定库

```
# myloader -u mydumper_user -p mydumper_pass -e -s db1 -d /data/backup/mydumper/
```

该命令会导入 db1 库的数据。

10. 将指定库中的数据导入另一个库

```
# myloader -u mydumper_user -p mydumper_pass -e -s db1 -B db1_new -d /data/backup/mydumper/
```

这里会将 db1 中的数据导入 db1_new。

5.2.6 总结

- 与 mysqldump 一样，mydumper 也是通过 SELECT * FROM TABLE_NAME 来实现备份的。
- mydumper 支持 chunk 级别的并行备份。在选择分区键时，优先选择主键，其次是唯一索引，最后是区分度最高的索引列。不仅如此，在 v0.12.6-2 之前，分区键只支持 4 种整数类型，不支持字符和时间类型。CHAR 和 VARCHAR 分别是从 v0.12.6-2 和 v0.13.0-3 开始支持的。
- 对于不支持并行备份的表，可指定 --chunk-filesize，将其备份到多个文件中。这样，myloader 在导入时，可使用多个线程进行并行恢复。

5.3 XtraBackup

XtraBackup 是 Percona 公司开源的 MySQL 物理备份工具，适用于 MySQL 和 Percona Server。

5.3.1 XtraBackup 的安装

XtraBackup 可从 Percona 官方网站下载。目前，XtraBackup 活跃的大版本有以下两个。

- XtraBackup 2.4，适用于 MySQL 5.6 和 MySQL 5.7。
- XtraBackup 8.0，适用于 MySQL 8.0。

注意，两者不可混用，因为 MySQL 8.0 的 redo log 和数据字典的格式发生了变化。

XtraBackup 只支持 Linux 系统。这里使用的是 Percona XtraBackup 8.0.27-19 (Linux - Generic)。以下是具体的安装步骤。

```
# tar xvf percona-xtrabackup-8.0.27-19-Linux-x86_64.glibc2.17.tar.gz -C /usr/local/
# cd percona-xtrabackup-8.0.27-19-Linux-x86_64.glibc2.17/
# yum install perl-Digest-MD5 perl-DBD-MySQL libev -y
# ls
bin  docs  include  lib  man  percona-xtrabackup-8.0-test
# ls bin/
xbcloud  xbcloud_osenv  xbcrypt  xbstream  xtrabackup
```

解压后，会有 6 个目录。bin 目录下均是二进制可执行文件，核心的备份恢复命令即位于此目录下。

下面来看看 bin 目录下各个文件的具体作用。

- xbcloud：与流式备份相结合，将备份存储到云服务的对象存储上。
- xbcloud_osenv：对 xbcloud 进行二次封装，可自动读取 OpenStack 环境中的 OS_*xxx* 变量。
- xbcrypt：用来加解密。
- xbstream：用来解压流式备份集。
- xtrabackup：常用的备份恢复工具。

在 XtraBackup 2.4 中，还有 innobackupex，不过它是 xtrabackup 的一个软链。

在 XtraBackup 2.3 之前，xtrabackup 只支持 InnoDB 表的备份，MyISAM 等非事务性存储引擎表的备份是通过 innobackupex 来实现的。从实现上看，innobackupex 是 Perl 脚本，而 xtrabackup 是用 C/C++ 编译的二进制文件。前者调用后者来备份 InnoDB 表，并在此基础上，实现非 InnoDB 表的备份、数据库的一致性备份及并行备份等。

这看上去似乎没问题，但既然是两个不同的工具协同处理同一个任务，就必然涉及这两个工具之间的交互。之前的解决方案是在不同的阶段，通过创建或删除不同的临时文件（xtrabackup_suspended_1、xtrabackup_suspended_2、xtrabackup_log_copied）与对方交互。但这种实现存在风险，例如在备份的过程中，临时文件可能被人误删。

于是从 XtraBackup 2.3 开始，Percona 用 C 语言重写了 innobackupex，并将其作为 xtrabackup 的一个软链。它依旧支持之前的语法，但不会增加新的特性，所有新的特性都只会集成在 xtrabackup 中。

建议从 XtraBackup 2.3 开始只使用 xtrabackup，而不是 innobackupex，后者在 XtraBackup 8.0 中被移除了。所以，下面的演示只使用 xtrabackup，没有使用 innobackupex。

5.3.2 基于源码分析 XtraBackup 的实现原理

XtraBackup 的 main 函数定义在 storage/innobase/xtrabackup/src/xtrabackup.cc 文件中。

从 main 函数中可以看到，对于 --backup 选项，会调用 xtrabackup_backup_func 函数。

```
int main(int argc, char **argv) {
  ...
  if (!xb_init()) {
    exit(EXIT_FAILURE);
  }
  ...
  /* --backup */
  if (xtrabackup_backup) {
    xtrabackup_backup_func();
  }

  /* --stats */
  if (xtrabackup_stats) {
    xtrabackup_stats_func(server_argc, server_defaults);
  }

  /* --prepare */
  if (xtrabackup_prepare) {
    xtrabackup_prepare_func(server_argc, server_defaults);
  }

  /* --copy-back 或 --move-back */
  if (xtrabackup_copy_back || xtrabackup_move_back) {
    if (!check_if_param_set("datadir")) {
      msg("Error: datadir must be specified.\n");
      exit(EXIT_FAILURE);
    }
    init_mysql_environment();
    if (!copy_back(server_argc, server_defaults)) {
      exit(EXIT_FAILURE);
    }
    cleanup_mysql_environment();
  }

  /* 解压压缩文件 */
  if (xtrabackup_decrypt_decompress && !decrypt_decompress()) {
    exit(EXIT_FAILURE);
  }

  backup_cleanup();

  xb_regex_end();

  msg_ts("completed OK!\n");

  exit(EXIT_SUCCESS);
}
```

下面重点看看 xtrabackup_backup_func 函数的处理逻辑。该函数同样位于 xtrabackup.cc 文件中。

```
void xtrabackup_backup_func(void) {
  ...
  /* 创建 redo log 拷贝线程 */
```

```
Redo_Log_Data_Manager redo_mgr;
...
/* 从最新的检查点开始拷贝 redo log */
if (!redo_mgr.start()) {
  exit(EXIT_FAILURE);
}

Tablespace_map::instance().scan(mysql_connection);

/* 获取 ibdata1、undo tablespaces 及所有的 ibd 文件 */
dberr_t err = xb_load_tablespaces();
if (err != DB_SUCCESS) {
  msg("xtrabackup: error: xb_load_tablespaces() failed with error code "
      "%u\n",
      err);
  exit(EXIT_FAILURE);
}
...
/* 创建数据拷贝线程 */
data_threads = (data_thread_ctxt_t *)ut::malloc_withkey(
    UT_NEW_THIS_FILE_PSI_KEY,
    sizeof(data_thread_ctxt_t) * xtrabackup_parallel);
count = xtrabackup_parallel;
mutex_create(LATCH_ID_XTRA_COUNT_MUTEX, &count_mutex);

/* 拷贝物理文件，其中 xtrabackup_parallel 是拷贝并发线程数，由 --parallel 参数指定 */
for (i = 0; i < (uint)xtrabackup_parallel; i++) {
  data_threads[i].it = it;
  data_threads[i].num = i + 1;
  data_threads[i].count = &count;
  data_threads[i].count_mutex = &count_mutex;
  data_threads[i].error = &data_copying_error;
  os_thread_create(PFS_NOT_INSTRUMENTED, i, data_copy_thread_func,
                   data_threads + i)
      .start();
}

/* 循环等待，直到拷贝结束 */
while (1) {
  std::this_thread::sleep_for(std::chrono::seconds(1));
  mutex_enter(&count_mutex);
  if (count == 0) {
    mutex_exit(&count_mutex);
    break;
  }
  if (redo_mgr.is_error()) {
    msg("xtrabackup: error: log copyiing failed.\n");
    exit(EXIT_FAILURE);
  }
  mutex_exit(&count_mutex);
}

mutex_free(&count_mutex);
ut::free(data_threads);
datafiles_iter_free(it);

if (data_copying_error) {
  exit(EXIT_FAILURE);
}
```

```
if (changed_page_bitmap) {
  msg("xtrabackup: WARNING: Incremental backup using page bitmap is "
      "deprecated and will be removed in a future release. Please use "
      "--page-tracking\n");
  xb_page_bitmap_deinit(changed_page_bitmap);
}
if (changed_page_tracking) {
  pagetracking::deinit(changed_page_tracking);
}

Backup_context backup_ctxt;

/* 调用 backup_start 函数，这个函数会加全局读锁，拷贝非 ibd 文件 */
if (!backup_start(backup_ctxt)) {
  exit(EXIT_FAILURE);
}

if (opt_debug_sleep_before_unlock) {
  msg_ts("Debug sleep for %u seconds\n", opt_debug_sleep_before_unlock);
  std::this_thread::sleep_for(std::chrono::seconds(opt_debug_sleep_before_unlock));
}

/* 停止 redo log 拷贝线程 */
if (!redo_mgr.stop_at(log_status.lsn, log_status.lsn_checkpoint)) {
  exit(EXIT_FAILURE);
}
if (redo_mgr.is_error()) {
  msg("xtrabackup: error: log copyiing failed.\n");
  exit(EXIT_FAILURE);
}
msg("\n");

io_watching_thread_stop = true;

if (!validate_missing_encryption_tablespaces()) {
  exit(EXIT_FAILURE);
}

if (!xtrabackup_incremental) {
  strcpy(metadata_type_str, "full-backuped");
  metadata_from_lsn = 0;
} else {
  strcpy(metadata_type_str, "incremental");
  metadata_from_lsn = incremental_lsn;
}
metadata_to_lsn = redo_mgr.get_last_checkpoint_lsn();
metadata_last_lsn = redo_mgr.get_stop_lsn();

if (!xtrabackup_stream_metadata(ds_meta)) {
  msg("xtrabackup: Error: failed to stream metadata.\n");
  exit(EXIT_FAILURE);
}
/* 调用 backup_finish 函数，这个函数会释放全局读锁或备份锁 */
if (!backup_finish(backup_ctxt)) {
  exit(EXIT_FAILURE);
}
...
}
```

该函数的处理流程如下。

- 在备份开始时，XtraBackup 首先会启动一个 redo log 拷贝线程，从最新的检查点（checkpoint）开始顺序拷贝 redo log。
- 创建数据文件拷贝线程，拷贝 ibdata1、undo tablespaces 及所有的 ibd 文件。

  命令行中可通过 --parallel 设置多个数据文件拷贝线程，提高物理文件的拷贝效率。若不设置，则默认为 1。

  在物理文件的拷贝过程中，redo log 拷贝线程会监听 redo log 的变化并持续拷贝。

  为什么 redo log 拷贝线程会先于 ibd 文件拷贝线程启动呢？这是因为 ibd 文件的拷贝只是物理拷贝，而物理拷贝的基本单位是页。在拷贝的过程中，有的页会先拷贝，有的则后拷贝。不仅如此，有的页在拷贝后又发生了变化。将 ibd 文件恢复到备份结束时的一致性状态，实际上是依赖 redo log 来实现的。如果 ibd 文件拷贝在前，则它的变化情况是无法得知的。
- ibd 文件拷贝完成后，调用 backup_start 函数。
- 停止 redo log 拷贝线程。
- 调用 backup_finish 函数。

接下来重点看看 backup_start 和 backup_finish 这两个函数的实现逻辑。

首先是 backup_start 函数。

```
bool backup_start(Backup_context &context) {
  // opt_no_lock 对应命令行中的 --no-lock 参数
  if (!opt_no_lock) {
    /*
       调用 backup_files 函数备份非 ibd 文件，加了全局读锁或备份锁后还会调用一次。
       这次调用只适用于 --rsync 方式
    */
    if (!backup_files(MySQL_datadir_path.path().c_str(), true)) {
      return (false);
    }

    history_lock_time = time(NULL);
    /*
       加全局读锁或备份锁。
       opt_backup_lock_timeout 对应命令行中的 --backup-lock-timeout。
       opt_backup_lock_retry_count 对应命令行中的 --backup-lock-retry-count
    */
    if (!lock_tables_maybe(mysql_connection, opt_backup_lock_timeout,
                           opt_backup_lock_retry_count)) {
      return (false);
    }
  }

  // 再次调用 backup_files 函数备份非 ibd 文件
  if (!backup_files(MySQL_datadir_path.path().c_str(), false)) {
    return (false);
  }
```

```cpp
if (opt_no_lock && opt_safe_slave_backup) {
  if (!wait_for_safe_slave(mysql_connection)) {
    return (false);
  }
}

if (have_rocksdb && opt_rocksdb_checkpoint_max_age > 0) {
  ...
}

if (have_rocksdb) {
  context.myrocks_checkpoint.create(mysql_connection, true);
}

/* 执行 FLUSH NO_WRITE_TO_BINLOG BINARY LOGS，切换 binlog */
msg_ts("Executing FLUSH NO_WRITE_TO_BINLOG BINARY LOGS\n");
xb_mysql_query(mysql_connection, "FLUSH NO_WRITE_TO_BINLOG BINARY LOGS",
               false);
/* 查询 performance_schema.log_status，获取 gtid_executed 及 binlog 的位置点信息 */
log_status_get(mysql_connection);

if (opt_page_tracking) {
  ...
}

debug_sync_point("xtrabackup_after_query_log_status");

/* 拷贝当前的 binlog */
if (!write_current_binlog_file(mysql_connection)) {
  return (false);
}

if (have_rocksdb) {
  ...
}

/* 如果设置了 --slave-info，则会将 SHOW SLAVE STATUS 的查询结果记录在 xtrabackup_slave_info 中 */
if (opt_slave_info) {
  if (!write_slave_info(mysql_connection)) {
    return (false);
  }
}
/*
    在 MySQL 8.0 之前，会将 SHOW MASTER STATUS 的查询结果记录在 xtrabackup_binlog_info 中。
    在 MySQL 8.0 中，记录的是 performance_schema.log_status 的查询结果
*/
write_binlog_info(mysql_connection);

/* 执行 FLUSH NO_WRITE_TO_BINLOG BINARY LOGS */
if (have_flush_engine_logs) {
  msg_ts("Executing FLUSH NO_WRITE_TO_BINLOG ENGINE LOGS...\n");
  xb_mysql_query(mysql_connection, "FLUSH NO_WRITE_TO_BINLOG ENGINE LOGS",
                 false);
}

return (true);
}
```

该函数的处理流程如下。

- 调用 lock_tables_maybe 函数加全局读锁或备份锁。
- 调用 backup_files 函数备份非 ibd 文件。具体来说，会备份以下面这些关键字作为扩展名的文件。

  ```
  const char *ext_list[] = {"MYD", "MYI", "MAD", "MAI", "MRG", "ARM",
                            "ARZ", "CSM", "CSV", "opt", "sdi", NULL};
  ```

- 执行 FLUSH NO_WRITE_TO_BINLOG BINARY LOGS，切换 binlog。
- 查询 performance_schema.log_status，获取 gtid_executed 及 binlog 的位置点信息。
- 拷贝当前的 binlog。
- 如果设置了 --slave-info，则会执行 SHOW SLAVE STATUS 获取从库复制的相关信息。
- 在 MySQL 8.0 中，会将 performance_schema.log_status 的查询结果记录在 xtrabackup_binlog_info 中。而在 MySQL 8.0 之前，记录的是 SHOW MASTER STATUS 的查询结果。

接下来分析 lock_tables_maybe 函数。

```
bool lock_tables_maybe(MYSQL *connection, int timeout, int retry_count) {
  bool force_ftwrl = opt_slave_info && !slave_auto_position &&
                     !(server_flavor == FLAVOR_PERCONA_SERVER);

  if (tables_locked || (opt_lock_ddl_per_table && !force_ftwrl)) {
    return (true);
  }

  if (!have_unsafe_ddl_tables && !force_ftwrl) {
    return (true);
  }

  if (have_backup_locks && !force_ftwrl) {
    return lock_tables_for_backup(connection, timeout, retry_count);
  }

  return lock_tables_ftwrl(connection);
}
```

该函数的处理流程如下。

- 通过 force_ftwrl 判断是否需要加全局读锁，它为 true 需满足以下 3 个条件。

  - 命令行中指定了 --slave-info。一般备份从库需要指定这个参数。
  - 从库 SHOW SLAVE STATUS 中的 Auto_Position 不为 1。可能的原因是：从库没有开启 GTID，或者虽然开启了 GTID 但未将 MASTER_AUTO_POSITION 设置为 1。
  - 备份实例不是 Percona Server。

- 如果 tables_locked 为 true，或者命令行中指定了 --lock-ddl-per-table 且无须加全局读锁，lock_tables_maybe 函数会直接返回 true。

  tables_locked 满足以下任意一个条件即为 true。

  - 已经执行过 LOCK TABLES FOR BACKUP。LOCK TABLES FOR BACKUP 是 Percona Server 中的备份锁。

- 已经执行过 FLUSH TABLES WITH READ LOCK。

□ 如果 have_unsafe_ddl_tables 为 false，且无须加全局读锁，lock_tables_maybe 函数会直接返回 true。

have_unsafe_ddl_tables 为 false 代表备份实例中不存在 MyISAM 或 RocksDB 表。

□ 如果 have_backup_locks 为 true，且无须加全局读锁，则会调用 lock_tables_for_backup。

have_backup_locks 同时满足以下两个条件才为 true。

- 备份实例存在参数 have_backup_locks。这个参数只在 Percona Server 中存在。
- 备份时没有指定 --no-backup-locks。

所以，如果备份实例是 Percona Server 且无须加全局读锁，则会直接调用 lock_tables_for_backup 加 Percona Server 中的备份锁。

□ 如果上述条件都不满足，则会调用 lock_tables_ftwrl 加全局读锁。

综上所述，如果备份的是 MySQL 社区版，以下两种情况都会加全局读锁。

□ 备份的是从库，指定了 --slave-info，而且从库的 Auto_Position 不为 1。
□ 备份实例中存在 MyISAM 表。

下面继续分析 lock_tables_for_backup 函数。

```
bool lock_tables_for_backup(MYSQL *connection, int timeout, int retry_count) {
  if (have_backup_locks) {
    execute_query_with_timeout(connection, "LOCK TABLES FOR BACKUP", timeout,
                               retry_count);
    tables_locked = true;

    return (true);
  }

  execute_query_with_timeout(connection, "LOCK INSTANCE FOR BACKUP", timeout,
                             retry_count);
  instance_locked = true;

  return (true);
}
```

这个函数比较简单，如果支持备份锁，则会直接执行 LOCK TABLES FOR BACKUP。如果不支持，则会执行 LOCK INSTANCE FOR BACKUP。

细心的读者可能会发现，对于 MySQL 社区版，似乎都没有机会执行 LOCK INSTANCE FOR BACKUP。毕竟，调用 lock_tables_for_backup 的前提条件之一是 have_backup_locks 必须为 true。

实际上，lock_tables_for_backup 这个函数在 xb_init() 中会调用一次，而 xb_init() 的执行时间早于 xtrabackup_backup_func()。

```
bool xb_init() {
  ...
  if (xtrabackup_backup) {
```

```cpp
...
if (opt_check_privileges) {
  check_all_privileges();
}

history_start_time = time(NULL);

/*
   opt_safe_slave_backup 对应命令行中的 --safe-slave-backup。
   如果指定了，则会在备份前停掉 SQL 线程
*/
if (!opt_no_lock && opt_safe_slave_backup) {
  if (!wait_for_safe_slave(mysql_connection)) {
    return (false);
  }
}

if (opt_lock_ddl &&
    !lock_tables_for_backup(mysql_connection, opt_lock_ddl_timeout, 0)) {
  return (false);
}

parse_show_engine_innodb_status(mysql_connection);
}

return (true);
}
```

接着继续分析 lock_tables_ftwrl 函数。

```cpp
bool lock_tables_ftwrl(MYSQL *connection) {
  if (have_lock_wait_timeout) {
    /* 设置全局读锁的超时时间 */
    xb_mysql_query(connection, "SET SESSION lock_wait_timeout=31536000", false);
  }

  /*
     这里的 opt_lock_wait_timeout 对应命令行中的 --ftwrl-wait-timeout,
     opt_kill_long_queries_timeout 对应命令行中的 --kill-long-queries-timeout,
     如果没有指定这两个参数，则会执行 FLUSH TABLES 操作，避免直接加全局读锁
  */
  if (!opt_lock_wait_timeout && !opt_kill_long_queries_timeout) {
    msg_ts("Executing FLUSH NO_WRITE_TO_BINLOG TABLES...\n");

    xb_mysql_query(connection, "FLUSH NO_WRITE_TO_BINLOG TABLES", false);
  }
  /* 这里的 opt_lock_wait_threshold 对应命令行中的 --ftwrl-wait-threshold   */
  if (opt_lock_wait_timeout) {
    if (!wait_for_no_updates(connection, opt_lock_wait_timeout,
                             opt_lock_wait_threshold)) {
      return (false);
    }
  }

  msg_ts("Executing FLUSH TABLES WITH READ LOCK...\n");
  /* 这里的 opt_kill_long_queries_timeout 对应命令行中的 --kill-long-queries-timeout */
  if (opt_kill_long_queries_timeout) {
```

```
    start_query_killer();
  }

  if (have_galera_enabled) {
    xb_mysql_query(connection, "SET SESSION wsrep_causal_reads=0", false);
  }
  /* 加全局读锁 */
  xb_mysql_query(connection, "FLUSH TABLES WITH READ LOCK", false);

  if (opt_kill_long_queries_timeout) {
    stop_query_killer();
  }

  tables_locked = true;

  return (true);
}
```

该函数的处理流程如下。

- 如果没有设置 --ftwrl-wait-timeout 和 --kill-long-queries-timeout，则会执行 FLUSH TABLES 操作，避免直接加全局读锁。
- 如果设置了 --ftwrl-wait-timeout，则会判断当前是否存在执行时间超过 --ftwrl-wait-threshold 的 SQL 语句。如果不存在，会进入下一步，加全局读锁。如果存在，则会在等待 1 秒后继续判断。判断的总时长由 --ftwrl-wait-timeout 决定。如果在这个时长内一直都有执行时间超过 --ftwrl-wait-threshold 的 SQL，xtrabackup 会报错并退出。
- 执行 FLUSH TABLES WITH READ LOCK 加全局读锁。
- 加了全局读锁后，如果有阻塞全局读锁的操作，正常情况下会等待这些操作结束。如果设置了 --kill-long-queries-timeout，则会在等待 --kill-long-queries-timeout 秒后，杀掉这些阻塞操作。

最后，看看 backup_finish 函数。

```
bool backup_finish(Backup_context &context) {
  /* 释放备份锁，如果加了全局读锁，还会释放全局读锁 */
  if (!opt_no_lock) {
    unlock_all(mysql_connection);
    history_lock_time = time(NULL) - history_lock_time;
  } else {
    history_lock_time = 0;
  }
  /* 如果设置了 --safe-slave-backup，且 SQL 线程停止了，则会开启 SQL 线程 */
  if (opt_safe_slave_backup && sql_thread_started) {
    msg("Starting slave SQL thread\n");
    xb_mysql_query(mysql_connection, "START SLAVE SQL_THREAD", false);
  }

  /* 拷贝 ib_buffer_pool 和 ib_lru_dump 文件 */
  if (!opt_rsync) {
    if (opt_dump_innodb_buffer_pool) {
      check_dump_innodb_buffer_pool(mysql_connection);
    }
```

```cpp
  if (buffer_pool_filename && file_exists(buffer_pool_filename)) {
    const char *dst_name;

    dst_name = trim_dotslash(buffer_pool_filename);
    copy_file(ds_data, buffer_pool_filename, dst_name, 0, FILE_PURPOSE_OTHER);
  }
  if (file_exists("ib_lru_dump")) {
    copy_file(ds_data, "ib_lru_dump", "ib_lru_dump", 0, FILE_PURPOSE_OTHER);
  }
}

if (have_rocksdb) {
  if (!backup_rocksdb_checkpoint(context, true)) {
    return (false);
  }
  context.myrocks_checkpoint.remove();
}

msg_ts("Backup created in directory '%s'\n", xtrabackup_target_dir);
if (!mysql_binlog_position.empty()) {
  msg("MySQL binlog position: %s\n", mysql_binlog_position.c_str());
}
if (!mysql_slave_position.empty() && opt_slave_info) {
  msg("MySQL slave binlog position: %s\n", mysql_slave_position.c_str());
}
if (xtrabackup::components::keyring_component_initialized &&
    !xtrabackup::components::write_component_config_file()) {
  msg("xtrabackup: write_component_config_file failed\n");
  return (false);
}
/* 生成配置文件 backup-my.cnf */
if (!write_backup_config_file()) {
  return (false);
}
/* 将备份的相关信息记录在 xtrabackup_info 文件中 */
if (!write_xtrabackup_info(mysql_connection)) {
  return (false);
}

return (true);
}
```

该函数的大致流程如下。

- 释放备份锁。如果加了全局读锁，还会释放全局读锁。
- 拷贝 ib_buffer_pool 和 ib_lru_dump 文件。
- 将备份的相关信息记录在 xtrabackup_info 文件中。如果设置了 --history，还会将备份信息记录在 PERCONA_SCHEMA.xtrabackup_history 中。

综合上面的分析，XtraBackup 8.0 的实现原理如图 5-1 所示。

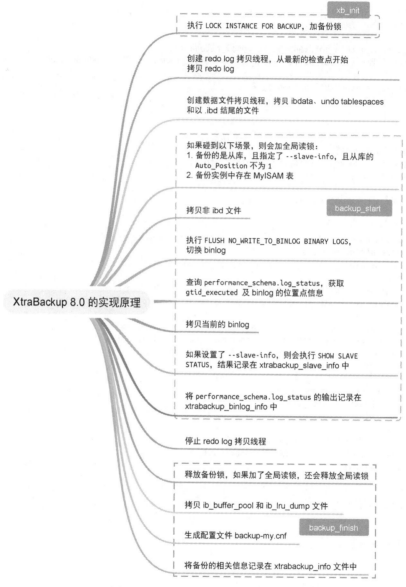

图 5-1　XtraBackup 8.0 的实现原理

5.3.3　XtraBackup 的常见用法

XtraBackup 的备份恢复一般包括以下 3 个阶段。

- 备份。

将物理文件拷贝到备份目录中。

❑ Prepare。

应用 redo log，将数据文件恢复到备份结束时的一致性状态。

❑ 恢复。

将备份文件恢复至数据库数据目录下。恢复阶段所做的，本质上就是拷贝文件，除了使用 XtraBackup，我们也可手动拷贝。

下面我们看看 XtraBackup 的常见用法。

1. 全库备份

首先创建一个全量备份。

```
# xtrabackup --user=backup_user --password=backup_pass --backup --parallel=10
--target-dir=/data/backup/full
```

备份时需指定 `--backup`。命令行中的 `--target-dir` 是备份目录，`--parallel` 指的是并发拷贝线程数。在线上，为了缩短备份时间，一般会指定 `parallel` 参数。

接下来看看备份目录的内容。

```
# ll /data/backup/full/
total 88196
-rw-r-----  1 root root      475 May  6 10:19 backup-my.cnf
drwxr-x---  2 root root     4096 May  6 10:19 db1
drwxr-x---  2 root root     4096 May  6 10:19 db2
-rw-r-----  1 root root     5575 May  6 10:19 ib_buffer_pool
-rw-r-----  1 root root 12582912 May  6 10:19 ibdata1
drwxr-x---  2 root root     4096 May  6 10:19 mysql
-rw-r-----  1 root root      501 May  6 10:19 mysql-bin.000009
-rw-r-----  1 root root       19 May  6 10:19 mysql-bin.index
-rw-r-----  1 root root 25165824 May  6 10:19 mysql.ibd
drwxr-x---  2 root root     4096 May  6 10:19 performance_schema
drwxr-x---  2 root root     4096 May  6 10:19 sys
-rw-r-----  1 root root 16777216 May  6 10:19 undo_001
-rw-r-----  1 root root 33554432 May  6 10:19 undo_002
-rw-r-----  1 root root       65 May  6 10:19 xtrabackup_binlog_info
-rw-r-----  1 root root      105 May  6 10:19 xtrabackup_checkpoints
-rw-r-----  1 root root      579 May  6 10:19 xtrabackup_info
-rw-r-----  1 root root  2174464 May  6 10:19 xtrabackup_logfile
-rw-r-----  1 root root       39 May  6 10:19 xtrabackup_tablespaces
```

除了数据库相关的文件，我们重点关注一下 xtrabackup 生成的备份文件。

- **backup-my.cnf**

注意，该文件不是参数文件的备份，它只是记录了 InnoDB 的相关参数，在 Prepare 阶段会用到。

```
# This MySQL options file was generated by innobackupex.

# The MySQL server
[mysqld]
innodb_checksum_algorithm=crc32
innodb_log_checksums=1
innodb_data_file_path=ibdata1:12M:autoextend
```

```
innodb_log_files_in_group=2
innodb_log_file_size=50331648
innodb_page_size=16384
innodb_undo_directory=./
innodb_undo_tablespaces=2
server_id=1
innodb_log_checksums=ON
innodb_redo_log_encrypt=OFF
innodb_undo_log_encrypt=OFF
server_uuid=a028d418-ccce-11ec-bf07-525400d51a16
master_key_id=0
```

- **xtrabackup_binlog_info**

该文件记录了备份结束时的一致性位置点信息,在搭建从库时会用到。

```
# cat xtrabackup_binlog_info
mysql-bin.000009  501    a028d418-ccce-11ec-bf07-525400d51a16:1-3692
```

- **xtrabackup_checkpoints**

该文件记录了备份的类型和 LSN 信息。

```
# cat xtrabackup_checkpoints
backup_type = full-backuped
from_lsn = 0
to_lsn = 170647793
last_lsn = 172718636
flushed_lsn = 172648890
```

部分项的具体含义如下。

- `backup_type`:备份的类型。`full-backuped` 代表全量备份,`incremental` 代表增量备份。
- `from_lsn`:上一次备份完成时的 LSN。如果是全量备份,则为 0;如果是增量备份,则为上一次备份的 `to_lsn`。
- `to_lsn`:备份完成时最近一个检查点的 LSN。
- `last_lsn`:redo log 停止拷贝时的 LSN。

- **xtrabackup_info**

该文件记录了更为详尽的备份信息。

```
# cat xtrabackup_info
uuid = f0541788-cce2-11ec-9255-525400d51a16
name =
tool_name = xtrabackup
tool_command = --user=backup_user --password=... --backup --parallel=10
--target-dir=/data/backup/full
tool_version = 8.0.27-19
ibbackup_version = 8.0.27-19
server_version = 8.0.27
start_time = 2022-05-06 10:19:18
end_time = 2022-05-06 10:19:20
lock_time = 1
binlog_pos = filename 'mysql-bin.000009', position '501', GTID of the last change
'a028d418-ccce-11ec-bf07-525400d51a16:1-3692'
innodb_from_lsn = 0
```

```
innodb_to_lsn = 170647793
partial = N
incremental = N
format = file
compressed = N
encrypted = N
```

部分项的具体含义如下。

- `uuid`：用于唯一标识这次备份，与备份实例的 `server_uuid` 无关。
- `name`：作用同 `uuid` 类似，可通过 `--history[=name]` 指定。
- `tool_name` 和 `tool_command`：备份工具及备份命令。
- `tool_version`、`ibbackup_version`、`server_version`：工具版本、xtrabackup 的版本、备份实例的版本。
- `start_time` 和 `end_time`：备份的开始时间和结束时间。
- `lock_time`：备份实例被锁定的时间，等于 `backup_finish` 中的 `history_lock_time` 减去 `backup_start` 中的 `history_lock_time`。
- `binlog_pos`：备份结束时的一致性位置点信息。
- `partial`：是否为部分库表备份。
- `incremental`：是否为增量备份。
- `format`：备份集的格式。`file` 是文件，`xbstream` 是流，用在流式备份中。
- `compressed`：是否为压缩备份。
- `encrypted`：是否为加密备份。

- xtrabackup_logfile

该文件用来保存拷贝的 redo log。

如果是对从库进行备份，还需要添加 `--slave-info` 参数，此时还会生成 xtrabackup_slave_info。

```
# cat xtrabackup_slave_info
CHANGE MASTER TO MASTER_LOG_FILE='mysql-bin.000007', MASTER_LOG_POS=2701634;
```

该文件记录了备份时对应主库的位置点信息。这里的 MASTER_LOG_FILE 和 MASTER_LOG_POS 取自 SHOW SLAVE STATUS 中的 Relay_Master_Log_File 和 Exec_Master_Log_Pos。注意，只有在从库上备份时，该文件才会有内容。在主库上备份时，也可指定 `--slave-info`，只不过这种场景下生成的 xtrabackup_slave_info 内容为空。

备份完毕后，下面进入 Prepare 阶段。

```
# xtrabackup --prepare --use-memory=2G --target-dir=/data/backup/full
```

在这个阶段，xtrabackup 会启动一个嵌入的 InnoDB 实例来进行 Crash Recovery。这个实例缓冲池的大小由 `--use-memory` 指定，默认是 100MB。如果想减少 Prepare 阶段的耗时，可适当增大 `--use-memory` 的值。

Prepare 完成后，备份目录下会生成 redo log（ib_logfile0、ib_logfile1）和临时表空间文件（ibtmp1）。

Prepare 阶段完成后，下面进入恢复阶段。

```
# xtrabackup --defaults-file=/etc/my_3307.cnf --copy-back --parallel=10
--target-dir=/data/backup/full
```

命令中的 `--copy-back` 表示将备份目录中的文件拷贝到数据目录下。

在恢复阶段，需要注意以下几点。

- `--copy-back` 拷贝备份集。在空间不足的情况下，也可使用 `--move-back` 移动备份集。除此之外，也可通过操作系统命令（cp 或 mv）手动拷贝备份集。
- 目标目录（一般是数据目录）必须为空，否则会报错。如果目标目录不为空，可指定 `--force-non-empty-directories` 忽略检查。但即便如此，如果要拷贝的文件在目标目录中存在，还是会报错。
- 拷贝完成后，检查数据目录的权限。必要时，修改数据目录的属主和属组。

```
# chown -R mysql.mysql /data/mysql/3307/data
```

恢复完毕后，就可以直接启动数据库实例了。

```
# /usr/local/mysql/bin/mysqld_safe --defaults-file=/etc/my_3307.cnf &
```

2. 增量备份

在前面介绍 XtraBackup 的备份原理时，提到过 LSN。LSN（log sequence number）就是日志序列号，是一个不断递增的 8 字节无符号整数，不仅存在于 redo log 中，也存在于数据页中。在每个数据页的头部，都会记录该页最后刷新时的 LSN。xtrabackup 在做增量备份时，会扫描每个数据页，检查它的 LSN 是否大于上次备份完成时的 LSN（xtrabackup_checkpoints 中的 to_lsn）。只有大于的数据页才会备份。

下面我们看看增量备份的具体用法。

首先，创建一个全量备份。

```
# xtrabackup --user=backup_user --password=backup_pass --backup --target-dir=/data/backup/base
```

然后，查看备份集中 xtrabackup_checkpoints 文件的内容。

```
# cat /data/backup/base/xtrabackup_checkpoints
backup_type = full-backuped
from_lsn = 0
to_lsn = 489275445
last_lsn = 497117118
flushed_lsn = 496948331
```

接下来，进行第一次增量备份。

```
# xtrabackup --user=backup_user --password=backup_pass --backup --target-dir=/data/backup/inc1
--incremental-basedir=/data/backup/base
```

命令中的 `--target-dir` 是这次的备份目录，`--incremental-basedir` 是上一次备份的备份目录。

xtrabackup 在进行增量备份时，对于每个 InnoDB 表，它会生成两个文件：delta 文件（差异文件，二进制格式）和 meta 文件（描述文件，文本格式）。而对于非 InnoDB 表，每次都会进行全量备份。

再来看看增量备份中 xtrabackup_checkpoints 文件的内容。

```
# cat /data/backup/inc1/xtrabackup_checkpoints
backup_type = incremental
from_lsn = 489275445
to_lsn = 501449835
last_lsn = 508799238
flushed_lsn = 508626988
```

可以看到，文件中的 from_lsn 等于全量备份中的 to_lsn。

在进行增量备份时，指定 --incremental-basedir 是为了获取 to_lsn，所以我们也可以通过 --incremental-lsn 显式指定 lsn。

```
# xtrabackup --user=backup_user --password=backup_pass --backup --target-dir=/data/backup/inc1 --incremental-lsn=489275445
```

注意，如果 --incremental-lsn 指定错误，xtrabackup 是不会识别出来的。

下面进行第二次增量备份。

```
# xtrabackup --user=backup_user --password=backup_pass --backup --target-dir=/data/backup/inc2 --incremental-basedir=/data/backup/inc1
```

再次查看 xtrabackup_checkpoints 的内容。

```
# cat /data/backup/inc2/xtrabackup_checkpoints
backup_type = incremental
from_lsn = 501449835
to_lsn = 529519812
last_lsn = 536093448
flushed_lsn = 535942235
```

文件中的 from_lsn 等于第一次增量备份中的 to_lsn。

接下来我们看看增量备份的恢复过程。

首先，对全量备份进行 Prepare。

```
# xtrabackup --prepare --apply-log-only --target-dir=/data/backup/base
```

注意，这里必须指定 --apply-log-only。为什么要指定这个参数呢？

在前面我们提到过，在 Prepare 阶段，InnoDB 会对备份集进行 Crash Recovery。Crash Recovery 首先会应用 redo log，然后回滚未提交的事务。指定 --apply-log-only，则只会应用 redo log，而不会回滚当前备份未提交的事务。毕竟当前备份未提交的事务很有可能会在下个备份中提交。

接下来，对第一个增量备份集进行 Prepare，同样需指定 --apply-log-only。

```
# xtrabackup --prepare --apply-log-only --target-dir=/data/backup/base --incremental-dir=/data/backup/inc1
```

接着，对最后一个增量备份进行 Prepare，此时无须指定 --apply-log-only。当然，也可以指定，这样的话，未提交事务的回滚操作将在数据库启动时进行。

```
# xtrabackup --prepare --target-dir=/data/backup/base --incremental-dir=/data/backup/inc2
```

Prepare 阶段完成后，就是恢复阶段。

```
# xtrabackup --defaults-file=/etc/my_3307.cnf --copy-back --target-dir=/data/backup/base/
```

上面介绍的这种增量备份方式会扫描所有的数据页。如果发生修改的数据页本身就很少，这么做的效率无疑会比较低。从 MySQL 8.0.17 开始，随着克隆插件的引入，MySQL 开始支持页面跟踪（Pge Tracking）。页面跟踪可用来跟踪指定 LSN 之后发生修改的数据页。

从 XtraBackup 8.0.27 开始，支持使用页面跟踪来进行增量备份。接下来，我们看看具体的使用方法。

首先安装 component_mysqlbackup。

```
mysql> install component "file://component_mysqlbackup";
Query OK, 0 rows affected (0.02 sec)

# 验证 component_mysqlbackup 是否安装成功
mysql> select count(*) from mysql.component where component_urn='file://component_mysqlbackup';
+----------+
| count(*) |
+----------+
|        1 |
+----------+
1 row in set (0.00 sec)
```

安装完 component_mysqlbackup，接下来就可以使用页面跟踪来进行增量备份了。

(1) 全量备份。

```
# xtrabackup --user=backup_user --password=backup_pass --backup --target-dir=/data/backup/base --page-tracking
```

开启页面跟踪后，MySQL 会在数据目录下生成一个 #ib_archive 目录，用来保存发生了修改的页的数据，这样即使实例重启了，也能继续使用页面跟踪功能。

(2) 第一次增量备份。

```
# xtrabackup --user=backup_user --password=backup_pass --backup --target-dir=/data/backup/inc1 --incremental-basedir=/data/backup/base --page-tracking
```

(3) 第二次增量备份。

```
# xtrabackup --user=backup_user --password=backup_pass --backup --target-dir=/data/backup/inc2 --incremental-basedir=/data/backup/inc1 --page-tracking
```

注意，无论是全量备份还是增量备份都需要指定 `--page-tracking`。

恢复过程与传统的增量备份方式一样。

3. 压缩备份

压缩备份需通过 `--compress` 指定压缩算法，具体命令如下。

```
# xtrabackup --user=backup_user --password=backup_pass --backup --compress --target-dir=/data/backup/compress
```

在 XtraBackup 8.0 中，支持两种压缩算法：quicklz 和 lz4，默认是 quicklz。而在 XtraBackup 2.4

中只支持 quicklz。

对于压缩备份集，在 Prepare 之前，必须先解压。

```
# xtrabackup --decompress --target-dir=/data/backup/compress/
```

解压时需注意以下几点。

- 如果使用的是 quicklz 算法，在执行解压操作之前，必须首先安装 qpress。在 quicklz 网站上下载安装包后解压，会在当前目录下生成一个可执行文件 qpress，接着将其移至 /usr/bin 下。具体命令如下。

  ```
  # tar xvf qpress-11-linux-x64.tar
  # mv qpress /usr/bin/
  ```

- 可同时指定 `--parallel` 进行并行解压。
- 解压后，默认不会删除原来的压缩文件。倘若要删除，需指定 `--remove-original`。
- 即使压缩文件没有被删除，在 `--copy-back` 时，也不会拷贝这些压缩文件。

解压备份集后，就可以进行 Prepare 和恢复了。

```
# xtrabackup --prepare --target-dir=/data/backup/compress/
# xtrabackup --defaults-file=/etc/my_3307.cnf --copy-back --target-dir=/data/backup/compress/
```

注意，使用 `--compress` 进行压缩备份，就只能使用 quicklz 和 lz4 这两种算法。如果要使用其他压缩算法，必须结合流式备份。

4. 指定库/表备份

在 XtraBackup 中进行指定库/表的备份有多种方式。

(1) 使用 `--tables` 参数。

备份指定表，支持正则表达式。比如下面两条命令会分别备份 db1 下的所有表和 db1.t1 表：

```
# xtrabackup --user=backup_user --password=backup_pass --backup --target-dir=/data/backup/partial --tables='^db1[.].*'
# xtrabackup --user=backup_user --password=backup_pass --backup --target-dir=/data/backup/partial --tables='^db1[.]t1$ '
```

(2) 使用 `--tables-file` 参数。

将要备份的表写到文件中。看下面这个示例。

```
# cat /tmp/tables.txt
db1.t1
db2.t3
# xtrabackup --user=backup_user --password=backup_pass --backup --target-dir=/data/backup/partial --tables-file=/tmp/tables.txt
```

注意，表必须是 `database.table` 的形式，一张表只能写一行，且不支持正则匹配。

(3) 使用 `--tables-exclude` 参数。

同 `--tables` 类似，跳过指定表的备份，同样支持正则匹配，优先级高于 `--tables`。

(4) 使用 `--databases` 参数。

`--databases` 比 `--tables` 更加灵活，既支持库的备份，也支持表的备份。比如下面这条命令会同时备份 db1.t1 表和 db2 库。多个备份对象之间需以空格隔开，不支持正则匹配。例如：

```
# xtrabackup --user=backup_user --password=backup_pass --backup --target-dir=/data/backup/partial
--databases="db1.t1 db2"
```

(5) 使用 `--databases-exclude` 参数。

同 `--databases` 类似，跳过指定库表的备份，优先级高于 `--databases`。例如：

```
# xtrabackup --user=backup_user --password=backup_pass --backup --target-dir=/data/backup/partial
--databases-exclude="sys performance_schema db1.t2"
```

(6) 使用 `--databases-file` 参数。

同 `--tables-file` 参数类似，将需要备份的库和表写入文件。例如：

```
# cat /tmp/databases.txt
db1
db2.t3
```

下面以 `--databases` 参数为例，看看指定库/表备份（Partial Backup）的原理。

```
# xtrabackup --user=backup_user --password=backup_pass --backup --target-dir=/data/backup/partial
--databases="db1.t1"
```

对于指定库/表备份，注意以下两点。

- 基于库/表的备份，只会备份系统表空间（ibdata1）、undo 表空间及需要备份的表，不会备份 mysql（MySQL 8.0 中会备份 mysql.ibd）、performance_schema 和 sys 库。
- 在 XtraBackup 2.4 中，即便只是备份一张表，也会加全局读锁。如无必要，可加 `--no-lock` 参数禁用它。

接下来进行 Prepare 操作。

```
# xtrabackup --prepare --export --target-dir=/data/backup/partial
```

命令中指定了 `--export` 参数，它会将普通的表空间导出为可传输表空间。该参数不仅适用于指定库/表备份，也适用于全库备份。

下面我们看看 db1.t1 表的变化。

Prepare 之前：

```
# ll /data/backup/partial/db1/
total 45056
-rw-r----- 1 root root 46137344 May  7 08:25 t1.ibd
```

Prepare 之后：

```
# ll /data/backup/partial/db1/
total 45060
-rw-r--r-- 1 root root      653 May  7 08:25 t1.cfg
-rw-r----- 1 root root 46137344 May  7 08:25 t1.ibd
```

如果备份的是 MySQL 5.6 和 MySQL 5.7，Prepare 之后会有 4 个文件：t1.cfg、t1.exp、t1.frm、t1.ibd。

Prepare 结束后，在目标实例中导入该表。具体的操作步骤如下。

(1) 在目标实例中创建一张空表，表名和表结构与要导入的表一致。

```
mysql> create database db1_new;
Query OK, 1 row affected (0.01 sec)

mysql> create table db1_new.t1(id int auto_increment primary key,c1 varchar(10));
Query OK, 0 rows affected (0.03 sec)
```

(2) 丢弃（discard）表空间。

```
mysql> alter table db1_new.t1 discard tablespace;
Query OK, 0 rows affected (0.12 sec)
```

(3) 将导出的文件拷贝到目标实例的 db1_new 目录下，修改文件属主和属组。

如果备份的是 MySQL 5.6 和 MySQL 5.7，则需拷贝 t1.cfg、t1.exp 和 t1.ibd。

```
# cd /data/backup/partial/db1/
# cp {t1.cfg,t1.ibd} /data/mysql/3306/data/db1_new/
# chown -R mysql.mysql /data/mysql/3306/data/db1_new/t1.cfg
# chown -R mysql.mysql /data/mysql/3306/data/db1_new/t1.ibd
```

(4) 导入表空间。

```
mysql> alter table db1_new.t1 import tablespace;
Query OK, 0 rows affected (0.36 sec)

mysql> select count(*) from db1_new.t1;
+----------+
| count(*) |
+----------+
|  1000000 |
+----------+
1 row in set (0.05 sec)
```

最后通过一个查询简单地判断一下表是否导入成功。这里的恢复利用了 MySQL 5.6 引入的可传输表空间特性。

很多时候，我们有对单表或多表进行备份恢复的需求，可传输表空间是个不错的选择，毕竟物理拷贝的效率要比逻辑恢复高。但在线上又不适合直接使用可传输表空间来拷贝数据，因为直接使用的话，必须首先对目标表执行 FLUSH TABLES t1 FOR EXPORT 操作，然后进行文件拷贝。在操作的这段时间内，目标表是不可用的。对于在线业务而言，这往往是不可接受的。所以，对于这种需求，更推荐使用 XtraBackup。

接下来，我们看看指定库/表备份的另外一种常见使用场景——基于部分库/表的拆分。

基于部分库/表的拆分是线上的一个高频需求。一般来说，业务在上线初期，由于数据量比较小，多个库通常会放到一个实例中。但随着业务量的增长，如果这个实例的性能或磁盘空间达到了瓶颈，就需要进行拆分了。比较常见的数据拆分是将部分库/表拆分到其他实例中。

为了保证业务的停机时间尽可能短，通常的做法是基于部分库/表搭建一个新的从库，同时利用复制中的过滤规则,实时同步这些库表的数据。

基于部分库/表搭建一个新的从库,就用到了这里的指定库/表备份。

这里以拆分 db1 库为例。

```
# xtrabackup --user=backup_user --password=backup_pass --backup --target-dir=/data/backup/partial
--databases="db1 mysql performance_schema sys"
```

指定的 --databases,除了 db1,还需要加上 mysql、performance_schema、sys 这 3 个库。如果不加,后续使用时还要单独初始化实例。

接下来,进行 Prepare 和恢复。

```
# xtrabackup --prepare --target-dir=/data/backup/partial
# xtrabackup --defaults-file=/etc/my_3307.cnf --copy-back --target-dir=/data/backup/partial
```

不过在 MySQL 8.0 中,没有备份的其他表的元数据信息依旧存在于数据字典中,实例启动后,依然能看到这些表,但不能访问。

```
mysql> show databases;
+--------------------+
| Database           |
+--------------------+
| db1                |
| db2                |
| information_schema |
| mysql              |
| performance_schema |
| sys                |
+--------------------+
6 rows in set (0.00 sec)

mysql> select * from db2.t3;
ERROR 1812 (HY000): Tablespace is missing for table `db2`.`t3`.

mysql> drop database db2;
ERROR 3679 (HY000): Schema directory './db2/' does not exist
```

对于这种情况,建议手动删除这些库表。表可以直接删除,但如果删除库,则会提示对应的目录不存在。

```
mysql> drop database db2;
ERROR 3679 (HY000): Schema directory './db2/' does not exist
```

这个时候,需创建一个目录,然后再执行删除操作,具体命令如下。

```
# mkdir -p /data/mysql/3307/data/db2
# chown -R mysql.mysql /data/mysql/3307/data/db2

mysql> drop database db2;
Query OK, 2 rows affected (0.04 sec)
```

5. 流式备份

流式备份指的是将备份数据通过流的方式输出到 STDOUT,而不是备份文件中。结合管道,可将多个功能组合在一起,如压缩、加密、流量控制等。

5.3 XtraBackup

XtraBackup 2.4 中支持的流格式有两种：tar 和 xbstream，不过 tar 不支持并行备份。而在 XtraBackup 8.0 中不再支持 tar，只支持 xbstream。

首先，我们看一个最简单的流式备份。

```
# xtrabackup --user=backup_user --password=backup_pass --backup --stream=xbstream --parallel=10 > /data/backup/stream/backup.xbstream
```

将备份数据写到一个文件中。接下来，进行 Prepare 和恢复。在 Prepare 之前，需使用 xbstream 解压备份集。

```
# xbstream -x -p 10 -C /data/backup/stream/extract < /data/backup/stream/backup.xbstream
# xtrabackup --prepare --target-dir=/data/backup/stream/extract/
# xtrabackup --defaults-file=/etc/my_3307.cnf --copy-back --target-dir=/data/backup/stream/extract/
```

xbstream 中的 -x 代表解压，-p 指定并行度，-C 指定解压目录。

接下来我们看看流式备份的其他常见场景。

(1) 流式备份 + 压缩。

这里以 lz4 压缩算法为例。

```
# xtrabackup --backup --stream=xbstream --parallel=10 | lz4 > full_backup.lz4
# lz4 -d full_backup.lz4 | xbstream -x -p 10 -C /data/backup/stream/extract
```

Prepare 和恢复步骤与之前一样。

(2) 直接备份到远程服务器。

```
# xtrabackup --backup --stream=xbstream --parallel=10 | lz4 | ssh mysql@192.168.244.20 "cat - > /data/backup/stream/full_backup.lz4"
```

这一点也可通过 nc 命令来实现。

```
# ssh mysql@192.168.244.20 "( nc -l 3366 > /data/backup/stream/full_backup.lz4 & )" \
&& xtrabackup --backup --stream=xbstream --parallel=10 | lz4 | nc 192.168.244.20 3366
```

注意，这里没有指定远程服务器的密码，是因为配置了免密码登录。

如果没有配置免密码登录，可通过 sshpass 命令指定密码，例如：

```
# xtrabackup --backup --stream=xbstream --parallel=10 | lz4 | sshpass -p 123456 ssh -o StrictHostKeyChecking=no mysql@192.168.244.20 "cat - > /data/backup/stream/full_backup.lz4"
```

(3) 直接备份到远程服务器并解压。

```
# xtrabackup --backup --stream=xbstream --parallel=10 | lz4 | \
ssh mysql@192.168.244.20 'cat - | lz4 -d | xbstream -x -p 10 -C /data/backup/stream/extract'
```

(4) 使用 pv 命令限速。

直接备份到远程服务器时，如果担心备份会占用较大的网络带宽，可使用 pv 命令来限速。

```
# xtrabackup --backup --stream=xbstream --parallel=10 | lz4 | \
pv -q -L20m | ssh mysql@192.168.244.20 "cat - > /data/backup/stream/full_backup.lz4"
```

pv 命令中的 -q 指的是 quiet，不输出进度信息；-L 指的是传输速率；20m 指的是 20MB。

(5) 对备份集进行校验

```
# xtrabackup --backup --stream=xbstream --parallel=10 | lz4 | \
tee >(sha1sum > /tmp/source_checksum) | ssh mysql@192.168.244.20 \
"cat - | tee >(sha1sum > /tmp/destination_checksum) > /data/backup/stream/full_backup.lz4"
```

这里的 sha1sum 其实对整个备份集进行了校验。

```
[mysql@node1 ~]$ cat /tmp/source_checksum
04dd8fb61a9aea7cc2616861cf9b2aabf9f12db3  -

[mysql@node2 ~]$ cat /tmp/destination_checksum
04dd8fb61a9aea7cc2616861cf9b2aabf9f12db3  -

[mysql@node2 ~]$ sha1sum /data/backup/stream/full_backup.lz4
04dd8fb61a9aea7cc2616861cf9b2aabf9f12db3  /data/backup/stream/full_backup.lz4
```

对比源端和目标端的校验和，就可知道备份集在传输过程中是否出现了问题。这一点可用来简单地判断备份集的有效性。

6. 加密备份

加密功能是基于 libgcrypt 库实现的，目前支持 AES128、AES192 和 AES256 这 3 种加密算法。

首先，创建一个加密备份。在此之前，需生成一个密钥。

```
# 通过 openssl 生成密钥。具体命令如下。
# 该命令会生成一个 24 位随机字节、以 Base64 编码的字符串
# openssl rand -base64 24
ynSEarw5ueDY+FMy031jIKnTlCCl0u4i

# 将生成的密钥写入文件
# echo -n "ynSEarw5ueDY+FMy031jIKnTlCCl0u4i" > /data/backup/keyfile

# xtrabackup --user=backup_user --password=backup_pass --backup --target-dir=/data/backup/encrypt
--encrypt=AES256 --encrypt-key-file=/data/backup/keyfile
```

命令中的 --encrypt 指定加密算法，--encrypt-key-file 指定密钥所在的文件。除此之外，加密备份还可指定以下参数。

- --encrypt-key=ENCRYPTION_KEY：指定密钥。--encrypt-key 和 --encrypt-key-file 是互斥的，只能指定其中一个。
- --encrypt-threads：加密线程数。该参数同样适用于解密。
- --encrypt-chunk-size：指定每个加密线程工作缓冲区的大小，单位字节。

加密备份集里面的文件会以 xbcrypt 结尾。

在 Prepare 之前，需使用 xtrabackup 解压备份集。

```
# xtrabackup --decrypt=AES256 --target-dir=/data/backup/encrypt --encrypt-key-file=/data/backup/keyfile
```

解密后，默认不会删除原来的加密文件，如果要删除，需指定 --remove-original。

最后进行 Prepare 和恢复。

```
# xtrabackup --prepare --target-dir=/data/backup/encrypt
# xtrabackup --defaults-file=/etc/my_3307.cnf --copy-back --target-dir=/data/backup/encrypt
```

5.3.4 Xtrabackup 的重要参数

- **--ftwrl-wait-timeout、--ftwrl-wait-threshold、--ftwrl-wait-query-type**

这 3 个参数与全局读锁有关。前两个参数在 5.3.2 节提过，简单来说，就是在加全局读锁之前，如果在指定的时间内（--ftwrl-wait-timeout）依然有执行时长超过 --ftwrl-wait-threshold（默认为 60 秒）的操作，xtrabackup 会报错并退出。

```
# xtrabackup --user=backup_user --password=backup_pass --backup --target-dir=/data/backup/full
--ftwrl-wait-timeout=5
...
220508 15:42:36 Waiting 5 seconds for queries running longer than 60 seconds to finish
220508 15:42:36 Waiting for query 523 (duration 64 sec): select id,sleep(65) from db1.t1 where id=1
220508 15:42:37 >> log scanned up to (2033178377)
220508 15:42:37 Waiting for query 523 (duration 65 sec): select id,sleep(65) from db1.t1 where id=1
220508 15:42:38 >> log scanned up to (2033178377)
220508 15:42:38 Waiting for query 626 (duration 62 sec): select sleep(65) from db1.t2 where id=1
220508 15:42:39 >> log scanned up to (2033178377)
220508 15:42:39 Waiting for query 626 (duration 63 sec): select sleep(65) from db1.t2 where id=1
220508 15:42:40 >> log scanned up to (2033178377)
220508 15:42:40 Waiting for query 626 (duration 64 sec): select sleep(65) from db1.t2 where id=1
220508 15:42:41 >> log scanned up to (2033178377)
220508 15:42:41 Waiting for query 626 (duration 65 sec): select sleep(65) from db1.t2 where id=1
220508 15:42:42 >> log scanned up to (2033178377)
220508 15:42:42 Unable to obtain lock. Please try again later.
```

--ftwrl-wait-timeout 默认为 0，不会判断实例当前是否存在长时间执行的操作，而是会直接加全局读锁。

--ftwrl-wait-query-type 是等待的操作类型，可设置为 ALL（默认值）、UPDATE 或 SELECT。若设置为 ALL，等待的操作类型有 INSERT、UPDATE、DELETE、REPLACE、ALTER、LOAD、SELECT、DO、HANDLER、CALL、EXECUTE、BEGIN。若设置为 UPDATE 或 SELECT，则等待的操作类型包括：INSERT、UPDATE、DELETE、REPLACE、ALTER、LOAD。

- **--kill-long-queries-timeout、--kill-long-query-type**

在执行 FTWRL 操作后，如果有阻塞全局读锁的操作，正常情况下会等待这些操作结束。如果设置了 --kill-long-queries-timeout，则会在等待 --kill-long-queries-timeout 秒后，杀掉这些阻塞操作。可被杀掉的操作的类型由 --kill-long-query-type 指定，可设置为 ALL、UPDATE 或 SELECT（默认值）。若设置为 ALL，对应的操作类型同 --ftwrl-wait-query-type 中的一样。若设置为 UPDATE 或 SELECT，对应的操作类型就只有 SELECT。--kill-long-queries-timeout 默认为 0，不会杀掉任何操作。

- **--no-lock**

在 XtraBackup 2.4 中，设置该参数则不执行 FTWRL 操作。在 XtraBackup 8.0 中，设置该参数则不执行 LOCK INSTANCE FOR BACKUP 和 FTWRL 操作。

- **--lock-ddl、--lock-ddl-per-table**

从 MySQL 5.7 开始，为了提高 DDL 的执行效率，对于部分 DDL 操作，在执行的过程中会禁用 redo log，而这会导致 XtraBackup 在 Prepare 的过程中出现问题。针对这个问题，一方面，XtraBackup

在备份的过程中，如果碰到 redo log 被禁用了，会直接报错。错误信息如下。

```
InnoDB: An optimized (without redo logging) DDL operation has been performed. All modified pages may
not have been flushed to the disk yet.
PXB will not be able to make a consistent backup. Retry the backup operation
```

另一方面，引入了 `--lock-ddl` 和 `--lock-ddl-per-table` 来阻塞备份过程中的 DDL。

在 MySQL 8.0 或者 Percona Server 5.7 中，可指定 `--lock-ddl`，它会在备份开始时添加备份锁，备份锁会阻塞备份过程中的 DDL。

在 MySQL 5.7 中，因为没有引入备份锁，如果要阻塞 DDL，只能设置 `--lock-ddl-per-table`。在备份开始时，XtraBackup 会单独启动一个线程，以 `SELECT 1 FROM db1.t1 LIMIT 0` 的方式获取表的元数据锁。但这种实现方式也有问题：第一，备份过程中的 DDL 会因为元数据锁而阻塞，从而导致后续针对该表的所有操作都会被阻塞；第二，指定了 `--lock-ddl-per-table`，就不会执行 FTWRL，进而无法保证非 InnoDB 表的备份一致性。

`--lock-ddl` 在 XtraBackup 8.0 中是默认开启的。

- `--throttle`

备份，是一个 I/O 密集型操作。如果担心备份占用太多的 I/O 资源，可通过 `--throttle` 限制 XtraBackup 的 I/O 吞吐量。XtraBackup 是以 chunk 为单位来拷贝文件的，每个 chunk 的大小是 10MB。`--throttle` 指的是 XtraBackup 每秒能拷贝的 chunk 的数量。

- `--history[=name]`

若指定了 `--history`，则会将备份信息记录在 PERCONA_SCHEMA.xtrabackup_history 中，记录的内容与 xtrabackup_info 中的一致。

如果我们将备份信息保存在表中，那么在进行增量备份时，无须指定 `--incremental-basedir` 或 `--incremental-lsn`，直接指定 `--incremental-history-name` 或 `--incremental-history-uuid` 即可，xtrabackup 会将对应 name 或 uuid 的最近一次备份信息的 innodb_to_lsn 作为下次增量备份的起点。

5.3.5 XtraBackup 的注意事项

(1) 如果待备份实例是 MySQL 5.7，则不要在备份过程中执行 DDL，否则很容易导致备份失败。

虽然可以通过 `--lock-ddl-per-table` 的方式来提前获取元数据锁，但不推荐这样做。

导致备份失败的基本上都是使用 INPLACE 算法的 DDL 操作（不包括只修改表的元数据信息的 DDL 操作），所以，在执行 DDL 时，我们可以将 DDL 的算法设置为 COPY，这样就不会出现 redo log 被禁用的情况。不过使用 COPY 算法的 DDL 会阻塞 DML。所以，这种方法不适用于 Online DDL，只适用于 pt-online-schema-change 和 gh-ost，因为这两个工具是对临时表（也是空表）进行 DDL，不会阻塞业务 SQL。使用时只需将 DDL 的算法设置为 COPY，例如：

```
--alter "add c2 varchar(10),algorithm=copy"
```

(2) XtraBackup 打开的文件描述符数过多。

在备份的过程中，如果出现以下错误信息，就意味着 XtraBackup 可打开的文件描述符数达到了上限。

```
InnoDB: Operating system error number 24 in a file operation.
InnoDB: Error number 24 means 'Too many open files'
InnoDB: Some operating system error numbers are described at
http://dev.mysql.com/doc/refman/5.7/en/operating-system-error-codes.html
InnoDB: File ./slowtech/t67.ibd: 'open' returned OS error 124. Cannot continue operation
InnoDB: Cannot continue operation.
```

XtraBackup 在备份的过程中会打开所有的 ibd 文件，之所以这么做，是为了避免备份过程中，由 `RENAME TABLE` 操作导致 ibd 文件拷贝错了。

解决方法是调整进程可打开的最大文件描述符数。具体步骤如下。

```
[mysql@slowtech ~]$ ulimit -Sn
65535
[mysql@slowtech ~]$ ulimit -Hn
100001
[mysql@slowtech ~]$ ulimit -Sn 80000
```

`ulimit -Sn` 查看的是软限制，`ulimit -Hn` 查看的是硬限制。如果软限制小于硬限制，我们可以直接通过 `ulimit -Sn xxx` 调整软限制的大小。设置多大比较合适呢？这里给个式子供参考：备份实例的 ibd 文件数 + 1000，式子中的备份实例的 ibd 文件数可根据以下命令获取。

```
# find /var/lib/mysql/ -name "*.ibd" | wc -l
```

命令中的 /var/lib/mysql/ 是备份实例的数据目录。

注意，上面这种方式只是临时调整，只对当前会话有效。如果要使调整永久有效，就只能修改 /etc/security/limits.conf，调整执行 xtrabackup 的系统用户的最大文件描述符数。

```
# vim /etc/security/limits.conf
mysql soft nofile 80000
mysql hard nofile 80000
```

5.3.6 备份用户需要的权限

备份用户需要以下权限。

- RELOAD：用于执行 `FLUSH TABLES WITH READ LOCK` 和 `FLUSH NO_WRITE_TO_BINLOG TABLES`，是必需权限。
- REPLICATION CLIENT：用于执行 `SHOW MASTER STATUS`，是必需权限。如果是对从库进行备份，还需要执行 `SHOW SLAVE STATUS`。
- BACKUP_ADMIN：用于执行 `LOCK INSTANCE FOR BACKUP`，是必需权限。
- PROCESS：用于执行 `SHOW ENGINE INNODB STATUS`、`SHOW PROCESSLIST`，是必需权限。
- SYSTEM_VARIABLES_ADMIN：非必需权限，用于在增量备份时执行 `SET GLOBAL mysqlbackup.backupid=xxx` 操作。

- SUPER：非必需权限。在以下两种场景下需要 SUPER 权限。
  - 指定了 --kill-long-queries-timeout，需要杀掉慢查询。
  - 从库备份指定了 --safe-slave-backup，需要重启复制。
- SHOW VIEW：在 XtraBackup 8.0 中，会通过下面这个 SQL 语句来判断备份实例中是否存在 MyISAM 表。如果存在，则加全局读锁。对于一个普通用户，查询 information_schema.tables，只能看到 information_schema 库中的表信息。如果要查看其他表，需要具有该表表级别的权限，常用的是 SELECT 权限，但 SELECT 权限能读取表中的数据，不满足合规要求，所以就选择了 SHOW VIEW 这个既能满足需求，又无实际影响的表级别权限。

  ```sql
  SELECT COUNT(*) FROM information_schema.tables WHERE engine = 'MyISAM' OR engine = 'RocksDB'
  ```

- 如果使用了 Page Tracking 来进行增量备份，还需要 mysql.component 的查询权限。
- 如果指定了 --history，还需要 PERCONA_SCHEMA.xtrabackup_history 的 SELECT、INSERT、CREATE、ALTER 权限。

具体的授权语句如下。

```sql
create user 'backup_user'@'localhost' identified by 'backup_pass';
grant reload, process, show databases, replication client, show view on *.* to 'backup_user'@'localhost';
grant backup_admin,system_variables_admin on *.* to 'backup_user'@'localhost';
grant select, insert, create, alter on PERCONA_SCHEMA.* to 'backup_user'@'localhost';
grant select on mysql.component to 'backup_user'@'localhost';
grant select on performance_schema.keyring_component_status to 'backup_user'@'localhost';
grant select on performance_schema.log_status to 'backup_user'@'localhost';
```

5.3.7 参考资料

- 数据库内核月报"MySQL·物理备份·Percona XtraBackup 备份原理"，作者：阿里云 PolarDB-数据库内核组。
- Percona 公司文档"Percona XtraBackup 8.0 Documentation"。
- 博客文章"Redesign of --lock-ddl-per-table in Percona XtraBackup"，作者：Marcelo Altmann。
- 博客文章"Avoiding the "An optimized (without redo logging) DDL operation has been performed" Error with Percona XtraBackup"，作者：Shahriyar Rzayev。

5.4 克隆插件

克隆插件（clone plugin）是 MySQL 8.0.17 引入的一个重大特性。为什么要实现这个特性呢？个人认为，主要还是为组复制服务。在组复制中，如果要添加一个新的节点，这个节点差异数据的补齐是通过分布式恢复（distributed recovery）来实现的。在 MySQL 8.0.17 之前，只支持一种恢复方式，即 binlog。如果新节点需要的 binlog 在集群中不存在，就只能先借助备份工具进行全量数据的备份恢复，再通过分布式恢复同步增量数据。这种方式虽然也能达到添加节点的目的，但总归还是要借助外部工具，相对来说有一定的使用门槛和工作量。

有了克隆插件，只用一条命令就能很方便地添加一个新的节点，无论是在组复制还是普通的主从环境中都是如此。

本节主要包括以下内容。

- 克隆插件的安装。
- 克隆插件的使用。
- 查看克隆操作的进度。
- 基于克隆数据搭建从库。
- 克隆插件的实现细节。
- 克隆插件的限制。
- 克隆插件与 XtraBackup 的异同。
- 克隆插件的参数解析。

5.4.1 克隆插件的安装

克隆插件可通过以下两种方式安装。

(1) 在配置文件中指定。

```
[mysqld]
plugin-load-add=mysql_clone.so
```

(2) 动态加载。

```
mysql> install plugin clone soname 'mysql_clone.so';
```

执行 SHOW PLUGINS 查看插件是否安装成功。

```
mysql> show plugins;
+------------------+----------+------+-----------------+---------+
| Name             | Status   | Type | Library         | License |
+------------------+----------+------+-----------------+---------+
...
| clone            | ACTIVE   | CLONE | mysql_clone.so | GPL     |
...
```

Status 为 ACTIVE 代表插件加载成功。

5.4.2 克隆插件的使用

克隆插件支持两种克隆方式：本地克隆和远程克隆。接下来我们分别看看它们的使用方法。

1. 本地克隆

本地克隆的原理如图 5-2 所示，它可将本地 MySQL 实例中的数据拷贝到本地服务器的一个目录中。本地克隆只能在实例本地发起。

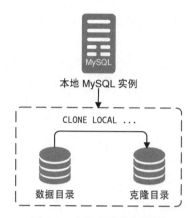

图 5-2　本地克隆的原理图

本地克隆命令的语法如下。

CLONE LOCAL DATA DIRECTORY [=] 'clone_dir';

这里的 clone_dir 是克隆目录。

下面看一个具体的示例。

(1) 创建克隆用户。

```
mysql> create user 'clone_user'@'%' identified by 'clone_pass';
mysql> grant backup_admin on *.* to 'clone_user'@'%';
```

这里的 BACKUP_ADMIN 是克隆操作必需的权限，它允许用户执行 LOCK INSTANCE FOR BACKUP 命令。

(2) 创建克隆目录。

```
# mkdir -p /data/backup
# chown -R mysql.mysql /data/backup/
```

(3) 执行本地克隆操作。

```
# mysql -uclone_user -pclone_pass
mysql> clone local data directory='/data/backup/3307';
```

这里的 /data/backup/3307 是克隆目录，它需要满足以下 3 个要求。

- 克隆目录必须是绝对路径。
- /data/backup 必须存在，且 MySQL 对其有写权限。
- 最后一级目录 3307 不能存在。

(4) 查看克隆目录的内容。

```
# ll /data/backup/3307
total 200644
drwxr-x--- 2 mysql mysql       89 Oct 23 22:09 #clone
-rw-r----- 1 mysql mysql     4049 Oct 23 22:09 ib_buffer_pool
```

```
-rw-r----- 1 mysql mysql 12582912 Oct 23 22:09 ibdata1
-rw-r----- 1 mysql mysql 50331648 Oct 23 22:09 ib_logfile0
-rw-r----- 1 mysql mysql 50331648 Oct 23 22:09 ib_logfile1
drwxr-x--- 2 mysql mysql        6 Oct 23 22:09 mysql
-rw-r----- 1 mysql mysql 25165824 Oct 23 22:09 mysql.ibd
drwxr-x--- 2 mysql mysql       20 Oct 23 22:09 slowtech
drwxr-x--- 2 mysql mysql       28 Oct 23 22:09 sys
-rw-r----- 1 mysql mysql 33554432 Oct 23 22:09 undo_001
-rw-r----- 1 mysql mysql 16777216 Oct 23 22:09 undo_002
```

(5) 可直接基于备份集启动实例。

```
# /usr/local/mysql/bin/mysqld --no-defaults --datadir=/data/backup/3307 --user mysql --port 3307 &
```

相较于 Xtrabackup，克隆插件无须 Prepare 阶段。

2. 远程克隆

远程克隆的原理如图 5-3 所示，涉及两个实例。被克隆的实例是 Donor，接受克隆数据的实例是 Recipient。克隆命令需在 Recipient 上发起。

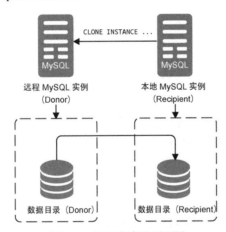

图 5-3　远程克隆的原理图

远程克隆命令的语法如下。

```
CLONE INSTANCE FROM 'user'@'host':port
IDENTIFIED BY 'password'
[DATA DIRECTORY [=] 'clone_dir']
[REQUIRE [NO] SSL];
```

其中各项解释如下。

- `host` 和 `port`：被克隆实例（Donor）的 IP 和端口。
- `user` 和 `password`：Donor 上的克隆用户和密码，需要 BACKUP_ADMIN 权限。
- `DATA DIRECTORY`：备份目录。不指定的话，则默认克隆到 Recipient 的数据目录下。
- `REQUIRE [NO] SSL`：是否开启 SSL 通信。

下面看一个具体的示例。

(1) 在 Donor 实例上创建克隆用户，加载克隆插件。

```
mysql> create user 'donor_user'@'%' identified by 'donor_pass';
mysql> grant backup_admin on *.* to 'donor_user'@'%';
mysql> install plugin clone soname 'mysql_clone.so';
```

(2) 在 Recipient 实例上创建克隆用户，加载克隆插件。

```
mysql> create user 'recipient_user'@'%' identified by 'recipient_pass';
mysql> grant clone_admin on *.* to 'recipient_user'@'%';
mysql> install plugin clone soname 'mysql_clone.so';
```

这里的 clone_admin 隐式含有 BACKUP_ADMIN 和 SHUTDOWN（重启实例）权限。

(3) 在 Recipient 实例上设置 Donor 白名单，Recipient 只能克隆白名单中的实例。

```
mysql> set global clone_valid_donor_list = '192.168.244.10:3306';
```

设置该参数需要 SYSTEM_VARIABLES_ADMIN 权限。

(4) 在 Recipient 上发起克隆命令。

```
# mysql -urecipient_user -precipient_pass
mysql> clone instance from 'donor_user'@'192.168.244.10':3306 identified by 'donor_pass';
```

远程克隆会依次执行以下操作。

- 获取备份锁（backup lock）。备份锁和 DDL 互斥。注意，获取的不仅是 Recipient 上的备份锁，Donor 上的也要获取。
- 用 DROP 操作删除用户表空间。注意，删除的只是用户表空间，不是数据目录，也不包括 ib_buffer_pool、ibdata 等系统文件。
- 从 Donor 实例拷贝数据。对于用户表空间，会直接拷贝。对于系统文件，则会重命名为 *xxx*.#clone，而不会直接替代原文件。例如：

```
# ll /data/mysql/3306/data/
...
-rw-r----- 1 mysql mysql     4049 Oct 24 09:11 ib_buffer_pool
-rw-r----- 1 mysql mysql     4049 Oct 24 10:54 ib_buffer_pool.#clone
-rw-r----- 1 mysql mysql 12582912 Oct 24 10:55 ibdata1
-rw-r----- 1 mysql mysql 12582912 Oct 24 10:55 ibdata1.#clone
...
-rw-r----- 1 mysql mysql 25165824 Oct 24 10:55 mysql.ibd
-rw-r----- 1 mysql mysql        0 Oct 24 10:54 mysql.ibd.#clone
...
```

- 重启实例。在启动的过程中，会用 *xxx*.#clone 替换原来的系统文件。

5.4.3 查看克隆操作的进度

查看克隆操作的进度，主要依托于 performance_schema 中的两张表：`clone_status` 和 `clone_progress`。下面我们看看这两张表的具体作用及各字段的含义。

首先看看 `clone_status` 表，该表记录了克隆操作的状态信息。

```
mysql> select * from performance_schema.clone_status\G
*************************** 1. row ***************************
             ID: 1
            PID: 0
          STATE: Completed
     BEGIN_TIME: 2021-10-24 10:54:35.565
       END_TIME: 2021-10-24 10:57:02.382
         SOURCE: 192.168.244.10:3306
    DESTINATION: LOCAL INSTANCE
       ERROR_NO: 0
  ERROR_MESSAGE:
    BINLOG_FILE: mysql-bin.000004
BINLOG_POSITION: 139952824
   GTID_EXECUTED: 453a5124-020e-11ec-8719-000c29f66609:1-17674
1 row in set (0.49 sec)
```

其中各字段的含义如下。

- PID：Processlist ID，对应 SHOW PROCESSLIST 中的 ID。如果要终止当前的克隆操作，可执行 KILL QUERY processlist_id。
- STATE：克隆操作的状态，包括 Not Started（尚未开始）、In Progress（进行中）、Completed（成功）、Failed（失败）。如果是 Failed 状态，ERROR_MESSAGE 会给出具体的错误信息。
- BEGIN_TIME 和 END_TIME：克隆操作的开始时间和结束时间。
- SOURCE：Donor 实例的地址。
- DESTINATION：克隆目录。LOCAL INSTANCE 代表当前实例的数据目录。
- GTID_EXECUTED 和 BINLOG_FILE（BINLOG_POSITION）：克隆操作对应的一致性位置点信息。可利用这些信息搭建从库。

接下来看看 clone_progress 表，该表记录了克隆操作的进度信息。

```
mysql> select * from performance_schema.clone_progress;
+----+-----------+-----------+----------------------------+----------------------------+---------+----------+----------+-----------+------------+---------------+
| ID | STAGE     | STATE     | BEGIN_TIME                 | END_TIME                   | THREADS | ESTIMATE | DATA     | NETWORK   | DATA_SPEED | NETWORK_SPEED |
+----+-----------+-----------+----------------------------+----------------------------+---------+----------+----------+-----------+------------+---------------+
|  1 | DROP DATA | Completed | 2021-10-24 10:54:48.395548 | 2021-10-24 10:54:58.352278 |       1 |        0 |        0 |         0 |          0 |             0 |
|  1 | FILE COPY | Completed | 2021-10-24 10:54:58.352527 | 2021-10-24 10:55:35.908919 |       2 | 616681425| 616681425| 616722587 |          0 |             0 |
|  1 | PAGE COPY | Completed | 2021-10-24 10:55:35.958036 | 2021-10-24 10:55:36.670272 |       2 |  7077888 |  7077888 |   7102277 |          0 |             0 |
|  1 | REDO COPY | Completed | 2021-10-24 10:55:36.671544 | 2021-10-24 10:55:37.160154 |       2 |  4372992 |  4372992 |   4373841 |          0 |             0 |
|  1 | FILE SYNC | Completed | 2021-10-24 10:55:37.160412 | 2021-10-24 10:55:39.724808 |       2 |        0 |        0 |         0 |          0 |             0 |
|  1 | RESTART   | Completed | 2021-10-24 10:55:39.724808 | 2021-10-24 10:56:55.207049 |       0 |        0 |        0 |         0 |          0 |             0 |
|  1 | RECOVERY  | Completed | 2021-10-24 10:56:55.207049 | 2021-10-24 10:57:02.382057 |       0 |        0 |        0 |         0 |          0 |             0 |
+----+-----------+-----------+----------------------------+----------------------------+---------+----------+----------+-----------+------------+---------------+
7 rows in set (0.00 sec)
```

其中各字段的含义如下。

- STAGE：一个克隆操作需要经过 DROP DATA、FILE COPY、PAGE COPY、REDO COPY、FILE SYNC、RESTART、RECOVERY 这 7 个阶段。当前阶段结束了才会开始下一个阶段。
- STATE：当前阶段的状态，包括 Not Started、In Progress、Completed。
- BEGIN_TIME 和 END_TIME：当前阶段的开始时间和结束时间。
- THREADS：当前阶段使用的并发线程数。

- ESTIMATE：预估的数据量。
- DATA：已经拷贝的数据量。
- NETWORK：通过网络传输的数据量。如果是本地克隆，该列的值为 0。
- DATA_SPEED 和 NETWORK_SPEED：当前数据拷贝速率和网络传输速率。注意，二者是当前值。

5.4.4 基于克隆数据搭建从库

这里基于两种场景（GTID 复制和基于位置点的复制）进行分析。

1. GTID 复制

对于 GTID 复制，无须关心具体的位置点信息，直接设置 MASTER_AUTO_POSITION = 1 即可。具体命令如下。

```
mysql> change master to master_host = 'master_host_name', master_port = master_port_num,
    ...
    master_auto_position = 1;
mysql> start slave;
```

通过 XtraBackup 搭建从库，在建立复制之前，必须执行 SET GLOBAL GTID_PURGED 操作。在克隆插件中则无须执行该操作。这是因为通过克隆数据启动的实例，GTID_PURGED 和 GTID_EXECUTED 默认已初始正确。

2. 基于位置点的复制

基于位置点的复制，需要的位置点信息可从 performance_schema.clone_status 中获取。具体命令如下。

```
mysql> select binlog_file, binlog_position from performance_schema.clone_status;
mysql> change master to master_host = 'master_host_name', master_port = master_port_num,
    ...
    master_log_file = 'master_log_name',
    master_log_pos = master_log_pos_num;
mysql> start slave;
```

其中，master_host_name 和 master_port_num 是 Donor 实例的 IP 和端口。master_log_name 和 master_log_pos_num 分别取自 performance_schema.clone_status 中的 binlog_file 和 binlog_position。

5.4.5 克隆插件的实现细节

克隆操作可细分为以下 5 个阶段：

[INIT] ---> [FILE COPY] ---> [PAGE COPY] ---> [REDO COPY] -> [Done]

下面我们看看各个阶段的具体作用。

- INIT：初始化一个克隆对象。
- FILE COPY：拷贝数据文件。

在拷贝之前，会将当前检查点的 LSN 记为 CLONE START LSN，同时启动 Page Tracking。

Page Tracking 会跟踪 CLONE START LSN 之后发生修改的页，记录这些页的 tablespace ID 和 page ID。数据文件拷贝结束后，会将当前检查点的 LSN 记为 CLONE FILE END LSN。

- PAGE COPY：拷贝 Page Tracking 中记录的页。在拷贝之前，会基于 tablespace ID 和 page ID 对这些页进行排序，以避免拷贝过程中的随机读写。同时，开启 Redo Archiving。

  Redo Archiving 会在后台开启一个归档线程，将 redo log 的内容按 chunk 拷贝到归档文件中。通常来说，归档线程的拷贝速度会快于 redo log 的生成速度。但即使慢于，在写入新的 redo log 时，也会等待归档线程完成拷贝，不会覆盖还未拷贝的 redo log。

  Page Tracking 中的页拷贝完毕后，会获取实例的一致性位置点信息，同时将此时的 LSN 记为 CLONE LSN。

- REDO COPY：拷贝归档文件中 CLONE FILE END LSN 与 CLONE LSN 之间的 redo log。
- Done：调用 `snapshot_end()` 销毁克隆对象。

5.4.6 克隆插件的限制

在使用克隆插件时，需注意以下限制。

- 克隆期间，会阻塞 DDL。同样，DDL 也会阻塞克隆命令的执行。不过从 MySQL 8.0.27 开始，克隆命令就不会阻塞 Donor 上的 DDL 了。
- 克隆插件不会拷贝 Donor 的配置参数。
- 克隆插件不会拷贝 Donor 的 binlog。
- 克隆插件只会拷贝 InnoDB 表的数据，对于其他存储引擎的表，只会拷贝表结构。
- Donor 实例中如果有表通过 `DATA DIRECTORY` 子句设置了绝对路径，在进行本地克隆时，会提示文件已存在。在进行远程克隆时，绝对路径必须存在且有写权限。
- 不允许通过 MySQL Router 连接 Donor 实例。
- 执行 `CLONE INSTANCE` 操作时，指定的 Donor 端口不能为 X Protocol 端口。

除此之外，在进行远程克隆时，还有如下限制。

- MySQL 版本必须一致。
- 主机的操作系统和位数（32 位/64 位）必须一致。两者可基于参数 `version_compile_os` 和 `version_compile_machine` 获取。
- Recipient 必须有足够的磁盘空间存储克隆数据。
- 字符集相关参数必须一致，具体包括：`character_set_server`（字符集）、`collation_server`（校验集）和 `character_set_filesystem`。
- 参数 `innodb_page_size` 必须一致。
- 无论是 Donor 还是 Recipient，同一时间只能执行一个克隆操作。
- Recipient 需要重启，所以它必须通过 mysqld_safe 或 systemd 等进程管理工具进行管理，否则在实例关闭后，需手动拉起。

5.4.7 克隆插件与 XtraBackup 的异同

下面看看克隆插件和 XtraBackup 的异同点。

- 在实现上，两者都有 FILE COPY 和 REDO COPY 阶段，但克隆插件比 XtraBackup 多了一个 PAGE COPY 阶段。由此，克隆插件的恢复速度比 XtraBackup 快。
- XtraBackup 没有 Redo Archiving 特性，有可能出现未拷贝的 redo log 被覆盖的情况。
- 在 GTID 下建立复制，克隆插件无须额外执行 SET GLOBAL GTID_PURGED 操作。

5.4.8 克隆插件的参数解析

- **clone_autotune_concurrency**

是否自动调节克隆过程中并发线程数的数量，默认为 ON，此时，最大线程数由参数 clone_max_concurrency 控制。若设置为 OFF，并发线程数将是固定的，等于 clone_max_concurrency。后者的默认值为 16。

- **clone_block_ddl**

克隆过程中是否对 Donor 实例加备份锁。如果加了，则会阻塞 DDL。默认为 OFF，即不加。该参数是 MySQL 8.0.27 引入的。

- **clone_delay_after_data_drop**

通过 DROP 操作删除完用户表空间，等待多久才执行数据拷贝操作。引入该参数的初衷是某些文件系统（如 VxFS）是异步释放空间的，如果删除用户表空间后，马上就执行数据拷贝操作，有可能会因为空间不足而导致克隆失败。该参数是 MySQL 8.0.29 引入的，默认为 0，表示不等待。

- **clone_buffer_size**

本地克隆时，中转缓冲区的大小，默认为 4MB。缓冲区越大，备份速度越快，相应地，对磁盘 I/O 的压力也越大。

- **clone_ddl_timeout**

克隆操作需要获取备份锁。在执行 CLONE 命令时，如果有 DDL 正在执行，则 CLONE 命令会被阻塞，等待获取备份锁（Waiting for backup lock）。等待的最大时长由 Recipient 实例上的 clone_ddl_timeout 决定，该参数默认为 300 秒。如果在这个时间内还没获取到备份锁，CLONE 命令会失败，并会提示 ERROR 3862 (HY000): Clone Donor Error: 1205 : Lock wait timeout exceeded; try restarting transaction。

需要注意的是，在执行 DDL 时，如果有 CLONE 命令在执行，DDL 也会因为备份锁而被阻塞，只不过，DDL 操作的等待时长由 lock_wait_timeout 决定，该参数默认为 31 536 000 秒，即 365 天。

- **clone_donor_timeout_after_network_failure**

在远程克隆时，如果出现了网络错误，克隆操作不会马上终止，而是会等待一段时间。如果在这个时间内，网络恢复了，操作会继续进行。在 MySQL 8.0.24 之前，等待时间是固定的 5 分钟。从 MySQL

8.0.24 开始，可通过 `clone_donor_timeout_after_network_failure` 设置这个时间，默认为 5 分钟。

- **`clone_enable_compression`**

远程克隆传输数据时，是否开启压缩。默认为 `OFF`。开启压缩能节省网络带宽，但会相应地增加 CPU 消耗。

- **`clone_max_data_bandwidth`**

在远程克隆时，可允许的最大数据拷贝速率，单位是 MiB/s。默认为 `0`，即不限制。如果 Donor 的磁盘 I/O 存在瓶颈，可通过该参数来限速。

注意，这里限制的只是单个线程的拷贝速率。如果是多个线程并行拷贝，实际最大拷贝速率等于 `clone_max_data_bandwidth` 乘以线程数。

- **`clone_max_network_bandwidth`**

在远程克隆时，可允许的最大网络传输速率，单位是 MiB/s。默认为 `0`，即不限制。如果网络带宽存在瓶颈，可通过该参数来限速。

- **`clone_valid_donor_list`**

在 Recipient 上设置 Donor 白名单，这样 Recipient 就只能克隆白名单中指定的实例。在执行克隆操作之前，必须设置该参数。

- **`clone_ssl_ca`、`clone_ssl_cert`、`clone_ssl_key`**

SSL 相关。

5.4.9 参考资料

- 官方工作日志 "InnoDB: Clone local replica"。
- 官方工作日志 "InnoDB: Clone remote replica"。
- 官方工作日志 "InnoDB: Clone Replication Coordinates"。
- 官方工作日志 "InnoDB: Clone Remote provisioning"。
- 知乎专栏 "MySQL/InnoDB 数据克隆插件（clone plugin）实现剖析"，作者：温正湖。

5.5　MySQL Shell Dump & Load

MySQL Shell 是 MySQL 的一个高级客户端和代码编辑器，是第二代 MySQL 客户端。第一代 MySQL 客户端即我们常用的 mysql。相比于 mysql，MySQL Shell 不仅支持 SQL，还具有以下关键特性。

- 支持 Python 和 JavaScript 两种语言模式。基于此，我们可以很容易地做一些脚本开发工作。
- 支持 AdminAPI。AdminAPI 可用来管理 InnoDB Cluster、InnoDB ClusterSet 和 InnoDB ReplicaSet。
- 支持 X DevAPI。X DevAPI 可对文档和表进行 CRUD 操作。

除此之外，MySQL Shell 还内置了很多实用工具。

- checkForServerUpgrade：检测目标实例能否升级到指定版本。
- dumpInstance：备份实例。
- dumpSchemas：备份指定库。
- dumpTables：备份指定表。
- loadDump：恢复通过上面 3 个工具生成的备份。
- exportTable：将指定的表导出到文本文件中。只支持单表，效果同 SELECT INTO OUTFILE 一样。
- importTable：将指定文本的数据导入到表中。

  在线上，如果有一个大文件需要导入，建议使用这个工具。它会将单个文件进行拆分，然后多线程并行执行 LOAD DATA LOCAL INFILE 操作。这样不仅提升了导入速度，还规避了大事务的问题。

- importJson：将 JSON 格式的数据导入 MySQL，比如将 MongoDB 中通过 mongoexport 导出的数据导入 MySQL。

在使用 MySQL Shell 时需注意以下两点。

- 通过 dumpInstance、dumpSchemas、dumpTables 生成的备份只能通过 loadDump 来恢复。
- 通过 exportTable 生成的备份只能通过 importTable 来恢复。

与 Dump & Load 相关的工具（包括 dumpInstance、dumpSchemas、dumpTables 和 loadDump）是本节介绍的重点。

本节主要包括以下内容。

- MySQL Shell Dump & Load 的用法。
- MySQL Shell Dump & Load 的关键特性。
- util.dumpInstance 的实现原理。
- util.dumpInstance 的参数解析。
- util.loadDump 的参数解析。
- 使用 MySQL Shell Dump & Load 时的注意事项。

5.5.1 MySQL Shell Dump & Load 的用法

- **util.dumpInstance(outputUrl[, options])**

备份实例。

outputUrl 是备份目录，必须为空。options 是可指定的选项。

首先，我们来看一个简单的示例。

```
# mysqlsh -h 10.0.20.4 -P3306 -uroot -p
mysql-js> util.dumpInstance('/data/backup/full',{compression: "none"})
Acquiring global read lock
Global read lock acquired
Initializing - done
```

```
1 out of 5 schemas will be dumped and within them 1 table, 0 views.
4 out of 7 users will be dumped.
Gathering information - done
All transactions have been started
Locking instance for backup
Global read lock has been released
Writing global DDL files
Writing users DDL
Running data dump using 4 threads.
NOTE: Progress information uses estimated values and may not be accurate.
Writing schema metadata - done
Writing DDL - done
Writing table metadata - done
Starting data dump
101% (650.00K rows / ~639.07K rows), 337.30K rows/s, 65.89 MB/s
Dump duration: 00:00:01s
Total duration: 00:00:01s
Schemas dumped: 1
Tables dumped: 1
Data size: 126.57 MB
Rows written: 650000
Bytes written: 126.57 MB
Average throughput: 65.30 MB/s
```

命令中的 /data/backup/full 是备份目录，compression: "none" 指的是不压缩，这里设置为不压缩主要是为了方便查看数据文件的内容。线上使用时，建议开启压缩。

接下来看看备份目录中的内容。

```
# ll /data/backup/full/
total 123652
-rw-r----- 1 root root      273 May 25 21:13 @.done.json
-rw-r----- 1 root root      854 May 25 21:13 @.json
-rw-r----- 1 root root      240 May 25 21:13 @.post.sql
-rw-r----- 1 root root      288 May 25 21:13 sbtest.json
-rw-r----- 1 root root 63227502 May 25 21:13 sbtest@sbtest1@0.tsv
-rw-r----- 1 root root      488 May 25 21:13 sbtest@sbtest1@0.tsv.idx
-rw-r----- 1 root root 63339214 May 25 21:13 sbtest@sbtest1@@1.tsv
-rw-r----- 1 root root      488 May 25 21:13 sbtest@sbtest1@@1.tsv.idx
-rw-r----- 1 root root      633 May 25 21:13 sbtest@sbtest1.json
-rw-r----- 1 root root      759 May 25 21:13 sbtest@sbtest1.sql
-rw-r----- 1 root root      535 May 25 21:13 sbtest.sql
-rw-r----- 1 root root      240 May 25 21:13 @.sql
-rw-r----- 1 root root     6045 May 25 21:13 @.users.sql
```

下面看看各个文件的具体作用。

- @.done.json：记录备份的结束时间和备份集的大小。在备份结束时生成。
- @.json：记录备份的一些元数据信息，包括备份时的一致性位置点信息：binlogFile、binlogPosition 和 gtidExecuted，这些信息可用来建立复制。
- @.sql 和 @.post.sql：这两个文件只有一些注释信息。不过在通过 util.loadDump 导入数据时，我们可以通过这两个文件自定义一些 SQL 语句，其中@.sql 在数据导入前执行，@.post.sql 在数据导入后执行。

- sbtest.json：记录 sbtest 中已经备份的表、视图、定时器、函数和存储过程。
- *.tsv：数据文件。我们看看数据文件的内容。

```
# head -2 sbtest@sbtest1@0.tsv
1	6461363	68487932199-96439406143-93774651418-41631865787-96406072701-20604855487-25459966574-28203206787-41238978918-19503783441	22195207048-70116052123-74140395089-76317954521-98694025897
2	1112248	13241531885-45658403807-79170748828-69419634012-13605813761-77983377181-01582588137-21344716829-87370944992-02457486289	28733802923-10548894641-11867531929-71265603657-36546888392
```

TSV 格式，每一行存储一条记录，字段与字段之间用制表符（\t）分隔。

- sbtest@sbtest1.json：记录了表相关的一些元数据信息，如列名、字段之间的分隔符（fieldsTerminatedBy）等。
- sbtest@sbtest1.sql：sbtest.sbtest1 的建表语句。
- sbtest.sql：建库语句。如果这个库中存在存储过程、函数、定时器，也会写到这个文件中。
- @.users.sql：创建账号及授权语句。默认不会备份 mysql.session、mysql.session、mysql.sys 这 3 个内部账号。

- **util.dumpSchemas(schemas, outputUrl[, options])**

备份指定库的数据。

用法同 util.dumpInstance 类似。第一个参数必须为数组，即使只需备份一个库，例如：

```
util.dumpSchemas(['sbtest'],'/data/backup/schema')
```

支持的配置大部分与 util.dumpInstance 相同。

从 MySQL Shell 8.0.28 开始，可直接使用 util.dumpInstance 中的 includeSchemas 选项进行指定库的备份。

- **util.dumpTables(schema, tables, outputUrl[, options])**

备份指定表的数据。

用法同 util.dumpInstance 类似。第二个参数必须为数组，例如：

```
util.dumpTables('sbtest',['sbtest1'],'/data/backup/table')
```

支持的配置大部分与 util.dumpInstance 相同。

从 MySQL Shell 8.0.28 开始，可直接使用 util.dumpInstance 中的 includeTables 选项进行指定表的备份。

- **util.loadDump(url[, options])**

导入通过 dump 命令生成的备份集。例如：

```
# mysqlsh -S /data/mysql/3307/data/mysql.sock
mysql-js> util.loadDump("/data/backup/full",{loadUsers: true})
Loading DDL, Data and Users from '/data/backup/full' using 4 threads.
Opening dump...
Target is MySQL 8.0.27. Dump was produced from MySQL 8.0.27
```

```
Scanning metadata - done
Checking for pre-existing objects...
Executing common preamble SQL
Executing DDL - done
Executing view DDL - done
Starting data load
2 thds loading - 100% (126.57 MB / 126.57 MB), 11.43 MB/s, 0 / 1 tables done
Recreating indexes - done
Executing user accounts SQL...
NOTE: Skipping CREATE/ALTER USER statements for user 'root'@'localhost'
NOTE: Skipping GRANT statements for user 'root'@'localhost'
Executing common postamble SQL
2 chunks (650.00K rows, 126.57 MB) for 1 tables in 1 schemas were loaded in 10 sec (avg throughput
13.96 MB/s)
0 warnings were reported during the load.
```

命令中的 /data/backup/full 是备份目录，loadUsers: true 是导入账号，默认不会导入。

5.5.2 MySQL Shell Dump & Load 的关键特性

util.dumpInstance 的关键特性如下。

- 多线程备份。并发线程数由 threads 决定，默认是 4。
- 支持单表 chunk 级别的并行备份，前提是表上存在主键或唯一索引。
- 默认是压缩备份。
- 支持备份限速。可通过 maxRate 限制单个线程的数据读取速率。

util.loadDump 的关键特性如下。

- 多线程恢复。并发线程数由 threads 决定，默认是 4。
- 支持断点续传功能。
- 支持延迟创建二级索引。
- 支持边备份边导入。
- 通过 LOAD DATA LOCAL INFILE 命令来导入数据。
- 如果单个文件过大，util.loadDump 在导入时会自动进行切割，以避免产生大事务。

5.5.3 util.dumpInstance 的实现原理

util.dumpInstance 的实现原理如图 5-4 所示。

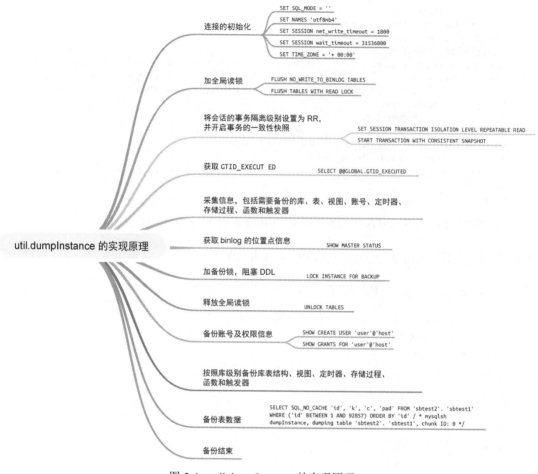

图 5-4　util.dumpInstance 的实现原理

不难看出，util.dumpInstance 的实现原理与 mysqldump 大致相同，主要有以下两个不同之处。

- util.dumpInstance 会加备份锁。备份锁可用来阻塞备份过程中的 DDL。
- util.dumpInstance 是并行备份，相较于 mysqldump 的单线程备份，备份效率更高。

5.5.4　util.dumpInstance 的参数解析

util.dumpInstance 的参数可分为如下几类。

1. 过滤相关

- excludeSchemas：忽略某些库的备份，多个库之间用逗号隔开。例如：

  ```
  excludeSchemas: ["db1", "db2"]
  ```

- includeSchemas：指定某些库的备份。
- excludeTables：忽略某些表的备份，表必须是 schema.table 的格式，多个表之间用逗号隔开。例如：

  excludeTables: ["sbtest.sbtest1", "sbtest.sbtest2"]
- includeTables：指定某些表的备份。
- events：是否备份定时器，默认为 true。
- excludeEvents：忽略某些定时器的备份。
- includeEvents：指定某些定时器的备份。
- routines：是否备份函数和存储过程，默认为 true。
- excludeRoutines：忽略某些函数和存储过程的备份。
- includeRoutines：指定某些函数和存储过程的备份。
- users：是否备份账号信息，默认为 true。
- excludeUsers：忽略某些账号的备份，可指定多个账号。
- includeUsers：指定某些账号的备份，可指定多个账号。
- triggers：是否备份触发器，默认为 true。
- excludeTriggers：忽略某些触发器的备份。
- includeTriggers：指定某些触发器的备份。
- ddlOnly：是否只备份表结构，默认为 false。
- dataOnly：是否只备份数据，默认为 false。

2. 并行备份相关

- chunking：是否开启 chunk 级别的并行备份功能，默认为 true。
- bytesPerChunk：每个 chunk 文件的大小，默认为 64MB。
- threads：并发线程数，默认为 4。

3. OCI[①]相关

- ocimds：是否检查备份集与 MySQL Database Service（MDS，是甲骨文云的 MySQL 云服务）的兼容性，默认为 false，即不检查。如果设置为 true，会输出所有的不兼容项及解决方法。不兼容项可通过下面的 compatibility 来解决。
- compatibility：如果要将备份数据导入 MDS，为了保证与后者的兼容性，可在导出的过程中进行相应的调整。

  - create_invisible_pks：对于没有主键的表，会创建一个隐藏主键，即 my_row_id BIGINT UNSIGNED AUTO_INCREMENT INVISIBLE PRIMARY KEY。隐藏列是 MySQL 8.0.23 引入的。
  - force_innodb：将表的引擎强制设置为 InnoDB。
  - ignore_missing_pks：忽略主键缺失导致的错误。与 create_invisible_pks 互斥，二者不能同时指定。

---

① 即甲骨文云基础设施（Oracle Cloud Infrastructure）。——编者注

- **skip_invalid_accounts**：忽略没有密码的账号，或者使用了 MDS 不支持的认证插件的账号。
- **strip_definers**：去掉视图、存储过程、函数、定时器、触发器中的 DEFINER=account 子句。
- **strip_restricted_grants**：去掉 MDS 中不允许授予的权限。
- **strip_tablespaces**：去掉建表语句中的 TABLESPACE=xxx 子句。

可设置其中一项或多项。例如：

```
compatibility: [ "create_invisible_pks", "force_innodb", "strip_definers", "strip_restricted_grants" ]
```

- **osBucketName**、**osNamespace**、**ociConfigFile**、**ociProfile**、**ociParManifest**、**ociParExpireTime**：与 OCI 对象存储相关。

4. 其他选项

- **tzUtc**：是否设置 TIME_ZONE = '+00:00'，默认为 true。
- **consistent**：是否开启一致性备份，默认为 true。若设置为 false，则不会加全局读锁，也不会开启事务的一致性快照。
- **dryRun**：试运行。此时只会打印备份信息，不会执行备份操作。
- **maxRate**：限制单个线程的数据读取速率，单位是字节每秒。默认为 0，即不限制。
- **showProgress**：是否打印进度信息，如果是 TTY 设备（命令行终端），则为 true，反之为 false。
- **defaultCharacterSet**：字符集，默认为 utf8mb4。
- **compression**：备份文件的压缩算法，默认为 zstd。也可设置为 gzip 或 none（不压缩）。

5.5.5 util.loadDump 的参数解析

util.loadDump 的参数可分为如下几类。

1. 过滤相关

- **excludeEvents**：忽略某些定时器的导入。
- **excludeRoutines**：忽略某些函数和存储过程的导入。
- **excludeSchemas**：忽略某些库的导入。
- **excludeTables**：忽略某些表的导入。
- **excludeTriggers**：忽略某些触发器的导入。
- **excludeUsers**：忽略某些账号的导入。
- **includeEvents**：导入指定定时器。
- **includeRoutines**：导入指定函数和存储过程。
- **includeSchemas**：导入指定库。
- **includeTables**：导入指定表。
- **includeTriggers**：导入指定触发器。
- **includeUsers**：导入指定账号。
- **loadData**：是否导入数据，默认为 true。

- loadDdl：是否导入 DDL 语句，默认为 true。
- loadUsers：是否导入账号，默认为 false。注意，即使将 loadUsers 设置为 true，也不会导入当前正在执行导入操作的用户。
- ignoreExistingObjects：是否忽略已经存在的对象，默认为 off。

2. 并行导入相关

- backgroundThreads：获取元数据和 DDL 文件内容的线程数。备份集如果存储在本地，则 backgroundThreads 默认和 threads 一致。
- threads：并发线程数，默认为 4。
- maxBytesPerTransaction：指定单个 LOAD DATA 操作可加载的最大字节数。默认与 bytesPerChunk 一致。这个参数可用来规避大事务。

3. 断点续传相关

- progressFile：在导入的过程中，会在备份目录中生成一个 progressFile，用于记录加载过程中的进度信息，这个进度信息可用来实现断点续传功能。默认为 load-progress.<server_uuid>.progress。
- resetProgress：如果备份目录中存在 progressFile，会默认从上次完成的地方继续执行。如果要从头开始执行，需将 resetProgress 设置为 true。该参数默认为 off。

4. OCI 相关

osBucketName、osNamespace、ociConfigFile 和 ociProfile。

5. 二级索引相关

- deferTableIndexes：是否延迟（数据加载完毕后）创建二级索引。可设置为 off（不延迟）、fulltext（只延迟创建全文索引，默认值）、all（延迟创建所有索引）。
- loadIndexes：与 deferTableIndexes 一起使用，用来决定数据加载完毕后，最后的二级索引是否创建，默认为 true。

6. 其他选项

- analyzeTables：表加载完毕后，是否执行 ANALYZE TABLE 操作。默认为 off（不执行），也可设置为 on 或 histogram（只对有直方图信息的表执行）。
- characterSet：字符集，无须显式设置，默认会从备份集中获取。
- createInvisiblePKs：是否创建隐式主键，默认从备份集中获取。这与备份时是否指定了 create_invisible_pks 有关，若指定了则为 true，反之为 false。
- dryRun：试运行。
- ignoreVersion：忽略 MySQL 的版本检测。
- schema：将表导入指定的 schema，适用于通过 util.dumpTables 创建的备份。
- showMetadata：导入时是否打印一致性备份时的位置点信息。
- showProgress：是否打印进度信息。

- **skipBinlog**：是否设置 sql_log_bin=0，默认为 false。这一点与 mysqldump、mydumper 不同。mydumper 默认会禁用 binlog，mysqldump 在开启 GTID，且备份时没有指定 --set-gtid-purged=OFF 的情况下，也会禁用 binlog。
- **updateGtidSet**：更新 GTID_PURGED。可设置为 off（不更新，默认值）、replace（替代目标实例的 GTID_PURGED）或 append（追加）。
- **waitDumpTimeout**：util.loadDump 可导入当前正在备份的备份集。处理完所有文件后，如果备份还没有结束（具体来说，是备份集中没有生成 @.done.json），util.loadDump 会报错并退出。可指定 waitDumpTimeout 等待一段时间，单位是秒。

5.5.6 MySQL Shell Dump & Load 的注意事项

- 表上存在主键或唯一索引时才能进行 chunk 级别的并行备份。字段的数据类型不限。
- 从 MySQL Shell 8.0.32 开始，对于不能进行并行备份的表，会将单表数据备份到多个文件中。在此之前，就只会备份到一个文件中。
- util.dumpInstance 只能保证 InnoDB 表的备份一致性。
- 默认不会备份 information_schema、mysql、ndbinfo、performance_schema、sys。
- 备份实例支持 MySQL 5.6 及以上版本，导入实例支持 MySQL 5.7 及以上版本。
- 备份的过程中，会将 BLOB 等非文本安全的列转换为 Base64，由此会导致转换后的数据大小超过原数据。导入时，要注意 max_allowed_packet 的限制。
- 导入之前，需将目标实例的 local_infile 设置为 ON。
- 在使用 MySQL Shell 时，应尽量选择最新版本。

5.5.7 参考资料

- 官方文档 "Instance Dump Utility, Schema Dump Utility, and Table Dump Utility"。
- 博客文章 "MySQL Shell Dump & Load part 1: Demo!"，作者：Kenny Gryp。
- 博客文章 "MySQL Shell Dump & Load part 2: Benchmarks"，作者：Kenny Gryp。
- 博客文章 "MySQL Shell Dump & Load part 3: Load Dump"。
- 博客文章 "MySQL Shell Dump & Load part 4: Dump Instance & Schemas"，作者：Alfredo Kojima。
- 博客文章 "Backup/Restore Performance Conclusion: mysqldump vs MySQL Shell Utilities vs mydumper vs mysqlpump vs XtraBackup"，作者：Vinicius Grippa。
- 官方文档 "Optimizing INSERT Statements"。

5.6 使用 XtraBackup 搭建从库

搭建从库，本质上需要的只是一个一致性备份集及这个备份集对应的位置点信息。之前介绍的几个备份工具均可满足需求。这里，我们重点看看如何基于 XtraBackup 搭建从库。

整个过程其实比较简单，无非是备份和还原。唯一需要注意的是建立复制时位置点的选择。

- 在基于位置点的复制中，注意 CHANGE MASTER TO 语句中 MASTER_LOG_FILE 和 MASTER_LOG_POS 的选择。
- 在 GTID 复制中，在执行 CHANGE MASTER TO 命令之前，必须设置 GTID_PURGED。

本节主要包括以下内容。

- 使用 XtraBackup 搭建从库的基本步骤。
- 基于从库备份搭建从库的注意事项。
- 设置 GTID_PURGED 的注意事项。
- 使用 XtraBackup 8.0 搭建从库的注意事项。

5.6.1 使用 XtraBackup 搭建从库的基本步骤

表 5-8 展示了测试环境信息。

表 5-8 测试环境信息

角色	IP 地址
主库	10.0.0.118
从库	10.0.0.195

下面我们看看具体的搭建步骤。

(1) 在主库上创建复制账号。

```
mysql> create user 'repl'@'%' identified by 'repl123';
Query OK, 0 rows affected (0.01 sec)

mysql> grant replication slave on *.* TO 'repl'@'%';
Query OK, 0 rows affected (0.00 sec)
```

(2) 对主库进行备份。

在 10.0.0.118 上执行备份命令。

```
# xtrabackup --user=backup_user --password=backup_pass --socket=/data/mysql/3306/data/mysql.sock --backup --parallel=10 --slave-info --target-dir=/data/backup/full
```

(3) 将备份文件传输到从库上。

```
# scp -r /data/backup/full/* root@10.0.0.195:/data/backup/full
```

(4) 在从库上准备好 MySQL 安装包及参数文件。

```
# tar xvf mysql-8.0.27-linux-glibc2.12-x86_64.tar.xz -C /usr/local/
# cd /usr/local/
# ln -s mysql-8.0.27-linux-glibc2.12-x86_64 mysql
```

(5) 在从库上进行 Prepare 和恢复。

```
# xtrabackup --prepare --target-dir=/data/backup/full
# xtrabackup --defaults-file=/etc/my.cnf --copy-back --parallel=10 --target-dir=/data/backup/full
```

恢复命令中的 /etc/my.cnf 是从库的配置文件。

步骤(2)、步骤(3)和步骤(5)可以简化为下面这两条命令。

```
# xtrabackup \
--user=backup_user --password=backup_pass --socket=/data/mysql/3306/data/mysql.sock \
--backup --stream=xbstream --slave-info --parallel=10 | lz4 | \
ssh mysql@10.0.0.195 'cat - | lz4 -d | xbstream -p10 -x -C /data/mysql/3306/data/'
# xtrabackup --prepare --target-dir=/data/mysql/3306/data/
```

第一条命令是在线上搭建从库时的一条常用命令，它将流式备份、管道结合在一起，具有以下优点。

- 边备份，边解压。相对于备份、传输再解压，花费的时间更短。
- 备份集是直接解压到从库服务器的，并不会保存到本地。这样，对于主库服务器，既可减少磁盘空间，又可减小磁盘的 I/O 压力。
- /data/mysql/3306/data/ 是从库的数据目录，在恢复时，无须 --copy-back，直接 Prepare 即可。

(6) 启动实例。

```
# chown -R mysql.mysql /data/mysql/3306/data/
# /usr/local/mysql/bin/mysqld_safe --defaults-file=/etc/my.cnf &
```

很多人误以为，要搭建从库，需要提前创建一个空白实例。对于逻辑备份确实如此，但对于物理备份，则无此必要，直接使用 mysqld_safe 启动还原后的备份文件即可。

(7) 建立复制。

这里需要区分两种场景：GTID 复制和基于位置点的复制。首先查看备份集中的 xtrabackup_binlog_info 文件的内容。

```
# cat xtrabackup_binlog_info
mysql-bin.000002    882880068       2cbdc21a-db11-11ec-83bf-020017003dc4:1-223148
```

如果 xtrabackup_binlog_info 中存在 GTID 信息，则代表备份实例开启了 GTID，这个时候就需要建立 GTID 复制。

- **GTID 复制**

对于 GTID 复制，在建立复制前，必须首先设置 `GTID_PURGED`。设置 `GTID_PURGED` 时，注意备份实例的版本。

在 MySQL 5.7 中，引入了 `mysql.gtid_executed`，从库实例启动后，会基于该表的值来初始化 `GTID_EXECUTED` 和 `GTID_PURGED`。

```
mysql> select * from mysql.gtid_executed;
+--------------------------------------+----------------+--------------+
| source_uuid                          | interval_start | interval_end |
+--------------------------------------+----------------+--------------+
| 2cbdc21a-db11-11ec-83bf-020017003dc4 |              1 |         2124 |
+--------------------------------------+----------------+--------------+
1 row in set (0.00 sec)

mysql> show global variables where variable_name in ('gtid_executed', 'gtid_purged');
+----------------+-------+
| Variable_name  | Value |
+----------------+-------+
```

```
| gtid_executed | 2cbdc21a-db11-11ec-83bf-020017003dc4:1-2124 |
| gtid_purged   | 2cbdc21a-db11-11ec-83bf-020017003dc4:1-2124 |
+---------------+----------------------------------------------+
2 rows in set (0.00 sec)
```

但很显然，GTID_PURGED 与 xtrabackup_binlog_info 中的 GTID 信息相差甚远。这一点不难理解，因为主库的 mysql.gtid_executed 在 MySQL 8.0.17 之前，只有在日志切换和实例关闭时才更新。

下面我们基于 xtrabackup_binlog_info 中的 GTID 信息重新设置 GTID_PURGED。

```
mysql> reset master;
Query OK, 0 rows affected (0.00 sec)

mysql> set global gtid_purged='2cbdc21a-db11-11ec-83bf-020017003dc4:1-223148';
Query OK, 0 rows affected (0.01 sec)
```

因为 GTID_EXECUTED 有值，所以在设置 GTID_PURGED 之前，必须通过 RESET MASTER 命令清空 GTID_EXECUTED。

在 MySQL 5.6 中，可直接基于 xtrabackup_binlog_info 中的 GTID 信息设置 GTID_PURGED。

```
mysql> set global gtid_purged='2cbdc21a-db11-11ec-83bf-020017003dc4:1-223148';
```

为什么在 MySQL 5.6 中无须执行 RESET MASTER 呢？因为 MySQL 5.6 中还没有引入 mysql.gtid_executed，实例恢复后，GTID_EXECUTED 和 GTID_PURGED 均为空。

在 MySQL 8.0 中，无须设置 GTID_PURGED。至于为什么不用设置，后面会详细介绍。

设置完 GTID_PURGED，接下来执行 CHANGE MASTER TO 命令。

```
CHANGE MASTER TO
  MASTER_HOST='10.0.0.118',
  MASTER_USER='repl',
  MASTER_PASSWORD='repl123',
  MASTER_PORT=3306,
  MASTER_AUTO_POSITION = 1;
```

对于 GTID 复制，需将 MASTER_AUTO_POSITION 设置为 1。在 MySQL 8.0 中，CHANGE MASTER TO 语句中还需添加 GET_MASTER_PUBLIC_KEY = 1。

- **基于位置点的复制**

如果 xtrabackup_binlog_info 没有 GTID 信息，则代表备份实例没有开启 GTID，这个时候就无须设置 GTID_PURGED，直接执行 CHANGE MASTER TO 命令即可。

```
CHANGE MASTER TO
  MASTER_HOST='10.0.0.118',
  MASTER_USER='repl',
  MASTER_PASSWORD='repl123',
  MASTER_PORT=3306,
  MASTER_LOG_FILE='mysql-bin.000002',
  MASTER_LOG_POS=882880068;
```

CHANGE MASTER TO 语句中的 MASTER_LOG_FILE 和 MASTER_LOG_POS 的值分别取自 xtrabackup_binlog_info 中的 filename 和 position。

(8) 开启复制。

- mysql> **start** slave;

(9) 检查主从复制是否正常。

- mysql> show slave status\G

Slave_IO_Running 和 Slave_SQL_Running 均为 Yes 代表复制正常。

以上就是使用 XtraBackup 搭建从库的基本步骤。

5.6.2 基于从库备份搭建从库的注意事项

在线上，我们很少备份主库，一般是备份从库。所以，基于从库的备份来搭建一个新的从库更为常见。

对于这种场景，上面的搭建步骤同样适用，不过有以下几点需要注意。

- 对从库进行备份时，需指定 --slave-info。这个时候备份集中会生成一个 xtrabackup_slave_info 文件，该文件记录了备份时备份实例对应主库的一致性位置点信息，例如：

```
# cat xtrabackup_slave_info
CHANGE MASTER TO MASTER_LOG_FILE='mysql-bin.000004', MASTER_LOG_POS=6263314;
```

如果从库开启了 GTID，则只会记录 GTID 信息，例如：

```
SET GLOBAL gtid_purged='2cbdc21a-db11-11ec-83bf-020017003dc4:1-2049780';
CHANGE MASTER TO MASTER_AUTO_POSITION=1;
```

其实，对主库备份，也可指定 --slave-info，只不过此时的 xtrabackup_slave_info 内容为空。所以，上面搭建步骤中的备份命令都带上了 --slave-info。

- 在基于位置点的复制中，CHANGE MASTER TO 语句中的 MASTER_LOG_FILE 和 MASTER_LOG_POS 必须取自 xtrabackup_slave_info，而不是 xtrabackup_binlog_info。

  对于 GTID 复制，则没关系，因为 xtrabackup_slave_info 和 xtrabackup_binlog_info 中的 GTID 信息是一致的。

- 只要是基于从库的备份来搭建从库，在执行 CHANGE MASTER TO 命令之前，都必须执行 RESET SLAVE ALL 清空 mysql.slave_master_info 和 mysql.slave_relay_log_info 表中的内容。

5.6.3 设置 GTID_PURGED 的注意事项

在 GTID 复制中，为什么需要设置 GTID_PURGED 呢？实际上，设置 GTID_PURGED 只是手段，最终目的还是设置 GTID_EXECUTED。

GTID_EXECUTED 代表了实例中已经执行过的 GTID 集。在建立复制后，从库会自动跳过 GTID_EXECUTED 的相关事务。如果这个值设置得不准确，会导致事务丢失，或者已经重放过的操作重复执行。但

GTID_EXECUTED 是一个只读参数，不能直接修改。

```
mysql> set global gtid_executed='411693c9-d512-11ec-9e11-525400d51a16:1-10369';
ERROR 1238 (HY000): Variable 'gtid_executed' is a read only variable
```

必须通过修改 GTID_PURGED 间接修改它。

GTID_PURGED 代表了实例中已经执行过，但 binlog 中不存在的 GTID 集。所以 GTID_PURGED 一定是 GTID_EXECUTED 的子集。

在 MySQL 8.0 之前，如果要修改 GTID_PURGED，则 GTID_EXECUTED 必须为空。

```
mysql> set global gtid_purged='411693c9-d512-11ec-9e11-525400d51a16:1-10369';
ERROR 1840 (HY000): @@GLOBAL.GTID_PURGED can only be set when @@GLOBAL.GTID_EXECUTED is empty.
```

而要 GTID_EXECUTED 为空，只能执行 RESET MASTER 操作。

```
mysql> reset master;
Query OK, 0 rows affected (0.02 sec)

mysql> show global variables where variable_name in ('gtid_executed', 'gtid_purged');
+----------------+-------+
| Variable_name  | Value |
+----------------+-------+
| gtid_executed  |       |
| gtid_purged    |       |
+----------------+-------+
2 rows in set (0.00 sec)

mysql> set global gtid_purged='411693c9-d512-11ec-9e11-525400d51a16:1-10369';
Query OK, 0 rows affected (0.00 sec)

mysql> show global variables where variable_name in ('gtid_executed', 'gtid_purged');
+----------------+------------------------------------------------+
| Variable_name  | Value                                          |
+----------------+------------------------------------------------+
| gtid_executed  | 411693c9-d512-11ec-9e11-525400d51a16:1-10369   |
| gtid_purged    | 411693c9-d512-11ec-9e11-525400d51a16:1-10369   |
+----------------+------------------------------------------------+
2 rows in set (0.00 sec)
```

可以看到，调整 GTID_PURGED 后，GTID_EXECUTED 也随之更改。

MySQL 8.0 中剔除了这一限制，即在设置 GTID_PURGED 时，无须 GTID_EXECUTED 为空。但也不能随便设置，需要满足一定的条件，具体可参考 2.2 节。

5.6.4 使用 XtraBackup 8.0 搭建从库的注意事项

在 MySQL 8.0 中，得益于 performance_schema.log_status 的引入，备份的过程中不再加全局读锁。对于备份结束时的位置点信息，查询的是 performance_schema.log_status。该表的内容如下。

```
mysql> select * from performance_schema.log_status\G
*************************** 1. row ***************************
   SERVER_UUID: d310871c-db0c-11ec-a557-020017003dc4
         LOCAL: {"gtid_executed": "d310871c-db0c-11ec-a557-020017003dc4:1-352559",
"binary_log_file": "mysql-bin.000022", "binary_log_position": 9698237}
```

```
    REPLICATION: {"channels": []}
STORAGE_ENGINES: {"InnoDB": {"LSN": 912297234, "LSN_checkpoint": 912297234}}
1 row in set (0.00 sec)
```

需要注意的是，LOCAL 中的 gtid_executed 和 binary_log_file + binary_log_position 对应的并不总是同一个事务。这一点很容易模拟出来，对一张表持续执行插入操作即可。

下面是一个具体的案例。

在备份过程中，持续对一张表执行插入操作。最后备份集中 xtrabackup_binlog_info 的内容如下。

```
# cat xtrabackup_binlog_info
mysql-bin.000024  507  d310871c-db0c-11ec-a557-020017003dc4:1-388482
```

接下来我们基于 binlog 的位置点信息 mysql-bin.000024 507 查找对应的事务。

```
# mysqlbinlog -v --base64-output=decode-rows --stop-position=507 mysql-bin.000024
# The proper term is pseudo_replica_mode, but we use this compatibility alias
# to make the statement usable on server versions 8.0.24 and older.
/*!50530 SET @@SESSION.PSEUDO_SLAVE_MODE=1*/;
/*!50003 SET @OLD_COMPLETION_TYPE=@@COMPLETION_TYPE,COMPLETION_TYPE=0*/;
DELIMITER /*!*/;
# at 4
#220529 11:19:07 server id 1  end_log_pos 126 CRC32 0xdcc54ec7  Start: binlog v 4, server v 8.0.28 created
220529 11:19:07
# at 126
#220529 11:19:07 server id 1  end_log_pos 197 CRC32 0x5d440f7c  Previous-GTIDs
# d310871c-db0c-11ec-a557-020017003dc4:1-388482
# at 197
#220529 11:19:07 server id 1  end_log_pos 276 CRC32 0x0dd893b5  GTID last_committed=0 sequence_number=1
rbr_only=yes original_committed_timestamp=1653823147539722
immediate_commit_timestamp=1653823147539722 transaction_length=310
/*!50718 SET TRANSACTION ISOLATION LEVEL READ COMMITTED*//*!*/;
# original_commit_timestamp=1653823147539722 (2022-05-29 11:19:07.539722 GMT)
# immediate_commit_timestamp=1653823147539722 (2022-05-29 11:19:07.539722 GMT)
/*!80001 SET @@session.original_commit_timestamp=1653823147539722*//*!*/;
/*!80014 SET @@session.original_server_version=80028*//*!*/;
/*!80014 SET @@session.immediate_server_version=80028*//*!*/;
SET @@SESSION.GTID_NEXT= 'd310871c-db0c-11ec-a557-020017003dc4:388483'/*!*/;
# at 276
#220529 11:19:07 server id 1  end_log_pos 365 CRC32 0xa49dc290  Query  thread_id=262 exec_time=0
error_code=0
...
BEGIN
/*!*/;
# at 365
#220529 11:19:07 server id 1  end_log_pos 425 CRC32 0x824f6309  Table_map: `slowtech`.`t1` mapped to number 157
# at 425
#220529 11:19:07 server id 1 end_log_pos 476 CRC32 0x5a6fe6ec  Write_rows: table id 157 flags: STMT_END_F
### INSERT INTO `slowtech`.`t1`
### SET
###   @1=1483132
###   @2='aaaaaaaaaa'
# at 476
#220529 11:19:07 server id 1  end_log_pos 507 CRC32 0x66a401f6  Xid = 4108904
COMMIT/*!*/;
SET @@SESSION.GTID_NEXT= 'AUTOMATIC' /* added by mysqlbinlog */ /*!*/;
DELIMITER ;
```

```
# End of log file
/*!50003 SET COMPLETION_TYPE=@OLD_COMPLETION_TYPE*/;
/*!50530 SET @@SESSION.PSEUDO_SLAVE_MODE=0*/;
```

可以看到，该事务对应的 GTID 是 d310871c-db0c-11ec-a557-020017003dc4:388483，不是 xtrabackup_binlog_info 中的 388482。

如果我们像在 MySQL 5.6 和 MySQL 5.7 中那样，基于 xtrabackup_binlog_info 中的 GTID 信息来设置 `GTID_PURGED`，在本例中会导致同一个 INSERT 操作被执行两次，进而出现主键冲突，导致主从复制中断。

问题看上去很严重，不过不用担心，XtraBackup 8.0 在查询 performance_schema.log_status 后，会基于查询到的 `binary_log_file` 和 `binary_log_position` 拷贝对应的 binlog。

下面是备份集中拷贝的 binlog。

```
# ll /data/backup/full/
...
-rw-r-----. 1 root root      507 May 29 11:19 mysql-bin.000024
-rw-r-----. 1 root root       19 May 29 11:19 mysql-bin.index
...
```

binlog 中记录了 mysql-bin.000024 507 这个位置点的事务所对应的 GTID 值。实例启动时，会自动基于 binlog 中的 GTID 信息来初始化 `GTID_EXECUTED` 和 `GTID_PURGED`。

```
mysql> show global variables where variable_name in ('gtid_executed', 'gtid_purged');
+----------------+----------------------------------------------+
| Variable_name  | Value                                        |
+----------------+----------------------------------------------+
| gtid_executed  | d310871c-db0c-11ec-a557-020017003dc4:1-388483 |
| gtid_purged    | d310871c-db0c-11ec-a557-020017003dc4:1-388482 |
+----------------+----------------------------------------------+
2 rows in set (0.00 sec)
```

所以，实例启动后，我们看到的 `GTID_EXECUTED` 就已经是正确的值，已经能正确反映备份结束时的一致性位置点信息了。这个时候，直接执行 CHANGE MASTER TO 操作就可以了。

5.6.5 总结

- 引入备份锁的初衷是阻塞备份过程中的 DDL，不是为了替代全局读锁。在 XtraBackup 8.0 中，我们可以指定 `--skip-lock-ddl` 禁用备份锁，这并不影响 XtraBackup 的正常使用。
- 基于物理备份搭建从库时，无须提前创建空白实例。
- 在基于位置点的复制中，注意 CHANGE MASTER TO 语句中 MASTER_LOG_FILE 和 MASTER_LOG_POS 的选择。

  以一个简单的主从复制拓扑为例：master -> slave1。

  - 如果基于 master 的备份添加一个 master 的从库，或者基于 slave1 的备份添加一个 slave1 的从库，在建立复制时，应使用 xtrabackup_binlog_info 的位置点信息。
  - 如果是基于 slave1 的备份添加 master 的一个从库，应使用 xtrabackup_slave_info 的位置点信息。

- 基于从库的备份搭建从库时,在执行 CHANGE MASTER TO 命令之前,必须执行 RESET SLAVE ALL。
- 无论是对主库还是从库进行备份,都可指定 --slave-info,此时会生成 xtrabackup_slave_info。只不过如果是对主库进行备份,则该文件会为空。
- 在 GTID 复制中设置 GTID_PURGED 时,注意备份实例的版本。如果是 MySQL 5.6 或 MySQL 5.7,可直接基于 xtrabackup_binlog_info 中的 GTID 信息设置 GTID_PURGED。如果是 MySQL 8.0,则无须再设置 GTID_PURGED。

5.7 指定时间点(位置点)的恢复

指定时间点(位置点)的恢复,常用于以下场景。

- 出现了人为的误操作,比如误删了线上的表。这个时候需要将表恢复到删除的前一个时刻。
- 需要查看数据库的历史状态。
- 业务代码存在 bug,导致数据被删除或污染了。这个时候需要结合历史数据来修复数据。

指定时间点(位置点)的恢复,主要包含如下两步。

- 恢复全量备份。这个备份具体来说是指定时间点的前一个全量备份。
- 应用增量数据。应用全量备份与指定时间点之间的 binlog。

下面通过一个具体的案例看看如何基于"备份 + binlog"进行指定时间点的恢复。

(1) 模拟故障。

手动删除 slowtech.t1 表。在 DROP 操作之前,我们看看该表的记录数及最新一条记录。

```
mysql> select count(*) from slowtech.t1;
+----------+
| count(*) |
+----------+
|   136424 |
+----------+
1 row in set (0.00 sec)

mysql> select * from slowtech.t1 order by insert_time desc limit 1;
+--------+----------------------------+
| id     | insert_time                |
+--------+----------------------------+
| 136424 | 2022-05-28 14:36:05.202220 |
+--------+----------------------------+
1 row in set (0.04 sec)

mysql> drop table slowtech.t1;
Query OK, 0 rows affected (0.02 sec)
```

(2) 恢复故障之前的最近一个全量备份集。

假设这个备份集是 /data/backup/20220528。在 DROP 操作之前通过以下命令进行备份。

```
# xtrabackup --user=backup_user --password=backup_pass --backup --parallel=10
--target-dir=/data/backup/20220528
```

5.7 指定时间点（位置点）的恢复

我们看看具体的恢复步骤。10.0.0.195 是新实例的地址。

```
# scp -r /data/backup/20220528/* mysql@10.0.0.195:/data/backup/20220528
# xtrabackup --prepare --target-dir=/data/backup/20220528
# xtrabackup --defaults-file=/etc/my.cnf --copy-back --target-dir=/data/backup/20220528
# chown -R mysql.mysql /data/mysql/3306/data/
# /usr/local/mysql/bin/mysqld_safe --defaults-file=/etc/my.cnf &
```

注意，是恢复到一个新的空白实例上，不是直接恢复到线上出故障的实例上。

(3) 确定 DROP 操作对应的 binlog 位置点。

首先查看这个备份对应的 binlog 位置点信息。

```
# cat /data/backup/20220528/xtrabackup_binlog_info
mysql-bin.000007    505    9b481834-de85-11ec-9045-020017003dc4:1-139225
```

接着，从 mysql-bin.000007 开始分析 binlog，找到主库 DROP 操作对应的 binlog 位置点。

```
# mysqlbinlog -v /data/mysql/3306/data/mysql-bin.000007 | grep -i -B 20 -A 2 "DROP"
# mysqlbinlog -v /data/mysql/3306/data/mysql-bin.000008 | grep -i -B 20 -A 2 "DROP"
# mysqlbinlog -v /data/mysql/3306/data/mysql-bin.000009 | grep -i -B 20 -A 2 "DROP"
### SET
###   @1=136424
###   @2='2022-05-28 14:36:05.202220'
# at 988347
#220528 14:36:05 server id 1  end_log_pos 988378 CRC32 0xc8973cbe     Xid = 302063
COMMIT/*!*/;
# at 988378
#220528 14:37:37 server id 1  end_log_pos 988455 CRC32 0xb3798f27     GTID    last_committed=3198
sequence_number=3199rbr_only=no    original_committed_timestamp=1653748657182976
...
SET @@SESSION.GTID_NEXT= '9b481834-de85-11ec-9045-020017003dc4:150967'/*!*/;
# at 988455
#220528 14:37:37 server id 1  end_log_pos 988593 CRC32 0x689a2493     Query   thread_id=28    exec_time=0
error_code=0Xid = 302095
SET TIMESTAMP=1653748657/*!*/;
SET @@session.pseudo_thread_id=28/*!*/;
/*!\C utf8mb4 *//*!*/;
SET
@@session.character_set_client=255,@@session.collation_connection=255,@@session.collation_server=2
55/*!*/;
DROP TABLE `slowtech`.`t1` /* generated by server */
/*!*/;
# at 988593
```

可以看到，DROP 操作对应的 binlog 位置点是 988593。我们要恢复到 DROP 操作的前一个事务，这个事务 COMMIT 操作对应的位置点是 988378。

(4) 开始应用 binlog。

```
# mysqlbinlog --start-position=505 --stop-position=988378 \
mysql-bin.000007 mysql-bin.000008 mysql-bin.000009 \
 | mysql -h 10.0.0.195 -uroot -p123456
```

注意，在使用 mysqlbinlog 时，如果指定了多个 binlog，则 --start-position 针对的是第一个 binlog，而 --stop-position 针对的是最后一个 binlog。第一个 binlog 和 --start-position 取的是备份集 xtrabackup_binlog_info 文件中的位置点。最后一个 binlog 和 --stop-position 取的是 DROP 操作的前

一个事务的位置点。

(5) 应用完 binlog，确认恢复的效果。

```
mysql> select count(*) from slowtech.t1;
+----------+
| count(*) |
+----------+
|   136424 |
+----------+
1 row in set (0.01 sec)

mysql> select * from slowtech.t1 order by insert_time desc limit 1;
+--------+---------------------+
| id     | insert_time         |
+--------+---------------------+
| 136424 | 2022-05-28 14:36:05.202220 |
+--------+---------------------+
1 row in set (0.03 sec)
```

与 DROP 前完全一致。确认没问题后，可先将该表从恢复实例中导出来，再导入故障实例。

5.8 搭建 binlog server

前面提到的几个备份工具，备份的都是实例在某个时间点的数据。如果我们要进行任意时间点的恢复，则必须借助于 binlog。

对于 binlog，常见的备份策略是在备份数据库时，同时备份 binlog。但这种方式不够及时，如果在实例运行期间，服务器宕机不能启动了，或者服务器硬盘损坏了，就会丢失上次备份之后的 binlog。即使通过脚本来定期备份，也无法对当前正在写入的 binlog 进行实时备份。

所幸从 MySQL 5.6 开始，mysqlbinlog 支持将远程服务器上的 binlog 实时复制到本地服务器上。它不是简单的文件拷贝，而是通过 Replication API 来实时获取 binlog 中的二进制日志事件。可将其简单理解为一个不会进行任何重放操作的从库。binlog server 就是基于这个特性来实现的。

5.8.1 基于 mysqlbinlog 搭建 binlog server

下面来看看如何基于 mysqlbinlog 搭建 binlog server。

核心命令如下。

```
# mysqlbinlog --read-from-remote-server --raw --host=10.0.0.118 --port=3306 --user=repl --password=repl123 --stop-never mysql-bin.000001
```

接着看看各个参数的具体含义。

- --read-from-remote-server（-R）：读取远程服务器的 binlog，是必需选项。如果不指定该选项，则读取的是本地的 binlog。
- --raw：以原生的二进制格式保存 binlog，是必需选项。如果不指定该选项，保存的则是 binlog 被 mysqlbinlog 解析后的文本内容。

- --user：复制用户，只需授予 REPLICATION SLAVE 权限。
- --stop-never：等待读取新的二进制日志事件，直到连接断开或命令终止。如果不指定该选项，则只会读取指定 binlog 的内容。
- mysql-bin.000001：在指定 --stop-never 的情况下，代表从哪个 binlog 开始读取。如果不指定 --stop-never，则只会读取指定的 binlog，此时可指定多个 binlog。

除了上述参数，其他相关的参数如下。

- --stop-never-slave-server-id：在指定了 --read-from-remote-server、--stop-never 后，mysqlbinlog 本质上就相当于一个从库。--stop-never-slave-server-id 用来指定从库的 server_id，默认为 1。在一个集群拓扑中，server_id 应全局唯一。

  如果从库的 server_id 出现冲突，主库会杀掉之前已经建立连接的从库的 binlog dump 线程，具体报错如下。

  ```
  ERROR: Got error reading packet from server: A slave with the same server_uuid/server_id as this slave
  has connected to the master; the first event 'mysql-bin.000001' at 4, the last event read from
  './mysql-bin.000009' at 989342, the last byte read from './mysql-bin.000009' at 989342.
  ```

- --to-last-log：读取到最后一个 binlog。指定 --stop-never 会隐式打开该选项。
- --result-file：设置 binlog 的存放路径。注意，如果将 --result-file 设置为目录，则目录末尾一定要带上目录分隔符 "/"，如 --result-file=/test/。如果将它设置为 /test，则保存到本地的文件名将为 /testmysql-bin.000001。

虽然命令很简单，也可以在终端直接运行，但这种方式有个弊端，就是 mysqlbinlog 断开后，不会自动重连。这不像复制，主从之间的连接断开后，有自动重连机制。

这个弊端可以通过 binlog_server.sh 脚本（已上传至 GitHub）来弥补。脚本内容如下。

```bash
#!/bin/bash

MBL=/usr/bin/mysqlbinlog
MYSQL_HOST=10.0.0.118
MYSQL_PORT=3306
MYSQL_USER=repl
MYSQL_PASS=repl123
BACKUP_DIR=/backup/binlogs/server1/
FIRST_BINLOG=mysql-bin.000001

# 重连的时间间隔
RESPAWN=10

# 创建备份目录
mkdir -p "$BACKUP_DIR"
cd "$BACKUP_DIR"

echo "Backup dir: $BACKUP_DIR"

while :
do
    if [ `ls -A "$BACKUP_DIR" |wc -l` -eq 0 ];then
        LAST_FILE="$FIRST_BINLOG"
    else
        LAST_FILE=`ls -l $BACKUP_DIR | tail -n 1 |awk '{print $9}'`
```

```
    fi
    echo "`date +"%Y/%m/%d %H:%M:%S"` starting binlog backup from $LAST_FILE"
    $MBL --read-from-remote-server --raw --stop-never --host=$MYSQL_HOST --port=$MYSQL_PORT
--user=$MYSQL_USER --password=$MYSQL_PASS $LAST_FILE
    echo "`date +"%Y/%m/%d %H:%M:%S"` mysqlbinlog exited with $? trying to reconnect in $RESPAWN seconds."
    sleep $RESPAWN
done
```

该脚本的逻辑比较简单，就是定义了一个死循环，如果 binlog 备份命令中途异常退出，10 秒后会自动重连。

最后让脚本在后台运行。

```
nohup sh binlog_server.sh 2>&1 > /var/log/binlog_server/server1.log &
```

5.8.2 参考资料

- 博客文章 "Backing up binary log files with mysqlbinlog"，作者：Tamas Kozak。

5.9 检测备份的有效性

为什么要检测备份的有效性？原因主要有两个。

- 验证整个备份环节的可靠性。

    包括备份参数是否准确完整、备份集是否有效、备份介质是否损坏等。

- 通过检查备份的有效性，搭建一套完整的自动化恢复体系。

    很多时候，影响数据库恢复时间的因素并不是备份集太老旧，而是在手动恢复过程中，因对命令、环境、流程不熟悉而产生的额外耗时。

检测备份的有效性，常用的方法有 3 个。

- 第一种方法是基于备份恢复实例，看实例能否恢复，并在此基础上进行随机查询。

    这种检测方法最简单。一般来说，实例能恢复，且随机查询也没问题，就意味着这个备份集是可用的。但备份集可用，并不意味着这个备份集能满足我们的需求，比如搭建从库的需求。而且一些常见的问题，如参数不准确，也无法通过这种方式检测出来。

- 第二种方法是在第一种方法的基础上，建立复制。

    如果从库在同步的过程中没有报错，大概率意味着主从数据是一致的。当然，只是大概率，并不是百分之百的。

- 第三种方法是在第二种方法的基础上，利用 pt-table-checksum 检查主从数据的一致性。

    如果检查结果没问题，则意味着主从数据是一致的，也就间接证明了备份的有效性。但因为 pt-table-checksum 在运行的过程中，会在 chunk 级别对表加 S 锁，这对更新频繁的业务还是有一定影响的。

一般来说，在线上使用第二种方法足矣。第三种方法因为要检查主从数据的一致性，耗时相对较久，如果要检测的备份集很多，反而会影响检测的效率。

5.10 本章总结

本章系统介绍了 MySQL 常用的备份工具。

- mysqldump。MySQL 官方自带的单线程逻辑备份工具。注意，对于 `--single-transaction`，只能保证事务表（InnoDB）的备份一致性，无法保证非事务表（MyISAM）的备份一致性。
- mydumper。业界开源的多线程逻辑备份工具。注意，只有指定 `--rows`，才能进行 chunk 级别的并行备份，否则只是进行表级别的并行备份。
- XtraBackup。Percona 开源的物理备份工具。
- 克隆插件。MySQL 8.0.17 引入。注意，在远程克隆中，如果没有指定 DATA DIRECTORY，数据会被拷贝到 Recipient 的数据目录下。此时，会通过 DROP 操作删除用户表空间。
- MySQL Shell Dump & Load。MySQL 官方新推出的多线程逻辑备份工具。备份速度高达 3GB/s，恢复速度高达 200MB/s。

在此基础上，本章还介绍了备份相关的两个高频操作。

- 搭建从库。

    如果是基于位置点的复制，要注意 CHANGE MASTER TO 语句中 MASTER_LOG_FILE 和 MASTER_LOG_POS 的选择。

    如果是 GTID 复制，在设置 GTID_PURGED 时，要注意备份实例的版本。在 MySQL 8.0 中，切记不要基于 xtrabackup_binlog_info 中的 GTID 信息来初始化 GTID_PURGED。

- 指定时间点（位置点）的恢复。注意 binlog 位置点的选择。

最后，本章介绍了如何基于 mysqlbinlog 搭建 binlog server，以及如何检测备份的有效性。在搭建 binlog server 时要注意，mysqlbinlog 早期版本存在一个 bug，会导致 binlog 中的最后一个二进制日志事件只会写入操作系统缓冲区，而不会持久化到文件中。该 bug 已经在 MySQL 5.7.19 和 MySQL 5.6.37 中修复。

重点问题回顾

- mysqldump 的常见用法及实现原理。
- 指定 `--master-data=2 --single-transaction`，mysqldump 可以保证非 InnoDB 表的备份一致性吗？
- START TRANSACTION WITH CONSISTENT SNAPSHOT 和 START TRANSACTION 的区别。
- mydumper 的常见用法及实现原理。
- mydumper 可以保证非 InnoDB 表的备份一致性吗？
- mydumper 基于 chunk 级别的适用场景。
- XtraBackup 的常见用法及实现原理。

- XtraBackup 中增量备份的实现原理。
- 如何使用 XtraBackup 进行实例拆分？
- 如何使用 XtraBackup 搭建从库？
- 克隆插件的常见用法及实现原理。
- 克隆插件与 XtraBackup 的异同点。
- util.dumpInstance 的实现原理。
- 如何进行指定时间点（位置点）的恢复？
- 如何搭建 binlog server？

第 6 章

监　控

与备份一样，监控的重要性也是毋庸置疑的。总的来说，监控的作用有以下两点。

- 事前告警。告警及时的话，能避免问题演变为故障，起到防患于未然的作用。
- 事后分析。很多时候，告警项只是表象，为了挖掘问题背后的根本原因，还要依赖其他监控项。如果监控项不丰富，或者采集不及时，我们常常很难挖掘出告警背后的真正原因。

所以，我们在选择监控方案时，一个核心的诉求就是监控项要足够丰富。

基于这个诉求，本章会介绍业内流行的两个开源监控方案。

- Zabbix。Zabbix 是一个企业级的开源分布式监控解决方案，由 Zabbix SIA 开发和维护，遵循 GNU 通用公共许可证（General Public License，GPL）V2 协议，百分之百免费和开源。Zabbix 简单，容易上手。除了需要将监控数据存储到外部数据库上，其他所有功能都已内置，包括告警和监控数据的可视化。它的发展时间较长，对于很多监控场景有现成的解决方案，成熟度相对更高。
- Prometheus。Prometheus 是一个开源的时间序列监控系统，灵感来自于 Google 的 Borgmon，由 Google 前员工于 2012 年在 SoundCloud 上创建。2016 年，它正式加入云原生计算基金会（Cloud Native Computing Foundation，CNCF），成为继 Kubernetes 之后的第二个托管项目。Prometheus 是目前最为热门的监控解决方案之一，尤其适合容器、云环境的监控。相对于 Zabbix，Prometheus 的学习曲线较为陡峭，告警需要依赖 Alertmanager，监控数据的可视化需要依赖 Grafana。但一旦熟悉了，你就会发现 Prometheus 的灵活性和可扩展性比 Zabbix 更高。

本章介绍的不是原生的 Prometheus，而是 PMM。PMM 的全称是 Percona Monitoring and Management，它是 Percona 公司基于 Prometheus 和 Grafana 开发的一个数据库监控管理平台，支持众多数据库组件，包括 MySQL、MongoDB、PostgreSQL、ProxySQL 等。这也是 Percona 公司目前大力推广的监控方案，更新迭代非常频繁。

本章主要包括以下内容。

- Zabbix。
- 安装 MySQL 监控插件 PMP。
- 深入理解 PMP。

- Zabbix 常见问题定位及性能优化。
- PMM。
- MySQL 中常用的监控指标。

6.1 Zabbix

Zabbix 主要由以下几个组件组成。

- Zabbix Server：Zabbix 服务端，是 Zabbix 的核心组件。它负责接收监控数据并触发告警，还负责将监控数据持久化到数据库中。
- Zabbix Agent：Zabbix 客户端，部署在被监控设备上，负责采集监控数据，并将采集后的数据发送给 Zabbix Server 处理。

  Zabbix Agent 目前有两个版本：Zabbix agent 和 Zabbix agent 2。前者是用 C 语言开发的，几乎支持所有的主流平台。后者是用 Go 语言开发的，优点包括：能有效降低 TCP 连接的数量，支持更高的并发，易于扩展。Zabbix agent 2 的目标是替代 Zabbix agent，目前只支持 Linux 和 Windows 两个平台。

- Zabbix Proxy：代替 Zabbix Server 接收监控数据并进行预处理，预处理后的数据将批量发送给 Zabbix Server，这样可减轻 Zabbix Server 的压力。
- Web 页面：可用来管理和维护被监控设备的配置信息、查看监控数据、配置告警等。
- 数据库：负责存储被监控设备的配置信息和监控数据。支持的数据库有：MySQL（Percona、MariaDB）、Oracle、PostgreSQL、TimescaleDB for PostgreSQL、SQLite。

6.1.1 安装 Zabbix Server

Zabbix 的演示环境如表 6-1 所示。

表 6-1 Zabbix 的演示环境

角　　色	IP	操作系统
Zabbix Server	192.168.244.128	CentOS 7.9
Zabbix Agent	192.168.244.10	CentOS 7.9

下面开始安装 Zabbix Server，具体步骤如下。

(1) 在 Zabbix 官网的下载页面（见图 6-1）上下载 Zabbix 软件包。

图 6-1　Zabbix 下载页面

(2) 解压源码包，建立软链接。

```
# cd /usr/src/
# tar xvf zabbix-6.0.2.tar.gz
# ln -s zabbix-6.0.2 zabbix
```

(3) 创建 zabbix 用户。

```
# groupadd --system zabbix
# useradd --system -g zabbix -d /usr/lib/zabbix -s /sbin/nologin -c "Zabbix Monitoring System" zabbix
```

(4) 创建 zabbix 数据库。

这里将 MySQL 作为 Zabbix 的后端数据库，使用的是 MySQL 8.0.27 二进制版本。

首先创建 zabbix 数据库和用户。

```
mysql> create database zabbix character set utf8 collate utf8_bin;
mysql> create user 'zabbix_admin'@'%' identified with mysql_native_password by 'zabbix_pass';
mysql> grant all privileges on zabbix.* to 'zabbix_admin'@'%';
```

接着初始化数据。

```
# cd /usr/src/zabbix/database/mysql/
# mysql -h127.0.0.1 -uzabbix_admin -pzabbix_pass zabbix < schema.sql
# mysql -h127.0.0.1 -uzabbix_admin -pzabbix_pass zabbix < images.sql
# mysql -h127.0.0.1 -uzabbix_admin -pzabbix_pass zabbix < data.sql
```

(5) 编译安装 Zabbix Server。

```
# yum install gcc mysql-devel libevent-devel libcurl-devel libxml2-devel net-snmp-devel
# cd /usr/src/zabbix
# ./configure --prefix=/usr/local/zabbix --enable-server --enable-agent
```

```
--with-mysql=/usr/local/mysql/bin/mysql_config --enable-ipv6 --with-net-snmp --with-libcurl
--with-libxml2
# make install
```

编译时需要注意以下两点。

- `configure` 命令中的 `--prefix` 是安装目录,如果不设置,则默认是 `/usr/local`。
- 在 `make install` 的过程中,如果出现如下错误:

```
/bin/ld: warning: libcrypto.so.1.1, needed by /usr/local/mysql/lib/libmysqlclient.so, not found (try using -rpath or -rpath-link)
/bin/ld: warning: libssl.so.1.1, needed by /usr/local/mysql/lib/libmysqlclient.so, not found (try using -rpath or -rpath-link)
```

则需要对依赖的两个库设置软链接。具体命令如下:

```
# find / -name libssl.so.1.1
/usr/local/mysql-8.0.27-linux-glibc2.12-x86_64/lib/private/libssl.so.1.1
# ln -s /usr/local/mysql-8.0.27-linux-glibc2.12-x86_64/lib/private/libssl.so.1.1 /usr/lib64
# ln -s /usr/local/mysql-8.0.27-linux-glibc2.12-x86_64/lib/private/libcrypto.so.1.1 /usr/lib64
```

编译完成后,我们看看 /usr/local/zabbix/ 目录中的内容。

```
# tree /usr/local/zabbix
/usr/local/zabbix
├── bin
│   ├── zabbix_get
│   ├── zabbix_js
│   └── zabbix_sender
├── etc
│   ├── zabbix_agentd.conf
│   ├── zabbix_agentd.conf.d
│   ├── zabbix_server.conf
│   └── zabbix_server.conf.d
├── lib
│   └── modules
├── sbin
│   ├── zabbix_agentd
│   └── zabbix_server
└── share
    ├── man
    │   ├── man1
    │   │   ├── zabbix_get.1
    │   │   └── zabbix_sender.1
    │   └── man8
    │       ├── zabbix_agentd.8
    │       └── zabbix_server.8
    └── zabbix
        ├── alertscripts
        └── externalscripts

14 directories, 11 files
```

这里面,etc 是配置文件目录。alertscripts 是告警脚本目录。externalscripts 是外部脚本目录。

(6) 修改配置文件。

首先,修改 Zabbix Server 的配置文件,修改后的配置如下所示。

```
# grep -Ev '^$|^#' /usr/local/zabbix/etc/zabbix_server.conf
LogFile=/tmp/zabbix_server.log
DBHost=127.0.0.1
DBName=zabbix
DBUser=zabbix_admin
DBPassword=zabbix_pass
DBPort=3306
Timeout=4
LogSlowQueries=3000
StatsAllowedIP=127.0.0.1
```

这里主要修改了连接数据库的相关参数。至于其他参数的调整，可参考 6.4 节。

其次，修改 Zabbix Agent 的配置文件，修改后的配置如下。

```
# grep -Ev '^$|^#' /usr/local/zabbix/etc/zabbix_agentd.conf
LogFile=/tmp/zabbix_agentd.log
Server=127.0.0.1
ServerActive=127.0.0.1
Hostname=Zabbix server
```

这里为什么也要修改 Zabbix Agent 的配置文件呢？因为 Zabbix Server 自身及所在服务器的监控数据也需要 Zabbix Agent 来采集，所以 Zabbix Agent 同样需要安装并启动。

(7) 配置服务管理脚本。

针对 Zabbix Server，主要修改 3 处：设置 zabbix_server 的路径，设置配置文件的路径，以及在启动时指定配置文件。

```
# vim /usr/src/zabbix/misc/init.d/fedora/core5/zabbix_server
...
ZABBIX_BIN="/usr/local/zabbix/sbin/zabbix_server"
CONFIG_FILE="/usr/local/zabbix/etc/zabbix_server.conf"
...
start() {
        echo -n $"Starting $prog: "
        daemon $ZABBIX_BIN -c $CONFIG_FILE
...

# cp /usr/src/zabbix/misc/init.d/fedora/core5/zabbix_server /etc/init.d/
```

针对 Zabbix Agent，同样调整这 3 处的内容。

```
# vim /usr/src/zabbix/misc/init.d/fedora/core5/zabbix_agentd
...
ZABBIX_BIN="/usr/local/zabbix/sbin/zabbix_agentd"
CONGIG_FILE="/usr/local/zabbix/etc/zabbix_agentd.conf"
...
start() {
        echo -n $"Starting $prog: "
        daemon $ZABBIX_BIN -c $CONGIG_FILE
...

# cp /usr/src/zabbix/misc/init.d/fedora/core5/zabbix_agentd /etc/init.d/
```

(8) 启动 Zabbix Server 和 Zabbix Agent。

```
# service zabbix_server start
# service zabbix_agentd start
```

启动 Zabbix Server 的过程中，如果提示以下错误：

```
Starting Zabbix Server: /usr/local/zabbix/sbin/zabbix_server: error while loading shared libraries:
libmysqlclient.so.21: cannot open shared object file: No such file or directory
```

同样可通过设置软链接来解决。

```
# ln -s /usr/local/mysql/lib/libmysqlclient.so.21 /usr/lib64
```

(9) 将 Zabbix Server 和 Zabbix Agent 设置为开机自启动。

```
# chkconfig zabbix_server on
# chkconfig zabbix_agentd on
```

安装完 Zabbix Server，接下来安装 Web 组件。

这里使用的 Web 服务器是 httpd，也可使用 Nginx。

(1) 安装 httpd 和 PHP。

方便起见，这里直接通过 yum 安装 httpd 和 PHP。从 Zabbix 5.0 开始，要求 PHP 的版本不低于 7.2。但在 CentOS 7 中，yum 源中默认的是 PHP 5.4.16。所以，这里需要安装额外的 epel 和 webtatic 源。

```
# rpm -Uvh https://dl.fedoraproject.org/pub/epel/epel-release-latest-7.noarch.rpm
# rpm -Uvh https://mirror.webtatic.com/yum/el7/webtatic-release.rpm
# yum install httpd php72w-cli php72w-common php72w-gd php72w-ldap php72w-mbstring php72w-mysqlnd
php72w-xml php72w-bcmath mod_php72w -y
```

(2) 将 Zabbix 源码包中的 PHP 文件拷贝到 httpd 的根目录下。

```
# mkdir -p /var/www/html/zabbix
# cp -r /usr/src/zabbix/ui/* /var/www/html/zabbix
# chown -R apache.apache /var/www/html/
```

(3) 修改 PHP 的配置文件。

```
# vim /etc/php.ini
max_execution_time = 300
max_input_time = 300
memory_limit = 128M
post_max_size = 16M
upload_max_filesize = 2M
date.timezone = Asia/Shanghai
```

注意，这里给出的只是最小需求值，可根据实际需要调大。

(4) 启动 httpd 服务。

```
# service httpd start
```

(5) 将 httpd 服务设置为开机自启动。

```
# chkconfig httpd on
```

安装完 Web 组件，接下来配置 Zabbix Web 页面。

(1) 登录 Zabbix Server Web 首页。

在浏览器中输入 http://192.168.244.128/zabbix，其中，192.168.244.128 是 Zabbix Server 的地址。

Zabbix Server Web 首页如图 6-2 所示。Default language 可用来设置安装过程中及 Zabbix 管理页面的语言，支持中文。

图 6-2　Zabbix Server Web 首页

(2) 进行依赖性检查。

如图 6-3 所示，依赖性检查主要检查 PHP 的插件、配置是否满足要求。如果不满足要求，最右边的 OK 将显示为 Fail。

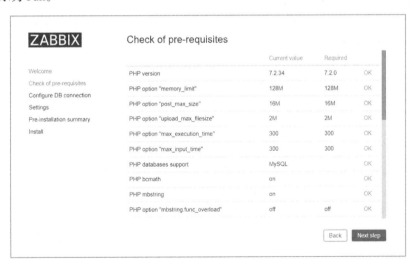

图 6-3　进行依赖性检查

(3) 配置数据库连接信息。

在我们的演示环境中，因为 Zabbix Server 和数据库部署在同一台主机上，所以这里的 Database host

设置为 127.0.0.1，如图 6-4 所示。在生产环境中建议分开部署。

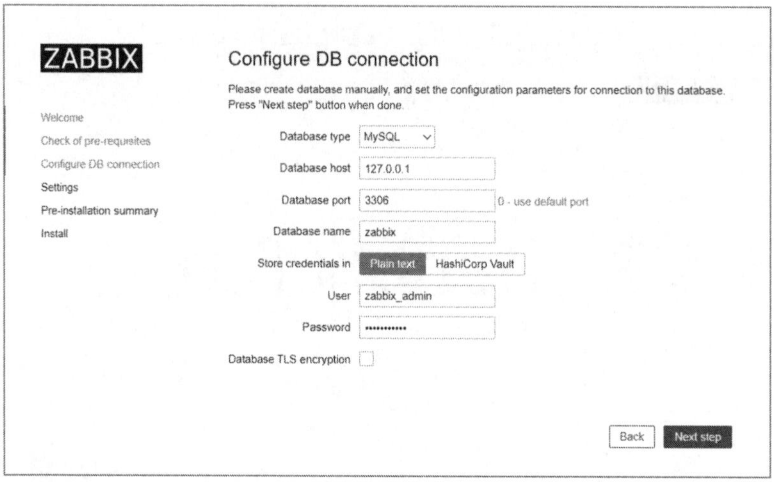

图 6-4　配置数据库连接信息

如果配置都正确，还是提示"Cannot connect to the database. Permission denied"，可检查服务器上的 SELinux 是否关闭。

(4) 设置 Zabbix Server 信息。

Zabbix Server 信息的设置如图 6-5 所示。

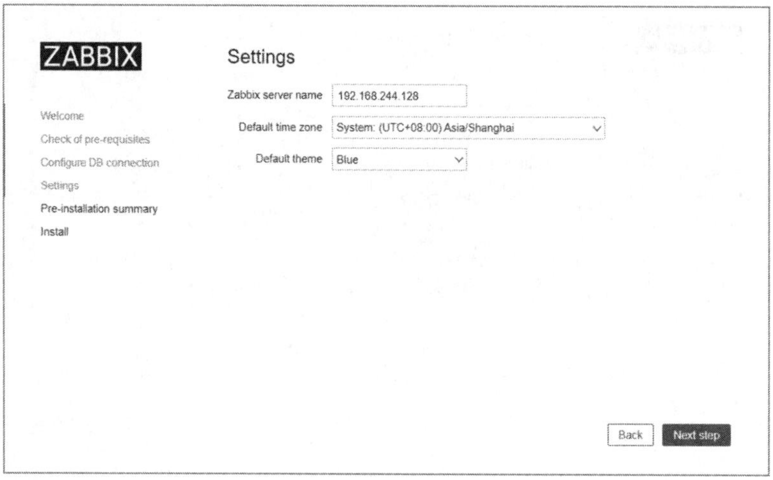

图 6-5　设置 Zabbix Server 信息

注意，这里的 Zabbix server name 会显示在管理页面导航栏的左上方，无其他作用。

Default theme 是 Zabbix 管理页面使用的默认主题。

(5) 检查配置是否正确。

如图 6-6 所示，如果配置不正确，可点击 Back 退回到之前的页面修改。

图 6-6　检查配置是否正确

(6) 安装并显示结果。

安装结果如图 6-7 所示。

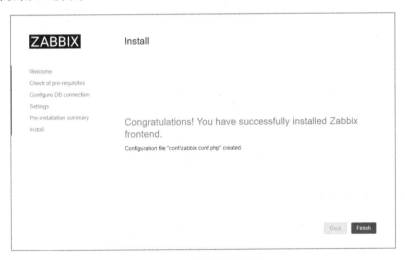

图 6-7　安装并显示结果

(7) 打开 Zabbix 登录页面。

图 6-8 是安装成功后 Zabbix 的登录页面。默认的用户名是 Admin，密码是 zabbix。Admin 是管理员账户。

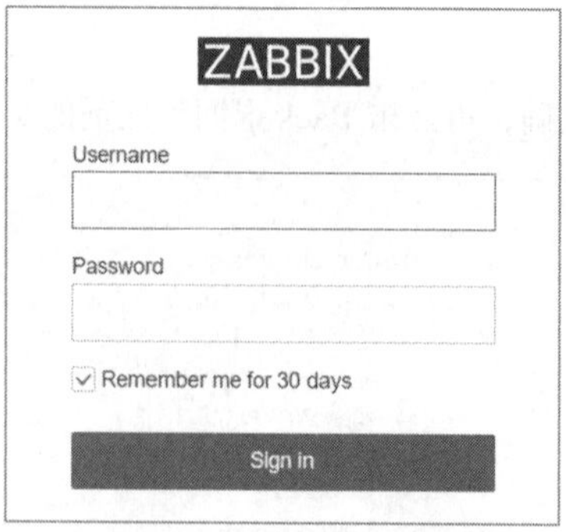

图 6-8　Zabbix 登录页面

(8) 进入 Zabbix 首页。

登录成功后的页面如图 6-9 所示。

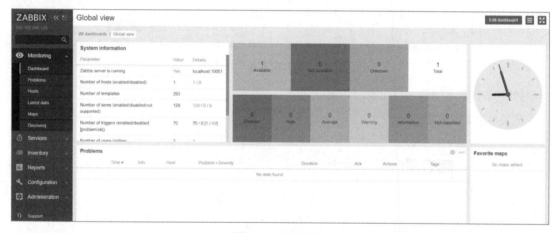

图 6-9　登录首页

如果登录后显示"Zabbix server is not running: the information displayed may not be current"，可通过 Zabbix Server 的日志（默认是 /tmp/zabbix_server.log）来定位问题。

6.1.2　安装 Zabbix Agent

在被监控主机（192.168.244.10）上部署同样的 Zabbix 源码包。

(1) 编译安装 Zabbix Agent。

```
# groupadd --system zabbix
# useradd --system -g zabbix -d /usr/lib/zabbix -s /sbin/nologin -c "Zabbix Monitoring System" zabbix
# cd /usr/src/
# tar xvf zabbix-6.0.2.tar.gz
# ln -s zabbix-6.0.2 zabbix
# yum install gcc pcre-devel -y
# cd zabbix
```

```
# ./configure --prefix=/usr/local/zabbix --enable-agent
# make install
```

(2) 修改配置文件。

```
# grep -Ev '^$|^#' /usr/local/zabbix/etc/zabbix_agentd.conf
LogFile=/tmp/zabbix_agentd.log
Server=192.168.244.128
ServerActive=127.0.0.1
Hostname=node1
```

部分配置项的具体如下。

- `Server`：被动模式下 Zabbix Server 的地址。在被动模式下，只有接受到 Zabbix Server 发送的请求，Zabbix Agent 才会响应数据。被动模式是 Zabbix Agent 默认的工作模式。
- `ServerActive`：主动模式下 Zabbix Server 的地址。在主动模式下，Zabbix Agent 会将采集到的数据主动发送给 Zabbix Server。
- `Hostname`：主机名。注意，这里的主机名并不一定是主机的主机名，也可以是 IP 或其他有标识性的字符串，只需保证这个配置在 Zabbix Server 监控的主机内全局唯一即可。`Hostname` 只适用于主动模式，在被动模式下无须设置。

(3) 配置服务管理脚本。

```
# vim /usr/src/zabbix/misc/init.d/fedora/core5/zabbix_agentd
...
ZABBIX_BIN="/usr/local/zabbix/sbin/zabbix_agentd"
CONGIG_FILE="/usr/local/zabbix/etc/zabbix_agentd.conf"
...
start() {
        echo -n $"Starting $prog: "
        daemon $ZABBIX_BIN -c $CONGIG_FILE
...
# cp /usr/src/zabbix/misc/init.d/fedora/core5/zabbix_agentd /etc/init.d/
```

(4) 启动 Zabbix Agent。

```
# service zabbix_agentd start
```

(5) 设置为开机自启动。

```
# chkconfig zabbix_agentd on
```

6.2 安装 MySQL 监控插件 PMP

PMP 的全称是 Percona Monitoring Plugins，是 Percona 公司为 MySQL 监控编写的插件，支持 Nagios、Cacti 和 Zabbix。

下面看看 PMP 的具体使用步骤。注意，下述操作都是在 Zabbix Agent 上执行的。

(1) 下载 PMP。

如图 6-10 所示，官方提供了多个平台的安装包，大家可基于线上环境自行选择。此处使用的是 CentOS 7 的 rpm 包。注意，官方已不提供安装包的下载。相关的 rpm 包（percona-zabbix-templates-

1.1.8-1.noarch.rpm）及压缩包（percona-zabbix-templates-1.1.8-1.tar.gz，适用于 Red Hat Enterprise Linux、CentOS 和 Oracle Linux 外的其他平台）已上传至 GitHub。

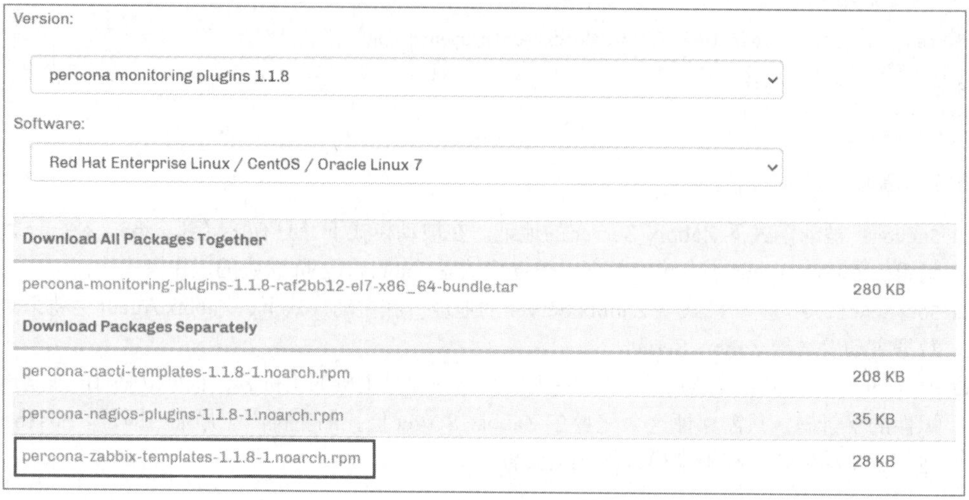

图 6-10　PMP 安装包

(2) 安装 PMP。

```
# rpm -ivh percona-zabbix-templates-1.1.8-1.noarch.rpm
warning: percona-zabbix-templates-1.1.8-1.noarch.rpm: Header V4 DSA/SHA1 Signature, key ID cd2efd2a: NOKEY
Preparing...                          ################################# [100%]
Updating / installing...
   1:percona-zabbix-templates-1.1.8-1 ################################# [100%]

Scripts are installed to /var/lib/zabbix/percona/scripts
Templates are installed to /var/lib/zabbix/percona/templates
```

基于安装过程中的提示，脚本和模板都安装在 /var/lib/zabbix/percona/ 目录下。我们看看这个目录下的内容。

```
# tree /var/lib/zabbix/percona/
/var/lib/zabbix/percona/
├── scripts
│   ├── get_mysql_stats_wrapper.sh
│   └── ss_get_mysql_stats.php
└── templates
    ├── userparameter_percona_mysql.conf
    └── zabbix_agent_template_percona_mysql_server_ht_2.0.9-sver1.1.8.xml

2 directories, 4 files
```

这里的 get_mysql_stats_wrapper.sh 和 ss_get_mysql_stats.php 用于采集 MySQL 数据。后者是核心，是用 PHP 开发的。前者是 shell 脚本，只不过对后者进行了一层封装。

userparameter_percona_mysql.conf 定义了监控项。例如：

```
UserParameter=MySQL.Sort-scan,/var/lib/zabbix/percona/scripts/get_mysql_stats_wrapper.sh kt
UserParameter=MySQL.slave-stopped,/var/lib/zabbix/percona/scripts/get_mysql_stats_wrapper.sh jh
```

zabbix_agent_template_percona_mysql_server_ht_2.0.9-sver1.1.8.xml 是 Zabbix 模板文件。

(3) 创建 MySQL 监控用户。

```
mysql> create user 'monitor_user'@'localhost' identified with mysql_native_password by 'monitor_pass';
mysql> grant process, replication client on *.* to 'monitor_user'@'localhost';
```

(4) 修改监控脚本。

```
# cd /var/lib/zabbix/percona/scripts/
# vim ss_get_mysql_stats.php
$mysql_user = 'monitor_user';
$mysql_pass = 'monitor_pass';
$mysql_port = 3306;
$mysql_socket = '/data/mysql/3306/data/mysql.sock';
```

这一步主要是调整数据库的连接信息。

(5) 测试监控脚本。

```
# yum install php-cli php-mysql -y
# php /var/lib/zabbix/percona/scripts/ss_get_mysql_stats.php --host localhost --items iu
iu:3
# rm -rf /tmp/localhost-mysql_cacti_stats.txt
```

在安装的依赖包中，php-cli 是 PHP 命令行工具，php-mysql 是 MySQL 的 PHP 驱动。测试命令中的 iu 对应的监控项是 Threads_connected，如果有值返回，则代表脚本能正常采集数据。采集结果默认缓存在 /tmp/localhost-mysql_cacti_stats.txt 中。为了避免对后续的测试造成影响，测试完后需删除该文件。

(6) 拷贝监控项文件。

将监控项文件拷贝到 Zabbix Agent 配置目录下。

```
# cp /var/lib/zabbix/percona/templates/userparameter_percona_mysql.conf
/usr/local/zabbix/etc/zabbix_agentd.conf.d/
```

(7) 编辑 Zabbix Agent 的配置文件。

允许 Zabbix Agent 使用 Include 目录下的配置文件。

```
# vim /usr/local/zabbix/etc/zabbix_agentd.conf
Include=/usr/local/zabbix/etc/zabbix_agentd.conf.d/
```

(8) 重启 Zabbix Agent。

```
# service zabbix_agentd restart
```

(9) 导入 PMP 模板。

这里要导入的 PMP 模板即 /var/lib/zabbix/percona/templates 下的 zabbix_agent_template_percona_mysql_server_ht_2.0.9-sver1.1.8.xml。

单击 Configuration -> Template -> Import，如图 6-11 所示。然后选择 PMP 模板并导入，如图 6-12 所示。

图 6-11　单击 Configuration -> Template -> Import

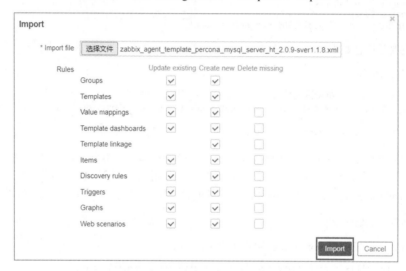

图 6-12　导入 PMP 模板

在导入的过程中却报错了，具体的错误信息如图 6-13 所示。

图 6-13　导入时报错

即使按照错误提示修改模板文件的时间戳，重新导入，还是会报错。这一次的错误信息如图 6-14 所示。

图 6-14　重新导入，再次报错

6.2 安装 MySQL 监控插件 PMP

按照官方文档的说法，这个模板只在 Zabbix 2.0.9 上测试过。很显然，这个模板并不适用于 Zabbix 6.x。如果要使用这个模板，难道就只能使用 Zabbix 2.x 版本？实际上，这里有个取巧的方法，即先将模板导入 Zabbix 2.x 再导出，这样导出的模板就可以在 Zabbix 6.x 中使用了。导出后的模板（zbx_export_templates.xml）已上传至 GitHub。

模板导入后，如图 6-15 所示。

图 6-15　PMP 模板

(10) 添加主机，关联 PMP 模板。

单击 Configuration -> Hosts -> Create host。

如图 6-16 所示，在 Host 页面上，常填的配置项有如下 4 个。

- Host name：主机名。在主动模式下，这里填的主机名应该与 Zabbix 客户端配置文件中的 Hostname 保持一致。在被动模式下，则无此要求。
- Templates：指定需要关联的模版。
- Groups：每个主机至少应属于一个主机组，下面选择的是 Percona Templates。
- Interfaces：填写客户端的 IP 和端口。

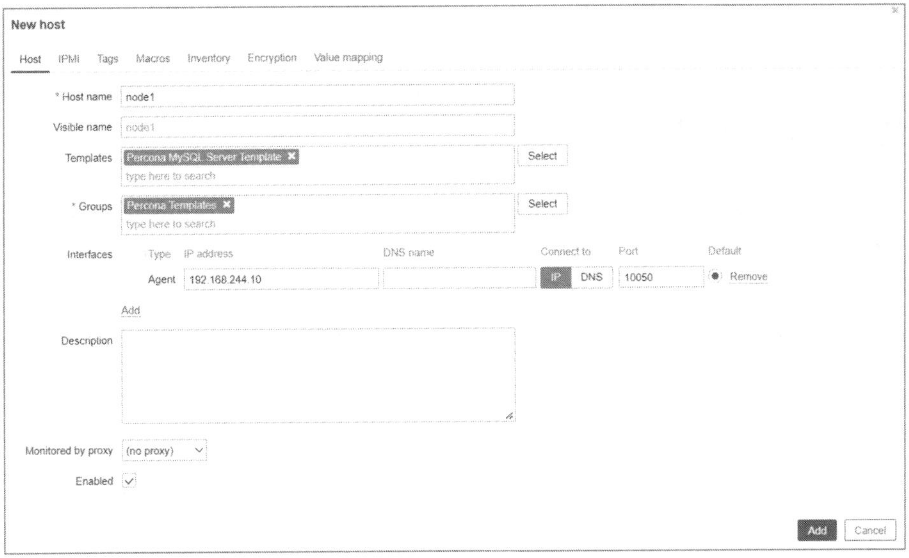

图 6-16　添加主机

(11) 确认模板是否关联成功。

单击 Configuration -> Hosts。主机关联的模板如图 6-17 所示。

图 6-17　查看主机关联模板

(12) 确认能否正常采集数据。

单击 Monitoring -> Latest data。在 Hosts 处选择主机，查看监控项的 Last value 列，如果 Last value 有值，则代表数据采集成功，如图 6-18 所示。

图 6-18　查看主机采集的数据

6.3　深入理解 PMP

本节主要从源码的角度分析 PMP 的实现逻辑。PMP 采集脚本由两部分组成：get_mysql_stats_wrapper.sh 和 ss_get_mysql_stats.php。

首先看看 get_mysql_stats_wrapper.sh 的内容。

```
# cat /var/lib/zabbix/percona/scripts/get_mysql_stats_wrapper.sh
#!/bin/sh

ITEM=$1
HOST=localhost
DIR=`dirname $0`
CMD="/usr/bin/php -q $DIR/ss_get_mysql_stats.php --host $HOST --items gg"
CACHEFILE="/tmp/$HOST-mysql_cacti_stats.txt"

if [ "$ITEM" = "running-slave" ]; then
    RES=`HOME=~zabbix mysql -e 'SHOW SLAVE STATUS\G' | egrep '(Slave_IO_Running|Slave_SQL_Running):'
```

```
        | awk -F: '{print $2}' | tr '\n' ','`
        if [ "$RES" = " Yes, Yes," ]; then
            echo 1
        else
            echo 0
        fi
        exit
elif [ -e $CACHEFILE ]; then
        TIMEFLM=`stat -c %Y /tmp/$HOST-mysql_cacti_stats.txt`
        TIMENOW=`date +%s`
        if [ `expr $TIMENOW - $TIMEFLM` -gt 300 ]; then
            rm -f $CACHEFILE
            $CMD 2>&1 > /dev/null
        fi
else
    $CMD 2>&1 > /dev/null
fi

if [ -e $CACHEFILE ]; then
    cat $CACHEFILE | sed 's/ /\n/g; s/-1/0/g'| grep $ITEM | awk -F: '{print $2}'
else
    echo "ERROR: run the command manually to investigate the problem: $CMD"
fi
```

该脚本接收一个参数，用作 Item 值。PMP 中支持的 Item 值及对应的监控项可在 userparameter_percona_mysql.conf 里查看，比如下面的 kt。

```
UserParameter=MySQL.Sort-scan,/var/lib/zabbix/percona/scripts/get_mysql_stats_wrapper.sh kt
```

接下来分析一下脚本的处理流程。

(1) 将接收到的参数赋值给 ITEM。

(2) 判断 ITEM 是否为 running-slave。如果是，则通过 SHOW SLAVE STATUS 检查从库的状态。

注意，这里有个"坑"：mysql -e 'SHOW SLAVE STATUS\G' 中没有指定用户名和密码，在执行时会提示 ERROR 1045 (28000): Access denied for user 'root'@'localhost' (using password: NO)，这样会导致 RES 的结果为空。虽然 RES 的结果为空，会执行 else 分支的内容，但这个错误信息会被 Zabbix Server 捕捉。

如果 ITEM 不是 running-slave，则会判断 CACHEFILE 是否存在。如果不存在，则会执行 CMD 命令，即 "/usr/bin/php -q $DIR/ss_get_mysql_stats.php --host $HOST --items gg"。这条命令会采集监控数据，并将其保存到 CACHEFILE 中。

(3) 如果 CACHEFILE 存在，则会判断 CACHEFILE 的修改时间是否在 300 秒之前。如果是，则意味着这个文件已经很"旧"了，这个时候会删除文件，执行 CMD 命令，重新采集数据。

将监控数据放到临时文件中，其实是一个很明智的做法。这就避免了每采集一个监控项就要调用一次脚本。

我们来简单评估一下这种方式的影响。

PMP 一共有 191 个监控项，剔除 running-slave 后，剩下 190 个。如果每采集一个监控项就调用一次脚本，那就意味着，在一个采集周期内（通常是 60 秒），会调用脚本 190 次。每调用一次，都会在数据库中执行如下操作：

```
SHOW GLOBAL STATUS
SHOW VARIABLES
SHOW SLAVE STATUS
SHOW MASTER LOGS
SHOW PROCESSLIST
SHOW INNODB STATUS
```

这么高频率的执行，无疑会增加数据库的负载。

如果 CACHEFILE 存在，且修改时间在 300 秒之内，则会直接从 CACHEFILE 中获取 Item 的采集值。

6.3.1 ss_get_mysql_stats.php 源码分析

接下来分析一下 ss_get_mysql_stats.php。分析的初衷很简单，就是为了自定义监控项。

下面对原脚本进行了精简处理，只保留了必要的配置项、主流程和相关函数名。如此，可以很容易地看出脚本的主要轮廓。

```php
# 配置
$mysql_user = 'monitor';
$mysql_pass = 'monitor_pass';
$mysql_port = 3306;
$mysql_socket = '/data/mysql/3306/data/mysql.sock';
$mysql_flags = 0;
$mysql_ssl = FALSE; # 是否使用 SSL 连接数据库
$mysql_ssl_key = '/etc/pki/tls/certs/mysql/client-key.pem';
$mysql_ssl_cert = '/etc/pki/tls/certs/mysql/client-cert.pem';
$mysql_ssl_ca = '/etc/pki/tls/certs/mysql/ca-cert.pem';
$mysql_connection_timeout = 5;

$heartbeat = FALSE; # 是否基于心跳表来计算主从延迟
$heartbeat_utc = FALSE; # pt-heartbeat 运行过程中是否指定了 --utc 选项
$heartbeat_server_id = 0; # 查询心跳表时指定的 server_id
$heartbeat_table = 'percona.heartbeat'; # 心跳表

$cache_dir = '/tmp'; # cache_file 的目录。cache_file 就是 get_mysql_stats_wrapper.sh 中的 CACHEFILE
$poll_time = 300; # 与 cache_file 的处理有关，具体可见对 ss_get_mysql_stats 函数的分析
$timezone = null;
# 是否开启额外的监控项，默认只会采集 variable (通过 SHOW VARIABLES 命令)
# 和 status (通过 SHOW GLOBAL STATUS 命令)
$chk_options = array (
    'innodb' => true, # 是否采集 InnoDB 的状态信息 (SHOW ENGINE INNODB STATUS)
    'master' => true, # 是否采集 binlog 的大小 (SHOW MASTER LOGS)
    'slave'  => true, # 是否采集从库的状态信息 (SHOW SLAVE STATUS)
    'procs'  => true, # 是否采集 SHOW PROCESSLIST 的输出
    'get_qrt' => true, # 采集 SQL 的响应时间，只适用于 Percona 和 MariaDB。如果使用的是 MySQL，可将该选项设置为 false
);

$use_ss = FALSE;
$debug = FALSE; # 是否开启 debug 模式
$debug_log = FALSE; # 指定 debug 日志的路径
...
```

```php
# 脚本的主流程
if (!isset($called_by_script_server)) {
   debug($_SERVER["argv"]);
   array_shift($_SERVER["argv"]);
   # 通过 parse_cmdline 函数解析命令行参数
   $options = parse_cmdline($_SERVER["argv"]);
   # 通过 validate_options 函数验证参数的有效性
   validate_options($options);
   # 通过 ss_get_mysql_stats 函数获取采集结果
   # 采集结果类似于 gg:0 gh:0 gi:0 gj:0 gk:0
   $result = ss_get_mysql_stats($options);
   debug($result);
   if ( !$debug ) {
      ob_end_clean();
   }
   else {
      ob_end_flush();
   }
   # $wanted 是命令行通过 --items 参数传入的采集项
   $wanted = explode(',', $options['items']);
   $output = array();
   # 循环遍历采集结果，判断采集项是否在采集结果中。如果在，则赋值给 $output，最后通过 print 函数输出
   foreach ( explode(' ', $result) as $item ) {
      if ( in_array(substr($item, 0, 2), $wanted) ) {
         $output[] = $item;
      }
   }
   debug(array("Final result", $output));
   print(implode(' ', $output));
}

# 验证命令行参数的有效性
function validate_options($options)

# 打印脚本的用法
function usage($message)

# 解析命令行参数
function parse_cmdline( $args )

# 脚本的主流程
function ss_get_mysql_stats( $options )

# 将 SHOW ENGINE INNODB STATUS 的结果以 key-value 的形式输出
function get_innodb_array($text, $mysql_version)

# 返回的是 big_add(big_multiply($hi, 4294967296), $lo)
function make_bigint ($hi, $lo)

# 从字符串中提取数字
function to_int ( $str )

# 封装了 mysqli_query
function run_query($sql, $conn)

# 加上指定的增量值
function increment(&$arr, $key, $howmuch)

# 两个数相乘
function big_multiply ($left, $right, $force = null)
```

```
# 两个数相减
function big_sub ($left, $right, $force = null)

# 两个数相加
function big_add ($left, $right, $force = null)

# 将 debug 信息写入日志
function debug($val)
```

我们具体分析一下 ss_get_mysql_stats 函数。该函数的代码量较多，这里只分析主要的处理逻辑。

- 处理 MySQL 连接选项。
- 创建 cache_file。

  cache_file 的生成规则如下：

  ```
  $sanitized_host = str_replace(array(":", "/"), array("", "_"), $host);
  $cache_file = "$cache_dir/$sanitized_host-mysql_cacti_stats.txt" . ($port != 3306 ? ":$port" : '');
  debug("Cache file is $cache_file");
  ```

  其中，$sanitized_host 是处理后的主机名，$cache_dir 是 cache_file 所在的目录，在脚本开头定义为 /tmp。如果端口不是 3306，则生成的 cache_file 后面还会带上端口号。这么做实际上考虑了单机多实例的情况。

  但在 get_mysql_stats_wrapper.sh 中，默认的 cache_file 为 /tmp/localhost-mysql_cacti_stats.txt。所以，如果实例的端口不是 3306，会导致两个脚本中的 cache_file 不一致。这个时候，就需要相应地调整 get_mysql_stats_wrapper.sh 中 CACHEFILE 的路径。

- 检查 cache_file。

  该部分的处理逻辑与 get_mysql_stats_wrapper.sh 脚本类似。

  ```
  $fp = null;
  if ( $cache_dir && !array_key_exists('nocache', $options) ) { # nocache 选项可在命令行中指定
    if ( $fp = fopen($cache_file, 'a+') ) {
       $locked = flock($fp, 1); # flock 中的 1 指的是 LOCK_SH。首先判断是否能获取到共享锁
       if ( $locked ) {
       # 如果能，则判断当前时间是否小于 cache_file 的修改时间加 poll_time/2。
       # 如果小于，则直接返回 cache_file 的内容作为这次的采集结果
         if ( filesize($cache_file) > 0
             && filectime($cache_file) + ($poll_time/2) > time()
             && ($arr = file($cache_file))
         ) {
            debug("Using the cache file");
            fclose($fp);
            return $arr[0];
         }
         else {
         # 如果大于，则代表文件比较"旧"，需要重新采集监控数据，这个时候，会尝试获取
         # 排他锁 debug("The cache file seems too small or stale");
            if ( flock($fp, 2) ) {
            # 在获取到排他锁的时候，会再次判断文件是否"旧"了。如果不旧，则直接返回文件的内容。
            # 为什么会再次判断呢？因为在执行加锁操作时，并不一定能马上获取到锁。
            # 有可能在获取到的时候，文件已经更新了
               if ( filesize($cache_file) > 0
  ```

```
                     && filectime($cache_file) + ($poll_time/2) > time()
                     && ($arr = file($cache_file))
                ) {
                    debug("Using the cache file");
                    fclose($fp);
                    return $arr[0];
                }
                # 如果文件确实"旧"了，则直接清空文件的内容
                ftruncate($fp, 0);
            }
        }
    }
    else {
      $fp = null;
      debug("Couldn't lock the cache file, ignoring it");
    }
  }
  else {
      $fp = null;
      debug("Couldn't open the cache file");
  }
}
else {
  debug("Caching is disabled.");
}
```

从这里可以看出，$poll_time 应设置为 get_mysql_stats_wrapper.sh 中时间的 2 倍。如果监控的采集频率为 60 秒，则 shell 脚本的时间应设置为 60，同时 $poll_time 设置为 120。

- 与 MySQL 建立连接。
- 执行 SHOW GLOBAL STATUS 和 SHOW VARIABLES，结果存储在字典中。

```
$result = run_query("SHOW /*!50002 GLOBAL */ STATUS", $conn);
foreach ( $result as $row ) {
    $status[$row[0]] = $row[1];
}

$result = run_query("SHOW VARIABLES", $conn);
foreach ( $result as $row ) {
    $status[$row[0]] = $row[1];
}
```

- 如果 $chk_options 中的 slave 为 true，会执行 SHOW SLAVE STATUS 采集从库的复制状态。

```
if ( $chk_options['slave'] ) {
  $result = run_query("SHOW SLAVE STATUS NONBLOCKING", $conn);
  if ( !$result ) {
      $result = run_query("SHOW SLAVE STATUS NOLOCK", $conn);
      if ( !$result ) {
        $result = run_query("SHOW SLAVE STATUS", $conn);
      }
  }
}
```

- 如果 $chk_options 中的 master 为 true，会执行 SHOW MASTER LOGS，获取 binlog 的总大小。

```
if ( $chk_options['master']
     && array_key_exists('log_bin', $status)
     && $status['log_bin'] == 'ON'
) {
```

```
    $binlogs = array(0);
    $result = run_query("SHOW MASTER LOGS", $conn);
    foreach ( $result as $row ) {
      $row = array_change_key_case($row, CASE_LOWER);
      if ( array_key_exists('file_size', $row) && $row['file_size'] > 0 ) {
          $binlogs[] = $row['file_size'];
      }
    }
    if (count($binlogs)) {
      $status['binary_log_space'] = to_int(array_sum($binlogs));
    }
}
```

- 如果 $chk_options 中的 procs 为 true，会执行 SHOW PROCESSLIST，汇总各个连接的状态。

```
if ( $chk_options['procs'] ) {
    $result = run_query('SHOW PROCESSLIST', $conn);
    foreach ( $result as $row ) {
        $state = $row['State'];
        if ( is_null($state) ) {
            $state = 'NULL';
        }
        if ( $state == '' ) {
            $state = 'none';
        }
        $state = preg_replace('/^(Table lock|Waiting for .*lock)$/', 'Locked', $state);
        $state = str_replace(' ', '_', strtolower($state));
        if ( array_key_exists("State_$state", $status) ) {
            increment($status, "State_$state", 1);
        }
        else {
            increment($status, "State_other", 1);
        }
    }
}
```

- 如果 $chk_options 中的 innodb 为 true，会执行 SHOW ENGINE INNODB STATUS，获取到的结果会通过 get_innodb_array 解析。

```
if ( $chk_options['innodb']
     && array_key_exists('InnoDB', $engines)
     && $engines['InnoDB'] == 'YES'
     || $engines['InnoDB'] == 'DEFAULT'
) {
    $result        = run_query("SHOW /*!50000 ENGINE*/ INNODB STATUS", $conn);
    $istatus_text = $result[0]['Status'];
    $istatus_vals = get_innodb_array($istatus_text, $mysql_version);
```

- 统计 SQL 的响应时间分布情况，这只适用于 Percona Server 和 MariaDB。

```
if ( $chk_options['get_qrt']
     && (( isset($status['have_response_time_distribution'])
     && $status['have_response_time_distribution'] == 'YES')
     || (isset($status['query_response_time_stats'])
     && $status['query_response_time_stats'] == 'ON')) )
{
    debug('Getting query time histogram');
    $i = 0;
    $result = run_query(
```

```
            "SELECT `count`, ROUND(total * 1000000) AS total "
          . "FROM INFORMATION_SCHEMA.QUERY_RESPONSE_TIME "
          . "WHERE `time` <> 'TOO LONG'",
        $conn);
        ...
    }
    else {
        debug('Not getting time histogram because it is not enabled');
    }
```

- 定义监控项，并为每个监控项设置一个简短的名字。之所以这么做，主要是为了规避 Cacti（一个老牌监控软件）中的限制。

```
$keys = array(
    'Key_read_requests'         => 'gg',
    'Key_reads'                 => 'gh',
    'Key_write_requests'        => 'gi',
    'Key_writes'                => 'gj',
    'history_list'              => 'gk',
    'innodb_transactions'       => 'gl',
    'read_views'                => 'gm',
    'current_transactions'      => 'gn',
    ...
    'pool_reads'                => 'qo',
    'pool_read_requests'        => 'qp',
);
```

- 将采集结果以 key:value 的形式写入 cache_file，其中，key 是重命名后的采集项。如果某些监控项没有采集到数据，则将其结果设置为 -1。

```
$output = array();
foreach ($keys as $key => $short ) {
    $val      = isset($status[$key]) ? $status[$key] : -1;
    $output[] = "$short:$val";
}
$result = implode(' ', $output);
if ( $fp ) {
    if ( fwrite($fp, $result) === FALSE ) {
        die("Can't write '$cache_file'");
    }
    fclose($fp);
}
return $result;
```

如此，我们已将 PMP 的关键流程梳理了一遍。

6.3.2 基于 ss_get_mysql_stats.php 自定义监控项

PMP 虽然提供了 191 个监控项，但还是缺少很多监控指标，比如我们平时比较关注的 TPS。

下面来看看如何基于 ss_get_mysql_stats.php 自定义监控项。

因为 TPS 等于 Com_commit 的每秒增量加上 Com_rollback 的每秒增量。所以首先必须采集 Com_commit 和 Com_rollback 的值，而这两个状态变量就存在于 SHOW GLOBAL STATUS 的输出中。

(1) 往 $keys 中添加监控项。

可以看到，监控项的映射名也比较有规则，按照字母顺序递增。

```
$keys = array(
    'Key_read_requests' => 'gg',
    'Key_reads' => 'gh',
    'Key_write_requests' => 'gi',
    'Key_writes' => 'gj',
    'history_list' => 'gk',
    'innodb_transactions' => 'gl',
    'read_views' => 'gm',
    'current_transactions' => 'gn',
    ...
    'pool_reads' => 'qo',
    'pool_read_requests' => 'qp',
    'Com_commit' => 'qq',
    'Com_rollback' => 'qr',
);
```

(2) 修改 PMP 配置文件，新增两个监控项。

```
# vim /usr/local/zabbix/etc/zabbix_agentd.conf.d/userparameter_percona_mysql.conf
UserParameter=MySQL.Com-commit,/var/lib/zabbix/percona/scripts/get_mysql_stats_wrapper.sh qq
UserParameter=MySQL.Com-rollback,/var/lib/zabbix/percona/scripts/get_mysql_stats_wrapper.sh qr
```

这里的 `MySQL.Com-commit` 是 key，在前端页面配置监控项时会用到。

`/var/lib/zabbix/percona/scripts/get_mysql_stats_wrapper.sh qq` 是命令，命令的结果即 key 的值。

(3) 重启 Zabbix Agent，让 PMP 配置文件生效。

```
# service zabbix_agentd restart
```

(4) 配置 Zabbix 监控项。

点击 Configuration -> Templates，选择 Percona MySQL Server Template，点击 Items，如图 6-19 所示。

图 6-19　配置监控项

要新增监控项，单击 Create item。

首先，配置 Com Commit 和 Com Rollback，两个配置项的 Preprocessing 需要配置为 Change per second，代表每秒增量。具体配置如图 6-20 和图 6-21 所示。

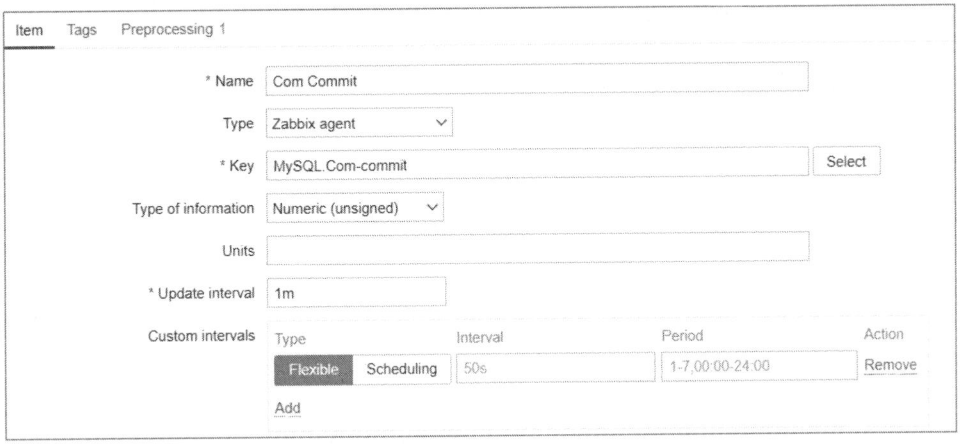

图 6-20　配置 Com Commit

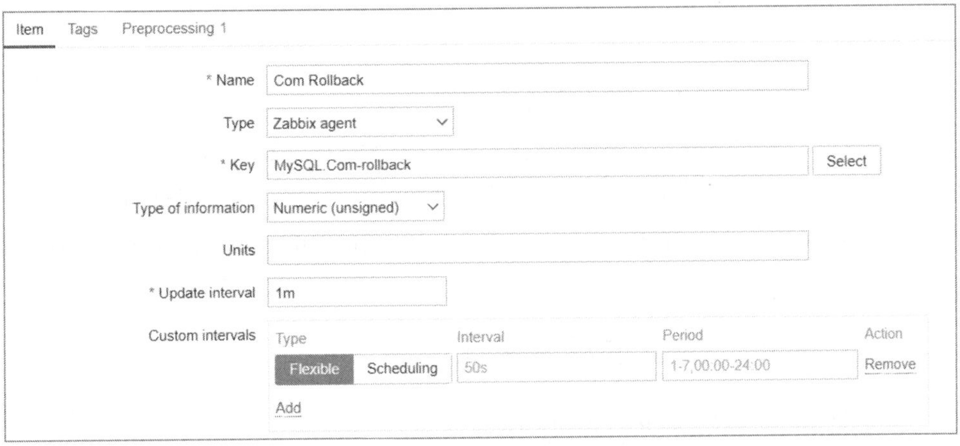

图 6-21　配置 Com Rollback

其次，配置 TPS。Type 需设置为 Calculated，代表该监控项的值是通过计算而不是采集得来的。具体的计算公式在 Formula 中设置，`last()` 指的是获取监控项的最近一个值。具体配置如图 6-22 所示。

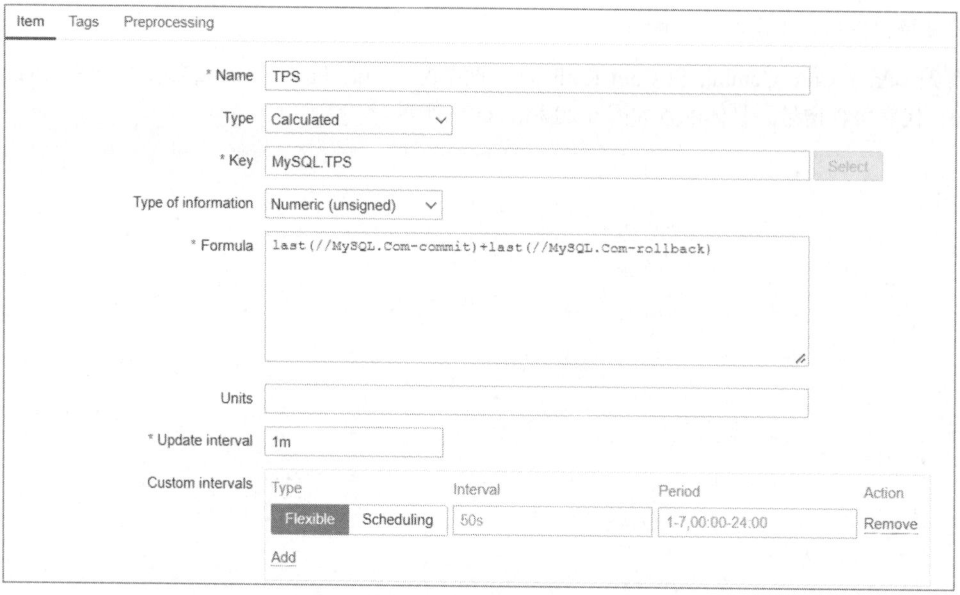

图 6-22　配置 TPS

通过上面的演示，可以看到基于 PMP 自定义监控项并不难。基本上，如果要获取其他 STATUS 或 VARIABLE，只需在 $keys 中添加相应的变量名，然后修改 Zabbix 的配置文件即可。

但在实际使用过程中，PMP 还有很多需要优化的地方。

- 很多采集项并不是很有意义，比如 Query Cache（在 MySQL 8.0 中，已从代码层完全移除）。
- 缺乏直观、可指导参数调整的性能指标，比如 InnoDB 缓存命中率、连接数使用率。

尽管如此，PMP 还是为我们监控 MySQL 提供了一个不错的起点，我们完全可以在其基础上进行二次开发。

6.4　Zabbix 常见问题定位及性能优化

6.4.1　定位监控项的状态 Not supported

Not supported 是我们在使用 Zabbix 的过程中经常会遇到的一个错误，尤其是在添加自定义监控项时。例如，在我们刚刚添加的 PMP 模板中，就有一个监控项的状态是 Not supported，如图 6-23 所示。

图 6-23　Not supported 的监控项

通过右边的 Info，可以看到该监控项失败的原因。但是，Info 并不总是会显示错误信息。

下面我们看看解决这类问题的一般思路。

(1) 在 Zabbix Server 端通过 `zabbix_get` 命令验证该 `Item` 的可用性。

```
# /usr/local/zabbix/bin/zabbix_get -s 192.168.244.10 -k MySQL.running-slave
/var/lib/zabbix/percona/scripts/get_mysql_stats_wrapper.sh: line 19: mysql: command not found
0
```

虽然有结果输出，但在 `0` 之前有一段错误信息。所以，Zabbix Server 获取到的结果是字符串，但是 `MySQL.running-slave` 定义的值的类型又是 `Numeric (unsigned)`，两者存在冲突。

(2) 在 Zabbix 客户端查找 `MySQL.running-slave` 对应的命令。

```
# grep "MySQL.running-slave"
/usr/local/zabbix/etc/zabbix_agentd.conf.d/userparameter_percona_mysql.conf
UserParameter=MySQL.running-slave,/var/lib/zabbix/percona/scripts/get_mysql_stats_wrapper.sh
running-slave
```

(3) 手动执行 `MySQL.running-slave` 对应的命令。

```
# sh /var/lib/zabbix/percona/scripts/get_mysql_stats_wrapper.sh running-slave
0
```

脚本执行的结果是 `0`，我们并没有看到上面的错误信息。

(4) 最后只能通过分析脚本来定位问题。

```
if [ "$ITEM" = "running-slave" ]; then
    RES=`HOME=~zabbix mysql -e 'SHOW SLAVE STATUS\G' | egrep '(Slave_IO_Running|Slave_SQL_Running):' | awk -F: '{print $2}' | tr '\n' ','`
    if [ "$RES" = " Yes, Yes," ]; then
        echo 1
    else
        echo 0
    fi
    exit
```

报错中提示没有找到 `mysql` 命令，这一点可通过显式指定 `mysql` 的绝对路径来解决。除此之外，还有另外一个问题，即命令行中没有指定用户名和密码。

我们已经通过分析脚本知道了问题所在，再来看看这个问题的解决方法。

将脚本中的 `mysql -e 'SHOW SLAVE STATUS\G'` 调整为

```
/usr/local/mysql/bin/mysql -umonitor_user -pmonitor_pass -S /data/mysql/3306/data/mysql.sock -e 'SHOW SLAVE STATUS\G' 2> /dev/null
```

在 Zabbix 5.2 之前，我们在很多时候会发现，虽然修改了脚本，通过 `zabbix_get` 能获取到数据，但监控项在前端页面中仍然显示为 Not supported。这实际上与 Refresh unsupported items（Administration -> General -> Other -> Refresh unsupported items）的配置有关，该配置默认为 10 分钟，即一个被判定为 Not supported 的监控项只有在等待 10 分钟后才能再次执行。个人觉得这个设置偏大，建议调整为 60 秒或与监控项的采集周期保持一致。不过这个配置在 Zabbix 5.2 中被移除了，刷新周期默认与监控项的采集周期保持一致。

6.4.2 分区表

Zabbix Server 采集的监控数据主要保存在与 history 和 trend 相关的两类表中。前者用于保存历史数据，会存储采集到的每个监控值，而后者用于保存趋势数据，会存储每个监控项每个小时内的最小值、平均值和最大值。

history 相关的表有以下 5 张。

- history：存储浮点类型的数据，对应 Item 中的 Numeric(float) 类型。
- history_uint：存储整数类型的数据，对应 Item 中的 Numeric(unsigned) 类型。
- history_log：存储日志类型的数据，一般用于日志监控，对应 Item 中的 Log 类型。
- history_str：存储字符串类型的数据，内容较短，对应 Item 中的 Character 类型。
- history_text：存储文本类型的数据，内容较长，对应 Item 中的 Text 类型。

trend 相关的表有以下两张。

- trend：存储浮点类型的趋势数据，对应 history 表。
- trend_uint：存储整数类型的趋势数据，对应 history_uint 表。

随着业务的发展，线上服务器会越来越多。同时，监控的精细化也会带来越来越多的监控项。这两方面的因素会导致监控数据越来越多，而监控数据一旦达到一定的量，数据库的操作也会越来越慢。一个直观的感受是，想在前端页面查看某个主机的 Latest data，往往要很久才能显示出来，在极端情况下甚至显示不出来。这时候的瓶颈往往在于数据库。

对于 Zabbix 数据库，业内较为常用的优化方案是分区表，它有两个显而易见的好处。

- 基于时间分区。无论是数据的插入还是查询，都可下推到一个或若干个分区内进行，相对于全表操作，性能更优。
- 默认情况下，过期数据是 Housekeeper 通过 DELETE 命令来删除的。在数据量很大的场景下，这种方式的效率较低。如果使用了分区表，数据的清理可通过删除分区来实现，简单高效。

下面看看如何将前面提到的 history 和 trend 两类表修改为分区表。

(1) 下载分区脚本。

这里用到的分区脚本是 zabbix_partition.sh，已上传至 GitHub。该脚本参考了 partitiontables.sh 的实现。

(2) 编辑分区脚本。

```
# head -5 zabbix_partition.sh
#!/bin/bash
SQL="/tmp/partition.sql"
HISTORY_KEEP_DAYS=30
TREND_KEEP_MONTHS=12
ZABBIX_VERSION=6

# sh zabbix_partition.sh
Bingo! Do not forget to set event_scheduler=on in my.cnf and disable Housekeeping
```

该脚本会对与 history 相关的表按天进行分区，对与 trend 相关的表按月进行分区。

该脚本有 4 个配置参数，各个参数的具体含义如下。

- SQL：执行完脚本，会生成一个 SQL 文件。该参数用来指定文件名。
- HISTORY_KEEP_DAYS：历史数据的保留时间。默认为 30，单位是天。
- TREND_KEEP_MONTHS：趋势数据的保留时间。默认为 12，单位是月。
- ZABBIX_VERSION：指定 Zabbix Server 的版本，如果要对 Zabbix 2.x 的表进行分区，需将该参数设置为 2。

(3) 应用生成的 SQL 文件。

```
# mysql zabbix < /tmp/partition.sql
```

这个文件会执行以下操作。

- 将 history 和 trend 两类表修改为分区表。
- 创建存储过程，用于添加和删除分区。
- 创建定时器，定期调用存储过程。

(4) 检查脚本执行的效果。

以 history 表为例，我们看看分区前后该表的结构。

分区前：

```
mysql> show create table history\G
*************************** 1. row ***************************
       Table: history
Create Table: CREATE TABLE `history` (
  `itemid` bigint unsigned NOT NULL,
  `clock` int NOT NULL DEFAULT '0',
  `value` double NOT NULL DEFAULT '0',
  `ns` int NOT NULL DEFAULT '0',
  PRIMARY KEY (`itemid`,`clock`,`ns`)
) ENGINE=InnoDB DEFAULT CHARSET=utf8mb3 COLLATE=utf8_bin
1 row in set (0.00 sec)
```

分区后：

```
mysql> show create table history\G
*************************** 1. row ***************************
       Table: history
Create Table: CREATE TABLE `history` (
  `itemid` bigint unsigned NOT NULL,
  `clock` int NOT NULL DEFAULT '0',
  `value` double NOT NULL DEFAULT '0',
  `ns` int NOT NULL DEFAULT '0',
  PRIMARY KEY (`itemid`,`clock`,`ns`)
) ENGINE=InnoDB DEFAULT CHARSET=utf8mb3 COLLATE=utf8_bin
/*!50100 PARTITION BY RANGE (`clock`)
(PARTITION p20220215 VALUES LESS THAN (1644940800) ENGINE = InnoDB,
 PARTITION p20220216 VALUES LESS THAN (1645027200) ENGINE = InnoDB,
 PARTITION p20220217 VALUES LESS THAN (1645113600) ENGINE = InnoDB,
 PARTITION p20220218 VALUES LESS THAN (1645200000) ENGINE = InnoDB,
```

```
PARTITION p20220219 VALUES LESS THAN (1645286400) ENGINE = InnoDB,
...
PARTITION p20220317 VALUES LESS THAN (1647532800) ENGINE = InnoDB,
PARTITION p20220318 VALUES LESS THAN (1647619200) ENGINE = InnoDB,
PARTITION p20220319 VALUES LESS THAN (1647705600) ENGINE = InnoDB) */
1 row in set (0.00 sec)
```

执行完 SQL 文件，还需注意以下两点。

- 将 event_scheduler=ON 添加到 MySQL 的参数文件中。在 MySQL 8.0 中无须添加，因为该参数默认为 ON。
- 单击 Administration -> General -> Housekeeping，禁用 history 和 trend 两类表的自动清理功能。具体如图 6-24 所示。

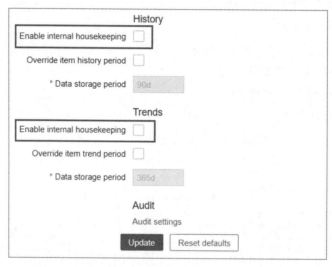

图 6-24　禁用 history 和 trend 两类表的自动清理功能

6.4.3　Zabbix Server 的参数优化

在 Zabbix 6.0 的配置文件中，一共提供了 107 个参数。从数量上看，还是相当多的。在这些参数中，与性能密切相关的有两类：进程类和缓存类。

进程类参数有：

- StartPollers
- StartIPMIPollers
- StartPreprocessors
- StartPollersUnreachable
- StartHistoryPollers
- StartTrappers
- StartPingers

6.4 Zabbix 常见问题定位及性能优化

- StartDiscoverers
- StartHTTPPollers
- StartTimers
- StartEscalators
- StartAlerters
- StartJavaPollers
- StartVMwareCollectors
- StartSNMPTrapper
- StartDBSyncers
- StartProxyPollers
- StartLLDProcessors
- StartReportWriters
- StartODBCPollers

缓存类参数有：

- VMwareCacheSize
- CacheSize
- HistoryCacheSize
- HistoryIndexCacheSize
- TrendCacheSize
- TrendFunctionCacheSize
- ValueCacheSize

很多 Zabbix 使用者很少调整这些参数，原因不外乎以下两点。

- 不是很明白这些参数的意义，即便配置文件中有简短的英文解释。
- 不确定应该何时调整这些参数。

针对第一个问题，我们看看各个参数的具体含义。首先是进程类参数。

- StartPollers：进程类参数中最重要的一个参数，用于配置 Poller 的数量。该进程负责收集通过 Zabbix Agent（被动模式）采集到的数据。注意，被动模式是 Zabbix Agent 的默认运行模式，如果监控的主机数较多，可适当增加这个进程的数量。默认为 5。
- StartIPMIPollers：配置 IPMI Poller 的数量。该进程通过 ipmitool 监控服务器的物理特征，如温度、电压、风扇工作状态等。
- StartPreprocessors：该进程对采集到的数据进行预处理。
- StartPollersUnreachable：同 Poller 的作用相同，只不过针对的是无法连接的（unreachable）主机。
- StartTrappers：配置 Trapper 的数量。该进程负责收集通过 Zabbix sender、Zabbix Agent（主动模式）、Proxy（主动模式）采集到的数据。如果有很多数据是通过上面这 3 种方式来采集的，可适当增加该进程的数量。默认为 5。

- StartPingers：配置 Pinger 的数量。该进程负责收集 ICMP ping 数据。
- StartDiscoverers：配置 Discoverer 的数量。该进程主要用来处理与 Network discovery 和 Low-level discovery 相关的任务，这两者是 Zabbix 的高级功能，主要用在机器数量较多、运维自动化的场景。
- StartHTTPPollers：配置 HTTP Poller 的数量。该进程负责收集通过 HTTP 请求采集到的 Web 可用性相关的数据。
- StartTimers：配置 Timer 的数量。该进程与维护周期及基于时间的触发器相关。
- StartEscalators：配置 Escalator 的数量。该进程用于告警升级。
- StartAlerters：配置 Alerter 的数量。该进程主要用于告警。
- StartJavaPollers：配置 Java Poller 的数量。该进程负责收集通过 JMX 采集到的数据。
- StartVMwareCollectors：配置 VMware Collector 的数量。该进程负责采集 VMware 虚拟机数据，默认为 0。如果要对 VMware ESXi 主机进行监控，该参数需大于 0。
- StartSNMPTrapper：配置 SNMP Trapper 的数量。该进程负责收集通过 SNMP 采集到的数据。
- StartDBSyncers：配置 DB Syncer 的数量。在 Zabbix 中，采集到的数据先放到自身的缓存中，再同步到数据库中。该进程即负责缓存数据的同步。
- StartProxyPollers：配置 Proxy Poller 的数量。该进程负责收集通过 Proxy（被动模式）采集到的数据。
- StartLLDProcessors：配置 Low-level discovery 工作进程的数量。Low-level discovery 可对网卡、磁盘等非固定的监控项进行自动发现。

然后是缓存类参数。

- VMwareCacheSize：缓存 VMware 相关的监控数据，默认为 8MB。
- CacheSize：缓存配置类信息，包括 host、item 和 trigger，默认为 8MB。
- HistoryCacheSize：缓存监控数据，默认为 16MB。
- HistoryIndexCacheSize：缓存监控数据的索引数据，默认为 4MB。
- TrendCacheSize：缓存趋势数据，默认为 4MB。

这里面比较重要的有 CacheSize、HistoryCacheSize 和 TrendCacheSize。如果监控的主机数及监控项较多，建议将这 3 个参数的大小调整到 1GB 以上。

接下来，我们看看第二个问题：如何判断上述参数需要调整呢？

实际上，Zabbix 提供了一个 Zabbix server health 模板，专门用来监控 Zabbix Server。该模板有 57 个监控项、42 个触发器，主要是监控各进程的繁忙程度、缓存的使用率等。

图 6-25 是模板中的一张监控图，直观地呈现了各个进程的繁忙程度。当进程的繁忙程度超过 75% 时，应适当增加对应进程的进程数。

图 6-26 是缓存类参数对应的监控图。同样，当缓存的使用率超过 75% 时，应适当增大对应缓存的大小。

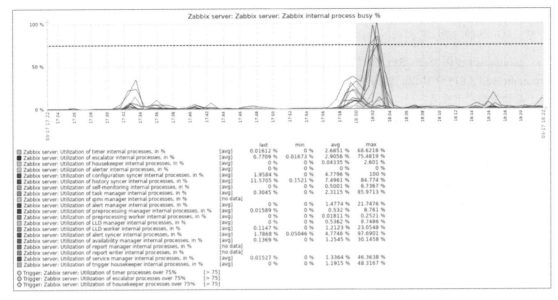

图 6-25　Zabbix 内部进程使用率

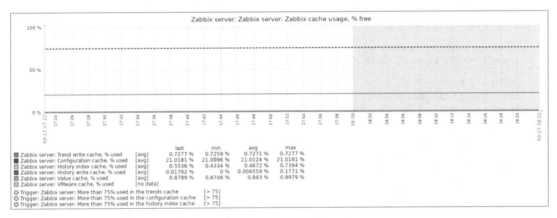

图 6-26　Zabbix 缓存使用率

我之前在生产中就碰到过缓存设置得过小，导致 Zabbix Server 因为 OOM（out of memory，内存不足）而崩溃。注意，这里的 OOM 与 Linux 的 OOM 不一样。后者指的是因为系统内存不足从而触发了 OOM Killer 杀掉内存占用较大的进程。而前者特指 Zabbix 缓存不足。在发生 OOM 时，Zabbix Server 的日志中会有相关记录（`zbx_mem_realloc(): out of memory`）。

6.4.4　Zabbix API

之前对于 node1 的监控是在 Web 页面上操作的。如果监控主机很少，这样做无可厚非。但如果需要管理的主机数很多，再通过页面来操作，就会显得烦琐且低效。在这种情况下，我们一般会使用

Zabbix API 来进行批量操作。除此之外，很多时候，我们也想利用 Zabbix 现有的监控数据来进行报表开发。对于类似需求，我们借助的往往也是 Zabbix API。

Zabbix API 使用 JSON-RPC 2.0 协议。JSON-RPC 是一个无状态、轻量级的远程过程调用（remote procedure call，RPC）协议，该协议定义了一些数据结构及相关的处理规则。具体来说，对于请求对象，必须包括以下成员。

- jsonrpc：JSON-RPC 协议版本。
- method：调用方法。
- params：参数。
- id：客户端分配的一个标识符。

对于响应对象，必须包含以下成员。

- jsonrpc：JSON-RPC 协议版本。
- result：成功响应时，需包含该字段。
- error：出现错误时，需包含该字段。注意，result 和 error 不能同时存在。
- id：与请求对象中的 ID 保持一致。

接下来，我们看看 Zabbix API 的使用方法，主要包括以下两个步骤。

(1) 基于用户名和密码，通过 user.login 方法获取认证 ID。例如：

```
# curl -i -X POST -H 'Content-Type: application/json' -d '{"jsonrpc":"2.0","method":"user.login","params":{"user":"Admin","password":"zabbix"},"auth":null,"id":0}' http://192.168.244.128/zabbix/api_jsonrpc.php

HTTP/1.1 200 OK
Date: Thu, 17 Mar 2022 09:21:12 GMT
Server: Apache/2.4.6 (CentOS) PHP/7.2.34
X-Powered-By: PHP/7.2.34
Access-Control-Allow-Origin: *
Access-Control-Allow-Headers: Content-Type
Access-Control-Allow-Methods: POST
Access-Control-Max-Age: 1000
Content-Length: 68
Content-Type: application/json

{"jsonrpc":"2.0","result":"6a6cf9decf8d93c73cc0c012cd8b6abe","id":0}
```

result 中的 `6a6cf9decf8d93c73cc0c012cd8b6abe` 即认证 ID。

(2) 基于认证 ID，调用其他方法。

这里通过 `host.get` 方法获取主机对应的 `hostid`。

```
# curl -i -X POST -H 'Content-Type: application/json' -d '{"jsonrpc": "2.0","method":"host.get","params":{"output":["hostid"],"filter": {"host":"node1"}},"auth": "6a6cf9decf8d93c73cc0c012cd8b6abe","id": 0}' http://192.168.244.128/zabbix/api_jsonrpc.php

HTTP/1.1 200 OK
Date: Thu, 17 Mar 2022 09:27:27 GMT
Server: Apache/2.4.6 (CentOS) PHP/7.2.34
X-Powered-By: PHP/7.2.34
```

```
Access-Control-Allow-Origin: *
Access-Control-Allow-Headers: Content-Type
Access-Control-Allow-Methods: POST
Access-Control-Max-Age: 1000
Content-Length: 54
Content-Type: application/json

{"jsonrpc":"2.0","result":[{"hostid":"10517"}],"id":0}
```

这里使用 CURL 命令只是为了演示 Zabbix API 怎么使用,在线上大规模使用显然不太现实。

下面具体看看如何使用 Python 封装 Zabbix API 来实现一些常用功能。

```python
#!/usr/bin/env python
# -*- coding: utf-8 -*-
import requests,json,time,datetime

API_URL = "http://192.168.244.128/zabbix/api_jsonrpc.php"
USER = "Admin"
PASSWORD = "zabbix"

class ZabbixAPI(object):
    def __init__(self, url, user, passwd, timeout=None):
        self.timeout = timeout
        self.url = url
        self.auth_id = self.get_auth_id(user, passwd)

    # HTTP 请求
    def http_request(self, data, timeout=None):
        headers = {"Content-Type": "application/json"}
        request = requests.post(self.url, data=data, headers=headers, timeout=timeout)
        r = request.json()
        if r.has_key("error"):
            raise ValueError(r["error"])
        else:
            return r["result"]

    # 获取认证 ID,所有的 API 请求都需要带上这个认证 ID
    def get_auth_id(self, user, passwd):
        data = json.dumps({
            "jsonrpc": "2.0",
            "method": "user.login",
            "params": {
                "user": user,
                "password": passwd
            },
            "id": 0
        })
        return self.http_request(data=data)

    # 基于主机名获取 hostid
    def host_get(self, host):
        data = json.dumps({
            "jsonrpc": "2.0",
            "method": "host.get",
            "params": {
                "output": ["hostid"],
                "filter": {"host": host}
            },
            "auth": self.auth_id,
```

```python
            "id": 1,
        })
        result = self.http_request(data=data)
        return result[0].get('hostid')

    # 设置维护策略,其中 duration 是维护周期,默认为 24 小时
    def maintenance_create(self, host, duration=24):
        data = json.dumps({
            "jsonrpc": "2.0",
            "method": "maintenance.create",
            "params": {
                "name": host,
                "active_since": int(time.time()),
                "active_till": int(time.time()) + duration * 3600,
                "hostids": [
                    self.host_get(host)
                ],
                "timeperiods": [
                    {
                        "timeperiod_type": 0,
                        "period": duration * 3600,
                    }
                ]
            },
            "auth": self.auth_id,
            "id": 1
        })
        return self.http_request(data=data)

    # 获取指定时间段的告警信息
    def problem_get(self,time_from):
        data = json.dumps({
            "jsonrpc": "2.0",
            "method": "problem.get",
            "params": {
                "output": ["eventid","objectid","clock","name","severity"],
                "sortfield": ["eventid"],
                "sortorder": "DESC",
                "time_from": time_from,
                #"limit": 1001
            },
            "auth": self.auth_id,
            "id": 1
        })
        return self.http_request(data=data)

    # 基于 triggerid 获取触发对象
    def trigger_get(self,trigger_ids):
        data = json.dumps({
            "jsonrpc": "2.0",
            "method": "trigger.get",
            "params": {
                "triggerids": trigger_ids,
                "output": ['triggerid'],
                "monitored": 1,
                "skipDependent":1,
                "selectHosts": ['name'],
                "filter": {
                    "value": 1
                }
```

```python
            },
            "auth": self.auth_id,
            "id": 1
        })
        return self.http_request(data=data)

#格式化秒数
def format_second(seconds):
    minutes, seconds = divmod(seconds, 60)
    hours, minutes = divmod(minutes, 60)
    days, hours = divmod(hours, 24)
    if days !=0:
      result="%dd %dh %dm %ds"%(days, hours, minutes, int(seconds))
    elif hours !=0:
      result="%dh %dm %ds"%(hours, minutes, int(seconds))
    elif minutes !=0:
      result="%dm %ds"%(minutes, int(seconds))
    else:
      result="%ds"%(int(seconds))
    return result


def main():
    zabbix_client=ZabbixAPI(API_URL,USER,PASSWORD)
    hostid=zabbix_client.host_get("node1")
    print hostid

    time_from=int(time.mktime((datetime.datetime.now() - datetime.timedelta(days=1)).timetuple()))
    problem_result=zabbix_client.problem_get(time_from)

    trigger_ids=[each_problem["objectid"]for each_problem in problem_result]

    trigger_info={}
    for each_trigger in zabbix_client.trigger_get(trigger_ids):
        triggerid=each_trigger['triggerid']
        hostname=each_trigger['hosts'][0]['name']
        trigger_info[triggerid]=hostname

    for each_problem in problem_result:
        problem_time=int(each_problem["clock"])
        trigger_id=each_problem["objectid"]
        problem_name=each_problem["name"]
        problem_time_format=datetime.datetime.fromtimestamp(problem_time).strftime('%Y-%m-%d %H:%M:%S')
        host=trigger_info[trigger_id]
        last_time=format_second(time.time()-problem_time)
        print problem_time_format,host,problem_name,last_time

    zabbix_client.maintenance_create("node1",12)

if __name__ == "__main__":
    main()
```

zabbix_api.py 脚本（已上传至 GitHub）实现了 3 个功能。

- 通过主机名获取 hostid。
- 给指定主机设置维护时间。
- 获取 1 天之内还没恢复的告警信息。

下面我们看看该脚本执行的效果。

```
# python zabbix_api.py
10517
2022-03-17 18:12:09 Zabbix server Load average is too high (per CPU load over 1.5 for 5m) 1m 25s
```

输出的告警信息实际上就是前端页面 Monitoring -> Problems 展示的内容，如图 6-27 所示。

图 6-27　前端页面的告警信息

既然知道了主机，知道了告警内容，就可以更进一步，通过 hostinterface.get 接口知道主机对应的 IP。基于这些信息，完全可以开发一个故障分析和故障自愈系统。

这里只是简单演示了几个接口。实际上，Zabbix API 提供了非常多的接口。只要能在前端页面操作的，基本上都有对应的 API 接口。例如，添加主机操作，就可以通过 host.create 接口实现。Zabbix 的官方文档也十分详尽，针对 API 的每个接口都给了一个示例。下面这个示例是通过 item.create 接口添加监控项的。

```
{
    "jsonrpc": "2.0",
    "method": "item.create",
    "params": {
        "name": "Free disk space on $1",
        "key_": "vfs.fs.size[/home/joe/,free]",
        "hostid": "30074",
        "type": 0,
        "value_type": 3,
        "interfaceid": "30084",
        "applications": [
            "609",
            "610"
        ],
        "delay": "30s"
    },
    "auth": "038e1d7b1735c6a5436ee9eae095879e",
    "id": 1
}
```

结合官方文档，大家可以尽情地扩展 zabbix_api.py 脚本，实现自己的个性化需求。

6.4.5　参考资料

- Zabbix 官方文档。
- GitHub 脚本"partitiontables.sh"，作者：吴兆松。

6.5　PMM

PMM（Percona Monitoring and Management）是 Percona 公司基于业界流行的 Prometheus 和 Grafana 开发的一个数据库监控管理平台，目前支持 MySQL、MongoDB、PostgreSQL、ProxySQL 等数据库组件的监控。

相对于原生的 Prometheus 和 Grafana，PMM 提供的是一体化的数据库监控解决方案，它具有以下特点。

- 在部署上，真正做到了开箱即用，极大地降低了新用户的使用门槛。
- 内置了多个 Exporter（数据采集组件）和 Dashboard（仪表盘）。无须额外下载 Exporter，也无须导入或自定义 Grafana 仪表盘。
- 监控指标更详尽。
- 提供 Query Analytics 功能，可对 MySQL、MongoDB、PostgreSQL 的查询进行分析。

6.5.1 PMM 的体系架构

图 6-28 是 PMM 的架构图。可以看到 PMM 是一个典型的客户端－服务端架构。PMM Client 主要由 pmm-admin、pmm-agent、Exporter 组成。PMM Server 则融合了多个流行的开源组件：Prometheus、Grafana、Nginx、PostgreSQL、ClickHouse，各个组件的具体作用后面会提到。

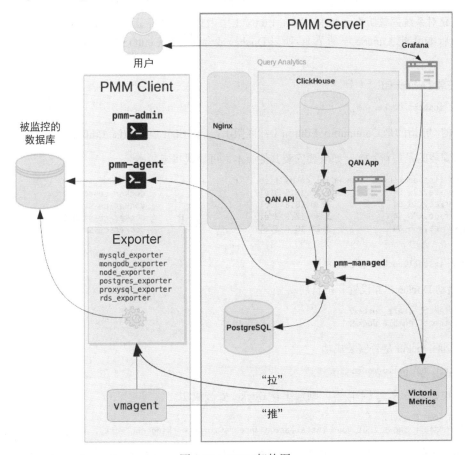

图 6-28　PMM 架构图

6.5.2　安装 PMM Server

对于 PMM Server，Percona 提供了 3 种安装方式。

- Docker。
- AWS Marketplace。
- Virtual Appliance（虚拟设备），比如 VMware、VirtualBox。

大家可根据自己的实际情况自行选择安装方式。这 3 种方式中，最为常用的是 Docker。

接下来看看如何基于 Docker 安装 PMM Server。

首先，安装 Docker，步骤如下。

(1) 安装 Docker yum 源。

```
# yum install -y yum-utils
# yum-config-manager --add-repo https://download.docker.com/linux/centos/docker-ce.repo
```

Docker 对系统内核版本有要求，需在 Linux 3.10 及以上版本上运行，而 CentOS 6 的内核版本是 2.6（内核版本可通过 uname -r 查看），所以 Docker 适合部署在 CentOS 7（内核版本是 3.10）及以上版本中。

(2) 安装 Docker CE。

```
# yum install docker-ce
```

这里的 CE 指的是 Community Edition（社区版），还有 EE（Enterprise Edition，企业版）。

默认安装的是最新版本。如果要安装其他版本，可显式指定版本号。

```
# yum list docker-ce --showduplicates | sort -r
docker-ce.x86_64            3:20.10.9-3.el7             docker-ce-stable
docker-ce.x86_64            3:20.10.8-3.el7             docker-ce-stable
docker-ce.x86_64            3:20.10.7-3.el7             docker-ce-stable
docker-ce.x86_64            3:20.10.6-3.el7             docker-ce-stable
...

# yum install docker-ce-20.10.6
```

(3) 启动 Docker，并设置开机自启动。

```
# systemctl start docker
# systemctl enable docker
```

(4) 验证 Docker 是否安装成功。

```
# docker run hello-world
```

输出中如果提示了以下内容，则意味着 Docker 安装成功。

```
Hello from Docker!
This message shows that your installation appears to be working correctly.
```

Docker 具体的安装过程，可参考 Docker 官方文档 "Install Docker Engine on CentOS"。

Docker 安装完毕，接下来安装 PMM Server，步骤如下。

(1) 下载 PMM Server 镜像。

```
# docker pull percona/pmm-server:2
```

这里下载的是 PMM 2.x 的最新版本。如果要下载其他版本的 PMM，可参考 Docker Hub percona/pmm-server 页面。

(2) 创建数据卷容器。

```
docker create --volume /srv \
--name pmm-data \
percona/pmm-server:2 /bin/true
```

单独创建数据卷容器有如下好处。

- 数据和应用分离。后续对应用进行调整（如升级、降级），不会影响到之前采集的数据。
- 方便数据的备份、恢复和迁移。

下面看看各个选项的具体含义。

- `--volume`：创建数据卷并将其挂载到容器中。/srv 是容器内目录，切记不要改变，因为 PMM 的各个组件默认都安装在这个路径下。如果设置为其他值，在升级 PMM 时会导致数据丢失。/srv 对应的宿主机目录可通过以下命令查看。

```
# docker inspect pmm-data | egrep "Source|Destination"
            "Source": "/var/lib/docker/volumes/c9f9964cd70014cff60078fd5c98e1e3a6125bc8d11e
89b89256540eefefedaa/_data",
            "Destination": "/srv",
```

- `--name`：容器名。后续其他容器可通过 `--volumes-from` 引用该容器的数据卷。
- `percona/pmm-server:2`：镜像名。

在使用数据卷容器时，需注意以下两点。

- 数据卷容器创建即可，无须运行。
- 在 Docker 中，数据卷一旦创建，它的生命周期就和创建它的数据卷容器无关，即使停止或删除了容器，数据卷也依然存在。如果要删除该数据卷，必须首先删除所有依赖它的容器，最后在删除数据卷容器时加上 `-v` 参数。

(3) 创建并启动 PMM Server 容器。

```
docker run --detach --restart always \
--publish 443:443 \
--volumes-from pmm-data \
--name pmm-server \
percona/pmm-server:2
```

下面看看各个选项的具体含义。

- `--detach`：让容器在后台运行。
- `--restart`：指定重启策略。

- `--publish`：端口映射。格式为 hostPort:containerPort，冒号前是宿主机端口，冒号后是容器内部端口。如果宿主机有多个地址，也可通过 ip:hostPort:containerPort 显式指定监听地址。这里只映射了 443 端口。除此之外，也可映射 80 端口。
- `--volumes-from`：引用 pmm-data 容器创建的数据卷。
- `--name`：容器名。
- `percona/pmm-server:2`：镜像名。

创建完毕后，可通过 `docker ps -a` 查看容器的运行情况。

```
# docker ps -a
CONTAINER ID   IMAGE                 COMMAND              CREATED          STATUS                     PORTS                                          NAMES
865710a17aa0   percona/pmm-server:2  "/opt/entrypoint.sh" About a minute ago  Up About a minute (healthy) 80/tcp, 0.0.0.0:443->443/tcp, :::443->443/tcp pmm-server
a81c19cd3ca1   percona/pmm-server:2  "/bin/true"          2 minutes ago    Created                                                                   pmm-data
e517b648e796   hello-world           "/hello"             3 minutes ago    Exited (0) 3 minutes ago                                                  vigorous_lehmann
```

STATUS 显示为 Up 代表容器处于运行状态。

(4) 登录 PMM Server。

登录地址为 https://10.0.0.160:443/。图 6-29 是 PMM Server 的登录页面。

图 6-29　PMM Server 的登录页面

默认的用户名和密码都是 admin。首次登录，会提示修改密码，这里我将其修改为 pmm_pass。登录后的页面如图 6-30 所示。

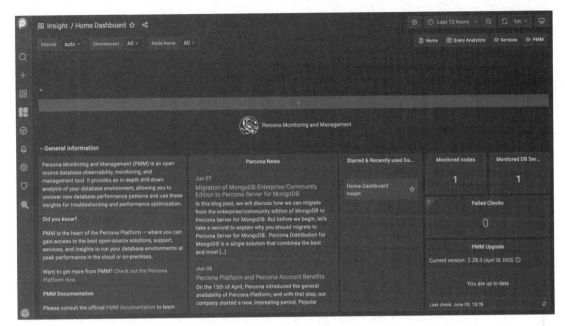

图 6-30　PMM Server 首页

至此，PMM Server 搭建完毕。接下来安装 PMM Client。

6.5.3　安装 PMM Client

图 6-31 是 PMM Client 的下载页面。

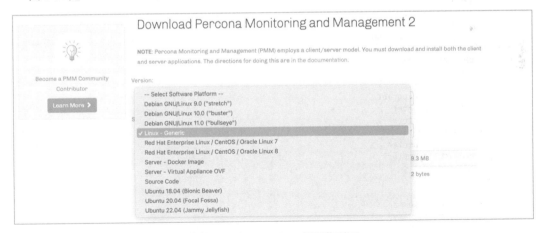

图 6-31　PMM Client 的下载页面

目前，支持 Debian、Linux 二进制包、RHEL、Docker、源码包、Ubuntu 等版本的下载。这里使用 Linux 二进制包，具体的安装步骤如下。

```
# cd /usr/local/
# tar xvf pmm2-client-2.28.0.tar.gz
# cd pmm2-client-2.28.0/
# ./install_tarball
# export PATH=$PATH:/usr/local/percona/pmm2/bin
```

二进制包默认会把文件安装在 /usr/local/percona/pmm2 目录下。我们不妨看看这个目录下的文件。

```
# tree /usr/local/percona/pmm2
/usr/local/percona/pmm2
├── bin
│   ├── pmm-admin
│   └── pmm-agent
├── collectors
│   ├── custom-queries
│   │   ├── mysql
│   │   │   ├── high-resolution
│   │   │   │   ├── queries-mysqld-group-replication.yml
│   │   │   │   └── queries-mysqld.yml
│   │   │   ├── low-resolution
│   │   │   │   └── queries-mysqld.yml
│   │   │   └── medium-resolution
│   │   │       └── queries-mysqld.yml
│   │   └── postgresql
│   │       ├── high-resolution
│   │       │   ├── example-queries-postgres.yml
│   │       │   └── queries-postgres-uptime.yml
│   │       ├── low-resolution
│   │       │   └── example-queries-postgres.yml
│   │       └── medium-resolution
│   │           ├── example-queries-postgres.yml
│   │           └── queries.yaml
│   └── textfile-collector
│       ├── high-resolution
│       │   └── example.prom
│       ├── low-resolution
│       │   └── example.prom
│       └── medium-resolution
│           └── example.prom
├── config
│   └── pmm-agent.yaml
├── exporters
│   ├── azure_exporter
│   ├── clickhouse_exporter
│   ├── mongodb_exporter
│   ├── mysqld_exporter
│   ├── node_exporter
│   ├── postgres_exporter
│   ├── proxysql_exporter
│   ├── rds_exporter
│   └── vmagent
└── tools
    ├── pt-mongodb-summary
    ├── pt-mysql-summary
    ├── pt-pg-summary
    └── pt-summary

18 directories, 28 files
```

其中各项解释如下。

- `exporters`：数据采集组件。可以看到，PMM 支持 Azure、ClickHouse、MongoDB、MySQL、OS、PostgreSQL、ProxySQL、RDS 的监控。
- `collectors`：采集器。可用来自定义采集项，是对 `exporters` 的一个补充。
- `textfile-collector` 是 `textfile` 采集器，是 Prometheus 自带的。这个采集器会扫描指定目录下以 .prom 结尾的文件，提取满足指定格式的指标数据，常与定时任务（如 crontab）一起使用。
- `custom-queries`：顾名思义，自定义查询，是 PMM 实现的。基于这个特性，可将表中的数据作为指标数据来展示，这就赋予了我们很大的灵活性，如果我们想展示某些应用数据，如日活、页面点击量等，就可将这些数据存储在表中，然后借助 `custom-queries` 来展示。具体使用方法可参考博客文章 "PMM's Custom Queries in Action: Adding a Graph for InnoDB mutex waits"，作者：Daniel Guzmán Burgos。
- `high-resolution`、`low-resolution`、`medium-resolution`：与指标的采集频率有关。PMM 默认的 Metrics resolution 是 Standard，对应的采集频率是 high: 5s, medium: 10s, low: 60s，意味着会分别以 5 秒一次、10 秒一次、60 秒一次的频率采集 `high-resolution`、`medium-resolution`、`low-resolution` 的指标。我们也可以在 Grafana 中自定义（Custom）采集频率，设置路径为：Configuration -> Settings -> Metrics Resolution。设置页面如图 6-32 所示。

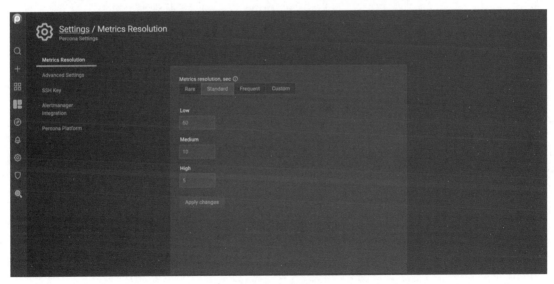

图 6-32　指标的采集频率设置页面

安装完 PMM Client，接下来看看具体的使用步骤。

(1) 注册 pmm-agent。

```
# pmm-agent setup --config-file=/usr/local/percona/pmm2/config/pmm-agent.yaml --server-address=
10.0.0.160 --server-insecure-tls --server-username=admin --server-password=pmm_pass 10.0.0.195
generic node1
...
```

```
Checking local pmm-agent status...
pmm-agent is not running.
Registering pmm-agent on PMM Server...
Registered.
Configuration file /usr/local/percona/pmm2/config/pmm-agent.yaml updated.
Please start pmm-agent: `pmm-agent --config-file=/usr/local/percona/pmm2/config/pmm-agent.yaml`.
```

其中各项解释如下。

- server-address、server-username、server-password 分别是 PMM Server 的地址、用户名、密码。
- 10.0.0.195：节点（客户端）地址。可不指定，默认是节点 IP。
- generic：节点类型，可设置为 generic 或 container。默认是 generic。
- node1：节点名。可不指定，默认是主机名。

(2) 启动 pmm-agent。

下面通过 systemd 来管理 pmm-agent。

首先编写 systemd 服务管理脚本。

```
# vim /usr/lib/systemd/system/pmm-agent.service
[Unit]
Description=pmm-agent
After=time-sync.target network.target

[Service]
Type=simple
ExecStart=/usr/local/percona/pmm2/bin/pmm-agent
--config-file=/usr/local/percona/pmm2/config/pmm-agent.yaml
Restart=always
RestartSec=2s

[Install]
WantedBy=multi-user.target
```

该脚本无须手动编写，二进制包的 config 目录下自带了 init 和 systemd 服务管理脚本。我们只需要在其基础上按照自己的实际环境修改即可。

```
# ls config/
pmm-agent.init  pmm-agent.logrotate  pmm-agent.service
```

加载服务配置文件。

```
# systemctl daemon-reload
```

启动 pmm-agent。

```
# systemctl start pmm-agent
```

查看服务的状态。如果服务启动失败，可通过 journalctl 查看具体的错误信息。

```
# systemctl status pmm-agent
# journalctl -u pmm-agent -n 20
```

journalctl 命令中的 -u 是服务名，-n 20 表示查看最近的 20 行日志。

将 pmm-agent 设置为开机自启动。

6.5 PMM

```
# systemctl enable pmm-agent
```

(3) 通过 pmm-admin 查看 agent 的状态。

```
# pmm-admin status
Agent ID: /agent_id/a075bdb3-f2ee-4584-b044-3caed9666243
Node ID : /node_id/49bfdd0b-6bf8-405c-8316-a8bffe7384f2

PMM Server:
        URL    : https://10.0.0.160:443/
        Version: 2.28.0

PMM Client:
        Connected        : true
        Time drift       : 129.381µs
        Latency          : 329.811µs
        pmm-admin version: 2.28.0
        pmm-agent version: 2.28.0
Agents:
        /agent_id/18f0f1b6-46e1-49a3-ace0-14930d4c31ba node_exporter Running
        /agent_id/7ae17c0d-af66-4b11-84fa-4efa7470528b vmagent Running
```

注意，对于不是通过二进制包，而是通过其他方式安装的 pmm-agent，在启动后，必须通过以下命令注册。

```
# pmm-admin config --server-insecure-tls --server-url=https://admin:pmm_pass@10.0.0.160:443
10.0.0.195 generic node1
```

最后登录 Grafana，看是否采集到了监控数据。

查看路径：PMM dashbords -> System (Node) -> Node Summary。图 6-33 是 node1 的监控页面。

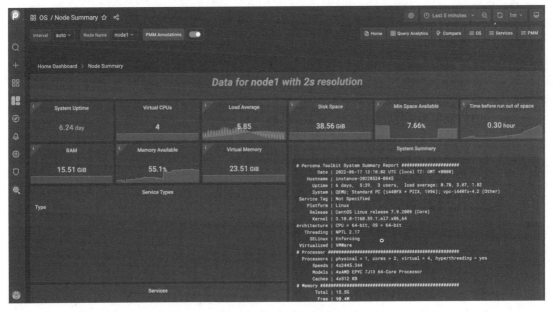

图 6-33　Node Summary 监控页面

6.5.4 添加 MySQL 服务

添加 MySQL 服务的具体步骤如下。

(1) 创建 MySQL 监控用户。

```
create user 'pmm_monitor'@'127.0.0.1' identified by 'monitor_pass' with max_user_connections 10;
grant select, process, replication client, reload, backup_admin on *.* to 'pmm_monitor'@'127.0.0.1';
```

BACKUP_ADMIN 权限是在 MySQL 8.0 中引入的，所以对于 MySQL 8.0 之前的版本，不要授予该权限。

(2) 确定 Query Analytics 的来源。

Query Analytics 的来源有两种：慢日志和 performance_schema。

- 慢日志

如果基于慢日志，则必须设置慢日志的相关参数，具体包括：

```
slow_query_log=ON
log_output=FILE
long_query_time=1
log_slow_admin_statements=ON
log_slow_slave_statements=ON
```

这些参数也可以动态设置，具体命令如下：

```
SET GLOBAL slow_query_log = ON;
SET GLOBAL log_output = 'FILE';
SET GLOBAL long_query_time = 1;
SET GLOBAL log_slow_admin_statements = ON;
SET GLOBAL log_slow_slave_statements = ON;
```

这里将慢日志的阈值定义为 1 秒，也可基于实际需要调整为其他值。

- performance_schema

如果基于 performance_schema，则必须开启与 SQL 操作相关的事件采集配置项（instrument），具体包括：

```
performance-schema-instrument='statement/%=ON'
performance-schema-consumer-events-statements-current=ON
performance-schema-consumer-events-statements-history=ON
performance-schema-consumer-events-statements-history-long=ON
performance-schema-consumer-statements-digest=ON
innodb_monitor_enable=all
```

这些事件采集配置项也可动态开启，具体命令如下：

```
UPDATE performance_schema.setup_instruments SET ENABLED = 'YES', TIMED = 'YES' WHERE NAME LIKE 'statement/%';
UPDATE performance_schema.setup_consumers SET ENABLED = 'YES' WHERE NAME LIKE '%statements%';
SET GLOBAL innodb_monitor_enable = all;
```

如果 pmm-agent 部署在实例本地，推荐使用慢日志。如果使用的是 performance_schema，PMM 会

基于 `performance_schema.events_statements_summary_by_digest` 来获取查询信息，而该表不会记录 Prepared Statement。所以当我们使用 sysbench 进行压测时，对于 sbtest 的增删改查操作就不会记录在 `performance_schema.events_statements_summary_by_digest` 中。

(3) 添加 MySQL 服务。

```
# pmm-admin add mysql --query-source=slowlog --username=pmm_monitor --password=monitor_pass
node1-mysql 127.0.0.1:3306
MySQL Service added.
Service ID  : /service_id/a0f7252e-3445-48e2-bf49-1f5b4048ecbc
Service name: node1-mysql

Table statistics collection enabled (the limit is 1000, the actual table count is 337).
```

其中各项解释如下。

- `--query-source`：SQL 的来源，可选值有 `slowlog`（慢查询，是默认值）、`perfschema`（performance_schema）、`none`（不采集）。
- `node1-mysql`：服务名。默认是"主机名-mysql"。
- `127.0.0.1:3306`：MySQL 实例地址。

默认会开启所有的采集器。采集器的名字可通过以下命令查看。

```
# ps -ef | grep mysqld_exporter
root      14317  1654  0 13:16 ?        00:00:00 /usr/local/percona/pmm2/exporters/mysqld_exporter
--collect.auto_increment.columns --collect.binlog_size --collect.custom_query.hr
--collect.custom_query.hr.directory=/usr/local/percona/pmm2/collectors/custom-queries/mysql/high-r
esolution --collect.custom_query.lr
--collect.custom_query.lr.directory=/usr/local/percona/pmm2/collectors/custom-queries/mysql/low-re
solution --collect.custom_query.mr
--collect.custom_query.mr.directory=/usr/local/percona/pmm2/collectors/custom-queries/mysql/medium
-resolution --collect.engine_innodb_status --collect.engine_tokudb_status --collect.global_status
--collect.global_variables --collect.heartbeat --collect.info_schema.clientstats
--collect.info_schema.innodb_cmp --collect.info_schema.innodb_cmpmem
--collect.info_schema.innodb_metrics --collect.info_schema.innodb_tablespaces
--collect.info_schema.processlist --collect.info_schema.query_response_time
--collect.info_schema.tables --collect.info_schema.tablestats --collect.info_schema.userstats
--collect.perf_schema.eventsstatements --collect.perf_schema.eventswaits
--collect.perf_schema.file_events --collect.perf_schema.file_instances
--collect.perf_schema.indexiowaits --collect.perf_schema.tableiowaits
--collect.perf_schema.tablelocks --collect.slave_status --collect.standard.go
--collect.standard.process --exporter.conn-max-lifetime=55s --exporter.global-conn-pool
--exporter.max-idle-conns=3 --exporter.max-open-conns=3 --web.listen-address=:42002
```

如果想禁用某些采集器，可指定 `--disable-collectors`，例如：

```
# pmm-admin add mysql --disable-collectors='engine_tokudb_status,info_schema.query_response_time,
info_schema.userstats' --username=pmm_monitor --password=monitor_pass node2-mysql 127.0.0.1:3306
```

`mysqld_exporter` 采集的指标信息可通过 "http://客户端 IP:42002/metrics" 来查看。URL 的 42002 是 `mysqld_exporter` 的 `--web.listen-address`。查看时需要输入用户名（默认是 pmm）和密码（默认是 Service 对应的 Agent ID）。Service 对应的 Agent ID 可通过 `pmm-admin list` 查看。具体来说，首先通过 `Address and port` 确认实例对应的 Service ID，然后根据 Service ID 确定其对应的 Agent ID。看下面的输出，127.0.0.1:3306 这个实例对应的 Agent ID 是 `/agent_id/9b840b6e-b62d-451d-9122-7809f03a9096`。

```
# pmm-admin list
Service type      Service name       Address and port    Service ID
MySQL             node1-mysql        127.0.0.1:3306      /service_id/a0f7252e-3445-48e2-bf49-1f5b4048ecbc

Agent type                                            Status        Metrics Mode    Agent ID
Service ID
pmm_agent                                             Connected                     /agent_id/a075bdb3-f2ee-4584-b044-3caed9666243
node_exporter                                         Running       push            /agent_id/18f0f1b6-46e1-49a3-ace0-14930d4c31ba
mysqld_exporter                                       Running       push            /agent_id/9b840b6e-b62d-451d-9122-7809f03a9096
/service_id/a0f7252e-3445-48e2-bf49-1f5b4048ecbc
mysql_slowlog_agent                                   Running                       /agent_id/4ed17036-6642-4ad8-9d9c-2b869d8a068f
/service_id/a0f7252e-3445-48e2-bf49-1f5b4048ecbc
vmagent                                               Running       push            /agent_id/7ae17c0d-af66-4b11-84fa-4efa7470528b
```

当然，这个密码也可以在执行 `pmm-admin add mysql` 时通过 `--agent-password` 来指定。

最后登录 Grafana，看是否采集到了监控数据。

查看路径是 PMM dashbords -> MySQL -> MySQL Summary。图 6-34 是 MySQL 的监控页面。

图 6-34　MySQL Instance Summary 监控页面

6.5.5　Query Analytics

Query Analytics（查询分析）可对 MySQL、MongoDB、PostgreSQL 中的查询进行分析。图 6-35 是 PMM Query Analytics 页面。

左边的 Filters 面板可按照不同的维度过滤实例。右边是汇总后的结果。与 `pt-query-digest` 一样，Query Analytics 会对采集到的所有 SQL 操作进行分类汇总。默认按照 Query 分类，Query 是同一类 SQL 操作的 fingerprint（指纹）。除了 Query，还可按照 Service Name、Database、Schema、User Name、Client Host 进行分类。Load 是平均活跃查询数，除了 Load，也可以选择其他指标，比如 Bytes Sent、Lock Time、Rows Sent、Rows Examined。Query Count 是执行次数，Query Time 是平均查询时间。除了 Load、Query

Count 和 Query Time，也可通过 Add column 添加其他统计指标。汇总后的结果默认按照 Load 降序排列。

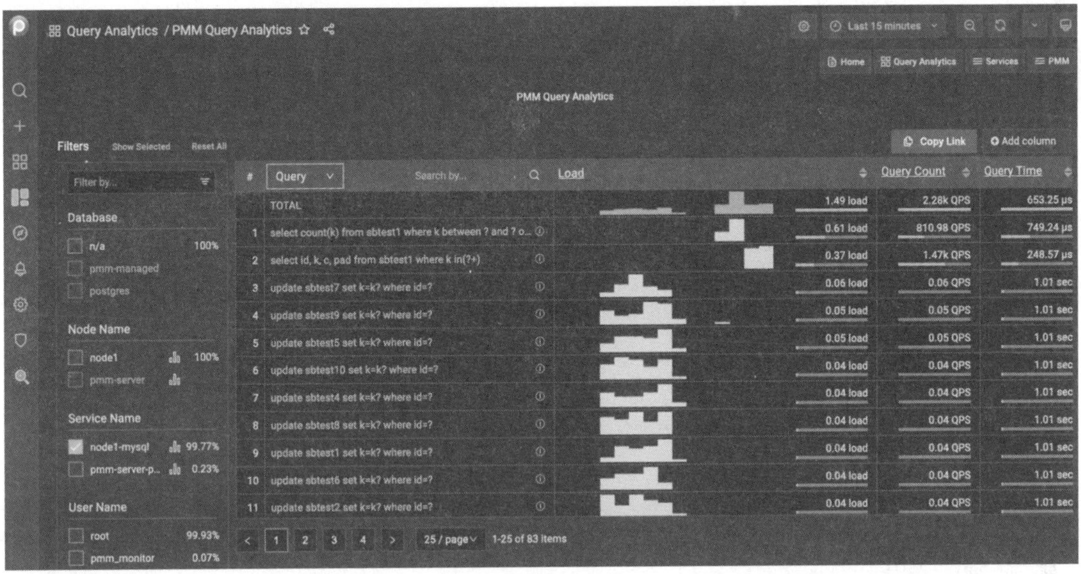

图 6-35　PMM Query Analytics 页面

单击 Query 这一列的 SQL，可以看到这类 SQL 更详细的执行信息，如图 6-36 所示。

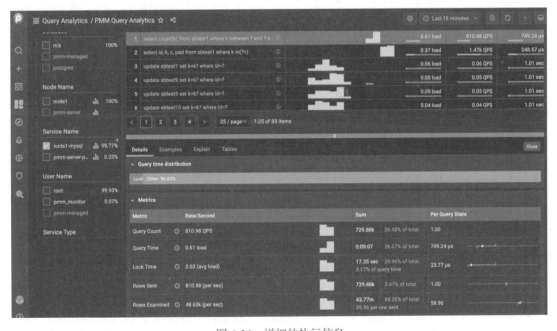

图 6-36　详细的执行信息

图中各项解释如下。

- Details：展示这类 SQL 操作更为详细的统计信息。
- Examples：给出一个具体的 SQL 语句。
- Explain：查询对应的执行计划。
- Tables：表及索引的相关信息。

6.5.6 深入理解 PMM Server

首先，我们分析一下 pmm-server:2.28 的 Dockerfile。

```
ADD file:b3ebbe8bd304723d43b7b44a6d990cd657b63d93d6a2a9293983a30bfc1dfa53 in /

LABEL org.label-schema.schema-version=1.0 org.label-schema.name=CentOS Base Image org.label-schema.
vendor=CentOS org.label-schema.license=GPLv2 org.label-schema.build-date=20201113 org.opencontainers.
image.title=CentOS Base Image org.opencontainers.image.vendor=CentOS org.opencontainers.image.
licenses=GPL-2.0-only org.opencontainers.image.created=2020-11-13 00:00:00+00:00

CMD ["/bin/bash"]

ARG VERSION

ARG BUILD_DATE

LABEL org.label-schema.build-date=2022-05-10 17:12:40+00:00

LABEL org.label-schema.license=AGPL-3.0

LABEL org.label-schema.name=Percona Monitoring and Management

LABEL org.label-schema.vendor=Percona

LABEL org.label-schema.version=202205101712

LABEL org.opencontainers.image.created=2022-05-10 17:12:40+00:00

LABEL org.opencontainers.image.licenses=AGPL-3.0

LABEL org.opencontainers.image.title=Percona Monitoring and Management

LABEL org.opencontainers.image.vendor=Percona

LABEL org.opencontainers.image.version=202205101712

EXPOSE 443 80

WORKDIR /opt

|2 BUILD_DATE=2022-05-10 17:12:40+00:00 VERSION=202205101712 /bin/sh -c yum -y install epel-release
&& yum -y install ansible

COPY dir:0642ba4f60532e85b3dc3617abd81ed5a8bc90398265242195163c0c2e0125de in /tmp/RPMS

COPY file:7d1f4654aac712b86eda9f9894ee6645cfa28663d1531d3fbb9be6b8e0060544 in /tmp/gitCommit

COPY dir:78212bd2b9ca8092e2482dff98b309181d280d8b0d28764488a19de1e5103072 in /opt/ansible
```

```
|2 BUILD_DATE=2022-05-10 17:12:40+00:00 VERSION=202205101712 /bin/sh -c cp -r /opt/ansible/roles
/opt/ansible/pmm2-docker/roles

|2 BUILD_DATE=2022-05-10 17:12:40+00:00 VERSION=202205101712 /bin/sh -c ansible-playbook -vvv -i
'localhost,' -c local /opt/ansible/pmm2-docker/main.yml     && ansible-playbook -vvv -i 'localhost,'
-c local /usr/share/pmm-update/ansible/playbook/tasks/update.yml     && ansible-playbook -vvv -i
'localhost,' -c local /opt/ansible/pmm2/post-build-actions.yml

|2 BUILD_DATE=2022-05-10 17:12:40+00:00 VERSION=202205101712 /bin/sh -c cp
/usr/share/pmm-server/entrypoint.sh /opt/entrypoint.sh

HEALTHCHECK &[["CMD-SHELL" "curl -f http://127.0.0.1/v1/readyz || exit 1"] "3s" "2s" "10s" '\x03'}

CMD ["/opt/entrypoint.sh"]
```

这里的关键步骤有两个。

第一个步骤如下。

```
/bin/sh -c ansible-playbook -vvv -i 'localhost,' -c local /opt/ansible/pmm2-docker/main.yml    &&
ansible-playbook -vvv -i 'localhost,' -c local
/usr/share/pmm-update/ansible/playbook/tasks/update.yml     && ansible-playbook -vvv -i 'localhost,'
-c local /opt/ansible/pmm2/post-build-actions.yml
```

可以看到，PMM Server 的各个组件是通过 Ansible 来安装的，具体的安装步骤可登录容器内部查看 /opt/ansible/pmm2-docker/main.yml 文件。

第二个步骤如下。

```
CMD ["/opt/entrypoint.sh"]
```

CMD 指定了容器启动时默认的执行命令。登录容器内部，查看该脚本的内容。

```
[root@slowtech ~]# docker exec -it pmm-server /bin/bash
[root@865710a17aa0 opt]# cat /opt/entrypoint.sh
#!/bin/bash

set -o errexit

# pmm-managed-init validates environment variables.
pmm-managed-init

# Start supervisor in foreground
exec supervisord -n -c /etc/supervisord.conf

[root@865710a17aa0 opt]# ls /etc/supervisord.d/
alertmanager.ini  dbaas-controller.ini  grafana.ini  pmm.ini  prometheus.ini  qan-api2.ini
victoriametrics.ini  vmalert.ini
```

可以看到，PMM Server 的各个组件是通过 supervisord 来管理的。各个组件的日志都在各自的 supervisord 配置文件中有定义。

接下来，我们具体看看 PMM Server 集成的开源组件。

- **Prometheus**

图 6-37 是 Prometheus 的架构图。

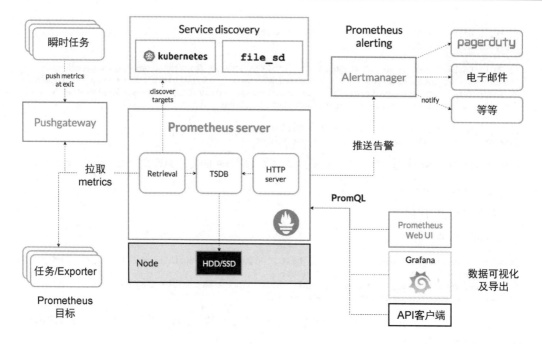

图 6-37　Prometheus 架构图

Prometheus 中各个组件的功能如下。

- Prometheus server：Retrieval 模块定时拉取数据，TSDB（Time Series Database，时序数据库）存储数据。Prometheus 提供了两种存储方式：本地存储和远端存储。本地存储使用的是 Prometheus 自带的时序数据库，非集群，存储容量有限。为了弥补这一不足，Prometheus 引入了远端存储，并定义了一套接口与之交互，目前支持的远端存储产品包括 InfluxDB、OpenTSDB、TiKV 等。
- Exporter：Prometheus 数据采集组件的统称。Prometheus 通过 Exporter 来采集主机及应用程序的性能指标。官方开源并维护了多个 Exporter，基本上覆盖了我们工作中的常用组件，具体列表可参考 Prometheus 官方文档"EXPORTERS AND INTEGRATIONS"。这里面，Node Exporter 可用来采集主机的各种指标数据，包括 CPU、内存、磁盘、网络等。
- Alertmanager：用于告警处理。告警规则是在 Prometheus 中定义的，当告警规则被触发后，Prometheus 会将告警信息推送到 Alertmanager。Alertmanager 会基于自身的配置，决定如何处理告警，包括告警内容、告警发送的频率、告警的介质等。
- Pushgateway：Prometheus 一般是通过 Pull 的方式拉取数据的，但在某些场景中这种方式就不太合适了，如批处理任务、目标作业的执行时间太短、防火墙限制等。这个时候，应用程序可将 metrics 数据主动推送给 Pushgateway。之后，由 Prometheus 定时去 Pushgateway 上拉取数据。
- Service discovery：服务发现。既然是通过 Pull 方式拉取数据，就需要事先定义拉取对象。目前，Prometheus 既支持静态配置，也支持动态配置。静态配置指的是在配置文件中通过 static_configs 显式指定监控对象。例如：

```
scrape_configs:
- job_name: prometheus
  scrape_interval: 5s
  scrape_timeout: 4s
  metrics_path: /prometheus/metrics
  static_configs:
  - targets:
    - 127.0.0.1:9090
    labels:
      instance: pmm-server
```

但这种方式的局限性也比较明显。

- 监控对象较多时，维护起来比较困难。
- 不适用于容器等资源动态变化的场景。

这个时候，可通过服务发现方式动态获取监控对象。官方支持的服务发现方式可归结为以下几类。

- 基于文件的服务发现（`file_sd_configs`）。
- 基于 API 的服务发现（`consul_sd_configs`、`azure_sd_configs`、`kubernetes_sd_configs`、`openstack_sd_configs` 等）。
- 基于 DNS 的服务发现（`dns_sd_configs`）。

❑ PromQL：Prometheus 内置的数据查询语言。

从 PMM 2.12 开始，PMM Server 使用 VictoriaMetrics 替换了 Prometheus。VictoriaMetrics 兼容 Prometheus 查询 API，除此之外，它还具有以下优点：支持 Pull 和 Push 两种数据采集方式；高压缩比；内存使用较少；有集群方案。

- **Grafana**

Grafana 是一个开源、跨平台的度量分析和可视化工具，可对数据源中的数据进行可视化展示，并提供告警功能。

Grafana 支持多个数据源，具体如下。

❑ 时序数据库，支持 Prometheus、Graphite、OpenTSDB、InfluxDB。
❑ 关系数据库，支持 MySQL、PostgreSQL、SQL Server。
❑ 云上，支持 AWS CloudWatch、Azure Monitor、Google Stackdriver。
❑ 其他数据源，如 Elasticsearch、ClickHouse。

除此之外，从 Grafana 3.0 开始，其他数据源可作为插件安装使用。

- **ClickHouse**

ClickHouse 是一个开源、面向 OLAP 场景的列式数据库。

在 PMM 中，ClickHouse 主要用在 Query Analytics 中，通过分析 SQL 性能数据，生成 SQL 分析报告。

```
[root@865710a17aa0 opt]# clickhouse-client --host 127.0.0.1
ClickHouse client version 21.3.20.1 (official build).
Connecting to 127.0.0.1:9000 as user default.
Connected to ClickHouse server version 21.3.20 revision 54447.

865710a17aa0 :) show databases;

SHOW DATABASES

Query id: bf52211c-4b84-4165-9974-a9ce6b679639

┌─name────┐
│ default │
│ pmm     │
│ system  │
└─────────┘

3 rows in set. Elapsed: 0.002 sec.

865710a17aa0 :) show tables from pmm;

SHOW TABLES FROM pmm

Query id: a20d01b9-e87f-4e8c-8e5b-9cfc4755c015

┌─name─────────────┐
│ metrics          │
│ schema_migrations│
└──────────────────┘

2 rows in set. Elapsed: 0.002 sec.
```

- **Nginx**

反向代理服务器。主要配置可参考 /etc/nginx/conf.d/pmm.conf。

- **PostgreSQL**

存储已注册的节点、服务等元数据信息。

```
[root@865710a17aa0 opt]# /usr/pgsql-11/bin/psql -U pmm-managed
psql (11.15)
Type "help" for help.

pmm-managed=> \d
                List of relations
 Schema |      Name       | Type  |   Owner
--------+-----------------+-------+-------------
 public | action_results  | table | pmm-managed
 public | agents          | table | pmm-managed
 public | artifacts       | table | pmm-managed
 public | backup_locations| table | pmm-managed
 public | check_settings  | table | pmm-managed
 public | ia_channels     | table | pmm-managed
 public | ia_rules        | table | pmm-managed
 public | ia_templates    | table | pmm-managed
 public | job_logs        | table | pmm-managed
 public | jobs            | table | pmm-managed
```

```
public    | kubernetes_clusters        | table | pmm-managed
public    | nodes                      | table | pmm-managed
public    | percona_sso_details        | table | pmm-managed
public    | restore_history            | table | pmm-managed
public    | scheduled_tasks            | table | pmm-managed
public    | schema_migrations          | table | pmm-managed
public    | service_software_versions  | table | pmm-managed
public    | services                   | table | pmm-managed
public    | settings                   | table | pmm-managed
(19 rows)
```

6.5.7 设置告警

PMM 在页面上集成了告警功能（Integrated Alerting），不过这个功能在本章完成时（PMM 2.28.0）尚处于技术预览（Technical Preview）阶段，还不成熟。这里介绍另外一种通用的告警方案——Alertmanager。

基于 Alertmanager 设置告警的具体步骤如下。

（1）下载 Alertmanager。

Alertmanager 的下载地址为 GitHub 的 prometheus/alertmanager 页面。

```
# cd /usr/local/
# tar xvf alertmanager-0.24.0.linux-amd64.tar.gz
# cd alertmanager-0.24.0.linux-amd64/
```

（2）修改 Alertmanager 的配置文件。

```
# vim alertmanager.yml
global:
    resolve_timeout: 5m  # 多长时间未收到告警才会视为告警恢复，此时会发送告警恢复信息
    smtp_smarthost: 'smtp.126.com:25' # 电子邮箱 SMTP 服务器地址
    smtp_from: 'slowtech@126.com' # 发送告警信息的电子邮件地址
    smtp_auth_username: 'slowtech' # 电子邮箱用户名
    smtp_auth_password: 'xxxxx' # 电子邮箱密码或授权码

route:
    group_by: ['alertname'] # 告警分组依据。这里指定了 alertname，意味着同一告警规则的告警会放在一起
    group_wait: 10s # 告警第一次触发时，会等待 10 秒，看 10 秒内组内是否有其他告警触发
    group_interval: 10s # 在发送新警报前的等待时间
    repeat_interval: 1m # 告警如果没有恢复，会持续发送，repeat_interval 定义了告警重复发送的时间间隔
    receiver: 'email' # 告警介质的名称，对应下面 receivers 中的 name

receivers:
- name: 'email'
    email_configs: # 电子邮箱配置
    - to: 'slowtech@126.com'   # 接收告警信息的电子邮箱地址

inhibit_rules:
  - source_match:
      severity: 'critical'
    target_match:
      severity: 'warning'
    equal: ['alertname', 'instance']
```

Alertmanager 的配置文件一般由以下几部分组成。

- global：定义一些全局的公共参数，如邮件告警中 SMTP 的配置。
- templates：定义告警通知时的模板。不同的 receiver 可设置不同的告警模板。
- route：定义路由规则。不同的路由规则可定义不同的 receiver。
- receivers：接收器。可理解为告警介质，目前支持 email、hipchat、pagerduty、pushover、slack、opsgenie、webhook、victorops、wechat 等告警方式。
- inhibit_rules：抑制规则。告警抑制指的是当告警发出后，就不再发送由此告警触发的其他告警。

(3) 启动 Alertmanager。

```
# nohup ./alertmanager --config.file alertmanager.yml &
```

Alertmanager 的默认监听端口是 9093。

注意，这里部署的只是一个单节点的 Alertmanager，没有高可用功能。在生产环境中，一般建议部署 Alertmanager Cluster，具体可参考 GitHub prometheus/alertmanager 页面上的 High Availability 部分。

(4) 在 PMM 上配置 Alertmanager 的相关信息。

设置路径：Configuration -> Settings -> Alertmanager Integration。设置页面如图 6-38 所示。

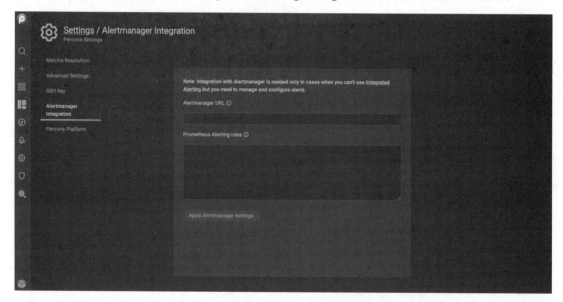

图 6-38　Alertmanager 设置页面

Alertmanager URL 是 Alertmanager 的地址，这里填写 http://10.0.0.118:9093/。

Prometheus Alerting rules 用来定义告警规则。告警规则一般定义在 Prometheus 的 rule_files（/srv/prometheus/rules/pmm.rules.yml）中。只不过 PMM 已经集成在页面中了，这样我们就可以在页面上直接定义告警规则。

下面添加一个 MySQL 端口告警规则。

```
groups:
- name: MySQLDown
  rules:
  - alert: MySQL is down
    expr: mysql_up == 0
    for: 1m
    labels:
      severity: critical
    annotations:
      summary: "Instance {{ $labels.instance }} MySQL is down"
      description: "MySQL is down."
```

这个规则中的 `MySQL is down` 是 alertname，`expr` 是触发告警的表达式。

(5) 验证告警能否触发。

关闭实例，看能否触发告警。告警信息可在两个地方查看：电子邮件和 Alertmanager 的 Web 控制台。控制台的地址是：http://10.0.0.118:9093/。图 6-39 是 Alertmanager Web 控制台的页面。

图 6-39　Alertmanager Web 控制台

可以看到，告警已成功触发。

(6) 告警维护。

通过 Alertmanager 触发的告警可通过以下两种方式来维护。

❑ Alertmanager Web 控制台中的 Silence。
❑ Alertmanager 安装包中自带的 amtool。

6.5.8　PMM 的常见问题

下面梳理了 PMM 使用过程中的一些常见问题及解决方法。

1. 如何自定义 pmm-agent 的日志文件

pmm-agent 的日志默认会输出到系统日志（/var/log/messages）中。不过我们可以在服务管理脚本里面自定义日志文件。

对于 systemd，可以在 /usr/lib/systemd/system/pmm-agent.service 中设置 `StandardError`。例如：

```
StandardError=file:/var/log/pmm-agent.log
```

不过 `StandardError` 这个配置项对 systemd 版本有要求，需 systemd 234 及以上版本。如果我们使用的是 CentOS 7，默认的 systemd 版本是 219，不支持这个配置项。

对于 init 脚本，可以在 /etc/init.d/pmm-agent 里面设置 `pmm_log`，例如：

```
pmm_log="/var/log/pmm-agent.log"
```

2. 如何删除一个服务

删除服务很简单，只需使用 `pmm-admin remove` 命令即可，例如：

```
# pmm-admin remove mysql node1-mysql
Service removed.
```

命令中的 `mysql` 是服务类型。除了 `mysql`，还支持 `mongodb`、`postgresql`、`proxysql`、`haproxy`、`external` 等类型。

`node1-mysql` 是服务名，可通过 `pmm-admin list` 查看。

需要注意的是，删除服务只是让 PMM Server 不再采集数据，并不会删除已经采集的数据。

如果要删除节点，直接停掉这个节点的 pmm-agent 服务即可。

```
# systemctl stop pmm-agent
```

当然，这么做的前提是节点可以登录。如果节点不能登录，可通过 `pmm-admin inventory remove` 来删除。该命令删除的是 PMM Server 中的元数据信息。支持删除 agent、node 和 service。具体语法如下：

```
inventory remove agent [<flags>] [<agent-id>]
  Remove agent from inventory

inventory remove node [<flags>] [<node-id>]
  Remove node from inventory

inventory remove service [<flags>] [<service-id>]
  Remove service from inventory
```

3. 如何离线安装 PMM Server

对于没有外网访问权限的服务器，可离线安装 PMM Server。具体步骤如下。

（1）下载 PMM Server Docker 镜像。

图 6-40 是 PMM Server Docker 镜像的下载页面。

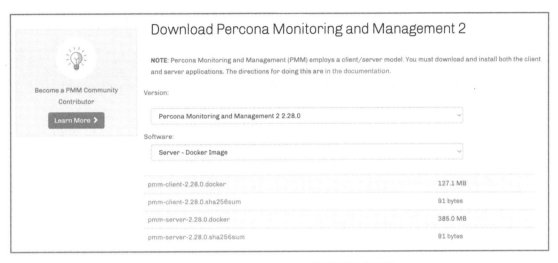

图 6-40　PMM Server Docker 镜像的下载页面

(2) 将下载的 Docker 镜像上传到服务器上。

(3) 通过 docker load 命令导入镜像。

```
# docker load < pmm-server-2.28.0.docker
174f56854903: Loading layer [==================================================>]  211.7MB/211.7MB
558def31e34a: Loading layer [==================================================>]  1.681GB/1.681GB
Loaded image: percona/pmm-server:2.28.0

# docker images
REPOSITORY            TAG      IMAGE ID        CREATED       SIZE
percona/pmm-server    2.28.0   15bdcd948d1e    5 weeks ago   1.85GB
```

镜像导入后，就可以创建数据卷容器和 PMM Server 容器了。

4. 如何设置不同的挂载目录

/srv 默认挂载到宿主机 /var/lib/docker 的一个目录下，而 /var/lib/docker 通常位于根目录下。根目录一般来说空间有限。所以通常来说，我们需要为 /srv 设置一个不同的挂载目录。

下面看看具体的操作步骤。

(1) 创建新的挂载目录。

```
# mkdir -p /data/pmm2/
```

(2) 创建数据卷容器。

```
docker create --volume /srv \
--name pmm-data \
percona/pmm-server:2 /bin/true
```

(3) 检查数据卷容器对应的宿主机目录。

```
# docker inspect pmm-data | egrep "Source|Destination"
                "Source":
```

```
"/var/lib/docker/volumes/8e10dfd3c9ae8d7356282d7ef7c4e28a99a89e6cac51b09be6dd3e75f6a3943f/_data",
        "Destination": "/srv",
```

(4) 将宿主机目录中的文件移动到新的挂载目录中。

```
# mv /var/lib/docker/volumes/8e10dfd3c9ae8d7356282d7ef7c4e28a99a89e6cac51b09be6dd3e75f6a3943f/_data/* /data/pmm2/
```

(5) 建立软链接。

```
# rm -rf /var/lib/docker/volumes/8e10dfd3c9ae8d7356282d7ef7c4e28a99a89e6cac51b09be6dd3e75f6a3943f/_data/
# ln -s /data/pmm2 /var/lib/docker/volumes/8e10dfd3c9ae8d7356282d7ef7c4e28a99a89e6cac51b09be6dd3e75f6a3943f/_data
```

(6) 创建 PMM Server 容器。

```
docker run --detach --restart always \
--publish 443:443 \
--volumes-from pmm-data \
--name pmm-server \
percona/pmm-server:2
```

除了上面这种方式，还有以下两种方式可供参考。

- 修改 Docker 的默认根目录。
- 将 /var/lib/docker 挂载到一个单独的物理盘上。

5. 如何查看监控项的来源

虽然 PMM 提供了这么多监控项，但很多人面临的一个问题是，不知道这些监控项的具体含义。如何知道监控项的具体含义呢？首先要知道这些监控项的来源。只有清楚了来源，才能通过官方文档了解这些监控项的含义。下面我们通过具体的示例来说明查看监控项来源的方法。

图 6-41 是 MySQL Instance Summary 中的一个面板，名为 MySQL Connections。

图 6-41　MySQL Connections 面板

单击面板名，选择 Inspect -> Panel JSON，查看面板的 JSON 定义。重点关注两项：legendFormat（监控项）和 expr（监控项表达式）。expr 里面有指标名。结合图 6-42 不难看出，Max Used Connections

对应的指标名是 `mysql_global_status_max_used_connections`，Max Connections 对应的指标名是 `mysql_global_variables_max_connections`。顾名思义，Max Used Connections 取自 GLOBAL STATUS 中的 `max_used_connections`，Max Connections 取自 GLOBAL VARIABLES 中的 `max_connections`。

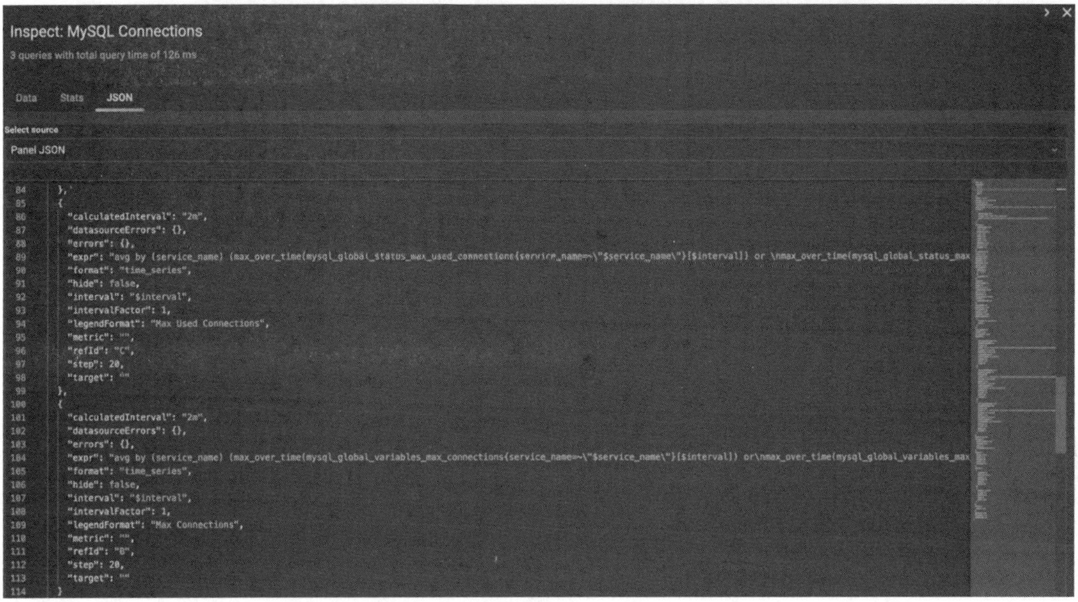

图 6-42 面板的 JSON 定义

`mysql_global_status_xxx` 和 `mysql_global_variables_xxx` 这两类指标名相对容易理解。下面看一个复杂一些的监控项。图 6-43 是 InnoDB IO Targets Write 面板。

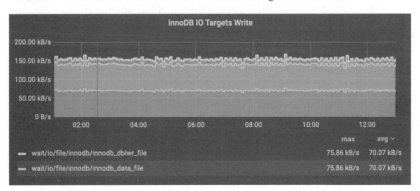

图 6-43 InnoDB IO Targets Write 面板

查看面板对应的 Panel JSON，监控项对应的指标名是 `mysql_perf_schema_file_events_bytes_total`。我们其实很难从这个指标名看出它的来源，只知道它跟 `performance_schema` 有关。这个时候，就只能分析源码包了。

以下是 mysqld_exporter 中的采集器（collector），MySQL 相关的指标都是在采集器中定义的。

```
# ls /usr/src/pmm2-client-2.28.0/mysqld_exporter-45c70da/collector
binlog.go                    info_schema_auto_increment.go              info_schema_tables_test.go
binlog_test.go               info_schema_clientstats.go                 info_schema_userstats.go
collector.go                 info_schema_clientstats_test.go            info_schema_userstats_test.go
collector_test.go            info_schema.go                             perf_schema_events_statements.go
custom_query.go              info_schema_innodb_cmp.go                  perf_schema_events_waits.go
custom_query_test.go         info_schema_innodb_cmpmem.go               perf_schema_file_events.go
engine_innodb.go             info_schema_innodb_cmpmem_test.go          perf_schema_file_instances.go
engine_innodb_test.go        info_schema_innodb_cmp_test.go             perf_schema_file_instances_test.go
engine_tokudb.go             info_schema_innodb_metrics.go              perf_schema.go
engine_tokudb_test.go        info_schema_innodb_metrics_test.go         perf_schema_index_io_waits.go
exporter.go                  info_schema_innodb_sys_tablespaces.go      perf_schema_index_io_waits_test.go
exporter_test.go             info_schema_innodb_sys_tablespaces_test.go perf_schema_table_io_waits.go
global_status.go             info_schema_processlist.go                 perf_schema_table_lock_waits.go
global_status_test.go        info_schema_query_response_time.go         scraper.go
global_variables.go          info_schema_query_response_time_test.go    slave_status.go
global_variables_test.go     info_schema_tables.go                      slave_status_test.go
heartbeat.go                 info_schema_tablestats.go                  standard.go
heartbeat_test.go            info_schema_tablestats_test.go
```

再来说说指标名，mysqld_exporter 中指标的命名还是有一定规则的。以 mysql_perf_schema_file_events_bytes_total 为例，mysql 是命名空间，一般用来避免库之间的冲突并指出指标出处，bytes 是指标单位，total 是指标后缀，perf_schema_file_events 是严格意义上的指标名。perf_schema_file_events 的相关指标是在 perf_schema_file_events.go 文件中定义的。

接下来，我们看看 perf_schema_file_events.go 这个文件的内容。

```
# vim perf_schema_file_events.go
...
const perfFileEventsQuery = `
    SELECT
        EVENT_NAME,
        COUNT_READ, SUM_TIMER_READ, SUM_NUMBER_OF_BYTES_READ,
        COUNT_WRITE, SUM_TIMER_WRITE, SUM_NUMBER_OF_BYTES_WRITE,
        COUNT_MISC, SUM_TIMER_MISC
      FROM performance_schema.file_summary_by_event_name
    `

var (
    performanceSchemaFileEventsDesc = prometheus.NewDesc(
        prometheus.BuildFQName(namespace, performanceSchema, "file_events_total"),
        "The total file events by event name/mode.",
        []string{"event_name", "mode"}, nil,
    )
    performanceSchemaFileEventsTimeDesc = prometheus.NewDesc(
        prometheus.BuildFQName(namespace, performanceSchema, "file_events_seconds_total"),
        "The total seconds of file events by event name/mode.",
        []string{"event_name", "mode"}, nil,
    )
    performanceSchemaFileEventsBytesDesc = prometheus.NewDesc(
        prometheus.BuildFQName(namespace, performanceSchema, "file_events_bytes_total"),
        "The total bytes of file events by event name/mode.",
        []string{"event_name", "mode"}, nil,
    )
)
...
```

这里只截取了两部分，先看第二部分：指标的定义。

在 Prometheus 中，指标（metric）通常是通过 prometheus.NewDesc 来定义的。而 prometheus.BuildFQName 则用来构造指标名，其语法如下：

```
func BuildFQName(namespace, subsystem, name string) string
```

其中，namespace 是命名空间，通常是应用程序名，subsystem 是子系统名，name 是一个简短的指标名。BuildFQName 的作用很简单，就是将 namespace、subsystem、name 这三个字符串通过下划线 "_" 连接到一起。

具体到 perf_schema_file_events.go 文件中的 BuildFQName，namespace 和 performanceSchema 是两个常量，分别在 collector.go 和 perf_schema.go 中定义。

```
# cat collector.go
...
const (
        namespace = "mysql"
        ...
)
# cat perf_schema.go
...
const performanceSchema = "perf_schema"
```

由此来看，mysql_perf_schema_file_events_bytes_total 这个指标确实是在 perf_schema_file_events.go 文件中定义的。

再来看看第一部分——perfFileEventsQuery。这个 SQL 实际上就是 mysql_perf_schema_file_events_bytes_total 这个指标的来源。

6.5.9 参考资料

- Prometheus 官方文档。
- 电子书 prometheus-book。
- 博客文章 "Deploying Percona Monitoring and Management 2 Without Access to the Internet"，作者：Ivan Groenewold。
- 博客文章 "How to Manually Remove Client Instances From Percona Monitoring and Management 2"，作者：Agustín。
- 博客文章 "Does Percona Monitoring and Management (PMM) Support External Monitoring Services? Yes It Does!"，作者：Borys Belinsky。
- 博客文章 "Running Custom MySQL Queries in Percona Monitoring and Management"，作者：Carlos Salguero。
- 博客文章 "Using Different Mount Points on Percona Monitoring and Management 2 Docker Deployments"，作者：Agustín。

6.6 MySQL 中常用的监控指标

MySQL 中绝大多数的监控指标来源于状态变量（status variable），下面我们按照不同的类别来看看这些常用状态变量的具体含义。

6.6.1 连接相关

- **Aborted_clients**

客户端已成功建立，但中途异常断开连接的次数。连接异常断开的常见原因有如下几种。

- 客户端程序在断开前，没有调用 `mysql_close()` 方法。
- 客户端连接的空闲时间超过 `wait_timeout` 的会话值，被服务端主动断开。
- 客户端程序在数据传输过程中突然结束。
- 数据包的大小超过 `max_allowed_packet` 的限制。

对于中途异常断开的连接，错误日志中通常会输出如下信息：

```
[Note] Aborted connection 42 to db: 'unconnected' user: 'root' host: 'localhost' (Got timeout reading communication packets)
```

错误日志中是否输出此类信息还与 `log_error_verbosity` 的设置有关，只有设置为 3 时才输出。该参数是 MySQL 5.7 引入的，默认值在 MySQL 5.7 中是 3，在 MySQL 8.0 中调整为了 2。所以，在 MySQL 5.7 中我们会经常看到此类信息。

- **Aborted_connects**

连接 MySQL 服务端失败的次数。连接失败的常见原因有如下几种。

- 客户端的账号或密码不准确。
- 没有指定库的访问权限。
- 连接包中没有包含正确的信息。
- 超过 `connect_timeout`（默认为 10 秒），服务端还没有收到客户端的连接包。

- **Threads_connected、Threads_running、Threads_cached、Threads_created**

MySQL 默认的线程调度方式是每个连接一个线程（one-thread-per-connection）。所以，在理解这 4 个变量时，基本上可将其视为连接。各个状态变量的具体含义如下。

- `Threads_connected`：当前连接（线程）数，这个值等于 SHOW PROCESSLIST 的总条数。
- `Threads_running`：当前处于活跃状态的线程数。如果该值过大，会导致系统频繁地进行上下文切换，CPU sys 使用率较高。
- `Threads_cached`：Thread Cache 缓存的线程数。在创建新的连接时，会首先检查 Thread Cache 中是否有缓存的线程。如果有，则直接复用；如果没有，则会创建新的线程。在线程池场景，会禁用 Thread Cache，此时，`Threads_cached` 会显示为 0。

- **Threads_created**：已创建的线程数。反映的是累计值。如果该值持续增大，则代表 Thread Cache 过小，此时可适当增大 thread_cache_size 的值。

- **Max_used_connections**、**Max_used_connections_time**、**Connection_errors_max_connections**

MySQL 中的最大连接数由参数 max_connections 控制，默认是 151。如果当前连接数达到了 max_connections 的限制，则新的连接会创建失败，并提示 Too many connections 错误，Connection_errors_max_connections 也会随之增大。

当我们观察到 Connection_errors_max_connections 大于 0 时，就已经晚了，最好是基于 Threads_connected / max_connections 做一个连接数使用率监控，如果达到 85% 则触发告警。

Max_used_connections 反映的是历史最大连接数。Max_used_connections_time 是连接数达到最大时的时间。

6.6.2 Com 相关

统计操作执行的次数。以下状态变量在监控中用得比较多，可以反映数据库的繁忙程度。

- **Com_select**、**Com_insert**、**Com_delete**、**Com_update**、**Com_commit**、**Com_rollback**

分别用来统计 SELECT、INSERT、DELETE、UPDATE、COMMIT 和 ROLLBACK 操作执行的次数。

以 Com_select 为例，看下面这个示例。

```
mysql> flush status;
Query OK, 0 rows affected (0.01 sec)

mysql> show status like 'com_select';
+---------------+-------+
| Variable_name | Value |
+---------------+-------+
| Com_select    | 0     |
+---------------+-------+
1 row in set (0.00 sec)

mysql> select * from t1 limit 1;
+----+------+
| id | c1   |
+----+------+
|  1 | a    |
+----+------+
1 row in set (0.00 sec)

mysql> show status like 'com_select';
+---------------+-------+
| Variable_name | Value |
+---------------+-------+
| Com_select    | 1     |
+---------------+-------+
1 row in set (0.00 sec)
```

在这几个变量中，可用 Com_commit 加上 Com_rollback 来衡量 TPS（transaction per second，每秒事务数）。在一些资料中，我们会看到下面这个计算公式：

TPS = (Com_commit + Com_rollback) / Uptime

不过通过这个公式得到的结果是平均值，参考意义不大。建议取 Com_commit 的每秒增量加上 Com_rollback 的每秒增量。

6.6.3 Handler 相关

Handler 是一个类，按不同的功能模块定义了若干接口。MySQL Server 在执行增删改查等操作时，不会直接与存储引擎层交互，而是调用 Handler 对象的相关方法。由此带来的好处是，实现了 Server 层与存储引擎层的解耦，方便了新引擎的引入。

- Handler_delete、Handler_update、Handler_write

分别对应 DELETE、UPDATE、INSERT 操作影响的记录数，这一点与 Com_xxx 不一样，后者反映的只是操作的次数。看下面这个示例。

```
mysql> select * from t;
+----+------+
| id | c1   |
+----+------+
|  1 | a    |
|  2 | b    |
|  3 | c    |
|  4 | d    |
|  5 | e    |
+----+------+
5 rows in set (0.00 sec)

mysql> flush status;
Query OK, 0 rows affected (0.02 sec)

mysql> show status where variable_name in ('Com_delete', 'Handler_delete');
+----------------+-------+
| Variable_name  | Value |
+----------------+-------+
| Com_delete     | 0     |
| Handler_delete | 0     |
+----------------+-------+
2 rows in set (0.00 sec)

mysql> delete from t limit 3;
Query OK, 3 rows affected (0.01 sec)

mysql> show status where variable_name in ('Com_delete', 'Handler_delete');
+----------------+-------+
| Variable_name  | Value |
+----------------+-------+
| Com_delete     | 1     |
| Handler_delete | 3     |
+----------------+-------+
2 rows in set (0.00 sec)
```

- `Handler_commit`、`Handler_prepare`、`Handler_rollback`、`Handler_savepoint`、`Handler_savepoint_rollback`

分别对应 COMMIT、PREPARE、ROLLBACK、SAVEPOINT、ROLLBACK TO SAVEPOINT 操作的数量。

需要注意的是，对于 InnoDB 存储引擎，在进行 DML 操作时，因为涉及事务的两阶段提交，所以对于任何一个操作，都会导致对应的 Handler 加 2（一个对应 binlog，一个对应 InnoDB）。

- `Handler_read_key`、`Handler_read_first`、`Handler_read_last`、`Handler_read_next`、`Handler_read_prev`、`Handler_read_rnd`、`Handler_read_rnd_next`

这些变量的具体含义如下。

- `Handler_read_key`：基于索引来定位记录。该值越大，代表基于索引的查询越多。
- `Handler_read_first`：读取索引的第一个值。该值越大，代表涉及索引全扫描的查询越多。
- `Handler_read_last`：和 `Handler_read_first` 相反，读取索引的最后一个值。若该值增大，基本上可以判定查询中使用了基于索引的 ORDER BY DESC 子句。
- `Handler_read_next`：根据索引的顺序来读取下一行的值。常用于基于索引的范围扫描和 ORDER BY LIMIT 子句中。
- `Handler_read_prev`：根据索引的顺序来读取上一行的值。一般用于基于索引的 ORDER BY DESC 子句中。
- `Handler_read_rnd`：将记录基于某种标准进行排序，然后再根据它们的位置信息来遍历排序后的结果，这往往会导致表的随机读。
- `Handler_read_rnd_next`：读取下一行记录的次数，常用于全表扫描。

在使用时需注意以下几点。

- `Handler_read_key` 的值越大越好，因为值越大，代表基于索引的查询越多。
- `Handler_read_first`、`Handler_read_last`、`Handler_read_next`、`Handler_read_prev` 都会利用索引。但查询是否高效还需要结合其他 `Handler_read` 的值来判断。
- `Handler_read_rnd` 不宜过大。
- `Handler_read_rnd_next` 不宜过大。过大的话，代表全表扫描过多，要引起足够的警惕。

6.6.4 临时表相关

- `Created_tmp_files`、`Created_tmp_tables`、`Created_tmp_disk_table`

这些变量的具体含义如下。

- `Created_tmp_files`：创建的临时文件的数量。
- `Created_tmp_tables`：创建的临时表的数量。
- `Created_tmp_disk_tables`：内存临时表转化为磁盘临时表的数量。

内存临时表如果超过一定的大小会转化为磁盘临时表。这个大小与内存临时表使用的引擎有关。

如果使用的是 MEMORY 引擎，这个大小由以下两个参数中的最小值决定。

- `tmp_table_size`：内存临时表的最大大小，默认为 16MB。
- `max_heap_table_size`：MEMORY 表的最大大小，默认为 16MB。

如果使用的是 TempTable 引擎，则上述大小由 `tmp_table_size` 决定。TempTable 引擎是 MySQL 8.0 引入的。

6.6.5 Table Cache 相关

为了提升表的访问效率，表使用完毕后，不会立即关闭，而是会缓存在 Table Cache 中。当 MySQL 访问一张表时，会首先检查该表的文件描述符是否在 Table Cache 中。如果在，则直接使用，同时增大变量 `Table_open_cache_hits` 的值。如果不在，则打开该表，此时会增大 `Opened_tables` 及 `Table_open_cache_misses` 的值。表使用完毕后，会将文件描述符缓存在 Table Cache 中。

如果缓存满了，Table Cache 达到了 `table_open_cache` 的限制，要考虑两种场景。

- 缓存中存在未使用的表。这个时候，会基于 LRU 算法关闭未使用的表，并将其从 Table Cache 中删除，同时会增大 `Table_open_cache_overflows` 的值。
- 缓存中的表都在使用中。这个时候，会临时扩容 Table Cache。一旦有表未使用，就会删除该表的缓存，让 Table Cache 的大小降到 `table_open_cache` 之下。

除此之外，FLUSH TABLES 操作也会删除 Table Cache 中的缓存。

Table Cache 相关的参数如下。

- `table_open_cache`：Table Cache 的大小。该参数在 MySQL 5.6 和 MySQL 5.7 中默认为 `2000`，在 MySQL 8.0 中默认为 `4000`。
- `table_open_cache_instances`：Table Cache 实例的数量。设置多个实例，每个实例可缓存的表的数量就等于 `table_open_cache` / `table_open_cache_instances`。在表数量较多的场景，能有效降低锁争用。如果主机 CPU 的核数大于或等于 16，建议将该参数设置为 `8` 或 `16`。
- `table_definition_cache`：限制可缓存的 frm 文件的数量。

Table Cache 相关的变量如下。

- `Open_tables`：当前打开的表的数量。
- `Open_table_definitions`：当前缓存的 frm 文件的数量。
- `Opened_tables`：打开过的表的数量。
- `Opened_table_definitions`：缓存过的 frm 文件的数量。
- `Table_open_cache_hits`：Table Cache 命中的次数。
- `Table_open_cache_misses`：Table Cache 没有命中的次数。
- `Table_open_cache_overflows`：表缓存被删除的次数。

如果我们观察到 Opened_tables 大于 table_open_cache 且还在持续增大，则意味着 table_open_cache 过小，此时可适当调大 table_open_cache 的值。

6.6.6 文件相关

MySQL 进程可打开的最大文件描述符数由 open_files_limit 决定，当文件描述符不足时，会提示 Too many open files 错误。需要注意的是，open_files_limit 并不是所设即所得，配置的和实际生效的并不完全一样。它的生成逻辑具体如下。

open_files_limit 首先会取以下 3 个值中的最大值。

- 10 + max_connections + table_open_cache * 2
- max_connections * 5
- open_files_limit ? open_files_limit : 5000。如果配置文件中指定了 open_files_limit，则取 open_files_limit，否则是 5000。

接着，会比较 open_files_limit 与 MySQL 进程可打开的最大文件描述符数的大小。

- 如果进程的最大文件描述符数超过 open_files_limit，则实际生效的是进程的最大文件描述符数。
- 如果进程的最大文件描述符数小于 open_files_limit，这个时候需区分启动用户。
  - 如果启动用户是 root，则实际生效的是 open_files_limit。
  - 如果启动用户是普通用户，则需要继续比较 open_files_limit 与硬限制（ulimit -Hn）的大小。
    - 如果 open_files_limit 小于等于硬限制，则实际生效的是 open_files_limit。
    - 如果 open_files_limit 大于硬限制，则实际生效的是进程的最大文件描述符数。

另外，如果进程是通过 mysqld_safe 启动的，mysqld_safe 会基于 open_files_limit 调整进程的最大文件描述符数。

```
ulimit -n $open_files
```

文件相关的状态变量如下。

- Innodb_num_open_files：InnoDB 当前打开的文件数。InnoDB 同时能打开的最大文件数由 innodb_open_files 参数决定，后者默认为 -1，基于 table_open_cache 自动调整。innodb_open_files 并不是一个硬限制，如果 InnoDB 打开的文件数达到了 innodb_open_files 的限制，会通过 LRU 算法关闭其他文件。
- Open_files：当前打开的文件数。注意，它不包括存储引擎使用自己内部函数打开的文件数。所以，在计算 MySQL 进程当前打开的文件数时，需加上 Innodb_num_open_files。
- Opened_files：打开过的文件数。

当 Open_files + Innodb_num_open_files 接近 open_files_limit 时，就需要调整进程的最大文件描述符数了。常规思路是修改 open_files_limit，但 open_files_limit 是个只读参数，重启实例才能

生效。实际上，不用这么麻烦。从 CentOS 6 开始，可在线调整进程的最大文件描述符数，无须重启进程。具体命令如下。

```
#CentOS 6
echo -n "Max open files=102400:102400" > /proc/${PID}/limits

#CentOS 7
prlimit --pid ${PID} --nofile=102400:102400
```

注意，CentOS 6 的调整方式在 CentOS 7 中不适用，同样，CentOS 6 中也不支持 prlimit 命令。

6.6.7 主从复制相关

主从复制重点关注以下 3 个指标。

- `Slave_IO_Running`：I/O 线程的状态。
- `Slave_SQL_Running`：SQL 线程的状态。
- `Seconds_Behind_Master`：主从延迟时间。

这 3 个指标均取自 SHOW SLAVE STATUS。

除此之外，还需关注从库 `read_only`（`super_read_only`）的值。为了避免对从库的误操作，建议将从库设置为只读。

6.6.8 缓冲池相关

缓冲池（buffer pool）中有 3 个比较核心的链表用来维护管理数据页。

- Free List：管理空闲页。空闲页指的是未被使用的页。在缓冲池初始化时，会将缓冲池中的所有页加入到 Free List 中。
- LRU List：管理干净页（clean page）和脏页（dirty page）。干净页指的是数据与磁盘中内容一致的数据页。脏页指的是被修改过，但还没有刷盘的数据页。
- FLU List：管理脏页。脏页按照 `oldest_modification`（最早修改时间）进行降序排列，最早修改的脏页会放到链表尾部。

当用户需要读取磁盘中的一个数据页时，会从 Free List 中获取空闲页。如果有，则直接返回；如果没有，则会尝试驱逐 LRU List 末尾的数据页。

以下是缓冲池中数据页相关的状态变量。

- `Innodb_buffer_pool_pages_data`：缓冲池中数据页的数量，包括干净页和脏页。
- `Innodb_buffer_pool_bytes_data`：数据页的大小，单位是字节。
- `Innodb_buffer_pool_pages_dirty`：脏页的数量。
- `Innodb_buffer_pool_bytes_dirty`：脏页的大小，单位是字节。
- `Innodb_buffer_pool_pages_free`：空闲页的数量。
- `Innodb_buffer_pool_pages_misc`：用于管理开销而分配的页的数量，比如行锁、自适应哈希等。

- Innodb_buffer_pool_pages_total：页的总数量。
- Innodb_buffer_pool_pages_flushed：脏页被刷盘的次数。
- Innodb_buffer_pool_wait_free：等待空闲页的次数。

这几个变量之间的关系如下。

- Innodb_buffer_pool_pages_total = Innodb_buffer_pool_pages_data + Innodb_buffer_pool_pages_free + Innodb_buffer_pool_pages_misc
- 在不使用表压缩的情况下，Innodb_buffer_pool_bytes_data = Innodb_buffer_pool_pages_data * 16KB，其中，16KB 为页的默认大小。
- 在不使用表压缩的情况下，innodb_buffer_pool_size = Innodb_buffer_pool_pages_total * 16KB。

- **Innodb_buffer_pool_read_requests、Innodb_buffer_pool_reads**

前者是逻辑读的数量，后者是物理读的数量。

InnoDB 缓冲池命中率 = (Innodb_buffer_pool_read_requests − Innodb_buffer_pool_reads) / Innodb_buffer_pool_read_requests。

可根据 InnoDB 缓冲池命中率来判断 innodb_buffer_pool_size 的设置是否合理。一般情况下，应保证该值大于 95%。

- **Innodb_rows_deleted、Innodb_rows_inserted、Innodb_rows_read、Innodb_rows_updated**

InnoDB 表被删除、插入、读取、更新的行数。

6.6.9 redo log 相关

以下是 redo log 相关的状态变量。

- Innodb_log_waits：因 redo log buffer 过小，导致 redo log buffer 刷盘的次数。
- Innodb_log_write_requests：写 redo log buffer 的次数。
- Innodb_log_writes：写 redo log 的次数。
- Innodb_os_log_fsyncs：对 redo log 调用 fsync 操作的次数。
- Innodb_os_log_pending_fsyncs：fsync 操作等待的次数。
- Innodb_os_log_pending_writes：写 redo log 等待的次数。
- Innodb_os_log_written：redo log 的写入量，单位是字节。

注意，不要使用 Innodb_os_log_written 来反映 redo log 的写入量。看下面这个示例。

- 事务 1 写入了 100 字节到 redo log buffer 中，相应地，LSN（日志序列号）会增加 100。
- 事务 1 提交，redo log buffer 会刷新到 redo log 中。因为 redo log 的基本存储单位是 block，且 block 的大小是 512 字节，所以，在刷新完毕后，Innodb_os_log_written 会增加 512。
- 接下来，事务 2 写入了 200 字节到 redo log buffer 中，相应地，LSN 会增加 200。
- 事务 2 提交，同样，Innodb_os_log_written 会增加 512。

redo log buffer 实际写入了 300 字节，但对应的 Innodb_os_log_written 却增加了 1024。

如果要评估 redo log 的写入量，推荐使用下面这种方式。

```
mysql> pager grep sequence
PAGER set to 'grep sequence'

mysql> show engine innodb status\G select sleep(60); show engine innodb status\G
Log sequence number 88396793310905
1 row in set (0.01 sec)

1 row in set (1 min 0.00 sec)

Log sequence number 88396813216350
1 row in set (0.01 sec)

mysql> nopager
PAGER set to stdout

mysql> select 88396813216350 - 88396793310905;
+---------------------------------+
| 88396813216350 - 88396793310905 |
+---------------------------------+
|                        19905445 |
+---------------------------------+
1 row in set (0.00 sec)
```

6.6.10 锁相关

以下是行锁相关的状态变量。

- `Innodb_row_lock_current_waits`：当前正在等待行锁的操作数。
- `Innodb_row_lock_time`、`Innodb_row_lock_time_avg`、`Innodb_row_lock_time_max`：获取行锁花费的总时间、平均时间、最大时间。单位是毫秒。
- `Innodb_row_lock_waits`：等待行锁的次数。

以下是表锁相关的状态变量。

- `Table_locks_immediate`：能立即获得表锁的次数。
- `Table_locks_waited`：等待表锁的次数。

在 MySQL 中，表锁只用于 MyISAM、MEMORY、MERGE 引擎，读写相互阻塞。

看下面这个示例。

```
session1> create table t1(id int primary key) engine=myisam;
Query OK, 0 rows affected (0.01 sec)

session1> insert into t1 values(1),(2);
Query OK, 2 rows affected (0.00 sec)
Records: 2  Duplicates: 0  Warnings: 0

session1> select id,sleep(100) from t1;
执行中……
```

```
session2> flush status;
Query OK, 0 rows affected (0.00 sec)

session2> show global status like 'Table_locks_waited';
+--------------------+-------+
| Variable_name      | Value |
+--------------------+-------+
| Table_locks_waited | 0     |
+--------------------+-------+
1 row in set (0.00 sec)

session2> delete from t1 where id=1;
阻塞中……

session3> show global status like 'Table_locks_waited';
+--------------------+-------+
| Variable_name      | Value |
+--------------------+-------+
| Table_locks_waited | 1     |
+--------------------+-------+
1 row in set (0.00 sec)

session3> show processlist;
+----+-----------------+-----------+----------+---------+------+-----------------------------+----------------------------------+
| Id | User            | Host      | db       | Command | Time | State                       | Info                             |
+----+-----------------+-----------+----------+---------+------+-----------------------------+----------------------------------+
| 5  | event_scheduler | localhost | NULL     | Daemon  | 1539 | Waiting on empty queue      | NULL                             |
| 18 | root            | localhost | slowtech | Query   | 39   | User sleep                  | select id,sleep(100) from t1     |
| 19 | root            | localhost | slowtech | Query   | 17   | Waiting for table level lock| delete from t1 where id=1        |
| 20 | root            | localhost | NULL     | Query   | 0    | init                        | show processlist                 |
+----+-----------------+-----------+----------+---------+------+-----------------------------+----------------------------------+
4 rows in set (0.00 sec)
```

6.6.11 排序相关

排序相关的变量有 Sort_merge_passes、Sort_range、Sort_rows、Sort_scan。

看下面这个示例。

```
mysql> create table t1(id int primary key,c1 varchar(10));
Query OK, 0 rows affected (0.03 sec)

mysql> insert into t1 values(1,'a'),(2,'b'),(3,'c');
Query OK, 3 rows affected (0.01 sec)
Records: 3  Duplicates: 0  Warnings: 0

mysql> flush status;
Query OK, 0 rows affected (0.01 sec)

mysql> select * from t1 order by c1;
...

mysql> show status like 'sort%';
+-------------------+-------+
| Variable_name     | Value |
+-------------------+-------+
| Sort_merge_passes | 0     |
| Sort_range        | 0     |
```

```
| Sort_rows            | 3     |
| Sort_scan            | 1     |
+----------------------+-------+
4 rows in set (0.00 sec)
```

结合实例来看看这 4 个变量的含义。

- Sort_merge_passes：在涉及排序（ORDER BY、GROUP BY、DISTINCT）操作时，如果无法使用索引，则会使用 filesort。使用 filesort 进行排序时，执行计划的 Extra 列会显示 Using filesort。在使用 filesort 时，MySQL 会分配单独的排序缓存区（sort buffer）。排序缓存区是需要时才分配的，并且是按需分配。它的最大大小由 sort_buffer_size 决定，默认是 256KB。如果要排序的记录数比较少，只用排序缓存区就能完成排序操作，这个时候处理的效率就会非常高。如果记录数较多，MySQL 会分批处理，每一批首先会在排序缓存区中排序，排序后的结果会存储到磁盘的临时文件中。每个排序缓存区对应临时文件中的一个 block。处理完毕后，最后再对临时文件中的 block 进行归并排序。Sort_merge_passes 反映的是归并操作的次数。如果 Sort_merge_passes 较大，可适当调大 sort_buffer_size 的值。
- Sort_range：对索引范围扫描的结果进行排序的次数。
- Sort_scan：对全表扫描的结果进行排序的次数。
- Sort_rows：排序的记录数。

6.6.12 查询相关

查询相关的变量有 Select_full_join、Select_full_range_join、Select_range、Select_range_check、Select_scan。

这些变量的具体含义如下。

- Select_scan：全表扫描。如果是多表关联查询，指的是最外层的驱动表执行了全表扫描。
- Select_full_join：同样是全表扫描，不过针对的是被驱动表。
- Select_range：范围查询。如果是多表关联查询，指的是最外层的驱动表执行了范围查询。
- Select_full_range_join：同样是范围查询，不过针对的是被驱动表。
- Select_range_check：常用在非等值的关联查询中。

6.6.13 其他重要指标

- **Uptime**

数据库的运行时间，单位是秒。可基于 Uptime 来判断数据库是否发生过重启。

- **Queries 和 Questions**

两者都是统计 MySQL 服务端执行的操作的数量，关系如下。

```
SUM(Com_xxx) + Qcache_hits
= Questions + statements executed within stored programs
= Queries
```

我们常用的 QPS 通常是基于 Questions 来计算的：QPS = Questions / Uptime。同 TPS 一样，建议取每秒增量。

- **Slow_queries**

慢查询的数量。注意，无论是否开启慢日志，只要操作的执行时间超过 `long_query_time`，就会导致 Slow_queries 增加。

- **Bytes_received 和 Bytes_sent**
  - Bytes_received：从客户端接收的流量大小，单位是字节。
  - Bytes_sent：发送给客户端的流量大小，单位是字节。

6.7 本章总结

本章介绍了两个监控方案。

一个是 Zabbix，依托的监控模板是 Percona 公司提供的 PMP。需要注意的是，从 2020 年 8 月 1 日开始，Percona 公司结束了对该产品的支持，后续不再维护、更新。其实 Zabbix 本身也内置了 MySQL 监控模板，只不过监控指标相对较少，不能很好地满足我们对于监控指标精细化及自定义监控指标的需求。

另一个是 PMM，PMM 是 Percona 公司基于 Prometheus 和 Grafana 开发的一个数据库监控管理平台，优点显而易见：开箱即用，使用简单；监控指标丰富；支持 Query Analytics，可对 MySQL 的慢查询进行分析。但同时我们也要注意到，PMM 融合了多个开源组件，虽然提升了易用性，但同时也带来了复杂性。如果我们对这些开源组件不熟悉，一旦出现问题，问题的定位也是一件棘手的事。

如果公司的监控平台本身使用的也是 Prometheus，且有独立的慢日志采集、展示平台，就没必要使用 PMM 了。好在 PMM 的 Dashboard 也可用在原生的 Prometheus 实例中。

本章最后介绍了 MySQL 中常用的监控指标。

重点问题回顾

- 如何使用 PMP 监控 MySQL。
- 如何基于 ss_get_mysql_stats.php 自定义监控项。
- Zabbix 分区表的使用及维护。
- Zabbix API。
- PMM 的安装及使用。
- Query Analytics 的来源有两种：慢日志和 `performance_schema`。两者有什么区别？
- PMM 如何设置告警。
- MySQL 中常用的监控指标。

第 7 章
DDL

DDL（data definition language，数据定义语言）主要是用来定义数据库对象的数据结构，主要包括以下几类操作：

- CREATE
- ALTER
- DROP
- TRUNCATE

上述定义其实是一种广义的说法。在实际工作中，当我们提起 DDL 时，更多的是指 ALTER（表结构变更）操作。

与之相对的是 DML（data manipulation language，数据操纵语言）。简单来说，DML 就是 INSERT、DELETE 和 UPDATE 操作的统称。

在线上进行表结构变更操作时，很多开发人员担心会锁表。这种担心其实不无道理，原因有二。

- 在 MySQL 5.6 引入 Online DDL 之前，所有的 ALTER 操作确实会阻塞 DML。
- 开发、测试环境都是直接通过 ALTER 语句修改表结构。即使引入了 Online DDL，很多高频操作，如调整字段的类型、增加字段的长度，还是会阻塞 DML。

但实际上，我们在线上很少直接通过 ALTER 语句修改表结构，更多的是使用第三方开源工具，如 pt-online-schema-change、gh-ost。为什么要使用这些工具呢？有以下 3 个原因。

- 锁定时间短。
- 对主库的性能影响较小。
- 不会造成主从延迟。严格来说，是当出现主从延迟时，这些工具能暂停当前的数据拷贝[①]操作。

本章主要包括以下内容。

- Online DDL。
- pt-online-schema-change。
- gh-ost。
- 元数据锁。

---

① 即 copy（复制），这里为了与"主从复制"中的"复制"区分而写成"拷贝"。——译者注

7.1 Online DDL

在 MySQL 5.6 之前，所有的 DDL 操作都会阻塞 DML，仅允许查询操作。此时的 DDL 支持两种算法：COPY 和 INPLACE。下面我们看看这两种算法的实现原理。

COPY 算法的实现原理如下。

- 新建临时表。
- 对原表加锁。这个时候会阻塞 DML，只允许查询操作。
- 将原表数据逐行拷贝到临时表中。
- 升级字典锁，禁止读写，执行 RENAME 操作。

INPLACE 算法是 MySQL 5.1 引入的，仅适用于索引的创建和删除。此种方式下，索引可直接在原表上创建，不会新建临时表，因此效率较高，但与 COPY 算法一样会阻塞 DML。

INPLACE 算法的实现原理如下。

- 新建索引的数据字典。
- 对原表加锁，这个时候会阻塞 DML，只允许查询操作。
- 读取聚集索引，构造新的索引项，排序并插入新索引。
- 等待当前表的所有只读事务提交。
- 创建索引结束。

Online DDL 是 MySQL 5.6.7 引入的。这里的 Online，指的是在执行 DDL 的过程中，不会阻塞 DML 操作。但注意，并非所有的 DDL 操作都不会阻塞 DML。

这里首先说说 ALTER TABLE 操作中 DDL 相关的两个子句：ALGORITHM 和 LOCK。

```
ALTER TABLE tbl_name ADD PRIMARY KEY (column), ALGORITHM=INPLACE, LOCK=NONE;
```

下面看看这两个子句的具体含义。

ALGORITHM 指定 DDL 操作使用的算法，支持如下选项。

- INSTANT：只会修改表的元数据信息。MySQL 8.0.12 引入，支持表的秒级加列特性。
- INPLACE：支持在原表上进行操作，包括 rebuild 方式和 no-rebuild 方式这两类。rebuild 方式会重建表，no-rebuild 方式只会修改表的元数据信息。
- COPY：会重建表且表的记录格式会发生变化。
- DEFAULT：如果不指定 ALGORITHM，则默认为 DEFAULT。此时，将由 MySQL 自行决定 DDL 使用的算法。优先使用 INSTANT，其次是 INPLACE，最后是 COPY。

LOCK 指定 DDL 操作的锁定级别，支持如下选项。

- NONE：允许查询和 DML 操作。
- SHARED：允许查询，但阻塞 DML 操作。
- EXCLUSIVE：阻塞查询和 DML 操作。

- DEFAULT：如果不指定 LOCK，则默认是 DEFAULT。由 MySQL 根据 DDL 的类型自行决定 LOCK 的级别。优先采用 NONE，其次是 SHARED，最后是 EXCLUSIVE。

当 ALGORITHM = COPY 时，LOCK 只能设置为 SHARED 和 EXCLUSIVE。如果不指定 LOCK 级别，则默认为 SHARED。

从上面两个子句的选项来看，ALGORITHM = INSTANT（INPLACE）且 LOCK = NONE 才算得上真正意义上的 Online DDL。

7.1.1 Online DDL 的分类

根据上面 ALGORITHM 和 LOCK 的选项，可将 DDL 分为 3 类。

1. 只修改表的元数据信息

只修改表的元数据信息的 DDL 操作有以下特点。

- 不会阻塞 DML。
- 耗时极短。如果表上没有正在执行的查询或事务，该类操作几乎会瞬间完成。
- 只会更新表结构文件。

这一点可从 DDL 前后 frm 文件的 inode 号看出。

```
# ll -ih /data/mysql/3307/data/sbtest/
total 2.3G
1062535 -rw-r----- 1 mysql mysql   65 Dec 11 10:24 db.opt
1062735 -rw-r----- 1 mysql mysql 8.5K Dec 11 10:41 sbtest1.frm
1062734 -rw-r----- 1 mysql mysql 2.3G Dec 11 10:42 sbtest1.ibd

mysql> alter table sbtest.sbtest1 drop index k_1;
Query OK, 0 rows affected (0.02 sec)
Records: 0  Duplicates: 0  Warnings: 0

# ll -ih /data/mysql/3307/data/sbtest/
total 2.3G
1062535 -rw-r----- 1 mysql mysql   65 Dec 11 10:24 db.opt
1062733 -rw-r----- 1 mysql mysql 8.5K Dec 11 10:44 sbtest1.frm
1062734 -rw-r----- 1 mysql mysql 2.3G Dec 11 10:44 sbtest1.ibd
```

注意，在 MySQL 8.0.12 引入 INSTANT 算法之前，只修改表的元数据信息的操作使用的是 INPLACE 算法，对应 no-rebuild 这种方式。引入了 INSTANT 算法之后，很多此类操作放到了 INSTANT 算法下面。虽然类别发生了变化，但本质没变。

属于此类的 DDL 操作如下。

- 删除二级索引。

    ```
    DROP INDEX index_name ON t1;
    ALTER TABLE t1 DROP INDEX index_name;
    ```

- 修改索引名（到 MySQL 5.7 才支持）。

    ```
    ALTER TABLE t1 RENAME INDEX old_index_name TO new_index_name;
    ```

在 MySQL 5.6 中，如果要修改索引名，只能重建索引。

- 修改字段名。

  ```
  ALTER TABLE t1 CHANGE old_col_name new_col_name datatype;
  ```

  注意，这里修改的只是字段名，不改变字段的其他属性。

- 设置（删除）字段的默认值。

  ```
  ALTER TABLE t1 ALTER COLUMN col SET DEFAULT literal;
  ALTER TABLE t1 ALTER COLUMN col DROP DEFAULT;
  ```

- 增加 VARCHAR 的长度。

  ```
  ALTER TABLE t1 CHANGE COLUMN c1 c1 VARCHAR(60);
  ```

  因为 VARCHAR 是变长字段，所以字符串在实际存储时，MySQL 需要 1~2 字节来表示它的长度。具体来说，如果字符串占用的字节数为 0~255，需 1 字节来表示；如果大于 255 字节，则需 2 字节来表示。这里，字符串占用的字节数与字符集有关。

  在增加 VARCHAR 的长度时，如果用来表示字符串长度的字节数不变，则只会修改表的元数据信息，否则只能使用 COPY 算法。看看下面这个示例。

  ```
  mysql> create table t1(c1 varchar(50)) charset=utf8;
  Query OK, 0 rows affected (0.01 sec)

  mysql> alter table t1 modify c1 varchar(85), ALGORITHM=INPLACE, LOCK=NONE;
  Query OK, 0 rows affected (0.00 sec)
  Records: 0  Duplicates: 0  Warnings: 0

  mysql> alter table t1 modify c1 varchar(86), ALGORITHM=INPLACE, LOCK=NONE;
  ERROR 1846 (0A000): ALGORITHM=INPLACE is not supported. Reason: Cannot change column type INPLACE.
  Try ALGORITHM=COPY.
  ```

  在 utf8 中，一个字符最多占用 3 字节。因为 85 × 3 = 255，所以使用 INPLACE 算法最多可将 c1 调整到 varchar(85)。

  注意，这个优化是到 MySQL 5.7 才支持的。在 MySQL 5.6 中，如果要增加 VARCHAR 的长度，就只能使用 COPY 算法。另外，如果要减小 VARCHAR 的长度，不管是在哪个版本中，都只能使用 COPY 算法。

- 修改自增主键的值。

  ```
  ALTER TABLE t1 AUTO_INCREMENT=next_value;
  ```

  其实，这里修改的是内存值。

- 修改表的统计信息相关选项。

  统计信息相关的选项有：STATS_PERSISTENT、STATS_AUTO_RECALC 和 STATS_SAMPLE_PAGES。

  ```
  ALTER TABLE t1 STATS_PERSISTENT=1, STATS_AUTO_RECALC=1, STATS_SAMPLE_PAGES=25;
  ```

- 重命名表。

```
ALTER TABLE old_tbl_name RENAME TO new_tbl_name;
RENAME TABLE old_tbl_name to new_tbl_name;
```

- 添加外键约束。

    区分以下两种场景:

    (1) 如果 foreign_key_checks 为 ON,则会使用 COPY 算法;

    (2) 如果 foreign_key_checks 为 OFF,则只会更新表的元数据信息。

    ```
    ALTER TABLE t1 ADD CONSTRAINT fk_name FOREIGN KEY (col1) REFERENCES t2(col2) referential_actions;
    ```

- 删除外键约束。

    无论 foreign_key_checks 是否开启,都只会更新表的元数据信息。

    ```
    ALTER TABLE t1 DROP FOREIGN KEY fk_name;
    ```

2. INPLACE

使用 INPLACE 算法 rebuild 这种方式的 DDL 操作有如下特点。

- 不会阻塞 DML。
- 执行时间与表的大小成正比。表越大,执行时间越长。
- 在执行过程中,会拷贝原表数据。具体来说,会在原表的当前目录创建一个临时 frm 文件和 ibd 文件。

    ```
    mysql> alter table sbtest.sbtest1 engine=innodb;

    # ll /data/mysql/3307/data/sbtest/
    total 4931620
    -rw-r----- 1 mysql mysql         65 Dec 11 10:24 db.opt
    -rw-r----- 1 mysql mysql       8632 Dec 11 10:44 sbtest1.frm
    -rw-r----- 1 mysql mysql 2441084928 Dec 11 10:44 sbtest1.ibd
    -rw-r----- 1 mysql mysql       8632 Dec 11 22:20 #sql-8d1_15.frm
    -rw-r----- 1 mysql mysql 2608857088 Dec 11 22:21 #sql-ib41-3512307169.ibd
    ```

    这两个临时文件同样以 #sql 开头。可以看到,临时 ibd 文件甚至超过了原表的大小,但基本上不会超出太多,可基于这一点来监控 Online DDL 的进度。不过既然会拷贝原表数据,该类操作对数据库的性能影响还是比较大的。

属于此类的 DDL 操作如下。

- 创建索引。

    实际上,创建索引属于 INPLACE 算法的 no-rebuild 方式,在执行过程中也不会创建临时 ibd 文件。之所以放到这里,主要是考虑到它的执行时间与表的大小有关。

    ```
    CREATE INDEX name ON t1 (col_list);
    ALTER TABLE t1 ADD INDEX name (col_list);
    ```

- 添加主键。

    ```
    ALTER TABLE t1 ADD PRIMARY KEY (id);
    ```

- 删除主键并添加另一个主键。

  ```
  ALTER TABLE t1 DROP PRIMARY KEY, ADD PRIMARY KEY(id1);
  ```

  如果只是删除主键，则只能使用 COPY 算法。

- 新增字段。

  ```
  ALTER TABLE t1 ADD [COLUMN] col_name column_definition [FIRST | AFTER col_name];
  ```

  注意，当添加的字段是自增列时，会阻塞 DML。

  从 MySQL 8.0.12 开始，新增字段可以使用 INSTANT 算法。

- 删除字段。

  ```
  ALTER TBALE t1 DROP [COLUMN] col_name;
  ```

  从 MySQL 8.0.29 开始，删除字段可以使用 INSTANT 算法。

- 调整字段的顺序。

  ```
  ALTER TABLE t1 MODIFY [COLUMN] col_name column_definition [FIRST | AFTER col_name];
  ```

- 修改表的属性（ROW_FORMAT 和 KEY_BLOCK_SIZE）。

  ```
  ALTER TABLE t1 ROW_FORMAT=COMPRESSED, KEY_BLOCK_SIZE=8;
  ```

- 修改字段的 NULL 属性。

  ```
  ALTER TABLE t1 MODIFY pad VARCHAR(60) NULL;
  ALTER TABLE t1 MODIFY c VARCHAR(255) NOT NULL;
  ```

  注意，对于后者，sql_mode 必须为 STRICT_ALL_TABLES 或 STRICT_TRANS_TABLES，否则，使用的还是 COPY 算法。

- OPTIMIZE TABLE。

  对于 InnoDB 表来说，OPTIMIZE TABLE 实际上对应的就是 ALTER TABLE ... FORCE 命令。该命令会重建表，回收空闲空间，并更新表的统计信息。在表碎片较多的情况下，推荐使用该命令。

  ```
  OPTIMIZE TABLE t1;
  ```

  注意，如果表中存在 FULLTEXT 索引，会使用 COPY 算法。

- ALTER TABLE ... FORCE。

  ```
  ALTER TABLE t1 FORCE;
  ```

- ALTER TABLE ... ENGINE=INNODB。

  作用同 ALTER TABLE ... FORCE 类似。

  ```
  ALTER TABLE t1 ENGINE=INNODB;
  ```

3. COPY

采用 COPY 算法，同样会在表的当前目录创建临时 frm 文件和 ibd 文件，但与 INPLACE 算法不一样的是，COPY 算法会阻塞 DML。

属于此类的 DDL 操作如下。

- 修改字段的定义。

    修改字段的定义，常见的有两种情况：调整字段的类型，例如将 int 调整为 bigint；调整字段的长度，例如在 latin1 字符集中，将 varchar(255) 调整为 varchar(256)。

    要修改字段的定义，既可通过 MODIFY，又可通过 CHANGE。

    ```
    ALTER TABLE t1 MODIFY [COLUMN] col_name column_definition [FIRST | AFTER col_name];
    ALTER TABLE t1 CHANGE c1 c1 BIGINT;
    ```

- 删除主键。

    ```
    ALTER TABLE t1 DROP PRIMARY KEY;
    ```

- 转换字符集。

    ```
    ALTER TABLE t CONVERT TO CHARACTER SET charset_name [COLLATE collation_name];
    ```

7.1.2 Online DDL 的实现原理

Online DDL 主要包括 3 个阶段：Prepare 阶段、DDL 执行阶段和 Commit 阶段。INPLACE 算法 rebuild 这种方式比仅更新表的元数据信息的方式多了一个 DDL 执行阶段，Prepare 阶段和 Commit 阶段一样。

下面看看这 3 个阶段的处理流程。

Prepare 阶段的处理流程如下。

- 创建新的临时 frm 文件。
- 获取 EXCLUSIVE-MDL，禁止读写。
- 根据 DDL 类型，确定执行的方式，是 copy（对应 COPY 算法）、online-rebuild（对应 INPLACE 算法 rebuild 这种方式），还是 online-norebuild（只修改表的元数据信息）。
- 更新数据字典的内存对象。
- 分配 row_log 对象记录 DDL 执行过程中产生的 DML 增量。
- 创建新的临时 ibd 文件。

DDL 执行阶段的处理流程如下。

- 降级 EXCLUSIVE-MDL，允许读写。
- 逐行扫描原表聚集索引中的记录 rec。
- 遍历新表的聚集索引和二级索引，逐一处理。
- 根据 rec 构造对应的索引项。
- 将构造好的索引项插入 sort_buffer 块。
- 将 sort_buffer 块插入新的索引。
- 处理 DDL 执行过程中产生的 DML 增量。

Commit 阶段的处理流程如下。

- 升级到 EXCLUSIVE-MDL，禁止读写。
- 重做 row_log 中最后一部分增量。
- 更新 InnoDB 的数据字典表。
- 提交事务。
- 更新统计信息。
- 重命名临时 idb 文件和 frm 文件。
- 变更完成。

在 Prepare 阶段，有个 row_log 对象用于保存 DDL 执行期间新增的 DML 操作，这个对象只在内存中创建，其大小由参数 innodb_online_alter_log_max_size 决定，默认为 128MB。如果在执行 DDL 操作的过程中，有大量的 DML 操作，可适当增大 innodb_online_alter_log_max_size 的值，否则会导致 DDL 失败，并提示如下错误：

```
ERROR 1799 (HY000): Creating index 'PRIMARY' required more than 'innodb_online_alter_log_max_size' bytes of modification log. Please try again.
```

7.1.3 如何检查 DDL 的进度

从 MySQL 5.7 开始，我们可以通过 performance_schema 中的相关表输出 DDL 的进度。

下面看看具体的操作步骤。

(1) 开启与 DDL 相关的事件采集配置项（instrument）。

```
mysql> update performance_schema.setup_instruments set enabled = 'yes' where name like 'stage/innodb/alter%';
Query OK, 0 rows affected (0.00 sec)
Rows matched: 7  Changed: 0  Warnings: 0

mysql> select * from performance_schema.setup_instruments WHERE NAME LIKE 'stage/innodb/alter%';
+--------------------------------------------------------------+---------+-------+
| NAME                                                         | ENABLED | TIMED |
+--------------------------------------------------------------+---------+-------+
| stage/innodb/alter table (end)                               | YES     | YES   |
| stage/innodb/alter table (flush)                             | YES     | YES   |
| stage/innodb/alter table (insert)                            | YES     | YES   |
| stage/innodb/alter table (log apply index)                   | YES     | YES   |
| stage/innodb/alter table (log apply table)                   | YES     | YES   |
| stage/innodb/alter table (merge sort)                        | YES     | YES   |
| stage/innodb/alter table (read PK and internal sort)         | YES     | YES   |
+--------------------------------------------------------------+---------+-------+
7 rows in set (0.00 sec)
```

相关的配置项有 7 个，其实代表了 Online DDL 的 7 个阶段，依次为：read PK and internal sort、merge sort、insert、log apply index、flush、log apply table 和 end。

(2) 开启事件状态表。

```
mysql> update performance_schema.setup_consumers set enabled = 'yes' where name like '%stages%';
Query OK, 3 rows affected (0.00 sec)
Rows matched: 3  Changed: 3  Warnings: 0
```

```
mysql> select * from performance_schema.setup_consumers where name like '%stages%';
+--------------------------------+---------+
| NAME                           | ENABLED |
+--------------------------------+---------+
| events_stages_current          | YES     |
| events_stages_history          | YES     |
| events_stages_history_long     | YES     |
+--------------------------------+---------+
3 rows in set (0.00 sec)
```

events_stages_current 只会记录当前事件的状态信息，之前事件的状态信息会保存到 events_stages_history 表中。默认情况下，events_stages_history 表最多保留 10 条记录，完整的记录会保存到 events_stages_history_long 中。

(3) 执行 DDL 操作。

```
mysql> alter table sbtest.sbtest1 engine=innodb;
```

(4) 查看事件状态表。

```
mysql> select event_name,work_completed,work_estimated from
performance_schema.events_stages_current;
+-------------------------------------------------+----------------+----------------+
| event_name                                      | work_completed | work_estimated |
+-------------------------------------------------+----------------+----------------+
| stage/innodb/alter table (read PK and internal sort) |          50903 |         205480 |
+-------------------------------------------------+----------------+----------------+
1 row in set (0.00 sec)
```

下面看看这 3 个字段的具体含义。

- event_name：当前正在执行的事件。
- work_estimated：整个 DDL 所需的工作量，它是预估值，会随着事件的执行而动态变化。
- work_completed：已经完成的工作量。

基于 work_estimated 和 work_completed 我们可以预估 DDL 的完成时间。

一个事件完成后，会执行下一个事件，而之前执行的事件会被移到 events_stages_history 表中。

需要注意的是，细分为 7 个事件只针对 INPLACE 算法的 DDL。对于 COPY 算法的 DDL，只有一个事件，即 copy to tmp table。

```
mysql> select event_name,work_completed,work_estimated from
performance_schema.events_stages_current;
+-----------------------------+----------------+----------------+
| event_name                  | work_completed | work_estimated |
+-----------------------------+----------------+----------------+
| stage/sql/copy to tmp table |        6484351 |        9859220 |
+-----------------------------+----------------+----------------+
1 row in set (0.00 sec)
```

上述方法适用于 MySQL 5.7 及以上版本。在 MySQL 5.6 中，我们有没有办法检查 DDL 的进度呢？当然有。之前提到过，如果 DDL 使用的是 INPLACE 算法 rebuild 这种方式或 COPY 算法，会在原表的当前目录创建一个临时 ibd 文件，而且这个文件的大小与原表相差不大，可基于这一点来计算 DDL

的进度。实际上，我也基于这个思路实现了一个简单的脚本 monitor_ddl_progress.sh，该脚本已上传至 GitHub。

7.1.4 MySQL 8.0.12 引入的秒级加列特性

在 MySQL 8.0.12 中，ALGORITHM 子句引入了一个新的选项 INSTANT，支持表的秒级加列特性。

下面看看 INSTANT 和 INPLACE 这两种算法在执行时间上的区别。

```
mysql> alter table sbtest.sbtest1 add column c1 varchar(10),ALGORITHM=INSTANT;
Query OK, 0 rows affected (0.02 sec)
Records: 0  Duplicates: 0  Warnings: 0

mysql> alter table sbtest.sbtest1 add column c2 varchar(10),ALGORITHM=INPLACE;
Query OK, 0 rows affected (37.35 sec)
Records: 0  Duplicates: 0  Warnings: 0
```

命令中的 sbtest.sbtest1 是通过 sysbench 生成的一张包含 100 万条数据的表。

可以看到，INSTANT 算法几乎瞬间就完成了，而 INPLACE 算法却用了 37.35 秒。表越大，效果越明显。

接下来，我们看看如何判断一张表中是否存在通过 INSTANT 算法添加的字段。

```
mysql> show create table sbtest.sbtest1\G
*************************** 1. row ***************************
       Table: sbtest1
Create Table: CREATE TABLE `sbtest1` (
  `id` int NOT NULL AUTO_INCREMENT,
  `k` int NOT NULL DEFAULT '0',
  `c` char(120) NOT NULL DEFAULT '',
  `pad` char(60) NOT NULL DEFAULT '',
  PRIMARY KEY (`id`),
  KEY `k_1` (`k`)
) ENGINE=InnoDB AUTO_INCREMENT=1000001 DEFAULT CHARSET=utf8mb4 COLLATE=utf8mb4_0900_ai_ci
1 row in set (0.00 sec)

mysql> select table_id,name,instant_cols,total_row_versions from information_schema.innodb_tables
where name like '%sbtest1%';
+----------+----------------+--------------+--------------------+
| table_id | name           | instant_cols | total_row_versions |
+----------+----------------+--------------+--------------------+
|     1652 | sbtest/sbtest1 |            0 |                  0 |
+----------+----------------+--------------+--------------------+
1 row in set (0.00 sec)

mysql> alter table sbtest.sbtest1 add column c1 varchar(10),ALGORITHM=INSTANT;
Query OK, 0 rows affected (0.03 sec)
Records: 0  Duplicates: 0  Warnings: 0

mysql> select table_id,name,instant_cols,total_row_versions from information_schema.innodb_tables
where name like '%sbtest1%';
+----------+----------------+--------------+--------------------+
| table_id | name           | instant_cols | total_row_versions |
+----------+----------------+--------------+--------------------+
|     1652 | sbtest/sbtest1 |            0 |                  1 |
+----------+----------------+--------------+--------------------+
```

```
1 row in set (0.00 sec)

mysql> select * from information_schema.innodb_columns where table_id = 1652;
+----------+------+-----+-------+----------+-----+-------------+---------------+
| TABLE_ID | NAME | POS | MTYPE | PRTYPE   | LEN | HAS_DEFAULT | DEFAULT_VALUE |
+----------+------+-----+-------+----------+-----+-------------+---------------+
|     1652 | id   |   0 |     6 |     1283 |   4 |           0 | NULL          |
|     1652 | k    |   1 |     6 |     1283 |   4 |           0 | NULL          |
|     1652 | c    |   2 |    13 | 16712190 | 480 |           0 | NULL          |
|     1652 | pad  |   3 |    13 | 16712190 | 240 |           0 | NULL          |
|     1652 | c1   |   4 |    12 | 16711695 |  40 |           1 | NULL          |
+----------+------+-----+-------+----------+-----+-------------+---------------+
5 rows in set (0.00 sec)
```

使用 INSTANT 算法添加字段后，表的 table_id 不会改变。如果要判断表上是否存在通过 INSTANT 算法添加的字段，可直接查看 information_schema.innodb_columns 表。如果 HAS_DEFAULT 为 1，则意味着对应字段是通过 INSTANT 算法添加的。除此之外，在 MySQL 8.0.29 之前，也可将 information_schema.innodb_tables 表中的 instant_cols 列作为判断的依据，该列会记录在执行第一次秒级加列操作时，字段的个数，默认为 0。不过从 MySQL 8.0.29 开始，该列就不再使用，与此同时，新增了 total_row_versions 列用来记录行的版本号。只要是使用 INSTANT 算法来添加或者删除字段，total_row_versions 都会递增 1。行的版本号允许的最大值是 64。当超过最大值限制时，MySQL 会提示以下错误信息。这个时候，可通过指定 INPLACE 算法来解决。

```
ERROR 4092 (HY000): Maximum row versions reached for table sbtest/sbtest1. No more columns
can be added or dropped instantly. Please use COPY/INPLACE.
```

同样是加列操作，如果指定的是 INPLACE 算法，不仅 table_id 会发生变化，total_row_versions 也会重置为 0。

```
mysql> alter table sbtest.sbtest1 add column c2 varchar(10),ALGORITHM=INPLACE;
Query OK, 0 rows affected (38.79 sec)
Records: 0  Duplicates: 0  Warnings: 0

mysql> select table_id,name,instant_cols,total_row_versions from information_schema.innodb_tables
where name like '%sbtest1%';
+----------+---------------+--------------+--------------------+
| table_id | name          | instant_cols | total_row_versions |
+----------+---------------+--------------+--------------------+
|     1653 | sbtest/sbtest1|            0 |                  0 |
+----------+---------------+--------------+--------------------+
1 row in set (0.00 sec)
```

最后再来看看秒级加列的局限性。

- 列只能添加到最后，不允许添加在其他位置。从 MySQL 8.0.29 开始，可以添加到任何位置。
- 不支持压缩表（ ROW_FORMAT=COMPRESSED ）。
- 不支持带有全文索引的表。
- 不支持存储在数据字典表空间中的表。
- 不支持临时表（ TEMPORARY TABLE ）。
- 多个 DDL 操作放到一条语句中执行，如果这些操作里面有不支持 INSTANT 算法的操作，同样无法进行秒级加列。

7.1.5 Online DDL 的优缺点

下面我们看看 Online DDL 的优缺点。

Online DDL 的优点如下。

- 原生支持，无须使用额外的工具。

Online DDL 的缺点如下。

- 虽然不会阻塞 DML，但 DDL 在刚开始和快结束的时候，都要获取元数据锁（metadata lock，MDL）。如果在这两个时间点上，表上存在未提交的事务或者未结束的查询，DDL 会因获取不到元数据锁被阻塞。DDL 一旦被阻塞，后续针对该表的其他操作也会被阻塞。
- "全量 + 增量"的变更方式，并不必然保证 DDL 的成功。
- 不能基于系统的负载自动调节 DDL 的进度。
- 在对大表进行 DDL 时，会导致较大的主从延迟。
- 如果大表 DDL 操作失败，回滚代价会比较高。

建议对于仅更新表的元数据信息的操作，使用 Online DDL。

7.1.6 Online DDL 的注意事项

Online DDL 在使用中的注意事项如下。

- 在 DDL 开始之初和结束之前，表上不允许有未提交的事务，事务读也不行，不然会阻塞 DDL 获取元数据锁。这一点从 DDL 的实现原理中也可以看出来。
- 在执行 DDL 的过程中，新增的 DML 操作都是先保存到 Online alter log 中，最后再应用到临时表中。所以有可能出现 DML 在执行的过程中没有问题，但在应用的过程中却因违反临时表的约束而导致 DDL 失败。

看看下面这个示例。在添加唯一键时，即使 session2 中的记录没有实际生成，DDL 仍出现了唯一键冲突。

```
session1> alter table t1 add unique key uk_c1 (c1);
执行中……

session2> begin;
Query OK, 0 rows affected (0.00 sec)

session2> insert into t1 values (10000001,1);
Query OK, 1 row affected (0.00 sec)

session2> delete from t1 where id=10000001 and c1=1;
Query OK, 1 row affected (0.00 sec)

session2> commit;
Query OK, 0 rows affected (0.03 sec)

session1> alter table t1 add unique key uk_c1 (c1);
ERROR 1062 (23000): Duplicate entry '1' for key 't1.uk_c1'
```

- 元数据锁的超时时间由参数 `lock_wait_timeout` 决定，默认为 31 536 000 秒，即 365 天。
- 无论是 INPLACE 算法还是 COPY 算法，生成的临时表都在原表目录下，而不是 /tmp 目录下（/tmp 是参数 `tmpdir` 的默认值）。`tmpdir` 只会用来存储临时排序文件。
- 一张表如果有多个 DDL 操作，建议将其合并在一起执行。例如：

  ```
  ALTER TABLE t1 ADD INDEX idx_c1(c1), ADD UNIQUE INDEX idx_c2(c2), CHANGE c3 c3_new_name INTEGER UNSIGNED;
  ```

- 如何判断一个 DDL 操作是使用 INPLACE 算法还是 COPY 算法呢？

  可创建一张测试表，插入少量的数据，然后执行 DDL 操作。如果影响的行数为 0，则代表该操作使用的是 INPLACE 算法，否则是 COPY 算法。

  ```
  Query OK, 0 rows affected (0.07 sec)
  ```

- INPLACE 算法 rebuild 方式和 COPY 算法的区别。

  两种算法的区别主要体现在以下几点。

  - 对于使用 INPLACE 算法的 DDL，`show processlist` 中的 State 会显示 `altering table`，而对于使用 COPY 算法的 DDL，则会显示 `copy to tmp table`。
  - 从性能上看，虽然都会创建临时 frm 文件和 ibd 文件，但 INPLACE 算法消耗的 CPU 和 I/O 资源更少，读取到缓冲池中的数据也更少。
  - 在执行时间上，INPLACE 算法更短。看看下面这个测试。

  ```
  mysql> show variables like 'sql_mode';
  +---------------+------------------+
  | Variable_name | Value            |
  +---------------+------------------+
  | sql_mode      | STRICT_ALL_TABLES |
  +---------------+------------------+
  1 row in set (0.00 sec)

  mysql> alter table sbtest1 modify pad varchar(60) not null default '';
  Query OK, 0 rows affected (39.80 sec)
  Records: 0  Duplicates: 0  Warnings: 0

  mysql> show variables like 'sql_mode';
  +---------------+-------+
  | Variable_name | Value |
  +---------------+-------+
  | sql_mode      |       |
  +---------------+-------+
  1 row in set (0.00 sec)

  mysql> alter table sbtest1 modify pad varchar(60) not null default '';
  Query OK, 10000000 rows affected (1 min 4.10 sec)
  Records: 10000000  Duplicates: 0  Warnings: 0
  ```

  同一个操作，因 `sql_mode` 不一样，导致使用的算法也不一样。很显然，使用 INPLACE 算法的 DDL 执行时间更短，差不多是 COPY 算法的三分之二。

- 在执行 DDL 时，若想使用 MySQL 5.6 之前的 Copy Table 方式，有两种方法。

  - 将参数 `old_alter_table` 设置为 ON。

- 在 ALTER TABLE 中指定 ALGORITHM=COPY。

如果两者发生冲突，后者将覆盖前者。

7.1.7 参考资料

- 博客文章"MySQL online ddl 原理"，作者：天士梦。
- 官方文档"InnoDB and Online DDL"。

7.2 pt-online-schema-change

pt-online-schema-change（OSC）是 Percona Toolkit 中的一个工具。顾名思义，它用于在线（真正意义上的 Online）变更表结构，是目前应用得最广泛的表结构变更工具。

虽然 MySQL 5.6 引入了 Online DDL，但通过上一节的分析，我们知道 Online DDL 还是有很大的局限性，包括：会造成主从延迟；无法限流；修改列的类型或调整列的长度甚至会阻塞 DML 操作。鉴于此，线上一般会使用 OSC 来修改表结构。

pt-online-schema-change 的安装方法可参考第 9 章。

7.2.1 pt-online-schema-change 的实现原理

下面，我们结合 general log 来分析 pt-online-schema-change 的实现原理。

首先，准备测试数据。

```
mysql> show create table sbtest.t1\G
*************************** 1. row ***************************
       Table: t1
Create Table: CREATE TABLE `t1` (
  `id` int NOT NULL AUTO_INCREMENT,
  `name` varchar(10) DEFAULT NULL,
  PRIMARY KEY (`id`)
) ENGINE=InnoDB AUTO_INCREMENT=10001 DEFAULT CHARSET=utf8mb4 COLLATE=utf8mb4_0900_ai_ci
1 row in set (0.00 sec)

mysql> select count(*) from sbtest.t1;
+----------+
| count(*) |
+----------+
|    10000 |
+----------+
1 row in set (0.00 sec)
```

其次，通过 OSC 进行加列操作。

```
# pt-online-schema-change h=192.168.244.10,P=3306,u=root,p=123456,D=sbtest,t=t1 --alter "add column c1 datetime" --execute
No slaves found.  See --recursion-method if host node1 has slaves.
Not checking slave lag because no slaves were found and --check-slave-lag was not specified.
Operation, tries, wait:
  analyze_table, 10, 1
```

```
    copy_rows, 10, 0.25
    create_triggers, 10, 1
    drop_triggers, 10, 1
    swap_tables, 10, 1
    update_foreign_keys, 10, 1
Altering `sbtest`.`t1`...
Creating new table...
Created new table sbtest._t1_new OK.
Altering new table...
Altered `sbtest`.`_t1_new` OK.
2021-12-15T21:13:14 Creating triggers...
2021-12-15T21:13:14 Created triggers OK.
2021-12-15T21:13:14 Copying approximately 10294 rows...
2021-12-15T21:13:14 Copied rows OK.
2021-12-15T21:13:14 Analyzing new table...
2021-12-15T21:13:14 Swapping tables...
2021-12-15T21:13:14 Swapped original and new tables OK.
2021-12-15T21:13:14 Dropping old table...
2021-12-15T21:13:14 Dropped old table `sbtest`.`_t1_old` OK.
2021-12-15T21:13:14 Dropping triggers...
2021-12-15T21:13:14 Dropped triggers OK.
Successfully altered `sbtest`.`t1`.
```

基于上面的输出，我们看看 OSC 的执行流程。

- 创建临时表，并修改临时表的表结构。注意，这里的临时表是指该表是临时中转的，不是只对当前会话可见的临时表（TEMPORARY TABLE）。
- 创建触发器。
- 将原表数据拷贝到临时表中。
- 拷贝完毕后，更新临时表的统计信息，将原表和临时表交换。
- 删除原表及触发器。

看似简单，这里其实涉及很多知识点。

下面，我们结合 general log 来具体看看。

```
14 Connect   root@192.168.244.10 on sbtest using TCP/IP
14 Query     SHOW VARIABLES LIKE 'innodb\_lock_wait_timeout'
14 Query     SET SESSION innodb_lock_wait_timeout=1
14 Query     SHOW VARIABLES LIKE 'lock\_wait_timeout'
14 Query     SET SESSION lock_wait_timeout=60
14 Query     SHOW VARIABLES LIKE 'wait\_timeout'
14 Query     SET SESSION wait_timeout=10000
14 Query     SELECT @@SQL_MODE
14 Query     SET @@SQL_QUOTE_SHOW_CREATE = 1/*!40101,
@@SQL_MODE='NO_AUTO_VALUE_ON_ZERO,ONLY_FULL_GROUP_BY,STRICT_TRANS_TABLES,NO_ZERO_IN_DATE,NO_ZERO_
DATE,ERROR_FOR_DIVISION_BY_ZERO,NO_ENGINE_SUBSTITUTION'*/
14 Query     SELECT VERSION()
14 Query     SHOW VARIABLES LIKE 'character_set_server'
14 Query     SET NAMES 'utf8mb4'
14 Query     SELECT @@server_id /*!50038 , @@hostname*/
15 Connect   root@192.168.244.10 on sbtest using TCP/IP
15 Query     SHOW VARIABLES LIKE 'innodb\_lock_wait_timeout'
...
15 Query     SELECT @@server_id /*!50038 , @@hostname*/
```

创建了两个连接，两个连接执行的操作一样，都是调整参数 innodb_lock_wait_timeout、lock_wait_ timeout、wait_timeout、sql_quote_show_create、sql_mode 的会话值。

在这几个参数中，需要注意的是 innodb_lock_wait_timeout 和 lock_wait_timeout。前者是行锁的等待时间，默认为 50 秒，后者是表锁的等待时间，默认为 1 年。之所以要将这两个参数的会话值设置得较小，是希望当发生锁争用时，尽量让 OSC 发起的 SQL 超时退出，而不是让业务 SQL 超时退出，以避免对线上业务造成较大的影响。

```
14 Query   SHOW VARIABLES LIKE 'wsrep_on'
14 Query   SHOW VARIABLES LIKE 'version%'
14 Query   SHOW ENGINES
14 Query   SHOW VARIABLES LIKE 'innodb_version'
14 Query   SHOW VARIABLES LIKE 'innodb_stats_persistent'
14 Query   SELECT @@SERVER_ID
14 Query   SHOW GRANTS FOR CURRENT_USER()
14 Query   SHOW FULL PROCESSLIST
14 Query   SHOW SLAVE HOSTS
14 Query   SHOW GLOBAL STATUS LIKE 'Threads_running'
14 Query   SHOW GLOBAL STATUS LIKE 'Threads_running'
14 Query   SELECT CONCAT(@@hostname, @@port)
14 Query   SHOW TABLES FROM `sbtest` LIKE 't1'
14 Query   SELECT VERSION()
14 Query   SHOW TRIGGERS FROM `sbtest` LIKE 't1'
14 Query   /*!40101 SET @OLD_SQL_MODE := @@SQL_MODE, @@SQL_MODE := '', @OLD_QUOTE :=
@@SQL_QUOTE_SHOW_CREATE, @@SQL_QUOTE_SHOW_CREATE := 1 */
```

通过 SHOW TABLES FROM `sbtest` LIKE 't1' 判断目标表是否存在。

通过 SHOW TRIGGERS FROM `sbtest` LIKE 't1' 查看目标表上的触发器信息。因为 OSC 会创建触发器，而在 MySQL 5.7 之前，针对同一触发事件（INSERT、DELETE、UPDATE），同一触发时间（BEFORE、AFTER）只能创建一个触发器，例如，对于每张表，我们只能创建一个 AFTER DELETE 触发器。所以，在 MySQL 5.7 之前，如果表上已经存在触发器，就不能通过 OSC 来修改表结构。MySQL 5.7 解除了这个限制。

```
14 Query   USE `sbtest`
14 Query   SHOW CREATE TABLE `sbtest`.`t1`
14 Query   /*!40101 SET @@SQL_MODE := @OLD_SQL_MODE, @@SQL_QUOTE_SHOW_CREATE := @OLD_QUOTE */
14 Query   EXPLAIN SELECT * FROM `sbtest`.`t1` WHERE 1=1
14 Query   SELECT table_schema, table_name FROM information_schema.key_column_usage WHERE
referenced_table_schema='sbtest' AND referenced_table_name='t1'
14 Query   SHOW VARIABLES LIKE 'version%'
14 Query   SHOW ENGINES
14 Query   SHOW VARIABLES LIKE 'innodb_version'
14 Query   SELECT table_schema, table_name FROM information_schema.key_column_usage WHERE
referenced_table_schema='sbtest' AND referenced_table_name='t1'
14 Query   SHOW VARIABLES LIKE 'wsrep_on'
14 Query   /*!40101 SET @OLD_SQL_MODE := @@SQL_MODE, @@SQL_MODE := '', @OLD_QUOTE :=
@@SQL_QUOTE_SHOW_CREATE, @@SQL_QUOTE_SHOW_CREATE := 1 */
14 Query   USE `sbtest`
14 Query   SHOW CREATE TABLE `sbtest`.`t1`
14 Query   /*!40101 SET @@SQL_MODE := @OLD_SQL_MODE, @@SQL_QUOTE_SHOW_CREATE := @OLD_QUOTE */
14 Query   CREATE TABLE `sbtest`.`_t1_new` (
  `id` int NOT NULL AUTO_INCREMENT,
  `name` varchar(10) DEFAULT NULL,
  PRIMARY KEY (`id`)
) ENGINE=InnoDB AUTO_INCREMENT=10001 DEFAULT CHARSET=utf8mb4 COLLATE=utf8mb4_0900_ai_ci
```

```
14 Query    ALTER TABLE `sbtest`.`_t1_new` add column c1 datetime
14 Query    /*!40101 SET @OLD_SQL_MODE := @@SQL_MODE, @@SQL_MODE := '', @OLD_QUOTE :=
@@SQL_QUOTE_SHOW_CREATE, @@SQL_QUOTE_SHOW_CREATE := 1 */
```

创建一张临时表 sbtest._t1_new，表结构与原表一致。

接着修改临时表的表结构，使用的 ALTER 语句是 OSC 命令行中的 alter 选项指定的。

如果对这张表有多个操作，建议将这些操作放到一个 ALTER 语句中。例如：

```
--alter "add column c1 datetime, modify column name varchar(20)"
```

```
14 Query    USE `sbtest`
14 Query    SHOW CREATE TABLE `sbtest`.`_t1_new`
14 Query    /*!40101 SET @@SQL_MODE := @OLD_SQL_MODE, @@SQL_QUOTE_SHOW_CREATE := @OLD_QUOTE */
14 Query    SELECT TRIGGER_SCHEMA, TRIGGER_NAME, DEFINER, ACTION_STATEMENT, SQL_MODE,
CHARACTER_SET_CLIENT, COLLATION_CONNECTION, EVENT_MANIPULATION, ACTION_TIMING    FROM
INFORMATION_SCHEMA.TRIGGERS  WHERE EVENT_MANIPULATION = 'DELETE'    AND ACTION_TIMING = 'AFTER'    AND
TRIGGER_SCHEMA = 'sbtest'      AND EVENT_OBJECT_TABLE = 't1'
14 Query    SELECT TRIGGER_SCHEMA, TRIGGER_NAME, DEFINER, ACTION_STATEMENT, SQL_MODE,
CHARACTER_SET_CLIENT, COLLATION_CONNECTION, EVENT_MANIPULATION, ACTION_TIMING    FROM
INFORMATION_SCHEMA.TRIGGERS  WHERE EVENT_MANIPULATION = 'UPDATE'    AND ACTION_TIMING = 'AFTER'    AND
TRIGGER_SCHEMA = 'sbtest'      AND EVENT_OBJECT_TABLE = 't1'
14 Query    SELECT TRIGGER_SCHEMA, TRIGGER_NAME, DEFINER, ACTION_STATEMENT, SQL_MODE,
CHARACTER_SET_CLIENT, COLLATION_CONNECTION, EVENT_MANIPULATION, ACTION_TIMING    FROM
INFORMATION_SCHEMA.TRIGGERS  WHERE EVENT_MANIPULATION = 'INSERT'    AND ACTION_TIMING = 'AFTER'    AND
TRIGGER_SCHEMA = 'sbtest'      AND EVENT_OBJECT_TABLE = 't1'
14 Query    SELECT TRIGGER_SCHEMA, TRIGGER_NAME, DEFINER, ACTION_STATEMENT, SQL_MODE,
CHARACTER_SET_CLIENT, COLLATION_CONNECTION, EVENT_MANIPULATION, ACTION_TIMING    FROM
INFORMATION_SCHEMA.TRIGGERS  WHERE EVENT_MANIPULATION = 'DELETE'    AND ACTION_TIMING = 'BEFORE'    AND
TRIGGER_SCHEMA = 'sbtest'      AND EVENT_OBJECT_TABLE = 't1'
14 Query    SELECT TRIGGER_SCHEMA, TRIGGER_NAME, DEFINER, ACTION_STATEMENT, SQL_MODE,
CHARACTER_SET_CLIENT, COLLATION_CONNECTION, EVENT_MANIPULATION, ACTION_TIMING    FROM
INFORMATION_SCHEMA.TRIGGERS  WHERE EVENT_MANIPULATION = 'UPDATE'    AND ACTION_TIMING = 'BEFORE'    AND
TRIGGER_SCHEMA = 'sbtest'      AND EVENT_OBJECT_TABLE = 't1'
14 Query    SELECT TRIGGER_SCHEMA, TRIGGER_NAME, DEFINER, ACTION_STATEMENT, SQL_MODE,
CHARACTER_SET_CLIENT, COLLATION_CONNECTION, EVENT_MANIPULATION, ACTION_TIMING    FROM
INFORMATION_SCHEMA.TRIGGERS  WHERE EVENT_MANIPULATION = 'INSERT'    AND ACTION_TIMING = 'BEFORE'    AND
TRIGGER_SCHEMA = 'sbtest'      AND EVENT_OBJECT_TABLE = 't1'

14 Query    CREATE TRIGGER `pt_osc_sbtest_t1_del` AFTER DELETE ON `sbtest`.`t1` FOR EACH ROW BEGIN DECLARE
CONTINUE HANDLER FOR 1146 begin end; DELETE IGNORE FROM `sbtest`.`_t1_new` WHERE `sbtest`.`_t1_new`.`id`
<=> OLD.`id`; END
14 Query    CREATE TRIGGER `pt_osc_sbtest_t1_upd` AFTER UPDATE ON `sbtest`.`t1` FOR EACH ROW BEGIN DECLARE
CONTINUE HANDLER FOR 1146 begin end; DELETE IGNORE FROM `sbtest`.`_t1_new` WHERE !(OLD.`id` <=> NEW.`id`)
AND `sbtest`.`_t1_new`.`id` <=> OLD.`id`; REPLACE INTO `sbtest`.`_t1_new` (`id`, `name`) VALUES
(NEW.`id`, NEW.`name`); END
14 Query    CREATE TRIGGER `pt_osc_sbtest_t1_ins` AFTER INSERT ON `sbtest`.`t1` FOR EACH ROW BEGIN DECLARE
CONTINUE HANDLER FOR 1146 begin end; REPLACE INTO `sbtest`.`_t1_new` (`id`, `name`) VALUES (NEW.`id`,
NEW.`name`);END
```

首先，会查询原表上是否存在触发器。如果存在，且指定了 preserve-triggers 参数，则会在临时表上创建这些触发器，以验证这些触发器在临时表中能否同样生效。例如：

```
Query    CREATE DEFINER=`root`@`localhost` TRIGGER `test`.`itjqbeetfswkwma` AFTER UPDATE ON _t1_new
FOR EACH ROW
INSERT INTO test.log VALUES (NOW(), CONCAT("updated row row with id ", OLD.id, " old f1:", OLD.f1,
" new f1: ", NEW.f1 ))
Query    DROP TRIGGER IF EXISTS `itjqbeetfswkwma`
```

如果没有问题，则会在创建完毕后删除触发器。

接着，会针对原表创建 3 个触发器，分别对应 DELETE、UPDATE 和 INSERT 操作。针对 UPDATE 和 INSERT 操作的触发器使用的是 REPLACE 语句，而 REPLACE 语句只有在主键或唯一索引存在的情况下才有意义，这也是为什么目标表上要有主键或唯一索引。

```
14 Query EXPLAIN SELECT * FROM `sbtest`.`t1` WHERE 1=1
14 Query SELECT /*!40001 SQL_NO_CACHE */ `id` FROM `sbtest`.`t1` FORCE INDEX(`PRIMARY`) ORDER BY `id` LIMIT 1 /*first lower boundary*/
14 Query SELECT /*!40001 SQL_NO_CACHE */ `id` FROM `sbtest`.`t1` FORCE INDEX (`PRIMARY`) WHERE `id` IS NOT NULL ORDER BY `id` LIMIT 1 /*key_len*/
14 Query EXPLAIN SELECT /*!40001 SQL_NO_CACHE */ * FROM `sbtest`.`t1` FORCE INDEX (`PRIMARY`) WHERE `id` >= '1' /*key_len*/
14 Query EXPLAIN SELECT /*!40001 SQL_NO_CACHE */ `id` FROM `sbtest`.`t1` FORCE INDEX(`PRIMARY`) WHERE ((`id` >= '1')) ORDER BY `id` LIMIT 999, 2 /*next chunk boundary*/
14 Query SELECT /*!40001 SQL_NO_CACHE */ `id` FROM `sbtest`.`t1` FORCE INDEX(`PRIMARY`) WHERE ((`id` >= '1')) ORDER BY `id` LIMIT 999, 2 /*next chunk boundary*/
14 Query EXPLAIN SELECT `id`, `name` FROM `sbtest`.`t1` FORCE INDEX(`PRIMARY`) WHERE ((`id` >= '1')) AND ((`id` <= '1000')) LOCK IN SHARE MODE /*explain pt-online-schema-change 51223 copy nibble*/
14 Query INSERT LOW_PRIORITY IGNORE INTO `sbtest`.`_t1_new` (`id`, `name`) SELECT `id`, `name` FROM `sbtest`.`t1` FORCE INDEX(`PRIMARY`) WHERE ((`id` >= '1')) AND ((`id` <= '1000')) LOCK IN SHARE MODE /*pt-online-schema-change 51223 copy nibble*/
14 Query SHOW WARNINGS
14 Query SELECT @@SERVER_ID
14 Query SHOW GRANTS FOR CURRENT_USER()
14 Query SHOW FULL PROCESSLIST
14 Query SHOW SLAVE HOSTS
14 Query SHOW GLOBAL STATUS LIKE 'Threads_running'
```

通过 EXPLAIN SELECT * FROM `sbtest`.`t1` WHERE 1=1 确定 t1 表的大概行数，我们在命令行的输出中可以看到这个行数。

```
Copying approximately 10294 rows...
```

接下来，会拷贝原表的全量数据。

可以看到，原表到临时表的数据拷贝是分 chunk 的，所以每次都会确认 chunk 的上下限。

下面这条 SQL 语句会输出第一个 chunk 的最小值。

```
SELECT /*!40001 SQL_NO_CACHE */ `id` FROM `sbtest`.`t1` FORCE INDEX(`PRIMARY`) ORDER BY `id` LIMIT 1 /*first lower boundary*/
```

下面这条 SQL 语句会输出两个值：1000 和 1001。第一个值将被作为当前 chunk 的最大值，第二个值则被作为下一个 chunk 的最小值。

```
SELECT /*!40001 SQL_NO_CACHE */ `id` FROM `sbtest`.`t1` FORCE INDEX(`PRIMARY`) WHERE ((`id` >= '1')) ORDER BY `id` LIMIT 999, 2 /*next chunk boundary*/
```

确定完 chunk 的上下限，会将原表的数据拷贝到临时表中，此时会对原表这个 chunk 加共享锁。

为什么要加共享锁呢？主要是为了避免这个 chunk 在拷贝的过程中发生修改。试想一下，如果不加任何锁，在拷贝的过程中，ID 为 1000 的这条记录被删除了。此时，原表的触发器会删除临时表中 ID 为 1000 的记录，但因为拷贝还未完成，此时临时表中并不存在这条记录。当拷贝操作完成时，ID 为 1000 的这条记录才会被插入临时表。这个时候，我们会发现，ID 为 1000 的这条记录，在原表中不存在，却存在于临时表中。这样，就出现了数据不一致。

当这个 chunk 的数据拷贝完成后，会通过 SHOW FULL PROCESSLIST、SHOW SLAVE HOSTS 获取从库信息，并检查从库的主从延迟情况。如果从库的 Seconds_Behind_Master 大于 1（由 max-lag 指定），则暂停下一个 chunk 的数据拷贝动作。

接着查看 Threads_running 的值，以判断系统当前负载是否过大。如果目标实例的 Threads_running 超过 25，则会暂停下一个 chunk 的拷贝动作，如果超过 50，工具会直接退出，并提示以下错误。

```
Error copying rows from `sbtest`.`t1` to `sbtest`.`_t1_new`: Threads_running=65 exceeds its critical threshold 50
```

这里的 Threads_running=25 和 Threads_running=50 分别是 OSC 中 max-load 和 critical-load 的默认值。

```
14 Query     EXPLAIN SELECT /*!40001 SQL_NO_CACHE */ `id` FROM `sbtest`.`t1` FORCE INDEX(`PRIMARY`) WHERE ((`id` >= '1001')) ORDER BY `id` LIMIT 7501, 2 /*next chunk boundary*/
14 Query     SELECT /*!40001 SQL_NO_CACHE */ `id` FROM `sbtest`.`t1` FORCE INDEX(`PRIMARY`) WHERE ((`id` >= '1001')) ORDER BY `id` LIMIT 7501, 2 /*next chunk boundary*/
14 Query     EXPLAIN SELECT `id`, `name` FROM `sbtest`.`t1` FORCE INDEX(`PRIMARY`) WHERE ((`id` >= '1001')) AND ((`id` <= '8502')) LOCK IN SHARE MODE /*explain pt-online-schema-change 51223 copy nibble*/
14 Query     INSERT LOW_PRIORITY IGNORE INTO `sbtest`.`_t1_new` (`id`, `name`) SELECT `id`, `name` FROM `sbtest`.`t1` FORCE INDEX(`PRIMARY`) WHERE ((`id` >= '1001')) AND ((`id` <= '8502')) LOCK IN SHARE MODE /*pt-online-schema-change 51223 copy nibble*/
14 Query     SHOW WARNINGS
14 Query     SELECT @@SERVER_ID
14 Query     SHOW GRANTS FOR CURRENT_USER()
14 Query     SHOW FULL PROCESSLIST
14 Query     SHOW SLAVE HOSTS
14 Query     SHOW GLOBAL STATUS LIKE 'Threads_running'

14 Query     EXPLAIN SELECT /*!40001 SQL_NO_CACHE */ `id` FROM `sbtest`.`t1` FORCE INDEX(`PRIMARY`) WHERE ((`id` >= '8503')) ORDER BY `id` LIMIT 29332, 2 /*next chunk boundary*/
14 Query     SELECT /*!40001 SQL_NO_CACHE */ `id` FROM `sbtest`.`t1` FORCE INDEX(`PRIMARY`) WHERE ((`id` >= '8503')) ORDER BY `id` LIMIT 29332, 2 /*next chunk boundary*/
14 Query     SELECT /*!40001 SQL_NO_CACHE */ `id` FROM `sbtest`.`t1` FORCE INDEX(`PRIMARY`) ORDER BY `id` DESC LIMIT 1 /*last upper boundary*/
14 Query     EXPLAIN SELECT `id`, `name` FROM `sbtest`.`t1` FORCE INDEX(`PRIMARY`) WHERE ((`id` >= '8503')) AND ((`id` <= '10000')) LOCK IN SHARE MODE /*explain pt-online-schema-change 51223 copy nibble*/
14 Query     INSERT LOW_PRIORITY IGNORE INTO `sbtest`.`_t1_new` (`id`, `name`) SELECT `id`, `name` FROM `sbtest`.`t1` FORCE INDEX(`PRIMARY`) WHERE ((`id` >= '8503')) AND ((`id` <= '10000')) LOCK IN SHARE MODE /*pt-online-schema-change 51223 copy nibble*/
14 Query     SHOW WARNINGS
14 Query     SELECT @@SERVER_ID
14 Query     SHOW GRANTS FOR CURRENT_USER()
14 Query     SHOW FULL PROCESSLIST
14 Query     SHOW SLAVE HOSTS
14 Query     SHOW GLOBAL STATUS LIKE 'Threads_running'
```

第二个和最后一个 chunk 的拷贝过程同第一个 chunk 的类似。

```
14 Query     ANALYZE TABLE `sbtest`.`_t1_new` /* pt-online-schema-change */
14 Query     RENAME TABLE `sbtest`.`t1` TO `sbtest`.`_t1_old`, `sbtest`.`_t1_new` TO `sbtest`.`t1`
14 Query     DROP TABLE IF EXISTS `sbtest`.`_t1_old`
14 Query     DROP TRIGGER IF EXISTS `sbtest`.`pt_osc_sbtest_t1_del`
14 Query     DROP TRIGGER IF EXISTS `sbtest`.`pt_osc_sbtest_t1_upd`
14 Query     DROP TRIGGER IF EXISTS `sbtest`.`pt_osc_sbtest_t1_ins`
```

```
14 Query    SHOW TABLES FROM `sbtest` LIKE '\_t1\_new'
15 Quit
14 Quit
```

在拷贝完原表的全量数据后,会对临时表执行 ANALYZE 操作。这样,可及时更新临时表的统计信息。

接着,通过 RENAME 操作对原表和临时表进行重命名,这个时候,同样需要获取原表的元数据锁。

最后,删除原表及创建的 3 个触发器。其实,删除触发器这个操作没太大必要,在对原表进行重命名后,触发器的触发对象也会随之改变。删除原表,这 3 个触发器同样也会被删除。

7.2.2　pt-online-schema-change 的参数解析

OSC 必需的参数有以下 3 个。

- `--alter`

  指定 DDL 子句。与一个完整的 DDL 语句相比,不用写前面的 `alter table table_name` 部分。

- `--execute`

  执行 OSC 操作。

  如果只是想验证参数的正确性,可指定 `--dry-run` 进行试运行,此时只会创建临时表及修改表结构,不会创建触发器、拷贝数据及执行 RENAME 操作。`dry-run` 和 `execute` 是互斥的。

- DSN

  DSN 在 Percona Toolkit 中比较常见,可理解为目标实例相关信息的缩写。支持的缩写及其含义如下。

```
缩写   含义
===    =========================================
A      默认的字符集
D      库名
F      只从给定文件读取配置信息,类似于MySQL中的--defaults-file
P      端口
S      用于连接的套接字文件
h      主机名
p      密码
t      表名
u      用户名
```

下面,我们看看 OSC 的其他常见参数。

1. 主从相关

前面提到过 Online DDL 很容易造成主从延迟,OSC 很好地解决了这个问题。在每次 chunk 拷贝完成后,都会检查从库的延迟情况。如果从库的 `Seconds_Behind_Master` 大于某个值(由 `--max-lag` 参数指定,默认为 1 秒),则会暂停下一个 chunk 的拷贝。在等待若干秒后(由 `--check-interval` 参数指定,默认为 1 秒),会再次检测从库的延迟情况。

下面再来看看主从相关的其他参数。

- **--check-slave-lag**

检测指定从库的延迟情况。

例如：

```
# pt-online-schema-change h=192.168.244.10,P=3306,u=pt_user,p=123456,D=sbtest,t=t1 --alter "modify column c1 varchar(10)" --check-slave-lag h=192.168.244.30,P=3308,u=pt_user,p=123456 –execute
```

注意，只能指定一个从库。如果要指定多个，可考虑使用 --skip-check-slave-lag 或 --recursion-method 参数。

- **--skip-check-slave-lag**

同 --check-slave-lag 相反，忽略某些从库的检测。与 --check-slave-lag 不一样的是，可指定多次。例如：

```
--skip-check-slave-lag h=192.168.244.20,P=3307 --skip-check-slave-lag h=192.168.244.30,P=3308
```

- **--recurse**

需检查的从库的层级。常用于级联复制场景，默认会检查所有从库的延迟情况。如果设置为 1，则只会检查一级从库。

- **--recursion-method**

发现从库的方式。支持的选项如下。

- processlist：通过 SHOW PROCESSLIST 发现从库。
- hosts：通过 SHOW SLAVE HOSTS 发现从库。
- dsn=DSN：从表中获取从库信息。
- none：不检查从库的延迟情况。

默认为 processlist,hosts，即优先选择 processlist，其次选择 hosts。

对于 SHOW PROCESSLIST，只会提供主机信息，没有端口信息。如果从库不是默认端口（3306），这个方法就不管用了。

同样，对于 SHOW SLAVE HOSTS，默认情况下，只会提供端口信息，没有主机信息。例如：

```
mysql> show slave hosts;
+-----------+------+------+-----------+--------------------------------------+
| Server_id | Host | Port | Master_id | Slave_UUID                           |
+-----------+------+------+-----------+--------------------------------------+
|         2 |      | 3307 |         1 | a78851ce-66c6-11ec-a023-000c297b8a24 |
|         3 |      | 3308 |         1 | c8de8958-6709-11ec-8053-000c299334ee |
+-----------+------+------+-----------+--------------------------------------+
2 rows in set, 1 warning (0.00 sec)
```

如果需要 SHOW SLAVE HOSTS 显示主机信息，需要从库配置 report_host 参数。但这个参数是静态参数，修改后需要重启实例才能生效。

```
mysql> show slave hosts;
+-----------+----------------+------+-----------+--------------------------------------+
| Server_id | Host           | Port | Master_id | Slave_UUID                           |
+-----------+----------------+------+-----------+--------------------------------------+
|         3 | 192.168.244.30 | 3308 |         1 | c8de8958-6709-11ec-8053-000c299334ee |
|         2 | 192.168.244.20 | 3307 |         1 | a78851ce-66c6-11ec-a023-000c297b8a24 |
+-----------+----------------+------+-----------+--------------------------------------+
2 rows in set, 1 warning (0.00 sec)
```

可以看到，上面这两种方式都不是很灵活，特别是 hosts 方式需要重启从库。

下面看看 dsn 方式。

这种方式需单独创建一张表，插入需要检测的从库信息。

```
mysql> create database percona;
Query OK, 1 row affected (0.00 sec)

mysql> create table percona.dsns (
    -> id int(11) not null auto_increment,
    -> parent_id int(11) default null,
    -> dsn varchar(255) not null,
    -> primary key (id)
    -> );
Query OK, 0 rows affected (0.02 sec)

mysql> insert into percona.dsns values(1,1,"h=192.168.244.20,P=3307");
Query OK, 1 row affected (0.00 sec)

mysql> insert into percona.dsns values(2,1,"h=192.168.244.30,P=3308");
Query OK, 1 row affected (0.00 sec)
```

该表虽然有 3 个字段，但实际起作用的只是 dsn。这里只指定了从库的 IP 和端口，也可指定用户名和密码。

此时，recursion-method 需指定为 dsn，例如：

```
--recursion-method dsn="h=192.168.244.128,P=3306,D=percona,t=dsns"
```

在 OSC 运行的过程中，如果我们不想检测某个从库的主从延迟，直接从表中删除对应记录即可，相当灵活。

> **注意**：不管通过何种方式获取从库信息，如果检测的从库在 OSC 运行过程停止复制了，OSC 会一直等待，直到复制恢复。

2. 性能相关

OSC 中全量数据的拷贝是分 chunk 来进行的，那么每个 chunk 的大小是如何确定的呢？这实际上与 chunk-size 和 chunk-time 有关。

- **--chunk-size 和 --chunk-time**

前者指定了 chunk 的大小，默认为 1000。后者指定了 chunk 的拷贝时间，默认为 0.5 秒。

在不指定 chunk-size 的情况下，1000 会被作为第一个 chunk 的大小。此后，会基于 chunk-time

自动调节 chunk 的大小。

下面，基于 OSC debug 模式下的输出，看看具体的实现算法。

```
# pt_online_schema_change:9950 11798 Average copy rate (rows/s): 99781
# WeightedAvgRate:5493 11798 Master op time: 1000 n / 0.0100219249725342 s
# WeightedAvgRate:5505 11798 Initial avg rate: 99781.2299274414 n/s
# WeightedAvgRate:5509 11798 Adjust n to 49890
# NibbleIterator:5958 11798 Set new chunk size (LIMIT): 49890
```

第一个 chunk 只有 1000 行，拷贝操作耗时 0.010 021 924 972 534 2 秒，所以拷贝的平均速率是每秒 99 781.229 927 441 4 行，若想下一次拷贝能在 0.5 秒内完成，则 chunk-size 应设置为 49890（即 99 781.229 927 441 4 × 0.5）。

如果显式指定了 chunk-size，则每次 chunk 的大小将是固定不变的，不会自动调节。

如果同时指定了 chunk-size 和 chunk-time，前者会覆盖后者。

在指定了 chunk-size 的情况下，还有另外一个参数需要注意：chunk-size-limit。如果 chunk index 不唯一，可能会导致单个 chunk 比较大，而 chunk-size-limit 限制了单个 chunk 的最大大小，其默认值为 4。

在执行的时候，如果通过 EXPLAIN 发现执行计划中的 rows > chunk-size * chunk-size-limit，工具会退出，并提示以下错误。

```
Cannot copy table `sbtest`.`sbtest1` because on the master it would be checksummed in one chunk but
on these replicas it has too many rows: 7874 rows on node2 The current chunk size limit is 4000 rows
(chunk size=1000 * chunk size limit=4.0).
```

- `--max-load` 和 `--critical-load`

这两个参数的作用在原理部分提到过，默认是监控 Threads_running 的。也可指定其他参数，如 `--max-load=Threads_connected=100`。也可指定多个参数，如 `--max-load=Threads_connected=500,Threads_running=50`。

- `--sleep`

每次 chunk 拷贝完等待的时间，默认为 0，单位是秒。如果在执行 OSC 时，业务较为繁忙，可通过这个参数减轻 OSC 对系统负载的影响。

3. 外键相关

- `--alter-foreign-keys-method`

指定外键被更新的方式。在目标表（父表）被其他表（子表）外键关联的情况下，必须指定该参数，否则会提示以下错误。

```
You did not specify --alter-foreign-keys-method, but there are foreign keys that reference the table.
Please read the tool's documentation carefully.
```

该参数有如下选项。

❑ auto。优先选择 rebuild_constraints 方式，其次是 drop_swap。

- rebuild_constraints。
- drop_swap。
- none。同 drop_swap 方式类似，只不过没有 swap 阶段。

以下面两张表为例。

```
create table t_parent (id int primary key);

create table t_child (
    id int primary key,
    parent_id int,
    foreign key (parent_id) references t_parent(id)
);
```

我们具体看看 rebuild_constraints、drop_swap 和 none 这 3 种方式的区别，其区别主要体现在最后的 RENAME 阶段。

(1) rebuild_constraints

RENAME 阶段对应的 SQL 如下：

```
RENAME TABLE `slowtech`.`t_parent` TO `slowtech`.`_t_parent_old`, `slowtech`.`_t_parent_new` TO `slowtech`.`t_parent`;
SET foreign_key_checks=0;
ALTER TABLE `slowtech`.`t_child` DROP FOREIGN KEY `_t_child_ibfk_1`, ADD CONSTRAINT `__t_child_ibfk_1` FOREIGN KEY (`parent_id`) REFERENCES `slowtech`.`t_parent` (`id`);
SET foreign_key_checks=1;
DROP TABLE IF EXISTS `slowtech`.`_t_parent_old`;
```

首先，重命名。注意，RENAME 操作完成后，t_child 的外键将指向 _t_parent_old，而不再是 t_parent 表。

其次，重建外键约束。DROP FOREIGN KEY 操作只会修改表的元数据信息，耗时极短。但 ADD FOREIGN KEY 操作的行为与 foreign_key_checks 参数有关，若该参数为 ON，则使用 COPY 算法。此时，会阻塞 t_child 的 DML 操作，而且 DDL 操作本身耗时也比较久。若该参数为 OFF，则只会修改表的元数据信息。注意，上面的 SET foreign_key_checks=0 操作是 Percona Toolkit v3.3.1 中才添加的，之前的版本是没有的。

最后，删除 _t_parent_old 表。

(2) drop_swap

RENAME 阶段对应的 SQL 如下：

```
SET foreign_key_checks=0;
DROP TABLE IF EXISTS `slowtech`.`t_parent`;
RENAME TABLE `slowtech`.`_t_parent_new` TO `slowtech`.`t_parent`;
```

首先，关闭外键检查，否则后面的 DROP TABLE 操作会失败。

其次，删除原表。

最后，将临时表重命名为原表。

相对于 rebuild_constraints，这种方式因为无须重建外键约束所以会更快，但存在以下风险。

- DROP TABLE 和 RENAME TABLE 是先后执行的。在这段时间内，t_parent 不存在，所有针对 t_parent 的操作都会失败。
- DROP TABLE 和 RENAME TABLE 不是一个原子操作，并不能保证 DROP TABLE 执行后，RENAME TABLE 操作就一定能执行成功。如果后者执行失败，影响太大。

(3) none

RENAME 阶段对应的 SQL 操作如下：

```
RENAME TABLE `slowtech`.`t_parent` TO `slowtech`.`_t_parent_old`, `slowtech`.`_t_parent_new` TO `slowtech`.`t_parent`;
SET foreign_key_checks=0;
DROP TABLE IF EXISTS `slowtech`.`_t_parent_old`;
```

首先，重命名。

其次，关闭外键检查。

最后，删除 _t_parent_old 表。删除后，t_child 的外键实际指向的是一张不存在的表。此时，针对 t_child 的新增和更新操作都会失败。

```
mysql> insert into t_child values(1,1);
ERROR 1452 (23000): Cannot add or update a child row: a foreign key constraint fails (`slowtech`.`t_child`, CONSTRAINT `t_child_ibfk_1` FOREIGN KEY (`parent_id`) REFERENCES `_t_parent_old` (`id`))
```

所以，在线上要慎用这种方式！

实际上，在执行的时候，也提示了风险，需要 DBA 手动确认。

```
WARNING! Using alter-foreign-keys-method = "none". This will typically cause foreign key violations! This method of handling foreign key constraints is only provided so that the database administrator can disable the tool's built-in functionality if desired.

Continue anyway? (y/N)
```

最后，再来看看 auto 选项。

设置为 auto，会自动选择外键更新方式，优先选择 rebuild_constraints，其次选择 drop_swap。那么什么情况下会使用 rebuild_constraints 这种方式呢？

工具在实际执行的过程中，会评估 rebuild_constraints 下的 ALTER 操作能否在 chunk-time 内完成。若能，则使用 rebuild_constraints，否则使用 drop_swap。这里假定：ALTER 操作的执行速度 = 数据拷贝速度 × chunk-size-limit。

如下面这个示例所示，因 t_child 的数据量较大（998 568 条），超过了可允许的最大值（595 382 条），最后使用了 drop_swap 方式。

```
2021-12-26T21:46:04 Creating triggers...
2021-12-26T21:46:04 Created triggers OK.
2021-12-26T21:46:04 Copying approximately 972000 rows...
2021-12-26T21:46:08 Copied rows OK.
```

```
2021-12-26T21:46:08 Max rows for the rebuild_constraints method: 595382
Determining the method to update foreign keys...
2021-12-26T21:46:08     `slowtech`.`t_child`: too many rows: 998568; must use drop_swap
2021-12-26T21:46:08 Drop-swapping tables...
2021-12-26T21:46:08 Analyzing new table...
2021-12-26T21:46:10 Dropped and swapped tables OK.
Not dropping old table because --no-drop-old-table was specified.
2021-12-26T21:46:10 Dropping triggers...
2021-12-26T21:46:10 Dropped triggers OK.
Successfully altered `slowtech`.`t_parent`.
```

考虑到 drop_swap 方式会删除原表，在高写入场景中，这对业务其实是有损的。

建议使用 rebuild_constraints 的方式，且使用 Percona Toolkit v3.3.1 及之后的版本，避免 ADD FOREIGN KEY 操作在 foreign_key_checks 为 ON 的情况下阻塞子表的 DML 操作。

4. 其他参数

- **--[no]check-replication-filters**

检查主从之间是否存在过滤规则。如果存在，OSC 默认会退出，此时可加 --no-check-replication-filters 跳过检测。

- **--default-engine**

默认情况下，创建的临时表和原表的表结构完全一样，包括存储引擎。如果指定了 --default-engine，则会使用系统默认的存储引擎。

- **--data-dir 和 --remove-data-dir**

默认情况下，表创建在数据目录下。从 MySQL 5.6 开始，支持通过 data directory 选项将表创建在其他目录下。例如：

```
mysql> create table slowtech.t1(id int primary key) data directory '/tmp';
Query OK, 0 rows affected (0.01 sec)

# ls /tmp/slowtech/
t1.ibd
```

如果单表过大，可通过 --data-dir 将表迁移到其他目录下。

--remove-data-dir 的作用与 --data-dir 的相反，是移除原表的 data directory 选项，将表迁回到数据目录下。

- **--[no]check-unique-key-change**

在使用 OSC 添加唯一索引或主键时，它默认不会执行，即使目标列中没有重复数据也是如此。

```
You are trying to add an unique key. This can result in data loss if the data is not unique.
Please read the documentation for the --check-unique-key-change parameter.
You can check if the column(s) contain duplicate content by running this/these query/queries:

SELECT IF(COUNT(DISTINCT id) = COUNT(*),
       'Yes, the desired unique index currently contains only unique values',
       'No, the desired unique index contains duplicated values. There will be data loss'
) AS IsThereUniqueness FROM `slowtech`.`t1`;
```

之所以不会执行与 OSC 的实现逻辑有关。

- 原表到临时表的数据拷贝使用的是 `INSERT IGNORE` 语句。如果原表该列存在重复数据，在拷贝的过程中，除了第一条，后续重复的数据都会被直接忽略。
- 在插入数据时，因为 INSERT 触发器使用了 REPLACE 操作，所以如果目标列存在重复数据，REPLACE 操作会导致后插入的记录覆盖之前的记录。

如果确定了目标列没有重复数据，可使用 --no-check-unique-key-change 参数跳过这个限制。

- `--print`

打印 OSC 执行过程中的 SQL 语句。

- `--pause-file`

如果指定的文件存在，会暂停当前的数据拷贝操作。

- `--[no]drop-new-table`

如果操作中途失败，是否删除临时表，默认为 yes。

- `--[no]drop-old-table`

在执行完 RENAME 操作后，是否删除原表，默认为 yes。

- `--[no]drop-triggers`

是否删除创建的 3 个触发器，默认为 yes。

- `--[no]swap-tables`

在完成数据拷贝后，是否对原表和临时表进行 RENAME 操作，默认为 yes。

- `--set-vars`

设置连接的会话值。默认会修改以下 3 个参数的值。

```
wait_timeout=10000
innodb_lock_wait_timeout=1
lock_wait_timeout=60
```

- `--tries`

定义关键操作的重试次数。

在执行 OSC 的过程中，可能会碰到以下报错：

```
Lock wait timeout
Deadlock found
Query is killed
Connection is killed
Lost connection to MySQL
```

针对这些报错，OSC 采取了重试策略。

关键操作及默认的重试次数如下：

```
操作                  重试次数    重试等待时间（单位是秒）
==================    =====     ====
create_triggers         10        1
drop_triggers           10        1
copy_rows               10        0.25
swap_tables             10        1
update_foreign_keys     10        1
analyze_table           10        1
```

在这些操作中，比较危险的是 swap_tables，即 RENAME 操作。因为这个操作需要获取原表的元数据锁。此时，如果原表有个慢查询还没结束，会导致 RENAME 操作被阻塞。不仅如此，在阻塞的这段时间内，后续针对该表的所有操作（包括查询操作）都会被 RENAME 操作阻塞。元数据锁的等待时间由参数 lock_wait_timeout 确定，默认为 365 天，只不过 OSC 将之调整为了 60 秒。但即使是 60 秒，还是较大。

建议进一步减小 lock_wait_timeout 的值，同时增加 swap_tables 操作的重试等待时间和重试次数。

最后，给出我在线上操作时的常用设置，供大家参考。

```
pt-online-schema-change h=192.168.244.10,P=3306,u=pt_user,p=123456,D=sbtest,t=t1 --alter="add column
c1 datetime" --charset utf8 --max-load=Threads_running=50 --critical-load=Threads_running=100
--set-vars lock_wait_timeout=5 --tries swap_tables:30:60 --execute
```

7.2.3　pt-online-schema-change 的优缺点

下面看看 pt-online-schema-change 的优缺点。

pt-online-schema-change 的优点如下。

- 使用广泛，久经考验。
- 由 Percona 公司开源并维护，质量相对有保证。

pt-online-schema-change 的缺点如下。

- 触发器是一种侵入式方案。

  触发器会将原表操作及触发器的触发操作放到一个事务空间内执行。如果触发动作失败，原表操作同样会失败。这也就是为什么如果 OSC 中途异常退出，在清理现场时，我们要首先删除触发器，其次才是临时表。

- 触发器是一种同步操作，并不能停止。

  前面提到过，主从延迟超过一定值，或通过设置 --pause-file，可暂停 OSC 的操作。这里的暂停，仅指数据拷贝操作。触发器的触发操作并不能停止。如果表本身的操作就很频繁，而且实例负载也比较高，那触发操作会进一步加重实例的负载。

7.2.4　pt-online-schema-change 的注意事项

下面，我们看看 OSC 在使用中的注意事项。

- 对于 OSC，不仅最后的 RENAME 操作会获取元数据锁。在开始阶段，创建触发器时，同样会获取元数据锁。如果此时原表上有个查询还没结束，同样会阻塞触发器的创建。
- 针对 UPDATE 操作的触发器，包含两个动作：DELETE 和 REPLACE。

  在 Percona Toolkit 3.0.2 之前，只有 REPLACE 这一个动作。在 OSC 执行的过程中，如果有对主键或唯一键的更新操作，会导致记录数增多。关于这个 bug 的更多信息可参考 Bug #1646713。
- 目标表上必须存在主键或唯一索引。唯一的例外是，当对表添加主键或唯一索引时，无此限制。
- 不能通过 OSC 添加自增主键。
- 不能通过 RENAME TO new_tbl_name 子句修改表名。

7.3 gh-ost

gh-ost 是著名的开源软件托管平台 GitHub 开源的 MySQL 表结构变更工具，由 Shlomi Noach 开发并维护。他也是 openark-kit（Python 开发的一套 MySQL 日常管理工具，类似于 Percona Toolkit）和 orchestrator（MySQL 高可用和复制管理工具）的作者。

这里还有个小故事。openark-kit 中有个名为 oak-online-alter-table 的工具，它是最早实现了基于触发器来进行表结构变更的工具，后来的很多工具（包括 pt-online-schema-change）借鉴了它的实现思路。Shlomi Noach 在 GitHub 工作期间，前期使用的也是 pt-online-schema-change，不过碰到了很多问题，于是他另辟蹊径，开发了 gh-ost 这个无须触发器的表结构变更工具。

gh-ost 的安装

这里直接使用二进制包进行安装。

```
# tar xvf gh-ost-binary-linux-20210617134741.tar.gz
```

解压完后，会在当前目录生成一个可执行文件 gh-ost。

7.3.1 gh-ost 的实现原理

在实现及处理细节上，gh-ost 与 OSC 有许多相似之处。

- 对于原表存量数据的处理，都是分 chunk 并加共享锁拷贝到临时表中的。
- 可自动调节 DDL 的进度，都基于 Threads_running 和主从延迟。

两者最大的不同是对增量数据的处理。OSC 基于触发器，可理解为实时处理。gh-ost 则基于 binlog，通过将自己"伪装"成一个从库，把捕捉到的 DML 操作异步应用到临时表中。异步处理赋予了 gh-ost 操作上极大的灵活性。例如，在业务高峰期可以暂停 gh-ost 的所有操作，而 OSC 只能暂停数据的拷贝动作，对于触发器的触发动作，无论你爱或不爱，它都在那里。

既然基于 binlog，那么理论上，无论是"伪装"成主库的从库还是从库的从库，都没有太大的区别。实际上，gh-ost 提供了 3 种工作模式，原理分别如图 7-1 所示。

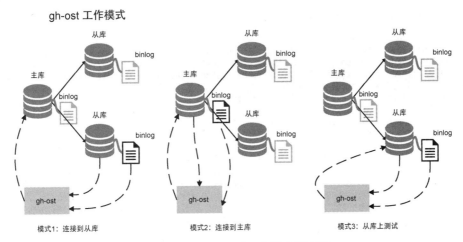

图 7-1 gh-ost 的 3 种工作模式原理图

接下来，我们看看这 3 种工作模式的区别。

(1) 读取从库的 binlog，修改主库的表结构。

这也是官方默认的工作模式，它会通过从库来获取主库的主机和端口信息。这种模式下，存量数据的拷贝是在主库上进行的，增量数据则是通过读取从库的 binlog，将其应用到主库上。

官方之所以推荐这种模式，主要还是考虑到这种模式对主库的侵入较小：既不用关心主库 binlog 的格式，也不用读取主库的 binlog。

(2) 读取主库的 binlog，修改主库的表结构。

与第一种模式不同的是，这种模式直接读取主库的 binlog，适用于没有从库或不想使用从库的场景。

使用这种模式，必须满足以下两个条件。

- 主库的 `binlog_format` 为 ROW。
- 指定 `-allow-on-master` 参数。

(3) 修改从库的表结构。

所有操作都在从库上进行。修改从库的表结构又包括以下两种模式。

- `-migrate-on-replica`：直接修改从库的表结构。
- `-test-on-replica`：与 `-migrate-on-replica` 类似，不一样的是在执行最终的 RENAME 操作之前，gh-ost 会停掉主从复制。之后，会执行两次 RENAME 操作，相当于原表还是原表。之所以要停掉复制，是为了方便大家对比两表的数据。

下面看看第一种工作模式的实现原理。还是采用老方法，打开 general log。

```
# gh-ost --host="192.168.244.20" --user="root" --password="123456" --database="sbtest" --table="t1"
--alter="add column c1 datetime" --chunk-size=3000 --max-load=Threads_running=50
--heartbeat-interval-millis=10000 --throttle-control-replicas="192.168.244.20" --execute
```

这里将 `--heartbeat-interval-millis` 设置为 10000（默认为 100）以避免心跳语句的频繁执行。

```
# Migrating `sbtest`.`t1`; Ghost table is `sbtest`.`_t1_gho`
# Migrating node1:3306; inspecting node2:3306; executing on node1
# Migration started at Mon Dec 27 21:42:20 +0800 2021
# chunk-size: 3000; max-lag-millis: 1500ms; dml-batch-size: 10; max-load: Threads_running=50;
critical-load: ; nice-ratio: 0.000000
# throttle-additional-flag-file: /tmp/gh-ost.throttle
# throttle-control-replicas count: 1
# Serving on unix socket: /tmp/gh-ost.sbtest.t1.sock
Copy: 0/10120 0.0%; Applied: 0; Backlog: 0/1000; Time: 3s(total), 0s(copy); streamer:
mysql-bin.000007:1269995; Lag: 0.09s, HeartbeatLag: 0.98s, State: migrating; ETA: N/A
Copy: 0/10120 0.0%; Applied: 0; Backlog: 0/1000; Time: 4s(total), 1s(copy); streamer:
mysql-bin.000007:1270913; Lag: 0.08s, HeartbeatLag: 1.00s, State: migrating; ETA: N/A
Copy: 3000/10120 29.6%; Applied: 0; Backlog: 0/1000; Time: 5s(total), 2s(copy); streamer:
mysql-bin.000007:1320295; Lag: 1.12s, HeartbeatLag: 0.99s, State: migrating; ETA: 4s
Copy: 9000/10120 88.9%; Applied: 0; Backlog: 0/1000; Time: 6s(total), 3s(copy); streamer:
mysql-bin.000007:1369903; Lag: 1.10s, HeartbeatLag: 0.98s, State: migrating; ETA: due
Copy: 10000/10000 100.0%; Applied: 0; Backlog: 0/1000; Time: 6s(total), 3s(copy); streamer:
mysql-bin.000007:1419066; Lag: 1.10s, HeartbeatLag: 0.15s, State: migrating; ETA: due
Copy: 10000/10000 100.0%; Applied: 0; Backlog: 1/1000; Time: 7s(total), 3s(copy); streamer:
mysql-bin.000007:1436991; Lag: 1.10s, HeartbeatLag: 0.96s, State: migrating; ETA: due
# Migrating `sbtest`.`t1`; Ghost table is `sbtest`.`_t1_gho`
# Migrating node1:3306; inspecting node2:3306; executing on node1
# Migration started at Mon Dec 27 21:42:20 +0800 2021
# chunk-size: 3000; max-lag-millis: 1500ms; dml-batch-size: 10; max-load: Threads_running=50;
critical-load: ; nice-ratio: 0.000000
# throttle-additional-flag-file: /tmp/gh-ost.throttle
# throttle-control-replicas count: 1
# Serving on unix socket: /tmp/gh-ost.sbtest.t1.sock
Copy: 10000/10000 100.0%; Applied: 0; Backlog: 0/1000; Time: 7s(total), 3s(copy); streamer:
mysql-bin.000007:1437914; Lag: 0.06s, HeartbeatLag: 0.16s, State: migrating; ETA: due
[2021/12/27 21:42:28] [info] binlogsyncer.go:164 syncer is closing...
[2021/12/27 21:42:28] [error] binlogstreamer.go:77 close sync with err: sync is been closing...
[2021/12/27 21:42:28] [info] binlogsyncer.go:179 syncer is closed
# Done
```

因为没有开启 `--verbose` 和 `--debug` 模式，所以打印的日志不是很多，主要是进度信息。

下面看看进度信息中各字段的具体含义。

- `Copy: 9000/10120 88.9%`：数据拷贝的进度，9000 是已经拷贝的行，10120 是原表的行数。
- `Applied: 0`：已经应用的 binlog 的数量。
- `Backlog: 0/1000`：gh-ost 将待应用的二进制日志事件放到队列中。前面的 0 代表队列中待应用的二进制日志事件的个数，后面的 1000 代表队列的长度。在 gh-ost 中，二进制日志事件应用的优先级大于数据的拷贝。
- `Time: 6s(total), 3s(copy)`：total 代表 gh-ost 执行的总时间，copy 代表数据拷贝花费的时间。
- `streamer: mysql-bin.000007:1369903`：当前读取的 binlog 的位置。如果上面的队列满了，该位置将不变。
- `Lag`：主从延迟时间。
- `State: migrating`：gh-ost 当前的状态。gh-ost 常见的状态如下。
  - `"migrating"`：正在迁移。

- "postponing cut-over"：在 cut-over 阶段暂停。
- "throttled, lag=23.509768s"：gh-ost 因主从延迟已暂停。
- "throttled, max-load Threads_running=4 >= 4"：gh-ost 因 max-load 超过阈值已暂停。

❑ ETA: due：预计操作还要多久才能完成，一般会给出具体的时间。due 代表不确定。

我们现在来看看 general log 的内容。

首先看看从库的 general log。

```
45 Connect   root@192.168.244.10 on sbtest using TCP/IP
45 Query   SET autocommit=true
45 Query   SET NAMES utf8mb4
45 Query   select @@global.version
45 Query   select @@global.port
45 Query   select @@global.hostname, @@global.port
45 Query   show /* gh-ost */ grants for current_user()
45 Query   select @@global.log_bin, @@global.binlog_format
45 Query   select @@global.binlog_row_image
46 Connect   root@192.168.244.10 on information_schema using TCP/IP
46 Query   SET autocommit=true
46 Query   SET NAMES utf8mb4
46 Query   show slave status
46 Quit
45 Query   stop slave
45 Query   start slave
```

检查从库是否开启 binlog、binlog 的格式以及 binlog_row_image。因为 gh-ost 是基于 binlog 来获取增量数据的，所以要求从库 binlog 的格式为 ROW，且 binlog_row_image 为 FULL。

如果当前从库 binlog 的格式为 STATEMENT，也可指定 --switch-to-rbr 参数，将其自动修改为 ROW。

接下来会重启复制。为什么要重启复制呢？如果当前从库的 binlog_format 为 STATEMENT，将其修改后，确实要重启复制让其生效。但有意思的是，即便检测到当前从库的 binlog_format 为 ROW，也还是会执行重启操作。大概是因为 gh-ost 的作者觉得看到的不一定就是真相，该参数可能是在复制启动后才修改的，所以还是重启比较保险。当然，如果可以确定从库就是 ROW 格式，可指定 --assume-rbr 参数，避免重启复制。

```
45 Query   show /* gh-ost */ table status from `sbtest` like 't1'
45 Query   SELECT SUM(REFERENCED_TABLE_NAME IS NOT NULL AND TABLE_SCHEMA='sbtest' AND TABLE_NAME='t1')
as num_child_side_fk, SUM(REFERENCED_TABLE_NAME IS NOT NULL AND REFERENCED_TABLE_SCHEMA='sbtest' AND
REFERENCED_TABLE_NAME='t1') as num_parent_side_fk FROM INFORMATION_SCHEMA.KEY_COLUMN_USAGE WHERE
REFERENCED_TABLE_NAME IS NOT NULL AND ((TABLE_SCHEMA='sbtest' AND TABLE_NAME='t1') OR
(REFERENCED_TABLE_SCHEMA='sbtest' AND REFERENCED_TABLE_NAME='t1') )
45 Query   SELECT COUNT(*) AS num_triggers FROM INFORMATION_SCHEMA.TRIGGERS WHERE
TRIGGER_SCHEMA='sbtest' AND EVENT_OBJECT_TABLE='t1'
45 Query   explain select /* gh-ost */ * from `sbtest`.`t1` where 1=1
45 Query   SELECT COLUMNS.TABLE_SCHEMA, COLUMNS.TABLE_NAME, COLUMNS.COLUMN_NAME, UNIQUES.INDEX_NAME,
UNIQUES.COLUMN_NAMES, UNIQUES.COUNT_COLUMN_IN_INDEX, COLUMNS.DATA_TYPE, COLUMNS.CHARACTER_SET_NAME,
LOCATE('auto_increment', EXTRA) > 0 as is_auto_increment, has_nullable FROM INFORMATION_SCHEMA.COLUMNS
INNER JOIN ( SELECT TABLE_SCHEMA, TABLE_NAME, INDEX_NAME, COUNT(*) AS COUNT_COLUMN_IN_INDEX,
GROUP_CONCAT(COLUMN_NAME ORDER BY SEQ_IN_INDEX ASC) AS COLUMN_NAMES,
SUBSTRING_INDEX(GROUP_CONCAT(COLUMN_NAME ORDER BY SEQ_IN_INDEX ASC), ',', 1) AS FIRST_COLUMN_NAME,
SUM(NULLABLE='YES') > 0 AS has_nullable FROM INFORMATION_SCHEMA.STATISTICS WHERE NON_UNIQUE=0 AND
TABLE_SCHEMA = 'sbtest' AND TABLE_NAME = 't1' GROUP BY TABLE_SCHEMA, TABLE_NAME, INDEX_NAME ) AS UNIQUES
```

```
ON ( COLUMNS.COLUMN_NAME = UNIQUES.FIRST_COLUMN_NAME ) WHERE COLUMNS.TABLE_SCHEMA = 'sbtest' AND
COLUMNS.TABLE_NAME = 't1' ORDER BY COLUMNS.TABLE_SCHEMA, COLUMNS.TABLE_NAME, CASE UNIQUES.INDEX_NAME
WHEN 'PRIMARY' THEN 0 ELSE 1 END, CASE has_nullable WHEN 0 THEN 0 ELSE 1 END, CASE
IFNULL(CHARACTER_SET_NAME, '') WHEN '' THEN 0 ELSE 1 END, CASE DATA_TYPE WHEN 'tinyint' THEN 0 WHEN
'smallint' THEN 1 WHEN 'int' THEN 2 WHEN 'bigint' THEN 3 ELSE 100 END, COUNT_COLUMN_IN_INDEX
45 Query   show columns from `sbtest`.`t1`
45 Query   SELECT AUTO_INCREMENT FROM INFORMATION_SCHEMA.TABLES WHERE TABLES.TABLE_SCHEMA = 'sbtest' AND
TABLES.TABLE_NAME = 't1' AND AUTO_INCREMENT IS NOT NULL
...
45 Query   select @@global.log_slave_updates
45 Query   select @@global.version
45 Query   select @@global.port
45 Query   show /* gh-ost readCurrentBinlogCoordinates */ master status
```

查看目标表是否存在外键关系，是否存在触发器。目前，gh-ost 不支持外键和触发器，这也是 gh-ost 的限制之一。

通过 `explain select /* gh-ost */ * from `sbtest`.`t1` where 1=1` 获取 sbtest.t1 表的大致行数。也可指定 `--exact-rowcount` 参数通过 `select count(*) from t` 的方式获取 t1 表的精确行数，从而输出更精确的进度信息。

接下来，会获取目标表上具有唯一性约束的列，包括主键和唯一键。如果没有，工具会退出。

```
FATAL No PRIMARY nor UNIQUE key found in table! Bailing out
```

接着，通过 `show master status` 获取 binlog 的位置点信息，后面会用到。

```
54 Connect    root@192.168.244.10 on  using TCP/IP
54 Query      SHOW GLOBAL VARIABLES LIKE 'BINLOG_CHECKSUM'
54 Query      SET @master_binlog_checksum='NONE'
54 Binlog Dump Log: 'mysql-bin.000007'  Pos: 1267523
```

以从库的身份连接到从库，开始读取从库的二进制日志事件，起始的位置点即上面 show master status 的输出。

```
45 Query   show columns from `sbtest`.`_t1_gho`
45 Query   select * from information_schema.columns where table_schema='sbtest' and table_name='t1'
45 Query   select * from information_schema.columns where table_schema='sbtest' and table_name='t1'
45 Query   select * from information_schema.columns where table_schema='sbtest' and
table_name='_t1_gho'
45 Query   select hint, value from `sbtest`.`_t1_ghc` where hint = 'heartbeat' and id <= 255
55 Query   select value from `sbtest`.`_t1_ghc` where hint = 'heartbeat' and id <= 255
...
45 Query   select hint, value from `sbtest`.`_t1_ghc` where hint = 'heartbeat' and id <= 255
55 Query   select value from `sbtest`.`_t1_ghc` where hint = 'heartbeat' and id <= 255
45 Quit
55 Quit
```

获取 sbtest._t1_ghc 表中心跳信息的值，用于判断主从延迟情况。

再来看看主库的 general log。

```
91 Connect    root@192.168.244.10 on information_schema using TCP/IP
91 Query      SET autocommit=true
91 Query      SET NAMES utf8mb4
91 Query      show slave status
91 Quit
```

```
92 Connect   root@192.168.244.10 on sbtest using TCP/IP
92 Query   SET autocommit=true
92 Query   SET NAMES utf8mb4
92 Query   select @@global.version
92 Query   select @@global.port
...
92 Query   select @@global.time_zone
92 Query   select @@global.hostname, @@global.port
92 Query   show columns from `sbtest`.`t1`
92 Query   show /* gh-ost */ table status from `sbtest` like '_t1_gho'
92 Query   show /* gh-ost */ table status from `sbtest` like '_t1_del'
92 Query   drop /* gh-ost */ table if exists `sbtest`.`_t1_ghc`
92 Query   create /* gh-ost */ table `sbtest`.`_t1_ghc` ( id bigint auto_increment, last_update timestamp
not null DEFAULT CURRENT_TIMESTAMP ON UPDATE CURRENT_TIMESTAMP, hint varchar(64) charset ascii not null,
value varchar(4096) charset ascii not null, primary key(id), unique key hint_uidx(hint) )
auto_increment=256
92 Query   create /* gh-ost */ table `sbtest`.`_t1_gho` like `sbtest`.`t1`
92 Query   alter /* gh-ost */ table `sbtest`.`_t1_gho` add column c1 datetime
92 Query   alter /* gh-ost */ table `sbtest`.`_t1_gho` AUTO_INCREMENT=10001

92 Query   insert /* gh-ost */ into `sbtest`.`_t1_ghc` (id, hint, value) values (NULLIF(2, 0), 'state',
'GhostTableMigrated') on duplicate key update last_update=NOW(), value=VALUES(value)
92 Query   insert /* gh-ost */ into `sbtest`.`_t1_ghc` (id, hint, value) values (NULLIF(0, 0), 'state
at 1640612543046884994', 'GhostTableMigrated') on duplicate key update last_update=NOW(),
value=VALUES(value)
92 Query       insert /* gh-ost */ into `sbtest`.`_t1_ghc` (id, hint, value) values (NULLIF(1, 0),
'heartbeat', '2021-12-27T21:42:23.127811028+08:00') on duplicate key update last_update=NOW(),
value=VALUES(value)
```

首先，执行 show slave status 判断主库是否为中间从库。

其次，判断 _t1_gho 和 _t1_del 表是否存在。

gh-ost 在执行过程中会创建 3 张表，分别以 gho、del 和 ghc 结尾。下面我们看看这 3 张表的具体作用。

- gho：临时表，DDL、数据拷贝及增量数据的应用都作用在该表上。
- del：辅助表，在后面的 RENAME 阶段会用到。
- ghc：信息记录表，会记录 3 类信息，分别对应上面的 3 个 insert into ... on duplicate key update 操作。

  (1) 标志位信息

  只有两个状态值：GhostTableMigrated 和 AllEventsUpToLockProcessed。在 gh-ost 读取 binlog 的过程中，如果碰到前者，则开始进行数据的拷贝。如果碰到后者，则代表数据拷贝的结束。

  (2) 工具执行的进度信息

  因为 NULLIF(0, 0) 的结果为 NULL，而 id 又为自增主键，所以每次执行的都是插入操作，不是更新操作。

  (3) 心跳信息

  因 NULLIF(1, 0) 的值为 1，所以每次执行的都是更新操作。

心跳信息用于判断主从的延迟情况，比 OSC 使用的 Seconds_Behind_Master 更可靠，而且粒度更小（可精确到毫秒）。

主从延迟的阈值由 `-max-lag-millis` 参数控制，默认为 1500，即 1.5 秒。

心跳信息的更新频率由 `-heartbeat-interval-millis` 参数控制，默认为 100，即 0.1 秒。

```
92 Query    select /* gh-ost `sbtest`.`t1` */ `id` from `sbtest`.`t1` order by `id` asc limit 1
92 Query    select /* gh-ost `sbtest`.`t1` */ `id` from `sbtest`.`t1` order by `id` desc limit 1
92 Query    show global status like 'Threads_running'

92 Query    insert /* gh-ost */ into `sbtest`.`_t1_ghc` (id, hint, value) values (NULLIF(0, 0), 'copy
iteration 0 at 1640612544', 'Copy: 0/10120 0.0%; Applied: 0; Backlog: 0/1000; Time: 3s(total), 0s(copy);
streamer: mysql-bin.000007:1269995; Lag: 0.09s, HeartbeatLag: 0.98s, State: migrating; ETA: N/A') on
duplicate key update last_update=NOW(), value=VALUES(value)
92 Query    insert /* gh-ost */ into `sbtest`.`_t1_ghc` (id, hint, value) values (NULLIF(1, 0),
'heartbeat', '2021-12-27T21:42:24.136033615+08:00') on duplicate key update last_update=NOW(),
value=VALUES(value)

92 Query    select  /* gh-ost `sbtest`.`t1` iteration:0 */ `id` from `sbtest`.`t1` where ((`id` >
_binary'1') or ((`id` = _binary'1'))) and ((`id` < _binary'10000') or ((`id` = _binary'10000'))) order
by `id` asc limit 1 offset 2999
92 Query    START TRANSACTION
92 Query    SET SESSION time_zone = 'SYSTEM', sql_mode = CONCAT(@@session.sql_mode,
',,NO_AUTO_VALUE_ON_ZERO,STRICT_ALL_TABLES')
92 Query    insert /* gh-ost `sbtest`.`t1` */ ignore into `sbtest`.`_t1_gho` (`id`, `name`) (select `id`,
`name` from `sbtest`.`t1` force index (`PRIMARY`) where (((`id` > _binary'1') or ((`id` = _binary'1')))
and ((`id` < _binary'3000') or ((`id` = _binary'3000')))) lock in share mode )
92 Query    COMMIT
```

开始进行数据的拷贝，实现方式与 OSC 一样，都是分 chunk 拷贝，且 chunk 上加共享锁。不一样的是，OSC 基于 `chunk-time` 自动调节 chunk 的大小，而 gh-ost 中 chunk 的大小是固定的，由参数 `-chunk-size` 决定，默认为 1000。

```
92 Query    START TRANSACTION
92 Query    select connection_id()
92 Query    select get_lock('gh-ost.92.lock', 0)
92 Query    set session lock_wait_timeout:=6
94 Query    show /* gh-ost */ table status from `sbtest` like '_t1_del'

94 Query    create /* gh-ost */ table `sbtest`.`_t1_del` ( id int auto_increment primary key )
engine=InnoDB comment='ghost-cut-over-sentry'

92 Query    lock /* gh-ost */ tables `sbtest`.`t1` write, `sbtest`.`_t1_del` write

95 Query    insert /* gh-ost */ into `sbtest`.`_t1_ghc` (id, hint, value) values (NULLIF(2, 0), 'state',
'AllEventsUpToLockProcessed:1640612547819766764') on duplicate key update last_update=NOW(),
value=VALUES(value)

94 Query    insert /* gh-ost */ into `sbtest`.`_t1_ghc` (id, hint, value) values (NULLIF(0, 0), 'state
at 1640612547821738591', 'AllEventsUpToLockProcessed:1640612547819766764') on duplicate key update
last_update=NOW(), value=VALUES(value)

95 Query    show global status like 'Threads_running'
```

```
95 Query    START TRANSACTION
95 Query    select connection_id()
95 Query    set session lock_wait_timeout:=3
95 Query    rename /* gh-ost */ table `sbtest`.`t1` to `sbtest`.`_t1_del`, `sbtest`.`_t1_gho` to
`sbtest`.`t1`

94 Query    select id from information_schema.processlist where id != connection_id() and 95 in (0, id)
and state like concat('%', 'metadata lock', '%') and info like concat('%', 'rename', '%')
94 Query    select is_used_lock('gh-ost.92.lock')

92 Query    drop /* gh-ost */ table if exists `sbtest`.`_t1_del`
92 Query    unlock tables
92 Query    ROLLBACK
95 Query    ROLLBACK
94 Query    drop /* gh-ost */ table if exists `sbtest`.`_t1_ghc`
92 Quit
95 Quit
94 Quit
```

在完成数据的拷贝后，开始进行 RENAME 操作，这个阶段在 gh-ost 中也称为 cut-over。处理过程如下。

- 创建辅助表 _t1_del。
- 对该表及原表执行 LOCK WRITE 操作。此时，会阻塞原表的所有操作，包括查询。
- 修改 _t1_ghc 表中 id 为 2 的那一行，将其 hint 修改为 AllEventsUpToLockProcessed:1640612547819766764。这里，重点是 AllEventsUpToLockProcessed。gh-ost 在读取 binlog 的过程中，如果碰到这个 hint，就意味着 binlog 中针对该表的操作已经被 gh-ost 执行完，此时可执行 RENAME 操作。
- 执行 RENAME 操作，该操作同样会被阻塞。
- 新建一个连接，用于监控 processlist id 为 95 的连接是否在执行 RENAME 操作。
- 一旦检测到，则通知之前执行 LOCK WRITE 的会话删除 _t1_del 表。此时，RENAME 操作依旧会被阻塞。
- 执行解锁操作，此时，RENAME 操作不再被阻塞。这里有个隐含的知识点，同样因为元数据锁被阻塞，RENAME 操作的优化级高于 DML。这就意味着，即使 DML 先于 RENAME 操作执行，但在解锁后，RENAME 操作将优先得到处理。

可能有人会好奇：为什么要这么处理？为什么不能像 OSC 那样直接对两张表进行 RENAME 操作？究其原因，还是与两者的实现方式有关：OSC 是基于触发器的，增量数据的处理是实时的；gh-ost 则基于 binlog，增量数据的处理是异步的，直接进行 RENAME 操作，很难保证 binlog 中的增量数据已处理完。

通过上面的分析，我们基本明白了 gh-ost 的实现原理，但上面的输出对应的只是一张"静态"表的变更，中间没有任何 DML 操作。对于增量数据，gh-ost 又是如何处理的呢？表 7-1 梳理了一些常见操作及 gh-ost 中对应的处理方式。

表 7-1 gh-ost 对于增量数据的处理方式

原表操作	gh-ost 的处理
insert into sbtest.t1 values(10001,'abc')	replace into `sbtest`.`_t1_gho` (`id`,`name`) values (10001, 'abc')
update sbtest.t1 set name='efg' where id=10003（基于主键进行更新）	update `sbtest`.`_t1_gho` set `id`=10003, `name`='efg' where ((`id` = 10003))
update sbtest.t1 set id=10002 where id=10001（对主键进行更新）	delete from `sbtest`.`_t1_gho` where ((`id` = 10001));replace into `sbtest`.`_t1_gho` (`id`,`name`) values (10002, 'abc')
delete from sbtest.t1 where id=10002	delete from `sbtest`.`_t1_gho` where ((`id` = 10002))

这里，id 是 sbtest.t1 的主键。

7.3.2 gh-ost 的参数解析

下面对常用参数进行分类说明。

1. 必需项

- **-alter**

指定 DDL 子句。

- **-database**

库名。

- **-table**

表名。

- **-execute**

执行表结构变更操作。如果不指定该参数，则默认为 noop（空操作）。noop 一般用于验证参数是否输入正确。

2. 工作模式相关

- **-allow-on-master**

直接读取主库的 binlog 来获取增量数据。对于没有从库或不想使用从库的场景，需指定该参数。

- **-migrate-on-replica**

修改从库的表结构。

- **-test-on-replica**

与 -migrate-on-replica 类似，也是修改从库的表结构，不一样的地方有两点。

- 会执行两次 RENAME 操作，相当于原表还是原表，没发生表结构的变更。
- 在进行 RENAME 操作之前，会关闭复制，即使整个过程执行完了，也不会开启复制。

这样，大家就可以在从库上验证两张表的数据是否一致。毕竟，对于一个新生工具，很多人还是持怀疑态度。如果不想 gh-ost 关闭复制，可指定 -test-on-replica-skip-replica-stop。

3. 性能相关

同 OSC 类似，gh-ost 也可根据当前的数据库负载来自动调整 DDL 的进度。

- `-max-load`

如果数据库负载超过 `-max-load` 指定的值，gh-ost 会暂停当前的所有操作，包括数据的拷贝和增量数据的应用，通常是设置 `Threads_running`，如 `-max-load=Threads_running=50`，每秒检测一次。

- `-critical-load, -critical-load-hibernate-seconds, -critical-load-interval-millis`

如果数据库负载超过 `-critical-load` 的设置，gh-ost 会直接退出。

可通过 `-critical-load-hibernate-seconds` 设置重试的间隔时间，重试期间会暂停当前的所有操作，没有重试次数限制。这一点比 OSC 更人性化，后者只要一达到 `--critical-load` 的阈值就会退出，不会重试。

也可设置 `-critical-load-interval-millis`。如果在该参数指定的时间内数据库负载依然超过 `-critical-load` 指定的值，则会退出。在这段时间内，依然是每秒检测一次数据库的负载情况。

- `-nice-ratio`

等待时间与 chunk 拷贝时间的比值。如果一个 chunk 的拷贝时间是 N 秒，则在这个 chunk 拷贝完成后，会等待 $N *$ `nice-ratio` 秒。默认为 `0`，即不等待。

- `-chunk-size`

chunk 的大小，默认为 `1000`，而 OSC 是基于 `chunk-time` 自动调节。

- `-dml-batch-size`

在应用 binlog event 时，将多少个事件作为一个事务提交，默认为 `10`。可选值为 1～100。

4. 主从相关

- `-max-lag-millis`

如果主从延迟超过 `-max-lag-millis` 指定的值，则暂停当前所有操作，默认为 `1500`，即 1.5 秒。

- `-throttle-control-replicas`

需要监控主从延迟的从库列表，如：

`-throttle-control-replicas="192.168.244.20:3306,192.168.244.30:3306"`

5. cut-over 相关

- `-cut-over`

指定 cut-over 的类型，默认是 atomic，即将原表和临时表的替换放到一个 RENAME 操作中完成。

另外一个选项是 two-step。

- **-cut-over-lock-timeout-seconds**

指定 cut-over 阶段 lock_wait_timeout 的值，默认为 3。对于 RENAME 操作，lock_wait_timeout 等于 cut-over-lock-timeout-seconds。对于 LOCK WRITE 操作，lock_wait_timeout 等于 cut-over-lock-timeout-seconds 的 2 倍，即 6 秒。

如果在给定的时间内，依然没有获取到元数据锁，则会间隔 1 秒后重试。间隔时间是固定的，无参数指定。重试次数由 -default-retries 参数指定，默认为 60。

注意，该参数可允许的最大值为 10，如果设置的值超过 10，则会重置为默认值 3。

6. 过程控制相关

- **-panic-flag-file**

如果指定文件存在，gh-ost 会直接退出，不做任何清理工作。

- **-throttle-flag-file**

如果指定文件存在，gh-ost 会暂停当前的所有操作。

- **-throttle-additional-flag-file**

作用同 -throttle-flag-file 一样。如果有暂停多个 gh-ost 操作的需求，可将 -throttle-additional-flag-file 设置为同一个文件。默认为 /tmp/gh-ost.throttle。

- **-postpone-cut-over-flag-file**

如果指定文件存在，gh-ost 会暂停 cut-over 阶段的操作。

有了这个选项，就可自定义 cut-over 的时间，对于 DBA 来说，DDL 操作将更加可控。

注意，设置了 -postpone-cut-over-flag-file，如果指定的文件不存在，gh-ost 会自动创建。

- **-throttle-query**

利用查询来判断是否需要暂停当前操作，如果查询的返回值大于 0，则暂停。

7. hook 相关

- **-hooks-path**

指定自定义脚本的路径。hook 可理解为插件。

gh-ost 针对不同阶段（具体可参考官方文档）开放了不同接口，使用者可自行开发脚本，实现邮件提醒、进度提醒等功能。脚本无语言要求，只要可执行且文件名以阶段名开头即可，比如 gh-ost-on-startup_123.sh、gh-ost-on-startup_456.py 等。

gh-ost 会以环境变量（具体可参考官方文档）的形式与脚本交互。

让 cut-over 在业务低峰期执行是一个比较常见的需求，但这个时间点很难控制，毕竟数据库负载和主从延迟都是动态变化的。这个时候,就可利用 postpone-cut-over-flag-file 加邮件提醒来自主决定 cut-over 的时间。

下面给出了一个简单的示例,实现了 cut-over 阶段进行邮件提醒的功能。该脚本已上传至 GitHub。

vim /gh-ost/scripts/gh-ost-on-before-cut-over.py

```python
#!/usr/bin/python3
# -*- coding: utf-8 -*-
import os,yagmail
GH_OST_MIGRATED_HOST=os.environ.get('GH_OST_MIGRATED_HOST')
GH_OST_DATABASE_NAME=os.environ.get('GH_OST_DATABASE_NAME')
GH_OST_TABLE_NAME=os.environ.get('GH_OST_TABLE_NAME')
GH_OST_DDL=os.environ.get('GH_OST_DDL')

message_head="Before cut-over: h=%s, D=%s, t=%s, DDL: %s"%(GH_OST_MIGRATED_HOST, GH_OST_DATABASE_NAME, GH_OST_TABLE_NAME, GH_OST_DDL)
yag = yagmail.SMTP(user='slowtech@126.com',password='******',host='smtp.126.com')
contents = [message_head]
yag.send('slowtech@126.com', message_head, contents)
```

当 gh-ost 执行到 cut-over 阶段时,会调用 -hooks-path 目录下以 gh-ost-on-before-cut-over 开头的脚本。

注意,gh-ost 只有执行到特定阶段,才会去执行相应阶段的脚本。如果该脚本存在问题(例如,在发送电子邮件时,电子邮件服务器存在问题),会导致脚本执行报错,gh-ost 也会异常退出。所以脚本中需做好相应的容错处理,建议无论是否出现问题,都将 0 作为脚本的返回值。

- **-hooks-hint**

对应 GH_OST_HOOKS_HINT 变量。

8. 其他选项

- **-default-retries**

操作的重试次数,默认为 60。

- **-allow-nullable-unique-key**

使用 gh-ost 的前提是目标表上必须存在主键或非空唯一键。如果表上没有主键,只有唯一键,且唯一键被定义为 NULL,此时 gh-ost 会直接退出。如果确定唯一键上没有 NULL 值,可指定该参数,让 gh-ost 正常处理。

- **-approve-renamed-columns**

如果 DDL 修改列名,则在执行 gh-ost 时会提示以下信息。此时,可指定该参数,允许修改列名。

```
FATAL gh-ost believes the ALTER statement renames columns, as follows: map[c1:c2]; as precaution, you
are asked to confirm gh-ost is correct, and provide with `--approve-renamed-columns`, and we're all happy.
Or you can skip renamed columns via `--skip-renamed-columns`, in which case column data may be lost
```

如果只是修改列名,个人更倾向于使用 Online DDL,毕竟它只会修改表的元数据信息。

- `-ok-to-drop-table`

默认情况下，在完成 DDL 后，gh-ost 不会删除原表，以避免删表操作给数据库的性能带来负面影响，尤其在缓冲池很大的情况下，可能会造成数据库"夯（hang）住"（MySQL 8.0.23 改善了 DROP/TRUNCATE 操作的性能），所以建议手动删除。如果需要自动删除，需指定 `-ok-to-drop-table`。

- `-initially-drop-old-table`

对同一张表再次执行 DDL 时，会检测之前的 del 表是否存在，如果存在，gh-ost 会直接退出。这个时候，可指定 `-initially-drop-old-table` 来删除它。

其他类似的选项还有 `-initially-drop-ghost-table`、`-initially-drop-socket-file`。正常情况下，gh-ost 在执行完毕后，会删除 DDL 执行过程中创建的 gho 表和套接字文件。但如果 gh-ost 中途异常退出，则不会删除。对同一张表再次执行 DDL 时，同样会检测 gho 表和套接字文件是否存在。这个时候，可指定 `-initially-drop-ghost-table`、`-initially-drop-socket-file` 来删除它们。

最后，给出官方的命令示例。当然，见仁见智，经过上面的讲解，相信大家对于参数的设置会有自己的判断和取舍。

```
gh-ost --max-load=Threads_running=25 --critical-load=Threads_running=1000 --chunk-size=1000
--throttle-control-replicas="myreplica.1.com,myreplica.2.com" --max-lag-millis=1500 --user="gh-ost"
--password="123456" --host=master.with.rbr.com --allow-on-master --database="my_schema"
--table="my_table" --verbose --alter="engine=innodb" --switch-to-rbr --allow-master-master
--cut-over=default --exact-rowcount --concurrent-rowcount --default-retries=120
--panic-flag-file=/tmp/ghost.panic.flag --postpone-cut-over-flag-file=/tmp/ghost.postpone.flag [--execute]
```

7.3.3 与 gh-ost 进行交互

上面介绍的可自定义 cut-over 时间足够吸引人吧？但 gh-ost 的惊艳之处并不只有这一点。实际上，在 gh-ost 运行期间，还能动态修改运行参数、发送控制命令等。

gh-ost 支持两种交互方式。

- 进程间通信。gh-ost 在启动时，默认会创建一个套接字文件（默认为 /tmp/gh-ost.dnname.tablename.sock），也可通过 `-serve-socket-file` 参数显式指定文件名。
- 网络通信。需通过 `-serve-tcp-port` 参数指定端口，默认不开启。

gh-ost 支持的操作可通过以下命令查看。

```
# 针对进程间通信
# echo help | nc -U /tmp/gh-ost.sbtest.t1.sock

# 针对网络通信
# echo help | nc 127.0.0.1 13306
```

下面列举了一些常见的使用场景。

- 查看 gh-ost 当前的执行状态。

```
# echo status | nc -U /tmp/gh-ost.sbtest.t1.sock
```

- 查看某个参数的值。

  ```
  # echo "max-load=?" | nc -U /tmp/gh-ost.sbtest.t1.sock
  Threads_running=50
  ```

- 修改某个参数的值。

  ```
  # echo "max-load=Threads_running=100" | nc -U /tmp/gh-ost.sbtest.t1.sock
  ```

- 让 gh-ost 暂停执行（throttle）、继续执行（no-throttle）、执行 cut-over 操作（unpostpone）、直接退出（panic）等。例如：

  ```
  # echo throttle | nc -U /tmp/gh-ost.sbtest.t1.sock
  ```

7.3.4 gh-ost 的优缺点

下面我们看看 gh-ost 的优缺点。

gh-ost 的优点如下。

- 无触发器。

  无触发器赋予了 gh-ost 极大的灵活性，主要体现在以下两点。

  - 所有操作都可暂停。无论是数据拷贝，还是增量数据的应用。
  - 对业务无侵入。不像触发器，会将业务操作和触发动作放到一个事务内执行。

- 可动态修改 gh-ost 的运行参数。

  动态可修改，意味着使用者对 gh-ost 有极大的控制权。

- 可在从库测试。

  这对很多对 gh-ost 好奇，但存在疑惑的人来说，是一个很好的切入点。

gh-ost 的限制如下。

- 目前不支持外键、触发器。
- 主键中不允许有 JSON 列。
- 表上必须存在主键或非空唯一键。

gh-ost 的缺点如下。

- 需消耗更多的网络流量。毕竟，二进制日志事件只有在读取后，才知道发生的修改是不是与目标表有关。这就意味着，在 DDL 执行期间，如果产生了 8GB 的 binlog，就有 8GB 流量的数据发送给 gh-ost 解析。从这一点来看，暂停 cut-over 并不是没有代价的。
- gh-ost 毕竟是个新工具，很多人对其准确性还是心存疑惑。如非必要，很多人不会轻易放弃自己熟悉的工具。所以从接触到建立信任，还需要一段时间。

7.4 元数据锁

在线上进行 DDL 操作时,相对于其可能带来的系统负载,其实我们最担心的还是元数据锁可能导致的阻塞问题。

一旦 DDL 操作因获取不到元数据锁而被阻塞,后续针对该表的所有操作都会被阻塞。下面是一个典型的例子,如果线上业务操作比较频繁,我们很快就会看到 Threads_running 飙升、CPU 告警。

```
mysql> show processlist;
+----+-----------------+-----------+------+---------+------+-------------------------------+---------------------------------------------+
| Id | User            | Host      | db   | Command | Time | State                         | Info                                        |
+----+-----------------+-----------+------+---------+------+-------------------------------+---------------------------------------------+
|  5 | event_scheduler | localhost | NULL | Daemon  |  273 | Waiting on empty queue        | NULL                                        |
|  9 | root            | localhost | NULL | Sleep   |  120 |                               | NULL                                        |
| 10 | root            | localhost | NULL | Query   |   70 | Waiting for table metadata lock | alter table sbtest.t1 add c1 datetime     |
| 11 | root            | localhost | NULL | Query   |   55 | Waiting for table metadata lock | select * from sbtest.t1 where id=1        |
| 12 | root            | localhost | NULL | Query   |    2 | Waiting for table metadata lock | select * from sbtest.t1 where id=2        |
| 13 | root            | localhost | NULL | Query   |    0 | init                          | show processlist                            |
+----+-----------------+-----------+------+---------+------+-------------------------------+---------------------------------------------+
6 rows in set (0.00 sec)
```

如果发生在线上,这无疑会严重影响业务。所以,一般建议将 DDL 操作放到业务低峰期来执行。这其实有 3 方面的考虑。

- 避免加重线上数据库的负载。
- 降低 DDL 被阻塞的概率。
- DDL 是个不可逆的操作。如果发布到线上后才暴露出问题,则很难回退。放到业务低峰期执行,即使出现了问题,影响也相对较小。

7.4.1 元数据锁引入的背景

元数据锁是 MySQL 5.5.3 引入的,主要用来解决下面两个问题。

1. RR 事务隔离级别下不可重复读的问题

看看下面这个示例,测试版本为 MySQL 5.5.0。

```
session1> create table sbtest.t1(id int primary key,name varchar(10)) engine=innodb;
Query OK, 0 rows affected (0.01 sec)

session1> insert into sbtest.t1 values(1,'a'),(2,'b');
Query OK, 2 rows affected (0.00 sec)
Records: 2  Duplicates: 0  Warnings: 0

session1> select @@tx_isolation;
+-----------------+
| @@tx_isolation  |
+-----------------+
| REPEATABLE-READ |
+-----------------+
1 row in set (0.00 sec)

session1> begin;
Query OK, 0 rows affected (0.00 sec)
```

```
session1> select * from t1;
+----+------+
| id | name |
+----+------+
|  1 | a    |
|  2 | b    |
+----+------+
2 rows in set (0.00 sec)

session2> alter table t1 add c1 datetime;
Query OK, 2 rows affected (0.02 sec)
Records: 2  Duplicates: 0  Warnings: 0

session1> select * from t1;
Empty set (0.00 sec)
```

虽然是 RR 隔离级别，但在开启事务的情况下，对于同一个查询，两次的结果却不一致。

2. 主从复制问题

看看下面这个示例。

```
(master) session1> create table t1(id int,name varchar(10)) engine=innodb;
Query OK, 0 rows affected (0.00 sec)

(master) session1> begin;
Query OK, 0 rows affected (0.00 sec)

(master) session1> insert into t1 values(1,'a');
Query OK, 1 row affected (0.00 sec)

(master) session2> truncate table t1;
Query OK, 0 rows affected (0.00 sec)

(master) session1> commit;
Query OK, 0 rows affected (0.00 sec)

(master) session1> select * from t1;
Empty set (0.00 sec)
```

主库的查询结果为空。

再来看看从库的查询结果，竟然有一条记录。

```
(slave) session1> select * from t1;
+------+------+
| id   | name |
+------+------+
|    1 | a    |
+------+------+
1 row in set (0.00 sec)
```

分析一下主库 binlog 的内容，可以看到，TRUNCATE 操作记录在前，INSERT 操作记录在后。

```
# at 279
#220101 22:24:48 server id 1  end_log_pos 398    Query    thread_id=10    exec_time=0    error_code=0
use sbtest/*!*/;
SET TIMESTAMP=1641047088/*!*/;
create table t1(id int,name varchar(10)) engine=innodb
```

```
/*!*/;
# at 398
#220101 22:25:15 server id 1  end_log_pos 468      Query    thread_id=11    exec_time=0    error_code=0
SET TIMESTAMP=1641047115/*!*/;
BEGIN
/*!*/;
# at 468
#220101 22:25:15 server id 1  end_log_pos 550      Query    thread_id=11    exec_time=0    error_code=0
SET TIMESTAMP=1641047115/*!*/;
truncate table t1
/*!*/;
# at 550
#220101 22:25:15 server id 1  end_log_pos 577      Xid = 35
COMMIT/*!*/;
# at 577
#220101 22:25:25 server id 1  end_log_pos 647      Query    thread_id=10    exec_time=0    error_code=0
SET TIMESTAMP=1641047125/*!*/;
BEGIN
/*!*/;
# at 647
#220101 22:25:01 server id 1  end_log_pos 740      Query    thread_id=10    exec_time=0    error_code=0
SET TIMESTAMP=1641047101/*!*/;
insert into t1 values(1,'a')
/*!*/;
# at 740
#220101 22:25:25 server id 1  end_log_pos 767      Xid = 28
COMMIT/*!*/;
```

7.4.2 元数据锁的基本概念

元数据锁出现的初衷是为了保护处于事务中的表，使其结构不被修改。这里提到的事务包括两类：显式事务和 AC-NL-RO（auto-commit non-locking read-only）事务。显式事务包括两类：关闭 autocommit（自动提交）下的操作，以及通过 START TRANSACTION 或 BEGIN 开启的事务。AC-NL-RO 可理解为开启 autocommit 下的 SELECT 操作。另外，元数据锁是事务级别的，只有在事务结束后才会释放。在此之前，其实也有类似的保护机制，只不过是语句级别的。

需要注意的是，元数据锁不仅适用于表，也适用于其他对象，如表 7-2 所示，表中的等待状态对应 SHOW PROCESSLIST 中的 State。

表 7-2　元数据锁的适用对象及对应 SHOW PROCESSLIST 中的状态

适用对象	等待状态
tablespace	Waiting for tablespace metadata lock
schema	Waiting for schema metadata lock
table	Waiting for table metadata lock
function	Waiting for stored function metadata lock
procedure	Waiting for stored procedure metadata lock
trigger	Waiting for trigger metadata lock
event	Waiting for event metadata lock
resource groups	Waiting for resource groups metadata lock
foreign key	Waiting for foreign key metadata lock
check constraint	Waiting for check constraint metadata lock

为了提高数据库的并发度，元数据锁被细分为了 11 种类型。

- `MDL_INTENTION_EXCLUSIVE`
- `MDL_SHARED`
- `MDL_SHARED_HIGH_PRIO`
- `MDL_SHARED_READ`
- `MDL_SHARED_WRITE`
- `MDL_SHARED_WRITE_LOW_PRIO`
- `MDL_SHARED_UPGRADABLE`
- `MDL_SHARED_READ_ONLY`
- `MDL_SHARED_NO_WRITE`
- `MDL_SHARED_NO_READ_WRITE`
- `MDL_EXCLUSIVE`

常用的有 `MDL_SHARED_READ`、`MDL_SHARED_WRITE` 和 `MDL_EXCLUSIVE`，分别对应 SELECT、DML 及 DDL 操作。其他类型的对应操作可参考源码 sql/mdl.h。

这里重点说说 `MDL_EXCLUSIVE`。`MDL_EXCLUSIVE` 是独占锁，持有此锁的连接可以修改表的表结构和数据。在其持有期间不允许授予其他类型的元数据库锁，自然也包括 SELECT 和 DML 操作。这也就是为什么 DDL 操作被阻塞时，后续针对该表的其他操作也会被阻塞。

需要注意的是，如果一条 SQL 语句在语法上有效，但执行时报错（如列名不存在），同样会获取元数据锁，直到事务结束才释放。

7.4.3 在 MySQL 5.7 和 8.0 中如何定位 DDL 被阻塞的问题

在 MySQL 5.7 和 8.0 中，如果我们要定位 DDL 被阻塞的问题，可直接使用 sys 中的 `schema_table_lock_waits` 视图。

看看下面这个示例。

```
session1> create table sbtest.t1(id int primary key,name varchar(10));
Query OK, 0 rows affected (0.02 sec)

session1> insert into sbtest.t1 values(1,'a');
Query OK, 1 row affected (0.01 sec)

session1> begin;
Query OK, 0 rows affected (0.00 sec)

session1> select * from sbtest.t1;
+----+------+
| id | name |
+----+------+
|  1 | a    |
+----+------+
1 row in set (0.00 sec)
```

```
session2> alter table sbtest.t1 add c1 datetime;

session3> show processlist;
+----+-----------------+-----------+------+---------+-------+--------------------------------+------------------------------------------+
| Id | User            | Host      | db   | Command | Time  | State                          | Info                                     |
+----+-----------------+-----------+------+---------+-------+--------------------------------+------------------------------------------+
|  5 | event_scheduler | localhost | NULL | Daemon  | 47628 | Waiting on empty queue         | NULL                                     |
| 24 | root            | localhost | NULL | Sleep   |    11 |                                | NULL                                     |
| 25 | root            | localhost | NULL | Query   |     5 | Waiting for table metadata lock | alter table sbtest.t1 add c1 datetime   |
| 26 | root            | localhost | NULL | Query   |     0 | init                           | show processlist                         |
+----+-----------------+-----------+------+---------+-------+--------------------------------+------------------------------------------+
4 rows in set (0.00 sec)
```

session2 中的 ALTER 操作被 session1 中的事务读操作所阻塞。

接下来，我们看看 sys.schema_table_lock_waits 的输出。

```
mysql> select * from sys.schema_table_lock_waits\G
*************************** 1. row ***************************
               object_schema: sbtest
                 object_name: t1
           waiting_thread_id: 62
                 waiting_pid: 25
             waiting_account: root@localhost
           waiting_lock_type: EXCLUSIVE
       waiting_lock_duration: TRANSACTION
               waiting_query: alter table sbtest.t1 add c1 datetime
          waiting_query_secs: 17
  waiting_query_rows_affected: 0
  waiting_query_rows_examined: 0
          blocking_thread_id: 61
                blocking_pid: 24
            blocking_account: root@localhost
          blocking_lock_type: SHARED_READ
      blocking_lock_duration: TRANSACTION
      sql_kill_blocking_query: KILL QUERY 24
 sql_kill_blocking_connection: KILL 24
*************************** 2. row ***************************
               object_schema: sbtest
                 object_name: t1
           waiting_thread_id: 62
                 waiting_pid: 25
             waiting_account: root@localhost
           waiting_lock_type: EXCLUSIVE
       waiting_lock_duration: TRANSACTION
               waiting_query: alter table sbtest.t1 add c1 datetime
          waiting_query_secs: 17
  waiting_query_rows_affected: 0
  waiting_query_rows_examined: 0
          blocking_thread_id: 62
                blocking_pid: 25
            blocking_account: root@localhost
          blocking_lock_type: SHARED_UPGRADABLE
      blocking_lock_duration: TRANSACTION
      sql_kill_blocking_query: KILL QUERY 25
 sql_kill_blocking_connection: KILL 25
2 rows in set (0.00 sec)
```

只有一个 ALTER 操作，却产生了两条记录，而且两条记录要杀掉的对象还不一样，其中一条记录要杀掉的对象还是 ALTER 操作本身。如果对表结构不熟悉或不仔细看记录内容，难免会杀错对象。

不仅如此，在 DDL 操作被阻塞后，如果后续有 N 个查询被 DDL 操作阻塞，还会产生 2N 条记录。在定位问题时，这 2N 条记录完全是噪声。

这个时候，就需要我们对上述记录进行过滤了。过滤的关键是 blocking_lock_type 不等于 SHARED_UPGRADABLE。SHARED_UPGRADABLE 是一个可升级的共享元数据锁，加锁期间，允许并发查询和更新，常用在 DDL 操作的第一阶段。所以，阻塞 DDL 的不会是 SHARED_UPGRADABLE。

所以，针对上面这个示例，我们可以通过下面这个查询来精确地定位需要 kill 的会话。

```
SELECT sql_kill_blocking_connection
FROM sys.schema_table_lock_waits
WHERE blocking_lock_type <> 'SHARED_UPGRADABLE'
    AND waiting_query = 'alter table sbtest.t1 add c1 datetime';
```

注意，sys.schema_table_lock_waits 视图依赖了一张数据锁相关的表——performance_schema.metadata_locks，该表是 MySQL 5.7 才引入的，会显示元数据锁的相关信息，包括作用对象、锁的类型及锁的状态等。

但在 MySQL 5.7 中，该表默认为空，因为与之相关的事件采集配置项默认没有开启。MySQL 8.0 才默认开启相关的事件采集配置项。

```
mysql> select * from performance_schema.setup_instruments where name='wait/lock/metadata/sql/mdl';
+----------------------------+---------+-------+
| NAME                       | ENABLED | TIMED |
+----------------------------+---------+-------+
| wait/lock/metadata/sql/mdl | NO      | NO    |
+----------------------------+---------+-------+
1 row in set (0.00 sec)
```

所以，在 MySQL 5.7 中，如果我们要使用 sys.schema_table_lock_waits，必须首先开启元数据锁相关的事件采集配置项。开启方式很简单，直接修改 performance_schema.setup_instruments 表即可，具体的 SQL 语句如下。

```
UPDATE performance_schema.setup_instruments SET ENABLED = 'YES', TIMED = 'YES'
WHERE NAME = 'wait/lock/metadata/sql/mdl';
```

但这种方式是临时生效的，实例重启后，又会恢复为默认值。

建议同步修改配置文件，修改方式如下。

```
[mysqld]
performance-schema-instrument='wait/lock/metadata/sql/mdl=ON'
```

7.4.4　在 MySQL 5.6 中如何定位 DDL 被阻塞的问题

sys.schema_table_lock_waits 是 MySQL 5.7 才引入的。但在实际生产环境中，MySQL 5.6 还是占有相当大的份额。如何解决 MySQL 5.6 的这个痛点呢？

细究下来，导致 DDL 被阻塞的操作无非以下两类。

- 表上有慢查询未结束。
- 表上有事务未提交。

第一类比较好定位，通过 SHOW PROCESSLIST 就能发现。而第二类仅凭 SHOW PROCESSLIST 很难定位，因为未提交事务的连接在 SHOW PROCESSLIST 中的命令（Command）类型同空闲连接一样，都是 Sleep。

所以，网上有杀掉空闲连接的说法，其实也不无道理，但这样做就太简单粗暴了，难免会"误杀"。

其实，既然是事务，在 information_schema.innodb_trx 中肯定会有记录，如 session1 中的事务在表中的记录如下：

```
mysql> select * from information_schema.innodb_trx\G
*************************** 1. row ***************************
                    trx_id: 421568246406360
                 trx_state: RUNNING
               trx_started: 2022-01-02 08:53:50
     trx_requested_lock_id: NULL
          trx_wait_started: NULL
                trx_weight: 0
       trx_mysql_thread_id: 24
                 trx_query: NULL
       trx_operation_state: NULL
         trx_tables_in_use: 0
         trx_tables_locked: 0
          trx_lock_structs: 0
     trx_lock_memory_bytes: 1128
           trx_rows_locked: 0
         trx_rows_modified: 0
   trx_concurrency_tickets: 0
       trx_isolation_level: REPEATABLE READ
         trx_unique_checks: 1
    trx_foreign_key_checks: 1
    trx_last_foreign_key_error: NULL
   trx_adaptive_hash_latched: 0
   trx_adaptive_hash_timeout: 0
          trx_is_read_only: 0
trx_autocommit_non_locking: 0
       trx_schedule_weight: NULL
1 row in set (0.00 sec)
```

其中 trx_mysql_thread_id 是线程 ID，结合 information_schema.processlist，可进一步缩小范围。

所以，我们可以通过下面这条 SQL 语句定位执行时间早于 DDL 的事务。

```
SELECT concat('kill ', i.trx_mysql_thread_id, ';')
FROM information_schema.innodb_trx i, (
    SELECT MAX(time) AS max_time
    FROM information_schema.processlist
    WHERE state = 'Waiting for table metadata lock'
      AND (info LIKE 'alter%'
      OR info LIKE 'create%'
      OR info LIKE 'drop%'
      OR info LIKE 'truncate%'
      OR info LIKE 'rename%'
)) p
WHERE timestampdiff(second, i.trx_started, now()) > p.max_time;
```

可喜的是，当前正在执行的查询也会显示在 information_schema.innodb_trx 中。

所以，上面这条 SQL 语句同样也适用于慢查询未结束的场景。

7.5 本章总结

本章围绕 DDL 展开，重点介绍了变更表结构的 3 种常用方式：Online DDL、pt-online-schema-change 和 gh-ost。可以看到，每种方式都有各自的适用场景。

对于以下操作，由于只涉及表元数据的变更，建议直接通过 ALTER 语句操作（Online DDL）。

- 删除二级索引
- 修改索引名（到 MySQL 5.7 才支持）
- 修改字段名
- 设置（删除）字段的默认值
- 增加 VARCHAR 的长度
- 修改自增主键的值
- 修改表的统计信息相关选项
- 重命名表

如果使用的是 MySQL 8.0，对于新增字段，也推荐使用 Online DDL。

对于其他操作，一律建议使用 pt-online-schema-change 或 gh-ost。

无论是 pt-online-schema-change 还是 gh-ost，在进行表结构变更时，都会涉及以下两个步骤。

- 全量数据的同步。两者都是以 chunk 为单位分批拷贝的。需要注意的是，在拷贝的过程中，会对这个 chunk 加共享锁，加锁期间，会阻塞针对该 chunk 的 DML 操作。
- 增量数据的同步。pt-online-schema-change 是通过触发器来实现的，而 gh-ost 则是通过 binlog 来实现的。相对而言，前者对业务有一定的侵入性，毕竟，触发器会将业务 SQL 和触发动作放到同一个事务空间内执行。不仅如此，在高负载环境中，触发器方案还会加重数据库的负载。

介绍完变更表结构的 3 种方式，我们重点介绍了元数据锁，包括其基本概念、引入的背景、类型及影响，并在此基础上分析了在 MySQL 5.6 和 MySQL 5.7（8.0）中如何定位 DDL 被阻塞的问题。

最后，我们看看 3 种表结构变更方式获取元数据锁的时间点。

- Online DDL：Prepare 阶段和 Commit 阶段。
- pt-online-schema-change：创建触发器及执行 RENAME 操作时。
- gh-ost：执行 LOCK WRITE 及 RENAME 操作时。

重点问题回顾

- Online DDL 的实现原理。
- 如何检查 Online DDL 的进度。

- innodb_online_alter_log_max_size 参数的作用。
- Online DDL 的适用场景。
- pt-online-schema-change 的实现原理。
- 如果 pt-online-schema-change 异常退出,在清理现场时,先删除触发器还是临时表?
- gh-ost 的实现原理。
- gh-ost 怎么保证,在执行 RENAME 操作时,binlog 中的增量数据已经处理完?
- 元数据锁的概念、类型、引入的背景及影响。
- 在 MySQL 5.7 和 8.0 中如何定位 DDL 被阻塞的问题。
- 在 MySQL 5.6 中如何定位 DDL 被阻塞的问题。

第 8 章

连接池和线程池

对于 MySQL 来说，传统的线程调度方式是每个连接一个线程（one-thread-per-connection）。连接和线程是紧密联系在一起的，以至于连接池和线程池这两个概念经常被混淆。

数据库连接池是在应用端（MySQL 客户端）实现的，它缓存了应用端到数据库服务端的连接。当应用端执行 SQL 操作，需要数据库连接时，直接从连接池中获取，无须新建。使用完毕后，应用端将连接归还给连接池，供其他操作复用。

使用连接池有如下显而易见的好处。

- 避免频繁创建和销毁连接带来的资源开销。
- 减少系统整体响应时间。在执行 SQL 操作时，只是复用之前的连接，避免了新建连接带来的时间开销。
- 可对连接进行统一管理，避免数据库连接泄漏。

而数据库线程池则是在 MySQL 服务端实现的，与传统的每个连接一个线程相比，它只会创建一定数量的线程来响应客户端的连接请求，每个线程对应多个客户端连接。

使用线程池同样有如下显而易见的好处。

- 避免线程的频繁创建和销毁，节省系统资源。
- 将服务端的线程数控制在一定范围内，避免了高并发场景下线程频繁的上下文切换。

所以，连接池和线程池没有必然的联系，在使用连接池的场景中也可使用线程池。

本章主要包括以下内容。

- 连接池。
- MySQL 线程池。
- MySQL server has gone away 深度解析。

8.1 连接池

8.1.1 连接池的运行原理

连接池在初始化时，会创建一定数量（`initialPoolSize`）的连接。当有数据库操作时，会从连接池中取出一个连接，如果连接池中没有空闲的连接，则会判断已创建的连接数是否达到了最大连接数（maxPoolSize）的限制。如果没有达到最大连接数的限制，连接池会创建新的连接，通常是一次创建多个（acquireIncrement）。如果已创建的连接数达到了最大连接数的限制，操作会等待其他操作释放连接。如果在一定时间（checkoutTimeout）内没有连接释放，操作会直接报错。

连接在使用完毕后，会被归还给连接池。如果连接池中连接的空闲时间超过一定的值（maxIdleTime），则会被销毁。如果连接池中连接的数量小于一定的值（minPoolSize），连接池会自动创建新的连接，以满足 minPoolSize 的要求。

8.1.2 常用的 JDBC 连接池

常用的 JDBC 连接池有：

- c3p0
- DBCP
- Druid
- Tomcat JDBC Pool
- HikariCP

DBCP 和 Tomcat JDBC Pool 是 Apache 开源的。Tomcat JDBC Pool 是 Tomcat 的默认连接池。

Druid 是阿里巴巴开源的，它不仅是一个数据库连接池，还可以监控数据库的访问性能，支持数据库密码加密等。

HikariCP 是目前风头最盛的 JDBC 连接池，号称性能最好。从 HikariCP 官网给出的压测结果（见图 8-1）来看，也确实如此，它在性能上远胜于 c3p0 和 DBCP2。Spring Boot 2.0 也将 HikariCP 作为默认的数据库连接池。

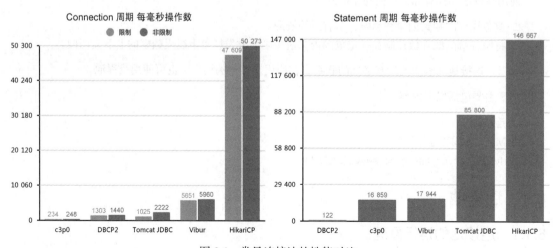

图 8-1 常见连接池的性能对比

下面会重点介绍 c3p0 和 DBCP 的参数，因为这两个连接池的参数最有代表性。Druid 和 Tomcat JDBC Pool 的参数设置基本上是基于 DBCP 的。

8.1.3　c3p0 连接池

本节分类阐述 c3p0 连接池中常见配置的具体含义。

1. 基本配置

- `minPoolSize`

最小连接数，默认为 3。

- `maxPoolSize`

最大连接数，默认为 15。

在使用 c3p0 时，一个常见的错误是 `An attempt by a client to checkout a Connection has timed out.`，具体错误信息如下，

```
Exception in thread "main" java.sql.SQLException: An attempt by a client to checkout a Connection has timed out.
    at com.mchange.v2.sql.SqlUtils.toSQLException(SqlUtils.java:118)
    at com.mchange.v2.sql.SqlUtils.toSQLException(SqlUtils.java:77)
    at com.mchange.v2.c3p0.impl.C3P0PooledConnectionPool.checkoutPooledConnection(C3P0PooledConnectionPool.java:690)
    at com.mchange.v2.c3p0.impl.AbstractPoolBackedDataSource.getConnection(AbstractPoolBackedDataSource.java:140)
    at com.victor.c3p0.test.main(test.java:41)
Caused by: com.mchange.v2.resourcepool.TimeoutException: A client timed out while waiting to acquire a resource from com.mchange.v2.resourcepool.BasicResourcePool@1554909b -- timeout at awaitAvailable()
    at com.mchange.v2.resourcepool.BasicResourcePool.awaitAvailable(BasicResourcePool.java:1467)
    at com.mchange.v2.resourcepool.BasicResourcePool.prelimCheckoutResource(BasicResourcePool.java:644)
    at com.mchange.v2.resourcepool.BasicResourcePool.checkoutResource(BasicResourcePool.java:554)
    at com.mchange.v2.c3p0.impl.C3P0PooledConnectionPool.checkoutAndMarkConnectionInUse(C3P0PooledConnectionPool.java:758)
    at com.mchange.v2.c3p0.impl.C3P0PooledConnectionPool.checkoutPooledConnection(C3P0PooledConnectionPool.java:685)
    ... 2 more
```

出现这个错误的常见原因是连接池没有空闲连接了。这个时候，可通过下面这条 SQL 语句统计各个主机的连接数。通常来说，一个应用在一台主机上只会部署一个节点，所以，一个 IP 对应的连接数基本上就反映了该应用节点已经创建的连接数的大小。通过对比这个连接数与 `maxPoolSize` 的大小，就能判断上述报错是否是连接池中无可用连接导致的。

```
mysql> select substring_index(host,':',1) ip,count(*) cnt from information_schema.processlist group by substring_index(host,':',1);
+-----------------+-----+
| ip              | cnt |
+-----------------+-----+
| localhost       |   2 |
| 192.168.244.20  |   3 |
+-----------------+-----+
2 rows in set (0.00 sec)
```

- `initialPoolSize`

连接池启动时创建的初始连接数，默认为 3，取值应在 `minPoolSize` 和 `maxPoolSize` 之间。

- **acquireIncrement**

当连接池中的连接耗尽时,连接池一次创建几个新的连接,默认为 3。

因为创建连接是一个成本相对高昂的操作,通常比较慢,所以为了避免应用端等待太久,连接池通常会批量创建新的连接。

- **checkoutTimeout**

获取一个连接的最大等待时间,单位是毫秒。默认为 0,代表一直等待。如果在指定时间内没有获取到连接,则会抛出 SQLException 异常。

2. 控制连接时长

- **maxIdleTime**

最大空闲时间。如果连接的空闲时间超过了该值,则连接会被连接池销毁。默认为 0,代表不销毁。单位是秒。

- **maxConnectionAge**

最大存活时间。如果连接的存活时间超过该值,则连接会被连接池销毁。默认为 0,代表不销毁。单位是秒。

与 maxIdleTime 不一样的是,它衡量的是连接的总时长,包括查询执行的时间,而 maxIdleTime 衡量的只是连接处于空闲状态的时长。

注意,当前正在使用的连接不会被销毁,而是会等到它调用 close() 后再销毁。

- **maxIdleTimeExcessConnections**

在并发量比较高的情况下,连接池中的连接数会飙升。如果希望在高峰期后能快速释放空闲连接,可设置此参数。默认为 0。单位是秒。

与 maxIdleTime 不一样的是,maxIdleTimeExcessConnections 只在连接池中连接的数量大于 minPoolSize 时触发,而 maxIdleTime 在释放连接时,通常不会考虑 minPoolSize 的大小,如果因为释放连接导致连接池中连接的数量小于 minPoolSize,连接池会自动创建新的连接,以满足 minPoolSize 的要求。

如果同时设置了 maxIdleTime 参数,maxIdleTimeExcessConnections 的值需小于 maxIdleTime,否则没有效果。

3. 检查连接的有效性

- **idleConnectionTestPeriod**

每隔多少秒检查一次连接池中空闲连接的有效性。默认为 0。

- **preferredTestQuery**

自定义探测语句。对于 MySQL 数据库,常用的是 SELECT 1。

- **automaticTestTable**

自定义探测表，假如我将该选项设置为 c3p0_test，则在启动连接池时，会创建一张空表。

```
CREATE TABLE `c3p0_test` ( a CHAR(1) )
```

对应的探测语句如下：

```
SELECT * FROM `c3p0_test`
```

如果同时配置了 automaticTestTable 和 preferredTestQuery 选项，则前者会覆盖后者。

- **testConnectionOnCheckout**

每次在通过 getConnection() 方法申请连接时，检测连接的可用性。如果不可用，则重新申请新的连接。这是最可靠的连接测试方法，但对应用端的性能影响较大，一般不建议线上开启。

- **testConnectionOnCheckin**

每次在通过 close() 方法关闭连接时，检测连接的可用性。不同于 testConnectionOnCheckout 的实时检测，关闭连接时的检测是异步处理的。所以，相对于 testConnectionOnCheckout，它对性能的影响没有那么大，线上可开启。

- **forceSynchronousCheckins**

默认情况下，关闭连接时的检测是异步处理的，但在连接池比较繁忙的情况下，可能会导致应用端在获取新的连接时，需要等待连接池中空闲连接的检测完成。为了减轻连接池的拥塞，可将关闭连接时的检测操作设置为同步处理。

在 c3p0 v0.9.5 之前，如果没有配置 preferredTestQuery 或 automaticTestTable，但又设置了 idleConnectionTestPeriod、testConnectionOnCheckout 或 testConnectionOnCheckin，则默认会调用 DatabaseMetaData.getTables() 方法来检测连接的可用性。具体在 MySQL 层面，会执行如下查询：

```
SHOW FULL TABLES FROM `sbtest` LIKE 'PROBABLYNOT'
```

这个查询通常会比 SELECT 1 和 SELECT * FROM c3p0_test 慢，在并发量较高的情况下，会对 MySQL 的性能产生较大的负面影响。

所以，在 c3p0 v0.9.5 之前的版本中，建议配置 preferredTestQuery。

从 v0.9.5 开始，c3p0 支持 JDBC 4 API，可通过 Connection.isValid() 来检测连接的有效性。这个检测是在驱动层实现的，相对来说更高效、更可靠，推荐使用。

4. 重试选项

- **acquireRetryAttempts**

从连接池中获取连接失败时，重试的次数。默认为 30。如果该值小于或等于 0，则会一直重试。

- **acquireRetryDelay**

重试的时间间隔，默认为 1000，单位是毫秒，即 1 秒。

- **breakAfterAcquireFailure**

在重试了 acquireRetryAttempts 次后，如果依旧无法获取连接，是否将连接池标记为损坏。标记为损坏的连接池，后续即使连接恢复正常，也不能从中获取连接。这个时候，就只能重启应用。默认为 false，建议保持默认值。

5. 配置 PreparedStatement 缓存

在介绍具体参数之前，首先看看 PreparedStatement 相对于 Statement 的优点。

- 对于 PreparedStatement 语句，数据库会进行预编译处理。这样，后续指定参数实际执行的时候，会省去词法分析、语法分析、选择执行计划等阶段，从而提升查询性能。
- 防止 SQL 注入。

- **maxStatements**

能缓存的 PreparedStatement 的总数量。默认为 0，代表不缓存。

- **maxStatementsPerConnection**

单个连接能缓存的 PreparedStatement 的总数量。默认为 0，代表不缓存。

- **statementCacheNumDeferredCloseThreads**

如果该值大于 0，则连接池会延迟释放缓存的 PreparedStatement，直到它所依附的父连接没被任何客户端使用。

默认为 0。如果在关闭连接时，日志中出现 APPARENT DEADLOCKS，可将该参数设置为 1。

6. 处理不确定事务

这里的不确定事务指的是应用端将 autoCommit（自动提交）设置为 false，在关闭连接前，没有显式调用 rollback() 或 commit() 方法，导致事务的状态不确定。

- **autoCommitOnClose**

关闭连接前，是提交还是回滚不确定事务。默认为 false，代表回滚不确定事务。

- **forceIgnoreUnresolvedTransactions**

如果不希望连接池提交或回滚不确定事务，可将该参数设置为 true（但不建议这样做）。默认为 false。

7. 其他配置

其他配置可参考官方文档"c3p0 - JDBC3 Connection and Statement Pooling"。

8.1.4 DBCP 连接池

下面分类阐述 DBCP 连接池中常见配置的具体含义。

1. 基本配置

- `initialSize`

启动连接池时的初始连接数,相当于 c3p0 中的 `initialPoolSize`。默认为 `0`。

- `maxTotal`

最大连接数,相当于 c3p0 中的 `maxPoolSize`。默认为 `8`。如果设置为 `0` 或负数,则代表没有限制。

- `maxIdle`

最大空闲连接数。多余的空闲连接会被连接池销毁。默认为 `8`。

- `minIdle`

最小空闲连接数,默认为 `0`。

注意,`maxIdle` 和 `minIdle` 生效的前提是必须激活工作线程,所以必须设置 `timeBetweenEvictionRunsMillis` 参数。

- `maxWaitMillis`

获取一个连接时的最大等待时间,单位是毫秒。默认为 `indefinitely`,代表一直等待。相当于 c3p0 中的 `checkoutTimeout`。

2. 控制连接时长

- `minEvictableIdleTimeMillis`

最大空闲时间,默认为 `1800000`,单位是毫秒,即 30 分钟。相当于 c3p0 中的 `maxIdleTime`。

注意,空闲连接是被工作线程销毁的,所以如果要让该参数有效,必须同时设置 `timeBetweenEvictionRunsMillis`。

- `maxConnLifetimeMillis`

最大存活时间,默认为 `-1`,即不限制。相当于 c3p0 中的 `maxConnectionAge`。

3. 检查连接的有效性

- `validationQuery`

自定义探测语句,相当于 c3p0 中的 `preferredTestQuery`。若不设置,则使用驱动层的 `isValid()` 方法来验证连接的有效性。

- `validationQueryTimeout`

自定义探测语句的查询时长,默认不限制。

- `testOnCreate`

在创建连接时验证其有效性,默认为 `false`。

- **testOnBorrow**

在申请连接时验证其有效性，默认为 true，建议设置为 false。相当于 c3p0 中的 testConnectionOnCheckout。

- **testOnReturn**

在归还连接时验证其有效性，默认为 false。相当于 c3p0 中的 testConnectionOnCheckin。

- **testWhileIdle**

是否通过工作线程来定时检测连接池中空闲连接的有效性，默认为 false。建议开启。

- **timeBetweenEvictionRunsMillis**

工作线程多久被唤醒一次，单位是毫秒。默认为-1，即不唤醒。该线程会检测空闲连接的有效性，控制空闲连接的数量。建议设置。

- **numTestsPerEvictionRun**

每次检查连接的数量，默认为 3。

4. 配置 PreparedStatement 缓存

- **poolPreparedStatements**

是否缓存 PreparedStatement。默认为 false，即不缓存。

- **maxOpenPreparedStatements**

缓存 PreparedStatement 的数量，默认为 unlimited，即不限制。

5. 销毁泄漏连接

有些低质量的代码，在执行完 SQL 后，没有显式关闭连接，导致这些连接不能释放回连接池，从而造成了数据库连接池的"泄漏"。在严重的情况下，这会导致应用在申请连接时，无连接可用。

这些内部状态活跃，但在一定时间内（阈值由 removeAbandonedTimeout 参数确定，默认为 300 秒）没有进行任何操作的连接，会被 DBCP 连接池定义为"泄漏"连接。泄漏连接可被工作线程销毁。

- **removeAbandonedOnMaintenance、removeAbandonedOnBorrow**

默认为 false，即不会销毁泄漏连接。若设置为 true，则会通过工作线程来销毁泄漏连接。

如果设置的是 removeAbandonedOnMaintenance，则只要是泄漏连接，就会被工作线程销毁。如果设置的是 removeAbandonedOnBorrow，则在应用申请连接且满足以下条件的情况下才会销毁泄漏连接。

- getNumActive() > getMaxTotal() – 3
- getNumIdle() < 2

注意，连接销毁后，如果再继续通过这个连接执行 SQL 操作，会抛出以下异常。

```
Exception in thread "main" java.sql.SQLException: org.apache.commons.dbcp2.DelegatingStatement with
address: "NULL" is closed.
    at org.apache.commons.dbcp2.DelegatingStatement.checkOpen(DelegatingStatement.java:122)
    at org.apache.commons.dbcp2.DelegatingStatement.executeUpdate(DelegatingStatement.java:229)
    at org.apache.commons.dbcp2.DelegatingStatement.executeUpdate(DelegatingStatement.java:234)
    at DBCP.DBCPTest.main(DBCPTest.java:79)
```

6. 其他重要配置

- **connectionInitSqls**

建立物理连接后执行的初始化语句,类似于 MySQL 中的 `init_connect` 参数。

- **defaultQueryTimeout**

设置 SQL 默认的超时时长。如果 SQL 在指定的时间内没有完成,则会被连接池中断,并提示以下错误。默认为 null。

```
Exception in thread "main" com.mysql.jdbc.exceptions.MySQLTimeoutException: Statement cancelled due
to timeout or client request
    at com.mysql.jdbc.StatementImpl.executeUpdate(StatementImpl.java:1626)
    at com.mysql.jdbc.StatementImpl.executeUpdate(StatementImpl.java:1524)
    at org.apache.commons.dbcp2.DelegatingStatement.executeUpdate(DelegatingStatement.java:234)
    at org.apache.commons.dbcp2.DelegatingStatement.executeUpdate(DelegatingStatement.java:234)
    at DBCP.DBCPTest.main(DBCPTest.java:60)
```

7. 其他配置

其他配置可参考官方文档"BasicDataSource Configuration Parameters"。

8.1.5　参考配置

下面是 Druid 在 GitHub 网站上给出的一个参考配置。可以看到,除了 `filters` 和 `asyncInit`,其他参数与 DBCP 中的参数名基本一致。这里列举出来仅供参考,最重要的是我们要了解这些配置项的具体含义,如此才能定制出最适合我们业务场景的参数配置。

```xml
<bean id="dataSource" class="com.alibaba.druid.pool.DruidDataSource" init-method="init"
destroy-method="close">
    <property name="url" value="${jdbc_url}" />
    <property name="username" value="${jdbc_user}" />
    <property name="password" value="${jdbc_password}" />

    <property name="filters" value="stat" />

    <property name="maxActive" value="20" />
    <property name="initialSize" value="1" />
    <property name="maxWait" value="6000" />
    <property name="minIdle" value="1" />

    <property name="timeBetweenEvictionRunsMillis" value="60000" />
    <property name="minEvictableIdleTimeMillis" value="300000" />

    <property name="testWhileIdle" value="true" />
    <property name="testOnBorrow" value="false" />
    <property name="testOnReturn" value="false" />
```

```xml
    <property name="poolPreparedStatements" value="true" />
    <property name="maxOpenPreparedStatements" value="20" />

    <property name="asyncInit" value="true" />
</bean>
```

8.1.6 总结

- 检测连接的有效性，最有效的方法是在申请连接时检测。但这种检测方式对应用端性能的影响比较大，不建议线上使用。
- 建议以一定的频率来检测连接池中空闲连接的有效性。对于 c3p0 连接池，是设置 `idleConnection-TestPeriod` 参数。对于 DBCP 连接池，是设置 `testWhileIdle` 和 `timeBetweenEvictionRunsMillis` 参数。同时，检测的频率需小于 MySQL 服务端 `wait_timeout` 的值，否则没有任何意义。
- 如果连接池支持 JDBC 4 API，建议通过 `isValid()` 方法来检测连接的有效性，这样无须自定义探测语句。
- 考虑到网络可能的故障，以一定的频率来检测连接池中空闲连接的有效性并不能百分之百检测出无效连接，所以应用端一定要做好容错处理。
- 从性能角度出发，建议线上配置 PreparedStatement 缓存。
- 注意泄漏连接的危害性。一个连接，如果没有被释放回连接池，连接池是不会检测它的有效性的。如果它在一段时间（阈值由 MySQL 服务端的 `wait_timeout` 参数决定，默认为 28800，单位是秒，即 8 小时）内没有执行任何操作，则会被 MySQL 服务端关闭。MySQL 服务端的关闭操作连接池是感知不到的。后续如果使用这些连接执行其他操作，应用端就会抛出 `Communications link failure Last packet sent to the server was xx ms ago` 异常。即使 DBCP 连接池提供了参数进行处理，但如果使用被销毁的连接执行其他操作，应用端同样会报错。所以，强烈建议应用端在执行完操作后尽快关闭连接，将其释放回连接池。

8.2 MySQL 线程池

在 MySQL 社区版中，线程调度方式（对应 `thread_handling` 参数）有以下两种。

- one-thread-per-connection：每个连接一个线程。
- no-threads：使用主线程来处理连接，一般用于 Linux 下的 debug 环境。

所以，在生产环境中，实际可用的基本上只有 one-thread-per-connection 这一种方式。在这种方式下，对于每一个数据库连接，MySQL 服务端都会创建一个单独的线程为其服务。当连接关闭后，线程要么被销毁，要么缓存起来供其他连接使用。

这种方式在并发量较低的情况下确实没什么问题。但如果并发量太高，会导致线程不断切换，MySQL 的处理能力急剧下降。极端情况下，还会导致整个实例"夯住"，无法正常响应客户端的请求。具体到操作系统层面，我们会观察到 CPU 上下文频繁切换，sys 使用率较高。在数据库层面，我们还会看到状态变量 `Threads_running` 的值较高。

8.2.1 线程池的实现原理

线程池由多个线程组组成，每个线程组管理一组客户端连接。连接建立后，线程池会以轮询的方式将它们分配给线程组处理。每个线程组包含一个任务队列、一个 Listener 线程和多个 Worker 线程。除了线程组，线程池还有一个 Timer 线程，这个线程会定期检查线程组是否处于超时或阻塞状态。

线程池初始化完毕后，每个线程组都会启动一个 Listener 线程，监听组内对应连接的网络事件。当有一个操作到来时，首先会判断任务队列中是否有待执行的操作。若有，则会将该操作放到任务队列中。倘若没有，则由 Listener 线程直接执行该操作，避免创建一个新的 Worker 线程，此时，Listener 线程扮演了 Worker 线程的角色。

这里的任务队列有两个：优先队列和普通队列。顾名思义，优先队列操作的优先级高于普通队列。如果一个操作是事务内操作，则会被放到优先队列中。这样，事务内的操作可以尽快执行，从而释放资源。如果一个操作针对的是非事务存储引擎或事务存储引擎，但开启了 autocommit，则会被放到普通队列中。普通队列中的操作会在一定时间（thread_pool_prio_kickup_timer）后移动到优先队列中，避免优先队列中的操作源源不断，导致普通队列中的操作被"饿死"。

线程池的目的是限制并发执行的操作数，所以，默认情况下，一个线程组内最多只能执行一个操作。如果这个操作执行得足够快，队列中的后续操作就能重用当前这个线程。线程池重用活跃的线程，能更好地利用 CPU 缓存。但如果这个操作执行得比较慢，实际上它会阻塞队列中的其他操作。为了避免这种情况的发生，当操作的执行时长超过一定时间（thread_pool_stall_limit）后，线程池会将该操作定义为超时，此时，会唤醒睡眠状态的线程或创建新的线程来处理队列中的其他操作。

图 8-2 是 MySQL 线程池的架构图。

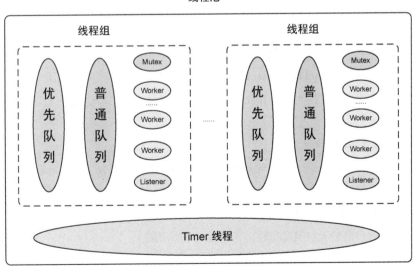

图 8-2　MySQL 线程池架构图

8.2.2 如何开启线程池功能

目前，线程池只存在于 MySQL 企业版、Percona 和 MariaDB 中。MySQL 社区版没有这个功能。

对于 MySQL 企业版，线程池是作为插件存在的，只需通过以下配置加载插件即可。

```
plugin-load-add=thread_pool.so
```

注意，thread_pool 只能通过配置文件配置，不能通过 INSTALL PLUGIN 动态加载。插件加载成功后，会自动将 thread_handling 参数设置为 loaded-dynamically。注意，如果在配置文件中同时设置了 thread_handling=loaded-dynamically，会导致 MySQL 服务启动失败。

对于 Percona 和 MariaDB，线程池已经集成到了服务端代码中，只需设置以下参数即可。

```
thread_handling=pool-of-threads
```

8.2.3 MySQL 企业版线程池参数解析

以 MySQL 企业版 8.0.25 为例，我们看看 MySQL 企业版各个参数的具体含义。

- **thread_pool_size**

线程组的个数，默认为 16。

- **thread_pool_stall_limit**

线程组内的操作经过多久可被视为超时。默认为 6，单位是 10 毫秒，即 60 毫秒。如果一个操作超时了，线程池会唤醒或创建另外一个线程，避免这个操作阻塞任务队列中的其他操作。

- **thread_pool_algorithm**

线程池算法。默认为 0，即使用传统的并行度较低的算法。也可设置为 1，此时，会使用一种并行度较高的激进算法，相对于传统算法，性能会有一定程度的提升，但如果连接数过大，会影响线程池的性能。该算法尚属于实验性质，不推荐线上使用。

- **thread_pool_high_priority_connection**

默认为 0，即会根据操作的类型将其放入优先队列或普通队列。若设置为 1，则所有操作都会被放入优先队列。

- **thread_pool_prio_kickup_timer**

普通队列中的操作被移动到优先队列需要等待的时间，默认为 1000，单位是毫秒，即 1 秒。为了避免短时间内移动过多，线程池限制了每个线程组每 10 毫秒最多可以移动一个语句。

- **thread_pool_max_unused_threads**

线程池中允许保留的最大空闲线程数，默认为 0，即不限制。

- **thread_pool_max_active_query_threads**

一个线程组内允许创建的最大活跃线程数，默认为 0，即使用尽可能多的线程。

- **thread_pool_max_transactions_limit**

线程池允许创建的最大事务数。默认为 0，即不限制。

如果当前事务数达到了 thread_pool_max_transactions_limit 的限制，则新的连接会被阻塞。不仅如此，已建立的连接在执行新的操作时也会被阻塞。此时，只允许具有 TP_CONNECTION_ADMIN 权限的用户访问。TP_CONNECTION_ADMIN 是 MySQL 8.0.31 引入的新权限。

- **thread_pool_dedicated_listeners**

若设置为 ON，则 Listener 线程只会监听网络事件，不会在优先队列和普通队列都为空时处理 SQL 操作。默认为 OFF。

8.2.4　Percona Server 线程池参数解析

以 Percona Server 8.0.25 为例，我们看看 Percona Server 各个参数的具体含义。

- **thread_handling**

默认为 one-thread-per-connection，即传统的每个连接一个线程模式。如果要开启线程池，需设置为 pool-of-threads。

- **thread_pool_size**

线程组的个数，默认为服务器 CPU 的核数。

- **thread_pool_stall_limit**

线程组内的操作经过多久可视为超时，默认为 500，单位是毫秒。

- **thread_pool_idle_timeout**

空闲线程的最大存活时间，默认为 60 秒。

- **thread_pool_high_prio_mode**

与 thread_pool_high_priority_connection 类似，控制任务队列的使用。可在全局和会话级别修改。该参数有 3 个选项。

- transactions：默认值。对于事务内操作且当事务所在连接对应的票数（high priority ticket）大于 0 时，会分配到优先队列。
- statements：操作放入优先队列。
- none：操作放入普通队列。

- **thread_pool_high_prio_tickets**

作用与 thread_pool_prio_kickup_timer 类似，都是为了避免普通队列中的操作被"饿死"。当 thread_pool_high_prio_mode 等于 transactions 时，每个新连接都会分配 thread_pool_high_prio_tickets 张票。连接内的操作只要进入优先队列，对应的票数就会减 1，直到票数为 0，进入普通队列。

而操作一旦进入普通队列，连接对应的票数将会重置为 thread_pool_high_prio_tickets。thread_pool_high_prio_tickets 默认为 4294967295。

- **thread_pool_max_threads**

  线程池的最大连接数，默认为 100000。

- **thread_pool_oversubscribe**

  一个线程组内允许额外增加几个活跃线程。默认为 3，此时，一个线程组内的最大活跃线程数为 4（即 thread_pool_oversubscribe + 1）。

8.2.5　MySQL 企业版线程池和 Percona Server 线程池的对比

- 线程组的个数

  对于 MySQL 企业版线程池，默认为 16；对于 Percona Server 线程池，默认为服务器 CPU 的核数。

- 对于普通队列的处理

  对于普通队列中的操作，MySQL 企业版线程池会在一定时间（由 thread_pool_prio_kickup_timer 决定，默认为 1 秒）后，将其移动到优先队列中。

  对于一个连接内的事务操作，Percona Server 线程池会在将其放入优先队列一定次数（由 thread_pool_high_prio_tickets 决定，默认为 4294967295）后放入普通队列。

- 对于空闲线程的处理。

  MySQL 企业版线程池用 thread_pool_max_unused_threads 参数来控制最大空闲线程数，默认不限制。

  而 Percona Server 线程池则指定了空闲线程的最大存活时间，超过存活时间则让其自动退出。默认为 60 秒。

- 一个线程组的最大活跃线程数。

  MySQL 企业版线程池通过 thread_pool_max_active_query_threads 来限制一个线程组的最大活跃线程数。默认为 0，即使用尽可能多的线程。

  Percona Server 线程池一个线程组内的最大活跃线程数默认为 4（即 thread_pool_oversubscribe + 1）。

下面，通过一个具体的案例来看看两者的区别。

模拟 20 个并发线程分别针对 MySQL 企业版 8.0.25 和 Percona Server 8.0.25 进行测试。因为是虚拟机环境，CPU 核数有限，为了保证测试环境的一致性，统一将两个测试实例的 thread_pool_size 设置为 2。测试脚本如下。

```python
#!/usr/bin/python
# -*- coding:UTF-8 -*-
import pymysql,time,threading

def run():
    try:
        conn = pymysql.connect(host="192.168.244.10",port=3306, user="root", password="123456", database="test")
        cursor = conn.cursor()
        cursor.execute('insert into t1 values(1,sleep(100))')
        conn.commit()
        conn.close()
    except Exception as e:
        print(e)

for i in range(20):
    t = threading.Thread(target=run, args=[])
    t.start()
    time.sleep(0.1)
```

首先看看 Percona Server 的执行结果：

```
mysql> select id,host,command,time,state,info from information_schema.processlist where host <> 'localhost' order by id;
+-----+----------------------+---------+------+------------+--------------------------------------------+
| id  | host                 | command | time | state      | info                                       |
+-----+----------------------+---------+------+------------+--------------------------------------------+
| 319 | 192.168.244.10:51106 | Query   |    9 | User sleep | insert into t1 values(1,sleep(100))        |
| 320 | 192.168.244.10:51108 | Query   |    9 | User sleep | insert into t1 values(1,sleep(100))        |
| 321 | 192.168.244.10:51110 | Query   |    9 | User sleep | insert into t1 values(1,sleep(100))        |
| 322 | 192.168.244.10:51112 | Query   |    9 | User sleep | insert into t1 values(1,sleep(100))        |
| 323 | 192.168.244.10:51114 | Query   |    9 | User sleep | insert into t1 values(1,sleep(100))        |
| 324 | 192.168.244.10:51116 | Query   |    9 | User sleep | insert into t1 values(1,sleep(100))        |
| 325 | 192.168.244.10:51118 | Query   |    9 | User sleep | insert into t1 values(1,sleep(100))        |
| 326 | 192.168.244.10:51120 | Query   |    8 | User sleep | insert into t1 values(1,sleep(100))        |
| 327 | connecting host      | Connect |    8 | login      | NULL                                       |
| 328 | connecting host      | Connect |    8 | login      | NULL                                       |
| 329 | connecting host      | Connect |    8 | login      | NULL                                       |
| 330 | connecting host      | Connect |    8 | login      | NULL                                       |
| 331 | connecting host      | Connect |    8 | login      | NULL                                       |
| 332 | connecting host      | Connect |    8 | login      | NULL                                       |
| 333 | connecting host      | Connect |    8 | login      | NULL                                       |
| 334 | connecting host      | Connect |    8 | login      | NULL                                       |
| 335 | connecting host      | Connect |    8 | login      | NULL                                       |
| 336 | connecting host      | Connect |    7 | login      | NULL                                       |
| 337 | connecting host      | Connect |    7 | login      | NULL                                       |
| 338 | connecting host      | Connect |    7 | login      | NULL                                       |
+-----+----------------------+---------+------+------------+--------------------------------------------+
20 rows in set (0.00 sec)
```

可以看到，20 个并发操作，只有 8 个在执行，其他的连接都处于登录状态。

因为 thread_pool_size 为 2，thread_pool_oversubscribe 为 3，所以最大活跃线程数为 $2 \times (3+1)=8$，与输出吻合。

在实际测试时发现，如果当前活跃线程数达到了线程池最大活跃线程数的限制，则新的客户端连接会被阻塞。不仅如此，已建立的连接，在执行新的操作时也会被阻塞。上面这个查询之所以没被阻塞，是因为在执行脚本之前，将当前连接的 thread_pool_high_prio_mode 的会话值调整为了 statements。注意，即使是通过管理端口（admin_port）连接也会存在上述问题。所以在碰到上述问题时，如果想杀掉当前的慢查询，实际上是无能为力的。除非一开始就将 thread_pool_high_prio_mode 的全局值设置为 statements。

接着我们看看 MySQL 企业版的执行结果：

```
mysql> select id,host,command,time,state,info from information_schema.processlist where host <>
'localhost' order by id;
+-----+---------------------+---------+------+------------+-------------------------------------+
| id  | host                | command | time | state      | info                                |
+-----+---------------------+---------+------+------------+-------------------------------------+
| 170 | 192.168.244.10:37496 | Query  |    8 | User sleep | insert into t1 values(1,sleep(100)) |
| 171 | 192.168.244.10:37498 | Query  |    8 | User sleep | insert into t1 values(1,sleep(100)) |
| 172 | 192.168.244.10:37500 | Query  |    8 | User sleep | insert into t1 values(1,sleep(100)) |
| 173 | 192.168.244.10:37502 | Query  |    8 | User sleep | insert into t1 values(1,sleep(100)) |
| 174 | 192.168.244.10:37504 | Query  |    8 | User sleep | insert into t1 values(1,sleep(100)) |
| 175 | 192.168.244.10:37506 | Query  |    8 | User sleep | insert into t1 values(1,sleep(100)) |
| 176 | 192.168.244.10:37508 | Query  |    8 | User sleep | insert into t1 values(1,sleep(100)) |
| 177 | 192.168.244.10:37510 | Query  |    8 | User sleep | insert into t1 values(1,sleep(100)) |
| 178 | 192.168.244.10:37512 | Query  |    7 | User sleep | insert into t1 values(1,sleep(100)) |
| 179 | 192.168.244.10:37514 | Query  |    7 | User sleep | insert into t1 values(1,sleep(100)) |
| 180 | 192.168.244.10:37516 | Query  |    7 | User sleep | insert into t1 values(1,sleep(100)) |
| 181 | 192.168.244.10:37518 | Query  |    7 | User sleep | insert into t1 values(1,sleep(100)) |
| 182 | 192.168.244.10:37520 | Query  |    7 | User sleep | insert into t1 values(1,sleep(100)) |
| 183 | 192.168.244.10:37522 | Query  |    7 | User sleep | insert into t1 values(1,sleep(100)) |
| 184 | 192.168.244.10:37524 | Query  |    7 | User sleep | insert into t1 values(1,sleep(100)) |
| 185 | 192.168.244.10:37526 | Query  |    7 | User sleep | insert into t1 values(1,sleep(100)) |
| 186 | 192.168.244.10:37528 | Query  |    7 | User sleep | insert into t1 values(1,sleep(100)) |
| 187 | 192.168.244.10:37530 | Query  |    6 | User sleep | insert into t1 values(1,sleep(100)) |
| 188 | 192.168.244.10:37532 | Query  |    6 | User sleep | insert into t1 values(1,sleep(100)) |
| 189 | 192.168.244.10:37534 | Query  |    6 | User sleep | insert into t1 values(1,sleep(100)) |
+-----+---------------------+---------+------+------------+-------------------------------------+
20 rows in set (0.00 sec)
```

虽然 thread_pool_size 也是 2，但 20 个并发操作可以同时执行。不存在 Percona Server 中的阻塞问题。

8.2.6 线程池的适用场景

线程池适用于有大量短连接或高并发的业务场景，不适合有大量慢查询的业务场景。尤其是 Percona Server 线程池限制了一个线程组的最大活跃线程数，如果一个线程组的活跃线程都被慢查询占用，则线程组内的后续操作都将被阻塞。

8.2.7 线程池的压测结果

图 8-3 和图 8-4 来自 MySQL 官方网站，展示了对线程池的压测结果，分别针对 TPCC（Transaction Processing Performance Council Benchmark C，针对 OLTP 的基准测试模型）和 OLTP 读写场景。两张图中的横坐标是用户并发数，纵坐标是 TPS。

以 Sysbench TPCC 的压测结果为例，可以看到，当并发数小于 256（实际上是一个介于 256 和

512 之间的值）时，随着并发数的增加，TPS 也会相应地增加。当并发数达到 256 时，无论是 MySQL 社区版还是企业版，TPS 都会达到最高。当并发数大于 256 时，随着并发数的进一步增加，MySQL 社区版的 TPS 反而越来越低。相反，MySQL 企业版因为有线程池的支撑，TPS 基本保持不变。

注意，这里的 256 不是一个绝对值，而是与被压测机器的 CPU 核数和测试模型有关。

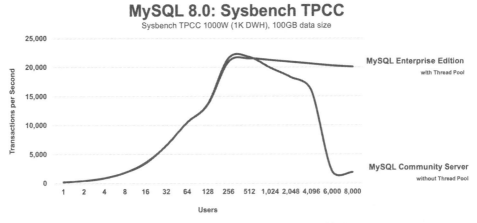

图 8-3　线程池 TPCC 场景的性能压测结果

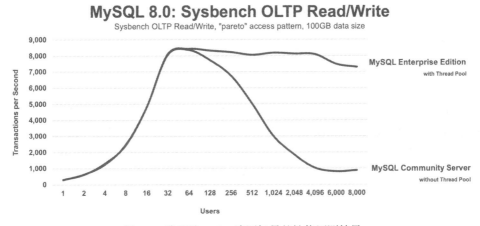

图 8-4　线程池 OLTP 读写场景的性能压测结果

8.2.8　线程池的监控

下面分别看看 MySQL 企业版和 Percona Server 是如何监控线程池性能的。

1. MySQL 企业版

MySQL 企业版提供了 3 张表来反映线程池的使用情况，这 3 张表均在 performance_schema 中。在 MySQL 8.0.14 之前，这 3 张表位于 information_schema 中。

- **TP_THREAD_GROUP_STATE**

该表反映了每个分组的状态信息，记录的是当前值。

```
mysql> desc tp_thread_group_state;
+-------------------------+---------------+------+-----+---------+-------+
| Field                   | Type          | Null | Key | Default | Extra |
+-------------------------+---------------+------+-----+---------+-------+
| TP_GROUP_ID             | int unsigned  | NO   | PRI | NULL    |       |
| CONSUMER_THREADS        | int unsigned  | NO   |     | NULL    |       |
| RESERVE_THREADS         | int unsigned  | NO   |     | NULL    |       |
| CONNECT_THREAD_COUNT    | int unsigned  | NO   |     | NULL    |       |
| CONNECTION_COUNT        | int unsigned  | NO   |     | NULL    |       |
| QUEUED_QUERIES          | int unsigned  | NO   |     | NULL    |       |
| QUEUED_TRANSACTIONS     | int unsigned  | NO   |     | NULL    |       |
| STALL_LIMIT             | int unsigned  | NO   |     | NULL    |       |
| PRIO_KICKUP_TIMER       | int unsigned  | NO   |     | NULL    |       |
| ALGORITHM               | varchar(20)   | NO   |     | NULL    |       |
| THREAD_COUNT            | int unsigned  | NO   |     | NULL    |       |
| ACTIVE_THREAD_COUNT     | int unsigned  | NO   |     | NULL    |       |
| STALLED_THREAD_COUNT    | int unsigned  | NO   |     | NULL    |       |
| WAITING_THREAD_NUMBER   | int unsigned  | YES  |     | NULL    |       |
| OLDEST_QUEUED           | bigint unsigned | YES |    | NULL    |       |
| MAX_THREAD_IDS_IN_GROUP | int unsigned  | NO   |     | NULL    |       |
+-------------------------+---------------+------+-----+---------+-------+
16 rows in set (0.00 sec)
```

表中各字段的含义如下。

- TP_GROUP_ID：线程组 ID。
- CONSUMER_THREADS：工作线程数。
- RESERVE_THREADS：空闲线程数。
- CONNECT_THREAD_COUNT：用来进行连接初始化和身份验证的线程数。
- CONNECTION_COUNT：该组对应的连接数。
- QUEUED_QUERIES：优先队列中等待执行的操作数。
- QUEUED_TRANSACTIONS：普通队列中等待执行的操作数。
- STALL_LIMIT：即参数 thread_pool_stall_limit。
- PRIO_KICKUP_TIMER：即参数 thread_pool_prio_kickup_timer。
- ALGORITHM：即参数 thread_pool_algorithm。
- THREAD_COUNT：线程数。
- ACTIVE_THREAD_COUNT：活跃线程数。
- STALLED_THREAD_COUNT：超时线程数。
- WAITING_THREAD_NUMBER：处理线程组中语句轮询的线程的线程 ID。
- OLDEST_QUEUED：队列中，等待时间最久的操作已经等待的时间，单位是 毫秒。
- MAX_THREAD_IDS_IN_GROUP：组内线程的最大线程 ID。

- **TP_THREAD_STATE**

该表反映了分组内每个线程的状态。

```
mysql> desc tp_thread_state;
+------------------+----------------+------+-----+---------+-------+
| Field            | Type           | Null | Key | Default | Extra |
+------------------+----------------+------+-----+---------+-------+
| TP_GROUP_ID      | int unsigned   | NO   | PRI | NULL    |       |
| TP_THREAD_NUMBER | int unsigned   | NO   | PRI | NULL    |       |
| PROCESS_COUNT    | bigint unsigned| NO   |     | NULL    |       |
| WAIT_TYPE        | varchar(30)    | YES  |     | NULL    |       |
+------------------+----------------+------+-----+---------+-------+
4 rows in set (0.00 sec)
```

表中各字段的含义如下。

- TP_GROUP_ID：线程组 ID。
- TP_THREAD_NUMBER：组内的线程 ID。
- PROCESS_COUNT：当前操作的执行时长，以 10 毫秒为单位。如果为 0，则代表当前线程没有操作。
- WAIT_TYPE：当前线程的等待事件类型。如果为 NULL，则代表当前线程没有被阻塞。具体的等待事件类型可参考下面的 TP_THREAD_GROUP_STATS 表。

- **TP_THREAD_GROUP_STATS**

该表反映了每个分组的性能信息，记录的是累计值。

```
mysql> desc tp_thread_group_stats;
+---------------------------+----------------+------+-----+---------+-------+
| Field                     | Type           | Null | Key | Default | Extra |
+---------------------------+----------------+------+-----+---------+-------+
| TP_GROUP_ID               | int unsigned   | NO   | PRI | NULL    |       |
| CONNECTIONS_STARTED       | bigint unsigned| NO   |     | NULL    |       |
| CONNECTIONS_CLOSED        | bigint unsigned| NO   |     | NULL    |       |
| QUERIES_EXECUTED          | bigint unsigned| NO   |     | NULL    |       |
| QUERIES_QUEUED            | bigint unsigned| NO   |     | NULL    |       |
| THREADS_STARTED           | bigint unsigned| NO   |     | NULL    |       |
| PRIO_KICKUPS              | bigint unsigned| NO   |     | NULL    |       |
| STALLED_QUERIES_EXECUTED  | bigint unsigned| NO   |     | NULL    |       |
| BECOME_CONSUMER_THREAD    | bigint unsigned| NO   |     | NULL    |       |
| BECOME_RESERVE_THREAD     | bigint unsigned| NO   |     | NULL    |       |
| BECOME_WAITING_THREAD     | bigint unsigned| NO   |     | NULL    |       |
| WAKE_THREAD_STALL_CHECKER | bigint unsigned| NO   |     | NULL    |       |
| SLEEP_WAITS               | bigint unsigned| NO   |     | NULL    |       |
| DISK_IO_WAITS             | bigint unsigned| NO   |     | NULL    |       |
| ROW_LOCK_WAITS            | bigint unsigned| NO   |     | NULL    |       |
| GLOBAL_LOCK_WAITS         | bigint unsigned| NO   |     | NULL    |       |
| META_DATA_LOCK_WAITS      | bigint unsigned| NO   |     | NULL    |       |
| TABLE_LOCK_WAITS          | bigint unsigned| NO   |     | NULL    |       |
| USER_LOCK_WAITS           | bigint unsigned| NO   |     | NULL    |       |
| BINLOG_WAITS              | bigint unsigned| NO   |     | NULL    |       |
| GROUP_COMMIT_WAITS        | bigint unsigned| NO   |     | NULL    |       |
| FSYNC_WAITS               | bigint unsigned| NO   |     | NULL    |       |
+---------------------------+----------------+------+-----+---------+-------+
22 rows in set (0.00 sec)
```

表中各字段的含义如下。

- TP_GROUP_ID：线程组 ID。
- CONNECTIONS_STARTED：创建的连接数。

- CONNECTIONS_CLOSED：关闭的连接数。
- QUERIES_EXECUTED：执行的操作数。
- QUERIES_QUEUED：放到队列中的操作的数量。
- THREADS_STARTED：启动的线程数。
- PRIO_KICKUPS：从普通队列移动到优先队列的操作数。
- STALLED_QUERIES_EXECUTED：因执行时间超过 thread_pool_stall_limit 而被线程池定义为超时的操作数。
- BECOME_CONSUMER_THREAD：线程被赋予 consumer 角色的次数。
- BECOME_RESERVE_THREAD：线程被赋予 reserve 角色的次数。
- BECOME_WAITING_THREAD：线程被赋予 waiter 角色的次数。
- WAKE_THREAD_STALL_CHECKER：因为检测到连接超时，线程池唤醒空闲连接或创建新的连接的次数。
- SLEEP_WAITS：线程进入休眠状态的次数。譬如在执行 sleep() 函数时。
- DISK_IO_WAITS：等待磁盘 I/O 的次数。
- ROW_LOCK_WAITS：因行级锁而等待的次数。
- GLOBAL_LOCK_WAITS：因全局锁而等待的次数。
- META_DATA_LOCK_WAITS：因元数据锁而等待的次数。
- TABLE_LOCK_WAITS：因表级锁而等待的次数。
- USER_LOCK_WAITS：因用户锁而等待的次数。
- BINLOG_WAITS：等待 binlog 空闲的次数。
- GROUP_COMMIT_WAITS：组提交相关的等待次数。
- FSYNC_WAITS：等待文件刷盘的次数。

可使用以下查询来衡量 thread_pool_stall_limit 参数设置得是否合理。结果越小越好。

```
SELECT SUM(STALLED_QUERIES_EXECUTED) / SUM(QUERIES_EXECUTED) FROM performance_schema.tp_thread_group_stats;
```

2. Percona Server

Percona Server 提供了两个状态变量来反映线程池的使用情况。

- **Threadpool_idle_threads**

线程池中空闲线程数。

- **Threadpool_threads**

线程池中总的线程数。

从监控的角度来看，MySQL 企业版线程池提供的信息较为详细，可基于这些信息来优化线程池的性能。

而 Percona Server 线程池只提供了两个状态变量，基于这两个变量其实很难评估线程池的性能、参数设置得是否合理等。

8.2.9 参考资料

- 官方文档"MySQL Enterprise Thread Pool"。
- 官网产品页"MySQL Thread Pool"。
- 官方文档"Thread Pool in MariaDB"。
- 博客文章"MySQL Thread Pool: Summary",作者:Mikael Ronstrom。

8.3 MySQL server has gone away 深度解析

对于 MySQL 使用者来说,MySQL server has gone away 是一个经常会遇到的错误信息。例如,在使用 mysql 客户端对数据库进行操作时,如果在一段时间内没有任何操作,那么再次操作时,就有可能出现这个错误。

```
mysql> select 1;
+---+
| 1 |
+---+
| 1 |
+---+
1 row in set (0.00 sec)

mysql> set session wait_timeout=10;
Query OK, 0 rows affected (0.00 sec)

# 等待 10 秒再操作

mysql> select 1;
ERROR 2006 (HY000): MySQL server has gone away
```

8.3.1 出现 MySQL server has gone away 错误的常见原因

本节我们看看出现 MySQL server has gone away 错误的常见原因。

- 会话的空闲时间超过服务端的设置,会话被服务端主动关闭了。
- 当前会话被杀掉了。
- 执行查询期间 MySQL 服务端宕机了。
- 客户端数据包的大小超过 max_allowed_packet(MySQL 8.0 默认为 64MB)的设置。例如:

```
mysql> select * from slowtech.t1;
ERROR 2006 (HY000): MySQL server has gone away
No connection. Trying to reconnect...
Connection id:    21
Current database: *** NONE ***

ERROR 2020 (HY000): Got packet bigger than 'max_allowed_packet' bytes
```

第一个原因最为常见,8.3 节开头提到的场景即可归为此类。

在 MySQL 8.0 中,8.3 节开头提到的场景不再报 MySQL server has gone away 错误,而是会提示以下错误。

```
ERROR 4031 (HY000): The client was disconnected by the server because of inactivity. See wait_timeout
and interactive_timeout for configuring this behavior.
```

错误提示很明显，客户端由于不活动被服务端断开连接了。与此相关的参数有两个：interactive_timeout 和 wait_timeout。这两个参数决定了连接的最大空闲时间。我们在下一节具体来看看这两个参数的含义及区别。

8.3.2 interactive_timeout 和 wait_timeout 的区别

首先看看官方文档对于这两个参数的定义。

- **interactive_timeout**

服务器在关闭交互式连接之前等待其活动的秒数。交互式客户端指的是 mysql_real_connect() 中使用了 CLIENT_INTERACTIVE 选项的客户端。

默认为 28800，单位是秒，即 8 小时。

- **wait_timeout**

服务器在关闭非交互式连接之前等待其活动的秒数。

在线程启动时，wait_timeout 的会话值继承自 wait_timeout 的全局值或 interactive_timeout 的全局值，具体取决于客户端的类型（由 mysql_real_connect() 中的 CLIENT_INTERACTIVE 选项决定）。

默认为 28800，单位是秒，即 8 小时。

根据上述定义，两者的区别显而易见。

interactive_timeout 适用于交互式连接，而 wait_timeout 适用于非交互式连接。交互式连接指的是客户端在使用 mysql_real_connect() 函数时指定了 CLIENT_INTERACTIVE 选项，这一般是在驱动层实现的，我们无须关心。

我们常用的 mysql 客户端即属于交互式连接，Java 程序中的 JDBC 连接则属于非交互式连接。

而且，从上面 wait_timeout 的定义中我们看到，在线程启动时， wait_timeout 的会话值既可能继承自 wait_timeout 的全局值，也可能继承自 interactive_timeout 的全局值。具体继承哪个参数，取决于客户端的类型是交互式连接还是非交互式连接。

下面具体测试一下，看看决定会话最大空闲时间的是哪个参数。是 wait_timeout 还是 interactive_timeout？

首先，测试 wait_timeout 会话级别的效果。

```
mysql> show session variables where variable_name in ('interactive_timeout','wait_timeout');
+---------------------+-------+
| Variable_name       | Value |
+---------------------+-------+
| interactive_timeout | 28800 |
| wait_timeout        | 28800 |
+---------------------+-------+
```

8.3 MySQL server has gone away 深度解析

```
2 rows in set (0.04 sec)

mysql> set session wait_timeout=10;
Query OK, 0 rows affected (0.00 sec)

# 等待 10 秒再操作

mysql> show session variables where variable_name in ('interactive_timeout','wait_timeout');
ERROR 2006 (HY000): MySQL server has gone away
```

其次，测试 interactive_timeout 会话级别的效果。

```
mysql> show session variables where variable_name in ('interactive_timeout','wait_timeout');
+---------------------+-------+
| Variable_name       | Value |
+---------------------+-------+
| interactive_timeout | 28800 |
| wait_timeout        | 28800 |
+---------------------+-------+
2 rows in set (0.02 sec)

mysql> set session interactive_timeout=10;
Query OK, 0 rows affected (0.03 sec)

# 同样等待 10 秒再操作

mysql> show session variables where variable_name in ('interactive_timeout','wait_timeout');
+---------------------+-------+
| Variable_name       | Value |
+---------------------+-------+
| interactive_timeout | 10    |
| wait_timeout        | 28800 |
+---------------------+-------+
2 rows in set (0.01 sec)
```

从上面的测试中可以看到，在 mysql 客户端中，决定会话最大空闲时间的是 wait_timeout 的会话值，与 interactive_timeout 无关。

其实，不仅仅是 mysql 客户端，对于 JDBC 连接同样如此，感兴趣的读者可自行测试一下。

既然我们知道了 wait_timeout 的会话值是会话最大空闲时间的决定因素，接着就来看看它的继承问题。

首先，我们看看 interactive_timeout 的测试结果。

(1) 调整 interactive_timeout 的全局值。

```
mysql> show global variables where variable_name in ('interactive_timeout','wait_timeout');
+---------------------+-------+
| Variable_name       | Value |
+---------------------+-------+
| interactive_timeout | 28800 |
| wait_timeout        | 28800 |
+---------------------+-------+
2 rows in set (0.02 sec)

mysql> set global interactive_timeout=10;
Query OK, 0 rows affected (0.01 sec)
```

(2) 查看 mysql 客户端的输出结果。注意，需开启一个新的会话。

```
mysql> show session variables like 'wait_timeout';
+---------------+-------+
| Variable_name | Value |
+---------------+-------+
| wait_timeout  | 10    |
+---------------+-------+
1 row in set (0.00 sec)
```

(3) 查看 JDBC 连接的输出结果。

```
wait_timeout: 28800
```

用来测试的 Java 程序如下。

```java
package com.victor_01;

import java.sql.Connection;
import java.sql.DriverManager;
import java.sql.ResultSet;
import java.sql.Statement;

public class JdbcTest {
    @SuppressWarnings("static-access")
    public static void main(String[] args) throws Exception {
        Class.forName("com.mysql.jdbc.Driver");
        Connection conn = DriverManager.getConnection("jdbc:mysql://192.168.244.10:3306/sbtest", "root", "123456");
        Statement stmt = conn.createStatement();
        String sql = "show variables where variable_name='wait_timeout'";
        ResultSet rs = stmt.executeQuery(sql);
        while (rs.next()) {
            System.out.println(rs.getString(1)+": "+rs.getString(2));
        }
        rs.close();
        stmt.close();
        conn.close();
    }
}
```

可以看到，当我们调整完 interactive_timeout 的全局值后，mysql 客户端中 wait_timeout 的会话值也随之改变。

接下来，我们看看 wait_timeout 的测试结果。

(1) 调整 wait_timeout 的全局值。

```
mysql> show global variables where variable_name in ('interactive_timeout','wait_timeout');
+---------------------+-------+
| Variable_name       | Value |
+---------------------+-------+
| interactive_timeout | 28800 |
| wait_timeout        | 28800 |
+---------------------+-------+
2 rows in set (0.01 sec)

mysql> set global wait_timeout=10;
Query OK, 0 rows affected (0.00 sec)
```

(2) 查看 mysql 客户端的输出结果。

```
mysql> show session variables like 'wait_timeout';
+---------------+-------+
| Variable_name | Value |
+---------------+-------+
| wait_timeout  | 28800 |
+---------------+-------+
1 row in set (0.00 sec)
```

(3) 查看 JDBC 连接的输出结果。

```
wait_timeout: 10
```

调整完 wait_timeout 的全局值，受影响的只有 JDBC 连接中的 wait_timeout，mysql 客户端中的 wait_timeout 保持不变。

基于上面的测试，我们可得出以下结论。

- 对于 mysql 客户端，wait_timeout 的会话值继承自 interactive_timeout 的全局值。
- 对于 JDBC 连接，wait_timeout 的会话值继承自 wait_timeout 的全局值。

8.3.3 wait_timeout 设置为多大比较合适

在回答这个问题之前，我们首先看一个测试。

对于 Java 程序，在执行数据库操作时，如果当前连接已经被 MySQL 服务端断开了，会出现什么问题？

首先调整 wait_timeout 的全局值。

```
mysql> set global wait_timeout=10;
Query OK, 0 rows affected (0.00 sec)
```

接着执行测试程序。

```java
package com.victor_01;

import java.sql.Connection;
import java.sql.DriverManager;
import java.sql.ResultSet;
import java.sql.Statement;

public class JdbcTest {
    @SuppressWarnings("static-access")
    public static void main(String[] args) throws Exception {
        Class.forName("com.mysql.jdbc.Driver");
        Connection conn = DriverManager.getConnection("jdbc:mysql://192.168.244.10:3306/sbtest",
            "root", "123456");
        Statement stmt = conn.createStatement();
        String sql = "show variables where variable_name='wait_timeout'";
        ResultSet rs = stmt.executeQuery(sql);
        while (rs.next()) {
            System.out.println(rs.getString(1)+": "+rs.getString(2));
        }
```

```java
        // 暂停 11 秒才执行下一个操作,这样可确保连接的空闲时间大于 wait_timeout 的设置
        Thread.currentThread().sleep(11000);

        sql = "select 1";
        rs=stmt.executeQuery(sql);
        while (rs.next()) {
            System.out.println(rs.getString(1));
        }

        rs.close();
        stmt.close();
        conn.close();
    }
}
```

测试程序的执行结果如下。

```
wait_timeout: 10
Exception in thread "main" com.mysql.jdbc.exceptions.jdbc4.CommunicationsException: Communications link failure

Last packet sent to the server was 26 ms ago.
    at sun.reflect.NativeConstructorAccessorImpl.newInstance0(Native Method)
    at sun.reflect.NativeConstructorAccessorImpl.newInstance(Unknown Source)
    at sun.reflect.DelegatingConstructorAccessorImpl.newInstance(Unknown Source)
    at java.lang.reflect.Constructor.newInstance(Unknown Source)
    at com.mysql.jdbc.Util.handleNewInstance(Util.java:406)
    at com.mysql.jdbc.SQLError.createCommunicationsException(SQLError.java:1074)
    at com.mysql.jdbc.MysqlIO.reuseAndReadPacket(MysqlIO.java:3009)
    at com.mysql.jdbc.MysqlIO.reuseAndReadPacket(MysqlIO.java:2895)
    at com.mysql.jdbc.MysqlIO.checkErrorPacket(MysqlIO.java:3438)
    at com.mysql.jdbc.MysqlIO.sendCommand(MysqlIO.java:1951)
    at com.mysql.jdbc.MysqlIO.sqlQueryDirect(MysqlIO.java:2101)
    at com.mysql.jdbc.ConnectionImpl.execSQL(ConnectionImpl.java:2548)
    at com.mysql.jdbc.ConnectionImpl.execSQL(ConnectionImpl.java:2477)
    at com.mysql.jdbc.StatementImpl.executeQuery(StatementImpl.java:1422)
    at com.victor_01.JdbcTest.main(JdbcTest.java:24)
Caused by: java.io.EOFException: Can not read response from server. Expected to read 4 bytes, read 0 bytes before connection was unexpectedly lost.
    at com.mysql.jdbc.MysqlIO.readFully(MysqlIO.java:2455)
    at com.mysql.jdbc.MysqlIO.reuseAndReadPacket(MysqlIO.java:2906)
    ... 8 more
```

执行结果中出现了大名鼎鼎的 Communications link failure. Last packet sent to the server was *xxx* ms ago. 错误。

回到本节开头提到的问题,即 wait_timeout 设置为多大比较合适。实际上,这个问题没有绝对的答案。之前我们的线上环境,有设置为默认值(8 小时)的,也有设置为 1800 秒的,区别不大,毕竟空闲连接占用不了多少系统资源。最重要的是,应用端在使用连接池的情况下,必须要确保空闲连接的检测周期小于 wait_timeout,否则很容易出现测试程序中的 Communications link failure. Last packet sent to the server was *xxx* ms ago. 错误。

8.4 本章总结

数据库连接池和线程池这两个概念很容易让人混淆。本质上，连接池是应用端实现的，而线程池是 MySQL 服务端实现的。为什么要实现它们呢？背后体现的其实是一种池化思想。池化思想的核心是"复用的成本"远远小于"创建的成本"。

在使用数据库连接池时，注意应用端在执行完 SQL 后应尽快关闭连接，将其释放回连接池，否则很容易出现连接泄漏问题。

而 MySQL 线程池呢？从官方的压测结果来看，它在高并发场景中确实有不可替代的优势。注意，线程池适用于有大量短连接或高并发的业务场景。如果业务中有大量慢查询，则不适用。

在使用 mysql 客户端时，我们经常会碰到 MySQL server has gone away 错误。出现这个错误的最常见原因是，会话的空闲时间超过了 wait_timeout 的会话值。而 wait_timeout 的会话值又与 interactive_timeout 和 wait_timeout 的全局设置有关。具体继承哪个参数，与连接的类型有关。如果连接是交互式的（如 mysql 客户端），则继承 interactive_timeout；如果连接是非交互式的（如 JDBC 连接），则继承 wait_timeout。

在 JDBC 连接中，如果连接的空闲时间超过了 wait_timeout，连接会被 MySQL 服务端断开，此时执行 SQL 会出现 Communications link failure. Last packet sent to the server was xxx ms ago. 错误，所以在配置连接池中空闲连接的检测周期（idleConnectionTestPeriod）时，切记不要超过 MySQL 服务端 wait_timeout 的值。

重点问题回顾

- 连接池的原理。
- 如何定位 An attempt by a client to checkout a Connection has timed out. 问题。
- 线程池的实现原理。
- 线程池的适用场景。
- 出现 MySQL server has gone away 错误的常见原因。
- interactive_timeout 和 wait_timeout 的区别。
- 出现 Communications link failure. Last packet sent to the server was xxx ms ago. 异常的原因。
- wait_timeout 设置为多大比较合适。

第 9 章

MySQL 的常用工具

本章会介绍 MySQL 中的一些常用工具，包括 sysbench 和 Percona Toolkit。sysbench 是目前用得最多的 MySQL 性能压测工具。Percona Toolkit 是 Percona 公司提供的一个工具包，里面包含 37 个工具。从功能上看，绝大部分工具是针对 MySQL 的，还有一部分工具是针对系统主机、MongoDB、K8s 和 PG 的。

从开发语言来看，工具集里较早出现的工具基本上都是用 Perl 和 shell 开发的，而最近几个工具则是用 Go 语言开发的。从这一点也可以看出，用 Go 语言做工具开发确实是一个趋势。毕竟，用 Go 语言开发出来的工具可以做到开箱即用，不像 Perl 和 Python，需要额外安装依赖包。

下面通过表 9-1 看看 Percona Toolkit 中都有哪些工具及各工具的作用。

表 9-1　Percona Toolkit 中的工具及其作用

工　　具	开发语言	功　　能
pt-align	Perl	将其他工具的输出按列对齐
pt-archiver	Perl	数据归档（删除）
pt-config-diff	Perl	找出不同参数源之间的参数差异
pt-deadlock-logger	Perl	采集死锁信息
pt-diskstats	Perl	打印磁盘读写信息
pt-duplicate-key-checker	Perl	查看数据库中的冗余索引
pt-fifo-split	Perl	通过管道方式实现文件切割功能
pt-find	Perl	基于条件查找数据库对象
pt-fingerprint	Perl	提取查询的一般形式
pt-fk-error-logger	Perl	采集外键错误信息
pt-heartbeat	Perl	检查主从延迟情况
pt-index-usage	Perl	通过慢日志分析哪些索引未使用
pt-ioprofile	Shell	统计指定进程的 I/O 操作
pt-kill	Perl	杀掉满足条件的连接
pt-k8s-debug-collector	Go	采集 K8s 的调试数据
pt-mext	shell	将 mysqladmin 输出的 status 的多次结果格式化到同一行
pt-mongodb-query-digest	Go	对 MongoDB 慢日志进行汇总分析
pt-mongodb-summary	Go	采集 MongoDB 实例的基本信息
pt-mysql-summary	shell	采集 MySQL 实例的基本信息
pt-online-schema-change	Perl	变更表结构

（续）

工具	开发语言	功能
pt-pg-summary	Go	采集 PostgreSQL 实例的基本信息
pt-pmp	shell	采集并汇总进程的栈信息
pt-query-digest	Perl	对慢日志进行汇总分析
pt-secure-collect	Go	采集主机及数据库的相关信息，并进行脱敏处理
pt-show-grants	Perl	打印实例的授权信息
pt-sift	shell	汇总 pt-stalk 的采集结果
pt-slave-delay	Perl	实现从库延迟复制功能
pt-slave-find	Perl	打印主从拓扑图
pt-slave-restart	Perl	复制中断时，重启复制
pt-stalk	shell	出现问题时，实时采集 MySQL 和服务器的现场数据
pt-summary	shell	采集服务器的基本信息
pt-table-checksum	Perl	校验主从数据一致性
pt-table-sync	Perl	修复主从不一致的数据
pt-table-usage	Perl	分析 SQL 语句对表的使用情况
pt-upgrade	Perl	数据库升级工具，可针对 SQL 语句进行功能测试和性能测试
pt-variable-advisor	Perl	分析 MySQL 的配置参数设置并输出建议
pt-visual-explain	Perl	将执行计划格式化为树状

不难看出，很多工具的功能在 MySQL 中已经实现了，下面是几个例子。

- pt-deadlock-logger：在 MySQL 5.6 中可通过 innodb_print_all_deadlocks 参数将死锁信息记录在错误日志中。
- pt-duplicate-key-checker：在 MySQL 5.7 中可通过 sys.schema_redundant_indexes 查看冗余索引。
- pt-slave-delay：MySQL 5.6 中原生支持了延迟从库的功能。

除此之外，还有一些工具用处并不大，或者说使用场景相对较窄，比如 pt-align 和 pt-table-usage。

所以，本章会重点介绍 Percona Toolkit 中那些使用频率较高的工具，具体如下：

- pt-archiver
- pt-config-diff
- pt-ioprofile
- pt-kill
- pt-pmp
- pt-query-digest
- pt-show-grants
- pt-slave-restart
- pt-stalk
- pt-table-checksum
- pt-table-sync
- pt-upgrade

9.1 sysbench

sysbench 是一个开源的、基于 LuaJIT（Lua 的即时编译器，可将代码直接翻译成机器码，性能比原生 Lua 要高）的、可自定义脚本的多线程基准测试工具，也是目前用得最多的 MySQL 性能压测工具。

基于 sysbench，我们可以对比 MySQL 在不同版本、不同硬件配置、不同参数（包括操作系统参数和数据库配置参数）下的性能差异。

本节会从 sysbench 的基本用法出发，逐渐延伸到 sysbench 的一些高级玩法。例如，如何阅读自带的测试脚本，如何自定义测试项，等等。除此之外，对于如何使用 sysbench 对 CPU 进行测试，网上的很多资料语焉不详，甚至是错误的，所以这里也会从源码的角度分析 CPU 测试的实现逻辑及 --cpu-max-prime 选项的具体含义。

本节主要包括以下内容。

- 安装 sysbench。
- sysbench 用法讲解。
- 对 MySQL 进行基准测试的基本步骤。
- 如何分析 MySQL 的基准测试结果。
- 如何使用 sysbench 对服务器性能进行测试。
- MySQL 常见的测试场景及对应的 SQL 语句。
- 如何自定义 sysbench 测试脚本。

9.1.1 安装 sysbench

下面是 sysbench 源码包的安装步骤。

```
# yum -y install make automake libtool pkgconfig libaio-devel openssl-devel mysql-devel
# cd /usr/src/
# wget https://github.com/akopytov/sysbench/archive/refs/tags/1.0.20.tar.gz
# tar xvf 1.0.20.tar.gz
# cd sysbench-1.0.20/
# ./autogen.sh
# ./configure
# make -j
# make install
```

安装完成后，压测脚本默认会安装在 /usr/local/share/sysbench 目录下。

我们看看该目录的内容。

```
# ls /usr/local/share/sysbench/
bulk_insert.lua      oltp_insert.lua        oltp_read_write.lua       oltp_write_only.lua        tests
oltp_common.lua      oltp_point_select.lua  oltp_update_index.lua     select_random_points.lua
oltp_delete.lua      oltp_read_only.lua     oltp_update_non_index.lua select_random_ranges.lua
```

除了 oltp_common.lua 是个公共模块，其他每个 Lua 脚本都对应一个测试场景。

9.1.2 sysbench 用法讲解

sysbench 命令的语法如下。

```
sysbench [options]... [testname] [command]
```

命令中的 `testname` 是测试项的名称。sysbench 支持的测试项如下。

- `*.lua`：数据库性能基准测试。
- `fileio`：磁盘 I/O 基准测试。
- `cpu`：CPU 性能基准测试。
- `memory`：内存访问基准测试。
- `threads`：基于线程的调度程序基准测试。
- `mutex`：POSIX 互斥量基准测试。

`command` 是 sysbench 要执行的命令，支持的选项有：`prepare`、`prewarm`、`run`、`cleanup` 和 `help`。注意，不是所有的测试项都支持这些选项。

`options` 是配置项。sysbench 中的配置项主要包括以下两部分。

- **通用配置项**

这部分配置项可通过 `sysbench --help` 查看。例如：

```
# sysbench --help
...
General options:
 --threads=N                number of threads to use [1]
 --events=N                 limit for total number of events [0]
 --time=N                   limit for total execution time in seconds [10]
...
```

- **测试项相关的配置项**

各个测试项支持的配置项可通过 `sysbench testname help` 查看。例如：

```
# sysbench memory help
sysbench 1.0.20 (using bundled LuaJIT 2.1.0-beta2)

memory options:
 --memory-block-size=SIZE      size of memory block for test [1K]
 --memory-total-size=SIZE      total size of data to transfer [100G]
 --memory-scope=STRING         memory access scope {global,local} [global]
 --memory-hugetlb[=on|off]     allocate memory from HugeTLB pool [off]
 --memory-oper=STRING          type of memory operations {read, write, none} [write]
 --memory-access-mode=STRING   memory access mode {seq,rnd} [seq]
```

9.1.3 对 MySQL 进行基准测试的基本步骤

下面以 `oltp_read_write` 为例，看看使用 sysbench 对 MySQL 进行基准测试的 4 个标准步骤。

1. prepare

生成压测数据。

```
# sysbench oltp_read_write --mysql-host=10.0.0.64 --mysql-port=3306 --mysql-user=admin
--mysql-password=Py@123456 --mysql-db=sbtest --tables=30 --table-size=1000000 --threads=30 prepare
```

命令中各个选项的具体含义如下：

- `oltp_read_write`：测试项，对应的是 /usr/local/share/sysbench/oltp_read_write.lua。这里也可指定脚本的绝对路径名。
- `--mysql-host`、`--mysql-port`、`--mysql-user` 和 `--mysql-password`：分别代表 MySQL 实例的主机名、端口、用户名和密码。
- `--mysql-db`：库名。不指定则默认 sbtest。
- `--tables`：表的数量，默认为 1。
- `--table-size`：单表的大小，默认为 10000。
- `--threads`：并发线程数，默认为 1。注意，导入时，单表只能使用一个线程。
- `prepare`：执行准备工作。

`oltp_read_write` 用来压测 OLTP 场景。在 sysbench 1.0 之前，该场景是通过 oltp.lua 这个脚本来测试的，不过该脚本在 sysbench 1.0 之后被废弃了。但为了兼容之前的版本，该脚本放到了 /usr/local/share/sysbench/tests/include/oltp_legacy/ 目录下。

鉴于 oltp_read_write.lua 和 oltp.lua 两者的压测内容完全一致。从 sysbench 1.0 开始，压测 OLTP 场景建议直接使用 `oltp_read_write`。

2. prewarm

预热。主要是将磁盘中的数据加载到内存中。

```
# sysbench oltp_read_write --mysql-host=10.0.0.64 --mysql-port=3306 --mysql-user=admin
--mysql-password=Py@123456 --mysql-db=sbtest --tables=30 --table-size=1000000 --threads=30 prewarm
```

除了需要将命令设置为 `prewarm`，其他配置与 `prepare` 中一样。

3. run

压测。

```
# sysbench oltp_read_write --mysql-host=10.0.0.64 --mysql-port=3306 --mysql-user=admin
--mysql-password=Py@123456 --mysql-db=sbtest --tables=30 --table-size=1000000 --threads=64 --time=60
--report-interval=10 run
```

命令中部分选项的具体含义如下。

- `--time`：压测时间。不指定则默认为 10 秒。除了 `--time`，也可通过 `--events` 限制需要执行的事件的数量。
- `--report-interval=10`：每 10 秒输出一次测试结果，默认为 0，即不输出。

4. cleanup

清理数据。

```
# sysbench oltp_read_write --mysql-host=10.0.0.64 --mysql-port=3306 --mysql-user=admin
--mysql-password=Py@123456 --mysql-db=sbtest --tables=30 cleanup
```

这里只需指定--tables，sysbench 会串行执行 DROP TABLE IF EXISTS sbtest 操作。

9.1.4 如何分析 MySQL 的基准测试结果

下面我们分析 oltp_read_write 场景下的压测结果。

```
Threads started!

[ 10s ] thds: 64 tps: 5028.08 qps: 100641.26 (r/w/o: 70457.59/20121.51/10062.16) lat (ms,95%): 17.32 err/s: 0.00 reconn/s: 0.00
# thds 是并发线程数。tps 是每秒事务数。qps 是每秒操作数，等于 r（读操作）加上 w（写操作）加上 o（其他
# 操作，主要包括 BEGIN 和 COMMIT）。lat 是延迟，(ms,95%)是 95%的查询时间小于或等于该值，单位为毫秒。
# err/s 是每秒错误数。reconn/s 是每秒重试的次数
[ 20s ] thds: 64 tps: 5108.93 qps: 102192.09 (r/w/o: 71533.28/20440.64/10218.17) lat (ms,95%): 17.32 err/s: 0.00 reconn/s: 0.00
[ 30s ] thds: 64 tps: 5126.50 qps: 102505.50 (r/w/o: 71756.30/20496.60/10252.60) lat (ms,95%): 17.32 err/s: 0.00 reconn/s: 0.00
[ 40s ] thds: 64 tps: 5144.50 qps: 102907.20 (r/w/o: 72034.07/20583.72/10289.41) lat (ms,95%): 17.01 err/s: 0.00 reconn/s: 0.00
[ 50s ] thds: 64 tps: 5137.29 qps: 102739.80 (r/w/o: 71916.99/20548.64/10274.17) lat (ms,95%): 17.01 err/s: 0.00 reconn/s: 0.00
[ 60s ] thds: 64 tps: 4995.38 qps: 99896.35 (r/w/o: 69925.98/19979.61/9990.75) lat (ms,95%): 17.95 err/s: 0.00 reconn/s: 0.00
SQL statistics:
    queries performed:
        read:                            4276622    # 读操作的数量
        write:                           1221592    # 写操作的数量
        other:                           610946     # 其他操作的数量
        total:                           6109460    # 总的操作数量, total = read + write + other
    transactions:                        305473 (5088.63 per sec.)    # 总的事务数（每秒事务数）
    queries:                             6109460 (101772.64 per sec.) # 总的操作数（每秒操作数）
    ignored errors:                      0      (0.00 per sec.)       # 忽略的错误数（每秒忽略的错误数）
    reconnects:                          0      (0.00 per sec.)       # 重试次数（每秒重试的次数）

General statistics:
    total time:                          60.0301s    # 总的执行时间
    total number of events:              305473      # 执行的事件的数量
                                                     # 在 oltp_read_write 中，默认参数下，一个事件其实
                                                     # 就是一个事务

Latency (ms):
         min:                                    5.81  # 最小耗时
         avg:                                   12.57  # 平均耗时
         max:                                  228.87  # 最大耗时
         95th percentile:                       17.32  # 95%事件的执行耗时
         sum:                              3840044.28  # 总耗时

Threads fairness:
    events (avg/stddev):           4773.0156/30.77    # 平均每个线程执行的事件的数量
                                                      # stddev 是标准差，值越小，代表结果越稳定
    execution time (avg/stddev):   60.0007/0.01       # 平均每个线程的执行时间
```

输出中，重点关注 3 个指标：

- 每秒事务数，即我们常说的 TPS；
- 每秒操作数，即我们常说的 QPS；
- 95%事件的执行耗时。

TPS 和 QPS 反映了系统的吞吐量，值越大越好。执行耗时代表了事务的执行时长，值越小越好。在一定范围内，并发线程数越大，TPS 和 QPS 也会越大。

9.1.5 如何使用 sysbench 对服务器性能进行测试

除了数据库基准测试，sysbench 还能对服务器的性能进行测试。服务器资源一般包括四大类：CPU、内存、磁盘 I/O 和网络。sysbench 可对 CPU、内存和磁盘 I/O 进行测试。下面具体来看看。

1. cpu

CPU 性能测试。支持的选项只有一个，即 --cpu-max-prime。

CPU 测试的命令如下。

```
# sysbench cpu --cpu-max-prime=20000 --threads=32 run
```

输出中，重点关注 events per second。值越大，代表 CPU 的计算性能越强。

```
CPU speed:
    events per second: 25058.08
```

下面是 CPU 测试相关的代码，可以看到，sysbench 是通过计算 --cpu-max-prime 范围内质数[1]的数量来衡量 CPU 的计算能力的。

```
int cpu_execute_event(sb_event_t *r, int thread_id)
{
  unsigned long long c;
  unsigned long long l;
  double t;
  unsigned long long n=0;

  (void)thread_id; /* unused */
  (void)r; /* unused */

  // max_prime 即命令行中指定的--cpu-max-prime
  for(c=3; c < max_prime; c++)
  {
    t = sqrt((double)c);
    for(l = 2; l <= t; l++)
      if (c % l == 0)
        break;
    if (l > t)
      n++;
  }
```

---

[1] 质数（prime number）又称素数，指的是大于 1 且只能被 1 和自身整除的自然数。在代码实现时，对于自然数 n，一般会用 2 和 $\sqrt{n}$ 之间的整数去除，如果都无法整除，则意味着 n 是个质数。

```
  return 0;
}
```

2. memory

内存测试。支持的选项如下。

- `--memory-block-size`：内存块的大小，默认为 1KB。测试时建议设置为 1MB。
- `--memory-total-size`：要传输的数据的总大小。默认为 100GB。
- `--memory-scope`：内存访问范围，可指定为 global 或 local，默认为 global。
- `--memory-hugetlb`：是否从 HugeTLB 池中分配内存，默认为 off。
- `--memory-oper`：内存操作类型，可指定为 read、write 或 none，默认为 write。
- `--memory-access-mode`：内存访问模式，可指定为 seq（顺序访问）或 rnd（随机访问），默认为 seq。

内存测试的命令如下。

```
# sysbench --test=memory --memory-block-size=1M --memory-total-size=100G --num-threads=1 run
```

输出中，重点关注以下部分。

```
102400.00 MiB transferred (23335.96 MiB/sec)
```

23335.96 MiB/sec 即数据在内存中的顺序写入速率。

3. fileio

磁盘 I/O 测试。支持的选项如下。

- `--file-num`：需要创建的文件数，默认为 128。
- `--file-block-size`：块的大小，默认为 16384，即 16KB。
- `--file-total-size`：需要创建的文件总大小，默认为 2GB。
- `--file-test-mode`：测试模式，可指定为 seqwr（顺序写）、seqrewr（顺序重写）、seqrd（顺序读）、rndrd（随机读）、rndwr（随机写）或 rndrw（随机读写）。
- `--file-io-mode`：文件的操作模式，可指定为 sync（同步 I/O）、async（异步 I/O）或 mmap，默认为 sync。
- `--file-async-backlog`：每个线程异步 I/O 队列的长度，默认为 128。
- `--file-extra-flags`：打开文件时指定的标志，可指定为 sync、dsync 或 direct，默认为空，即没有指定。
- `--file-fsync-freq`：指定持久化操作的频率，默认为 100，即每执行 100 个 I/O 请求会进行一次持久化操作。
- `--file-fsync-all`：在每次写入操作后执行持久化操作，默认为 off。
- `--file-fsync-end`：在测试结束时执行持久化操作，默认为 on。
- `--file-fsync-mode`：持久化操作的模式，可指定为 fsync 或 fdatasync，默认为 fsync。fdatasync 和 fsync 类似，只不过 fdatasync 只会更新数据，而 fsync 还会同步更新文件的属性。

- --file-merged-requests：允许合并的最多 I/O 请求数，默认为 0，即不合并。
- --file-rw-ratio：混合测试中的读写比例，默认为 1.5。

磁盘 I/O 测试主要分为以下三步。

```
# 准备测试文件
# sysbench fileio --file-num=1 --file-total-size=10G --file-test-mode=rndrw prepare

# 测试
# sysbench fileio --file-num=1 --file-total-size=10G --file-test-mode=rndrw run

# 删除测试文件
# sysbench fileio --file-num=1 --file-total-size=10G --file-test-mode=rndrw cleanup
```

输出中，重点关注以下两部分。

```
File operations:
    reads/s:                      4978.26
    writes/s:                     3318.84
    fsyncs/s:                     83.07

Throughput:
    read, MiB/s:                  77.79
    written, MiB/s:               51.86
```

输出中，reads/s 加上 writes/s 即我们常说的 IOPS。read, MiB/s 加上 written, MiB/s 即我们常说的吞吐量。

9.1.6 MySQL 常见的测试场景及对应的 SQL 语句

接下来会列举 MySQL 常见的测试场景及各个场景对应的 SQL 语句。

为了让大家清晰地知道 SQL 语句的含义，我们首先看看测试表的表结构。除了 bulk_insert 会创建单独的测试表，其他场景都会使用下面的表结构。

```
mysql> show create table sbtest.sbtest1\G
*************************** 1. row ***************************
       Table: sbtest1
Create Table: CREATE TABLE `sbtest1` (
  `id` int NOT NULL AUTO_INCREMENT,
  `k` int NOT NULL DEFAULT '0',
  `c` char(120) NOT NULL DEFAULT '',
  `pad` char(60) NOT NULL DEFAULT '',
  PRIMARY KEY (`id`),
  KEY `k_1` (`k`)
) ENGINE=InnoDB AUTO_INCREMENT=1000001 DEFAULT CHARSET=utf8mb4 COLLATE=utf8mb4_0900_ai_ci
1 row in set (0.00 sec)
```

1. bulk_insert

批量插入测试。

```
# sysbench bulk_insert --mysql-host=10.0.0.64 --mysql-port=3306 --mysql-user=admin
--mysql-password=Py@123456 --mysql-db=sbtest --tables=30 --table-size=1000000 --threads=64 --time=60
--report-interval=10 run
```

下面是 bulk_insert 场景下创建的测试表。

```
mysql> show create table sbtest.sbtest1\G
*************************** 1. row ***************************
       Table: sbtest1
Create Table: CREATE TABLE `sbtest1` (
  `id` int NOT NULL,
  `k` int NOT NULL DEFAULT '0',
  PRIMARY KEY (`id`)
) ENGINE=InnoDB DEFAULT CHARSET=utf8mb4 COLLATE=utf8mb4_0900_ai_ci
1 row in set (0.01 sec)
```

测试对应的 SQL 语句如下。

```
INSERT INTO sbtest1 VALUES(?, ?),(?, ?),(?, ?),(?, ?)...
```

2. oltp_delete

删除测试。

```
# sysbench oltp_delete --mysql-host=10.0.0.64 --mysql-port=3306 --mysql-user=admin
--mysql-password=Py@123456 --mysql-db=sbtest --tables=30 --table-size=1000000 --threads=64 --time=60
--report-interval=10 run
```

基于主键进行删除。测试对应的 SQL 语句如下。

```
DELETE FROM sbtest1 WHERE id=?
```

3. oltp_insert

插入测试。

```
# sysbench oltp_insert --mysql-host=10.0.0.64 --mysql-port=3306 --mysql-user=admin
--mysql-password=Py@123456 --mysql-db=sbtest --tables=30 --table-size=1000000 --threads=64 --time=60
--report-interval=10 run
```

测试对应的 SQL 语句如下。

```
INSERT INTO sbtest1 (id, k, c, pad) VALUES (?, ?, ?, ?)
```

4. oltp_point_select

基于主键进行查询。

```
# sysbench oltp_point_select --mysql-host=10.0.0.64 --mysql-port=3306 --mysql-user=admin
--mysql-password=Py@123456 --mysql-db=sbtest --tables=30 --table-size=1000000 --threads=64 --time=60
--report-interval=10 run
```

测试对应的 SQL 语句如下。

```
SELECT c FROM sbtest1 WHERE id=?
```

5. oltp_read_only

只读测试。

```
# sysbench oltp_read_only --mysql-host=10.0.0.64 --mysql-port=3306 --mysql-user=admin
--mysql-password=Py@123456 --mysql-db=sbtest --tables=30 --table-size=1000000 --threads=64 --time=60
--report-interval=10 run
```

测试对应的 SQL 语句如下。

```
SELECT c FROM sbtest1 WHERE id=? # 默认会执行 10 次,由--point_selects 选项控制
SELECT c FROM sbtest1 WHERE id BETWEEN ? AND ?
SELECT SUM(k) FROM sbtest1 WHERE id BETWEEN ? AND ?
SELECT c FROM sbtest1 WHERE id BETWEEN ? AND ? ORDER BY c
SELECT DISTINCT c FROM sbtest1 WHERE id BETWEEN ? AND ? ORDER BY c
```

6. oltp_read_write

读写测试。

测试对应的 SQL 语句如下。

```
SELECT c FROM sbtest1 WHERE id=? # 默认会执行 10 次,由--point_selects 选项控制
SELECT c FROM sbtest1 WHERE id BETWEEN ? AND ?
SELECT SUM(k) FROM sbtest1 WHERE id BETWEEN ? AND ?
SELECT c FROM sbtest1 WHERE id BETWEEN ? AND ? ORDER BY c
SELECT DISTINCT c FROM sbtest1 WHERE id BETWEEN ? AND ? ORDER BY c
UPDATE sbtest1 SET k=k+1 WHERE id=?
UPDATE sbtest1 SET c=? WHERE id=?
DELETE FROM sbtest1 WHERE id=?
INSERT INTO sbtest1 (id, k, c, pad) VALUES (?, ?, ?, ?)
```

7. oltp_update_index

基于主键进行更新,更新的是索引字段。

```
# sysbench oltp_update_index --mysql-host=10.0.0.64 --mysql-port=3306 --mysql-user=admin
--mysql-password=Py@123456 --mysql-db=sbtest --tables=30 --table-size=1000000 --threads=64 --time=60
--report-interval=10 run
```

测试对应的 SQL 语句如下。

```
UPDATE sbtest1 SET k=k+1 WHERE id=?
```

8. oltp_update_non_index

基于主键进行更新,更新的是非索引字段。

```
# sysbench oltp_update_non_index --mysql-host=10.0.0.64 --mysql-port=3306 --mysql-user=admin
--mysql-password=Py@123456 --mysql-db=sbtest --tables=30 --table-size=1000000 --threads=64 --time=60
--report-interval=10 run
```

测试对应的 SQL 语句如下。

```
UPDATE sbtest1 SET c=? WHERE id=?
```

9. oltp_write_only

只写测试。

```
# sysbench oltp_write_only --mysql-host=10.0.0.64 --mysql-port=3306 --mysql-user=admin
--mysql-password=Py@123456 --mysql-db=sbtest --tables=30 --table-size=1000000 --threads=64 --time=60
--report-interval=10 run
```

测试对应的 SQL 语句如下。

```
UPDATE sbtest1 SET k=k+1 WHERE id=?
UPDATE sbtest1 SET c=? WHERE id=?
```

```
DELETE FROM sbtest1 WHERE id=?
INSERT INTO sbtest1 (id, k, c, pad) VALUES (?, ?, ?, ?)
```

10. select_random_points

基于索引进行随机查询。

```
# sysbench select_random_points --mysql-host=10.0.0.64 --mysql-port=3306 --mysql-user=admin
--mysql-password=Py@123456 --mysql-db=sbtest --tables=30 --table-size=1000000 --threads=64 --time=60
--report-interval=10 run
```

测试对应的 SQL 语句如下。

```
SELECT id, k, c, pad
       FROM sbtest1
       WHERE k IN (?, ?, ?, ?, ?, ?, ?, ?, ?, ?)
```

11. select_random_ranges

基于索引进行随机范围查询。

```
# sysbench select_random_ranges --mysql-host=10.0.0.64 --mysql-port=3306 --mysql-user=admin
--mysql-password=Py@123456 --mysql-db=sbtest --tables=30 --table-size=1000000 --threads=64 --time=60
--report-interval=10 run
```

测试对应的 SQL 语句如下。

```
SELECT count(k)
       FROM sbtest1
       WHERE k BETWEEN ? AND ? OR k BETWEEN ? AND ? OR k BETWEEN ? AND ? OR k BETWEEN ? AND ? OR
k BETWEEN ? AND ? OR k BETWEEN ? AND ? OR k BETWEEN ? AND ? OR k BETWEEN ? AND ? OR k BETWEEN ? AND ?
OR k BETWEEN ? AND ?
```

9.1.7 如何自定义 sysbench 测试脚本

下面通过 bulk_insert.lua 和 oltp_point_select.lua 这两个脚本分析下 sysbench 测试脚本的实现逻辑。

首先看看 bulk_insert.lua 脚本的内容。

```
# cat bulk_insert.lua
#!/usr/bin/env sysbench

cursize=0
function thread_init()
   drv = sysbench.sql.driver()
   con = drv:connect()
end

function prepare()
   local i

   local drv = sysbench.sql.driver()
   local con = drv:connect()

   for i = 1, sysbench.opt.threads do
      print("Creating table 'sbtest" .. i .. "'...")
      con:query(string.format([[
        CREATE TABLE IF NOT EXISTS sbtest%d (
```

```
         id INTEGER NOT NULL,
         k INTEGER DEFAULT '0' NOT NULL,
         PRIMARY KEY (id))]], i))
   end
end

function event()
   if (cursize == 0) then
      con:bulk_insert_init("INSERT INTO sbtest" .. thread_id+1 .. " VALUES")
   end

   cursize = cursize + 1

   con:bulk_insert_next("(" .. cursize .. "," .. cursize .. ")")
end

function thread_done(thread_9d)
   con:bulk_insert_done()
   con:disconnect()
end

function cleanup()
   local i

   local drv = sysbench.sql.driver()
   local con = drv:connect()

   for i = 1, sysbench.opt.threads do
      print("Dropping table 'sbtest" .. i .. "'...")
      con:query("DROP TABLE IF EXISTS sbtest" .. i )
   end
end
```

下面,我们看看这几个函数的具体作用。

- thread_init():线程初始化时调用。这个函数常用来创建数据库连接。
- prepare():指定 prepare 时调用。这个函数常用来创建测试表,生成测试数据。
- event():指定 run 时调用。这个函数会定义需要测试的 SQL 语句。
- thread_done():线程退出时调用。这个函数常用来关闭 Prepared Statement 和数据库连接。
- cleanup():指定 cleanup 时调用。这个函数常用来删除测试表。

如果我们要自定义测试脚本,只需实现这几个函数即可。

如果我们要基于 sbtest 表自定义测试项,就要分析 oltp*.lua 脚本的实现逻辑。

下面,以 oltp_point_select.lua 脚本为例。

```
#!/usr/bin/env sysbench
...
require("oltp_common")

function prepare_statements()
   -- point_selects 是 oltp_point_select 中支持的选项,默认为 10,这里调整为了 1
   sysbench.opt.point_selects=1

   prepare_point_selects()
end
```

```
function event()
    execute_point_selects()
end
```

与 bulk_insert.lua 不一样的是，oltp_point_select.lua 只简单地定义了两个函数：prepare_statements() 和 event()。实际上，不仅仅是 oltp_point_select.lua，其他 oltp*.lua 脚本也只定义了这两个函数。

虽然只定义了这两个函数，但脚本导入了 oltp_common 模块，所以实际上，脚本中的 prepare_point_selects()、execute_point_selects() 和 bulk_insert.lua 中的 thread_init()、prepare()、thread_done()、cleanup() 都是在 oltp_common.lua 这个公共模块中定义的。

接下来，我们看看 prepare_point_selects() 和 execute_point_selects() 这两个函数的实现逻辑。

首先看看 prepare_point_selects()。它调用的是 prepare_for_each_table()。prepare_for_each_table() 是一个基础函数。所有 prepare 相关的函数都会调用 prepare_for_each_table()，只不过不同的 prepare 函数会传入不同的参数名。

prepare_for_each_table() 会填充两张表（Lua 中的表既可用来表示数组，也可用来表示集合）：stmt 和 param，其中，stmt 用来存储 Prepared Statement，param 用来存储 Prepared Statement 相关的参数类型。

填充完毕后，再通过 bind_param 函数将两者绑定在一起。

可以看到，无论是 Prepared Statement 还是相关的参数类型，都是在 stmt_defs 中定义的。

```
function prepare_point_selects()
    prepare_for_each_table("point_selects")
end

function prepare_for_each_table(key)
    for t = 1, sysbench.opt.tables do
        -- t 是表的序号，key 是测试项的名字
        stmt[t][key] = con:prepare(string.format(stmt_defs[key][1], t))

        local nparam = #stmt_defs[key] - 1

        if nparam > 0 then
            param[t][key] = {}
        end

        for p = 1, nparam do
            local btype = stmt_defs[key][p+1]
            local len

            if type(btype) == "table" then
                len = btype[2]
                btype = btype[1]
            end
            if btype == sysbench.sql.type.VARCHAR or
               btype == sysbench.sql.type.CHAR then
                   param[t][key][p] = stmt[t][key]:bind_create(btype, len)
            else
               param[t][key][p] = stmt[t][key]:bind_create(btype)
            end
```

```
        end

        if nparam > 0 then
            stmt[t][key]:bind_param(unpack(param[t][key]))
        end
    end
end
```

接下来,看看 stmt_defs 的内容。

```
local stmt_defs = {
    point_selects = {
        "SELECT c FROM sbtest%u WHERE id=?",
        t.INT},
    simple_ranges = {
        "SELECT c FROM sbtest%u WHERE id BETWEEN ? AND ?",
        t.INT, t.INT},
    sum_ranges = {
        "SELECT SUM(k) FROM sbtest%u WHERE id BETWEEN ? AND ?",
        t.INT, t.INT},
    order_ranges = {
        "SELECT c FROM sbtest%u WHERE id BETWEEN ? AND ? ORDER BY c",
        t.INT, t.INT},
    distinct_ranges = {
        "SELECT DISTINCT c FROM sbtest%u WHERE id BETWEEN ? AND ? ORDER BY c",
        t.INT, t.INT},
    index_updates = {
        "UPDATE sbtest%u SET k=k+1 WHERE id=?",
        t.INT},
    non_index_updates = {
        "UPDATE sbtest%u SET c=? WHERE id=?",
        {t.CHAR, 120}, t.INT},
    deletes = {
        "DELETE FROM sbtest%u WHERE id=?",
        t.INT},
    inserts = {
        "INSERT INTO sbtest%u (id, k, c, pad) VALUES (?, ?, ?, ?)",
        t.INT, t.INT, {t.CHAR, 120}, {t.CHAR, 60}},
}
```

可以看到,stmt_defs 是一张表,里面定义了不同测试项对应的 Prepared Statement 和参数类型。

具体到 point_selects 这个测试项,它对应的 Prepared Statement 是"SELECT c FROM sbtest%u WHERE id=?",对应的参数类型是 t.INT。

梳理完 prepare_point_selects()函数的实现逻辑,最后我们看看 execute_point_selects()函数的实现逻辑。

```
function execute_point_selects()
    local tnum = get_table_num()
    local i
    -- point_selects 对应命令行中的--point_selects 选项,默认为 10。
    for i = 1, sysbench.opt.point_selects do
        param[tnum].point_selects[1]:set(get_id())

        stmt[tnum].point_selects:execute()
    end
end
```

逻辑也非常简单，先赋值，最后执行。

所以如果我们要基于 sbtest 表自定义测试项，最关键的一步其实就是在 `stmt_defs` 中定义 Prepared Statement 和相关的参数类型。至于 `prepare_xxx` 和 `execute_xxx` 函数，实现起来都非常简单。

9.1.8 总结

- 基准测试一般会关注 3 个指标：TPS/QPS、执行耗时和并发量。
- 只有进行全链路压测，我们才知道系统的瓶颈在哪里。不能想当然地以为，数据库不容易横向扩展，系统瓶颈就一定会出在数据库层。事实上，很多系统在设计之初就引入了缓存，而缓存会分担很大一部分读流量，这种架构下的数据库压力其实并不大。
- 不能简单地将 sysbench 的测试结果（TPS/QPS）作为业务系统的吞吐量指标，因为两者的业务模型并不一致。
- 如果要自定义测试脚本，实现的方式有以下两种。
  - 自己实现测试相关的所有函数，具体实现细节可参考 bulk_insert.lua。
  - 基于 `sbtest` 表自定义测试项。实现过程中最关键的一步是在 `stmt_defs` 中定义 Prepared Statement 和相关的参数类型。

9.2 pt-archiver

顾名思义，pt-archiver 用于数据归档。

9.2.1 安装

我们首先来看看 Percona Toolkit 的安装。

官方提供了 Debian、Linux - Generic（Linux 通用二进制包）、RHEL、源码包和 Ubuntu 这 5 个版本的安装包。较为常用的是 Linux - Generic 版本的安装包。

下面是 Linux - Generic 包的安装步骤。

```
# yum install perl-ExtUtils-MakeMaker perl-DBD-MySQL perl-Digest-MD5
# cd /usr/src
# tar xvf percona-toolkit-3.3.1_x86_64.tar.gz
# cd percona-toolkit-3.3.1
# perl Makefile.PL
# make
# make install
```

9.2.2 实现原理

首先，通过一个简单的示例看看 pt-archiver 的实现原理。

```
pt-archiver --source h=192.168.244.10,P=3306,u=pt_user,p=pt_pass,D=employees,t=departments --dest h=192.168.244.128,P=3306,u=pt_user,p=pt_pass,D=employees,t=departments --where "1=1"
```

上述命令会将 192.168.244.10 中 employees.departments 表的数据归档到 192.168.244.128 的同名表中。

命令行中指定了 3 个参数，含义分别如下。

- `--source`：源库（业务实例）的 DSN。
- `--dest`：目标库（归档实例）的 DSN。
- `--where`：归档条件，其中"1=1"代表归档全表。

接下来看看 employees.departments 表的内容。

```
mysql> show create table employees.departments\G
*************************** 1. row ***************************
       Table: departments
Create Table: CREATE TABLE `departments` (
  `dept_no` char(4) NOT NULL,
  `dept_name` varchar(40) NOT NULL,
  PRIMARY KEY (`dept_no`),
  UNIQUE KEY `dept_name` (`dept_name`)
) ENGINE=InnoDB DEFAULT CHARSET=utf8mb4 COLLATE=utf8mb4_0900_ai_ci
1 row in set (0.00 sec)

mysql> select * from employees.departments;
+---------+--------------------+
| dept_no | dept_name          |
+---------+--------------------+
| d009    | Customer Service   |
| d005    | Development        |
| d002    | Finance            |
| d003    | Human Resources    |
| d001    | Marketing          |
| d004    | Production         |
| d006    | Quality Management |
| d008    | Research           |
| d007    | Sales              |
+---------+--------------------+
9 rows in set (0.00 sec)
```

结合 general log 的输出，我们看看 pt-archiver 的实现原理。

源库日志：

```
2022-01-05T16:00:34.768495+08:00         14 Query     SELECT /*!40001 SQL_NO_CACHE */ `dept_no`,`dept_name` FROM `employees`.`departments` FORCE INDEX(`PRIMARY`) WHERE (1=1) ORDER BY `dept_no` LIMIT 1
2022-01-05T16:00:34.769007+08:00         14 Query     DELETE FROM `employees`.`departments` WHERE (`dept_no` = 'd001')
2022-01-05T16:00:35.123870+08:00         14 Query     commit

2022-01-05T16:00:35.128063+08:00         14 Query     SELECT /*!40001 SQL_NO_CACHE */ `dept_no`,`dept_name` FROM `employees`.`departments` FORCE INDEX(`PRIMARY`) WHERE (1=1) AND ((`dept_no` >= 'd001')) ORDER BY `dept_no` LIMIT 1
2022-01-05T16:00:35.128729+08:00         14 Query     DELETE FROM `employees`.`departments` WHERE (`dept_no` = 'd002')
2022-01-05T16:00:35.342320+08:00         14 Query     commit
...
```

目标库日志：

```
2022-01-05T16:00:34.768747+08:00         18 Query     INSERT INTO `employees`.`departments`(`dept_no`,`dept_name`) VALUES ('d001','Marketing')
```

```
2022-01-05T16:00:35.120320+08:00         18 Query    commit
2022-01-05T16:00:35.128420+08:00         18 Query    INSERT INTO `employees`.`departments`(`dept_no`,
`dept_name`) VALUES ('d002','Finance')
2022-01-05T16:00:35.336487+08:00         18 Query    commit
```

结合源库和目标库的日志，可以看出以下 5 点。

- pt-archiver 首先会从源库查询一条记录，然后插入目标库，只有目标库插入成功，才会从源库中删除这条记录。这样就能确保数据在删除之前，一定是归档成功的。
- 仔细观察这几个操作的执行时间，其先后顺序如下。

    (1) 源库查询记录。
    (2) 目标库插入记录。
    (3) 源库删除记录。
    (4) 目标库 COMMIT。
    (5) 源库 COMMIT。

    这种实现借鉴了分布式事务中的两阶段提交算法。

- --where 参数中的 "1=1" 会传递到 SELECT 操作中。"1=1" 代表归档全表。也可指定其他条件，如常用的时间。
- 每次查询都是使用主键索引，这样即使归档条件中没有索引，也不会产生全表扫描。
- 每次删除都是基于主键，这样可避免归档条件没有索引导致全表被锁的风险。

9.2.3 常见用法

1. 批量归档

如果使用示例中的参数进行归档，在数据量比较大的情况下，效率会非常低，毕竟 COMMIT 是一个昂贵的操作。所以在线上，我们通常会进行批量操作。具体命令如下。

```
pt-archiver --source h=192.168.244.10,P=3306,u=pt_user,p=pt_pass,D=employees,t=departments --dest
h=192.168.244.128,P=3306,u=pt_user,p=pt_pass,D=employees,t=departments --where "1=1" --bulk-delete
--limit 1000 --commit-each --bulk-insert
```

相对于之前的归档命令，这条命令额外指定了如下 4 个参数。

- --bulk-delete：批量删除。
- --limit：每批归档的记录数。
- --commit-each：对于每一批记录，只会提交一次。
- --bulk-insert：归档数据以 LOAD DATA INFILE 的方式导入到归档库中。

看看上述命令对应的 general log。

源库：

```
2022-01-05T16:42:36.090574+08:00         18 Query    SELECT /*!40001 SQL_NO_CACHE */ `dept_no`,`dept_name`
FROM `employees`.`departments` FORCE INDEX(`PRIMARY`) WHERE (1=1) ORDER BY `dept_no` LIMIT 1000
...
```

```
2022-01-05T16:42:36.099878+08:00        18 Query    DELETE FROM `employees`.`departments` WHERE (((`dept_no`
>= 'd001'))) AND (((`dept_no` <= 'd009'))) AND (1=1) LIMIT 1000
2022-01-05T16:42:39.301175+08:00        18 Query    SELECT /*!40001 SQL_NO_CACHE */ `dept_no`,`dept_name`
FROM `employees`.`departments` FORCE INDEX(`PRIMARY`) WHERE (1=1) AND ((`dept_no` >= 'd009')) ORDER BY `dept_no`
LIMIT 1000
2022-01-05T16:42:39.305939+08:00        18 Query    commit
```

目标库：

```
2022-01-05T16:42:36.098438+08:00             25 Query    LOAD DATA LOCAL INFILE '/tmp/H5jUQKXdgapt-
archiver' INTO TABLE `employees`.`departments`(`dept_no`,`dept_name`)
2022-01-05T16:42:39.301848+08:00             25 Query    commit
```

在执行批量归档操作时，注意以下 4 点。

- 如果要执行 LOAD DATA LOCAL INFILE 操作，需将目标库的 `local_infile` 参数设置为 ON。
- 如果不指定 `--bulk-insert` 且没指定 `--commit-each`，则目标库的插入还是会像示例中显示的那样，逐行提交。
- 如果不指定 `--commit-each`，即使表中的 9 条记录是通过一条 DELETE 命令删除的，但因为涉及了 9 条记录，pt-archiver 也会执行 COMMIT 操作 9 次。目标库同样如此。
- 在使用 `--bulk-insert` 归档时要注意，如果导入的过程中出现问题，如主键冲突，pt-archiver 是不会提示任何错误的。

接下来我们通过表 9-2 看看归档 20 万条数据，不同归档参数之间的执行时间对比。

表 9-2　不同归档参数之间的执行时间对比

归档参数	执行时间（单位是秒）
不指定任何批量相关参数	850.040
--bulk-delete --limit 1000	422.352
--bulk-delete --limit 1000 --commit-each	46.646
--bulk-delete --limit 5000 --commit-each	46.111
--bulk-delete --limit 1000 --commit-each --bulk-insert	7.650
--bulk-delete --limit 5000 --commit-each --bulk-insert	6.540
--bulk-delete --limit 1000 --bulk-insert	47.273

通过表格中的数据，我们可以得出以下 5 点：

- 第一种方式是最慢的。这种情况下，无论是源库还是归档库，都是逐行操作并提交的。
- 只指定 `--bulk-delete --limit 1000` 依然很慢。这种情况下，源库虽然是批量删除，但 COMMIT 次数并没有减少。归档库依然是逐行插入并提交的。
- `--bulk-delete --limit 1000 --commit-each` 相当于第二种归档方式，源库和目标库都是批量提交的。
- `--limit 1000` 和 `--limit 5000` 归档性能相差不大。
- `--bulk-delete --limit 1000 --bulk-insert` 与 `--bulk-delete --limit 1000 --commit-each --bulk-insert` 相比，没有设置 `--commit-each`。虽然都是批量操作，但前者会执行 COMMIT 操作 1000 次。由此来看，空事务并不是没有代价的。

2. 删除数据

删除数据是 pt-archiver 另外一个常见的使用场景。具体命令如下。

```
pt-archiver --source h=192.168.244.10,P=3306,u=pt_user,p=pt_pass,D=employees,t=departments --where
"1=1" --bulk-delete --limit 1000 --purge
```

命令行中的 `--purge` 代表只删除，不归档。除了 `--purge`，建议加上 `--primary-key-only`。这样，在执行 SELECT 操作时，就只会查询主键，不会查询所有列。

接下来，我们看看删除命令相关的 general log。

为了直观地展示 pt-archiver 删除数据的实现逻辑，实际测试时将 `--limit` 设置为了 3。

```
# 开启事务
set autocommit=0;

# 查看表结构，获取主键
SHOW CREATE TABLE `employees`.`departments`;

# 开始删除第一批数据
# 通过 FORCE INDEX(`PRIMARY`) 强制使用主键
# 指定了 --primary-key-only，所以只会查询主键
# 这里其实无须获取所有满足条件的主键值，只取一个最小值和最大值即可
SELECT /*!40001 SQL_NO_CACHE */ `dept_no` FROM `employees`.`departments` FORCE INDEX(`PRIMARY`) WHERE
(1=1) ORDER BY `dept_no` LIMIT 3;

# 基于主键进行删除，删除的时候同时带上了 --where 指定的删除条件，以避免误删
DELETE FROM `employees`.`departments` WHERE (((`dept_no` >= 'd001'))) AND (((`dept_no` <= 'd003')))
AND (1=1) LIMIT 3;

# 提交
commit;

# 删除第二批数据
SELECT /*!40001 SQL_NO_CACHE */ `dept_no` FROM `employees`.`departments` FORCE INDEX(`PRIMARY`) WHERE
(1=1) AND ((`dept_no` >= 'd003')) ORDER BY `dept_no` LIMIT 3;
DELETE FROM `employees`.`departments` WHERE (((`dept_no` >= 'd004'))) AND (((`dept_no` <= 'd006')))
AND (1=1); LIMIT 3
commit;

# 删除第三批数据
SELECT /*!40001 SQL_NO_CACHE */ `dept_no` FROM `employees`.`departments` FORCE INDEX(`PRIMARY`) WHERE
(1=1) AND ((`dept_no` >= 'd006')) ORDER BY `dept_no` LIMIT 3;
DELETE FROM `employees`.`departments` WHERE (((`dept_no` >= 'd007'))) AND (((`dept_no` <= 'd009')))
AND (1=1) LIMIT 3;
commit;

# 删除最后一批数据
SELECT /*!40001 SQL_NO_CACHE */ `dept_no` FROM `employees`.`departments` FORCE INDEX(`PRIMARY`) WHERE
(1=1) AND ((`dept_no` >= 'd009')) ORDER BY `dept_no` LIMIT 3;
commit;
```

在业务代码中，如果我们有类似的删除需求，不妨借鉴 pt-archiver 的实现方式。

3. 将数据归档到文件中

数据除了能归档到数据库，也可归档到文件中。

```
pt-archiver --source h=192.168.244.10,P=3306,u=pt_user,p=pt_pass,D=employees,t=departments --where
"1=1" --bulk-delete --limit 1000 --file '/tmp/%Y-%m-%d-%D.%t'
```

命令中指定的是 `--file`，而不是 `--dest`。生成的文件是 CSV 格式，后续可通过 LOAD DATA INFILE 命令加载到数据库中。

无论是数据归档还是删除，对于源库，都需要执行 DELETE 操作。很多人担心，如果删除的记录数太多，会造成主从延迟。事实上，pt-archiver 本身就具备了基于主从延迟来自动调节归档（删除）操作的能力。如果从库的延迟超过 1 秒（由 `--max-lag` 指定）或复制状态不正常，则会暂停归档（删除）操作，直到其恢复。默认情况下，pt-archiver 不会检查从库的延迟情况，如果要检查，需通过 `--check-slave-lag` 设置从库的地址，例如：

```
pt-archiver --source h=192.168.244.10,P=3306,u=pt_user,p=pt_pass,D=employees,t=departments --where
"1=1" --bulk-delete --limit 1000 --commit-each --primary-key-only --purge --check-slave-lag
h=192.168.244.20,P=3306,u=pt_user,p=pt_pass
```

这里只会检查 192.168.244.20 的延迟情况。如果有多个从库需要检查，需多次指定 `--check-slave-lag`，每次对应一个从库。

9.2.4 常用参数

下面看看 pt-archiver 中的其他常用参数。

- `--analyze`

在执行完归档操作后，执行 ANALYZE TABLE 操作。

后面可接任意字符串，如果字符串中含有 s，则会在源库中执行 ANALYZE 操作。如果字符串中含有 d，则会在目标库中执行 ANALYZE 操作。如果字符串同时含有 d 和 s，则在源库和目标库中均会执行 ANALYZE 操作。例如：

```
--analyze ds
```

- `--optimize`

在执行完归档操作后，执行 OPTIMIZE TABLE 操作。用法与 `--analyze` 类似。

- `--charset`

指定连接（Connection）字符集。在 MySQL 8.0 之前，默认是 latin1。在 MySQL 8.0 中，默认是 utf8mb4。注意，这里的默认值与 MySQL 服务端字符集 character_set_server 无关。

若显式设置了该值，pt-archiver 在建立连接后，会首先执行 SET NAMES charset_name 操作。

- `--[no]check-charset`

检查源库及目标库中连接（Connection）字符集和表的字符集是否一致。

如果不一致，会提示以下错误：

```
Character set mismatch: --source DSN uses latin1, table uses gbk. You can disable this check by specifying
--no-check-charset.
```

这个时候，切记不要按照提示指定 `--no-check-charset` 忽略检查，否则很容易导致乱码。针对上述报错，可将 `--charset` 指定为表的字符集。

注意，该选项并不是比较源库和目标库的字符集是否一致。

- `--[no]check-columns`

检查源表和目标表列名是否一致。注意，只会检查列名，不会检查列的顺序、列的数据类型是否一致。

- `--columns`

归档指定列。在有自增列的情况下，如果源表和目标表的自增列存在交集，可不归档自增列。这个时候，需要使用 `--columns` 显式指定归档列。

- `--dry-run`

只打印待执行的 SQL，而不实际执行。常用于实际操作之前，校验待执行的 SQL 是否符合自己的预期。

- `--ignore`

使用 INSERT IGNORE 归档数据。

- `--no-delete`

不删除源库的数据。

- `--replace`

使用 REPLACE 操作归档数据。

- `--[no]safe-auto-increment`

在归档有自增主键的表时，默认不会删除自增主键最大的那一行。这样做，主要是为了规避在 MySQL 8.0 之前自增主键不能持久化的问题。在对全表进行归档时，需要注意这一点。如果确实需要删除，需指定 `--no-safe-auto-increment`。

- `--source`

给出源库实例的信息。

除了常用的选项，它还支持如下选项。

(1) a：指定连接的默认数据库。
(2) b：设置 SQL_LOG_BIN=0。如果是在源库指定，则 DELETE 操作不会写入到 binlog 中；如果是在目标库指定，则 INSERT 操作不会写入 binlog。
(3) i：设置归档操作使用的索引，默认是主键。

- `--progress`

显示进度信息，如 `--progress 10000` 代表每归档（删除）10 000 行就打印一次进度信息。

```
TIME                    ELAPSED    COUNT
2022-01-05T18:24:19           0        0
2022-01-05T18:24:20           0    10000
2022-01-05T18:24:21           1    20000
```

第一列是当前时间，第二列是已经消耗的时间，第三列是已归档（删除）的行数。

另外，Percona Toolkit 中用 Perl 语言开发的工具，可通过环境变量 PTDEBUG 开启调试模式。例如：

```
PTDEBUG=1 pt-archiver ... > pt_archiver_debug.log 2>&1
```

输出的日志会记录在 pt_archiver_debug.log 文件中。如果不指定，则默认输出到终端。

9.2.5 总结

前面对比了归档操作中不同参数的执行时间，其中 --bulk-delete --limit 1000 --commit-each --bulk-insert 是最快的。不指定任何批量操作参数是最慢的。

但在使用 --bulk-insert 时要注意，如果导入的过程中出现问题，pt-archiver 是不会提示任何错误的。常见的错误有主键冲突，数据和目标列的数据类型不一致。

如果不使用 --bulk-insert，而是通过默认的 INSERT 操作来归档，大部分错误是可以识别出来的。例如，主键冲突，会提示以下错误。

```
DBD::mysql::st execute failed: Duplicate entry 'd001' for key 'PRIMARY' [for Statement
"INSERT INTO `employees`.`departments`(`dept_no`,`dept_name`) VALUES (?,?)" with
ParamValues: 0='d001', 1='Marketing'] at /usr/local/bin/pt-archiver line 6772.
```

导入的数据和目标列的数据类型不一致，会提示以下错误。

```
DBD::mysql::st execute failed: Incorrect integer value: 'Marketing' for column 'dept_name'
at row 1 [for Statement "INSERT INTO `employees`.`departments`(`dept_no`,`dept_name`) VALUES
(?,?)" with ParamValues: 0='d001', 1='Marketing'] at /usr/local/bin/pt-archiver line 6772.
```

当然，数据和类型不一致，能被识别出来的前提是归档实例的 SQL_MODE 为严格模式。如果待归档的实例中有 MySQL 5.6，我们其实很难将归档实例的 SQL_MODE 开启为严格模式。因为 MySQL 5.6 的 SQL_MODE 默认为非严格模式，所以难免会产生很多无效数据，比如时间字段中的 0000-00-00 00:00:00。如果把这种无效数据插入开启了严格模式的归档实例中，会直接报错。

从数据安全的角度出发，最推荐的归档方式如下。

(1) 先归档，但不删除源库的数据。
(2) 比对源库和归档库的数据是否一致。
(3) 如果比对结果一致，再删除源库的归档数据。

步骤(1)和步骤(3)可通过 pt-archiver 实现，步骤(2)可通过 pt-table-sync 实现。

虽然这种归档方式相对于边归档边删除的方式麻烦不少，但更安全。

9.3 pt-config-diff

pt-config-diff 能找出不同参数源之间的参数差异，其语法如下。

```
pt-config-diff [OPTIONS] CONFIG CONFIG [CONFIG...]
```

这里的 CONFIG 是参数源，既可以是 DSN，也可以是文件名。文件既可以是配置文件，也可以是 mysqld --help --verbose、my_print_defaults 或 SHOW VARIABLES 的输出。

pt-config-diff 常见的使用场景有 3 个。

- 配置文件和配置文件之间的参数对比。

  ```
  # pt-config-diff /etc/my_3306.cnf /etc/my_3307.cnf
  ```

- 实例和实例之间的参数对比。

  ```
  # pt-config-diff h=127.0.0.1,P=3306,u=root,p=123456 h=127.0.0.1,P=3307,u=root,p=123456
  ```

- 配置文件和实例之间的参数对比。

  ```
  # pt-config-diff /etc/my.cnf h=127.0.0.1,P=3307,u=root,p=123456
  ```

不足之处是，pt-config-diff 只会输出在两个参数源中都存在但值不同的参数。如果某个参数只在一个参数源中存在，则不会输出。

9.4 pt-ioprofile

pt-ioprofile 的作用是统计指定进程的 I/O 操作。

pt-ioprofile 是用 Shell 语言开发的，实际依赖的是 lsof 和 strace 命令。strace 命令会对进程的性能造成较大的影响，所以在线上应慎用。

pt-ioprofile 的使用较为简单，直接执行 pt-ioprofile 即可，它默认统计 mysqld 进程的 I/O 操作。

```
# pt-ioprofile
Wed Jan 12 11:06:40 CST 2022
Tracing process ID 7950
     total      pread64     pwrite64        write    fdatasync        fsync filename
 18.771912     0.000000     1.587137     0.000000     0.000000    17.184775 /data/mysql/3306/data/ib_logfile1
  6.202627     5.300540     0.000000     0.000000     0.000000     0.902087 /data/mysql/3306/data/sbtest/sbtest3.ibd
  6.060029     5.329841     0.000000     0.000000     0.000000     0.730188 /data/mysql/3306/data/sbtest/sbtest5.ibd
  5.608060     4.960819     0.000000     0.000000     0.000000     0.647241 /data/mysql/3306/data/sbtest/sbtest9.ibd
  5.392277     4.683582     0.000000     0.000000     0.000000     0.708695 /data/mysql/3306/data/sbtest/sbtest7.ibd
  5.384356     4.519996     0.000000     0.000000     0.000000     0.864360 /data/mysql/3306/data/sbtest/sbtest4.ibd
  5.330371     4.554545     0.000000     0.000000     0.000000     0.775826 /data/mysql/3306/data/sbtest/sbtest6.ibd
  5.276978     0.000000     0.000000     0.113769     5.163209     0.000000 /data/mysql/3306/data/mysql-bin.000027
  5.149105     4.211879     0.000000     0.000000     0.000000     0.937226 /data/mysql/3306/data/sbtest/sbtest2.ibd
  ...
```

输出中，pread64、pwrite64、write、fdatasync、fsync 是 I/O 操作相关的系统调用，默认统计 I/O 操作的时间。

下面看看 I/O 操作相关的系统调用的具体含义。

(1) read：把文件中的数据读取到缓冲区中。

(2) write：把缓冲区中的数据写入文件。

(3) lseek：设置文件的读写指针，可用来设置文件的下一个读写位置。

(4) pread64：相当于顺序执行了 lseek 和 read 操作。

(5) pwrite64：相当于顺序执行了 lseek 和 write 操作。

(6) fsync：将数据同步刷新到磁盘上，该操作会等待硬盘 I/O 完成再返回。除了数据，fsync 操作每次还会更新文件的元数据信息，如 size（文件大小）、atime（访问时间）、mtime（修改时间）。

(7) fdatasync：功能和 fsync 类似，只同步数据，仅在必要的时候才会更新文件的元数据信息。

(8) open：打开一个文件，并返回这个文件的描述符。

(9) close：关闭文件。

(10) fcntl：基于文件描述符来进行文件控制和 I/O 控制。

在使用时，需要注意以下几点。

- pt-ioprofile 不仅可以统计 mysqld 进程的 I/O 操作，也可统计其他进程的 I/O 操作。可通过 --profile-process 指定进程名，或通过 --profile-pid 指定进程 ID。
- 可通过 --cell 参数定义 I/O 操作的统计维度，默认是 times，即统计 I/O 操作的时间。也可设置为 sizes，统计 I/O 操作的大小，或设置为 count，统计 I/O 操作的次数。
- 采集时间由 --run-time 参数指定，默认为 30 秒。

在 CentOS 7 中，pt-ioprofile 存在 bug，不会输出任何 I/O 统计结果。

```
# pt-ioprofile
Wed Jan  5 21:01:55 CST 2022
Tracing process ID 1122
     total filename
```

究其原因，是 strace 的输出从 CentOS 6.x 到 CentOS 7.x 发生了变化，导致 awk 在聚合分析时出现了问题，相关 bug 可参考 "pt-ioprofile doesn't work in CentOS 7.5 box"。截止到 Percona Toolkit 3.3.1，该 bug 还没被修复。

该 bug 的修复方法如下。

```
574c574
<     /^Process/ { mode = "strace"; }
---
>     /^(strace: )?Process/ { mode = "strace"; }
```

也就是将 /^Process/ { mode = "strace"; } 替换为 /^(strace:)?Process/ { mode = "strace"; }。

9.5　pt-kill

杀掉指定连接。

9.5.1　实现原理

首先，看看线上的一个高频需求，即杀掉执行时间超过 30 秒的慢查询。具体命令如下。

```
pt-kill h=192.168.244.10,P=3306,u=pt_user,p=pt_pass --busy-time 30 --interval 10 --print --kill
--match-info "(?i-xsm:select)"
```

命令行中的 --busy-time 定义了慢查询的阈值，--interval 指的是检测时间间隔，这里 pt-kill 会每隔 10 秒执行一次 SHOW FULL PROCESSLIST 操作，看看是否有执行时间超过 30 秒的查询。如果有，则执行 KILL 操作（由 --kill 参数决定），并将执行的 KILL 操作及被杀掉的 SQL 语句打印出来（--print）。

注意，--busy-time 针对的是 Command 列为 Query 的操作，而 SHOW PROCESSLIST 中 Command 列为 Query 的操作不仅仅包括 SELECT，同样也包括 DELECT、INSERT、UPDATE 和 ALTER 操作。所以为了保证杀掉的一定是 SELECT 操作，这里使用了 --match-info 进行过滤。--match-info 匹配的是 SHOW PROCESSLIST 中 Info 列的内容。?i-xsm:^select 是正则表达式，匹配以 select 开头的操作，不区分大小写。

看看该命令的输出及对应的 general log。

```
# 2022-01-05T21:28:57 KILL 103 (Query 39 sec) select sleep(100)
# 2022-01-05T21:29:07 KILL 105 (Query 47 sec) select sleep(200)

2022-01-05T21:28:47.348592+08:00      106 Query     SHOW FULL PROCESSLIST
2022-01-05T21:28:57.349148+08:00      106 Query     SHOW FULL PROCESSLIST
2022-01-05T21:28:57.349763+08:00      106 Query     KILL '103'
2022-01-05T21:29:07.350167+08:00      106 Query     SHOW FULL PROCESSLIST
2022-01-05T21:29:07.350651+08:00      106 Query     KILL '105'
2022-01-05T21:29:17.352402+08:00      106 Query     SHOW FULL PROCESSLIST
```

可以看到，在杀掉第一个查询的时候，第二个查询其实也满足条件，但没被杀掉，而是等到下一轮检测才被杀掉。这个行为实际上是由 --victims 参数控制的，--victims 取值如下。

(1) oldest：每次只会杀掉执行时间最长的那个查询，是默认值。
(2) all：杀掉所有符合条件的查询。
(3) all-but-oldest：杀掉所有符合条件的查询，除了执行时间最长的那个。

既然是基于 SHOW PROCESSLIST 的输出，pt-kill 就可从多个维度进行过滤，具体的过滤参数如下。

- **--ignore-user**、**--match-user**

基于 USER 列的输出进行过滤。

- **--ignore-host**、**--match-host**

基于 HOST 列的输出进行过滤。

- **--ignore-db**、**--match-db**

基于 db 列的输出进行过滤。

- **--ignore-command**、**--match-command**

基于 command 列的输出进行过滤。

- **--ignore-state**、**--match-state**

基于 State 列的输出进行过滤。

- `--ignore-info`、`--match-info`

  基于 Info 列的输出进行过滤。

以上过滤参数均支持正则匹配。

需要注意的是，如果同时指定了 `--busy-time` 和过滤参数，对于 Command 列不为 Query 的操作，此时起作用的将只有过滤参数，没有`--busy-time`。看下面这个示例。

```
# mysql -h 192.168.244.10 -uu1 -p123456
mysql> select connection_id();
+-----------------+
| connection_id() |
+-----------------+
|             113 |
+-----------------+
1 row in set (0.00 sec)

mysql> begin;
Query OK, 0 rows affected (0.00 sec)

mysql> delete from slowtech.t1 limit 1;
Query OK, 1 row affected (0.00 sec)
```

执行 pt-kill 操作。

```
pt-kill h=192.168.244.10,P=3306,u=pt_user,p=pt_pass --busy-time 30 --interval 10 --print --kill --match-user u1
# 2022-01-05T21:40:43 KILL 113 (Sleep 9 sec) NULL
```

这本来是要将来自 u1 的执行时间超过 30 秒的操作杀掉，却意外地杀掉了一个未提交的事务。

究其原因，是 `--busy-time` 只对 Command 列为 Query 的操作才有效果，而这个事务对应的 Command 列是 Sleep。

9.5.2 过滤逻辑

下面从源码的角度分析 pt-kill 的过滤逻辑，这样我们才能更加清晰地知道`--busy-time` 和过滤参数之间的关系。

```perl
sub find {
   my ( $self, $proclist, %find_spec ) = @_;
   PTDEBUG && _d('find specs:', Dumper(\%find_spec));
   my $ms  = $self->{MasterSlave};
   # 定义一个数组，用来存储需要杀掉的操作
   my @matches;
   $self->{_reasons_for_matching} = undef;
   QUERY:
   # 遍历 SHOW FULL PROCESSLIST 的输出
   foreach my $query ( @$proclist ) {
      PTDEBUG && _d('Checking query', Dumper($query));
      my $matched = 0;
      # 如果命令行中不指定--replication-threads，则默认会跳过复制相关线程
      if (   !$find_spec{replication_threads}
          && $ms->is_replication_thread($query) ) {
         PTDEBUG && _d('Skipping replication thread');
```

```perl
            next QUERY;
        }
        # $self->{kill_busy_commands}是一张哈希表，exists 用来判断哈希表中是否有指定键
        # $self->{kill_busy_commands}中的键由--kill-busy-commands 指定，不指定则默认为 Query。
        if ( $find_spec{busy_time} && exists($self->{kill_busy_commands}->{$query->{Command} || ''}) ) {
            next QUERY unless defined($query->{Time});
            # 如果操作的执行时间小于--busy-time，则会跳过当前操作，不会进行其他判断
            if ( $query->{Time} < $find_spec{busy_time} ) {
                PTDEBUG && _d("Query isn't running long enough");
                next QUERY;
            }
            my $reason = 'Exceeds busy time';
            PTDEBUG && _d($reason);
            push @{$self->{_reasons_for_matching}->{$query} ||= []}, $reason;
            $matched++;
        }

        # 如果命令行中指定了--idle-time，则只会匹配 Command 为 Sleep 类型的操作
        if ( $find_spec{idle_time} && ($query->{Command} || '') eq 'Sleep' ) {
            next QUERY unless defined($query->{Time});
            # 如果操作的执行时间小于--idle-time，则会跳过当前操作，不会进行其他判断
            if ( $query->{Time} < $find_spec{idle_time} ) {
                PTDEBUG && _d("Query isn't idle long enough");
                next QUERY;
            }
            my $reason = 'Exceeds idle time';
            PTDEBUG && _d($reason);
            push @{$self->{_reasons_for_matching}->{$query} ||= []}, $reason;
            $matched++;
        }

    PROPERTY:
        # 判断操作是否满足--ignore-user，--match-user 之类参数指定的条件
        foreach my $property ( qw(Id User Host db State Command Info) ) {
            my $filter = "_find_match_$property";
            # 如果设置了 ignore 相关的参数，且操作满足 ignore 参数指定的条件，则会跳过当前操作
            if ( defined $find_spec{ignore}->{$property}
                && $self->$filter($query, $find_spec{ignore}->{$property}) ) {
                PTDEBUG && _d('Query matches ignore', $property, 'spec');
                next QUERY;
            }
            # 如果设置了 match 相关的参数，且操作不满足 match 参数指定的条件，则会跳过当前操作
            if ( defined $find_spec{match}->{$property} ) {
                if ( !$self->$filter($query, $find_spec{match}->{$property}) ) {
                    PTDEBUG && _d('Query does not match', $property, 'spec');
                    next QUERY;
                }
                my $reason = 'Query matches ' . $property . ' spec';
                PTDEBUG && _d($reason);
                push @{$self->{_reasons_for_matching}->{$query} ||= []}, $reason;
                $matched++;
            }
        }
        # 将满足条件、需要杀掉的操作添加到@matches
        # $find_spec{all}对应命令行中的--match-all 参数
        if ( $matched || $find_spec{all} ) {
            PTDEBUG && _d("Query matched one or more specs, adding");
            push @matches, $query;
            next QUERY;
        }
        PTDEBUG && _d('Query does not match any specs, ignoring');
```

```
    } # QUERY
    return @matches;
}
```

从源码中可以得出以下几点。

(1) --busy-time 只适用于 Command 列为 Query 的操作。

(2) --idle-time 只适用于 Command 列为 Sleep 的操作。

(3) --idle-time 和--busy-time 的处理逻辑相同。

(4) 对于 Command 列不为 Query 的操作，只能通过 --ignore-user、--match-user 之类的参数进行过滤。

(5) 对于 Command 列为 Query 的操作，当执行时长超过 --busy-time 时，将进一步通过 --ignore-user、--match-user 之类的参数进行过滤。

(6) --match-all 参数用来匹配所有未被忽略的操作，可用来实现否定匹配的功能。

9.5.3 常见用法

1. 将 KILL 操作记录在数据库中

具体命令如下。

```
pt-kill h=192.168.244.10,P=3306,u=pt_user,p=pt_pass --busy-time 30 --interval 10 --print --kill
--log-dsn h=192.168.244.10,P=3306,u=pt_user,p=pt_pass,D=percona,t=kill_log --create-log-table
```

KILL 操作会记录在 --log-dsn 指定的实例中，如果表不存在，可指定 --create-log-table 创建。表中记录如下。

```
mysql> select *  from percona.kill_log limit 1\G
*************************** 1. row ***************************
   kill_id: 1
 server_id: 1
 timestamp: 2022-01-05 22:00:11
    reason: Exceeds busy time
kill_error:
        Id: 128
      User: root
      Host: localhost
        db: NULL
   Command: Query
      Time: 35
     State: User sleep
      Info: select sleep(120)
   Time_ms: NULL
1 row in set (0.00 sec)
```

2. 将 pt-kill 作为守护进程运行

具体命令如下。

```
pt-kill h=192.168.244.10,P=3306,u=pt_user,p=pt_pass --busy-time 30 --interval 10 --print --kill --log
/tmp/pt-kill.log --daemonize
```

执行的 kill 操作会记录在 `--log` 指定的文件中。

默认情况下，pt-kill 不会杀掉复制相关的连接。

上述命令都指定了 `--kill`，此时会杀掉连接。如果只想杀掉查询，而不是连接，可指定 `--kill-query`。如果只是打印，而不是实际执行 KILL 操作，只需指定 `--print`。

9.6 pt-pmp

pt-pmp 有两方面的作用：一是采集进程的栈信息，二是对这些栈信息进行汇总。

进程的栈信息是通过 gdb 来获取的，在采集的过程中，会对进程的性能造成一定的影响。如非必要，在线上应慎用。

pt-pmp 的使用较为简单，直接执行 `pt-pmp` 即可，默认的采集对象是 mysqld 进程。

```
# pt-pmp
Thu Jan  6 10:47:29 CST 2022
      7 __io_getevents_0_4(libaio.so.1),LinuxAIOHandler::collect,LinuxAIOHandler::poll,
os_aio_linux_handler,os_aio_handler,fil_aio_wait,io_handler_thread,void,std::__invoke<void,
std::_Bind<void,std::_Bind<void,Detached_thread::operator()<void,std::__invoke_impl<void,,
std::__invoke<Detached_thread,,std::thread::_Invoker<std::tuple<Detached_thread,,std::thread::
_Invoker<std::tuple<Detached_thread,,std::thread::_State_impl<std::thread::_Invoker<std::tuple
<Detached_thread,,execute_native_thread_routine,start_thread(libpthread.so.0),clone(libc.so.6)
      3 pthread_cond_wait,os_event::wait,os_event::wait_low,os_event_wait_low,sync_array_wait_
event,TTASEventMutex::wait,TTASEventMutex::spin_and_try_lock,TTASEventMutex::enter,PolicyMutex<TTA
SEventMutex<GenericPolicy>,mutex_enter_inline<PolicyMutex<TTASEventMutex<GenericPolicy>>>,buf_page
_io_complete,fil_aio_wait,io_handler_thread,void,std::__invoke<void,std::_Bind<void,std::_Bind<voi
d,Detached_thread::operator()<void,std::__invoke_impl<void,,std::__invoke<Detached_thread,,std::th
read::_Invoker<std::tuple<Detached_thread,,std::thread::_Invoker<std::tuple<Detached_thread,,std::
thread::_State_impl<std::thread::_Invoker<std::tuple<Detached_thread,,execute_native_thread_routin
e,start_thread(libpthread.so.0),clone(libc.so.6)
      2 pthread_cond_wait,os_event::wait,os_event::wait_low,os_event_wait_low,sync_array_wait_
event,TTASEventMutex::wait,TTASEventMutex::spin_and_try_lock,TTASEventMutex::enter,PolicyMutex<TTA
SEventMutex<GenericPolicy>,mutex_enter_inline<PolicyMutex<TTASEventMutex<GenericPolicy>>>,buf_pool
_watch_set,Buf_fetch<Buf_fetch_other>::is_on_watch,Buf_fetch_other::get,Buf_fetch_other::get,Buf_f
etch::single_page,buf_page_get_gen,btr_cur_search_to_nth_level,btr_pcur_t::open,row_search_index_e
ntry,row_purge_remove_sec_if_poss_leaf,row_purge_remove_sec_if_poss,row_purge_del_mark,row_purge_r
ecord_func,row_purge,row_purge_step,que_thr_step,que_run_threads_low,que_run_threads,srv_task_exec
ute,srv_worker_thread,void,std::__invoke<void,std::_Bind<void,std::_Bind<void,Detached_thread::ope
rator()<void,std::__invoke_impl<void,,std::__invoke<Detached_thread,,std::thread::_Invoker<std::tu
ple<Detached_thread,,std::thread::_Invoker<std::tuple<Detached_thread,,std::thread::_State_impl<st
d::thread::_Invoker<std::tuple<Detached_thread,,execute_native_thread_routine,start_thread(libpthr
ead.so.0),clone(libc.so.6)
```

常见用法

1. 汇总 pstack 获取的结果

```
# ps -ef |grep mysqld
# pstack 10230 > 10230.info
# pt-pmp 10230.info
```

这里的 10230 是 mysqld 的进程 ID。

2. 打印指定进程的栈信息

要打印指定进程的栈信息，既可指定进程名：

```
# pt-pmp --binary mysqld
```

也可指定 pid：

```
# pt-pmp --pid 20291
```

上述命令只会迭代一次。也可迭代多次，每次迭代间隔若干时间。例如：

```
# pt-pmp --binary mysqld --iterations 2 --interval 1
```

如果要同时保留汇总前的栈信息，可指定 `--save-samples` 参数。

```
# pt-pmp --binary mysqld --save-samples mysqld.txt
```

9.7 pt-query-digest

pt-query-digest 可对多种日志进行汇总分析，如 binlog、general log、tcpdump、slowlog、SQL 文件，最常见的对慢日志进行分析。

9.7.1 常见用法

1. 分析慢日志

首先看看慢日志的内容。

```
# Time: 2022-01-06T11:19:37.097139+08:00
# User@Host: root[root] @ localhost []  Id:      18
# Query_time: 0.075636  Lock_time: 0.000115 Rows_sent: 2  Rows_examined: 19580
SET timestamp=1641439177;
select * from dept_emp where dept_no='d001' and from_date='1986-06-26';
```

第一行的 Time 是 SQL 的结束时间。

第二行记录了 SQL 的执行用户及 Processlist ID。

第三行给出了 SQL 的查询时间、锁等待时间、返回给客户端的行数，以及存储引擎中检索的行数。一般而言，Rows_sent 和 Rows_examined 越小越好。

需要注意的是，这里的 Query_time 包括了 Lock_time。如果一个 SQL 的 Lock_time 大于等于 long_query_time 但 Query_time 减去 Lock_time 小于 long_query_time，则这个 SQL 是不会记录在慢日志中的。不过从 MySQL 8.0.28 开始，这个计算方式发生了变化，在新的计算方式中，只要 SQL 的 Query_time 大于等于 long_query_time 就会记录在慢日志中。这里的 long_query_time 是慢日志的阈值。

最后给出了 SQL 开始执行的时间，及具体的 SQL。注意，在 MySQL 8.0.14 之前，SET timestamp 记录的是 SQL 语句的结束时间。

接下来，我们使用 pt-query-digest 分析一下慢日志。使用方法很简单，直接指定慢日志即可。例如：

```
# pt-query-digest /data/mysql/3306/log/slow.log
```

下面我们分析一下输出结果。

```
# 110ms user time, 10ms system time, 22.29M rss, 185.92M vsz
# Current date: Thu Jan  6 11:21:49 2022
# Hostname: slowtech
# Files: /data/mysql/3306/data/slowtech-slow.log
# Overall: 7 total, 6 unique, 0.05 QPS, 0.01x concurrency _____
# Time range: 2022-01-06T11:19:13 to 2022-01-06T11:21:25
# Attribute          total     min     max     avg     95%  stddev  median
# ============     =======  ======  ======  ======  ======  ======  ======
# Exec time           904ms   222us   438ms   129ms   433ms   137ms    75ms
# Lock time           750us    91us   117us   107us   113us     8us   103us
# Rows sent          244.81k      0 225.65k  34.97k 222.42k  77.01k    1.96
# Rows examine       674.19k      1 313.04k  96.31k 312.96k 129.25k   18.47k
# Query size             391     31      85   55.86   84.10   18.00   49.17

# Profile
# Rank Query ID                           Response time Calls R/Call V/M
# ==== ================================== ============= ===== ====== ====
#    1 0xFFCB5B203CE3B95911C5FC9FDE5E2DAF  0.4384 48.5%     1 0.4384  0.00 DELETE dept_emp
#    2 0x7A467E69C8DF893A40D350695FC57CB7  0.1995 22.1%     1 0.1995  0.00 SELECT employees
#    3 0xE073FD6E336BC64DC3EB6F7076A83DE0  0.1093 12.1%     2 0.0547  0.02 SELECT dept_emp
#    4 0x7CE88895C3EBE38E4CB690F31E139ACD  0.0941 10.4%     1 0.0941  0.00 SELECT dept_emp
#    5 0xAC3538779FD7FC6571905C13C844D3D8  0.0628  6.9%     1 0.0628  0.00 SELECT dept_emp
# MISC 0xMISC                              1.4141  3.2%     3 0.4714   0.0 <3 ITEMS>
```

首先，输出工具在执行过程中消耗的 CPU 及内存。

接着，打印系统当前时间、主机名，以及被分析的慢日志文件。

接下来，是慢日志的查询分析结果。

- Overall：一共分析了 7 个 SQL，形式不一样的有 6 个。QPS 等于被分析 SQL 的数量除以时间范围，concurrency 等于被分析 SQL 总的 Query_time 除以时间范围。
- Time range：被分析 SQL 的时间范围。
- Attribute：分别给出了 Exec time（执行时间）、Lock time（锁等待时间）、Rows sent（返回给客户端的行数）、Rows examine（存储引擎中检索的行数）、Query size 这 5 个维度的 total （汇总值）、min（最小值）、max（最大值）、avg（平均值）、95%（95% 的查询时间小于或等于该值）、stddev（标准差）、median（中位数，一组数据中居于中间位置的数）。
- Profile：对每一类 SQL 按照总的执行时间进行了排序。各项解释如下。
  - Rank：排名。排名越靠前，代表总的执行时间越长。
  - Query ID：每一类 SQL 的指纹，查看具体 SQL 时会用到。
  - Response time：总的执行时长及执行时长占比。
  - Calls：执行次数。
  - R/Call：平均执行时间。
  - V/M：平均方差。一般而言，平均方差越大，代表 SQL 执行时间的波动越大。

最后看看每一类 SQL 的具体执行情况。

```
# Query 1: 0 QPS, 0x concurrency, ID 0xFFCB5B203CE3B95911C5FC9FDE5E2DAF at byte 0
# Scores: V/M = 0.00
# Time range: all events occurred at 2022-01-06T11:19:13
# Attribute    pct   total    min     max     avg     95%  stddev  median
# ============ ===   =======  =======  =======  =======  =======  =======  =======
# Count         14      1
# Exec time     48    438ms   438ms   438ms   438ms   438ms       0   438ms
# Lock time     13    104us   104us   104us   104us   104us       0   104us
# Rows sent      0        0       0       0       0       0       0       0
# Rows examine   1   10.79k  10.79k  10.79k  10.79k  10.79k       0  10.79k
# Query size    10       41      41      41      41      41       0      41
# String:
# Hosts        localhost
# Users        root
# Query_time distribution
#   1us
#  10us
# 100us
#   1ms
#  10ms
# 100ms  ################################################################
#    1s
#  10s+
# Tables
#    SHOW TABLE STATUS LIKE 'dept_emp'\G
#    SHOW CREATE TABLE `dept_emp`\G
delete from dept_emp where emp_no <=20000\G
# Converted for EXPLAIN
# EXPLAIN /*!50100 PARTITIONS*/
select * from  dept_emp where emp_no <=20000\G
```

第一行的 ID 即 Profile 中的 Query ID。Attribute 是针对当前这类 SQL 的执行耗时统计。Query_time distribution 给出了当前这类 SQL 查询耗时的分布情况，1us（即 1μs）是查询时间小于 1 微秒的区间，10us 是查询时间大于等于 1 微秒但小于 10 微秒的区间，依此类推。# 越多，代表落在这个区间的 SQL 越多。从这里可以直观地看出 SQL 的执行时间是否稳定。

一般我们优化一个 SQL 会查看表的统计信息及表结构，所以，这里直接给出了对应的 SQL。

最后，给出了这类 SQL 中执行时间最久的那个 SQL。

在 MySQL 5.6 之前，只能查看 SELECT 语句的执行计划，如果要查看 DELETE、UPDATE 语句的执行计划，只能先将其转化为等价的 SELECT 语句。

2. 对实例当前正在执行的 SQL 操作进行分析

需要指定 --processlist 和目标实例的 DSN。具体命令如下。

```
# pt-query-digest --processlist h=127.0.0.1,P=3306,u=root,p=123456 --run-time 60s
Reading from STDIN ...

# 230ms user time, 30ms system time, 25.35M rss, 226.02M vsz
# Current date: Thu Jan  6 15:04:46 2022
# Hostname: slowtech
# Files: STDIN
# Overall: 1 total, 1 unique, 0 QPS, 0x concurrency _____
```

```
# Time range: all events occurred at 2022-01-06T15:03:46
# Attribute          total     min     max     avg     95%  stddev  median
# ============     =======  ======  ======  ======  ======  ======  ======
# Exec time            37s     37s     37s     37s     37s       0     37s
# Lock time              0       0       0       0       0       0       0
# Query size            16      16      16      16      16       0      16
...
select sleep(40)\G
```

命令行中的 `--run-time` 是采集周期。指定了 `--processlist`，这样 pt-query-digest 会定时采集目标实例 SHOW FULL PROCESSLIST 的输出，采集间隔由 `--interval` 参数指定，默认为 0.1 秒。

注意，pt-query-digest 只会分析采集周期内结束的 SQL，不会分析采集结束时仍在执行的 SQL。

9.7.2 常用参数

下面看看 pt-query-digest 中的常用参数。

- `--since`、`--until`

分析指定时间段的慢 SQL，推荐的时间格式是 YYYY-MM-DD [HH:MM:SS]。

- `--type`

指定分析的类型。可选值有 `binlog`、`genlog`、`slowlog`、`tcpdump`、`rawlog`。默认为 `slowlog`。

- `--limit`

按总的查询时间进行排序，只输出指定比例或指定数量的分析结果，默认为 95%。

- `--filter`

对事件进行过滤。例如给定一个指纹，输出慢日志中所有这一类 SQL。

```
pt-query-digest slow.log
--no-report
--output slowlog
--filter '$event->{fingerprint} \
&& make_checksum($event->{fingerprint}) eq "FDEA8D2993C9CAF3"'
```

又例如，只对来自 pt_user 的慢查询进行分析汇总。

```
--filter '($event->{user} || "") =~ m/pt_user/'
```

- `--order-by`

默认值为 `Query_time:sum`，按照总的查询时间输出查询分析结果。

- `--explain`

打印慢 SQL 的执行计划，这里需指定目标库的 DSN。

- `--output`

指定查询分析结果输出的格式，默认为 `report`，其他选项有：`slowlog`、`json`、`json-anon`、`secure-slowlog`。

- **--review、--history**

将查询分析结果保存到数据库中，方便后续的审核和分析。

- **--[no]report**

是否输出查询分析结果。当指定 --history 或 --review 时，可通过 --no-report 屏蔽输出。

- **--report-all**

当指定 --review 时，对于已经检查（reviewed_by 不为空）过的 SQL 语句，不再输出。如果想让其输出，可指定 --report-all。

9.8 pt-show-grants

pt-show-grants 的作用是打印实例的授权信息。它常见的使用场景是将某个实例的用户权限复制到其他实例中。

pt-show-grants 的使用也较为简单，只需指定目标库的 DSN。

```
# pt-show-grants h=192.168.244.10,P=3306,u=pt_user,p=pt_pass
-- Grants dumped by pt-show-grants
-- Dumped from server 192.168.244.10 via TCP/IP, MySQL 8.0.27 at 2022-01-06 15:27:10
-- Grants for 'mysql.infoschema'@'localhost'
CREATE USER IF NOT EXISTS `mysql.infoschema`@`localhost`;
ALTER USER `mysql.infoschema`@`localhost` IDENTIFIED WITH 'caching_sha2_password' AS
'$A$005$THISISACOMBINATIONOFINVALIDSALTANDPASSWORDTHATMUSTNEVERBRBEUSED' REQUIRE NONE PASSWORD
 EXPIRE DEFAULT ACCOUNT LOCK PASSWORD HISTORY DEFAULT PASSWORD REUSE INTERVAL DEFAULT PASSWORD
 REQUIRE CURRENT DEFAULT;
GRANT SELECT ON *.* TO `mysql.infoschema`@`localhost`;
GRANT SYSTEM_USER ON *.* TO `mysql.infoschema`@`localhost`;
...
-- Grants for 'u1'@'%'
CREATE USER IF NOT EXISTS `u1`@`%`;
ALTER USER `u1`@`%` IDENTIFIED WITH 'mysql_native_password' AS
'*6BB4837EB74329105EE4568DDA7DC67ED2CA2AD9' REQUIRE NONE PASSWORD EXPIRE DEFAULT ACCOUNT UNLOCK
PASSWORD HISTORY DEFAULT PASSWORD REUSE INTERVAL DEFAULT PASSWORD REQUIRE CURRENT DEFAULT;
GRANT DELETE, INSERT, SELECT, UPDATE ON *.* TO `u1`@`%`;
-- Grants for 'u2'@'%'
CREATE USER IF NOT EXISTS `u2`@`%`;
ALTER USER `u2`@`%` IDENTIFIED WITH 'mysql_native_password' AS
'*6BB4837EB74329105EE4568DDA7DC67ED2CA2AD9' REQUIRE NONE PASSWORD EXPIRE DEFAULT ACCOUNT UNLOCK
PASSWORD HISTORY DEFAULT PASSWORD REUSE INTERVAL DEFAULT PASSWORD REQUIRE CURRENT DEFAULT;
GRANT DELETE, INSERT, SELECT, UPDATE ON *.* TO `u2`@`%`;
```

常用参数

下面看看 pt-show-grants 中的常用参数。

- **--drop**

在 CREATE USER 之前打印 DROP USER 操作。

- **--flush**

在末尾打印 FLUSH PRIVILEGES 操作。

- **--revoke**

在 CREATE USER 之前打印 REVOKE 操作。

- **--only**

只打印指定用户的授权信息，多个用户之间用逗号隔开。例如：

```
# pt-show-grants h=192.168.244.10,P=3306,u=pt_user,p=pt_pass --only 'u1'@'%','u1'@'localhost'
```

也可使用 --ignore 参数，忽略某些用户的授权信息。

9.9　pt-slave-restart

pt-slave-restart 的作用是监控从库的复制状态。如果复制因 SQL 线程重放失败（如主键冲突）而中断，pt-slave-restart 会自动跳过当前重放失败的事务，恢复复制。

下面看一个简单的示例。

```
# pt-slave-restart h=192.168.244.10,P=3306,u=pt_user,p=pt_pass -vv --recurse 1
P=3306,h=192.168.244.20,p=...,u=pt_user delayed 0 sec
P=3306,h=192.168.244.20,p=...,u=pt_user sleeping 2.000000
P=3306,h=192.168.244.20,p=...,u=pt_user delayed 0 sec
P=3306,h=192.168.244.20,p=...,u=pt_user sleeping 4.000000
P=3306,h=192.168.244.20,p=...,u=pt_user delayed 0 sec
P=3306,h=192.168.244.20,p=...,u=pt_user sleeping 8.000000
P=3306,h=192.168.244.20,p=...,u=pt_user delayed 0 sec
P=3306,h=192.168.244.20,p=...,u=pt_user sleeping 16.000000
P=3306,h=192.168.244.20,p=...,u=pt_user delayed 0 sec
P=3306,h=192.168.244.20,p=...,u=pt_user sleeping 32.000000
P=3306,h=192.168.244.20,p=...,u=pt_user delayed 0 sec
P=3306,h=192.168.244.20,p=...,u=pt_user sleeping 64.000000
P=3306,h=192.168.244.20,p=...,u=pt_user delayed 0 sec
P=3306,h=192.168.244.20,p=...,u=pt_user sleeping 64.000000
2022-01-06T16:38:16 P=3306,h=192.168.244.20,p=...,u=pt_user node2-relay-bin.000046      458 1032
Could not execute Delete_rows event on table slowtech.t1;Can't find record in 't1', Error_code: 1032;
handler error HA_ERR_KEY_NOT_FOUND; the event's master log mysql-bin.000008, end_log_pos
49892355P=3306,h=192.168.244.20,p=...,u=pt_user sleeping 1.000000
2022-01-06T16:38:17 P=3306,h=192.168.244.20,p=...,u=pt_user node2-relay-bin.000046      750 1032
Could not execute Delete_rows event on table slowtech.t1; Can't find record in 't1', Error_code: 1032;
handler error HA_ERR_KEY_NOT_FOUND; the event's master log mysql-bin.000008, end_log_pos
49892647P=3306,h=192.168.244.20,p=...,u=pt_user sleeping 1.000000
2022-01-06T16:38:18 P=3306,h=192.168.244.20,p=...,u=pt_user node2-relay-bin.000046     1042 1032
Could not execute Delete_rows event on table slowtech.t1; Can't find record in 't1', Error_code: 1032;
handler error HA_ERR_KEY_NOT_FOUND; the event's master log mysql-bin.000008, end_log_pos
49892939P=3306,h=192.168.244.20,p=...,u=pt_user sleeping 1.000000
2022-01-06T16:38:19 P=3306,h=192.168.244.20,p=...,u=pt_user node2-relay-bin.000046     1334 1032
Could not execute Delete_rows event on table slowtech.t1; Can't find record in 't1', Error_code: 1032;
handler error HA_ERR_KEY_NOT_FOUND; the event's master log mysql-bin.000008, end_log_pos
49893231P=3306,h=192.168.244.20,p=...,u=pt_user sleeping 1.000000
2022-01-06T16:38:20 P=3306,h=192.168.244.20,p=...,u=pt_user node2-relay-bin.000046     1626 1032
Could not execute Delete_rows event on table slowtech.t1; Can't find record in 't1', Error_code: 1032;
```

```
handler error HA_ERR_KEY_NOT_FOUND; the event's master log mysql-bin.000008, end_log_pos
49893523P=3306,h=192.168.244.20,p=...,u=pt_user sleeping 1.000000
P=3306,h=192.168.244.20,p=...,u=pt_user delayed 0 sec
P=3306,h=192.168.244.20,p=...,u=pt_user sleeping 4.000000
P=3306,h=192.168.244.20,p=...,u=pt_user delayed 0 sec
P=3306,h=192.168.244.20,p=...,u=pt_user sleeping 8.000000
P=3306,h=192.168.244.20,p=...,u=pt_user delayed 0 sec
P=3306,h=192.168.244.20,p=...,u=pt_user sleeping 16.000000
```

这里的 192.168.244.10 是主库地址，-vv 表示打印详细信息，--recurse 1 指的是监控给定实例一级从库的复制状态。

从示例打印的日志来看，对于从库复制状态的检查，pt-slave-restart 会使用一种启发式算法。该算法的具体规则如下。

❑ 每次检测都会等待一段时间，如果当前检测到主从复制状态正常，则下次等待的时间会是上次等待时间的 2 倍。

❑ 等待时间，最长是 64 秒，由 --max-sleep 参数控制；最短是 0.015 625 秒，由 --min-sleep 参数控制。

常用参数

- **--daemonize**

让 pt-slave-restart 以后台进程的形式运行，同时指定 --log，记录操作日志。

```
# pt-slave-restart h=192.168.244.10,P=3306,u=pt_user,p=pt_pass -v --recurse 1 --daemonize --log
/tmp/pt-slave-restart.log
```

- **--error-numbers**

跳过指定错误码的复制错误，常见的错误码有 1062（主键冲突）、1032（指定记录不存在）。

- **--error-text**

与 --error-numbers 参数类似，都是跳过指定错误，只不过这里给的是错误文本。

- **--skip-count**

跳过的事务的数量，默认为 1。

- **--master-uuid**

在 GTID 复制中，跳过事务是通过注入空事务来实现的。默认情况下，空事务的 UUID 取自 SHOW SLAVE STATUS 中的 Master_UUID，但是在级联复制中，这样做会有问题，例如：

```
master -> slave1 -> slave2
```

对于 slave2，默认注入的是 slave1 的空事务，而不是 master 的空事务。这个时候，可显式指定 --master-uuid。

- **--always**

如果复制正常关闭，pt-slave-restart 不会自动开启。如果需要开启，需设置 --always 参数。

注意，pt-slave-restart 无法用于 GTID 且从库开启了并行复制的场景。

```
# pt-slave-restart h=192.168.244.10,P=3306,u=pt_user,p=pt_pass -vv --recurse 1
Cannot skip transactions properly because GTID is enabled and slave_parallel_workers > 0. See 'GLOBAL
TRANSACTION IDS' in the tool's documentation.
```

9.10　pt-stalk

很多时候，当我们收到线上告警，准备处理的时候，会发现告警已经恢复。我们试图定位问题的原因，却苦于没有现场数据，监控数据也不够详尽。针对这个痛点，可使用 pt-stalk，在问题出现时，实时采集主机及数据库的相关信息。

首先，看一个简单的示例。

```
# pt-stalk -h 192.168.244.10 -p 3306 -u pt_user -p pt_pass
2022_01_06_20_01_23 Starting /usr/local/bin/pt-stalk --function=status --variable=Threads_running
--threshold=25 --match= --cycles=5 --interval=1 --iterations= --run-time=30 --sleep=300
--dest=/var/lib/pt-stalk --prefix= --notify-by-email= --log=/var/log/pt-stalk.log
--pid=/var/run/pt-stalk.pid --plugin=
2022_01_06_20_01_33 Check results: status(Threads_running)=32, matched=yes, cycles_true=1
2022_01_06_20_01_34 Check results: status(Threads_running)=32, matched=yes, cycles_true=2
2022_01_06_20_01_35 Check results: status(Threads_running)=32, matched=yes, cycles_true=3
2022_01_06_20_01_36 Check results: status(Threads_running)=32, matched=yes, cycles_true=4
2022_01_06_20_01_37 Check results: status(Threads_running)=32, matched=yes, cycles_true=5
2022_01_06_20_01_37 Collect 1 triggered
2022_01_06_20_01_37 MYSQL_ONLY:
2022_01_06_20_01_37 Collect 1 PID 22190
2022_01_06_20_01_37 Collect 1 done
2022_01_06_20_01_37 Sleeping 300 seconds after collect
```

这里只指定了实例的连接信息，但还是能正常工作。从第二行的输出来看，它使用的是参数的默认值。

下面看看这些参数的具体含义。

(1) --function=status --variable=Threads_running --threshold=25：定义了触发条件，默认是 Threads_running 大于 25。

(2) --cycles=5 和 --interval=1：连续检测 5 次，每次间隔 1 秒，只有 5 次的检测结果都为 True，才会触发实际的采集动作。

(3) --run-time=30：采集的时长为 30 秒。

(4) --sleep=300：采集一次后，会等待 300 秒。这段时间内，即使满足触发条件，也不会再次采集。

(5) --dest=/var/lib/pt-stalk：采集数据的存储目录。

接下来，我们看看 pt-stalk 都采集了哪些信息。

```
# ls /var/lib/pt-stalk/
2022_01_06_20_01_37-df          2022_01_06_20_01_37-iostat-overall  2022_01_06_20_01_37-opentables2
```

```
2022_01_06_20_01_37-slave-status
2022_01_06_20_01_37-disk-space        2022_01_06_20_01_37-log_error          2022_01_06_20_01_37-output
2022_01_06_20_01_37-sysctl
2022_01_06_20_01_37-diskstats         2022_01_06_20_01_37-lsof               2022_01_06_20_01_37-pmap
2022_01_06_20_01_37-top
2022_01_06_20_01_37-hostname          2022_01_06_20_01_37-meminfo            2022_01_06_20_01_37-processlist
2022_01_06_20_01_37-trigger
2022_01_06_20_01_37-innodbstatus1     2022_01_06_20_01_37-mpstat             2022_01_06_20_01_37-procstat
2022_01_06_20_01_37-variables
2022_01_06_20_01_37-innodbstatus2     2022_01_06_20_01_37-mpstat-overall     2022_01_06_20_01_37-procvmstat
2022_01_06_20_01_37-vmstat
2022_01_06_20_01_37-interrupts        2022_01_06_20_01_37-mysqladmin         2022_01_06_20_01_37-ps
2022_01_06_20_01_37-vmstat-overall
2022_01_06_20_01_37-iostat            2022_01_06_20_01_37-opentables1        2022_01_06_20_01_37-slabinfo
```

从文件名可以看出来，pt-stalk 采集了服务器和 MySQL 的性能指标。在服务器层面，采集了磁盘、CPU、I/O、内存的使用情况，在 MySQL 层面采集了 SHOW INNODB STATUS、SHOW PROCESSLIST、SHOW SLAVE STATUS、SHOW VARIABLES、SHOW STATUS 的内容。

文件名以采集时的时间戳开头，这样，不同时间点的采集数据才不至于混淆。

常用参数

- `--collect`

满足触发条件时是否采集数据，默认为 yes。如果设置为 `--no-collect`，则不采集数据。

- `--stalk`

默认情况下，只有满足触发条件才会采集数据。如果设置为 `--no-stalk`，则会立即进行数据的采集操作，无论条件是否满足。假如我们想立即采集性能数据，采集时长为 1 分钟，且只采集一次，可像下面这样指定。

```
--no-stalk --run-time 60 --iterations 1
```

- `--collect-gdb`、`--collect-oprofile`、`--collect-strace`、`--collect-tcpdump`

是否采集 gdb、oprofile、strace、tcpdump 数据。默认不会采集，尤其是 gdb 和 strace 对数据库性能影响比较大，一般不建议开启。

- `--mysql-only`

只会采集 MySQL 相关的内容，适用于 MySQL 云服务。

- `--retention-time`

采集数据的保留时间，默认值为 30，单位是天。

- `--disk-bytes-free`、`--disk-pct-free`

前者默认为 100MB，后者默认为 5。如果数据存储目录的剩余空间小于 100MB 或 5%，则不进行数据采集。

- **--function**

定义触发条件，默认为 status，即基于 SHOW GLOBAL STATUS 的输出。--variable 定义了具体的变量值，默认为 Threads_running。

触发条件也可指定为 processlist，即基于 SHOW FULL PROCESSLIST 的输出，此时的 --variable 指的是列名，具体的监控项由 --match 指定。看下面这个示例基于元数据锁进行触发。

```
--function processlist --variable State --match "Waiting for table metadata lock" --threshold 0
```

除此之外，还可通过脚本自定义触发条件，脚本只需定义一个 trg_plugin 函数，函数的输出值必须为数字。

9.11 pt-table-checksum

pt-table-checksum 主要用来校验主从数据的一致性。

9.11.1 实现原理

首先，看一个简单的示例。

为了简化 general log 的输出，这里忽略了 mysql 和 sys 库的检测。

```
# pt-table-checksum h=192.168.244.10,P=3306,u=pt_user,p=pt_pass --no-check-binlog-format
--ignore-databases mysql,sys
Checking if all tables can be checksummed ...
Starting checksum ...
            TS ERRORS  DIFFS     ROWS  DIFF_ROWS  CHUNKS  SKIPPED    TIME TABLE
01-08T11:50:14      0      0    20000          0       5        0   1.822 sbtest.sbtest1
```

下面看看输出中各个字段的具体含义。

(1) TS：校验完后的时间戳。

(2) ERRORS：校验过程中出现 error 和 warning 的总次数。

(3) DIFFS：校验和（checksum）不相同的 chunk 的数量。

(4) ROWS：被校验的记录数。通常情况下为表的总行数。如果指定了 --where 选项，则为满足条件的记录数。

(5) CHUNKS：chunk 的个数。

(6) SKIPPED：跳过的 chunk 的个数。跳过的常见原因有：没有使用 --chunk-index 指定的索引；没有使用完整的索引，具体来说，是指 chunk 使用的索引的 key_len 比历史上最大的 key_len 小；chunk 的大小超过 --chunk-size * --chunk-size-limit；锁等待超时；校验操作被杀死了。

(7) TIME：校验表所花费的时间。

(8) TABLE：表名。

重点关注 DIFFS 这一列，如果 DIFFS 不为 0，则意味着该表主从数据不一致了。

接下来，我们看看主库 general log 的内容。

```
17 Connect   pt_user@192.168.244.10 on  using TCP/IP
17 Query    SHOW VARIABLES LIKE 'innodb\_lock_wait_timeout'
17 Query    SET SESSION innodb_lock_wait_timeout=1
17 Query    SHOW VARIABLES LIKE 'wait\_timeout'
17 Query    SET SESSION wait_timeout=10000
17 Query    SELECT @@SQL_MODE
17 Query    SET @@SQL_QUOTE_SHOW_CREATE = 1/*!40101, @@SQL_MODE='NO_AUTO_VALUE_ON_ZERO,ONLY_FULL_GROUP_BY,
STRICT_TRANS_TABLES,NO_ZERO_IN_DATE,NO_ZERO_DATE,ERROR_FOR_DIVISION_BY_ZERO,NO_ENGINE_SUBSTITUTION'*/
17 Query    SELECT VERSION()
17 Query    SHOW VARIABLES LIKE 'character_set_server'
17 Query    SET NAMES 'utf8mb4'
17 Query    SELECT @@server_id /*!50038 , @@hostname*/
17 Query    SELECT @@SQL_MODE
17 Query    SET SQL_MODE=',NO_AUTO_VALUE_ON_ZERO,STRICT_TRANS_TABLES,NO_ZERO_IN_DATE,NO_ZERO_DATE,
ERROR_FOR_DIVISION_BY_ZERO,NO_ENGINE_SUBSTITUTION'
17 Query    SHOW VARIABLES LIKE 'version%'
17 Query    SHOW ENGINES
17 Query    SHOW VARIABLES LIKE 'innodb_version'
17 Query    SELECT @@binlog_format
17 Query    /*!50108 SET @@binlog_format := 'STATEMENT'*/
17 Query    SET SESSION TRANSACTION ISOLATION LEVEL REPEATABLE READ
17 Query    SHOW /*!40103 GLOBAL*/ VARIABLES
17 Query    SELECT VERSION()
17 Query    SHOW ENGINES
17 Query    SHOW VARIABLES LIKE 'wsrep_on'
17 Query    SELECT @@SERVER_ID
17 Query    SHOW GRANTS FOR CURRENT_USER()
17 Query    SHOW FULL PROCESSLIST
17 Query    SHOW VARIABLES LIKE 'wsrep_on'
17 Query    SELECT @@SERVER_ID
17 Query    SHOW VARIABLES LIKE 'wsrep_on'
17 Query    SELECT @@SERVER_ID
```

这里需要关注以下两点。

- 将 innodb_lock_wait_timeout 的会话值设置为 1。这样，如果校验操作与业务正常 SQL 发生了锁争用，校验操作会首先超时退出，避免对线上业务造成影响。
- 将 binlog_format 的会话值设置为 STATEMENT。这样，校验操作才会复制到从库执行。

```
17 Query    SHOW DATABASES LIKE 'percona'
17 Query    CREATE DATABASE IF NOT EXISTS `percona` /* pt-table-checksum */
17 Query    USE `percona`
17 Query    SHOW TABLES FROM `percona` LIKE 'checksums'
17 Query    CREATE TABLE IF NOT EXISTS `percona`.`checksums` (
       db              CHAR(64)       NOT NULL,
       tbl             CHAR(64)       NOT NULL,
       chunk           INT            NOT NULL,
       chunk_time      FLOAT          NULL,
       chunk_index     VARCHAR(200)   NULL,
       lower_boundary  TEXT           NULL,
       upper_boundary  TEXT           NULL,
       this_crc        CHAR(40)       NOT NULL,
       this_cnt        INT            NOT NULL,
       master_crc      CHAR(40)       NULL,
       master_cnt      INT            NULL,
       ts              TIMESTAMP      NOT NULL DEFAULT CURRENT_TIMESTAMP ON UPDATE CURRENT_TIMESTAMP,
       PRIMARY KEY (db, tbl, chunk),
       INDEX ts_db_tbl (ts, db, tbl)
```

```
             ) ENGINE=InnoDB DEFAULT CHARSET=utf8
17 Query     SHOW GLOBAL STATUS LIKE 'Threads_running'
17 Query     SELECT CONCAT(@@hostname, @@port)
17 Query     SELECT CRC32('test-string')
17 Query     SELECT CRC32('a')
17 Query     SELECT CRC32('a')
17 Query     SHOW VARIABLES LIKE 'wsrep_on'
17 Query     SHOW DATABASES
17 Query     SHOW /*!50002 FULL*/ TABLES FROM `percona`
```

创建 percona.checksums 表，用来保存校验的结果。

```
17 Query     SHOW /*!50002 FULL*/ TABLES FROM `sbtest`
17 Query     /*!40101 SET @OLD_SQL_MODE := @@SQL_MODE, @@SQL_MODE := '', @OLD_QUOTE := @@SQL_QUOTE_SHOW_
CREATE, @@SQL_QUOTE_SHOW_CREATE := 1 */
17 Query     USE `sbtest`
17 Query     SHOW CREATE TABLE `sbtest`.`sbtest1`
17 Query     /*!40101 SET @@SQL_MODE := @OLD_SQL_MODE, @@SQL_QUOTE_SHOW_CREATE := @OLD_QUOTE */
```

查看 sbtest.sbtest1 的表结构，选取分片键，一般为主键或唯一索引。

```
17 Query     EXPLAIN SELECT * FROM `sbtest`.`sbtest1` WHERE 1=1
17 Query     SELECT /*!40001 SQL_NO_CACHE */ `id` FROM `sbtest`.`sbtest1` FORCE INDEX(`PRIMARY`) ORDER BY
`id` LIMIT 1 /*first lower boundary*/
17 Query     SELECT /*!40001 SQL_NO_CACHE */ `id` FROM `sbtest`.`sbtest1` FORCE INDEX (`PRIMARY`) WHERE
`id` IS NOT NULL ORDER BY `id` LIMIT 1 /*key_len*/
17 Query     EXPLAIN SELECT /*!40001 SQL_NO_CACHE */ * FROM `sbtest`.`sbtest1` FORCE INDEX (`PRIMARY`)
WHERE `id` >= '1' /*key_len*/
17 Query     USE `percona`
17 Query     DELETE FROM `percona`.`checksums` WHERE db = 'sbtest' AND tbl = 'sbtest1'
17 Query     USE `sbtest`
17 Query     EXPLAIN SELECT /*!40001 SQL_NO_CACHE */ `id` FROM `sbtest`.`sbtest1` FORCE INDEX(`PRIMARY`)
WHERE ((`id` >= '1')) ORDER BY `id` LIMIT 999, 2 /*next chunk boundary*/
17 Query     SELECT /*!40001 SQL_NO_CACHE */ `id` FROM `sbtest`.`sbtest1` FORCE INDEX(`PRIMARY`) WHERE
((`id` >= '1')) ORDER BY `id` LIMIT 999, 2 /*next chunk boundary*/
17 Query     EXPLAIN SELECT COUNT(*) AS cnt, COALESCE(LOWER(CONV(BIT_XOR(CAST(CRC32(CONCAT_WS('#', `id`,
`k`, convert(`c` using utf8mb4), convert(`pad` using utf8mb4))) AS UNSIGNED)), 10, 16)), 0) AS crc FROM
`sbtest`.`sbtest1` FORCE INDEX(`PRIMARY`) WHERE ((`id` >= '1')) AND ((`id` <= '1000')) /*explain
checksum chunk*/
17 Query     REPLACE INTO `percona`.`checksums` (db, tbl, chunk, chunk_index, lower_boundary,
upper_boundary, this_cnt, this_crc) SELECT 'sbtest', 'sbtest1', '1', 'PRIMARY', '1', '1000', COUNT(*)
AS cnt, COALESCE(LOWER(CONV(BIT_XOR(CAST(CRC32(CONCAT_WS('#', `id`, `k`, convert(`c` using utf8mb4),
convert(`pad` using utf8mb4))) AS UNSIGNED)), 10, 16)), 0) AS crc FROM `sbtest`.`sbtest1` FORCE
INDEX(`PRIMARY`) WHERE ((`id` >= '1')) AND ((`id` <= '1000')) /*checksum chunk*/
17 Query     SHOW WARNINGS
17 Query     SELECT this_crc, this_cnt FROM `percona`.`checksums` WHERE db = 'sbtest' AND tbl = 'sbtest1'
AND chunk = '1'
17 Query     UPDATE `percona`.`checksums` SET chunk_time = '0.090402', master_crc = '9d9b481f', master_cnt
= '1000' WHERE db = 'sbtest' AND tbl = 'sbtest1' AND chunk = '1'
17 Query     SHOW GLOBAL STATUS LIKE 'Threads_running'
```

通过 EXPLAIN SELECT * FROM `sbtest`.`sbtest1` WHERE 1=1 确定 sbtest1 的大致行数。

获取第一个 chunk 的最小值。

删除 percona.checksums 表中 sbtest.sbtest1 之前的校验记录。

获取第一个 chunk 的最大值和下一个 chunk 的最小值。

执行 REPLACE INTO 操作，将第一个 chunk 的校验和写入 percona.checksums 表。在执行具体的 REPLACE INTO 操作之前，会查看对应 SELECT 操作的执行计划，检查读取的行数、使用的索引是否合理。可以看到，这里计算 chunk 的校验和使用的是 CRC32 算法。注意，在执行校验操作的过程中，会在 chunk 级别对表加 S 锁。此时，针对该 chunk 的所有 DML 操作都会被阻塞。

最后，会将这个 chunk 对应的校验和、涉及的行数更新到 master_cnt 和 master_crc 中，否则，REPLACE INTO 操作在从库重放时会覆盖 this_crc 和 this_cnt 中的数据。

每次执行完校验操作，都会查看实例 Threads_running 的值，如果该值大于 25，则会暂停下一个 chunk 的校验操作。

```
17 Query  EXPLAIN SELECT /*!40001 SQL_NO_CACHE */ `id` FROM `sbtest`.`sbtest1` FORCE INDEX(`PRIMARY`)
WHERE ((`id` >= '1001')) ORDER BY `id` LIMIT 5529, 2 /*next chunk boundary*/
17 Query  SELECT /*!40001 SQL_NO_CACHE */ `id` FROM `sbtest`.`sbtest1` FORCE INDEX(`PRIMARY`) WHERE
((`id` >= '1001')) ORDER BY `id` LIMIT 5529, 2 /*next chunk boundary*/
17 Query  EXPLAIN SELECT COUNT(*) AS cnt, COALESCE(LOWER(CONV(BIT_XOR(CAST(CRC32(CONCAT_WS('#', `id`,
`k`, convert(`c` using utf8mb4), convert(`pad` using utf8mb4))) AS UNSIGNED)), 10, 16)), 0) AS crc FROM
`sbtest`.`sbtest1` FORCE INDEX(`PRIMARY`) WHERE ((`id` >= '1001')) AND ((`id` <= '6530')) /*explain
checksum chunk*/
17 Query  REPLACE INTO `percona`.`checksums` (db, tbl, chunk, chunk_index, lower_boundary,
upper_boundary, this_cnt, this_crc) SELECT 'sbtest', 'sbtest1', '2', 'PRIMARY', '1001', '6530',
COUNT(*) AS cnt, COALESCE(LOWER(CONV(BIT_XOR(CAST(CRC32(CONCAT_WS('#', `id`, `k`, convert(`c` using
utf8mb4), convert(`pad` using utf8mb4))) AS UNSIGNED)), 10, 16)), 0) AS crc FROM `sbtest`.`sbtest1`
FORCE INDEX(`PRIMARY`) WHERE ((`id` >= '1001')) AND ((`id` <= '6530')) /*checksum chunk*/
17 Query  SHOW WARNINGS
17 Query  SELECT this_crc, this_cnt FROM `percona`.`checksums` WHERE db = 'sbtest' AND tbl = 'sbtest1'
AND chunk = '2'
17 Query  UPDATE `percona`.`checksums` SET chunk_time = '0.116826', master_crc = 'e9d54858', master_cnt
= '5530' WHERE db = 'sbtest' AND tbl = 'sbtest1' AND chunk = '2'
17 Query  SHOW GLOBAL STATUS LIKE 'Threads_running'

17 Query  EXPLAIN SELECT /*!40001 SQL_NO_CACHE */ `id` FROM `sbtest`.`sbtest1` FORCE INDEX(`PRIMARY`)
WHERE ((`id` >= '6531')) ORDER BY `id` LIMIT 17006, 2 /*next chunk boundary*/
17 Query  SELECT /*!40001 SQL_NO_CACHE */ `id` FROM `sbtest`.`sbtest1` FORCE INDEX(`PRIMARY`) WHERE
((`id` >= '6531')) ORDER BY `id` LIMIT 17006, 2 /*next chunk boundary*/
17 Query  SELECT /*!40001 SQL_NO_CACHE */ `id` FROM `sbtest`.`sbtest1` FORCE INDEX(`PRIMARY`) ORDER BY
`id` DESC LIMIT 1 /*last upper boundary*/
17 Query  EXPLAIN SELECT COUNT(*) AS cnt, COALESCE(LOWER(CONV(BIT_XOR(CAST(CRC32(CONCAT_WS('#', `id`,
`k`, convert(`c` using utf8mb4), convert(`pad` using utf8mb4))) AS UNSIGNED)), 10, 16)), 0) AS crc FROM
`sbtest`.`sbtest1` FORCE INDEX(`PRIMARY`) WHERE ((`id` >= '6531')) AND ((`id` <= '20000')) /*explain
checksum chunk*/
17 Query  REPLACE INTO `percona`.`checksums` (db, tbl, chunk, chunk_index, lower_boundary,
upper_boundary, this_cnt, this_crc) SELECT 'sbtest', 'sbtest1', '3', 'PRIMARY', '6531', '20000',
COUNT(*) AS cnt, COALESCE(LOWER(CONV(BIT_XOR(CAST(CRC32(CONCAT_WS('#', `id`, `k`, convert(`c` using
utf8mb4), convert(`pad` using utf8mb4))) AS UNSIGNED)), 10, 16)), 0) AS crc FROM `sbtest`.`sbtest1`
FORCE INDEX(`PRIMARY`) WHERE ((`id` >= '6531')) AND ((`id` <= '20000')) /*checksum chunk*/
17 Query  SHOW WARNINGS
17 Query  SELECT this_crc, this_cnt FROM `percona`.`checksums` WHERE db = 'sbtest' AND tbl = 'sbtest1'
AND chunk = '3'
17 Query  UPDATE `percona`.`checksums` SET chunk_time = '0.149785', master_crc = 'b729aa46', master_cnt
= '13470' WHERE db = 'sbtest' AND tbl = 'sbtest1' AND chunk = '3'
17 Query  SHOW GLOBAL STATUS LIKE 'Threads_running'
```

第二个 chunk 和第三个 chunk 同样如此。sbtest1 的 20 000 行数据分 3 个 chunk 处理完了。

```
17 Query     EXPLAIN SELECT  COUNT(*), '0' FROM `sbtest`.`sbtest1` FORCE INDEX(`PRIMARY`) WHERE ((`id` <
'1')) ORDER BY `id` /*explain past lower chunk*/
17 Query     REPLACE INTO `percona`.`checksums` (db, tbl, chunk, chunk_index, lower_boundary,
upper_boundary, this_cnt, this_crc) SELECT 'sbtest', 'sbtest1', '4', 'PRIMARY', NULL, '1', COUNT(*),
'0' FROM `sbtest`.`sbtest1` FORCE INDEX(`PRIMARY`) WHERE ((`id` < '1')) ORDER BY `id` /*past lower
chunk*/
17 Query     SHOW WARNINGS
17 Query     SELECT this_crc, this_cnt FROM `percona`.`checksums` WHERE db = 'sbtest' AND tbl = 'sbtest1'
AND chunk = '4'
17 Query     UPDATE `percona`.`checksums` SET chunk_time = '0.090021', master_crc = '0', master_cnt = '0'
WHERE db = 'sbtest' AND tbl = 'sbtest1' AND chunk = '4'
17 Query     SHOW GLOBAL STATUS LIKE 'Threads_running'

17 Query     EXPLAIN SELECT  COUNT(*), '0' FROM `sbtest`.`sbtest1` FORCE INDEX(`PRIMARY`) WHERE ((`id` >
'20000')) ORDER BY `id` /*explain past upper chunk*/
17 Query     REPLACE INTO `percona`.`checksums` (db, tbl, chunk, chunk_index, lower_boundary,
upper_boundary, this_cnt, this_crc) SELECT 'sbtest', 'sbtest1', '5', 'PRIMARY', '20000', NULL,
COUNT(*), '0' FROM `sbtest`.`sbtest1` FORCE INDEX(`PRIMARY`) WHERE ((`id` > '20000')) ORDER BY `id`
/*past upper chunk*/
17 Query     SHOW WARNINGS
17 Query     SELECT this_crc, this_cnt FROM `percona`.`checksums` WHERE db = 'sbtest' AND tbl = 'sbtest1'
AND chunk = '5'
17 Query     UPDATE `percona`.`checksums` SET chunk_time = '0.001914', master_crc = '0', master_cnt = '0'
WHERE db = 'sbtest' AND tbl = 'sbtest1' AND chunk = '5'
17 Query     SHOW GLOBAL STATUS LIKE 'Threads_running'
17 Query     SHOW VARIABLES LIKE 'wsrep_on'
17 Query     SHOW MASTER STATUS
17 Quit
```

不仅如此，pt-table-checksum 还额外校验了两个 chunk：id < 1 和 id > 20000。之所以这么做，是考虑到从库的这两个区间内有可能会存在数据。

接下来看看从库的 general log。

```
25 Query     SHOW SLAVE STATUS
25 Query     SHOW VARIABLES LIKE 'version%'
25 Query     SHOW ENGINES
25 Query     SHOW VARIABLES LIKE 'innodb_version'
25 Query     SELECT MASTER_POS_WAIT('mysql-bin.000001', 3826550, 60 )
25 Query     SELECT MAX(chunk) FROM `percona`.`checksums` WHERE db='sbtest' AND tbl='sbtest1' AND
master_crc IS NOT NULL
25 Query     SELECT CONCAT(db, '.', tbl) AS `table`, chunk, chunk_index, lower_boundary, upper_boundary,
COALESCE(this_cnt-master_cnt, 0) AS cnt_diff, COALESCE(this_crc <> master_crc OR ISNULL(master_crc)
<> ISNULL(this_crc), 0) AS crc_diff, this_cnt, master_cnt, this_crc, master_crc FROM
`percona`.`checksums` WHERE (master_cnt <> this_cnt OR master_crc <> this_crc OR ISNULL(master_crc)
<> ISNULL(this_crc)) AND (db='sbtest' AND tbl='sbtest1')
```

通过 SELECT MASTER_POS_WAIT() 等待从库应用到指定位置点，这个位置点是主库在执行完 REPLACE INTO 操作后，执行 SHOW MASTER STATUS 的结果。应用到这个位置点，代表从库已经执行完了该表所有的 REPLACE INTO 操作。最后，执行上面的最后一个查询，输出该表的主从一致性检测结果。

综合上面的分析，我们可以看到，在实现上，pt-table-checksum 会将一张大表分成多个 chunk 来检验，每次校验一个 chunk，校验的结果通过 REPLACE INTO 语句写入 percona.checksums 表。该语句通过主从复制后，会在从库同样执行一次，执行的结果同样会被写入 percona.checksums 表。

9.11.2 常见用法

1. 实例校验

```
# pt-table-checksum h=192.168.244.10,P=3306,u=pt_user,p=pt_pass --no-check-binlog-format
```

校验整个实例。

2. 基于指定库的校验

```
# pt-table-checksum h=192.168.244.10,P=3306,u=pt_user,p=pt_pass --no-check-binlog-format --databases sbtest,slowtech
```

校验指定库，多个库之间用逗号隔开。

3. 基于指定表的校验

```
# pt-table-checksum h=192.168.244.10,P=3306,u=pt_user,p=pt_pass --no-check-binlog-format --tables slowtech.t1,sbtest.sbtest1
```

校验指定表，多个表之间用逗号隔开。

9.11.3 常用参数

下面看看 pt-table-checksum 中的常用参数。

首先看看 `--replicate-check-only` 参数。它会根据 `percona.checksums` 表中的内容输出每个从库的主从差异结果，不做任何校验。

```
# pt-table-checksum h=192.168.244.10,P=3306,u=pt_user,p=pt_pass --no-check-binlog-format --replicate-check-only
Checking if all tables can be checksummed ...
Starting checksum ...
Differences on node2
TABLE CHUNK CNT_DIFF CRC_DIFF CHUNK_INDEX LOWER_BOUNDARY UPPER_BOUNDARY
mysql.engine_cost 1 0 1
mysql.server_cost 1 0 1
mysql.tables_priv 1 0 1
mysql.user 1 0 1

Differences on node3
TABLE CHUNK CNT_DIFF CRC_DIFF CHUNK_INDEX LOWER_BOUNDARY UPPER_BOUNDARY
slowtech.t1 1 -2 1
```

下面看看输出中各个字段的具体含义。

(1) `TABLE`：表名。
(2) `CHUNK`：校验和不相同的 chunk 的序号。
(3) `CNT_DIFF`：该 chunk 从库的记录数减去主库的记录数。
(4) `CRC_DIFF`：如果校验和相同，则 `CRC_DIFF` 为 0，否则为 1。
(5) `CHUNK_INDEX`：分片索引。
(6) `LOWER_BOUNDARY`、`UPPER_BOUNDARY`：该 chunk 的最小值和最大值。

1. 过滤相关

- `--databases`、`--ignore-databases`

校验指定库，多个库之间用逗号隔开。`--ignore-databases` 是忽略指定库的校验。

- `--databases-regex`、`--ignore-databases-regex`

基于正则匹配校验指定库。

- `--tables`、`--ignore-tables`

校验指定表，多个表之间用逗号隔开。

- `--tables-regex`、`--ignore-tables-regex`

基于正则匹配校验指定表。

- `--columns`、`--ignore-columns`

只校验指定列。

- `--where`

只校验满足 where 条件的行。

2. 主从复制相关

pt-table-checksum 在运行的过程中，会以固定频率（由 `--check-interval` 决定，默认为 1 秒）检测从库的复制情况。如果从库的复制停止或延迟超过一定的阈值（由 `--max-lag` 决定，默认为 1 秒），则会暂停当前的 REPLACE INTO 操作，直到复制或延迟恢复。

主从复制相关的参数包括 `--recurse`、`--recursion-method`、`--check-slave-lag`、`--skip-check-slave-lag`，具体用法可参考 7.2 节的讲解。

- `--[no]check-binlog-format`

检查主库的 binlog_format 是否为 ROW。

考虑下面这个场景：级联复制（master -> slave1 -> slave2），且 binlog_format 为 ROW。虽然可将 master 的 binlog_format 的会话值调整为 STATEMENT，但因为 slave1 的 binlog_format 依然为 ROW，所以 slave1 执行的 REPLACE INTO 操作还是没办法复制到 slave2 中。

针对这个场景，如果要检查主从一致性，只能先比较 master 和 slave1，然后再比较 slave1 和 slave2。不过这种方式也有缺陷，如果是 GTID 复制，直接对 slave1 和 slave2 进行校验会生成额外的 GTID。

- `--[no]check-replication-filters`

是否检查过滤规则。

- `--fail-on-stopped-replication`

pt-table-checksum 在运行的过程中，如果从库复制停止，默认它会一直等待，直到从库恢复。如果设置了该参数，则直接退出。

3. 性能相关

- `--chunk-size`、`--chunk-time`

前者指定 chunk 的大小，默认为 1000。后者指定 chunk 的校验时间，默认为 0.5 秒。默认使用的是后者，此时会基于校验时间自动调节 chunk 的大小。如果显式设置 `--chunk-size`，则会覆盖 `--chunk-time`。此时，每个 chunk 的大小将是固定不变的。

- `--chunk-size-limit`

虽然默认会基于 `chunk-time` 自动调节 chunk 的大小，但第一个 chunk 的大小总是固定不变的（`chunk-size`）。如果表上没有索引或者没有合适的索引，导致 EXPLAIN 预估的行数超过 `chunk-size * chunk-size-limit`（默认为 2）的大小，会提示如下错误。

```
# pt-table-checksum h=192.168.244.10,P=3306,u=pt_user,p=pt_pass --no-check-binlog-format --tables sbtest.sbtest2
Checking if all tables can be checksummed ...
Starting checksum ...
01-08T19:06:51 Cannot checksum table sbtest.sbtest2: There is no good index and the table is oversized. at /usr/local/bin/pt-table-checksum line 6811.
```

这个时候，可将 `chunk-size-limit` 设置为 0 关闭检测，或者增大 `chunk-size` 或 `chunk-size-limit` 的值，例如：

```
# pt-table-checksum h=192.168.244.10,P=3306,u=pt_user,p=pt_pass --no-check-binlog-format --tables sbtest.sbtest2 --chunk-size=10000
Checking if all tables can be checksummed ...
Starting checksum ...
            TS ERRORS  DIFFS     ROWS DIFF_ROWS CHUNKS SKIPPED    TIME TABLE
01-08T19:07:26      0      0    20000         0      1       0   2.450 sbtest.sbtest2
# pt-table-checksum h=192.168.244.10,P=3306,u=pt_user,p=pt_pass --no-check-binlog-format --tables sbtest.sbtest2 --chunk-size-limit=20
Checking if all tables can be checksummed ...
Starting checksum ...
            TS ERRORS  DIFFS     ROWS DIFF_ROWS CHUNKS SKIPPED    TIME TABLE
01-08T19:08:06      0      0    20000         0      1       0   0.837 sbtest.sbtest2
```

- `--max-load`

基于数据库的负载情况来自动调节校验操作。默认为 `Threads_running=25`。也可指定其他状态变量，如 `--max-load=Threads_connected=1000`。

4. checksum 表相关

- `--replicate`

指定 checksum 表，默认为 `percona.checksums`。

- `--[no]create-replicate-table`

是否自动创建 `--replicate` 指定的表。

- `--[no]empty-replicate-table`

在校验每个表之前，是否删除 checksum 表中该表之前的校验记录。

- `--truncate-replicate-table`

在执行校验操作之前，是否用 TRUNCATE 操作清空 checksum 表。

5. 输出相关

- `--explain`

只打印校验操作，并不实际执行。如果将 explain 指定两次，还会输出每个 chunk 的最小值和最大值。

- `--quiet`

只会输出 error、warning 以及主从数据不一致的信息。

6. 其他参数

- `--[no]check-plan`

在执行具体的校验操作之前，会先查看执行计划。如果执行计划不合理，会跳过该 chunk 的检测。

- `--chunk-index`

指定分片索引。如果指定索引不存在，pt-table-checksum 会使用默认的分片索引。

- `--chunk-index-columns`

指定使用复合索引中的最左侧的几个索引。常用于复合索引列数过多，导致校验操作执行计划不合理的场景。

- `--resume`

对于 pt-table-checksum 还没检测完，就提前终止的情况，可在下次执行的时候指定 `--resume`，这样它就会从上次终止的地方继续执行。

9.12 pt-table-sync

pt-table-checksum 只能用于校验主从数据一致性。如果要修复不一致的数据，需使用 pt-table-sync。很多人有个误解，以为在执行 pt-table-sync 之前，必须先执行 pt-table-checksum。实际上不用，因为 pt-table-sync 自带主从数据一致性校验功能。

9.12.1 实现原理

首先，看个简单的示例，分析其实现原理。

主库：192.168.244.10。从库：192.168.244.20，192.168.244.30。

测试表：sbtest.sbtest1，有 3000 条记录，从库相对于主库少一条 ID 为 3000 的记录。

```
# pt-table-sync --execute --sync-to-master h=192.168.244.20,P=3306,u=pt_user,p=pt_pass
--tables=sbtest.sbtest1
```

命令中的 `--sync-to-master` 指的是修复指定从库的数据。h=192.168.244.20,P=3306,u=pt_user,p=pt_pass 是从库的 DSN 。

接下来看看主库的 general log。

```
...
55 Query    SELECT @@binlog_format
55 Query    /*!50108 SET @@binlog_format := 'STATEMENT'*/
...
55 Query    SELECT MIN(`id`), MAX(`id`) FROM `sbtest`.`sbtest1` FORCE INDEX (`PRIMARY`)
55 Query    EXPLAIN SELECT * FROM `sbtest`.`sbtest1` FORCE INDEX (`PRIMARY`)
55 Query    SELECT CRC32('test-string')
55 Query    SELECT CRC32('a')
55 Query    SELECT CRC32('a')
55 Query    USE `sbtest`
55 Query    SET @crc := '', @cnt := 0
55 Query    commit

55 Query    START TRANSACTION /*!40108 WITH CONSISTENT SNAPSHOT */
55 Query    SELECT /*sbtest.sbtest1:1/4*/ 0 AS chunk_num, COUNT(*) AS cnt,
COALESCE(LOWER(CONV(BIT_XOR(CAST(CRC32(CONCAT_WS('#', `id`, `k`, `c`, `pad`)) AS UNSIGNED)), 10, 16)),
0) AS crc FROM `sbtest`.`sbtest1` FORCE INDEX (`PRIMARY`) WHERE (`id` = 0) FOR UPDATE
56 Query    SHOW MASTER STATUS
55 Query    SET @crc := '', @cnt := 0
55 Query    commit

55 Query    START TRANSACTION /*!40108 WITH CONSISTENT SNAPSHOT */
55 Query    SELECT /*sbtest.sbtest1:2/4*/ 1 AS chunk_num, COUNT(*) AS cnt,
COALESCE(LOWER(CONV(BIT_XOR(CAST(CRC32(CONCAT_WS('#', `id`, `k`, `c`, `pad`)) AS UNSIGNED)), 10, 16)),
0) AS crc FROM `sbtest`.`sbtest1` FORCE INDEX (`PRIMARY`) WHERE (`id` > 0 AND `id` < '1019') FOR UPDATE
56 Query    SHOW MASTER STATUS
55 Query    SET @crc := '', @cnt := 0
55 Query    commit
...
55 Query    START TRANSACTION /*!40108 WITH CONSISTENT SNAPSHOT */
55 Query    SELECT /*sbtest.sbtest1:4/4*/ 3 AS chunk_num, COUNT(*) AS cnt,
COALESCE(LOWER(CONV(BIT_XOR(CAST(CRC32(CONCAT_WS('#', `id`, `k`, `c`, `pad`)) AS UNSIGNED)), 10, 16)),
0) AS crc FROM `sbtest`.`sbtest1` FORCE INDEX (`PRIMARY`) WHERE (`id` >= '2037') FOR UPDATE
56 Query    SHOW MASTER STATUS
55 Query    SET @crc := '', @cnt := 0

55 Query    SELECT /*rows in chunk*/ `id`, `k`, `c`, `pad`, CRC32(CONCAT_WS('#', `id`, `k`, `c`, `pad`))
AS __crc FROM `sbtest`.`sbtest1` FORCE INDEX (`PRIMARY`) WHERE (`id` >= '2037') ORDER BY `id` FOR UPDATE
55 Query    SELECT `id`, `k`, `c`, `pad` FROM `sbtest`.`sbtest1` WHERE `id`='3000' LIMIT 1
55 Query    REPLACE INTO `sbtest`.`sbtest1`(`id`, `k`, `c`, `pad`) VALUES ('3000', '9969',
'79434169942-59769757845-37520311867-03925425301-14673511735-51426543096-37890133679-03683925749-3
8148672309-26666656854', '01283653869-84678229063-94668819233-88581759678-29055933492')
/*percona-toolkit src_db:sbtest src_tbl:sbtest1 src_dsn:P=3306,h=192.168.244.10,p=...,u=pt_user
dst_db:sbtest dst_tbl:sbtest1 dst_dsn:P=3306,h=192.168.244.20,p=...,u=pt_user lock:1 transaction:1
changing_src:1 replicate:0 bidirectional:0 pid:65817 user:root host:node1*/
55 Query    commit
55 Query    commit
55 Quit
```

对比看一下从库的 general log。

```
...
51 Query    SHOW SLAVE STATUS
51 Query    SHOW VARIABLES LIKE 'version%'
51 Query    SHOW ENGINES
51 Query    SHOW VARIABLES LIKE 'innodb_version'
51 Query    SELECT MASTER_POS_WAIT('mysql-bin.000002', 10924000, 60 )
51 Query    SELECT /*sbtest.sbtest1:1/4*/ 0 AS chunk_num, COUNT(*) AS cnt,
COALESCE(LOWER(CONV(BIT_XOR(CAST(CRC32(CONCAT_WS('#', `id`, `k`, `c`, `pad`)) AS UNSIGNED)), 10, 16)),
0) AS crc FROM `sbtest`.`sbtest1` FORCE INDEX (`PRIMARY`) WHERE (`id` = 0) LOCK IN SHARE MODE
51 Query    SET @crc := '', @cnt := 0
51 Query    commit

51 Query    SHOW SLAVE STATUS
51 Query    SHOW VARIABLES LIKE 'version%'
51 Query    SHOW ENGINES
51 Query    SHOW VARIABLES LIKE 'innodb_version'
51 Query    SELECT MASTER_POS_WAIT('mysql-bin.000002', 10924000, 60 )
51 Query    SELECT /*sbtest.sbtest1:2/4*/ 1 AS chunk_num, COUNT(*) AS cnt,
COALESCE(LOWER(CONV(BIT_XOR(CAST(CRC32(CONCAT_WS('#', `id`, `k`, `c`, `pad`)) AS UNSIGNED)), 10, 16)),
0) AS crc FROM `sbtest`.`sbtest1` FORCE INDEX (`PRIMARY`) WHERE (`id` > 0 AND `id` < '1019') LOCK IN
SHARE MODE
51 Query    SET @crc := '', @cnt := 0
51 Query    commit
...
```

结合主从的 general log，我们看看 pt-table-sync 的实现原理。

- pt-table-sync 对于单表的校验也是分 chunk 来进行的。chunk 的大小由 --chunk-size 指定，默认为 1000。
- 以第一个 chunk 为例。首先开启事务，接着计算第一个 chunk 的校验和，此时，会对这个 chunk 加排它锁，避免其他会话修改这个 chunk 对应记录的值。查看主库当前 binlog 的位置点，获取到的位置点信息将用于从库执行 SELECT MASTER_POS_WAIT() 操作，该操作是为了确保从库已经应用到了主库对 chunk 加排它锁的位置点。可以看到，主从延迟越大，这个 chunk 被锁定的时间就越久，不过因为 MASTER_POS_WAIT 指定了超时时间，所以最多只会等待 60 秒。等 MASTER_POS_WAIT 执行完，再计算从库相同 chunk 的校验和。
- 如果校验和一致，则意味着这个 chunk 在主从库中的数据是一致的，此时会进行下一个 chunk 的对比。
- 如果校验和不一致，则会对比这个 chunk 每一行的校验和。

    ```
    SELECT /*rows in chunk*/ `id`, `k`, `c`, `pad`, CRC32(CONCAT_WS('#', `id`, `k`, `c`, `pad`)) AS __crc
    FROM `sbtest`.`sbtest1` FORCE INDEX (`PRIMARY`) WHERE (`id` >= '2037') ORDER BY `id` FOR UPDATE
    ```

- 确定完不一致后，会在主库而不是从库进行修复操作。因为是 STATEMENT 格式，且使用的是 REPLACE INTO 语句，所以只会修改从库不一致的数据，不会实际修改主库的数据。

从 general log 的内容来看，pt-table-sync 少比较了一个 chunk 的值，即 id < 0 的区间。这里使用了默认的分片算法 Chunk，若使用 Nibble 算法就不会有这个问题。

9.12.2 常见用法

1. 修复所有从库的数据

```
# pt-table-checksum h=192.168.244.10,P=3306,u=pt_user,p=pt_pass --no-check-binlog-format
# pt-table-sync --execute --replicate percona.checksums h=192.168.244.10,P=3306,u=pt_user,p=pt_pass
```

首先执行 pt-table-checksum，将主从校验结果保存到 percona.checksums 表中。然后执行 pt-table-sync。此时需指定 `--replicate` 和主库的 DSN。

注意，要修复所有从库的数据，只有这一种方法。

2. 修复某个从库的数据

第一种方法是直接修复。具体命令如下。

```
# pt-table-sync --execute --sync-to-master h=192.168.244.20,P=3306,u=pt_user,p=pt_pass
```

必须指定 `--sync-to-master` 和需要修复的从库的 DSN。

第二种方法是直接使用 percona.checksums 的校验结果。

```
# pt-table-sync --execute --replicate percona.checksums --sync-to-master h=192.168.244.20,P=3306,u=pt_user,p=pt_pass
```

3. 独立环境下的数据同步

```
# pt-table-sync --execute host1 host2 host3
```

将 host1 中的数据同步到 host2、host3 中。注意，这里的 host1、host2、host3 必须是独立的，没有主从复制关系，否则会提示错误。

9.12.3 常用参数

1. 过滤相关

- `--databases`、`--ignore-databases`、`--engines`、`--ignore-engines`、`--tables`、`--ignore-tables`、`--ignore-tables-regex`、`--columns`、`--ignore-columns`、`--where`

具体可参考 pt-table-checksum。

2. 输出相关

- `--execute`

执行修复操作。

- `--print`

打印修复语句。

- `--dry-run`

只用于验证命令是否可行。

注意：以上三者之间并不是互斥关系，可搭配使用。规则如下。

(1) 在有 --dry-run 的情况下，只会验证命令是否可行。即便搭配了--execute 或 --print 参数，也不会校验数据。
(2) 如果只指定--print，则 pt-table-sync 会校验表，并将修复语句打印出来。对于不太信任该工具或者想手动执行修复语句的人来说，可指定该参数。
(3) --execute 会执行修复操作，默认情况下，执行是不会输出任何结果的。如果要同时打印修复操作，需指定 --print 选项。

3. 双向同步

- --bidirectional、--conflict-column、--conflict-comparison、--conflict-error、--conflict-threshold、--conflict-value

双向同步需指定 --bidirectional，后面几个参数是冲突出现时的解决规则。需要注意的是，双向同步是一个处于实验阶段的新特性，在线上应慎用。

4. 分片相关

- --chunk-column

分片列。

- --chunk-index

指定分片索引。

- --chunk-size

指定 chunk 的大小，默认为 1000。

5. 锁相关

锁相关的参数主要有两个：--[no]transaction 和 --lock。

如果不显式指定 --transaction，则加锁规则由 --lock 参数控制，该参数值默认为 1。在这种情况下，会区分 InnoDB 表和非 InnoDB 表。对于 InnoDB 表，会加排它锁（FOR UPDATE），对于非 InnoDB 表，会加表锁（LOCK TABLES slowtech.t1 WRITE）。

如果显式指定了 --transaction，则不区分 InnoDB 表和非 InnoDB 表，两者都会加排它锁。

如果显式指定了 --no-transaction，则无论是 InnoDB 表还是非 InnoDB 表，都会加表锁。

6. 主从复制相关

- --sync-to-master

修复指定从库的数据。

在设置 --sync-to-master 的情况下，只能设置一个 DSN，且这个 DSN 必须是从库。设置该参数，会隐式将 --wait 设置为 60，将 --lock 设置为 1。

- `--wait`、`--timeout-ok`

指定 `SELECT MASTER_POS_WAIT('mysql-bin.000002', 10924000, 60)` 中的第三个参数，如果从库在给定的时间内没有应用到主库的指定位置，工具会退出。如果要让工具继续执行，可指定 `--timeout-ok` 参数，但此时主从数据的一致性就很难保证了。

7. 其他参数

- `--function`

指定校验函数，默认为 `CRC32`，还可选择 `MD5` 和 `SHA1`。

- `--algorithms`

指定分片算法。有 Chunk、Nibble、GroupBy、Stream 这 4 种分片算法，按优先级从高到低排列。Chunk 和 Nibble 类似，都是将一张表分成多个 chunk 进行校验，适用于有主键或唯一索引的表。而 GroupBy 和 Stream 则是将整个表作为一个 chunk 进行校验，适用于没有主键或唯一索引的表。

- `--[no]check-master`

当指定 `--sync-to-master` 时，会检测 DSN 对应的主库是否是真正的主库，而不是级联复制中的中间节点。

- `--replicate`

基于 `--replicate` 表中的对比结果进行修复。如果要使用该选项，必须首先执行 `pt-table-checksum`，将主从的校验结果保存到该表中。

9.13 pt-upgrade

数据库升级是一项让人喜忧参半的工程。喜的是，通过升级，可以享受新版本带来的新特性及性能提升；忧的是，新版本可能与旧版本不兼容。不兼容主要体现在 3 个方面：一是语法不兼容；二是语义不兼容，同一条 SQL 语句在新旧版本中的执行结果不一致；三是新版本的查询性能可能更差。所以，在对线上数据库进行升级之前，一般会在测试环境进行大量的测试，包括功能测试和性能测试。

pt-upgrade 可帮我们从 SQL 层面检查新旧版本是否兼容。它的检测思路很简单，给定一条 SQL 语句，分别在两个不同版本的实例上执行，看看是否存在不一致的情况。具体来说，它会检查以下几项。

(1) Row count：查询返回的行数是否一致。
(2) Row data：查询的结果是否一致。
(3) Warnings：是否提示 warning。正常来说，要么都提示 warning，要么都不提示 warning。
(4) Query time：查询时间是否在同一个量级，或者新版本的执行时间是否更短。
(5) Query errors：查询如果在一个实例中出现语法错误，会提示 "Query errors"。
(6) SQL errors：查询如果在两个实例中同时出现语法错误，会提示 "SQL errors"。

看下面这个示例，pt_upgrade_test.sql 包含了若干条测试语句。

```
# cat /tmp/pt_upgrade_test.sql
select "a word a" REGEXP "[[:<:]]word[[:>:]]";
select dept_no,count(*) from employees.dept_emp group by dept_no desc;
grant select on employees.* to 'u1'@'%' identified by '123456';
create table employees.t1(id int primary key,c1 text not null default (''));
select * from employees.dept_emp group by dept_no;
```

看看这些语句在 MySQL 5.7 和 MySQL 8.0 中的执行结果有何不同。

```
# pt-upgrade h=127.0.0.1,P=3307,u=pt_user,p=pt_pass h=127.0.0.1,P=3306,u=pt_user,p=pt_pass --type
rawlog /tmp/pt_upgrade_test.sql --no-read-only

#-----------------------------------------------------------------
# Logs
#-----------------------------------------------------------------

File: /tmp/pt_upgrade_test.sql
Size: 311

#-----------------------------------------------------------------
# Hosts
#-----------------------------------------------------------------

host1:

  DSN:       h=127.0.0.1,P=3307
  hostname:  slowtech
  MySQL:     MySQL Community Server (GPL) 5.7.36

host2:

  DSN:       h=127.0.0.1,P=3306
  hostname:  slowtech
  MySQL:     MySQL Community Server - GPL 8.0.27

##################################################################
# Query class 00A13DD81BF65D41
##################################################################

Reporting class because it has diffs, but hasn't been reported yet.

Total queries      1
Unique queries     1
Discarded queries  0

grant select on employees.* to ?@? identified by ?;

##
## Query errors diffs: 1
##

-- 1.

No error

vs.

DBD::mysql::st execute failed: You have an error in your SQL syntax; check the manual that corresponds
to your MySQL server version for the right syntax to use near 'identified by '123456'' at line 1 [for
Statement "grant select on employees.* to 'u1'@'%' identified by '123456';"]
```

```
grant select on employees.* to 'u1'@'%' identified by '123456';

#######################################################################
# Query class 296E46FE3AEE9B6C
#######################################################################

Reporting class because it has SQL errors, but hasn't been reported yet.

Total queries      1
Unique queries     1
Discarded queries  0

select * from employees.dept_emp group by dept_no;

##
## SQL errors: 1
##

-- 1.

On both hosts:

DBD::mysql::st execute failed: Expression #1 of SELECT list is not in GROUP BY clause and contains
nonaggregated column 'employees.dept_emp.emp_no' which is not functionally dependent on columns in GROUP
BY clause; this is incompatible with sql_mode=only_full_group_by [for Statement "select * from
employees.dept_emp group by dept_no;"]

select * from employees.dept_emp group by dept_no;

#######################################################################
# Query class 8B81ACF1E68DE066
#######################################################################

Reporting class because it has diffs, but hasn't been reported yet.

Total queries      1
Unique queries     1
Discarded queries  0

create table employees.t?(id int primary key,c? text not ? default (?));

##
## Query errors diffs: 1
##

-- 1.

DBD::mysql::st execute failed: You have an error in your SQL syntax; check the manual that corresponds
to your MySQL server version for the right syntax to use near '('')' at line 1 [for Statement "create
table employees.t1(id int primary key,c1 text not null default (''));  "]

vs.

No error

create table employees.t1(id int primary key,c1 text not null default (''));

#######################################################################
# Query class 92E8E91AB47593A5
```

```
######################################################################

Reporting class because it has diffs, but hasn't been reported yet.

Total queries       1
Unique queries      1
Discarded queries   0

select ? regexp ?;

##
## Query errors diffs: 1
##

-- 1.

No error

vs.

DBD::mysql::st execute failed: Illegal argument to a regular expression. [for Statement "select "a word
a" REGEXP "[[:<:]]word[[:>:]]";"]

select "a word a" REGEXP "[[:<:]]word[[:>:]]";

######################################################################
# Query class D3F390B1B46CF9EA
######################################################################

Reporting class because it has diffs, but hasn't been reported yet.

Total queries       1
Unique queries      1
Discarded queries   0

select dept_no,count(*) from employees.dept_emp group by dept_no desc;

##
## Query errors diffs: 1
##

-- 1.

No error

vs.

DBD::mysql::st execute failed: You have an error in your SQL syntax; check the manual that corresponds
to your MySQL server version for the right syntax to use near 'desc' at line 1 [for Statement "select
dept_no,count(*) from employees.dept_emp group by dept_no desc;"]

select dept_no,count(*) from employees.dept_emp group by dept_no desc;

#-----------------------------------------------------------------
# Stats
#-----------------------------------------------------------------

failed_queries      1
not_select          0
queries_filtered    0
```

```
queries_no_diffs      0
queries_read          5
queries_with_diffs    0
queries_with_errors   4
```

3307 和 3306 端口分别对应 MySQL 5.7 和 MySQL 8.0 实例。

文件中的每一条 SQL 语句都会在这两个实例中执行。如果将每一条存在差异的 SQL 语句的结果都打印出来的话，最后的输出将十分庞杂。为了简化最后的输出结果，pt-upgrade 会对 SQL 进行分类，同一类 SQL 的输出次数受到 --max-class-size 和 --max-examples 的限制。

结合执行的 SQL，我们看看最后的分析结果。

- **grant select on** employees.* **to** 'u1'@'%' **identified by** '123456';

在 MySQL 8.0 之前，对一个用户进行授权（grant）操作时，如果该用户不存在，会隐式创建。而在 MySQL 8.0 中，该命令会直接报错。

必须先创建用户，再授权。

所以，上面这条 SQL 需拆分为以下两条 SQL 来执行。

```
create user 'u1'@'%' identified by '123456';
grant select on employees.* to 'u1'@'%';
```

一个查询如果只在一个实例中出现语法错误，pt-upgrade 会将其归类为 Query errors。

- **select** * **from** employees.dept_emp **group by** dept_no;

从 MySQL 5.7 开始，SQL_MODE 的默认值发生了变化，包含了 ONLY_FULL_GROUP_BY。ONLY_FULL_GROUP_BY 要求，对于 GROUP BY 操作，SELECT 列表中只能出现分组列（即 GROUP BY 后面的列）和聚合函数（SUM、AVG、MAX 等），不能出现其他非分组列。很明显，上面这条 SQL 语句违背了这一要求。所以，无论是在 MySQL 5.7 还是 MySQL 8.0 中，该 SQL 语句都会报错。

一个查询如果在两个实例中都出现语法错误，pt-upgrade 会将其归类为 SQL errors。

- **create table** employees.t1(**id** int **primary key**,c1 text **not null default** ('''));

从 MySQL 8.0.13 开始，允许对 BLOB、TEXT、GEOMETRY 和 JSON 字段设置默认值。之前的版本中则不允许。

- **select** "a word a" REGEXP "[[:<:]]word[[:>:]]";

在 MySQL 8.0 中，正则表达式底层库由 Henry Spencer 调整为了 International Components for Unicode（ICU）。在 Henry Spencer 库中，[[:<:]]和[[:>:]] 用来表示一个单词的开头和结尾。但在 ICU 库中，则不能这样表示，类似的功能要通过 \b 来实现。所以，上面这条 SQL 语句在 MySQL 8.0 中的写法如下。

```
select "a word a" REGEXP "\\bword\\b";
```

- **select** dept_no,count(*) **from** employees.dept_emp **group by** dept_no **desc**;

在 MySQL 8.0 之前，如果要对分组后的结果进行排序，可使用 GROUP BY ASC/DESC，命令中没有指定排序列，默认是对分组列进行排序。在 MySQL 8.0 中，不再支持这一语法，如果要进行排序，需显式指定排序列。所以，上面这个 SQL 在 MySQL 8.0 中的写法如下。

```
select dept_no,count(*) from employees.dept_emp group by dept_no order by dept_no desc;
```

常见参数

- --type

指定文件的类型，可选值有 slowlog（慢日志）、genlog（general log）、binlog、rawlog（SQL 语句）、tcpdump。若不指定，则默认为 slowlog。

- --[no]read-only

默认情况下，pt-upgrade 只会执行 SELECT 和 SET 操作。如果要执行其他操作，必须指定 --no-read-only。

- --[no]create-upgrade-table、--upgrade-table

默认情况下，pt-upgrade 会在目标实例上创建一张 percona_schema.pt_upgrade 表（由 --upgrade-table 参数指定），每执行完一个 SQL，都会执行一次 SELECT * FROM percona_schema.pt_upgrade LIMIT 1 以清除上一个 SQL 中有可能出现的 warning。

- --save-results

该参数涉及 pt-upgrade 的第二个使用场景，即先生成一个基准测试结果，再基于这个结果测试其他环境的兼容性。

```
# pt-upgrade h=127.0.0.1,P=3307,u=pt_user,p=pt_pass --save-results /tmp/pt_upgrade_result --type rawlog /tmp/pt_upgrade_test.sql --no-read-only
# pt-upgrade /tmp/pt_upgrade_result/ h=127.0.0.1,P=3306,u=pt_user,p=pt_pass
```

9.14 本章总结

本章主要介绍了 sysbench 和 Percona Toolkit 中一些常用工具的常见用法及实现原理。虽然 Percona 工具集里有 37 个工具，但很多工具的用处并不大，还有一部分工具的功能在 MySQL 中已经实现了。需要注意的是，有些工具虽然实现了功能，但在使用上还是有一定的局限性，以下面这两个工具为例。

(1) pt-config-diff：只能比较两个参数源都有的参数。如果某个参数只在一个参数源中存在，就不会输出。例如，比较 MySQL 5.7 和 MySQL 8.0，就不会输出 MySQL 8.0 中新增的参数。

(2) pt-slave-find：在一主多从的场景中，如果从库的端口不一致，就无法正确输出集群拓扑图。

尽管如此，但终究瑕不掩瑜，并不影响 Percona Toolkit 在 MySQL 社区的影响力。

建议重点掌握 pt-archiver、pt-query-digest、pt-table-checksum、pt-table-sync、pt-online-schema-change 等线上使用频率比较高的工具。

重点问题回顾

- 如何使用 sysbench 对 MySQL 进行基准测试？
- sysbench 支持的 MySQL 测试场景及对应的 SQL 语句。
- 如何自定义 sysbench 测试脚本？
- pt-archiver 的实现原理。
- 如何使用 pt-archiver 删除数据？
- 为什么不建议在线上直接使用 pt-ioprofile 和 pt-pmp？
- 如何使用 pt-kill 杀掉查询时间超过 30 秒的 SELECT 操作？
- pt-slave-restart 的适用场景。
- pt-table-checksum 的实现原理。
- pt-table-sync 的实现原理。
- 主从延迟对 pt-table-sync 的影响。
- 在使用 pt-table-sync 修复数据时，需要首先执行 pt-table-checksum 吗？
- 数据库升级的注意事项。

第 10 章
中 间 件

中间件这个概念应用得比较广泛。我们首先看看维基百科里对中间件的解释。

中间件是一类提供系统软件和应用软件之间连接、便于软件各部件之间沟通的软件，应用软件可以借助中间件在不同的技术架构之间共享信息与资源。

中间件技术建立在对应用软件部分常用功能的抽象上，将常用且重要的过程调用、分布式组件、消息队列、事务、安全、连结器、商业流程、网络并发、HTTP 服务器、Web Service 等功能集于一身或者分别在不同品牌的不同产品中分别完成。

具体到数据库中间件，简单来说，就是介于应用和数据库之间、进行数据处理与交互的中间服务。

什么场景下会用到数据库中间件呢？

- 要扩展读能力。扩展读能力，一般就需要进行读写分离。实现读写分离，常见的方式有两种：(1) 开发人员自己梳理业务代理逻辑，为需要发往从库查询的 SQL 语句单独配置一个数据源 (DataSource)；(2) 中间件实现，数据库中间件会分析数据包的内容，只将以 SELECT 开头的 SQL 语句发往从库查询。
- 要扩展写能力。扩展写能力，一般就需要进行分库分表。
- 需要一些安全能力，例如防火墙、审计。
- 为了管理上的便利性。例如，后端节点的平滑上下线、SQL 黑名单等。

当然，不是所有的数据库中间件都具备上述能力。在选择数据库产品时，我们要从自己的实际需求出发。

具体到 MySQL 中，目前常用的数据库中间件有两种模式。

- 代理（proxy）模式

图 10-1 是代理模式的架构图。在这种模式下，我们会部署一个单独的代理服务。应用会直接连接代理节点。代理节点收到应用发送的 SQL 请求后，会进行分析处理（有的中间件会跳过这一步，直接转发），然后转发给后端 MySQL 数据库。虽然应用连接的是中间件，但从实际的效果来看，应用访问的就是 MySQL 数据库。属于这一模式的产品有：MySQL Router、ProxySQL、Vitess、MaxScale、Atlas、DBProxy、MyCAT、DBLE、OneProxy、Cetus 和 kingshard。虽然数量众多，不过这里提到的很多产品现在已经不再维护了。

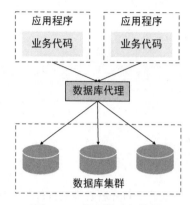

图 10-1　代理模式架构图

下面我们看看代理模式的优缺点。

优点：

- 对开发语言没有限制，适用于任何兼容 MySQL 协议的客户端；
- 对应用程序完全透明，应用程序可直接将代理节点作为 MySQL 使用；
- 功能全面。

缺点：

- 实现复杂，因为要实现 MySQL 的二进制通信协议；
- 代理节点存在单点风险，因此需要考虑代理节点的高可用；
- 如果代理节点和应用节点不在同一台机器上，还会多一层网络开销。

- 客户端（client）模式

图 10-2 是客户端模式的架构图。在这种模式下，客户端以 SDK 的方式提供给应用使用。SDK 会对外暴露一套接口，应用直接使用这些接口就可以进行读写分离、分库分表操作。属于这一模式的产品有：Sharding-JDBC、Zebra 和 TDDL。

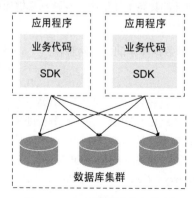

图 10-2　客户端模式的架构图

下面我们看看客户端模式的优缺点。

优点：

- 实现简单，一般是在驱动层上进行封装的；
- 去中心化，因为没有代理层，所以不用考虑高可用问题；
- 直连数据库，性能损耗低。

缺点：

- 只支持特定语言，例如 Sharding-JDBC 就只支持 Java 语言；
- 对业务有一定的侵入——无论是 SDK 的引入，还是 SDK 本身的限制，都需要业务代码进行兼容适配；
- 功能有限——除了读写分离、分库分表、分布式事务等业务相关的功能，一般不会实现 SQL 黑名单、审计等管理功能；
- 主要面向开发人员，在出现问题时，只能由开发人员来定位和分析。

表 10-1 选取了读者比较熟悉的部分 MySQL 中间件进行了对比。在选择中间件产品时，除了产品功能以外，我们也要考虑产品的活跃度。事实上，上面提到的很多中间件产品现在已经不再维护了。

表 10-1 MySQL 中间件功能对比

	MySQL Router	ProxySQL	Vitess	Sharding-JDBC	MaxScale	Atlas	MyCAT
开发语言	C++	C++	Go	Java	C	C	Java
部署形式	代理	代理	代理	客户端	代理	代理	代理
读写分离	支持	支持	支持	支持	支持	支持	支持
分库分表	不支持	支持简单的分片功能	支持	支持	支持简单的分片功能	支持	支持
支持的数据库	MySQL	MySQL	MySQL	虽然理论上支持任意实现 JDBC 规范的数据库，但用得较多的还是 MySQL	MariaDB	MySQL	虽然支持多种数据库，但用得较多的还是 MySQL
产品活跃度	活跃	活跃	活跃	活跃	活跃	基本停止更新	一般
产品架构复杂度	简单	简单	复杂	简单	简单	简单	简单

本书会介绍两个中间件：一个是将在第 12 章介绍的 MySQL Router，另一个就是本章即将介绍的 ProxySQL。

ProxySQL 使用 C++ 开发，于 2013 年基于 GPLv3 协议开源，支持的功能包括读写分离、连接池、负载均衡、审计、SQL 黑名单、查询重写、缓存、简单的分库分表等。除此之外，它还支持多层配置系统。基于这个系统，我们可以动态地调整配置，而无须重启 ProxySQL，可谓相当灵活。

国内外开源的 MySQL 中间件不少，但 ProxySQL 是业界为数不多仍在持续迭代的中间件产品。虽然它也有 bug，但一般都能在很短的时间内发版解决。

本章主要包括以下内容。

- ProxySQL 的安装。
- ProxySQL 入门。
- 多层配置系统。
- 读写分离。
- 深入理解 ProxySQL 表。
- ProxySQL 的高级特性。
- ProxySQL 连接池。
- ProxySQL Cluster。
- ProxySQL 的常见参数。
- ProxySQL 中的常见问题。

10.1 ProxySQL 的安装

首先在 ProxySQL 的 GitHub 页面上下载安装包。

以下是 ProxySQL rpm 包的安装步骤。

```
# yum install proxysql-2.4.1-1-centos7.x86_64.rpm
# service proxysql start
# netstat -ntlup | grep proxysql
tcp        0      0 0.0.0.0:6032            0.0.0.0:*               LISTEN      10046/proxysql
tcp        0      0 0.0.0.0:6033            0.0.0.0:*               LISTEN      10046/proxysql
```

服务启动后，默认会开启两个端口：6032 和 6033。6032 是管理端口，我们所有的管理操作都是在管理端口上进行的，默认的管理用户名和密码都是 admin（admin 用户只允许在本地登录）。6033 是 ProxySQL 的对外服务端口。

10.2 ProxySQL 入门

下面通过一个简单的示例来看看如何使用 ProxySQL。

(1) 配置后端节点。

主要是配置 mysql_servers 表。

```
# mysql -h127.0.0.1 -uadmin -padmin -P6032 --prompt 'ProxySQL> '
ProxySQL> insert into mysql_servers(hostgroup_id,hostname,port) values(1,'192.168.244.10',3306);
Query OK, 1 row affected (0.00 sec)

ProxySQL> load mysql servers to runtime;
Query OK, 0 rows affected (0.01 sec)

ProxySQL> save mysql servers to disk;
Query OK, 0 rows affected (0.02 sec)
```

插入后端节点的 IP 和端口，并将其主机组 ID 设置为 1。

主机组是 ProxySQL 中的一个核心概念，所有路由规则的作用对象都是主机组，而不是后端节点。`load mysql servers to runtime` 将 `mysql_servers` 中的内容加载到 RUNTIME 中，至于什么是 RUNTIME，后面会详细介绍，这里可简单理解为将表中的数据刷新到内存中。至于为什么要这样做，涉及 ProxySQL 的多层配置系统，后面会详细介绍。`save mysql servers to disk` 则将 `mysql_servers` 中的内容持久化到磁盘中。

(2) 配置访问用户。

这里的用户有以下两层含义。

- 前端用户（frontend user）：客户端用来访问 ProxySQL 的用户。
- 后端用户（backend user）：ProxySQL 用来访问后端 MySQL 节点的用户。

在目前的版本中，前后端用户没有分离。所以在配置时，前端用户的账号和密码必须与后端一致。

首先在 MySQL 中配置后端用户。

```
# mysql -h192.168.244.10 -P3306 -uroot -p123456
mysql> create user proxy_user@'%' identified with mysql_native_password by 'proxy_pass';
Query OK, 0 rows affected (0.15 sec)

mysql> grant select,delete,insert,update on *.* to proxy_user@'%';
Query OK, 0 rows affected (0.05 sec)
```

这里创建用户时使用了 `mysql_native_password`。如果我们要使用 MySQL 8.0 引入的 `caching_sha2_password`，需将 ProxySQL 中的 `admin-hash_passwords` 设置为 `false`，否则即使密码正确，也会提示 "Access denied"。

接着在 ProxySQL 中配置前端用户，主要是配置 `mysql_users` 表。

```
# mysql -h127.0.0.1 -uadmin -padmin -P6032 --prompt 'ProxySQL> '
ProxySQL> insert into mysql_users(username,password,default_hostgroup)
values('proxy_user','proxy_pass',1);
Query OK, 1 row affected (0.00 sec)

ProxySQL> load mysql users to runtime;
Query OK, 0 rows affected (0.01 sec)

ProxySQL> save mysql users to disk;
Query OK, 0 rows affected (0.01 sec)
```

配置完后，同样需要将更新的内容刷新到 RUNTIME 及持久化到磁盘中。配置用户时，同时指定了 `default_hostgroup`（默认主机组）。如果没有匹配的路由规则，则默认将请求转发给 `default_hostgroup`。

(3) 验证配置的效果。

执行几个简单 SQL，看能否达到预期的效果。

```
# mysql -h192.168.244.128 -uproxy_user -pproxy_pass -P 6033 -e "select @@hostname"
+------------+
| @@hostname |
+------------+
| node1      |
+------------+
```

```
# mysql -h192.168.244.128 -uproxy_user -pproxy_pass -P 6033 -e "insert into slowtech.t1 values(1,'a')"
# mysql -h192.168.244.128 -uproxy_user -pproxy_pass -P 6033 -e "select * from slowtech.t1"
+----+------+
| id | c1   |
+----+------+
| 1  | a    |
+----+------+
```

至此，一个简单的 ProxySQL 环境搭建完毕，后续针对 ProxySQL 6033 端口的所有请求都会转发给 192.168.244.10:3306 这个 MySQL 实例。

这只是一个简单的示例，如果我们要将它部署到线上，实际上还是有很多问题需要考虑的。

- 如果后端节点宕机了怎么办？ProxySQL 能探测到吗？
- 这里只有一个节点，如果我部署的是一套主从环境，这套主从环境的复制关系如何体现出来？
- 常用的读写分离如何配置？
- ProxySQL 自身的高可用性如何保障？

带着这些问题，下面我们深入学习 ProxySQL。

10.3 多层配置系统

在上面的示例中可以看到，无论是插入 mysql_servers 表还是 mysql_users 表，都会执行 LOAD *xxx* TO RUNTIME 和 SAVE *xxx* TO DISK 这两个操作，这实际上与多层配置系统（multi layer configuration system）有关。

在 ProxySQL 中，配置信息分为如图 10-3 所示的 3 层。

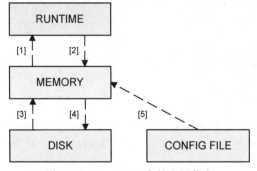

图 10-3　ProxySQL 中的配置信息

最上面一层是 RUNTIME。RUNTIME 是一种基于内存的数据结构，保存着 ProxySQL 当前正在使用的配置，无法直接修改，需要通过第二层的 MEMORY 来间接修改。

第二层的 MEMORY 是基于内存的 SQLite3 数据库。如果我们要修改 RUNTIME 的配置，只能先通过 SQL 修改 MEMORY 层的配置，再将其加载到 RUNTIME 中。既然是内存中的配置，一旦 ProxySQL

发生重启，配置就会丢失。如果不想丢失，需将其持久化到第三层的 DISK 中。

第三层的 DISK 是基于磁盘的 SQLite3 数据库。

第三层的 CONFIG FILE 指的是配置文件。如果是通过 rpm 包安装的，则默认的配置文件是 /etc/proxysql.cnf。

ProxySQL 在启动的时候，会首先通过配置文件确认数据目录（datadir，默认为 /var/lib/proxysql）的位置，接着判断数据目录中是否存在 proxysql.db（SQLite3 数据库）。如果存在，则从 proxysql.db 中加载 ProxySQL 的配置信息。如果不存在，则从配置文件中加载 ProxySQL 的配置信息。

3 层的配置系统看上去有点复杂，但实际上赋予了我们极大的便利性。

- 可通过 SQL 命令修改配置，简单方便。
- 可动态修改配置，无须重启 ProxySQL 进程。
- 可及时回滚不合适的配置。

接下来看看这 3 层分别包含了哪些配置信息。

首先是 ProxySQL 中的库信息。

```
# mysql -h127.0.0.1 -uadmin -padmin -P6032 --prompt 'ProxySQL> '
ProxySQL> show databases;
+-----+---------------+-------------------------------------+
| seq | name          | file                                |
+-----+---------------+-------------------------------------+
| 0   | main          |                                     |
| 2   | disk          | /var/lib/proxysql/proxysql.db       |
| 3   | stats         |                                     |
| 4   | monitor       |                                     |
| 5   | stats_history | /var/lib/proxysql/proxysql_stats.db |
+-----+---------------+-------------------------------------+
5 rows in set (0.00 sec)
```

一共有 5 个库，其中与配置相关的有两个：main 和 disk。前者包含了 RUNTIME 和 MEMORY 层的配置信息，后者包含了 DISK 层的配置信息。

接下来是 disk 库中的表信息。

在 ProxySQL 中，如果要查看某个库中的表，只能通过 SHOW TABLES FROM DBNAME 命令查看，不能通过 MySQL 中的 USE DBNAME; SHOW TABLES 命令查看。在执行 SHOW TABLES 命令时，如果没指定 DBNAME，则默认是 main 库。

```
ProxySQL> show tables from disk;
+-----------------------------------------+
| tables                                  |
+-----------------------------------------+
| global_settings                         |
| global_variables                        |
| mysql_aws_aurora_hostgroups             |
| mysql_collations                        |
| mysql_firewall_whitelist_rules          |
| mysql_firewall_whitelist_sqli_fingerprints |
| mysql_firewall_whitelist_users          |
```

```
| mysql_galera_hostgroups                  |
| mysql_group_replication_hostgroups       |
| mysql_query_rules                        |
| mysql_query_rules_fast_routing           |
| mysql_replication_hostgroups             |
| mysql_servers                            |
| mysql_users                              |
| proxysql_servers                         |
| restapi_routes                           |
| scheduler                                |
+------------------------------------------+
17 rows in set (0.00 sec)
```

`disk` 库中一共有 17 张表，常用的有以下几张。

- `global_variables`：配置 ProxySQL 的参数。
- `mysql_servers`：配置后端 MySQL 节点信息。
- `mysql_users`：配置用户信息。
- `mysql_replication_hostgroups`、`mysql_galera_hostgroups`、`mysql_group_replication_hostgroups`、`mysql_aws_aurora_hostgroups`：定义主机组之间的复制关系，这 4 张表依次对应主从复制、Galera Cluster、组复制和 AWS Aurora。
- `mysql_query_rules`、`mysql_query_rules_fast_routing`：定义路由规则。

除了 `global_settings`，`disk` 库中的其他 16 张表在 main 库中同样存在，表名也一致，对应 MEMORY 层的配置信息。

main 库中除了属于 MEMORY 层的 16 张表，还有 16 张表属于 RUNTIME 层。RUNTIME 层和 MEMORY 层的表名基本一致，只不过是以 `runtime_` 开头的。RUNTIME 层的表也可修改，但不会生效。

RUNTIME、MEMORY 和 DISK 这三层的配置信息都是相互独立的。如果要将下层的配置信息加载到上层（如从 DISK 到 MEMORY），需使用 LOAD 命令。相反，如果要将上层的配置信息持久化到下层，需使用 SAVE 命令。

在使用这两个命令时，既可以通过 FROM 子句指定源层，也可以通过 TO 子句指定目标层。例如，要将 MYSQL USERS 中的配置信息从 MEMORY 层加载到 RUNTIME 层，以下两个命令是等价的。

```
LOAD MYSQL USERS FROM MEMORY;
LOAD MYSQL USERS TO RUNTIME;
```

表 10-2 梳理了与 MYSQL USERS 相关的 LOAD 和 SAVE 命令。图 10-3 中的箭头对应的其实就是这 5 条命令。

表 10-2 MYSQL USERS 相关的 LOAD 和 SAVE 命令

命令	作用
LOAD MYSQL USERS FROM MEMORY / LOAD MYSQL USERS TO RUNTIME	将配置信息从 MEMORY 加载到 RUNTIME
SAVE MYSQL USERS TO MEMORY / SAVE MYSQL USERS FROM RUNTIME	将配置信息从 RUNTIME 持久化到 MEMORY
LOAD MYSQL USERS TO MEMORY / LOAD MYSQL USERS FROM DISK	将配置信息从 DISK 加载到 MEMORY
SAVE MYSQL USERS FROM MEMORY / SAVE MYSQL USERS TO DISK	将配置信息从 MEMORY 持久化到 DISK
LOAD MYSQL USERS FROM CONFIG	将配置信息从配置文件加载到 MEMORY

除了 MYSQL USERS，其他常用的配置信息有：MYSQL SERVERS、MYSQL QUERY RULES、MYSQL VARIABLES、ADMIN VARIABLES。

10.4 读写分离

读写分离是中间件的一个高频使用场景。下面我们看看在 ProxySQL 中实现读写分离的具体步骤。

(1) 配置后端节点。

```
# mysql -h127.0.0.1 -uadmin -padmin -P6032 --prompt 'ProxySQL> '
ProxySQL> insert into mysql_servers(hostgroup_id,hostname,port) values(10,'192.168.244.10',3306);
Query OK, 1 row affected (0.00 sec)

ProxySQL> insert into mysql_servers(hostgroup_id,hostname,port) values(20,'192.168.244.20',3306);
Query OK, 1 row affected (0.00 sec)

ProxySQL> load mysql servers to runtime;
Query OK, 0 rows affected (0.00 sec)

ProxySQL> save mysql servers to disk;
Query OK, 0 rows affected (0.02 sec)
```

在配置的两个节点中，192.168.244.10:3306 是主库，192.168.244.20:3306 是从库，对应的 `hostgroup_id` 分别是 10 和 20。`hostgroup_id` 在后面通过 `mysql_replication_hostgroups` 定义主机组之间的复制关系时会用到。

(2) 配置用户。

这里需要配置两个用户：监控用户和业务访问用户。

首先，创建业务访问用户。该用户在 MySQL 和 ProxySQL 中都要配置。

```
# mysql -h192.168.244.10 -P3306 -uroot -p123456
mysql> create user split_user@'%' identified with mysql_native_password by 'split_pass';
Query OK, 0 rows affected (0.01 sec)

mysql> grant create,select,insert,delete,update on *.* to split_user@'%';
Query OK, 0 rows affected (0.00 sec)

# mysql -h127.0.0.1 -uadmin -padmin -P6032 --prompt 'ProxySQL> '
ProxySQL> insert into mysql_users(username, password, default_hostgroup)
values('split_user','split_pass',10);
Query OK, 1 row affected (0.00 sec)

ProxySQL> load mysql users to runtime;
Query OK, 0 rows affected (0.00 sec)

ProxySQL> save mysql users to disk;
Query OK, 0 rows affected (0.00 sec)
```

接着，创建监控用户。该用户用于检测 MySQL 节点的健康状态，监控的范围如下。

- 后端节点的连通性。
- 后端节点 `read_only` 的值。

❑ 后端节点的主从延迟情况。

监控的结果会记录在 monitor 库中。

在 ProxySQL 中，监控用户名和密码分别由参数 mysql-monitor_username 和 mysql-monitor_password 决定，这两个参数的默认值都为 monitor。

这里，保持 ProxySQL 中默认的账号和密码不变，只在 MySQL 中创建对应的用户。

```
mysql> create user 'monitor'@'%' identified with mysql_native_password by 'monitor';
Query OK, 0 rows affected (0.01 sec)

mysql> grant replication client on *.* to 'monitor'@'%';
Query OK, 0 rows affected (0.00 sec)
```

(3) 定义主机组之间的复制关系。

主机组之间的复制关系是在 mysql_replication_hostgroups 中配置的。

```
ProxySQL> insert into mysql_replication_hostgroups (writer_hostgroup, reader_hostgroup, check_type, comment) values (10,20,'super_read_only','repl_group_1');
Query OK, 1 row affected (0.00 sec)

ProxySQL> load mysql servers to runtime;
Query OK, 0 rows affected (0.00 sec)

ProxySQL> save mysql servers to disk;
Query OK, 0 rows affected (0.03 sec)
```

其中各项解释如下。

❑ writer_hostgroup：写节点（主库）所在的主机组。
❑ reader_hostgroup：读节点（从库）所在的主机组。
❑ check_type：判断读节点还是写节点的标准，可以取 read_only、innodb_read_only、super_read_only、read_only|innodb_read_only（read_only 或 innodb_read_only）、read_only&innodb_read_only（read_only 与 innodb_read_only）这 5 个值。
❑ comment：注释。

因为 ProxySQL 会基于后端节点 check_type 的值来动态调整它所属的主机组，所以我们必须确保读节点（从库）的 super_read_only 处于开启状态。

```
# mysql -h192.168.244.20 -P3306 -uroot -p123456
mysql> show variables like 'super_read_only';
+-----------------+-------+
| Variable_name   | Value |
+-----------------+-------+
| super_read_only | ON    |
+-----------------+-------+
1 row in set (0.00 sec)
```

注意，如果我们不想让 ProxySQL 自动调整节点所属的主机组，不配置 mysql_replication_hostgroups 即可。

(4) 配置路由规则。

路由规则是在 mysql_query_rules 中配置的。

```
ProxySQL> insert into mysql_query_rules(rule_id,active,match_digest,destination_hostgroup,apply)
values (1,1,'^SELECT.*FOR UPDATE$',10,1);
Query OK, 1 row affected (0.00 sec)

ProxySQL> insert into mysql_query_rules(rule_id,active,match_digest,destination_hostgroup,apply)
values (2,1,'^SELECT',20,1);
Query OK, 1 row affected (0.00 sec)

ProxySQL> load mysql query rules to runtime;
Query OK, 0 rows affected (0.00 sec)

ProxySQL> save mysql query rules to disk;
Query OK, 0 rows affected (0.01 sec)
```

其中各项解释如下。

- rule_id：规则 ID。在进行规则匹配时，会按照规则 ID 从小到大的顺序依次匹配。
- active：是否启用规则。
- match_digest：指定匹配规则，支持正则匹配。
- destination_hostgroup：指定目标主机组。匹配上规则的操作将发往该主机组处理。
- apply：设置为 1，如果当前的规则匹配，则不再进行后续其他规则的匹配。

这里虽然只配置了两个规则，但我们在创建用户时指定了 default_hostgroup，所以整个路由规则如下。

- 所有 SELECT FOR UPDATE 操作将发往主机组 10（主库）处理。
- 其他所有 SELECT 操作将发往主机组 20（从库）处理。
- 除此之外的所有操作将发往默认主机组 10（主库）处理。

(5) 验证读写分离的效果。

这里测试了几条常用的 SQL 语句。

```
# mysql -h192.168.244.128 -usplit_user -psplit_pass -P6033
mysql> create table slowtech.t1(id int, c1 varchar(10));
Query OK, 0 rows affected (0.01 sec)

mysql> insert into slowtech.t1 values(1,'a'),(2,'b');
Query OK, 2 rows affected (0.01 sec)
Records: 2  Duplicates: 0  Warnings: 0

mysql> update slowtech.t1 set c1='c' where id=1;
Query OK, 1 row affected (0.00 sec)
Rows matched: 1  Changed: 1  Warnings: 0

mysql> delete from slowtech.t1 where id=1;
Query OK, 1 row affected (0.00 sec)

mysql> select * from slowtech.t1 for update;
+------+------+
| id   | c1   |
```

```
+------+------+
|  2   |  b   |
+------+------+
1 row in set (0.00 sec)

mysql> select * from slowtech.t1;
+------+------+
|  id  |  c1  |
+------+------+
|  2   |  b   |
+------+------+
1 row in set (0.00 sec)
```

接下来,我们验证一下这些操作分别是在哪个节点上执行的。

```
ProxySQL> select hostgroup,digest_text from stats.stats_mysql_query_digest order by first_seen;
+-----------+-------------------------------------------------+
| hostgroup | digest_text                                     |
+-----------+-------------------------------------------------+
| 10        | select @@version_comment limit ?                |
| 10        | create table slowtech.t1(id int,c1 varchar(?))  |
| 10        | insert into slowtech.t1 values(?,?),(?,?)       |
| 10        | update slowtech.t1 set c1=? where id=?          |
| 10        | delete from slowtech.t1 where id=?              |
| 10        | select * from slowtech.t1 for update            |
| 20        | select * from slowtech.t1                       |
+-----------+-------------------------------------------------+
7 rows in set (0.00 sec)
```

可以看到,除了最后一个 SELECT 操作是在从库上执行的,其他所有操作都是在主库上执行的。

下面再验证另外的一个常见操作——事务读。事务读指的是显式开启事务后执行的 SELECT 操作。

在验证之前,我们先清空 stats.stats_mysql_query_digest 表中的数据。

```
ProxySQL> select * from stats.stats_mysql_query_digest_reset limit 1;
```

对 stats.stats_mysql_query_digest_reset 执行查询操作就能清空 stats.stats_mysql_query_digest 表的内容。

清空完 stats.stats_mysql_query_digest 表中的数据,我们继续来验证。测试 SQL 如下。

```
mysql> begin;
Query OK, 0 rows affected (0.00 sec)

mysql> select * from slowtech.t1 where id=1;
Empty set (0.00 sec)

mysql> commit;
Query OK, 0 rows affected (0.00 sec)
```

接下来重点看看上面这个 SELECT 操作是在哪个节点上执行的。

```
ProxySQL> select hostgroup,digest_text from stats.stats_mysql_query_digest order by first_seen;
+-----------+-------------------------------------------+
| hostgroup | digest_text                               |
+-----------+-------------------------------------------+
| 10        | begin                                     |
| 10        | select * from slowtech.t1 where id=?      |
```

```
| 10         | commit                                |
+------------+---------------------------------------+
3 rows in set (0.01 sec)
```

奇怪的是，这个 SELECT 操作竟然是在主库上执行的，不应该匹配第二个规则吗？

这实际上与账号创建时 transaction_persistent 的值有关。

```
ProxySQL> select username,password,active,default_hostgroup,transaction_persistent from mysql_users;
+------------+------------+--------+-------------------+------------------------+
| username   | password   | active | default_hostgroup | transaction_persistent |
+------------+------------+--------+-------------------+------------------------+
| split_user | split_pass | 1      | 10                | 1                      |
+------------+------------+--------+-------------------+------------------------+
1 row in set (0.00 sec)
```

在创建账号时，如果不显式设置 transaction_persistent，则其默认为 1。transaction_persistent 为 1 意味着，如果一个事务开始时的第一个操作是在某个主机组内执行的，则这个事务后续的其他操作都将转发到这个主机组内执行，此时会忽略任何路由规则。

具体到上面这个查询，因为 BEGIN 操作是在主库上执行的，所以后续的其他操作也会转发到主库上执行。

接下来看看将 transaction_persistent 设置为 0 的效果。

```
ProxySQL> update mysql_users set transaction_persistent=0 where username='split_user';
Query OK, 1 row affected (0.36 sec)

ProxySQL> load mysql users to runtime;
Query OK, 0 rows affected (0.00 sec)
```

断开客户端连接，重新执行上面的操作。

```
ProxySQL> select hostgroup,digest_text from stats.stats_mysql_query_digest order by first_seen;
+-----------+---------------------------------------+
| hostgroup | digest_text                           |
+-----------+---------------------------------------+
| 10        | begin                                 |
| 20        | select * from slowtech.t1 where id=?  |
| 10        | commit                                |
+-----------+---------------------------------------+
3 rows in set (0.00 sec)
```

同样的 SELECT 操作，这次转发到从库上执行了。很显然，transaction_persistent 为 1 更符合我们对事务的认知习惯。

以上是 ProxySQL 中实现读写分离的基本步骤。

虽然实现了读写分离，但这种方式还是过于简单，没有考虑以下两种常见场景。

- 从库宕机。从库如果宕机了，所有转发到从库的查询都会失败。
- 主从延迟。基于上面配置的路由规则，只要是非 SELECT FOR UPDATE 的查询操作都会转发到从库上执行，无论从库是否延迟。但实际上，很多做了读写分离的业务，只能容忍短时间内的主从延迟。如果主从延迟过大，也会影响业务。

针对上面这两种常见场景，通常的解决思路如下。

- 从库如果宕机了，将读请求转发到主库上执行。
- 为主从延迟设置阈值。将超过阈值的读请求转发到主库上执行。

对于上面这两个需求，ProxySQL 同样也能满足，具体的 SQL 语句如下。

```
ProxySQL> insert into mysql_servers(hostgroup_id,hostname,port) values(20,'192.168.244.10',3306);
Query OK, 1 row affected (0.00 sec)

ProxySQL> update mysql_servers set weight=100,max_replication_lag=10 where hostgroup_id=20 and hostname='192.168.244.20';
Query OK, 1 row affected (0.00 sec)

ProxySQL> load mysql servers to runtime;
Query OK, 0 rows affected (0.01 sec)

ProxySQL> save mysql servers to disk;
Query OK, 0 rows affected (0.02 sec)
```

这几条 SQL 语句的作用如下。

- 将主库添加到 reader_hostgroup 中。这样当从库出现故障时，读请求能转发到主库执行。
- 提高从库的权重（weight）。这样即使主库也在 reader_hostgroup 中，它也不会承担太多的读请求。
- 设置从库可允许的最大延迟时间（max_replication_lag）。若主从延迟超过这个值，读请求会转发到主库执行。

最后我们看看调整后的配置。

```
ProxySQL> select hostgroup_id,hostname,port,status,weight,max_replication_lag from mysql_servers;
+--------------+----------------+------+--------+--------+---------------------+
| hostgroup_id | hostname       | port | status | weight | max_replication_lag |
+--------------+----------------+------+--------+--------+---------------------+
| 10           | 192.168.244.10 | 3306 | ONLINE | 1      | 0                   |
| 20           | 192.168.244.20 | 3306 | ONLINE | 100    | 10                  |
| 20           | 192.168.244.10 | 3306 | ONLINE | 1      | 0                   |
+--------------+----------------+------+--------+--------+---------------------+
3 rows in set (0.00 sec)
```

至此，一个生产可用的 ProxySQL 读写分离环境搭建完毕。

但实际上，ProxySQL 的作者并不推荐这种读写分离方案。他主张基于 stats.stats_mysql_query_digest 中的统计信息，只将耗时较久的查询放到从库上执行。

下面就来看看 ProxySQL 作者推崇的读写分库方案。该方案的实现思路如下。

- 将所有的读写请求都转发到主库上执行。
- 执行一段时间后，统计 stats.stats_mysql_query_digest 中耗时较久的 SELECT 操作。
- 配置路由规则，将这些耗时较久的 SELECT 操作转发到从库上执行。

还是基于之前的环境，我们看看具体的操作步骤。

10.4 读写分离

(1) 清除之前的路由规则。

```
ProxySQL> delete from mysql_query_rules;
Query OK, 2 rows affected (0.00 sec)

ProxySQL> load mysql query rules to runtime;
Query OK, 0 rows affected (0.00 sec)
```

因为没有设置路由规则，所以所有的操作都会转发到默认主机组，即主库上执行。

(2) 统计 stats.stats_mysql_query_digest 中耗时较久的 SELECT 操作。

```
ProxySQL> select hostgroup,digest_text,digest,count_star,sum_time from stats.stats_mysql_query_digest
          where digest_text like 'select%' order by sum_time desc limit 5;
+-----------+-------------------------------------------------+--------------------+------------+----------+
| hostgroup | digest_text                                     | digest             | count_star | sum_time |
+-----------+-------------------------------------------------+--------------------+------------+----------+
| 10        | select distinct c1 from slowtech.t1             | 0xE39877A1DD28E8E2 | 7          | 2163801  |
| 10        | select count(*) from slowtech.t1 where id<=?    | 0x7247DC0F2A77831D | 4          | 569432   |
| 10        | select id,count(*) from slowtech.t1 group by id | 0xE5F2E6176D25C1DF | 1          | 247629   |
| 10        | select count(*) from slowtech.t1                | 0x67336BFB39A5ADD3 | 4          | 129104   |
+-----------+-------------------------------------------------+--------------------+------------+----------+
4 rows in set (0.00 sec)
```

这里的 sum_time 是总的执行时间，单位是微秒。

(3) 将耗时较久的查询放到从库上执行。

以第一条 SQL 为例。

```
ProxySQL> insert into mysql_query_rules (rule_id,active,digest,destination_hostgroup,apply) values
(1,1,'0xE39877A1DD28E8E2',20,1);
Query OK, 1 row affected (0.00 sec)

ProxySQL> load mysql query rules to runtime;
Query OK, 0 rows affected (0.00 sec)

ProxySQL> select * from stats.stats_mysql_query_digest_reset limit 1;
```

这里用到了 digest，它是对规范后的 SQL 取的哈希值。一个 digest 代表一类 SQL。

再次执行第一条 SQL，并查看统计信息。

```
ProxySQL> select hostgroup,digest_text,digest,count_star,sum_time from
stats.stats_mysql_query_digest;
+-----------+-------------------------------------+--------------------+------------+----------+
| hostgroup | digest_text                         | digest             | count_star | sum_time |
+-----------+-------------------------------------+--------------------+------------+----------+
| 20        | select distinct c1 from slowtech.t1 | 0xE39877A1DD28E8E2 | 2          | 644596   |
+-----------+-------------------------------------+--------------------+------------+----------+
1 row in set (0.00 sec)
```

可以看到，这条 SQL 被转发到了从库上执行。

总结

在过往的工作中，我看到很多业务在上线之初就做了读写分离。这些做了读写分离的业务中，有很多业务受业务类型所限,在未来几年里体量不会太大,请求量也不会太高。对于这些业务,其实没必要做读写分离。毕竟读写分离并不是万能的"银弹",有时候带来的问题反而比预想的收益还多。在使用读写分离时,需注意以下几点。

- 读写分离适合"读多写少"的场景。如果读操作不多,完全没有必要做读写分离。
- 对于数据实时性要求很高的读操作,不建议放到从库上执行。只要是主从复制,即使使用的是半同步复制,也有读取到脏数据的风险。

10.5 深入理解 ProxySQL 表

本节介绍的几张表是 ProxySQL 中最常用的表。熟悉这些表的具体含义,有助于我们更好地使用 ProxySQL 的各种功能。

- **mysql_users**

该表用来配置用户信息。

```
ProxySQL> show create table mysql_users\G
*************************** 1. row ***************************
       table: mysql_users
Create Table: CREATE TABLE mysql_users (
    username VARCHAR NOT NULL,
    password VARCHAR,
    active INT CHECK (active IN (0,1)) NOT NULL DEFAULT 1,
    use_ssl INT CHECK (use_ssl IN (0,1)) NOT NULL DEFAULT 0,
    default_hostgroup INT NOT NULL DEFAULT 0,
    default_schema VARCHAR,
    schema_locked INT CHECK (schema_locked IN (0,1)) NOT NULL DEFAULT 0,
    transaction_persistent INT CHECK (transaction_persistent IN (0,1)) NOT NULL DEFAULT 1,
    fast_forward INT CHECK (fast_forward IN (0,1)) NOT NULL DEFAULT 0,
    backend INT CHECK (backend IN (0,1)) NOT NULL DEFAULT 1,
    frontend INT CHECK (frontend IN (0,1)) NOT NULL DEFAULT 1,
    max_connections INT CHECK (max_connections >=0) NOT NULL DEFAULT 10000,
    attributes VARCHAR CHECK (JSON_VALID(attributes) OR attributes = '') NOT NULL DEFAULT '',
    comment VARCHAR NOT NULL DEFAULT '',
    PRIMARY KEY (username, backend),
    UNIQUE (username, frontend))
1 row in set (0.00 sec)
```

表中各个字段的具体含义如下。

- username 和 password:用户名和密码。
- active:是否激活账号。
- use_ssl:设置为 1,则强制用户使用 SSL 证书进行身份验证。
- default_hostgroup:默认主机组。如果没有匹配的路由规则或路由规则没配置,请求会转发到默认主机组。
- default_schema:默认 schema。如果不设置,则由 mysql-default_schema(默认为 information_schema)决定。
- schema_locked:目前暂未实现。

- transaction_persistent：设置为 1，则代表一个事务内的所有操作都将转发到一个主机组内执行。具体是哪个主机组，由事务刚开始的那个操作决定，如 BEGIN、START TRANSACTION。
- fast_forward：设置为 1，则该用户发起的 SQL 会跳过重写（rewriting）、缓存（caching）等查询处理层，直接转发给后端节点。
- backend：是否为后端账号。
- frontend：是否为前端账号。当前，前后端账号没有分离，所以 backend 和 frontend 都必须设置为 1。
- max_connections：设置账号的最大连接数，默认为 10000。
- attributes：目前暂未实现。
- comment：注释。

● mysql_servers

该表用来配置后端节点信息。

```
ProxySQL> show create table mysql_servers\G
*************************** 1. row ***************************
       table: mysql_servers
Create Table: CREATE TABLE mysql_servers (
    hostgroup_id INT CHECK (hostgroup_id>=0) NOT NULL DEFAULT 0,
    hostname VARCHAR NOT NULL,
    port INT CHECK (port >= 0 AND port <= 65535) NOT NULL DEFAULT 3306,
    gtid_port INT CHECK ((gtid_port <> port OR gtid_port=0) AND gtid_port >= 0 AND gtid_port <= 65535) NOT NULL DEFAULT 0,
    status VARCHAR CHECK (UPPER(status) IN ('ONLINE','SHUNNED','OFFLINE_SOFT', 'OFFLINE_HARD')) NOT NULL DEFAULT 'ONLINE',
    weight INT CHECK (weight >= 0 AND weight <=10000000) NOT NULL DEFAULT 1,
    compression INT CHECK (compression IN(0,1)) NOT NULL DEFAULT 0,
    max_connections INT CHECK (max_connections >=0) NOT NULL DEFAULT 1000,
    max_replication_lag INT CHECK (max_replication_lag >= 0 AND max_replication_lag <= 126144000) NOT NULL DEFAULT 0,
    use_ssl INT CHECK (use_ssl IN(0,1)) NOT NULL DEFAULT 0,
    max_latency_ms INT UNSIGNED CHECK (max_latency_ms>=0) NOT NULL DEFAULT 0,
    comment VARCHAR NOT NULL DEFAULT '',
    PRIMARY KEY (hostgroup_id, hostname, port) )
1 row in set (0.00 sec)
```

表中各个字段的具体含义如下。

- hostgroup_id：主机组 ID。
- hostname：后端节点的主机名或 IP。
- port：后端节点的端口。
- status：节点的状态。可配置以下值。
    - ONLINE：节点功能正常，可对外提供服务。
    - SHUNNED：节点暂时不可用。
    - OFFLINE_SOFT 和 OFFLINE_HARD：两者功能类似，都是将节点置为离线状态。置为离线状态的节点，将不再接受新的请求。两者的主要区别是前者会等待当前正在执行的事务完成，后者会直接杀掉连接。

    将节点从 mysql_servers 表中删除，效果等同于 OFFLINE_HARD。

- weight：权重。权重越高，被分发的请求越多。
- compression：是否开启压缩。
- max_connections：限制 ProxySQL 到后端节点的最大连接数。
- max_replication_lag：主从延迟的阈值。一旦超过该值，ProxySQL 就会将该节点的状态置为 SHUNNED，直到延迟恢复。
- use_ssl：是否开启 SSL。针对的是 ProxySQL 与后端节点之间的连接。注意，后端连接是否使用 SSL 与前端连接是否使用 SSL 没有任何关系，两者是相互独立的。
- max_latency_ms：ProxySQL 会定期对后端节点进行 PING 操作，如果 PING 操作的响应时间超过 max_latency_ms，则会将该节点从连接池中剔除，尽管该节点的状态依然是 ONLINE。

- **mysql_replication_hostgroups**

该表用来定义主从复制中主机组之间的关系。

```
ProxySQL> show create table mysql_replication_hostgroups\G
*************************** 1. row ***************************
       table: mysql_replication_hostgroups
Create Table: CREATE TABLE mysql_replication_hostgroups (
    writer_hostgroup INT CHECK (writer_hostgroup>=0) NOT NULL PRIMARY KEY,
    reader_hostgroup INT NOT NULL CHECK (reader_hostgroup<>writer_hostgroup AND reader_hostgroup>=0),
    check_type VARCHAR CHECK (LOWER(check_type) IN ('read_only','innodb_read_only','super_read_only',
'read_only|innodb_read_only','read_only&innodb_read_only')) NOT NULL DEFAULT 'read_only',
    comment VARCHAR NOT NULL DEFAULT '', UNIQUE (reader_hostgroup))
1 row in set (0.00 sec)
```

表中各个字段的具体含义如下。

- writer_hostgroup：定义可写的主机组。read_only 为 OFF 的节点会置于 writer_hostgroup 中。
- reader_hostgroup：定义只读的主机组。read_only 为 ON 的节点会置于 reader_hostgroup 中。
- check_type：检测类型。ProxySQL 会定期检测后端节点 check_type 的值。如果 check_type 的值从 OFF 变为 ON，则 ProxySQL 会将这个节点调整到 reader_hostgroup 中。反之，若 check_type 的值从 ON 变为 OFF，则 ProxySQL 会将其调整到 writer_hostgroup 中。注意，这个调整是 ProxySQL 自动进行的，涉及 mysql_servers 和 runtime_mysql_servers 表。

  由此可见，在 ProxySQL 中，一定要将从库的 check_type 值设置为 ON。如果不小心将其设置为了 OFF，ProxySQL 会自动将这个节点调整到 writer_hostgroup 中，进而出现双写。

  与此相关的一个参数是 mysql-monitor_writer_is_also_reader，该参数决定了后端节点在调整到 writer_hostgroup 后，是否依然保留在 reader_hostgroup 中。默认为 true，即保留在 reader_hostgroup 中。

- **mysql_group_replication_hostgroups 和 mysql_galera_hostgroups**

这两张表的表结构相同，只不过前者针对 MySQL 的组复制，后者针对 Galera Cluster 和 Percona XtraDB Cluster。

```
ProxySQL> show create table mysql_group_replication_hostgroups\G
*************************** 1. row ***************************
       table: mysql_group_replication_hostgroups
```

```
Create Table: CREATE TABLE mysql_group_replication_hostgroups (
    writer_hostgroup INT CHECK (writer_hostgroup>=0) NOT NULL PRIMARY KEY,
    backup_writer_hostgroup INT CHECK (backup_writer_hostgroup>=0 AND
backup_writer_hostgroup<>writer_hostgroup) NOT NULL,
    reader_hostgroup INT NOT NULL CHECK (reader_hostgroup<>writer_hostgroup AND
backup_writer_hostgroup<>reader_hostgroup AND reader_hostgroup>0),
    offline_hostgroup INT NOT NULL CHECK (offline_hostgroup<>writer_hostgroup AND
offline_hostgroup<>reader_hostgroup AND backup_writer_hostgroup<>offline_hostgroup AND
offline_hostgroup>=0),
    active INT CHECK (active IN (0,1)) NOT NULL DEFAULT 1,
    max_writers INT NOT NULL CHECK (max_writers >= 0) DEFAULT 1,
    writer_is_also_reader INT CHECK (writer_is_also_reader IN (0,1,2)) NOT NULL DEFAULT 0,
    max_transactions_behind INT CHECK (max_transactions_behind>=0) NOT NULL DEFAULT 0,
    comment VARCHAR,
    UNIQUE (reader_hostgroup),
    UNIQUE (offline_hostgroup),
    UNIQUE (backup_writer_hostgroup))
1 row in set (0.00 sec)
```

表中各个字段的具体含义如下。

- writer_hostgroup：定义可写的主机组。
- backup_writer_hostgroup：与 max_writers 有关，max_writers 限制了 writer_hostgroup 中写节点的个数。超过 max_writers 后，写节点会放到 backup_writer_hostgroup 中。一般用在组复制的多主模式下，限制可写的节点数。
- reader_hostgroup：定义只读的主机组。
- offline_hostgroup：定义 OFFLINE 的主机组。非 ONLINE 节点会置于 offline_hostgroup 中。
- active：启用后，ProxySQL 会检测各节点的状态，并自动调整节点所属的主机组。
- writer_is_also_reader：是否将 writer_hostgroup 中的节点放到 reader_hostgroup 中。若设置为 2，则会将 backup_writer_hostgroup 中的节点放到 reader_hostgroup 中。
- max_transactions_behind：节点可允许的最大延迟事务数。一旦超过该值，ProxySQL 会将节点的状态置为 SHUNNED。延迟事务数在组复制中由 sys.gr_member_routing_candidate_status（注意，这个视图需单独创建，组复制中没有这张视图）中的 transactions_behind 列决定；在 Galera Cluster 中，则由变量 wsrep_local_recv_queue 决定。

- mysql_query_rules

该表用来定义路由规则。

```
ProxySQL> show create table mysql_query_rules\G
*************************** 1. row ***************************
       table: mysql_query_rules
Create Table: CREATE TABLE mysql_query_rules (
    rule_id INTEGER PRIMARY KEY AUTOINCREMENT NOT NULL,
    active INT CHECK (active IN (0,1)) NOT NULL DEFAULT 0,
    username VARCHAR,
    schemaname VARCHAR,
    flagIN INT CHECK (flagIN >= 0) NOT NULL DEFAULT 0,
    client_addr VARCHAR,
    proxy_addr VARCHAR,
    proxy_port INT CHECK (proxy_port >= 0 AND proxy_port <= 65535),
    digest VARCHAR,
    match_digest VARCHAR,
```

```
        match_pattern VARCHAR,
        negate_match_pattern INT CHECK (negate_match_pattern IN (0,1)) NOT NULL DEFAULT 0,
        re_modifiers VARCHAR DEFAULT 'CASELESS',
        flagOUT INT CHECK (flagOUT >= 0),
        replace_pattern VARCHAR CHECK(CASE WHEN replace_pattern IS NULL THEN 1 WHEN replace_pattern IS NOT
NULL AND match_pattern IS NOT NULL THEN 1 ELSE 0 END),
        destination_hostgroup INT DEFAULT NULL,
        cache_ttl INT CHECK(cache_ttl > 0),
        cache_empty_result INT CHECK (cache_empty_result IN (0,1)) DEFAULT NULL,
        cache_timeout INT CHECK(cache_timeout >= 0),
        reconnect INT CHECK (reconnect IN (0,1)) DEFAULT NULL,
        timeout INT UNSIGNED CHECK (timeout >= 0),
        retries INT CHECK (retries>=0 AND retries <=1000),
        delay INT UNSIGNED CHECK (delay >=0),
        next_query_flagIN INT UNSIGNED,
        mirror_flagOUT INT UNSIGNED,
        mirror_hostgroup INT UNSIGNED,
        error_msg VARCHAR,
        OK_msg VARCHAR,
        sticky_conn INT CHECK (sticky_conn IN (0,1)),
        multiplex INT CHECK (multiplex IN (0,1,2)),
        gtid_from_hostgroup INT UNSIGNED,
        log INT CHECK (log IN (0,1)),
        apply INT CHECK(apply IN (0,1)) NOT NULL DEFAULT 0,
        attributes VARCHAR CHECK (JSON_VALID(attributes) OR attributes = '') NOT NULL DEFAULT '',
        comment VARCHAR)
1 row in set (0.00 sec)
```

表中各个字段的具体含义如下。

- rule_id：指定规则 ID。规则 ID 越小，越先匹配。
- active：是否启用规则。
- username：基于用户名进行匹配。
- schemaname：基于 schema 名进行匹配。
- flagIN 和 flagOUT：定义规则的入口和出口，可用来实现链式规则。如果一条规则中定义了 flagOUT，则在继续匹配规则时，只会匹配 flagIN 等于 flagOUT 的规则。
- client_addr：基于客户端地址进行匹配。通过 client_addr 可实现简单的 IP 白名单功能。
- proxy_addr 和 proxy_port：如果 ProxySQL 所在的主机有多个 IP，可匹配来自指定 IP 的流量。
- digest：基于 Query ID 进行匹配。
- match_digest 和 match_pattern：基于正则表达式进行匹配，只不过前者匹配规范化后的 SQL，后者匹配原生 SQL。

下面看看 match_pattern、match_digest 和 digest 这三者的区别。

SQL（match_pattern）	digest_text（match_digest）	Query ID（digest）
select * from slowtech.t1 where id=1	select * from slowtech.t1 where id=?	0x21CF9688079F6E21

这里的 digest_text 和 digest 均取自 stats_mysql_query_digest。

ProxySQL 内置了两个正则引擎：PCRE（Perl Compatible Regular Expressions，Perl 兼容正则表达式）和 RE2（Google 开源的正则表达式库）。具体使用哪个引擎由参数 mysql-query_processor_regex 决定，默认使用前者。

- negate_match_pattern：设置为 1，相当于没匹配上 match_digest 或 match_pattern 才为真。
- re_modifiers：设置正则引擎的修饰符。目前支持的修饰符有 CASELESS 和 GLOBAL。前者表示忽略大小写，后者表示全局替换。默认开启 CASELESS。
- replace_pattern：替换后的文本，与查询重写有关，具体如何使用后面会提到。
- destination_hostgroup：指定目标主机组。
- cache_ttl：设置结果集的缓存时长，单位是毫秒。
- cache_empty_result：是否缓存空的结果集。
- cache_timeout：暂未实现。
- reconnect：暂未实现。
- timeout：定义查询的超时时长，单位是毫秒。如果查询在指定时间内没有执行完，则会被 ProxySQL 杀掉。不指定此字段的话，则超时时长由 mysql-default_query_timeout 决定，该参数默认为 36000000 毫秒，即 10 小时。
- retries：查询失败重试的次数。不指定此字段的话，则查询失败重试的次数由 mysql-query_retries_on_failure 决定，该参数默认为 1。
- delay：定义查询延迟执行的时长，单位是毫秒。可用于流控，或调整某些 SQL 的优先级。不指定此字段的话，则查询延迟执行的时长由 mysql-default_query_delay 决定，该参数默认为 0。
- mirror_flagOUT、mirror_hostgroup：与 Mirroring 功能有关，具体什么是 mirroring 后面会提到。
- error_msg：匹配规则的操作将返回 error_msg。可用这个字段来实现 SQL 黑名单功能，具体如何实现后面会提到。
- sticky_conn：暂未实现。
- multiplex：是否开启连接复用功能，具体什么是连接复用功能后面会提到。
- gtid_from_hostgroup：与 GTID Consistent Reads 有关。
- log：是否将匹配规则的查询记录到审计日志中。不指定的话，则由 mysql-eventslog_default_log 决定，该参数默认为 0。
- apply：若设置为 1，则操作在匹配上当前规则时，会被直接转发给后端节点处理，不再进行其他规则的匹配。

10.6　ProxySQL 的高级特性

10.6.1　定时器

ProxSQL 实现这个功能的初衷是，基于外部事件的检测结果来自动调整 ProxySQL 的配置。定时任务是在 scheduler 表中配置的。下面看看 scheduler 表的表结构。

```
ProxySQL> show create table scheduler\G
*************************** 1. row ***************************
       table: scheduler
Create Table: CREATE TABLE scheduler (
```

```
        id INTEGER PRIMARY KEY AUTOINCREMENT NOT NULL,
        active INT CHECK (active IN (0,1)) NOT NULL DEFAULT 1,
        interval_ms INTEGER CHECK (interval_ms>=100 AND interval_ms<=100000000) NOT NULL,
        filename VARCHAR NOT NULL,
        arg1 VARCHAR,
        arg2 VARCHAR,
        arg3 VARCHAR,
        arg4 VARCHAR,
        arg5 VARCHAR,
        comment VARCHAR NOT NULL DEFAULT '')
1 row in set (0.00 sec)
```

表中部分字段的具体含义如下。

- interval_ms：脚本执行的时间间隔，单位是毫秒。
- filename：脚本名，必须是绝对路径，且有可执行权限。
- arg1… arg5：最多可设置 5 个参数。

例如：

```
ProxySQL> insert into scheduler(active,interval_ms,filename,arg1)
    -> values (1,10000,'/data/scripts/check_node_health.sh','192.168.244.10');
Query OK, 1 row affected (0.00 sec)

ProxySQL> load scheduler to runtime;
Query OK, 0 rows affected (0.00 sec)
```

脚本能否正常运行，可通过错误日志（/var/lib/proxysql/proxysql.log）来判断。

10.6.2 SQL 审计

ProxySQL 可将流经它的 SQL 都记录下来，方便后续的审计和问题定位。以下是 SQL 审计的相关参数。

```
ProxySQL> show variables like 'mysql-eventslog%';
+-----------------------------+-----------+
| Variable_name               | Value     |
+-----------------------------+-----------+
| mysql-eventslog_filename    |           |
| mysql-eventslog_filesize    | 104857600 |
| mysql-eventslog_default_log | 0         |
| mysql-eventslog_format      | 1         |
+-----------------------------+-----------+
4 rows in set (0.00 sec)
```

各参数的具体含义如下。

- mysql-eventslog_filename：日志前缀名。默认为空，代表 SQL 审计没有开启。
- mysql-eventslog_filesize：日志的最大大小。超过此限制，则会进行日志切割，默认为 100MB。
- mysql-eventslog_default_log：是否将操作记录在审计日志中。默认为 0，即不记录。
- mysql-eventslog_format：日志的格式。设置为 1 代表二进制格式，设置为 2 代表 JSON 格式。
  如果要查看二进制格式日志的内容，必须使用专门的解析工具 eventslog_reader_sample。

下面看看如何记录所有 SQL。

```
ProxySQL> set mysql-eventslog_filename='/var/lib/proxysql/proxysql_query';
Query OK, 1 row affected (0.00 sec)

ProxySQL> set mysql-eventslog_default_log=1;
Query OK, 1 row affected (0.00 sec)

ProxySQL> set mysql-eventslog_format=2;
Query OK, 1 row affected (0.00 sec)

ProxySQL> load mysql variables to runtime;
Query OK, 0 rows affected (0.00 sec)
```

审计日志的内容如下。

```
# tailf  /var/lib/proxysql/proxysql_query.00000001 | jq
{
  "client": "192.168.244.10:60808",
  "digest": "0xE39877A1DD28E8E2",
  "duration_us": 474449,
  "endtime": "2022-06-30 11:58:52.785517",
  "endtime_timestamp_us": 1656590332785517,
  "event": "COM_QUERY",
  "hostgroup_id": 20,
  "query": "select distinct c1 from slowtech.t1",
  "rows_affected": 0,
  "rows_sent": 1,
  "schemaname": "information_schema",
  "server": "192.168.244.20:3306",
  "starttime": "2022-06-30 11:58:52.311068",
  "starttime_timestamp_us": 1656590332311068,
  "thread_id": 10,
  "username": "split_user"
}
```

10.6.3 查询重写

在线上有时候会遇到这种场景：由于优化器不完善，某个 SQL 没有用上合适的索引，导致查询速度不理想。此时，通常的做法是使用 Hint（FORCE INDEX）强制使用某个索引。强制使用某个索引就意味着 SQL 需要重写。

在哪里重写比较合适呢？如果在代码层重写，会存在两方面的问题：其一，发布新版本需要一定的时间；其二，表的数据量是动态变化的，这个索引在另一个数据量下很可能就不太合适。基于上述痛点，对于这类重写，我们更推荐放到中间件层或 MySQL 服务端来实现。毕竟后者一般支持动态修改，相对比较灵活。

以下面这个 SQL 为例，我们看看在 ProxySQL 中如何进行查询重写。

```
select * from employees.dept_emp where dept_no='d008' and from_date>='1989-02-10'
```

查询重写还是基于路由规则表来实现的。

```
ProxySQL> select * from stats_mysql_query_digest_reset limit 1;
Empty set (0.00 sec)

ProxySQL> insert into mysql_query_rules(rule_id,active,match_digest,match_pattern,replace_pattern,
apply) values
```

```
    -> (1,1,'^select \* from employees\.dept_emp where dept_no=(.*)and from_date>=(.*)$','dept_emp',
'dept_emp force index (idx_from_date)',0);
Query OK, 1 row affected (0.00 sec)

ProxySQL> load mysql query rules to runtime;
Query OK, 0 rows affected (0.00 sec)
```

部分字段的具体含义如下。

- match_digest：指定匹配规则。只有满足规则的 SQL 才能重写。
- match_pattern：旧字符串。
- replace_pattern：新字符串。
- apply：设置为 0，则继续匹配其他路由规则。建议设置为 0，这样的话，就无须关心重写后的 SQL 会转发到哪个后端节点中。

接下来看看重写的效果。

```
# mysql -h192.168.244.128 -usplit_user -psplit_pass -P6033
mysql> select * from employees.dept_emp where dept_no='d008' and from_date>='1989-02-10';

# mysql -h127.0.0.1 -uadmin -padmin -P6032 --prompt 'ProxySQL> '
ProxySQL> select hostgroup,digest,digest_text,count_star from stats_mysql_query_digest;
+-----------+--------------------+--------------------------------------------------------------------------------+------------+
| hostgroup | digest             | digest_text                                                                    | count_star |
+-----------+--------------------+--------------------------------------------------------------------------------+------------+
| 10        | 0x3DB0DE1907BFEC06 | select * from employees.dept_emp force index (idx_from_date) where dept_no=? and from_date>=? | 1 |
+-----------+--------------------+--------------------------------------------------------------------------------+------------+
1 row in set (0.00 sec)
```

从 digest_text 的内容来看，SQL 已重写成功。

上面虽然指定了 match_digest，但对于 ProxySQL 的查询重写来说，必需的字段其实只有两个：match_pattern 和 replace_pattern。上面的 match_digest 只是用来过滤的。

下面看看另外一个示例，重写 C3P0 0.9.5 版本之前的连接探测语句 SHOW FULL TABLES FROM `slowtech` LIKE 'PROBABLYNOT'。

```
ProxySQL> insert into mysql_query_rules(rule_id,active,match_pattern,replace_pattern,apply) values
    -> (100,1,"^SHOW FULL TABLES FROM(.*)LIKE 'PROBABLYNOT'$",'SELECT 1',0);
Query OK, 1 row affected (0.00 sec)

ProxySQL> load mysql query rules to runtime;
Query OK, 0 rows affected (0.06 sec)

mysql> SHOW FULL TABLES FROM `slowtech` LIKE 'PROBABLYNOT';
+---+
| 1 |
+---+
| 1 |
+---+
1 row in set (0.00 sec)
```

最后，我们看看如何在 Server 端进行查询重写。Server 端的查询重写依赖于查询重写插件（query rewrite plugin），这个插件是 MySQL 5.7 引入的。在 Server 端进行查询重写的基本步骤如下。

(1) 加载 Query Rewrite Plugin。

```
# cd /usr/local/mysql/share/
# mysql < install_rewriter.sql
```

(2) 配置重写规则。

```
insert into query_rewrite.rewrite_rules ( pattern, replacement )
values
        ( 'select * from employees.dept_emp where dept_no=? and from_date>=?', 'select * from
employees.dept_emp force index (idx_from_date) where dept_no=? and from_date>=?' );
```

(3) 将规则刷新到内存中。

```
call query_rewrite.flush_rewrite_rules();
```

(4) 检查重写的效果。

```
mysql> select * from employees.dept_emp where dept_no='d008' and from_date>='1989-02-10';

mysql> show warnings\G
*************************** 1. row ***************************
  Level: Note
   Code: 1105
Message: Query 'select * from employees.dept_emp where dept_no='d008' and from_date>='1989-02-10''
rewritten to 'select * from employees.dept_emp force index (idx_from_date) where dept_no='d008' and
from_date>='1989-02-10'' by a query rewrite plugin
1 row in set (0.00 sec)
```

10.6.4　mirroring

mirroring 从字面上看，是镜像的意思。具体在 ProxySQL 中，可理解为流量复制。

首先，看一个简单的示例。

```
ProxySQL> delete from mysql_query_rules;
Query OK, 0 rows affected (0.05 sec)

ProxySQL> insert into mysql_query_rules
(rule_id,active,match_pattern,destination_hostgroup,mirror_hostgroup,apply)
    -> values (101,1,'^SELECT',20,30,1);
Query OK, 1 row affected (0.05 sec)

ProxySQL> load mysql query rules to runtime;
Query OK, 0 rows affected (0.05 sec)

ProxySQL> save mysql query rules to disk;
Query OK, 0 rows affected (0.46 sec)
```

这条规则的作用是，所有的查询操作除了发送给 destination_hostgroup，还会同时发送给 mirror_hostgroup。

下面看看这条规则的实际效果。

```
# 清空 stats.stats_mysql_query_digest_reset 表的内容
ProxySQL> select count(*) from stats.stats_mysql_query_digest_reset;
```

```
# mysql -h192.168.244.128 -usplit_user -psplit_pass -P6033
mysql> select * from slowtech.t1;

ProxySQL> select hostgroup,digest_text,count_star,sum_time from stats_mysql_query_digest order by
digest;
+-----------+------------------------+------------+----------+
| hostgroup | digest_text            | count_star | sum_time |
+-----------+------------------------+------------+----------+
| 30        | select * from slowtech.t1 | 1       | 525      |
| 20        | select * from slowtech.t1 | 1       | 525      |
+-----------+------------------------+------------+----------+
2 rows in set (0.01 sec)
```

可以看到，同一个查询被分发到两个主机组处理了。

实际上，一个 SQL 匹配的路由规则中只要指定了 mirror_flagOUT 或 mirror_hostgroup，ProxySQL 就会自动开启 mirroring 功能。此时 ProxySQL 会创建一个新的会话（这个会话会复制原始会话的所有属性，包括用户名和密码，以及 schema）来执行镜像 SQL。注意，这个 SQL 不一定是原始 SQL，如果路由规则中指定了 replace_pattern，发送给 mirror_hostgroup 的将是重写后的 SQL。

看下面这个示例。

```
ProxySQL> update mysql_query_rules set match_pattern="^SELECT.*$",replace_pattern='select "Hello World"';
Query OK, 1 row affected (0.00 sec)

ProxySQL> load mysql query rules to runtime;
Query OK, 0 rows affected (0.00 sec)

ProxySQL> select count(*) from stats.stats_mysql_query_digest_reset;

mysql> select * from slowtech.t1;
+-------------+
| Hello World |
+-------------+
| Hello World |
+-------------+
1 row in set (0.00 sec)

ProxySQL> select hostgroup,digest_text,count_star,sum_time from stats_mysql_query_digest order by digest;
+-----------+-------------+------------+----------+
| hostgroup | digest_text | count_star | sum_time |
+-----------+-------------+------------+----------+
| 30        | select ?    | 1          | 442      |
| 20        | select ?    | 1          | 442      |
+-----------+-------------+------------+----------+
2 rows in set (0.00 sec)
```

下面结合另一个示例来理解 mirror_flagOUT 的用法，看看如何在不改变原始 SQL 的情况下，只镜像重写后的 SQL。

```
ProxySQL> delete from mysql_query_rules;
Query OK, 1 row affected (0.00 sec)

ProxySQL> insert into mysql_query_rules (rule_id,active,match_pattern,destination_hostgroup,
mirror_flagOUT,apply) VALUES (102,1,'from employees.dept_emp',20,300,1);
Query OK, 1 row affected (0.00 sec)
```

```
ProxySQL> insert into mysql_query_rules (rule_id,active,flagIN,match_pattern,destination_hostgroup,
replace_pattern,apply) VALUES (103,1,300,'from employees.dept_emp',30,'from employees.dept_emp ignore
index(dept_no)',1);
Query OK, 1 row affected (0.00 sec)

ProxySQL> load mysql query rules to runtime;
Query OK, 0 rows affected (0.00 sec)
```

这里配置了两条规则。

- 102 号规则设置了 `mirror_flagOUT`。此时，被镜像的 SQL 将继续匹配 flagIN 等于 mirror_flagOUT 的规则，即 103 号规则。
- 103 号规则会对镜像 SQL 进行重写，重写后的结果会被发送给 destination_hostgroup 处理。注意，103 号规则设置的是 destination_hostgroup，不是 mirror_hostgroup。

下面看看这两条规则的效果。

```
ProxySQL> select count(*) from stats.stats_mysql_query_digest_reset;

mysql> select * from employees.dept_emp where dept_no='d008';

ProxySQL> select hostgroup,digest_text,count_star,sum_time from stats_mysql_query_digest order by digest;
+-----------+--------------------------------------------------------------------+------------+----------+
| hostgroup | digest_text                                                        | count_star | sum_time |
+-----------+--------------------------------------------------------------------+------------+----------+
| 30        | select * from employees.dept_emp ignore index(dept_no) where dept_no=? | 1          | 157573   |
| 20        | select * from employees.dept_emp where dept_no=?                   | 1          | 92475    |
+-----------+--------------------------------------------------------------------+------------+----------+
2 rows in set (0.00 sec)
```

可以看到，原始 SQL 没有变，但镜像 SQL 被重写了。

注意，mirroring 不支持 Prepared Statement。如果要使用 sysbench 进行测试，需要在测试命令中加上 `--db-ps-mode=disable` 来禁用 Prepared Statement。

10.6.5　SQL 黑名单

SQL 黑名单，有时候也称为 SQL 限流或 SQL 防火墙。考虑以下两种场景。

- 在线上发布新版本，但某些慢 SQL 未能提前识别（毕竟线上环境和测试环境的数据量不一致），导致数据库响应变慢。
- 业务流量突增，或某个对外服务接口被恶意调用，导致某些 SQL 请求量变高，执行速度变慢。

如果碰到以上情况，通常的做法有两种。

- DBA 杀掉慢查询。

  优点：见效快。

  缺点有两个。

  - 适用于耗时较久的慢查询。对于那些执行不慢但并发量较高的 SQL，通过 KILL 也不能有效解决问题。

- 查询在被杀掉之前，实际上已经消耗了部分系统资源。从这个角度来看，KILL 其实是一种后知后觉的方案。
- 通知开发人员，由开发人员定位并屏蔽相关接口。

  优点：可从根本上解决问题。

  缺点：一般耗时较久。毕竟开发人员定位接口需要一定的时间。

以上两种方案的痛点实际上可以通过 SQL 黑名单来解决。SQL 黑名单其实类似于接口屏蔽，只不过是在中间件层实现的，针对的是具体的 SQL。凡是黑名单中定义的 SQL，都会在中间件层直接被屏蔽，不会发往后端节点。

在 ProxySQL 中，实现 SQL 黑名单的方法有两种。

- mysql_query_rules 中的 error_msg 字段。
- 防火墙。这个功能是 ProxySQL 2.0.9 引入的。

看下面这个示例，即基于第一种方法实现的 SQL 黑名单功能。

```
ProxySQL> insert into mysql_query_rules(rule_id,active,match_pattern,error_msg,apply) values
    -> (100,1,'^select \* from employees\.employees$','query not allowed',1);
Query OK, 1 row affected (0.00 sec)

ProxySQL> load mysql query rules to runtime;
Query OK, 0 rows affected (0.00 sec)

mysql> select * from employees.employees;
ERROR 1148 (42000): query not allowed
```

10.7　ProxySQL 连接池

连接池的一个核心功能是连接复用（multiplexing）。multiplexing 即多路复用技术，最开始是一个通信术语，指的是一个通信信道可以传输多路信号。这样做的好处显而易见：可以充分利用信道的带宽。在服务更多用户的同时，还无须搭建更多的线路。在 ProxySQL 的连接池中，连接复用指的是一个后端连接可以被多个查询复用。

下面通过一个简单的示例直观地看看什么是连接复用。

```
# mysql -h192.168.244.128 -usplit_user -psplit_pass -P6033 --prompt 'session1> '
session1> select 1;
+---+
| 1 |
+---+
| 1 |
+---+
1 row in set (0.00 sec)
```

查看对应节点的 general log。

```
25 Query     select 1
```

这里的 25 是 MySQL 中执行这个 SQL 的线程 ID。

再来看看连接池中空闲连接的情况

```
ProxySQL> select * from stats_mysql_free_connections\G
*************************** 1. row ***************************
          fd: 44
    hostgroup: 20
     srv_host: 192.168.244.20
     srv_port: 3306
         user: split_user
       schema: information_schema
 init_connect: NULL
    time_zone: NULL
     sql_mode: NULL
   autocommit: 1
      idle_ms: 84761
   statistics: {"address":"0x7fb5675b0f00","age_ms":84761,"bytes_recv":6,"bytes_sent":8,
"myconnpoll_get":1,"myconnpoll_put":1,"questions":1}
   mysql_info: {"address":"0x7fb562db0000","affected_rows":0,"charset":255,"charset_name":"utf8mb4",
"client_flag":{"client_found_rows":0,"client_multi_results":1,"client_multi_statements":1},"db":
"information_schema","host":"192.168.244.20","host_info":"192.168.244.20 via
TCP/IP","insert_id":0,"net":{"fd":44,"last_errno":0,"max_packet_size":1073741824,"sqlstate":
"00000"},"options":{"charset_name":"","use_ssl":0},"port":3306,"server_status":2,"server_version":
"8.0.27","thread_id":25,"unix_socket":"","user":"split_user"}
1 row in set (0.00 sec)
```

不难发现，mysql_info 中的 thread_id 与 general log 中的线程 ID 一样，都是 25。虽然 session1 还没有断开，但后端连接已经被放回连接池了。

打开一个新窗口，执行 session1 中的类似查询。

```
# mysql -h192.168.244.128 -usplit_user -psplit_pass -P6033 --prompt 'session2> '
session2> select 2;
+---+
| 2 |
+---+
| 2 |
+---+
1 row in set (0.04 sec)
```

还是查看该节点的 general log。

```
25 Query    select 2
```

两个不同的客户端连接都没有断开，却使用了同一个后端连接来执行查询操作。这就意味着，ProxySQL 中的连接复用是语句级别的。这一点和应用层的连接池（如 C3P0）不一样。对于应用层的连接池，连接只有在释放后才能放回连接池。相对来说，语句级别的连接复用更高效，但并不是所有连接都可以复用。还是基于上面的两个会话，考虑下面这个场景：session1 设置了一个会话变量，接着 session2 也设置了这个同名变量。因为是同一个后端连接，所以 session2 的设置势必会覆盖 session1 的，这样就违背了隔离性的原则。因此，在 ProxySQL 中，一个连接能否复用是有很多限制条件的，具体条件后面会讲到。

结合上面的示例，我们总结一下 ProxySQL 连接池的运行流程。

- 当会话发起查询，需要一个后端连接时，它会首先检查连接池中是否有对应后端节点的空闲连接。
- 如果有，则复用其中一个连接，否则新建一个连接。

- 查询完成后，会将连接发送给 Hostgroup Manager，由它来判断这个连接能否放回连接池。
  - 如果连接可复用，且池中可允许的最大空闲连接数还未满，则放回。对于每个后端节点，在连接池中最多允许保留的空闲连接数等于 mysql_servers.max_connections * mysql-free_connections_pct / 100。
  - 如果连接不可复用，则将其销毁。

下面重点说说连接不可复用的场景。连接不可复用的场景主要分为两类。

- 连接只是暂时不可复用，只要满足指定条件，就可恢复复用。属于这一类的操作如下。
  - 事务。当一个连接开启了事务时，连接会暂停复用，直到事务提交或回滚。
  - 锁表操作。具体来说，如果执行了 LOCK TABLE、LOCK TABLES、FLUSH TABLES WITH READ LOCK 操作，连接会暂停复用，直到执行 UNLOCK TABLES 操作。
  - SQL_LOG_BIN 设置为 0，连接会暂停复用，直到 SQL_LOG_BIN 设置为 1。
- 连接不会恢复复用。属于这一类的操作如下。
  - 使用了会话变量。具体来说，命令中包含 @ 字符。但有 3 个变量除外：tx_isolation、transaction_isolation、version。这 3 个变量是在 mysql-keep_multiplexing_variables 中定义的。之所以排除它们，是因为很多应用程序在建立连接时会查询它们。
  - 创建临时表（CREATE TEMPORARY TABLE）。
  - GET_LOCK() 操作。
  - 显式设置如下参数：
    - SET SQL_SAFE_UPDATES=?,SQL_SELECT_LIMIT=?,MAX_JOIN_SIZE=?
    - SET FOREIGN_KEY_CHECKS
    - SET UNIQUE_CHECKS
    - SET AUTO_INCREMENT_INCREMENT
    - SET AUTO_INCREMENT_OFFSET
    - SET GROUP_CONCAT_MAX_LEN
  - 使用了 SQL_CALC_FOUND_ROWS。
  - 使用了 PREPARE。

除此之外，连接能否复用还与参数的设置有关。连接复用的相关参数如下。

```
ProxySQL> show variables like '%multiple%';
+----------------------------------------------+------------------------------------------+
| Variable_name                                | Value                                    |
+----------------------------------------------+------------------------------------------+
| mysql-connection_delay_multiplex_ms          | 0                                        |
| mysql-multiplexing                           | true                                     |
| mysql-auto_increment_delay_multiplex         | 5                                        |
| mysql-auto_increment_delay_multiplex_timeout_ms | 10000                                 |
| mysql-keep_multiplexing_variables            | tx_isolation,transaction_isolation,version |
+----------------------------------------------+------------------------------------------+
5 rows in set (0.00 sec)
```

各个参数的具体含义如下。

- `mysql-connection_delay_multiplex_ms`：默认情况下，在查询结束后，如果连接可被复用，连接会马上被返回连接池。也可设置 `mysql-connection_delay_multiplex_ms` 等待一段时间再返回。
- `mysql-multiplexing`：是否开启连接复用功能，默认开启。如果关闭，则所有连接都不会被复用。
- `mysql-auto_increment_delay_multiplex`：如果使用了 LAST_INSERT_ID() 函数，接下来的若干个查询会直接复用当前的连接。`mysql-auto_increment_delay_multiplex` 决定了查询的数量，默认为 5。
- `mysql-auto_increment_delay_multiplex_timeout_ms`：作用与 `mysql-auto_increment_delay_multiplex` 类似，只不过指定的是时间。
- `mysql-keep_multiplexing_variables`：不会禁用连接复用的变量列表。

除了上面的 5 个全局参数，还可通过 `mysql_query_rules` 中的 `multiplexing` 字段，在路由规则层个性化设置复用开关。例如：

```
insert into mysql_query_rules(active,match_digest,multiplex) values ('1','^SELECT
@@max_allowed_packet',2);
```

该字段默认为 NULL，即继承全局配置。除此之外，还可设置以下值。

- `0`：禁用连接复用。
- `1`：开启连接复用。
- `2`：即使命令中包含 @ 字符，也不禁用连接复用。

需要注意的是，连接不能复用，指的是这个连接不会被其他会话的查询使用，并不意味着当前会话的所有查询都会使用这个后端连接。毕竟，查询分发到哪个后端节点与路由规则有关。

下面看两个比较有代表性的示例，还是读写分离的配置。

```
ProxySQL> select hostgroup_id,hostname,port,status from mysql_servers;
+--------------+----------------+------+--------+
| hostgroup_id | hostname       | port | status |
+--------------+----------------+------+--------+
| 10           | 192.168.244.10 | 3306 | ONLINE |
| 20           | 192.168.244.20 | 3306 | ONLINE |
+--------------+----------------+------+--------+
2 rows in set (0.00 sec)

ProxySQL> select rule_id,active,match_digest,destination_hostgroup,apply from mysql_query_rules;
+---------+--------+-------------------+-----------------------+-------+
| rule_id | active | match_digest      | destination_hostgroup | apply |
+---------+--------+-------------------+-----------------------+-------+
| 1       | 1      | ^SELECT.*FOR UPDATE$ | 10                 | 1     |
| 2       | 1      | ^SELECT           | 20                    | 1     |
+---------+--------+-------------------+-----------------------+-------+
2 rows in set (0.00 sec)

ProxySQL> select * from mysql_replication_hostgroups;
+------------------+------------------+------------+---------+
| writer_hostgroup | reader_hostgroup | check_type | comment |
```

```
+-----------------+-----------------+------------------+-----------------+
| 10              | 20              | super_read_only  | repl_group_1    |
+-----------------+-----------------+------------------+-----------------+
1 row in set (0.00 sec)

# mysql -h192.168.244.128 -usplit_user -psplit_pass -P6033
mysql> set @name='slowtech';
Query OK, 0 rows affected (0.00 sec)

mysql> select @name;
ERROR 9006 (Y0000): ProxySQL Error: connection is locked to hostgroup 10 but trying to reach hostgroup 20

# mysql -h192.168.244.128 -usplit_user -psplit_pass -P6033
mysql> create temporary table slowtech.tmp(id int primary key);
Query OK, 0 rows affected (0.00 sec)

mysql> insert into slowtech.tmp values(1);
Query OK, 1 row affected (0.00 sec)

mysql> select * from slowtech.tmp;
ERROR 1146 (42S02): Table 'slowtech.tmp' doesn't exist
```

是不是有点出乎意料？这也是为什么我们在配置路由规则时，尤其要小心，一定要进行充分的测试。

讲解完连接复用，接下来看看与连接相关的知识点。

(1) 与连接相关的配置。

```
ProxySQL> select username,max_connections from mysql_users;
+------------+-----------------+
| username   | max_connections |
+------------+-----------------+
| split_user | 10000           |
+------------+-----------------+
1 row in set (0.00 sec)

ProxySQL> select hostgroup_id,hostname,port,max_connections from mysql_servers;
+--------------+----------------+------+-----------------+
| hostgroup_id | hostname       | port | max_connections |
+--------------+----------------+------+-----------------+
| 10           | 192.168.244.10 | 3306 | 1000            |
| 20           | 192.168.244.20 | 3306 | 1000            |
+--------------+----------------+------+-----------------+
2 rows in set (0.00 sec)
```

mysql_users 中的 max_connections 指的是用户可以创建的最大连接数，而 mysql_servers 中的 max_connections 指的是 ProxySQL 到后端节点的最大连接数。

(2) 与连接相关的参数。

```
+-------------------------------------+-------+
| Variable_name                       | Value |
+-------------------------------------+-------+
| mysql-connect_retries_delay         | 1     |
| mysql-connect_retries_on_failure    | 10    |
| mysql-connect_timeout_server        | 3000  |
```

```
| mysql-connect_timeout_server_max               | 10000  |
| mysql-connection_max_age_ms                    | 0      |
| mysql-free_connections_pct                     | 10     |
| mysql-connection_warming                       | false  |
| mysql-kill_backend_connection_when_disconnect  | true   |
| mysql-log_unhealthy_connections                | true   |
| mysql-max_connections                          | 2048   |
| mysql-ping_interval_server_msec                | 120000 |
| mysql-ping_timeout_server                      | 500    |
| mysql-shun_on_failures                         | 5      |
| mysql-shun_recovery_time_sec                   | 10     |
+------------------------------------------------+--------+
```

上述参数的具体含义如下。

- `mysql-connect_retries_delay`：如果在建立连接时失败，ProxySQL 会不断重试，重试的时间间隔由 `mysql-connect_retries_delay` 决定，默认为 1 毫秒。重试的次数由 `mysql-connect_retries_on_failure` 决定，默认为 10。除了次数，重试还有最大超时时长的限制，最大超时时长由 `mysql-connect_timeout_server_max` 决定，默认为 10000，即 10 秒。如果达到了最大超时时长，ProxySQL 会提示以下错误：

  ```
  ERROR 9001 (HY000): Max connect timeout reached while reaching hostgroup 10 after 10000ms
  ```

- `mysql-connect_timeout_server`：连接超时时长，默认为 3000，即 3 秒。
- `mysql-connection_max_age_ms`：连接池中，连接可允许的最大空闲时间。
- `mysql-free_connections_pct`：连接池每个后端节点的最大空闲连接比，所以每个后端节点允许保留的空闲连接等于 `mysql_servers.max_connections` * `mysql-free_connections_pct` / 100。
- `mysql-connection_warming`：当会话需要一个连接时，默认情况下，会直接复用当前的空闲连接。如果该参数为 true，则会优先创建一个新的连接，直到每个节点的连接数等于 `mysql_servers.max_connections` * `mysql-free_connections_pct` / 100，才开始复用空闲连接。
- `mysql-kill_backend_connection_when_disconnect`：客户端连接断开时，是否同时断开对应的后端连接。
- `mysql-log_unhealthy_connections`：若为 true，则当客户端连接异常断开时，ProxySQL 会在日志中打印 `Closing unhealthy client connection IP:port`。
- `mysql-max_connections`：ProxySQL 的最大连接数。
- `mysql-ping_timeout_server`：PING 操作的超时时长。ProxySQL 会定期（由 `mysql-ping_interval_server_msec` 决定，默认为 120000，即 120 秒）对创建的连接进行 PING 操作，以确保这些连接处于活跃状态。
- `mysql-shun_on_failures`：1 秒内如果连续 `mysql-shun_on_failures` 次（如果 `mysql-connect_retries_on_failure` 小于 `mysql-shun_on_failures`，则取 `mysql-connect_retries_on_failure`）连接创建失败，ProxySQL 会将节点的状态置为 SHUNNED，并在一段时间内不使用这个节点。具体时间由 `mysql-shun_recovery_time_sec` 决定，默认为 10 秒。10 秒后会再次判断节点是否可用。

  ```
  MySQL_HostGroups_Manager.cpp:939:connect_error(): [ERROR] Shunning server 192.168.244.10:3306 with 5 errors/sec. Shunning for 10 seconds
  ```

(3) 与连接池相关的状态表。

通过这些表，我们可以直观地看到连接池当前的运行状况。

- **stats_mysql_users**

查看前端用户的连接使用情况。

```
ProxySQL> select * from stats_mysql_users;
+------------+---------------------+-------------------------+
| username   | frontend_connections | frontend_max_connections |
+------------+---------------------+-------------------------+
| split_user | 2                   | 10000                   |
+------------+---------------------+-------------------------+
1 row in set (0.00 sec)
```

其中各项解释如下。

- `frontend_connections`：用户当前使用的连接数。
- `frontend_max_connections`：用户允许创建的最大连接数，对应 `mysql_users` 表中的 `max_connections`。

- **stats_mysql_processlist**

查看当前的连接信息。除此之外，也可使用 SHOW PROCESSLIST 和 SHOW FULL PROCESSLIST。

```
ProxySQL> select * from stats_mysql_processlist\G
*************************** 1. row ***************************
      ThreadID: 0
     SessionID: 16
          user: split_user
            db: information_schema
      cli_host: 192.168.244.10
      cli_port: 44288
      hostgroup: 20
    l_srv_host: 192.168.244.128
    l_srv_port: 35284
      srv_host: 192.168.244.20
      srv_port: 3306
       command: Query
       time_ms: 12012
          info: select sleep(1000)
  status_flags: 0
 extended_info: NULL
1 row in set (0.00 sec)
```

其中各项解释如下。

- `ThreadID`：ProxySQL 线程的内部 ID。
- `SessionID`：会话 ID。可通过 KILL CONNECTION SessionID 命令杀掉指定连接。
- `cli_host` 和 `cli_port`：前端连接（应用程序与 ProxySQL 之间的连接）中应用程序端的 IP 和端口。
- `l_srv_host` 和 `l_srv_port`：后端连接（ProxySQL 与后端 MySQL 节点之间的连接）中 ProxySQL 端的 IP 和端口。
- `srv_host` 和 `srv_port`：后端节点的 IP 和端口。

- time_ms：SQL 执行的时长。
- info：执行的 SQL。

表中的 extended_info 默认为 NULL，如果要输出更详细的会话信息，需将参数 mysql-show_processlist_extended 设置为 1。在输出的详细信息中，status 部分与连接复用相关。只要有一项为 true，连接就不能复用。

```
"status": {
            "compression": false,
            "found_rows": false,
            "get_lock": false,
            "has_savepoint": false,
            "lock_tables": false,
            "no_multiplex": true,
            "prepared_statement": false,
            "temporary_table": false,
            "user_variable": true
        }
```

- **stats_mysql_connection_pool**

查看各个主机组每个后端节点的连接和流量使用情况。

```
ProxySQL> select hostgroup,srv_host,srv_port,status,ConnUsed,ConnFree,ConnOK,ConnERR,MaxConnUsed,
Queries from stats_mysql_connection_pool;
+-----------+----------------+----------+--------+----------+----------+--------+---------+-------------+---------+
| hostgroup | srv_host       | srv_port | status | ConnUsed | ConnFree | ConnOK | ConnERR | MaxConnUsed | Queries |
+-----------+----------------+----------+--------+----------+----------+--------+---------+-------------+---------+
| 20        | 192.168.244.20 | 3306     | ONLINE | 1        | 0        | 1      | 0       | 1           | 3       |
| 10        | 192.168.244.10 | 3306     | ONLINE | 1        | 0        | 2      | 0       | 1           | 5       |
+-----------+----------------+----------+--------+----------+----------+--------+---------+-------------+---------+
2 rows in set (0.00 sec)
```

其中各项解释如下。

- status：后端节点的状态，可以是 ONLINE、SHUNNED、OFFLINE_SOFT、OFFLINE_HARD。
- ConnUsed：当前正在使用的连接数。
- ConnFree：空闲连接数。空闲连接的具体信息可通过 stats_mysql_free_connections 表查看。
- ConnOK：创建成功的连接数。
- ConnERR：创建失败的连接数。
- MaxConnUsed：曾经使用的最大连接数。
- Queries：发送给后端节点的查询的数量。

除此之外，stats_mysql_connection_pool 还提供了以下 3 列信息。

- Bytes_data_sent：发送给后端节点的流量大小。
- Bytes_data_recv：从后端节点接收的流量大小。
- Latency_us：通过 PING 请求监测到的后端节点的网络延迟情况。

与 stats_mysql_query_digest 一样，stats_mysql_connection_pool 的内容可通过查询 stats_mysql_connection_pool_reset 来清空。

- **stats_mysql_free_connections**

查看连接池中空闲连接的情况。

```
ProxySQL> select * from stats_mysql_free_connections\G
*************************** 1. row ***************************
           fd: 44
     hostgroup: 20
      srv_host: 192.168.244.20
      srv_port: 3306
          user: split_user
        schema: information_schema
  init_connect: NULL
     time_zone: NULL
      sql_mode: NULL
    autocommit: 1
       idle_ms: 1873
    statistics: {"address":"0x7fb5651b9600","age_ms":1873,"bytes_recv":6,"bytes_sent":8,"myconnpoll_
get":1,"myconnpoll_put":1,"questions":1}
    mysql_info: {"address":"0x7fb5675b1900","affected_rows":0,"charset":255,"charset_name":"utf8mb4"
,"client_flag":{"client_found_rows":0,"client_multi_results":1,"client_multi_statements":1},"db":
"information_schema","host":"192.168.244.20","host_info":"192.168.244.20 via TCP/IP","insert_id":0,
"net":{"fd":44,"last_errno":0,"max_packet_size":1073741824,"sqlstate":"00000"},"options":{"charset
_name":"","use_ssl":0},"port":3306,"server_status":2,"server_version":"8.0.27","thread_id":177,
"unix_socket":"","user":"split_user"}
1 row in set (0.00 sec)
```

输出中，init_connect、time_zone、sql_mode、autocommit 都是连接的会话值，若为 NULL，代表客户端没有设置这些参数的会话值。

10.8 ProxySQL Cluster

ProxySQL 作为一个中间件，本身是无状态的，所以线上一般会部署多个节点，前端通过 LVS 或 HAProxy 来进行负载均衡和故障节点的自动移除。既然涉及多个节点，就存在节点与节点之间配置信息如何同步的问题。假如我要增加一条路由规则，怎么保证这个规则在很短的时间内在所有节点上都生效呢？通常会使用 Ansible 等配置管理工具或 ZooKeeper 等服务发现工具，但这些工具也有缺点，比如配置的收敛时间不可预测。基于此，官方推出了 ProxySQL Cluster。图 10-4 是 ProxySQL Cluster 的架构图。

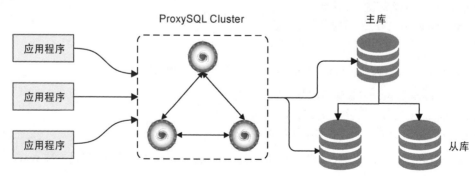

图 10-4　ProxySQL Cluster 架构图

ProxySQL Cluster 能保证，只要在任意一个节点上对 mysql_users、mysql_servers、mysql_query_rules、proxysql_servers、global_variables 这 5 张表进行操作，修改的内容就会很快（近实时）被同步到其他节点上。

10.8.1 搭建 ProxySQL Cluster

下面以两个节点（192.168.244.128 和 192.168.244.129）为例，搭建一个 ProxySQL Cluster。

(1) 安装 ProxySQL。

```
# yum install proxysql-2.4.1-1-centos7.x86_64.rpm
# service proxysql start
```

(2) 配置 ProxySQL Cluster 检测用户。

只有配置了这个用户，才能检测其他 Cluster 节点的状态。所有节点上都要配置。

```
# mysql -h127.0.0.1 -uadmin -padmin -P6032 --prompt 'ProxySQL> '
ProxySQL> set admin-admin_credentials='admin:admin;cluster_check:cluster_check_pass';
Query OK, 1 row affected (0.00 sec)

ProxySQL> set admin-cluster_username='cluster_check';
Query OK, 1 row affected (0.00 sec)

ProxySQL> set admin-cluster_password='cluster_check_pass';
Query OK, 1 row affected (0.00 sec)

ProxySQL> load admin variables to runtime;
Query OK, 0 rows affected (0.00 sec)

ProxySQL> save admin variables to disk;
Query OK, 46 rows affected (0.00 sec)
```

(3) 配置 ProxySQL Cluster 节点信息。

所有节点上都要进行此项配置。

```
ProxySQL> insert into proxysql_servers (hostname,port,weight,comment) values 
('192.168.244.128',6032,100,'proxysql1');
Query OK, 1 row affected (0.00 sec)

ProxySQL> insert into proxysql_servers (hostname,port,weight,comment) values 
('192.168.244.129',6032,100,'proxysql2');
Query OK, 1 row affected (0.00 sec)

ProxySQL> load proxysql servers to runtime;
Query OK, 0 rows affected (0.00 sec)

ProxySQL> save proxysql servers to disk;
Query OK, 0 rows affected (0.00 sec)
```

`proxysql_servers` 表里的 `weight` 指的是节点的权重，目前暂未使用。

至此，ProxySQL Cluster 搭建完毕。

接下来，我们验证 ProxySQL Cluster 的效果。

以 `mysql_users` 表为例,新建一个用户,看看这个用户能否被同步到其他节点上。

```
ProxySQL> insert into mysql_users(username, password, default_hostgroup)
values('test_user','123456',10);
Query OK, 1 row affected (0.01 sec)

ProxySQL> load mysql users to runtime;
Query OK, 0 rows affected (0.01 sec)

ProxySQL> save mysql users to disk;
Query OK, 0 rows affected (0.01 sec)
```

查看另外一个节点上 `mysql_users` 表的内容。

```
proxysql2> select username,password,active,default_hostgroup from mysql_users;
+-----------+-------------------------------------------+--------+-------------------+
| username  | password                                  | active | default_hostgroup |
+-----------+-------------------------------------------+--------+-------------------+
| test_user | *6BB4837EB74329105EE4568DDA7DC67ED2CA2AD9 | 1      | 10                |
| test_user | *6BB4837EB74329105EE4568DDA7DC67ED2CA2AD9 | 1      | 10                |
+-----------+-------------------------------------------+--------+-------------------+
2 rows in set (0.00 sec)
```

确实同步过来了。再来看看该节点对应的日志。

```
# tailf /var/lib/proxysql/proxysql.log
2022-07-02 10:18:59 [INFO] Cluster: detected a new checksum for mysql_users from peer
192.168.244.128:6032, version 2, epoch 1656757138, checksum 0x6217F366FC18A167 . Not syncing yet ...
2022-07-02 10:19:01 [INFO] Cluster: detected a peer 192.168.244.128:6032 with mysql_users version 2,
epoch 1656757138, diff_check 3. Own version: 1, epoch: 1656756838. Proceeding with remote sync
2022-07-02 10:19:02 [INFO] Cluster: detected a peer 192.168.244.128:6032 with mysql_users version 2,
epoch 1656757138, diff_check 4. Own version: 1, epoch: 1656756838. Proceeding with remote sync
2022-07-02 10:19:02 [INFO] Cluster: detected peer 192.168.244.128:6032 with mysql_users version 2, epoch
1656757138
2022-07-02 10:19:02 [INFO] Cluster: Fetching MySQL Users from peer 192.168.244.128:6032 started
2022-07-02 10:19:02 [INFO] Cluster: Fetching MySQL Users from peer 192.168.244.128:6032 completed
2022-07-02 10:19:02 [INFO] Cluster: Loading to runtime MySQL Users from peer 192.168.244.128:6032
2022-07-02 10:19:02 [INFO] Cluster: Saving to disk MySQL Users from peer 192.168.244.128:6032
```

结合日志,我们总结一下 ProxySQL Cluster 的同步流程。

在分析具体的同步流程之前,首先介绍一下 `runtime_checksums_values` 表。

```
ProxySQL> select * from runtime_checksums_values;
+-------------------+---------+------------+--------------------+
| name              | version | epoch      | checksum           |
+-------------------+---------+------------+--------------------+
| admin_variables   | 2       | 1656756897 | 0x1BCAF50BF53FDFB0 |
| mysql_query_rules | 1       | 1656756854 | 0x0000000000000000 |
| mysql_servers     | 1       | 1656756854 | 0x0000000000000000 |
| mysql_users       | 2       | 1656757138 | 0x6217F366FC18A167 |
| mysql_variables   | 1       | 1656756854 | 0x3735B7F44B73D294 |
| proxysql_servers  | 2       | 1656756990 | 0xAEA04FADB2A49710 |
+-------------------+---------+------------+--------------------+
6 rows in set (0.00 sec)
```

该表会记录各个模块的版本号(version)、修改时间(epoch)以及内容的校验和(checksum)。

表中的 `name` 是模块名。`version` 的初始值为 1,每执行一次 LOAD xxx TO RUNTIME 操作,无论表的内容是否发生变化，都会将对应模块的版本号加1。epoch是执行LOAD xxx TO RUNTIME操作时的时间戳。

需要注意的是，对于proxysql_servers表，只要执行LOAD PROXYSQL SERVERS TO RUNTIME操作，就会更新runtime_checksums_values表的内容，而对于其他几个模块，是否更新与对应模块的admin-checksum_xxx参数有关。admin-checksum_xxx参数默认都为true，所以只要执行LOAD xxx TO RUNTIME操作，就会更新runtime_checksums_values表的内容；若对应模块的该参数为false，则不会更新，也不会从其他节点同步数据。

接下来，我们看看ProxySQL Cluster的同步流程。

- ProxySQL 会定期检查其他节点各个模块的校验和值。检查的频率由 admin-cluster_check_interval_ms 决定，默认为 1000，即 1 秒。
- 如果发现某个模块的校验和值发生了变化，这个时候会判断当前节点该模块的版本号。
  - 如果版本号等于 1，则会立即触发同步。同步的对象是版本号大于 1 且时间戳最大的那个节点。
  - 如果版本号大于 1，则会进行多轮检测。只有不一致的次数超过了指定值（由对应模块的 admin-cluster_xxx_diffs_before_sync 决定，默认为 3），才会触发同步。同步的对象同样是版本号大于 1 且时间戳最大的那个节点。

(4) 同步会进行以下操作，这里以mysql_users表为例。

```
# 查询同步对象 mysql_users 表的内容
SELECT username, password, active, use_ssl, default_hostgroup, default_schema, schema_locked,
transaction_persistent, fast_forward, backend, frontend, max_connections, attributes, comment FROM
runtime_mysql_users;

# 删除当前节点 mysql_users 表的内容
DELETE FROM mysql_users;

# 将第一步的查询结果插入 mysql_users 表
INSERT INTO mysql_users (username, password, active, use_ssl, default_hostgroup, default_schema,
schema_locked, transaction_persistent, fast_forward, backend, frontend, max_connections, attributes,
comment) VALUES (...)

# 将更新加载到内存中
LOAD MYSQL USERS TO RUNTIME;

# 将更新持久化到磁盘中
SAVE MYSQL USERS TO DISK;
```

最后一步持久化操作是否执行是由 admin-cluster_mysql_users_save_to_disk 决定的，该参数默认为 true。

10.8.2 添加一个新的节点

在现有的 ProxySQL Cluster 上添加一个新的节点，也是一个相当常见的需求。下面我们看看具体的操作步骤。还是在上面的集群基础上添加一个新的节点——192.168.244.130。

(1) 在 192.168.244.130 上配置 ProxySQL Cluster 检测用户。

```
proxysql3> set admin-admin_credentials='admin:admin;cluster_check:cluster_check_pass';
Query OK, 1 row affected (0.00 sec)

proxysql3> set admin-cluster_username='cluster_check';
Query OK, 1 row affected (0.00 sec)

proxysql3> set admin-cluster_password='cluster_check_pass';
Query OK, 1 row affected (0.00 sec)

proxysql3> load admin variables to runtime;
Query OK, 0 rows affected (0.00 sec)

proxysql3> save admin variables to disk;
Query OK, 46 rows affected (0.00 sec)
```

(2) 在 192.168.244.130 上添加集群任意一个节点的配置信息。

```
proxysql3> insert into proxysql_servers (hostname,port,weight,comment) values
('192.168.244.128',6032,100,'proxysql1');
Query OK, 1 row affected (0.00 sec)

proxysql3> load proxysql servers to runtime;
Query OK, 0 rows affected (0.00 sec)
```

这里添加的是 192.168.244.128。此时，192.168.244.130 只能接受 192.168.244.128 的单向同步，它的修改不会同步到集群中。

(3) 在 192.168.244.128 上将 192.168.244.130 添加进来。

```
proxysql1> insert into proxysql_servers (hostname,port,weight,comment) values
('192.168.244.130',6032,100,'proxysql3');
Query OK, 1 row affected (0.05 sec)

proxysql1> load proxysql servers to runtime;
```

这样，192.168.244.130 就完全加入到集群中了。

在实际测试过程中，我们发现步骤(2)和步骤(3)的顺序不能颠倒。如果先将 192.168.244.130 添加到集群，然后再在 192.168.244.130 中配置集群其他节点的信息，会导致整个集群的 proxysql_servers 信息被 192.168.244.130 中的覆盖。

之所以会存在这样的问题，是因为当前版本的 ProxySQL Cluster 是一个去中心化的架构。每个节点的角色、功能都是对等的。节点与节点之间的配置冲突，是通过版本号来解决的。谁的版本号最高，谁就有权利将自身的配置同步给其他节点。不过，在官方的路线图中，会实现只有主节点可写、主从复制等功能。让我们拭目以待吧！

10.9 ProxySQL 的常见参数

作为一个功能丰富的中间件，ProxySQL 自身提供了多个参数来调整其行为。这些参数大致可分为以下几类。

10.9.1 管理参数

管理参数以 admin- 开头,主要用来控制管理接口的相关行为。相关参数如下。

```
ProxySQL> show variables like 'admin%';
+-------------------------------------------+--------------------+
| Variable_name                             | Value              |
+-------------------------------------------+--------------------+
| admin-stats_credentials                   | stats:stats        |
| admin-stats_mysql_connections             | 60                 |
| admin-stats_mysql_connection_pool         | 60                 |
| admin-stats_mysql_query_cache             | 60                 |
| admin-stats_mysql_query_digest_to_disk    | 0                  |
| admin-stats_system_cpu                    | 60                 |
| admin-stats_system_memory                 | 60                 |
| admin-telnet_admin_ifaces                 | (null)             |
| admin-telnet_stats_ifaces                 | (null)             |
| admin-refresh_interval                    | 2000               |
| admin-read_only                           | false              |
| admin-hash_passwords                      | true               |
| admin-vacuum_stats                        | true               |
| admin-version                             | 2.4.1-1-g1ea371d   |
| admin-cluster_username                    |                    |
| admin-cluster_password                    |                    |
| admin-cluster_check_interval_ms           | 1000               |
...
| admin-checksum_admin_variables            | true               |
| admin-checksum_ldap_variables             | true               |
| admin-restapi_enabled                     | false              |
| admin-restapi_port                        | 6070               |
| admin-web_enabled                         | false              |
| admin-web_port                            | 6080               |
| admin-web_verbosity                       | 0                  |
| admin-prometheus_memory_metrics_interval  | 61                 |
| admin-admin_credentials                   | admin:admin        |
| admin-mysql_ifaces                        | 0.0.0.0:6032       |
+-------------------------------------------+--------------------+
46 rows in set (0.00 sec)
```

上述参数的具体含义如下。

- admin-admin_credentials:管理用户及密码,默认为 admin:admin。需要注意的是, admin 用户只能本地登录。如果需要远程登录,必须设置其他用户。例如:

  admin-admin_credentials="admin:admin;proxy_admin:proxy_pass"

- admin-mysql_ifaces:管理端口,默认为 0.0.0.0:6032。如果要指定多个,中间需用分号(;)隔开。
- admin-stats_credentials:只读用户及密码。该用户只能查看 stats 库的内容。对于其他库,没有读写权限。同 admin 用户一样,只能本地登录。
- admin-stats_xxx:指定连接、连接池、Query Cache、CPU、MEMORY 等信息的更新周期,默认为 60 秒。
- admin-read_only:将 ProxySQL 管理模块设置为只读。只读状态下的 ProxySQL 只能查询,不能修改。如果要恢复为可写状态,需执行 PROXYSQL READWRITE 命令,不过 PROXYSQL READWRITE 只在当前会话中有效。

- admin-hash_passwords：在执行 LOAD MYSQL USERS TO RUNTIME 时，是否将 mysql_users 中的明文密码加密保存到 runtime_mysql_users 中，默认为 true。如果 mysql_users 表中有使用 caching_sha2_password 的用户，需将该参数设置为 false。
- admin-checksum_xxx、admin-cluster_xxx：ProxySQL Cluster 的相关参数。
- admin-restapi_enabled、admin-restapi_port：ProxySQL 内置了 Prometheus Exporter。admin-restapi_enabled 用来开启 Metrics Endpoint，admin-restapi_port 是 Exporter 的监听端口，admin-prometheus_memory_metrics_interval 定义了 memory_metrics 的采集周期。
- admin-web_enabled、admin-web_port：是否启用 ProxySQL 内嵌的 Web Server。默认禁用，可在线开启。该 Web Server 会展示 Query Cache、连接数等信息。

10.9.2 监控参数

监控参数以 mysql-monitor_ 开头，主要是控制 Monitor 模块的相关行为。Monitor 模块在 ProxySQL 中用来检查后端节点的状态。相关参数如下。

```
ProxySQL> show variables like 'mysql-monitor%';
+-----------------------------------------------------------------+--------+
| Variable_name                                                   | Value  |
+-----------------------------------------------------------------+--------+
| mysql-monitor_enabled                                           | true   |
| mysql-monitor_connect_timeout                                   | 600    |
| mysql-monitor_ping_max_failures                                 | 3      |
| mysql-monitor_ping_timeout                                      | 1000   |
| mysql-monitor_read_only_max_timeout_count                       | 3      |
| mysql-monitor_replication_lag_group_by_host                     | false  |
| mysql-monitor_replication_lag_interval                          | 10000  |
| mysql-monitor_replication_lag_timeout                           | 1000   |
| mysql-monitor_replication_lag_count                             | 1      |
| mysql-monitor_groupreplication_healthcheck_interval             | 5000   |
| mysql-monitor_groupreplication_healthcheck_timeout              | 800    |
| mysql-monitor_groupreplication_healthcheck_max_timeout_count    | 3      |
| mysql-monitor_groupreplication_max_transactions_behind_count    | 3      |
| mysql-monitor_groupreplication_max_transactions_behind_for_read_only | 1 |
| mysql-monitor_galera_healthcheck_interval                       | 5000   |
| mysql-monitor_galera_healthcheck_timeout                        | 800    |
| mysql-monitor_galera_healthcheck_max_timeout_count              | 3      |
| mysql-monitor_replication_lag_use_percona_heartbeat             |        |
| mysql-monitor_query_interval                                    | 60000  |
| mysql-monitor_query_timeout                                     | 100    |
| mysql-monitor_slave_lag_when_null                               | 60     |
| mysql-monitor_threads_min                                       | 8      |
| mysql-monitor_threads_max                                       | 128    |
| mysql-monitor_threads_queue_maxsize                             | 128    |
| mysql-monitor_wait_timeout                                      | true   |
| mysql-monitor_writer_is_also_reader                             | true   |
| mysql-monitor_username                                          | monitor|
| mysql-monitor_password                                          | monitor|
| mysql-monitor_history                                           | 600000 |
| mysql-monitor_connect_interval                                  | 60000  |
| mysql-monitor_ping_interval                                     | 10000  |
| mysql-monitor_read_only_interval                                | 1500   |
| mysql-monitor_read_only_timeout                                 | 500    |
+-----------------------------------------------------------------+--------+
33 rows in set (0.00 sec)
```

看起来没有规律，实际上有迹可循。Monitor 模块主要会进行以下 4 个方面的检查。

❑ CONNECT

连接性检查。检查用户及密码由 `mysql-monitor_username`、`mysql-monitor_password` 决定，默认都是 `monitor`。检查频率由 `mysql-monitor_connect_interval` 决定，默认为 60000，即 60 秒。检查的超时时长由 `mysql-monitor_connect_timeout` 决定，默认为 600 毫秒。连接动作会记录在 general log 中。

```
12 Connect   monitor@192.168.244.128 on   using TCP/IP
12 Quit
```

❑ PING

PING 检查。检查频率由 `mysql-monitor_ping_interval` 决定，默认为 10000，即 10 秒。检查的超时时长由 `mysql-monitor_ping_timeout` 参数决定，默认为 1000，即 1 秒。检查失败或超时会重试，重试的次数由 `mysql-monitor_ping_max_failures` 决定，默认为 3。如果重试还是失败，Monitor 模块会通知 `MySQL_Hostgroups_Manager` 将该节点的状态置为 `SHUNNED` 并杀掉该节点的所有连接。

注意，这里的 PING 检查使用的是 `mysql_ping()` 函数，不会记录在 general log 中。

❑ READ_ONLY

只读检查。这个检查非常关键，ProxySQL 判断一个节点是属于 `reader_hostgroup` 还是 `writer_hostgroup` 的依据就是 `read_only` 的值。注意，这里的 `read_only` 不是特指 `read_only` 这个参数，而是泛指 `read_only`、`innodb_read_only`、`super_read_only` 等参数。具体检查哪些参数，由 `mysql_replication_hostgroups` 表中的 `check_type` 决定。`read_only` 的检查频率由 `mysql-monitor_read_only_interval` 参数决定，默认为 1500，即 1.5 秒一次，检查的超时时长由 `mysql-monitor_read_only_timeout` 决定，默认为 500 毫秒。检查超时会重试，重试的次数由 `mysql-monitor_read_only_max_timeout_count` 决定，默认为 3。如果重试还是失败，ProxySQL 会将该节点的状态置为 `OFFLINE_HARD`。

❑ 主从延迟

主从延迟在默认情况下是通过 SHOW SLAVE STATUS 中的 `Seconds_Behind_Master` 来判断的。检查频率由 `mysql-monitor_replication_lag_interval` 决定，默认为 10000，即 10 秒一次。检查的超时时长由 `mysql-monitor_replication_lag_timeout` 决定，默认为 1000 毫秒。如果主从延迟过大，超过 `mysql_servers` 表中 `max_replication_lag` 的设置，ProxySQL 会将该节点的状态置为 `SHUNNED`。`mysql-monitor_replication_lag_count` 是检查次数，默认为 1，即第一次发现主从延迟大于 `max_replication_lag` 时，就调整节点的状态。

当从库的 I/O 线程或 SQL 线程停止时，`Seconds_Behind_Master` 会显示为 `NULL`，此时，ProxySQL 会将其转换为一个具体的数值，这个值由 `mysql-monitor_slave_lag_when_null` 决定，默认为 60 秒。

除了 Seconds_Behind_Master，也可通过 pt-heartbeat 来获取主从延迟情况，此时，需通过 mysql-monitor_replication_lag_use_percona_heartbeat 指定心跳表，一般是 percona.heartbeat。

注意，主从延迟检测只针对主从复制。如果后端节点是组复制，则对应的检查参数如下。

- mysql-monitor_groupreplication_healthcheck_interval：检查频率。
- mysql-monitor_groupreplication_healthcheck_timeout：检查的超时时长。
- mysql-monitor_groupreplication_healthcheck_max_timeout_count：超时次数。达到次数限制才将节点状态置为 OFFLINE。
- mysql-monitor_groupreplication_max_transactions_behind_count：如果节点的延迟事务数超过 mysql_group_replication_hostgroups 表中 max_transactions_behind 的设置，ProxySQL 会将节点的状态置为 SHUNNED。mysql-monitor_groupreplication_max_transactions_behind_count 是检查次数，默认为 3，即连续 3 次超过才调整节点的状态。节点是否被调整，还与 mysql-monitor_groupreplication_max_transactions_behind_for_read_only 的设置有关。
- mysql-monitor_groupreplication_max_transactions_behind_for_read_only：设置为 0，只将 read_only=0 的延迟节点置为 SHUNNED；设置为 1，只将 read_only=1 的延迟节点置为 SHUNNED；设置为 2，则不区分 read_only 的值，只要是延迟节点，就置为 SHUNNED。默认为 1。

其他需要注意的参数有。

- mysql-monitor_enabled：是否开启 Monitor 模块。
- mysql-monitor_username、mysql-monitor_password：监控用户和密码。
- mysql-monitor_history：Monitor 模块的检查结果会保存在 monitor 库中。mysql-monitor_history 用来指定检查结果的保留时长，默认为 600 000 毫秒，即 600 秒。
- mysql-monitor_writer_is_also_reader：当后端节点 read_only 的值由 1 变为 0 时，该节点会从 reader_hostgroup 移到 writer_hostgroup 中。参数若为 true，则该节点会同时保留在 reader_hostgroup 中；若为 false，则会从 reader_hostgroup 中移除。

10.9.3　MySQL 参数

除了管理参数和监控参数，其他参数都属于 MySQL 参数。MySQL 参数以 mysql- 开头，主要用来调整 MySQL 相关的功能，可分为以下几类。

1. 查询和事务的执行时长

```
+-------------------------------+----------+
| Variable_name                 | Value    |
+-------------------------------+----------+
| mysql-default_query_timeout   | 36000000 |
| mysql-max_transaction_idle_time | 14400000 |
| mysql-max_transaction_time    | 14400000 |
+-------------------------------+----------+
```

上述参数的具体含义如下。

- mysql-default_query_timeout：查询的超时时长，默认为 36 000 000 毫秒，即 10 小时。

- mysql-max_transaction_idle_time：事务的最大空闲时间，默认为 14 400 000 毫秒，即 4 小时。
- mysql-max_transaction_time：事务的最大持续时间，等于事务执行时间加上事务空闲时间，默认为 14 400 000 毫秒，即 4 小时。

无论是查询还是事务，只要超过指定时长，就会被 ProxySQL 杀掉。注意，因超时被杀掉的查询（事务）不会自动重试。

2. 审计日志

```
+------------------------+-----------+
| Variable_name          | Value     |
+------------------------+-----------+
| mysql-auditlog_filename |           |
| mysql-auditlog_filesize | 104857600 |
+------------------------+-----------+
2 rows in set (0.00 sec)
```

上述参数的具体含义如下。

- mysql-auditlog_filename：日志前缀名。
- mysql-auditlog_filesize：日志的最大大小。

需要注意的是，auditlog 记录的是与连接相关的事件，不是具体的 SQL 语句。

```
# tailf /var/lib/proxysql/proxysql_audit.00000001 | jq
{
  "client_addr": "192.168.244.10:53416",
  "event": "MySQL_Client_Connect_OK",
  "proxy_addr": "0.0.0.0:6033",
  "schemaname": "information_schema",
  "ssl": false,
  "thread_id": 26,
  "time": "2022-07-05 08:45:55.847",
  "timestamp": 1657010755847,
  "username": "split_user"
}
```

3. 防火墙

```
+--------------------------------+-----------------------------+
| Variable_name                  | Value                       |
+--------------------------------+-----------------------------+
| mysql-firewall_whitelist_enabled  | false                    |
| mysql-firewall_whitelist_errormsg | Firewall blocked this query |
+--------------------------------+-----------------------------+
2 rows in set (0.00 sec)
```

上述参数的具体含义如下。

- mysql-firewall_whitelist_enabled：开启 ProxySQL 2.0.9 引入的防火墙功能。具体的防火墙规则需在 mysql_firewall_whitelist_rules 和 mysql_firewall_whitelist_users 中定义。
- mysql-firewall_whitelist_errormsg：因不满足防火墙规则，而返回给客户端的错误信息。

4. 流量控制

```
+------------------------------------------------+---------+
| Variable_name                                  | Value   |
+------------------------------------------------+---------+
| mysql-throttle_connections_per_sec_to_hostgroup| 1000000 |
| mysql-throttle_max_bytes_per_second_to_client  | 0       |
| mysql-throttle_ratio_server_to_client          | 0       |
| mysql-default_query_delay                      | 0       |
+------------------------------------------------+---------+
```

上述参数的具体含义如下。

- `mysql-throttle_connections_per_sec_to_hostgroup`：限制单个主机组每秒新增连接的数量。注意，是主机组，不是节点。
- `mysql-throttle_max_bytes_per_second_to_client`：限制每秒可以发送给客户端的最大字节数。
- `mysql-throttle_ratio_server_to_client`：暂未实现。
- `mysql-default_query_delay`：查询延迟执行的时间，类似于 `mysql_query_rules` 表中的 `delay`。在紧急情况下，可以使用这个参数进行简单的限流操作。

5. MySQL 中的同名参数

```
+-------------------------+----------+
| Variable_name           | Value    |
+-------------------------+----------+
| mysql-init_connect      |          |
| mysql-long_query_time   | 1000     |
| mysql-max_allowed_packet| 67108864 |
| mysql-max_connections   | 2048     |
| mysql-wait_timeout      | 28800000 |
+-------------------------+----------+
```

上述参数的具体含义如下。

- `mysql-init_connect`：ProxySQL 在创建或初始化后端连接时执行的 SQL。可指定多个 SQL，中间用分号隔开。
- `mysql-long_query_time`：慢查询的阈值，默认为 1 秒。慢查询的次数会记录在 `stats_mysql_global` 表的 `Slow_queries` 变量中。
- `mysql-max_allowed_packet`：限制单个数据包的最大大小。
- `mysql-max_connections`：ProxySQL 的最大连接数。
- `mysql-wait_timeout`：如果前端连接的空闲时间超过 `mysql-wait_timeout`，则会被 ProxySQL 杀掉。

6. 其他需要注意的参数

```
+-----------------------------------+---------------------+
| Variable_name                     | Value               |
+-----------------------------------+---------------------+
| mysql-default_max_latency_ms      | 1000                |
| mysql-default_schema              | information_schema  |
| mysql-enable_load_data_local_infile| false              |
| mysql-have_ssl                    | false               |
| mysql-interfaces                  | 0.0.0.0:6033        |
+-----------------------------------+---------------------+
```

```
| mysql-log_mysql_warnings_enabled   | false    |
| mysql-max_stmts_cache              | 10000    |
| mysql-server_version               | 5.5.30   |
| mysql-threads                      | 4        |
+------------------------------------+----------+
```

上述参数的具体含义如下。

- `mysql-default_max_latency_ms`：最大延迟时长。ProxySQL 在转发请求时，会自动忽略 PING 请求超过最大延迟时长的节点。
- `mysql-default_schema`：默认的 schema。
- `mysql-enable_load_data_local_infile`：是否支持 LOAD DATA LOCAL INFILE 命令，默认为 `false`，即不支持。如果要支持，导入的文件只能放到 ProxySQL 本地，不能放到客户端本地。
- `mysql-have_ssl`：是否允许前端连接开启 SSL 通信。
- `mysql-interfaces`：指定 ProxySQL 的对外服务端口。修改后，需重启 ProxySQL（PROXYSQL RESTART）才能生效。
- `mysql-log_mysql_warnings_enabled`：是否将 MySQL 产生的 warning 记录在日志中。
- `mysql-max_stmts_cache`：使用 Prepared Statement 时，可以缓存的 Statement 的数量。
- `mysql-server_version`：ProxySQL 用来响应客户端的服务端版本号。这个版本号与后端节点的实际版本没有任何关系。
- `mysql-threads`：ProxySQL 用来处理客户端请求的线程数。修改后，需重启 ProxySQL 才能生效。

10.9.4 如何修改参数

除了 `mysql-interfaces`、`mysql-threads` 和 `mysql-stacksize`，ProxySQL 中的其他参数均可动态修改。支持的修改方式如下。

- 直接使用 SET 命令修改参数，例如：

  ```
  set mysql-long_query_time=2000;
  ```

- 使用 UPDATE 命令修改 GLOBAL_VARIABLES，例如：

  ```
  update global_variables set variable_value=2000 where variable_name='mysql-long_query_time';
  ```

参数修改后需加载到 RUNTIME 中才会生效。对于管理参数，对应的加载命令是 LOAD ADMIN VARIABLES TO RUNTIME。对于监控参数及 MySQL 参数，对应的加载命令是 LOAD MYSQL VARIABLES TO RUNTIME。

10.10 ProxySQL 中的常见问题

10.10.1 如何自定义 ProxySQL 的数据目录

(1) 修改配置文件。

```
# vim /etc/proxysql.cnf
datadir="/data/proxysql/6033/data"
errorlog="/data/proxysql/6033/log/proxysql.log"
```

(2) 初始化配置。

```
# /usr/bin/proxysql --initial -f -c /etc/proxysql.cnf
```

初始化时，需指定 `--initial`。

(3) 修改数据目录和日志目录的属主。

```
# chown -R proxysql.proxysql /data/proxysql/6033/
```

(4) 修改服务启动脚本 PIDFile 的路径。

```
# vim /etc/systemd/system/proxysql.service
PIDFile=/data/proxysql/6033/data/proxysql.pid
```

(5) 启动服务。

```
# systemctl daemon-reload
# service proxysql start
```

10.10.2 通过 USE DBNAME 切换数据库

在 ProxySQL 中，USE DBNAME 命令总是会执行成功，即使指定的数据库在后端节点不存在也是如此。

```
# mysql -h192.168.244.128 -usplit_user -psplit_pass -P6033
mysql> use db1
Database changed
mysql> show tables;
ERROR 1049 (42000): Unknown database 'db1'
```

10.10.3 ProxySQL 的高可用性

上面提到的 ProxySQL Cluster 只是用来同步配置的，并不是一个高可用方案。如果要实现 ProxySQL 的高可用性，可考虑组合使用 Keepalived、VIP 和负载均衡组件（LVS、HAproxy、F5）。

10.11 本章总结

在使用 ProxySQL 时，需注意以下几点。

- 如果我们要做读写分离，推荐的方案是将那些耗时较久的查询而不是所有查询放到从库上执行。
- ProxySQL 会基于后端节点 `check_type` 的值来动态调整它所属的主机组。如果一个从库的 `check_type` 发生了变化，如 `super_read_only` 不小心关闭了，ProxySQL 会自动将这个从库调整到 `writer_hostgroup` 中。而 `writer_hostgroup` 中的节点会承担写流量。这一点在线上尤其需要注意。
- ProxySQL 只是中间件，并不是 MySQL 高可用方案。如果要实现 MySQL 的高可用，需借助于组复制、MHA、Orchestrator 等高可用方案。

重点问题回顾

- ProxySQL 的多层配置系统。
- 官方推荐的读写分离方案。
- 如何开启 SQL 审计功能？
- 如何进行查询重写？
- 如何使用 SQL 黑名单功能进行限流？
- 如何实现 IP 白名单功能？
- ProxySQL 连接池的运行流程。
- 连接在哪些场景下不可复用？
- ProxySQL Cluster 引入的背景及实现的功能。

第 11 章
组 复 制

MySQL 组复制（MySQL Group Replication，MGR）是官方在 MySQL 5.7.17 中推出的一个基于 Paxos 协议的高可用方案。在 MGR 推出之前，MySQL 只支持主从复制。传统的主从复制包括异步复制和半同步复制。无论是异步复制还是半同步复制，都存在主从数据不一致的风险。下面我们结合异步复制和半同步复制的原理图分析一下两者的风险。图 11-1 是异步复制的原理图。

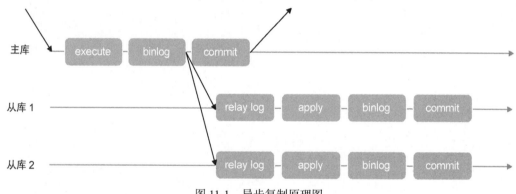

图 11-1　异步复制原理图

从图 11-1 中可以看到，事务只要写入 binlog，就可以发送到从库了，至于从库有没有接受或者主从复制有没有中断，主库并不关心。事务写入 binlog 后，直接在引擎层提交，然后返回给客户端。如果因为网络问题或者其他问题导致主库的 binlog 没有及时发送到从库，而主库此时又出现故障，那么当切换到从库后，就会发现主库之前写入的数据丢失了。所以，单从原理上来说，异步复制很难保证主从数据的一致性。

图 11-2 是半同步复制的原理图。

相比于异步复制，半同步复制中的事务在引擎层提交之前，需要得到从库的反馈。这一点确实能保证事务在返回给客户端之后，对应的二进制日志事件一定写入 relay log 了。但考虑下面这个场景：事务在写入 binlog 之后、写入 relay log 之前，主库宕机了。当切换到从库后，虽然从客户端的角度看，主从数据是一致的，但主库恢复后，因为事务已经写入 binlog 了，所以它在实例恢复的过程中会在引擎层提交，这样就有可能导致主从数据不一致了。

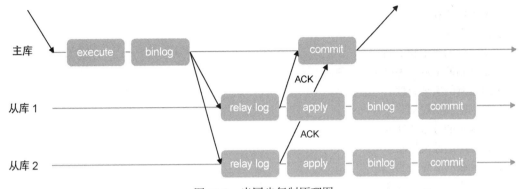

图 11-2 半同步复制原理图

当然，上面说的都是特殊场景，正常情况下，异步复制和半同步复制能够满足大部分系统的高可用需求。但不能百分之百保证主从数据的一致性，自然无法满足那些对数据一致性要求很高的系统的需求。

基于上述背景，MySQL 5.7 引入了基于 Paxos 协议实现的组复制。图 11-3 是组复制的原理图。

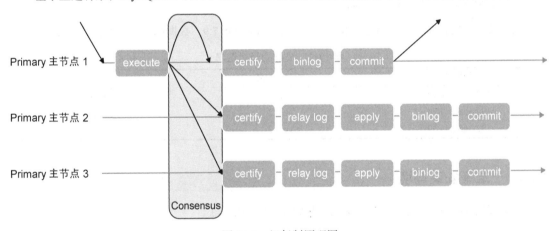

图 11-3 组复制原理图

相对于传统的主从复制，组复制主要新增了两个模块。

- Consensus：共识。共识是基于 Paxos 实现的，可以保证消息的全局有序和消息会被半数以上节点确认接受。
- certify：认证。认证可以保证所有节点会以确定的行为来处理同一个事务，要么全部认证通过，要么全部认证失败。

从原理上看，组复制能百分之百保证各个在线节点的数据一致性。

本章主要包括以下内容。

- 部署组复制。

- 单主模式和多主模式。
- 监控组复制。
- 组复制的要求和限制。
- 组复制的常见管理操作。
- 组复制的实现原理。
- 组复制的实现细节。
- 组复制的分布式恢复。
- 组复制的冲突检测。
- 组复制的故障检测。
- 组复制的事务一致性。
- 组复制的流量控制机制。
- 组复制的重点参数。

11.1 部署组复制

我们来看看组复制单主模式的部署方法。

集群的节点信息如表 11-1 所示。操作系统版本是 CentOS 7.9，数据库版本是 MySQL 8.0.27。

表 11-1 集群的节点信息

IP 地址	主 机 名	角 色
192.168.244.10	node1	Primary
192.168.244.20	node2	Secondary
192.168.244.30	node3	Secondary

11.1.1 准备安装环境

安装环境的准备工作主要是清空防火墙规则和关闭 SELinux。

清空防火墙规则：

```
# iptables -F
```

关闭 SELinux：

```
# setenforce 0
```

这只会临时生效，系统重启后又会恢复为默认值。如果要让这永久生效，需设置 /etc/sysconfig/selinux。

```
# vim /etc/sysconfig/selinux
SELINUX=disabled
```

11.1.2 初始化 MySQL 实例

集群的 3 个节点都要执行以下操作。

(1) 准备安装包。

```
# useradd mysql
# cd /usr/local/
# tar xvf mysql-8.0.27-linux-glibc2.12-x86_64.tar.xz
# ln -s mysql-8.0.27-linux-glibc2.12-x86_64 mysql
```

(2) 编辑 MySQL 配置文件。

以下是 node1 的配置文件。

```
[client]
socket = /data/mysql/3306/data/mysql.sock

[mysqld]
# Server
user = mysql
datadir = /data/mysql/3306/data
basedir = /usr/local/mysql
port = 3306
socket = /data/mysql/3306/data/mysql.sock
log_timestamps = system
log_error = /data/mysql/3306/data/mysqld.err
skip_name_resolve
report_host = "192.168.244.10"
disabled_storage_engines = "MyISAM,BLACKHOLE,FEDERATED,ARCHIVE,MEMORY"
sql_require_primary_key = ON

# Replication
server_id = 1
log_bin = mysql-bin
binlog_format = ROW
log_slave_updates = ON
gtid_mode = ON
enforce_gtid_consistency = ON
master_info_repository = TABLE
relay_log_info_repository = TABLE
super_read_only = ON
binlog_transaction_dependency_tracking = WRITESET
transaction_write_set_extraction = XXHASH64

# Multi-threaded Replication
slave_parallel_type = LOGICAL_CLOCK
slave_preserve_commit_order = ON
slave_parallel_workers = 4

# Group Replication Settings
plugin_load_add = "group_replication.so"
loose_group_replication_group_name = "4d8dd73b-018c-11ed-8463-525400d51a16"
loose_group_replication_start_on_boot = OFF
loose_group_replication_local_address = "192.168.244.10:33061"
loose_group_replication_group_seeds =
"192.168.244.10:33061,192.168.244.20:33061,192.168.244.30:33061"
loose_group_replication_bootstrap_group = OFF
loose_group_replication_recovery_get_public_key = ON
```

下面看看各个参数的具体含义。

首先是 Server 部分。因为本章的重点是组复制，所以 Server 部分只列举了一些基本参数，需要注意的参数如下。

- report_host：显式指定主机 IP。如果不设置该参数，会默认将主机名作为 performance_schema.replication_group_members 中的 member_host。如果主机名没有绑定或 DNS 解析不到，则其他待加入的节点将无法启动 group_replication_recovery 通道。
- disabled_storage_engines：禁用了除 InnoDB 之外的其他存储引擎，因为组复制要求存储引擎必须支持事务。
- sql_require_primary_key：强制要求设置主键。该参数是 MySQL 8.0.13 引入的。

其次是 Replication 部分，需要注意的参数如下。

- gtid_mode = ON、enforce_gtid_consistency = ON：开启 GTID 复制。组复制依赖 GTID 来简化事务的冲突检测。
- master_info_repository = TABLE、relay_log_info_repository = TABLE：将复制相关的元数据信息记录在表中。
- super_read_only = ON：设置为只读，这样在实例启动之后、组复制开启之前，都会保持只读状态，避免写入脏数据。
- binlog_transaction_dependency_tracking = WRITESET：组复制的冲突检测是基于 WRITESET 的。
- transaction_write_set_extraction = XXHASH64：将采集的主键信息、唯一索引信息、外键信息提取为 WRITESET 使用的哈希算法。

除此之外，在 MySQL 8.0.21 之前，还必须设置 binlog_checksum = NONE，因为在这个版本之前，组复制还不支持带有校验和的 binlog。

接着是 Multi-threaded Replication 部分。考虑到从库重放的效率，建议开启多线程复制。

```
slave_parallel_type = LOGICAL_CLOCK
slave_preserve_commit_order = ON
slave_parallel_workers = 4
```

从 MySQL 8.0.27 开始，多线程复制就默认开启了。

最后是组复制的相关参数。

- plugin_load_add：加载组复制插件。也可在实例启动后，通过以下命令手动加载。

    ```
    mysql> install plugin group_replication soname 'group_replication.so';
    ```

- group_replication_group_name：集群名。用于唯一标识一个集群，必须是个有效的 UUID 值。UUID 可通过 SELECT UUID() 生成。
- group_replication_start_on_boot：是否在实例启动时自动开启组复制，默认为 ON。
- group_replication_local_address：当前节点的内部通信地址。注意，这个地址只用于集群的内部通信，不可用于客户端连接。

- group_replication_group_seeds：种子节点地址。当有新的节点加入时，它会首先与种子节点建立连接。种子节点是组节点的一个子集。
- group_replication_bootstrap_group：是否由当前节点初始化集群，默认为 OFF。在配置文件中，切记不要将该参数设置为 ON，否则，任何一个节点重启都会创建一个新的集群。
- group_replication_recovery_get_public_key：如果复制用户使用了 caching_sha2_password，那么在分布式恢复阶段若没有使用 SSL（group_replication_recovery_use_ssl=ON），就必须使用 RSA 密钥对进行密码交换。将 group_replication_recovery_get_public_key 设置为 ON，则允许该节点直接从源节点获取公钥。

注意，组复制的每个参数前都加了一个 loose。如果不加 loose，实例在初始化的过程中会报错，错误信息如下。

```
[ERROR] [MY-000067] [Server] unknown variable 'group_replication_group_name=4d8dd73b-018c-11ed-8463-525400d51a16'.
[ERROR] [MY-013236] [Server] The designated data directory /data/mysql/3306/data/ is unusable. You can remove all files that the server added to it.
[ERROR] [MY-010119] [Server] Aborting
```

提示有未知参数。这是因为实例在初始化的过程中不会加载插件，而组复制的相关参数又只有在插件加载的情况下才有效。加了 loose，对于未知的参数，就只会提示 warning，并不影响实例的初始化和正常启动。

node2 的配置文件如下。

参数与 node1 基本相同，只需修改 report_host、server_id 和 loose-group_replication_local_address。

```
report_host = "192.168.244.20"
server_id = 2
group_replication_local_address = "192.168.244.20:33061"
```

node3 的配置文件如下。

```
report_host = "192.168.244.30"
server_id = 3
group_replication_local_address = "192.168.244.30:33061"
```

(3) 创建数据目录。

```
# mkdir -p /data/mysql/3306/data
```

(4) 初始化实例。

```
# /usr/local/mysql/bin/mysqld --defaults-file=/etc/my.cnf --initialize-insecure
```

方便起见，这里直接使用 --initialize-insecure 进行初始化。

(5) 启动实例。

```
# /usr/local/mysql/bin/mysqld_safe --defaults-file=/etc/my.cnf &
```

11.1.3　启动组复制

(1) 查看插件是否加载成功。

```
mysql> show plugins;
...
| group_replication              | ACTIVE  | GROUP REPLICATION | group_replication.so | GPL |
```

(2) 初始化组复制。

只在 node1 上执行初始化。

```
mysql> set global group_replication_bootstrap_group=on;
mysql> start group_replication;
mysql> set global group_replication_bootstrap_group=off;
```

通过 performance_schema.replication_group_members 查看集群的节点信息。

```
mysql> select member_id,member_host,member_port,member_state,member_role from
performance_schema.replication_group_members;
+--------------------------------------+----------------+-------------+--------------+-------------+
| member_id                            | member_host    | member_port | member_state | member_role |
+--------------------------------------+----------------+-------------+--------------+-------------+
| 207db264-0192-11ed-92c9-02001700754e | 192.168.244.10 |        3306 | ONLINE       | PRIMARY     |
+--------------------------------------+----------------+-------------+--------------+-------------+
1 row in set (0.00 sec)
```

ONLINE 代表节点状态正常。

(3) 创建复制用户。

```
create user 'rpl_user'@'%' identified by 'rpl_password';
grant replication slave on *.* to 'rpl_user'@'%';
grant backup_admin on *.* to 'rpl_user'@'%';
```

在新节点加入时，为了让它的数据与组内其他节点保持一致，会首先让它经历一个分布式恢复阶段。在这个阶段，新节点会随机选择组内的一个在线节点作为 Donor 同步差异数据。

在 MySQL 8.0.17 之前，同步的方式只有一种，即 binlog。这种方式依赖于传统的异步复制，所以必须创建复制用户，且授予其 REPLICATION SLAVE 权限。从 MySQL 8.0.17 开始，新增了一种同步方式——物理备份恢复，这种方式依赖于 MySQL 8.0.17 引入的克隆插件。使用克隆插件需要 BACKUP_ADMIN 权限。

从 MySQL 8.0.21 开始，如果通过 group_replication_advertise_recovery_endpoints 指定了 admin_port 来进行分布式恢复操作，还需要授予用户 SERVICE_CONNECTION_ADMIN 权限。

从 MySQL 8.0.27 开始，如果组成员之间的连接要使用 MySQL 通信栈（group_replication_communication_stack），还需要授予用户 GROUP_REPLICATION_STREAM 和 CONNECTION_ADMIN 权限。

如果使用 MySQL 5.7 搭建组复制，只需授予复制用户 REPLICATION SLAVE 权限。

官方文档中，在所有节点上都创建了复制用户。为了避免冲突，在创建之前，会将 SQL_LOG_BIN 设置为 0。但这实际上没有必要，因为新节点加入后，这个操作会自动在新节点上重放。

(4) 配置恢复通道。

```
mysql> change master to master_user='rpl_user', master_password='rpl_password' for channel
'group_replication_recovery';
```

在这里，CHANGE MASTER TO 语句只是指定了 master_user 和 master_password，没有指定 master_host 和 master_port。为什么不用指定呢？因为当一个新的节点加入时，组会随机选择一个在线节点作为 Donor，所以主机和端口是不固定的。

对于 Primary 节点（主节点），这一步其实是非必需的，但考虑到节点的角色有可能会发生变化，比如从 Primary 节点切换为 Secondary 节点（从节点），这个时候同样要经历分布式恢复阶段，所以，建议为所有节点都配置恢复通道。

(5) 构造测试数据。

在这里构造测试数据是为了方便验证组复制是否搭建成功。

```
mysql> create database slowtech;
mysql> create table slowtech.t1(id int primary key,c1 varchar(10));
mysql> insert into slowtech.t1 values(1,'abc');
```

11.1.4 添加节点

添加节点比较简单，只需两步。

- 配置恢复通道。
- 启动组复制。

依次在 node2 和 node3 上执行以下命令。

```
mysql> change master to master_user='rpl_user', master_password='rpl_password' for channel
'group_replication_recovery';
mysql> start group_replication;
```

如果组复制启动失败，可通过错误日志查看具体的原因。

```
mysql> start group_replication;
ERROR 3092 (HY000): The server is not configured properly to be an active member of the group. Please
see more details on error log.
```

节点添加完毕后，通过 performance_schema.replication_group_members 查看集群的节点信息。

```
mysql> select member_id,member_host,member_port,member_state,member_role from
performance_schema.replication_group_members;
+--------------------------------------+----------------+-------------+--------------+-------------+
| member_id                            | member_host    | member_port | member_state | member_role |
+--------------------------------------+----------------+-------------+--------------+-------------+
| 207db264-0192-11ed-92c9-02001700754e | 192.168.244.10 |        3306 | ONLINE       | PRIMARY     |
| 2cee229d-0192-11ed-8eff-02001700f110 | 192.168.244.20 |        3306 | ONLINE       | SECONDARY   |
| 4cbfdc79-0192-11ed-8b01-02001701bd0a | 192.168.244.30 |        3306 | ONLINE       | SECONDARY   |
+--------------------------------------+----------------+-------------+--------------+-------------+
3 rows in set (0.01 sec)
```

最后分别在 node2 和 node3 上执行以下操作，如果能读取到在 node1 中写入的数据，则意味着从应用的角度来说，组复制已经搭建成功了。

```
mysql> select * from slowtech.t1;
+----+------+
| id | c1   |
+----+------+
|  1 | abc  |
+----+------+
1 row in set (0.00 sec)
```

11.2 单主模式和多主模式

在上一节中，部署的是单主模式（single-primary mode）的组复制，这也是官方推荐的部署模式。在单主模式下，只有一个 Primary 节点（可读可写），其他节点都是 Secondary 节点（只能读，不能写）。除了单主模式，组复制还支持多主模式（multi-primary mode）。在多主模式下，所有节点都是 Primary 节点。

下面我们从 4 个方面看看单主模式和多主模式的区别。

11.2.1 单主模式和多主模式的区别

单主模式和多主模式的本质区别是多主模式会开启冲突检测，而单主模式不会。除此之外，还有如下几方面的区别。

1. 部署方式

两者的部署方式基本相同。相对于单主模式，多主模式需额外设置以下两个参数。

```
group_replication_single_primary_mode = OFF
group_replication_enforce_update_everywhere_checks = ON
```

前者会关闭单主模式，后者会开启严格的一致性检查，检查的内容如下。

- 不允许将事务隔离级别设置为 SERIALIZABLE。
- 不支持外键的级联操作。

2. read_only 的设置

对于单主模式，组复制会自动将 Secondary 节点的 super_read_only 和 read_only 设置为 ON。

3. 自增主键

在 MySQL 中，自增主键的生成受两个参数的影响：auto_increment_offset（初始值）和 auto_increment_increment（步长）。两个参数的默认值均为 1。

在组复制中，如果我们显式设置了这两个参数，则以实际设置的为主。如果没有设置，那么在单主模式下，这两个参数还是默认为 1，但在多主模式下，这两个参数则分别取自 server_id 和 group_replication_auto_increment_increment。

group_replication_auto_increment_increment 默认为 7。之所以是 7，实际上是考虑到组复制的最大节点数是 9。如果设置得过小，小于集群的节点数，会造成自增主键冲突。而如果设置得过大，又会造成自增主键的浪费。

既然 auto_increment_offset 是由 server_id 决定的，这里就会有一个潜在的问题。是什么问题呢？

首先，我们看看自增主键的生成公式。

auto_increment_offset + N × auto_increment_increment

因为 auto_increment_increment 取自 group_replication_auto_increment_increment，是固定的，所以如果某个节点的 server_id 等于另外一个节点的 server_id + N × auto_increment_increment，就会造成自增主键的冲突。

表 11-2 列举了当 group_replication_auto_increment_increment（auto_increment_increment）为 7 时，不同 server_id（auto_increment_offset）下自增主键的生成情况。

表 11-2　不同 server_id 下自增主键的生成情况

server_id（auto_increment_offset）	自增主键值（第一次）	自增主键值（第二次）	自增主键值（第三次）
1	1	8	15
2	2	9	16
3	3	10	17
4	4	11	18
5	5	12	19
6	6	13	20
7	7	14	21
8	8	15	22
9	9	16	23
10	1	10	17
11	2	11	18
12	3	12	19
13	4	13	20

可以看到，server_id 为 1 和 8 的会发生冲突，为 2 和 9 的会发生冲突，依此类推。看下面这个示例，node1 和 node3 中的 server_id 分别是 1 和 8。

```
# 查看两个节点自增主键的相关参数
node1> show global variables like '%auto_increment%';
+-------------------------------------------+-------+
| Variable_name                             | Value |
+-------------------------------------------+-------+
| auto_increment_increment                  | 7     |
| auto_increment_offset                     | 1     |
| group_replication_auto_increment_increment| 7     |
+-------------------------------------------+-------+
3 rows in set (0.00 sec)

node3> show global variables like '%auto_increment%';
+-------------------------------------------+-------+
| Variable_name                             | Value |
```

```
+-----------------------------------------------+-------+
| auto_increment_increment                      | 7     |
| auto_increment_offset                         | 8     |
| group_replication_auto_increment_increment    | 7     |
+-----------------------------------------------+-------+
3 rows in set (0.00 sec)

# 新建一张测试表并插入两条记录
node1> create table t1(id int auto_increment primary key,c1 varchar(10));
Query OK, 0 rows affected (0.01 sec)

node1> insert into t1(c1) values('a');
Query OK, 1 row affected (0.00 sec)

node1> insert into t1(c1) values('a');
Query OK, 1 row affected (0.01 sec)

node1> select * from t1;
+----+------+
| id | c1   |
+----+------+
|  1 | a    |
|  8 | a    |
+----+------+
2 rows in set (0.00 sec)

# node1 开启事务并插入一条记录，但没有提交
node1> begin;
Query OK, 0 rows affected (0.00 sec)

node1> insert into t1(c1) values('a');
Query OK, 1 row affected (0.00 sec)

node1> select * from t1;
+----+------+
| id | c1   |
+----+------+
|  1 | a    |
|  8 | a    |
| 15 | a    |
+----+------+
3 rows in set (0.00 sec)

# node3 同样插入一条记录，自动提交
node3> insert into t1(c1) values('a');
Query OK, 1 row affected (0.00 sec)

node3> select * from t1;
+----+------+
| id | c1   |
+----+------+
|  1 | a    |
|  8 | a    |
| 15 | a    |
+----+------+
3 rows in set (0.00 sec)

# 因为 id=15 这个主键值已经在 node3 中生成了，所以 node1 中的事务在提交时，会因为主键冲突而回滚
node1> commit;
ERROR 1180 (HY000): Got error 149 - 'Lock deadlock; Retry transaction' during COMMIT
```

所以，在多主模式下，如果没有显式设置 auto_increment_offset 和 auto_increment_increment 的话，server_id 的设置尤其需要注意。特别是在很多自动化部署环境中，server_id 都是随机生成的。

4. 组复制的限制

组复制本身就有一定的要求和限制，具体后面会提到，其中很多限制其实是针对多主模式的，比如以下几点。

- 乐观事务模型。
- 验证阶段不会考虑间隙锁（gap lock）、表锁（lock table）和 GET_LOCK 操作。
- DDL 与 DML 的并发执行问题。

在单主模式下，因为所有的操作都是在一个节点上执行的，所以就不存在上述限制。

11.2.2　单主模式和多主模式的在线切换

在组复制中，单主模式和多主模式不能混合部署。在添加一个新的节点时，如果它的模式和集群中其他节点的模式不一样，则在执行 START GROUP_REPLICATION 时会报错。同时，错误日志中会提示以下信息。

```
[ERROR] [MY-011529] [Repl] Plugin group_replication reported: 'The member configuration is not
compatible with the group configuration. Variables such as group_replication_single_primary_mode or
group_replication_enforce_update_everywhere_checks must have the same value on every server in the
group. (member configuration option: [group_replication_single_primary_mode], group configuration
option: [group_replication_enforce_update_everywhere_checks]).'
```

在 MySQL 8.0.13 之前，不支持在线调整集群模式。如果要调整，只能重启整个组复制。从 MySQL 8.0.13 开始，可通过以下命令在线调整集群模式。

```
# 单主模式切换为多主模式
select group_replication_switch_to_multi_primary_mode();

# 多主模式切换为单主模式
select group_replication_switch_to_single_primary_mode(member_uuid);
```

对于多主模式切换为单主模式，函数中的 member_uuid 是单主模式下 Primary 节点的 member_id，可不指定，此时基于默认的选举算法选择 Primary 节点。

在执行这两个命令时，会同步修改 group_replication_single_primary_mode 和 group_replication_enforce_update_everywhere_checks 这两个参数的值，并持久化到 mysqld-auto.cnf 文件中。

可通过以下命令查看切换进度。

```
mysql> select event_name, work_completed, work_estimated from
performance_schema.events_stages_current where event_name like "%stage/group_rpl%"\G
*************************** 1. row ***************************
    event_name: stage/group_rpl/Multi-primary Switch: applying buffered transactions
work_completed: 0
work_estimated: 1
1 row in set (0.00 sec)
```

注意，切换时，集群所有节点的状态必须是 ONLINE。

```
mysql> select group_replication_switch_to_single_primary_mode('207db264-0192-11ed-92c9-02001700754e');
ERROR 1123 (HY000): Can't initialize function 'group_replication_switch_to_single_primary_mode';
A member is joining the group, wait for it to be ONLINE.
```

除了能在线调整集群模式，从 MySQL 8.0.13 开始，还可通过以下命令切换单主模式下的 Primary 节点。

select group_replication_set_as_primary(member_uuid);

注意，该命令只适用于单主模式，且必须指定新的 Primary 节点的 member_id。

有意思的是，在单主模式下，将 Secondary 节点的 super_read_only 和 read_only 设置为 OFF，同样允许执行写操作，并且写操作也会同步到其他节点上。从这一点来看，单主模式和多主模式似乎没有严格的区别。但实际上，单主模式下不会开启冲突检测。当针对同一行数据的 DML 操作在不同节点并发执行时，有可能会导致节点之间的数据不一致。这一点很容易通过 conflict_detection_test.py 脚本（已上传至 GitHub）模拟出来。该脚本的实现思路很简单，就是在两个节点上分别对同一行数据执行 UPDATE 操作，只不过为了模拟并行的效果，首先执行了 UPDATE 操作，然后开启两个线程并发执行 COMMIT 操作，最后对比这两个节点更新后的值。

11.3 监控组复制

对组复制的监控主要是基于 performance_schema 中的 6 张表实现的：

- replication_group_members
- replication_group_member_stats
- replication_connection_status
- replication_applier_status
- replication_applier_status_by_coordinator
- replication_applier_status_by_worker

在这 6 张表中，只有 replication_group_members 和 replication_group_member_stats 是组复制特有的。其他 4 张表适用于所有的复制场景。关于这 4 张表的具体作用，可参考第 4 章。

下面重点说说 replication_group_members 和 replication_group_members_stats 这两张表的具体作用及含义。

11.3.1 replication_group_members

该表显示了集群各节点的状态信息。

```
mysql> select * from performance_schema.replication_group_members\G
*************************** 1. row ***************************
  CHANNEL_NAME: group_replication_applier
     MEMBER_ID: 207db264-0192-11ed-92c9-02001700754e
   MEMBER_HOST: 192.168.244.10
   MEMBER_PORT: 3306
  MEMBER_STATE: ONLINE
   MEMBER_ROLE: PRIMARY
```

```
                MEMBER_VERSION: 8.0.27
MEMBER_COMMUNICATION_STACK: XCom
...
3 rows in set (0.00 sec)
```

其中各项解释如下。

- CHANNEL_NAME：channel 名。组复制引入了两个 channel：group_replication_recovery 和 group_replication_applier。前者用于新成员加入时的分布式恢复阶段，后者用来执行组复制中的事务。
- MEMBER_ID：节点的 UUID，由 @@server_uuid 决定。实例的 server_uuid 可在实例启动之前通过 auto.cnf 文件指定。
- MEMBER_HOST：节点的主机名。默认由 @@hostname 决定，建议通过 report_host 参数显式指定。
- MEMBER_PORT：节点的对外服务端口，由 @@port 决定。
- MEMBER_STATE：节点的状态，取值如下。
    - ONLINE：节点状态正常。
    - RECOVERING：节点处于分布式恢复阶段。
    - OFFLINE：已加载组复制插件，但还未开启组复制。
    - ERROR：错误状态。这个时候，可通过错误日志定位处于 ERROR 状态的具体原因。节点处于 ERROR 状态的行为由 group_replication_exit_state_action 参数决定，默认为 READ_ONLY，也可设置为 OFFLINE_MODE 或 ABORT_SERVER。
    - UNREACHABLE：发送给目标节点的组消息超时。常见原因包括目标节点无法访问，当前节点和目标节点之间的网络中断。
- MEMBER_ROLE：节点角色，为 PRIMARY 或 SECONDARY。
- MEMBER_VERSION：实例版本。
- MEMBER_COMMUNICATION_STACK：节点之间使用的通信栈。这个列是 MySQL 8.0.27 新增的。

11.3.2 replication_group_member_stats

该表显示了组复制运行过程中集群各节点的状态信息。

```
mysql> select * from performance_schema.replication_group_member_stats\G
*************************** 1. row ***************************
                              CHANNEL_NAME: group_replication_applier
                                   VIEW_ID: 16575963794856729:3
                                 MEMBER_ID: 207db264-0192-11ed-92c9-02001700754e
               COUNT_TRANSACTIONS_IN_QUEUE: 0
                COUNT_TRANSACTIONS_CHECKED: 27617
                  COUNT_CONFLICTS_DETECTED: 0
        COUNT_TRANSACTIONS_ROWS_VALIDATING: 35481
        TRANSACTIONS_COMMITTED_ALL_MEMBERS: 4d8dd73b-018c-11ed-8463-525400d51a16:1-15573
            LAST_CONFLICT_FREE_TRANSACTION: 4d8dd73b-018c-11ed-8463-525400d51a16:27620
COUNT_TRANSACTIONS_REMOTE_IN_APPLIER_QUEUE: 0
         COUNT_TRANSACTIONS_REMOTE_APPLIED: 3
         COUNT_TRANSACTIONS_LOCAL_PROPOSED: 27617
         COUNT_TRANSACTIONS_LOCAL_ROLLBACK: 0
...
3 rows in set (0.00 sec)
```

其中各项解释如下。

- CHANNEL_NAME：channel 名。
- VIEW_ID：组视图 ID。
- MEMBER_ID：节点的 UUID。
- COUNT_TRANSACTIONS_IN_QUEUE：队列中等待冲突检测的事务数。
- COUNT_TRANSACTIONS_CHECKED：已经进行过冲突检测的事务数，包括通过和没有通过的。
- COUNT_CONFLICTS_DETECTED：冲突检测失败的事务数。
- COUNT_TRANSACTIONS_ROWS_VALIDATING：冲突检测数据库当前的记录数。
- TRANSACTIONS_COMMITTED_ALL_MEMBERS：所有成员都执行过的事务的 GTID 集合，每 60 秒更新一次。组复制会启动一个 GC（Garbage Collector）线程，定期（每 60 秒）将该集合涉及的事务信息从冲突检测数据库中清理掉。
- LAST_CONFLICT_FREE_TRANSACTION：最近一个通过冲突检测的事务的 GTID。
- COUNT_TRANSACTIONS_REMOTE_IN_APPLIER_QUEUE：从组接受，等待应用的事务数。
- COUNT_TRANSACTIONS_REMOTE_APPLIED：从组接受，已经应用的事务数。
- COUNT_TRANSACTIONS_LOCAL_PROPOSED：由当前节点发起，发送给组的事务数。
- COUNT_TRANSACTIONS_LOCAL_ROLLBACK：由当前节点发起，但被组回滚的事务数。

11.4 组复制的要求和限制

在部署组复制时，有以下要求和限制。

(1) 只支持 InnoDB 存储引擎。

在关系数据库中，一般有两种并发控制机制：Pessimistic Concurrency Control（悲观并发控制，又称"悲观锁"）和 Optimistic Concurrency Control（乐观并发控制，又称"乐观锁"）。

首先，看看乐观锁的特点。

- 在修改阶段不加锁，占用资源较少。
- 只在 COMMIT 阶段加锁，锁定时间较短。

它的缺点也是显而易见的：如果写入量比较大，并发度比较高，就容易出现事务冲突。冲突就意味着事务的回滚和重试，这反而会影响数据库的性能。

与乐观锁不一样的是，悲观锁在操作之初就开始加锁，并持续到最后事务提交。绝大多数关系数据库使用的是悲观锁，包括 MySQL。

组复制在多主模式下使用的是乐观锁，只在 COMMIT 阶段进行冲突检测。如果冲突检测失败，会把事务回滚。既然涉及回滚，就必然要求存储引擎支持事务。目前，只有 InnoDB 符合这个条件。

注意，在组复制的搭建过程中，如果没有使用 disabled_storage_engines 参数显式禁用非 InnoDB 存储引擎，组复制本身是允许创建非 InnoDB 表的，只不过在执行 DML 时会报错。

```
mysql> create table slowtech.t1(id int primary key,c1 varchar(10)) engine=myisam;
Query OK, 0 rows affected (0.01 sec)

mysql> insert into slowtech.t1 values(1,'a');
ERROR 3098 (HY000): The table does not comply with the requirements by an external plugin.
```

(2) 表上必须存在主键或唯一非空索引。

因为组复制是基于主键来进行冲突检测的，所以表上必须存在主键或唯一非空索引。

(3) 在 MySQL 8.0.14 之前，只支持 IPv4，不支持 IPv6。

从 MySQL 8.0.14 开始，支持 IPv6，以及 IPv4 和 IPv6 的混合部署。

(4) 在 MySQL 8.0.21 之前，不支持带有校验和的二进制日志事件。

在 MySQL 8.0.21 之前，binlog_checksum 只能设置为 NONE。从 MySQL 8.0.21 开始，可将 binlog_checksum 设置为 CRC32。

(5) 组复制节点数的限制。

组复制可允许的最大节点数是 9。

(6) 不允许对 group_replication_applier 或 group_replication_recovery 通道设置过滤规则。

```
mysql> change replication filter replicate_ignore_table=(db1.t1) for channel 'group_replication_applier';
ERROR 3139 (HY000): CHANGE REPLICATION FILTER cannot be performed on channel 'group_replication_applier'.
```

(7) 在 MySQL 8.0.20 之前，不允许手动执行克隆操作。

从 MySQL 8.0.20 开始，允许手动执行克隆操作，但必须指定备份目录。

```
mysql> clone local data directory='/data/local_backup/full';
Query OK, 0 rows affected (1.74 sec)

# 不指定备份目录会报错
mysql> clone instance from 'root'@'192.168.244.10':3306 identified by '123456';
ERROR 3875 (HY000): The clone operation cannot be executed when Group Replication is running.

mysql> clone instance from 'root'@'192.168.244.10':3306 identified by '123456' data directory = '/data/remote_backup/full';
Query OK, 0 rows affected (2.61 sec)
```

上述要求和限制，适用于单主模式和多主模式。除此之外，多主模式还存在其他限制，下一节会具体介绍。

多主模式下的额外限制

在多主模式下，组复制还有以下额外限制。

(1) 验证阶段不会考虑间隙锁、表锁和 GET_LOCK 操作。

多主模式下，加锁操作不能在多个实例间共享。所以，对于多主模式，官方建议将事务隔离级别设置为 RC。

(2) 不能通过 SELECT ... FOR UPDATE、SELECT ... FOR SHARE 锁定数据。

看下面这个示例。

node1	node2
`mysql> begin;` `Query OK, 0 rows affected (0.00 sec)` `mysql> select * from slowtech.t1 where id=1 for update;` `+----+------+` `| id	c1 |` `+----+------+` `| 1	abc |` `+----+------+` `1 row in set (0.00 sec)`	
	`mysql> delete from slowtech.t1 where id=1;` `Query OK, 1 row affected (0.01 sec)`

`mysql> select * from slowtech.t1 where id=1 for update;`
`ERROR 1213 (40001): Deadlock found when trying to get lock;`
`try restarting transaction`	

(3) DDL 与 DML 的并发执行问题。

组复制中没有对 DDL 进行冲突检测。当针对同一个对象的 DDL 和 DML 操作在不同节点并发执行时，有可能会导致节点数据不一致。看下面这个示例。

node1	node2

`mysql> create database slowtech;`
`Query OK, 1 row affected (0.00 sec)`
 	
`mysql> create table slowtech.t1(id int primary key,c1 varchar(10));`
`Query OK, 0 rows affected (0.04 sec)`
 	
`mysql> begin;`
`Query OK, 0 rows affected (0.00 sec)`
 	
`mysql> insert into slowtech.t1 values(1,'a');`
`Query OK, 1 row affected (0.01 sec)`	

	`mysql> truncate table slowtech.t1;` `Query OK, 0 rows affected (0.03 sec)`
`mysql> commit;` `Query OK, 0 rows affected (0.00 sec)` `mysql> select * from slowtech.t1;` `Empty set (0.00 sec)`	`mysql> select * from slowtech.t1;` `+----+------+` `| id	c1 |` `+----+------+` `| 1	a |` `+----+------+` `1 row in set (0.01 sec)`

从上面的输出中可以看到，node1 和 node2 中 t1 表的数据不一致了。

如果分析 binlog，可以看到，node1 中的 INSERT 操作在前、TRUNCATE 操作在后，而 node2 中则恰恰相反。于是就导致了两个节点的数据不一致。

(4) 不允许将事务隔离级别设置为 SERIALIZABLE。

```
mysql> set session transaction isolation level serializable;
Query OK, 0 rows affected (0.00 sec)

mysql> insert into slowtech.t1 values(1,'a');
ERROR 3098 (HY000): The table does not comply with the requirements by an external plugin.
```

同时，错误日志中会提示以下信息。

```
[ERROR] [MY-011598] [Repl] Plugin group_replication reported: 'Transaction isolation level
(tx_isolation) is set to SERIALIZABLE, which is not compatible with Group Replication'
```

(5) 不允许外键的级联删除。

```
mysql> create table slowtech.t_parent (id int primary key);
Query OK, 0 rows affected (0.01 sec)

mysql> create table slowtech.t_child (
    -> id int primary key,
    -> parent_id int,
    -> foreign key (parent_id) references t_parent(id) on delete cascade
    -> );
Query OK, 0 rows affected (0.02 sec)

mysql> insert into slowtech.t_parent values(1);
Query OK, 1 row affected (0.01 sec)

mysql> insert into slowtech.t_child values(1,1);
ERROR 3098 (HY000): The table does not comply with the requirements by an external plugin.
```

同时，错误日志中会提示以下信息。

```
[ERROR] [MY-011543] [Repl] Plugin group_replication reported: 'Table t_child has a foreign key with
'CASCADE', 'SET NULL' or 'SET DEFAULT' clause. This is not compatible with Group Replication.'
```

上面两个限制实际上是由 group_replication_enforce_update_everywhere_checks 参数控制的。在多主模式下，一般会将其设置为 ON。

注意，以上限制只针对多主模式。为什么单主模式没有这些限制呢？因为在单主模式下，所有的写操作都是在一个实例中执行的，而实例本身的锁机制能规避这些问题。

11.5 组复制的常见管理操作

本节将介绍组复制中的一些常见管理操作。

11.5.1 强制组成员的重新配置

组复制是基于 Paxos 协议实现的，传输的任何数据都需要半数以上的节点确认接受才算成功。表 11-3 梳理了集群的节点数及可容忍的故障节点数。

表 11-3　集群节点数及可容忍的故障节点数

节点数	半数以上节点所对应的节点数	可容忍的故障节点数
1	1	0
2	2	0
3	2	1
4	3	1
5	3	2
6	4	2
7	4	3
8	5	3
9	5	4

所以，集群要想具备容灾能力，至少应部署 3 个节点。另外，如果是通过 STOP GROUP_REPLICATION 命令手动让节点离开集群的，那么这个节点并不能视为故障节点。这也就是为什么在一个三节点的集群中，当通过 STOP GROUP_REPLICATION 命令关闭其中两个节点的组复制时，剩下的一个节点仍能正常提供服务。

如果出现网络分区或节点异常宕机，导致组内大多数节点的状态是 UNREACHABLE，剩下的少数 ONLINE 节点是不能对外提供写服务的。具体来说，所有的写操作都会被阻塞。

```
session1> select member_id,member_host,member_port,member_state,member_role from
performance_schema.replication_group_members;
+--------------------------------------+----------------+-------------+--------------+-------------+
| member_id                            | member_host    | member_port | member_state | member_role |
+--------------------------------------+----------------+-------------+--------------+-------------+
| 207db264-0192-11ed-92c9-02001700754e | 192.168.244.10 |        3306 | UNREACHABLE  | PRIMARY     |
| 2cee229d-0192-11ed-8eff-02001700f110 | 192.168.244.20 |        3306 | UNREACHABLE  | PRIMARY     |
| 4cbfdc79-0192-11ed-8b01-02001701bd0a | 192.168.244.30 |        3306 | ONLINE       | PRIMARY     |
+--------------------------------------+----------------+-------------+--------------+-------------+
3 rows in set (0.00 sec)

session1> delete from sbtest.sbtest1 limit 1;
……阻塞中

session2> select * from performance_schema.processlist where info like 'delete%';
+----+------+-----------+------+---------+------+-----------------------------+-------------------------------------+
| ID | USER | HOST      | DB   | COMMAND | TIME | STATE                       | INFO                                |
+----+------+-----------+------+---------+------+-----------------------------+-------------------------------------+
| 84 | root | localhost | NULL | Query   |   55 | waiting for handler commit  | delete from sbtest.sbtest1 limit 1  |
+----+------+-----------+------+---------+------+-----------------------------+-------------------------------------+
1 row in set (0.00 sec)

session2> show variables like '%read_only%';
+-----------------------+-------+
| Variable_name         | Value |
+-----------------------+-------+
| innodb_read_only      | OFF   |
| read_only             | OFF   |
| super_read_only       | OFF   |
| transaction_read_only | OFF   |
+-----------------------+-------+
4 rows in set (0.00 sec)
```

当节点不能对外提供写服务时,错误日志中会提示以下信息。

```
[ERROR] [MY-011495] [Repl] Plugin group_replication reported: 'This server is not able to reach a
majority of members in the group. This server will now block all updates. The server will remain blocked
until contact with the majority is restored. It is possible to use group_replication_force_members to
force a new group membership.'
```

需要注意的是,这里的"不能对外提供写服务",并不是将 read_only 和 super_read_only 设置为 ON,而是指组复制的通信协议基于 Paxos 的一个多数派协议,因此无论是事务提交,还是节点的状态变化,都需要经过大多数节点的同意。

如果希望剩下的少数 ONLINE 节点能立即对外提供写服务,可通过 group_replication_force_members 强制重新配置组成员。例如:

```
session2> set global group_replication_force_members='192.168.244.30:33061';
Query OK, 0 rows affected (1 min 21.92 sec)

session2> select member_id,member_host,member_port,member_state,member_role from
performance_schema.replication_group_members;
+--------------------------------------+----------------+-------------+--------------+-------------+
| member_id                            | member_host    | member_port | member_state | member_role |
+--------------------------------------+----------------+-------------+--------------+-------------+
| 4cbfdc79-0192-11ed-8b01-02001701bd0a | 192.168.244.30 |        3306 | ONLINE       | PRIMARY     |
+--------------------------------------+----------------+-------------+--------------+-------------+
1 row in set (0.00 sec)
```

执行该操作时需谨慎。如果少数 ONLINE 节点不能对外提供写服务是由网络分区导致的,那么强制执行该操作会人为造成"脑裂"(split brain),即使节点之间的网络恢复,两个集群依旧相互独立。所以,如非必要,切忌执行该操作。如果一定要执行该操作,务必确保其他节点已关闭。

关于 group_replication_force_members,还有一点需要注意,即在执行 START GROUP_REPLICATION 操作时,它必须为空。

```
[ERROR] [MY-011634] [Repl] Plugin group_replication reported: 'group_replication_force_members must
be empty on group start. Current value: '192.168.244.30:33061''
```

11.5.2 如何设置 IP 白名单

IP 白名单是组复制从安全角度提供的一个特性。只有白名单中的实例,才可以加入组中。

IP 白名单由参数 group_replication_ip_whitelist(从 MySQL 8.0.22 开始,该参数被弃用,取而代之的是 group_replication_ip_allowlist)指定。不显式指定的话,则默认为 AUTOMATIC,它会将当前节点私有子网的 IP 自动加入白名单。这一点可以从组复制启动过程中的错误日志(需将 log_error_verbosity 设置为 3)看出。

```
[Note] [MY-011735] [Repl] Plugin group_replication reported: '[GCS] Added automatically IP ranges
127.0.0.1/8,192.168.244.10/24,::1/128,fe80::17ff:fe00:754e/64 to the allowlist'
```

192.168.244.10/24 的子网掩码是 24,这就意味着 192.168.244 这个网段的实例都可加入组中。另外,无论是否显式指定,本地地址(127.0.0.1)都会加入白名单。

在 MySQL 5.7.21 之前,IP 白名单只能是 IP,不能为主机名。从 MySQL 5.7.21 开始,支持主机名。

多个 IP 和主机名之间需用逗号隔开，例如：

```
group_replication_ip_allowlist="192.168.244.10/24,example.org,www.example.com/24"
```

从 MySQL 8.0.14 开始，支持 IPv6。

group_replication_ip_whitelist 可动态修改。在 MySQL 8.0.24 之前，需重启组复制才能生效。例如：

```
mysql> stop group_replication;
mysql> set global group_replication_ip_allowlist="192.168.244.10/24,example.org,www.example.com/24";
mysql> start group_replication;
```

从 MySQL 8.0.24 开始，无须重启组复制即可生效。建议所有节点的 group_replication_ip_whitelist 参数保持一致。

需要注意的是，IP 白名单只适用于 XCom 通信栈。如果我们使用的是 MySQL 通信栈（group_replication_communication_stack=MYSQL），则会通过账号自身的权限来进行访问控制，此时 IP 白名单是无效的。

11.5.3　如何查找单主模式下的 Primary 节点

在 MySQL 8.0 中，通过 performance_schema.replication_group_members 可直接看到当前集群的 Primary 节点。

```
mysql> select member_id,member_host,member_port,member_state,member_role from
performance_schema.replication_group_members;
+--------------------------------------+----------------+-------------+--------------+-------------+
| member_id                            | member_host    | member_port | member_state | member_role |
+--------------------------------------+----------------+-------------+--------------+-------------+
| 207db264-0192-11ed-92c9-02001700754e | 192.168.244.10 |        3306 | ONLINE       | PRIMARY     |
| 2cee229d-0192-11ed-8eff-02001700f110 | 192.168.244.20 |        3306 | ONLINE       | SECONDARY   |
| 4cbfdc79-0192-11ed-8b01-02001701bd0a | 192.168.244.30 |        3306 | ONLINE       | SECONDARY   |
+--------------------------------------+----------------+-------------+--------------+-------------+
3 rows in set (0.00 sec)
```

但在 MySQL 5.7 中，replication_group_members 表没有 MEMBER_ROLE 这一列。这个时候，只能查询 group_replication_primary_member 变量。在单主模式下，该变量会输出 Primary 节点的 member_id。如果是多主模式，这个变量的值会为空。

```
mysql> show status like 'group_replication_primary_member';
+----------------------------------+--------------------------------------+
| Variable_name                    | Value                                |
+----------------------------------+--------------------------------------+
| group_replication_primary_member | ade82c77-0a61-11ed-b3e4-02001700754e |
+----------------------------------+--------------------------------------+
1 row in set (0.00 sec)
```

11.5.4　新主选举算法

MGR 的新主选举算法，在节点版本一致的情况下，其实挺简单的。首先比较权重，权重越高，选为新主（新的 Primary 节点）的优先级越高。如果权重一致，则会进一步比较节点的 server_uuid。

server_uuid 越小，被选为新主的优先级越高。所以，在节点版本一致的情况下，会选择权重最高、server_uuid 最小的节点作为新主。

节点的权重由 group_replication_member_weight 决定，该参数是 MySQL 5.7.20 引入的，可设置为 0 和 100 之间的任意整数值，默认为 50。

但如果集群节点版本不一致，实际的新主选举算法就没这么简单了。我们结合源码具体分析一下。

新主选举算法主要涉及 3 个函数：

- pick_primary_member
- sort_and_get_lowest_version_member_position
- sort_members_for_election

这 3 个函数都是在 primary_election_invocation_handler.cc 中定义的。pick_primary_member 是主函数，会基于其他两个函数的结果选择 Primary 节点。

下面从 pick_primary_member 出发，看看这 3 个函数的具体实现逻辑。

```
bool Primary_election_handler::pick_primary_member(
    std::string &primary_uuid,
    std::vector<Group_member_info *> *all_members_info) {
  DBUG_TRACE;

  bool am_i_leaving = true;
#ifndef NDEBUG
  int n = 0;
#endif
  Group_member_info *the_primary = nullptr;

  std::vector<Group_member_info *>::iterator it;
  std::vector<Group_member_info *>::iterator lowest_version_end;

  // 基于 member_version 选择候选节点
  lowest_version_end =
      sort_and_get_lowest_version_member_position(all_members_info);

  // 基于节点权重和 server_uuid 对候选节点进行排序
  sort_members_for_election(all_members_info, lowest_version_end);

  // 循环遍历所有节点，判断 Primary 节点是否已定义
  for (it = all_members_info->begin(); it != all_members_info->end(); it++) {
#ifndef NDEBUG
    assert(n <= 1);
#endif

    Group_member_info *member = *it;
    // 如果当前节点是单主模式且遍历的节点中有 Primary 节点，则将该节点赋值给 the_primary
    if (local_member_info->in_primary_mode() && the_primary == nullptr &&
        member->get_role() == Group_member_info::MEMBER_ROLE_PRIMARY) {
      the_primary = member;
#ifndef NDEBUG
      n++;
#endif
    }
```

```
// 检查当前节点的状态是否为 OFFLINE
if (!member->get_uuid().compare(local_member_info->get_uuid())) {
  am_i_leaving =
      member->get_recovery_status() == Group_member_info::MEMBER_OFFLINE;
}
}

// 如果当前节点的状态不是 OFFLINE 且 the_primary 仍为空，则选择一个 Primary 节点
if (!am_i_leaving) {
  if (the_primary == nullptr) {
    // 因为循环的结束条件是 it != lowest_version_end 且 the_primary 为空，
    // 所以基本上会将候选节点中的第一个节点作为 Primary 节点
    for (it = all_members_info->begin();
         it != lowest_version_end && the_primary == nullptr; it++) {
      Group_member_info *member_info = *it;

      assert(member_info);
      if (member_info && member_info->get_recovery_status() ==
                             Group_member_info::MEMBER_ONLINE)
        the_primary = member_info;
    }
  }
}

if (the_primary == nullptr) return true;

primary_uuid.assign(the_primary->get_uuid());
return false;
}
```

这个函数中比较关键的地方有 3 个。

- 调用 sort_and_get_lowest_version_member_position。这个函数会基于 member_version（节点版本）选择候选节点。只有候选节点才有资格被选为 Primary 节点。
- 调用 sort_members_for_election。这个函数会基于节点权重和 server_uuid 对候选节点进行排序。
- 基于排序后的候选节点选择 Primary 节点。因为候选节点是从头开始遍历的，所以基本上，只要第一个节点是 ONLINE 状态，就会把这个节点作为 Primary 节点。

接下来看看 sort_and_get_lowest_version_member_position 函数的实现逻辑。

```
sort_and_get_lowest_version_member_position(
    std::vector<Group_member_info *> *all_members_info) {
std::vector<Group_member_info *>::iterator it;

// 按照版本对 all_members_info 从小到大排序
std::sort(all_members_info->begin(), all_members_info->end(),
          Group_member_info::comparator_group_member_version);

// std::vector::end 会返回一个迭代器，该迭代器引用 vector（向量容器）中的末尾元素。
// 注意，这个元素指向的是 vector 最后一个元素的下一个位置，不是最后一个元素
std::vector<Group_member_info *>::iterator lowest_version_end =
    all_members_info->end();

// 获取排序后的第一个节点，这个节点版本最低
it = all_members_info->begin();
```

```
    Group_member_info *first_member = *it;
    // 获取第一个节点的 major_version
    // 对于 MySQL 5.7, major_version 是 5; 对于 MySQL 8.0, major_version 是 8
    uint32 lowest_major_version =
        first_member->get_member_version().get_major_version();

    // 循环遍历剩下的节点，注意 it 是从 all_members_info->begin() + 1 开始的
    for (it = all_members_info->begin() + 1; it != all_members_info->end();
         it++) {
      // 如果第一个节点的版本号高于 MySQL 8.0.17，且节点的版本号不等于第一个节点的版本号，
      // 则将该节点赋值给 lowest_version_end，并退出循环
      if (first_member->get_member_version() >=
              PRIMARY_ELECTION_PATCH_CONSIDERATION &&
          (first_member->get_member_version() != (*it)->get_member_version())) {
        lowest_version_end = it;
        break;
      }
      // 如果节点的 major_version 不等于第一个节点的 major_version，
      // 则将该节点赋值给 lowest_version_end，并退出循环
      if (lowest_major_version !=
          (*it)->get_member_version().get_major_version()) {
        lowest_version_end = it;
        break;
      }
    }
    return lowest_version_end;
}
```

函数中的 PRIMARY_ELECTION_PATCH_CONSIDERATION 是 0x080017，即 MySQL 8.0.17。在 MySQL 8.0.17 中，组复制引入了兼容性策略。引入兼容性策略的初衷是为了避免集群中出现节点不兼容的情况。

该函数首先会对 all_members_info 按照版本从小到大排序。接着会基于第一个节点的版本（最小版本）确定 lowest_version_end。组复制用 lowest_version_end 标记最低版本的结束点。只有 lowest_version_end 之前的节点才是候选节点。

lowest_version_end 的取值逻辑如下。

- 如果最小版本高于等于 MySQL 8.0.17，则会将最小版本之后的第一个节点设置为 lowest_version_end。
- 如果集群中既有 MySQL 5.7，又有 MySQL 8.0，则会将 MySQL 8.0 的第一个节点设置为 lowest_version_end。
- 如果最小版本低于 MySQL 8.0.17，且只有一个大版本（major_version），则会取 all_members_info->end()。此时，所有节点都是候选节点。

最后，我们看看 sort_members_for_election 函数的实现逻辑。

```
void sort_members_for_election(
    std::vector<Group_member_info *> *all_members_info,
    std::vector<Group_member_info *>::iterator lowest_version_end) {
  Group_member_info *first_member = *(all_members_info->begin());
  // 获取第一个节点的版本，这个节点版本最低
  Member_version lowest_version = first_member->get_member_version();

  // 如果最小版本高于等于 MySQL 5.7.20，则根据节点的权重来排序。权重越高，在 vector 中的位置越靠前
  // 注意，这里只会对 [all_members_info->begin(), lowest_version_end) 这个区间内的元素进行排序，
```

```
    // 不包括 lowest_version_end
    if (lowest_version >= PRIMARY_ELECTION_MEMBER_WEIGHT_VERSION)
      std::sort(all_members_info->begin(), lowest_version_end,
                Group_member_info::comparator_group_member_weight);
    else
      // 如果最小版本低于 MySQL 5.7.20,则根据节点的 server_uuid 来排序。server_uuid 越小,
      // 在 vector 中的位置越靠前
      std::sort(all_members_info->begin(), lowest_version_end,
                Group_member_info::comparator_group_member_uuid);
}
```

函数中的 PRIMARY_ELECTION_MEMBER_WEIGHT_VERSION 是 0x050720,即 MySQL 5.7.20。如果最小节点的版本高于等于 MySQL 5.7.20,则会基于权重来排序。权重越高,在 all_members_info 中的位置越靠前。如果最小节点的版本低于 MySQL 5.7.20,则会基于节点的 server_uuid 来排序。server_uuid 越小,在 all_members_info 中的位置越靠前。

注意,std::sort 中的结束位置是 lowest_version_end,所以 lowest_version_end 这个节点不会参与排序。

在基于权重进行排序时,如果两个节点的权重一致,还会进一步比较这两个节点的 server_uuid。这个逻辑是在 comparator_group_member_weight 中定义的。如果权重一致,则节点的 server_uuid 越小,在 all_members_info 中的位置越靠前。

```
bool Group_member_info::comparator_group_member_weight(Group_member_info *m1,
                                                        Group_member_info *m2) {
  return m1->has_greater_weight(m2);
}

bool Group_member_info::has_greater_weight(Group_member_info *other) {
  MUTEX_LOCK(lock, &update_lock);
  if (member_weight > other->get_member_weight()) return true;
  // 如果权重一致,会按照节点的 server_uuid 来排序
  if (member_weight == other->get_member_weight())
    return has_lower_uuid_internal(other);

  return false;
}
```

基于上面代码的逻辑,接下来分析 sort_and_get_lowest_version_member_position 函数注释部分列举的 4 个案例。

案例 1:5.7.18、5.7.18、5.7.19、5.7.20、5.7.21、8.0.2

- 在这几个节点中,最小版本号是 5.7.18,小于 MySQL 8.0.17,所以会比较各个节点的 major_version。因为最后一个节点(8.0.2)的 major_version 和第一个节点不一致,所以会将 8.0.2 作为 lowest_version_end。此时,除了 8.0.2,其他节点都是候选节点。
- 最小版本号 5.7.18 低于 MySQL 5.7.20,所以 5.7.18、5.7.18、5.7.19、5.7.20、5.7.21 这几个节点会根据 server_uuid 进行排序。注意,lowest_version_end 的节点不会参与排序。
- 选择 server_uuid 最小的节点作为 Primary 节点。

案例 2：5.7.20、5.7.21、8.0.2、8.0.2

- 和案例 1 一样，会将 8.0.2 作为 lowest_version_end。此时，候选节点只有 5.7.20 和 5.7.21。
- 最小版本号 5.7.20 等于 MySQL 5.7.20，所以，5.7.20、5.7.21 这两个节点会根据节点的权重进行排序。如果权重一致，则会基于 server_uuid 进行进一步的排序。
- 选择权重最高、server_uuid 最小的节点作为 Primary 节点。

案例 3：8.0.17、8.0.18、8.0.19

- 最小版本号是 8.0.17，等于 MySQL 8.0.17，所以会判断其他节点的版本号是否与第一个节点相同。若不相同，则会将该节点的版本号赋值给 lowest_version_end。所以，会将 8.0.18 作为 lowest_version_end。此时，候选节点只有 8.0.17。
- 选择 8.0.17 这个节点作为 Primary 节点。

案例 4：8.0.13、8.0.17、8.0.18

- 最小版本号是 8.0.13，低于 MySQL 8.0.17，而且各个节点的 major_version 一致，所以最后返回的 lowest_version_end 实际上是 all_members_info->end()。此时，这 3 个节点都是候选节点。
- MySQL 8.0.13 高于 MySQL 5.7.20，所以这三个节点会根据权重进行排序。如果权重一致，则会基于 server_uuid 进行进一步的排序。
- 选择权重最高、server_uuid 最小的节点作为 Primary 节点。

结合代码和上面 4 个案例的分析，最后总结一下 MGR 的新主选举算法。

- 如果集群中存在 MySQL 5.7 的节点，则会将 MySQL 5.7 的节点作为候选节点。
- 如果集群节点的版本都是 MySQL 8.0，需要区分两种情况。
  - 如果最小版本低于 MySQL 8.0.17，则所有的节点都可作为候选节点。
  - 如果最小版本高于等于 MySQL 8.0.17，则只有最小版本的节点会作为候选节点。
- 在候选节点的基础上，会进一步根据候选节点的权重和 server_uuid 选择 Primary 节点。具体来说，
  - 如果候选节点中存在 MySQL 5.7.20 之前版本的节点，则会选择 server_uuid 最小的节点作为 Primary 节点；
  - 如果候选节点都大于等于 MySQL 5.7.20，则会选择权重最高、server_uuid 最小的节点作为 Primary 节点。

11.5.5 如何查看 Secondary 节点的延迟情况

如果我们使用 MySQL Shell 来管理组复制，cluster.status() 的输出中有个 replicationLag 字段，可以用它直接查看 Secondary 节点的延迟情况。例如：

```
cluster.status()
{
    "clusterName": "myCluster",
    "defaultReplicaSet": {
```

```
    "name": "default",
    "primary": "192.168.244.10:3306",
    "ssl": "DISABLED",
    "status": "OK",
    "statusText": "Cluster is ONLINE and can tolerate up to ONE failure.",
    "topology": {
        ...
        "192.168.244.20:3306": {
            "address": "192.168.244.20:3306",
            "memberRole": "SECONDARY",
            "mode": "R/O",
            "readReplicas": {},
            "replicationLag": "00:00:20.235056",
            "role": "HA",
            "status": "ONLINE",
            "version": "8.0.27"
        },
        ...
    },
    "topologyMode": "Single-Primary"
},
"groupInformationSourceMember": "192.168.244.10:3306"
}
```

如果没有使用 MySQL Shell，可通过以下 SQL 查看 Secondary 节点的延迟情况。该 SQL 同样适用于主从复制。

```
SELECT CASE
       WHEN min_commit_timestamp IS NULL THEN 0
       ELSE unix_timestamp(now(6)) - unix_timestamp(min_commit_timestamp)
       END AS seconds_behind_master
FROM (
    SELECT MIN(applying_transaction_original_commit_timestamp) AS min_commit_timestamp
    FROM performance_schema.replication_applier_status_by_worker
    WHERE applying_transaction <> ''
) t;
```

11.5.6 大事务

在 MySQL 中，大事务的危害性显而易见。

- 会造成主从延迟，而主从延迟又会影响数据库的高可用切换。
- 锁定的记录多，相对而言更容易导致锁等待。
- 回滚表空间不断膨胀。在 MySQL 8.0 之前，回滚表空间默认会放到系统表空间中，而系统表空间一旦膨胀，就不会收缩。

除此之外，在组复制中，大事务还会导致以下问题。

- 因为内存分配问题，大事务可能导致系统变慢。
- 大事务在通过网络传播的过程中，如果在 5 秒的窗口期内还未拷贝完，则节点会被其他节点标记为可疑节点，驱逐出集群。

所以，对于线上的大规模操作，建议分而治之，每次只操作一部分数据，分多次执行。

在传统的主从复制中,我们无法限制单个事务的大小,但在组复制中,可以通过 group_replication_transaction_size_limit 限制单个事务的大小。该参数默认为 150 000 000 字节,约 143MB。

看下面这个示例,单个事务因为过大而被回滚。

```
mysql> select count(*) from sbtest.sbtest1;
+----------+
| count(*) |
+----------+
|   800000 |
+----------+
1 row in set (0.04 sec)

mysql> delete from sbtest.sbtest1;
ERROR 3100 (HY000): Error on observer while running replication hook 'before_commit'.
```

同时,错误日志中会提示以下信息。

```
[ERROR] [MY-011608] [Repl] Plugin group_replication reported: 'Error on session 105. Transaction of size 163777114 exceeds specified limit 150000000. To increase the limit please adjust group_replication_transaction_size_limit option.'
[ERROR] [MY-010207] [Repl] Run function 'before_commit' in plugin 'group_replication' failed
```

在 MySQL 8.0.16 中,组复制引入了消息分片特性。基于这个特性,任何大于 group_replication_communication_max_message_size(默认为 10 485 760 字节,即 10MB)的消息将被自动拆分为多个分片进行传输,这样可以在很大程度上规避(注意,不能完全规避)大事务导致节点被驱逐出集群的问题。除此之外,group_replication_communication_max_message_size 还限制了单个分片的最大大小。

11.5.7 查看组复制的内存使用

在 MySQL 8.0.30 之前,我们只能看到 XCom Cache 的内存使用情况。

```
mysql> SELECT NAME,ENABLED FROM performance_schema.setup_instruments
    ->        WHERE NAME LIKE 'memory/group_rpl/%';
+----------------------------------------+---------+
| NAME                                   | ENABLED |
+----------------------------------------+---------+
| memory/group_rpl/GCS_XCom::xcom_cache  | YES     |
+----------------------------------------+---------+
1 row in set (0.00 sec)
```

从 MySQL 8.0.30 开始,支持查看以下事件采集配置项(instrument)的内存使用情况。

```
mysql> SELECT NAME,ENABLED FROM performance_schema.setup_instruments
    ->        WHERE NAME LIKE 'memory/group_rpl/%';
+-------------------------------------------------+---------+
| NAME                                            | ENABLED |
+-------------------------------------------------+---------+
| memory/group_rpl/write_set_encoded              | YES     |
| memory/group_rpl/certification_data             | YES     |
| memory/group_rpl/certification_data_gc          | YES     |
| memory/group_rpl/certification_info             | YES     |
```

```
| memory/group_rpl/transaction_data                                  | YES |
| memory/group_rpl/sql_service_command_data                          | YES |
| memory/group_rpl/mysql_thread_queued_task                          | YES |
| memory/group_rpl/message_service_queue                             | YES |
| memory/group_rpl/message_service_received_message                  | YES |
| memory/group_rpl/group_member_info                                 | YES |
| memory/group_rpl/consistent_members_that_must_prepare_transaction  | YES |
| memory/group_rpl/consistent_transactions                           | YES |
| memory/group_rpl/consistent_transactions_prepared                  | YES |
| memory/group_rpl/consistent_transactions_waiting                   | YES |
| memory/group_rpl/consistent_transactions_delayed_view_change       | YES |
| memory/group_rpl/GCS_XCom::xcom_cache                              | YES |
| memory/group_rpl/Gcs_message_data::m_buffer                        | YES |
+--------------------------------------------------------------------+-----+
17 rows in set (0.00 sec)
```

以上基本涵盖了组复制中会消耗内存的一些常见场景。至于各个事件采集配置项的具体含义，可查阅官方文档。

官方文档中还贴心地提供了一些示例查询，方便我们分析组复制的内存使用情况。

例如，利用下面这个查询可以查看组复制使用的总内存。

```
SELECT * FROM (
              SELECT
                (CASE
                    WHEN EVENT_NAME LIKE 'memory/group_rpl/%'
                    THEN 'memory/group_rpl/memory_gr'
                    ELSE 'memory_gr_rest'
                END) AS EVENT_NAME, SUM(COUNT_ALLOC), SUM(COUNT_FREE),
                SUM(SUM_NUMBER_OF_BYTES_ALLOC),
                SUM(SUM_NUMBER_OF_BYTES_FREE), SUM(LOW_COUNT_USED),
                SUM(CURRENT_COUNT_USED), SUM(HIGH_COUNT_USED),
                SUM(LOW_NUMBER_OF_BYTES_USED), SUM(CURRENT_NUMBER_OF_BYTES_USED),
                SUM(HIGH_NUMBER_OF_BYTES_USED)
              FROM performance_schema.memory_summary_global_by_event_name
              GROUP BY (CASE
                        WHEN EVENT_NAME LIKE 'memory/group_rpl/%'
                        THEN 'memory/group_rpl/memory_gr'
                        ELSE 'memory_gr_rest'
                      END)
     ) f
     WHERE f.EVENT_NAME != 'memory_gr_rest'\G
*************************** 1. row ***************************
                     EVENT_NAME: memory/group_rpl/memory_gr
               SUM(COUNT_ALLOC): 6568
                SUM(COUNT_FREE): 4422
 SUM(SUM_NUMBER_OF_BYTES_ALLOC): 2199060
  SUM(SUM_NUMBER_OF_BYTES_FREE): 905599
            SUM(LOW_COUNT_USED): 0
        SUM(CURRENT_COUNT_USED): 2146
           SUM(HIGH_COUNT_USED): 2164
    SUM(LOW_NUMBER_OF_BYTES_USED): 0
SUM(CURRENT_NUMBER_OF_BYTES_USED): 1293461
   SUM(HIGH_NUMBER_OF_BYTES_USED): 1297584
1 row in set (0.02 sec)
```

11.6 组复制的实现原理

组复制是基于数据库状态机（database state machine）实现的，接下来我们从数据库状态机的基本概念及事务在组复制中的处理流程这两个方面来看看组复制的实现原理。

11.6.1 数据库状态机

数据库状态机是 Fernando Pedone、Rachid Guerraoui 和 Andre Schiper 在 1999 年发表的论文 "The Database State Machine Approach" 中提出的。论文中描述了状态机方法及相关的两个主要概念，简要介绍如下。

- **状态机方法**

状态机方法，也称为主动复制，是一种非集中式复制协调技术。它的关键概念是所有节点都会接收和处理相同的请求序列。不仅如此，节点还应具有确定性行为，这就意味着当提供相同的输入（例如，外部请求）时，每个节点将产生相同的输出（例如，状态更改）。

- **原子广播**

原子广播可以将消息发送到多个节点，并保证所有节点都同意交付的消息集和消息交付的顺序。原子广播需确保以下属性。

- 共识（agreement）：如果一个节点传递消息 m，那么每个节点都会传递消息 m。
- 有序性（order）：没有两个节点会以不同的顺序传递任何两条消息。
- 终止（termination）：如果一个节点广播消息 m 并且没有失败，那么每个节点最终都会发送消息 m。

- **延迟更新复制技术**

在延迟更新复制技术中，事务在一个数据库节点本地执行，在执行期间，该节点不会和其他数据库节点之间发生交互。当客户端请求事务提交时，事务的更新（例如，重做日志记录）和一些控制结构会传播到所有数据库节点。在这些节点中，事务将被认证并在可能的情况下提交。认证是为了确保单个副本的可串行化。如果事务的提交会导致数据库进入不一致状态（即不可序列化），则节点会中止事务。

11.6.2 事务在组复制中的处理流程

结合数据库状态机的实现原理，我们看看事务在组复制中的处理流程，如图 11-4 所示。

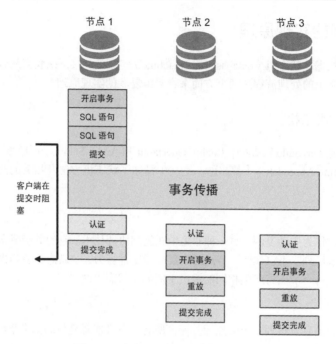

图 11-4 事务在组复制中的处理流程

在介绍具体流程之前，首先回顾一下 MySQL 事务的两阶段提交协议。

☐ 第一阶段

在存储引擎内部进行事务的 PREPARE 操作（InnoDB Prepare），写 redo log 并刷盘（write/fsync Redo log），此时 binlog 不执行任何操作。

☐ 第二阶段

- 写 binlog 并刷盘（write/fsync Binlog）。
- 在存储引擎内部进行事务的 COMMIT 操作（InnoDB Commit）。

接下来分析事务在组复制中的处理流程。

☐ 事务在 node1 上发起，COMMIT 之前的所有操作都不会和其他节点交互。所以在多主模式下，组复制使用的是乐观事务模型。
☐ 客户端发起 COMMIT 操作。事务开始执行提交操作。
☐ 事务在生成 binlog cache 之后，写入 binlog 之前，开始进入组复制的处理流程。
☐ 组复制会打包事务的相关信息，并通过 GCS（Group Communication System，组通信系统）发送。GCS 是组复制的消息通信层，具体实现是 XCom。XCom 是基于 Paxos 实现的组通信引擎。
☐ 当消息被大多数节点（包括自身节点）确认接受后，开始进行冲突检测。因为每个节点都维护了一个冲突检测数据库，所以各个节点的冲突检测都是独立进行的。
☐ 如果没有通过冲突检测（冲突检测失败），则回滚事务。

- 如果通过了冲突检测，则提交事务。这个时候，会区分本地事务和远程事务。
- 对于本地事务，会直接写 binlog 提交，然后反馈给客户端。
- 对于远程事务，则会将 binlog event 写入 relay log，然后由 group_replication_applier 通道重放 relay log 中的事务。

由此可见，事务在传播阶段（基于 Paxos 协议）是强一致的，但在应用阶段（重放 relay log 中的事务）又是异步的，这也就是为什么说 MGR 是一个最终一致性的系统。

图 11-5 是组复制的整体架构图。可以看到，组复制只是 MySQL 中的一个插件。

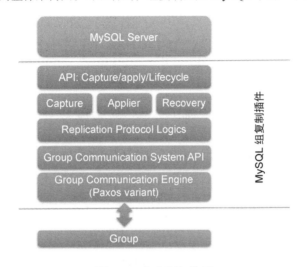

图 11-5　组复制架构图

从图 11-5 中可以看到组复制由多个组件构成，各个组件的主要功能如下。

- Capture：负责跟踪当前事务的上下文。
- Applier：负责执行远程事务。对应的复制通道是 group_replication_applier。
- Recovey：负责分布式恢复。当有新的节点加入时，会首先通过分布式恢复补齐差异数据。对应的复制通道是 group_replication_recovery。
- Replication Protocol Logics：复制协议模块。这个模块会处理冲突检测，接受事务并将事务传播到组中。
- Group Communication System API：组通信系统（GCS）API。这个 API 抽象了构建复制状态机所必需的属性。
- Group Communication Engine：基于 Paxos 实现的组通信引擎（Xcom）。

11.6.3　参考资料

- 论文 "The Database State Machine Approach"，作者：Fernando Pedone、Rachid Guerraoui 和 André Schiper。

第 11 章　组复制

❑ 博客文章 "'The king is dead, long live the king'：Our Paxos-based consensus"，作者：Alfranio Correia。

11.7　组复制的实现细节

组复制实现了几个钩子函数，这些函数的调用时机从函数的名称中就可以看出来。

```
Trans_observer trans_observer = {
    sizeof(Trans_observer),

    group_replication_trans_before_dml,
    group_replication_trans_before_commit,
    group_replication_trans_before_rollback,
    group_replication_trans_after_commit,
    group_replication_trans_after_rollback,
    group_replication_trans_begin,
};
```

在这些钩子函数里，用得最多的是 `group_replication_trans_before_commit`。只要事务进行了提交，就会调用这个函数。

图 11-6 展示了这个函数调用的具体时间点。

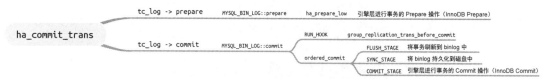

图 11-6　group_replication_trans_before_commit 的调用时间点

`group_replication_trans_before_commit` 是在 RUN_HOOK 中调用的。

```
if (stmt_stuff_logged || trx_stuff_logged) {
    CONDITIONAL_SYNC_POINT_FOR_TIMESTAMP("before_invoke_before_commit_hook");
    if (RUN_HOOK(
            transaction, before_commit,
            (thd, all, thd_get_cache_mngr(thd)->get_trx_cache(),
             thd_get_cache_mngr(thd)->get_stmt_cache(),
             max<my_off_t>(max_binlog_cache_size, max_binlog_stmt_cache_size),
             is_atomic_ddl)) ||
        DBUG_EVALUATE_IF("simulate_failure_in_before_commit_hook", true,
                         false)) {
        trx_coordinator::rollback_in_engines(thd, all);
        gtid_state->update_on_rollback(thd);
        thd_get_cache_mngr(thd)->reset();
        if (thd->get_stmt_da()->is_ok())
            thd->get_stmt_da()->reset_diagnostics_area();
        my_error(ER_RUN_HOOK_ERROR, MYF(0), "before_commit");
        return RESULT_ABORTED;
    }
```

当 RUN_HOOK 被调用的时候，事务所有的事件都写入了 binlog cache，只待调用 ordered_commit 进行组提交。

接下来，我们重点分析 group_replication_trans_before_commit 函数的实现逻辑。

```
int group_replication_trans_before_commit(Trans_param *param) {
  DBUG_TRACE;
  int error = 0;
  const int pre_wait_error = 1;
  const int post_wait_error = 2;
  int64 sequence_number = 1;

  ...
  // 判断复制通道的类型，如果是 Applier 线程，则意味着当前正在执行的事务是个远程事务，不是本地事务
  Replication_thread_api channel_interface;
  if (GR_APPLIER_CHANNEL == param->rpl_channel_type) {
    if (nullptr == local_member_info) {
      return 0;
    }

    // 如果插件没有停止，则会更新统计信息，这些统计信息会用在流控中
    bool fail_to_lock = shared_plugin_stop_lock->try_grab_read_lock();
    if (!fail_to_lock) {
      const Group_member_info::Group_member_status member_status =
          local_member_info->get_recovery_status();
      if (Group_member_info::MEMBER_ONLINE == member_status ||
          Group_member_info::MEMBER_IN_RECOVERY == member_status) {
        applier_module->get_pipeline_stats_member_collector()
            ->decrement_transactions_waiting_apply();
        applier_module->get_pipeline_stats_member_collector()
            ->increment_transactions_applied();
      }
      if (Group_member_info::MEMBER_IN_RECOVERY == member_status) {
        applier_module->get_pipeline_stats_member_collector()
            ->increment_transactions_applied_during_recovery();
      }
      shared_plugin_stop_lock->release_read_lock();

      // 一般情况下，after_applier_prepare 能很快返回。
      // 但如果事务的 consistency_level 包含 AFTER，则会在此处等待其他节点的响应并向其他节点发送
      // 一个 ACK 包
      if ((Group_member_info::MEMBER_ONLINE == member_status ||
          Group_member_info::MEMBER_IN_RECOVERY == member_status) &&
          transaction_consistency_manager->after_applier_prepare(
              param->gtid_info.sidno, param->gtid_info.gno, param->thread_id,
              member_status)) {
        return 1; /* purecov: inspected */
      }
    }
    // 直接返回
    // 可以看到，对于远程事务，虽然也会调用 group_replication_trans_before_commit，但主要是更新统计信息
    return 0;
  }
  // 下面是本地事务的处理逻辑

  ...
  assert(applier_module != nullptr && recovery_module != nullptr);
  // 获取事务的最大大小，这个大小由 group_replication_transaction_size_limit 参数设置
  const ulong transaction_size_limit = get_transaction_size_limit();

  ...
  // 判断事务是否已经指定 GTID
  // 什么情况下会指定 GTID 呢？比较常见的是当前实例是个从库
  const bool is_gtid_specified = param->gtid_info.type == ASSIGNED_GTID;
```

```cpp
Gtid gtid = {param->gtid_info.sidno, param->gtid_info.gno};
if (!is_gtid_specified) {
  // 设置一个初始值, 冲突检测完成后更新
  gtid.sidno = 1;
  gtid.gno = 1;
}
...
// 获取事务 group_replication_consistency 的会话值
const enum_group_replication_consistency_level consistency_level =
    static_cast<enum_group_replication_consistency_level>(
        param->group_replication_consistency);
...
// 更新统计信息, 将本地执行的事务数加 1
applier_module->get_pipeline_stats_member_collector()
    ->increment_transactions_local();

// 创建一个 Transaction_context_log_event, 里面包括 server_uuid、是否是 dml、线程 id 和是否已经指定 GTID
tcle = new Transaction_context_log_event(param->server_uuid,
                                         is_dml || param->is_atomic_ddl,
                                         param->thread_id, is_gtid_specified);
...
if (is_dml) {
  // 如果是 DML, 则获取事务的 write_set
  Transaction_write_set *write_set =
      get_transaction_write_set(param->thread_id);
  ...
  // 将事务的 write_set 添加到 Transaction_context_log_event 中
  if (write_set != nullptr) {
    if (add_write_set(tcle, write_set)) {
      cleanup_transaction_write_set(write_set);
      LogPluginErr(ERROR_LEVEL, ER_GRP_RPL_FAILED_TO_GATHER_TRANS_WRITE_SET,
                   param->thread_id);
      error = pre_wait_error;
      goto err;
    }
    cleanup_transaction_write_set(write_set);
    assert(is_gtid_specified || (tcle->get_write_set()->size() > 0));
  }
  ...
}
...

// 创建一个 Gtid_log_event
gle = new Gtid_log_event(
    param->server_id, is_dml || param->is_atomic_ddl, 0, sequence_number,
    may_have_sbr_stmts, *(param->original_commit_timestamp),
    immediate_commit_timestamp, gtid_specification,
    *(param->original_server_version), *(param->immediate_server_version));
...

// 获取事务的大小
transaction_size =
    cache_log_position + tcle->get_event_length() + gle->get_event_length();
// 如果事务的大小超过 group_replication_transaction_size_limit, 则回滚事务
if (is_dml && transaction_size_limit &&
    transaction_size > transaction_size_limit) {
  LogPluginErr(ERROR_LEVEL, ER_GRP_RPL_TRANS_SIZE_EXCEEDS_LIMIT,
               param->thread_id, transaction_size, transaction_size_limit);
```

```cpp
    error = pre_wait_error;
    goto err;
}

// 将 Transaction_context_log_event, Gtid_log_event 和 binlog cache 封装到一个事务消息中
try {
    // 如果事务的 consistency_level 不包含 AFTER，会创建一个 Transaction_message，否则会创建
    // 一个 Transaction_with_guarantee_message
    // Transaction_with_guarantee_message 带有 consistency_level 信息
    if (consistency_level < GROUP_REPLICATION_CONSISTENCY_AFTER) {
        transaction_msg = new Transaction_message(transaction_size);
    } else {
        transaction_msg = new Transaction_with_guarantee_message(
            transaction_size, consistency_level);
    }

    // 将 Transaction_context_log_event 和 Gtid_log_event 中的内容写到 transaction_msg 中
    if (binary_event_serialize(tcle, transaction_msg) ||
        binary_event_serialize(gle, transaction_msg)) {
        LogPluginErr(ERROR_LEVEL, ER_GRP_RPL_WRITE_TO_TRANSACTION_MESSAGE_FAILED,
                     param->thread_id);
        error = pre_wait_error;
        goto err;
    }
    delete tcle;
    tcle = nullptr;
    delete gle;
    gle = nullptr;
    // 将 binlog cache 中的内容写到 transaction_msg 中
    if (cache_log->copy_to(transaction_msg)) {
        LogPluginErr(ERROR_LEVEL, ER_GRP_RPL_WRITE_TO_TRANSACTION_MESSAGE_FAILED,
                     param->thread_id);
        error = pre_wait_error;
        goto err;
    }
} catch (const std::bad_alloc &) {
    LogPluginErr(ERROR_LEVEL, ER_OUT_OF_RESOURCES);
    error = pre_wait_error;
    goto err;
}
// 注册 Ticket，后面会通过 waitTicket 暂停事务
if (transactions_latch->registerTicket(param->thread_id)) {
    LogPluginErr(ERROR_LEVEL,
                 ER_GRP_RPL_FAILED_TO_REGISTER_TRANS_OUTCOME_NOTIFICTION,
                 param->thread_id);
    error = pre_wait_error;
    goto err;
}

// 判断是否需要进行流控
applier_module->get_flow_control_module()->do_wait();

// 通过 GCS 模块广播事务消息
send_error = gcs_module->send_transaction_message(*transaction_msg);

// 删除 transaction_msg，释放内存
delete transaction_msg;
transaction_msg = nullptr;

if (send_error == GCS_MESSAGE_TOO_BIG) {
```

```
          LogPluginErr(ERROR_LEVEL, ER_GRP_RPL_MSG_TOO_LONG_BROADCASTING_TRANS_FAILED,
                      param->thread_id);
          error = pre_wait_error;
          goto err;
      } else if (send_error == GCS_NOK) {
          LogPluginErr(ERROR_LEVEL, ER_GRP_RPL_BROADCASTING_TRANS_TO_GRP_FAILED,
                      param->thread_id);
          error = pre_wait_error;
          goto err;
      }

      shared_plugin_stop_lock->release_read_lock();

      // 暂停事务
      if (transactions_latch->waitTicket(param->thread_id)) {
          LogPluginErr(ERROR_LEVEL,
                      ER_GRP_RPL_ERROR_WHILE_WAITING_FOR_CONFLICT_DETECTION,
                      param->thread_id);
          error = post_wait_error;
          goto err;
      }

err:
      delete gle;
      delete tcle;
      delete transaction_msg;
      ...
      return error;
  }
```

组复制在收到广播的消息后,会针对不同类型的消息,调用不同的函数来处理。图 11-7 梳理了 GCS 的消息类型及对应的处理函数。

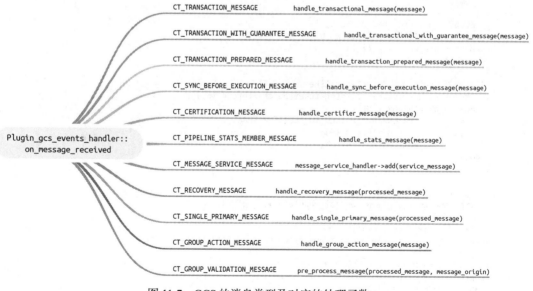

图 11-7 GCS 的消息类型及对应的处理函数

GCS 的消息类型解释如下。

- CT_TRANSACTION_MESSAGE：普通的事务消息。
- CT_TRANSACTION_WITH_GUARANTEE_MESSAGE：consistency_level 中包括 AFTER 的事务消息。
- CT_TRANSACTION_PREPARED_MESSAGE：consistency_level 中包含 AFTER 的事务在执行到 group_replication_trans_before_commit 阶段时，会发送一个 TRANSACTION_PREPARED_MESSAGE。
- CT_SYNC_BEFORE_EXECUTION_MESSAGE：consistency_level 中包含 BEFORE 的事务在执行到 group_replication_trans_before_commit 阶段时，会发送一个 SYNC_BEFORE_EXECUTION_MESSAGE。

这里重点说说普通事务消息的处理函数 handle_transactional_message。它的主要作用是将事务消息推送到 Applier 模块的 incoming 队列中。

```
int handle(const uchar *data, ulong len,
           enum_group_replication_consistency_level consistency_level,
           std::list<Gcs_member_identifier> *online_members) override {
  this->incoming->push(
      new Data_packet(data, len, consistency_level, online_members));
  return 0;
}
```

消息被推送到队列后，Applier 线程就会消费它。同样，针对不同的包（packet），会调用不同的函数来处理。图 11-8 梳理了 Applier 模块的包的类型及对应的处理函数。

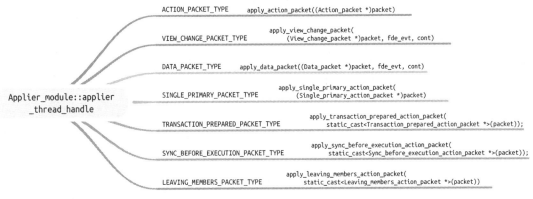

图 11-8　Applier 模块的包的类型及对应的处理函数

这里重点说说 DATA_PACKET_TYPE（数据包）的处理逻辑。图 11-9 是数据包的处理流程。

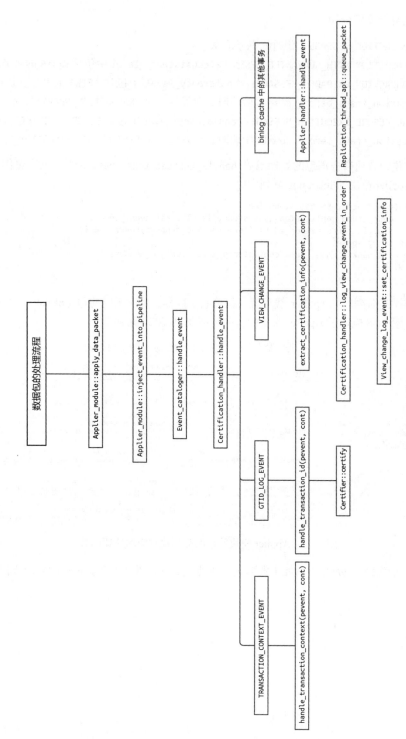

图 11-9 数据包的处理流程

之前，我们分析过一个普通的事务消息会封装 3 部分内容：Transaction_context_log_event、Gtid_log_event 和 binlog cache。这其中，binlog cache 又会包含 QUERY_EVENT、TABLE_MAP_EVENT、xxx_ROWS_EVENT 和 XID_EVENT。对于不同的 EVENT，实际上也是调用不同的函数来处理的。

下面看看数据包处理流程中一些关键函数的具体作用。

- **Applier_module::inject_event_into_pipeline**

将事件注入流水线（pipeline）中处理。流水线是计算机中普遍使用的一种技术，核心思想是将一个任务拆分为多个处理阶段，每个阶段由一个专门的功能模块来实现，并且会把上一个阶段任务的处理结果交给下一个阶段来处理。当任务足够多的时候，可以保证多个阶段之间的处理是并行的，这样就能充分利用 CPU 资源，提升系统吞吐量。

- **Event_cataloger::handle_event**

基于 transaction_discarded 判断 EVENT 是否需要继续执行。

transaction_discarded 是一个布尔值，用来标记事务是否已被丢弃，是在冲突检测后设置的。transaction_discarded 会在以下两种场景中设置为 true：本地事务；事务冲突检测失败。对于新事务，Event_cataloger::handle_event 又会将 transaction_discarded 重置为 false。

```
int Event_cataloger::handle_event(Pipeline_event *pevent, Continuation *cont) {
  Log_event_type event_type = pevent->get_event_type();
  // 判断 EVENT 的类型是否是 TRANSACTION_CONTEXT_EVENT。如果是，则代表它是一个新事务。
  // 这个时候，会将 EVENT 标记为 TRANSACTION_BEGIN
  if (event_type == binary_log::TRANSACTION_CONTEXT_EVENT) {
    pevent->mark_event(TRANSACTION_BEGIN);
  } else if (pevent->get_event_context() != SINGLE_VIEW_EVENT) {
    pevent->mark_event(UNMARKED_EVENT);
  }

  if (cont->is_transaction_discarded()) {
    // 对于新事务，会将 transaction_discarded 重置为 false
    if ((pevent->get_event_context() == TRANSACTION_BEGIN) ||
        (pevent->get_event_context() == SINGLE_VIEW_EVENT)) {
      cont->set_transation_discarded(false);
    } else {
      // 函数直接返回 0，不会继续处理
      cont->signal(0, true);
      return 0;
    }
  }
  // 将 EVENT 传递给流水线的下一个阶段继续处理
  next(pevent, cont);
  return 0;
}
```

- **Certification_handler::handle_event**

基于 EVENT 的类型，调用不同的函数进行处理。

```
int Certification_handler::handle_event(Pipeline_event *pevent,
                                        Continuation *cont) {
  DBUG_TRACE;
```

```
    Log_event_type ev_type = pevent->get_event_type();
    switch (ev_type) {
      case binary_log::TRANSACTION_CONTEXT_EVENT:
        return handle_transaction_context(pevent, cont);
      // 对于 GTID_LOG_EVENT,会调用 handle_transaction_id 进行冲突检测
      case binary_log::GTID_LOG_EVENT:
        return handle_transaction_id(pevent, cont);
      // 对于 VIEW_CHANGE_EVENT,会调用 extract_certification_info 构建冲突检测数据库
      case binary_log::VIEW_CHANGE_EVENT:
        return extract_certification_info(pevent, cont);
      default:
        // 对于远程事务 binlog cache 中的 event,会通过 Applier_handler::handle_event 写入 relay log
        next(pevent, cont);
        return 0;
    }
  }
```

- **Certification_handler::handle_transaction_id**

这个函数非常关键,会进行冲突检测并定义冲突检测后的处理逻辑。

```
seq_number =
    cert_module->certify(tcle->get_snapshot_version(), tcle->get_write_set(),
                         !tcle->is_gtid_specified(), tcle->get_server_uuid(),
                         gle, local_transaction);
```

冲突检测完成后,对于本地事务,会释放 group_replication_trans_before_commit 结尾处的 transactions_latch->waitTicket,让事务继续处理。

```
// seq_number <=0 代表冲突检测失败
if ((seq_number <= 0 || pevent->get_consistency_level() <
                       GROUP_REPLICATION_CONSISTENCY_AFTER) &&
    transactions_latch->releaseTicket(tcle->get_thread_id())) {
  LogPluginErr(ERROR_LEVEL, ER_GRP_RPL_NOTIFY_CERTIFICATION_OUTCOME_FAILED);
  cont->signal(1, true);
  error = 1;
  goto end;
}

cont->signal(0, true);
```

同时,通过 cont->signal(0, true) 将事务的 transaction_discarded 设置为 true。这样, Event_cataloger::handle_event 在处理的时候,对于 binlog cache 中的其他事件,会直接返回 0,不再继续处理。

对于远程事务,如果冲突检测通过,则会生成一个新的 Gtid_log_event 并写入 relay log。为什么要生成一个新的 Gtid_log_event 呢?

```
Gtid_log_event *gle_generated = new Gtid_log_event(
    gle->server_id, gle->is_using_trans_cache(), gle->last_committed,
    gle->sequence_number, gle->may_have_sbr_stmts,
    gle->original_commit_timestamp, gle->immediate_commit_timestamp,
    gtid_specification, gle->original_server_version,
    gle->immediate_server_version);
```

因为这个 EVENT 中的 last_committed、sequence_number、immediate_commit_timestamp、immediate_server_version 都发生了变化。

11.8 组复制的分布式恢复

分布式恢复（distributed recovery）是组复制中的一个核心功能模块。前面提到过，组复制是基于数据库状态机来实现的，而数据库状态机的一个核心要求是当提供相同的输入时，每个副本会产生相同的输出。要产生相同的输出，其实有个隐含的前提，即我们操作的对象（数据）必须是一致的。怎么保证数据一致呢？这就要求操作前各个节点的初始数据一致。对于新集群，因为本来就没有数据，所以我们无须担心数据一致性的问题。但对于已经上线且运行很久的集群，如果我们要添加一个新的节点，怎么保证这个节点的数据和集群数据一致呢？在组复制中，这实际上是通过分布式恢复来实现的。

在介绍分布式恢复的具体实现原理之前，首先看看分布式恢复中的一个关键概念——视图。视图指的是组在一段时间内的成员状态。当组的成员状态发生变化时（例如，有节点加入或离开），视图也会随之变化。

视图用 VIEW_ID 来唯一标识。VIEW_ID 可通过以下 SQL 语句查看。

```
mysql> select distinct view_id from performance_schema.replication_group_member_stats;
+---------------------+
| view_id             |
+---------------------+
| 16599266502075199:3 |
+---------------------+
1 row in set (0.00 sec)
```

以冒号为分隔符，VIEW_ID 由两部分组成。

- 第一部分是一个随机数，在组初始化（group_replication_bootstrap_group=ON）时产生。只要集群没有重新初始化（集群完全重启时需要重新初始化），这个值就不会发生变化。
- 第二部分是一个整数，初始值为 1。当组的成员状态发生变化时，该整数会递增 1。

11.8.1 分布式恢复的实现原理

当一个新的节点（Joiner）加入时，它会首先与 group_replication_group_seeds 中的第一个种子节点建立连接。如果连接被拒绝（如 Joiner 不在白名单中），则 Joiner 会依次与 group_replication_group_seeds 中的其他节点建立连接。连接成功建立后，开始恢复数据。恢复的主要流程是在 handle_joining_members 中定义的。

下面分析一下 handle_joining_members 函数的处理逻辑。

```
// mysql-8.0.27/plugin/group_replication/src/gcs_event_handlers.cc
void Plugin_gcs_events_handler::handle_joining_members(const Gcs_view &new_view,
                                                       bool is_joining,
                                                       bool is_leaving) const {
  size_t number_of_members = new_view.get_members().size();
  if (number_of_members == 0 || is_leaving) {
    return;
  }
  size_t number_of_joining_members = new_view.get_joined_members().size();
  size_t number_of_leaving_members = new_view.get_leaving_members().size();
```

```cpp
if (is_joining) {
  int error = 0;
  // 兼容性检查
  if ((error = check_group_compatibility(number_of_members))) {
    gcs_module->notify_of_view_change_cancellation(error);
    return;
  }
  gcs_module->notify_of_view_change_end();

  // 将节点的状态置为 RECOVERING
  update_member_status(
      new_view.get_joined_members(), Group_member_info::MEMBER_IN_RECOVERY,
      Group_member_info::MEMBER_OFFLINE, Group_member_info::MEMBER_END);

  // 判断节点加入期间是否正在选举
  primary_election_handler->set_election_running(
      is_group_running_a_primary_election());

  // 将节点设置为只读
  if (enable_server_read_mode(PSESSION_DEDICATED_THREAD)) {
    leave_group_on_failure::mask leave_actions;
    leave_actions.set(leave_group_on_failure::SKIP_SET_READ_ONLY, true);
    leave_actions.set(leave_group_on_failure::SKIP_LEAVE_VIEW_WAIT, true);
    leave_group_on_failure::leave(
        leave_actions, ER_GRP_RPL_SUPER_READ_ONLY_ACTIVATE_ERROR,
        PSESSION_DEDICATED_THREAD, &m_notification_ctx, "");
    set_plugin_is_setting_read_mode(false);

    return;
  } else {
    set_plugin_is_setting_read_mode(false);
  }

  /*
    获取节点 auto_increment_increment 的值
    在多主模式下，如果集群的大小超过 auto_increment_increment，容易出现主键冲突
  */
  ulong auto_increment_increment = get_auto_increment_increment();

  if (!local_member_info->in_primary_mode() &&
      new_view.get_members().size() > auto_increment_increment) {
    LogPluginErr(ERROR_LEVEL, ER_GRP_RPL_EXCEEDS_AUTO_INC_VALUE,
                 new_view.get_members().size(), auto_increment_increment);
  }

  // 发送一个暂停包（SUSPENSION_PACKET）给 Applier 模块
  applier_module->add_suspension_packet();

  // 发送一个 View_change_packet 给 Applier 模块
  std::string view_id = new_view.get_view_id().get_representation();
  View_change_packet *view_change_packet = new View_change_packet(view_id);
  applier_module->add_view_change_packet(view_change_packet);

  // 选择恢复的方式
  Remote_clone_handler::enum_clone_check_result recovery_strategy =
      Remote_clone_handler::DO_RECOVERY;

  // 使用克隆插件进行恢复需满足一定的条件
  if (number_of_members > 1)
    recovery_strategy = remote_clone_handler->check_clone_preconditions();
```

```cpp
    if (Remote_clone_handler::DO_CLONE == recovery_strategy) {
      LogPluginErr(SYSTEM_LEVEL, ER_GRP_RPL_RECOVERY_STRAT_CHOICE,
                   "Cloning from a remote group donor.");

      // 启动克隆线程，通过克隆插件进行恢复
      if (remote_clone_handler->clone_server(
              new_view.get_group_id().get_group_id(),
              new_view.get_view_id().get_representation()))  {
        /* purecov: begin inspected */
        LogPluginErr(WARNING_LEVEL, ER_GRP_RPL_RECOVERY_STRAT_FALLBACK,
                     "Incremental Recovery.");
        recovery_strategy = Remote_clone_handler::DO_RECOVERY;
        /* purecov: end */
      }
    }
    // 如果不满足克隆条件，且能通过binlog来恢复，则基于binlog来恢复
    if (Remote_clone_handler::DO_RECOVERY == recovery_strategy) {
      LogPluginErr(SYSTEM_LEVEL, ER_GRP_RPL_RECOVERY_STRAT_CHOICE,
                   "Incremental recovery from a group donor");

      // 启动恢复线程
      recovery_module->start_recovery(
          new_view.get_group_id().get_group_id(),
          new_view.get_view_id().get_representation());
    } else if (Remote_clone_handler::CHECK_ERROR == recovery_strategy ||
               Remote_clone_handler::NO_RECOVERY_POSSIBLE ==
                   recovery_strategy) {
      // 如果克隆插件不存在，且不能通过binlog来恢复，则意味着无法通过分布式恢复来恢复
      if (Remote_clone_handler::NO_RECOVERY_POSSIBLE == recovery_strategy)
        LogPluginErr(ERROR_LEVEL, ER_GRP_RPL_NO_POSSIBLE_RECOVERY);
      else {
        /* purecov: begin inspected */
        LogPluginErr(ERROR_LEVEL, ER_GRP_RPL_RECOVERY_EVAL_ERROR, "");
        /* purecov: end */
      }
      leave_group_on_failure::mask leave_actions;
      leave_actions.set(leave_group_on_failure::SKIP_LEAVE_VIEW_WAIT, true);
      leave_group_on_failure::leave(leave_actions, 0, PSESSION_DEDICATED_THREAD,
                                    &m_notification_ctx, "");
      return;
    }
  }
  ...
}
```

handle_joining_members 函数的处理流程如下。

(1) 进行兼容性检查，检查的内容如下。

❏ 集群的数量：算上新节点，集群总的节点数不能超过 9。

❏ 新节点的版本需与集群中的其他节点兼容。

❏ 新节点的参数需与集群中的其他节点一致。具体包括：group_replication_gtid_assignment_block_size、transaction_write_set_extraction、lower_case_table_names、default_table_encryption、group_replication_view_change_uuid、group_replication_single_primary_mode、group_replication_enforce_update_everywhere_checks。

❑ 新节点的 gtid_executed 必须是集群 gtid_executed 的子集。换言之，新节点中不能存在集群中没有的事务。

(2) 将节点的状态置为 RECOVERING 并设置为只读。

(3) 获取节点 auto_increment_increment 的值。

在多主模式下，如果集群的大小超过 auto_increment_increment，则容易出现主键冲突。此时，错误日志中会提示以下信息。

```
[ERROR] [MY-011514] [Repl] Plugin group_replication reported: 'Group contains 3 members which is greater than group_replication_auto_increment_increment value of 1. This can lead to a higher transactional abort rate.'
```

(4) 发送一个暂停包（SUSPENSION_PACKET）给 Applier 模块。

Applier 模块接收到这个暂停包后将只会接受，而不会处理新的数据包。接受的数据包会堆积在 incoming 队列中，直到节点恢复完成再处理。

(5) 发送一个 View_change_packet 给 Applier 模块。

Applier 模块接收到这个包后会创建一个 View_change_log_event。这个事件会被写入 relay log，然后通过重放写入 binlog。在基于 binlog 进行恢复时会用到这个事件。

(6) 选择恢复的方式。

恢复的方式有两种：Remote_clone_handler::DO_CLONE 和 Remote_clone_handler::DO_RECOVERY。前者基于克隆插件进行恢复，后者则基于 binlog。在 MySQL 8.0.17 之前，只支持 binlog 这一种恢复方式。但考虑一下以下两种场景。

❑ Joiner 需要的 binlog 被集群中的其他节点删除了。
❑ 差异事务太多。

在第一种场景中，无法通过 binlog 来恢复，只能使用备份工具将 Joiner 的数据恢复到一个较近的时间点。在第二种场景中，虽然可以通过 binlog 来恢复，但效率会比较低，恢复时间也比较久。

为了解决上述痛点，MySQL 8.0.17 引入了克隆插件，而组复制也支持通过克隆插件来执行恢复操作。使用克隆插件执行恢复操作，需满足以下条件。

❑ MySQL 的版本大于等于 8.0.17。
❑ Joiner 需要安装克隆插件。注意，代码中没有判断 Donor 是否安装了克隆插件。虽然没有判断，但为了保证克隆操作能成功执行，Donor 中其实也需要安装克隆插件。
❑ 差异事务的数量大于 group_replication_clone_threshold 或者无法通过 binlog 来恢复。

如果不满足克隆条件且差异事务对应的 binlog 在其他节点也存在，则会通过 binlog 的方式来恢复。

需要注意的是，通过克隆插件恢复后，实例会重启。实例重启之后，我们需要再次执行 START GROUP_REPLICATION 命令，启动组复制。这个时候会从头开始执行恢复操作，只不过这一次因为差异事务的数量小于 group_replication_clone_threshold，所以会选择 binlog 而不是克隆插件来恢复。

现在就来分析一下 binlog 这种恢复方式。

```cpp
int Recovery_module::recovery_thread_handle() {
  DBUG_TRACE;

  // 初始化恢复线程
  int error = 0;
  Plugin_stage_monitor_handler stage_handler;
  if (stage_handler.initialize_stage_monitor())
    LogPluginErr(ERROR_LEVEL, ER_GRP_RPL_NO_STAGE_SERVICE);

  set_recovery_thread_context();
  mysql_mutex_lock(&run_lock);
  recovery_thd_state.set_initialized();
  mysql_mutex_unlock(&run_lock);

  size_t number_of_members = group_member_mgr->get_number_of_members();
  // 构造 Donor 列表
  recovery_state_transfer.initialize_group_info();

  mysql_mutex_lock(&run_lock);
  // 将恢复线程的状态置为 running
  recovery_thd_state.set_running();
  stage_handler.set_stage(info_GR_STAGE_module_executing.m_key, __FILE__,
                         __LINE__, 0, 0);
  mysql_cond_broadcast(&run_cond);
  mysql_mutex_unlock(&run_lock);

  // 在本地恢复阶段, 会启动 group_replication_applier 重放 relay log 中还没执行完的事件
  error =
      applier_module->wait_for_applier_complete_suspension(&recovery_aborted);

  if (error == APPLIER_THREAD_ABORTED) {
    error = 0;
    recovery_aborted = true;
    goto cleanup;
  }

  if (!recovery_aborted && error) {
    LogPluginErr(ERROR_LEVEL, ER_GRP_RPL_UNABLE_TO_EVALUATE_APPLIER_STATUS);
    goto cleanup;
  }

  // 如果集群的节点数为 1, 则意味着它无须从其他节点同步数据, 这个时候会跳过下面的全局恢复阶段
  if (number_of_members == 1) {
    if (!recovery_aborted) {
      LogPluginErr(INFORMATION_LEVEL, ER_GRP_RPL_ONLY_ONE_SERVER_ALIVE);
    }
    goto single_member_online;
  }

  // 全局恢复阶段, 会从其他节点同步差异事务
  m_state_transfer_return =
      recovery_state_transfer.state_transfer(stage_handler);
  error = m_state_transfer_return;

  stage_handler.set_stage(info_GR_STAGE_module_executing.m_key, __FILE__,
                         __LINE__, 0, 0);

  if (error) {
    goto cleanup;
  }
```

```
single_member_online:

  // 唤醒 Applier 模块，开始应用缓存事务
  if (!recovery_aborted) applier_module->awake_applier_module();

  error = wait_for_applier_module_recovery();

cleanup:

  // 如果没有报错，则会通知集群将新节点的状态置为 ONLINE
  if (!recovery_aborted && !error) {
    notify_group_recovery_end();
  }

  // 如果报错，则新节点会离开集群，并执行 auto-rejoin 操作
  if (!recovery_aborted && error) {
    leave_group_on_recovery_failure();
  }

  stage_handler.end_stage();
  stage_handler.terminate_stage_monitor();

  // 销毁恢复线程
  clean_recovery_thread_context();

  mysql_mutex_lock(&run_lock);

  recovery_aborted = true;
  delete recovery_thd;

  Gcs_interface_factory::cleanup_thread_communication_resources(
      Gcs_operations::get_gcs_engine());

  my_thread_end();
  recovery_thd_state.set_terminated();
  mysql_cond_broadcast(&run_cond);
  mysql_mutex_unlock(&run_lock);
  my_thread_exit(nullptr);

  return error; /* purecov: inspected */
}
```

一共包括8步，下面看看各个步骤的具体作用。

(1) 初始化恢复线程。

在这个阶段，会构造 Donor 列表，并通过随机函数打散该列表。

```
if (suitable_donors.size() > 1) {
  std::random_device rng;
  std::mt19937 urng(rng());
  std::shuffle(suitable_donors.begin(), suitable_donors.end(), urng);
}
```

(2) 本地恢复阶段。

如果这个节点之前加入过集群，它的 group_replication_applier 通道的 relay log 中可能还有一部分事件没有执行。这个时候，会启动 group_replication_applier 重放这些还没有执行的事件。怎么判

断这些事件是否已经执行完了呢？监控 GTID 的变化，当 retrieved_gtid_set 是 gtid_executed 的子集时，就意味着 relay log 中的事务应用完了。

(3) 如果集群的节点数为 1，则意味着这个节点是初始化（Bootstrap）节点，无须从其他节点同步数据。这个时候会跳过下面的全局恢复阶段。

(4) 全局恢复阶段。

在这个阶段，Joiner 会与 Donor 建立一个主从复制关系。注意，这里的复制是异步复制，与组复制无关。这个复制使用的通道是 group_replication_recovery。所以，在开启组复制之前，必须通过以下命令配置该通道的账号密码。

```
mysql> change master to master_user='rpl_user', master_password='rpl_password' for channel
'group_replication_recovery';
```

不用指定 master_host 和 master_port，因为 Donor 是随机选择的，这两部分信息由组复制来填充。复制开启时会同时指定 UNTIL_SQL_VIEW_ID，因此复制会在 SQL 线程应用到指定 view_id 时停止。注意，UNTIL_SQL_VIEW_ID 仅供组复制内部使用，无法在 START SLAVE 命令中指定。

在指定 UNTIL_SQL_VIEW_ID 的情况下，在应用每个 EVENT 之前，都会调用下面这个函数。

```
// sql/rpl_replica_until_options.cc
bool Until_view_id::check_before_dispatching_event(const Log_event *ev) {
  if (ev->get_type_code() == binary_log::VIEW_CHANGE_EVENT) {
    View_change_log_event *view_event = const_cast<View_change_log_event *>(
        static_cast<const View_change_log_event *>(ev));

    if (m_view_id.compare(view_event->get_view_id()) == 0) {
      set_group_replication_retrieved_certification_info(view_event);
      until_view_id_found = true;
      return false;
    }
  }

  if (until_view_id_found && ev->ends_group()) {
    until_view_id_commit_found = true;
    return false;
  }

  return false;
}
```

该函数会判断当前待应用的 EVENT 是否是 VIEW_CHANGE_EVENT。如果是，会继续判断 View_change_log_event 中的 view_id 是否是复制开启时指定的 view_id。如果是，则会基于 View_change_log_event 初始化冲突检测数据库。

(5) 唤醒 Applier 模块，开始应用缓存事务。

在应用的过程中，会基于步骤(4)构建的冲突检测数据库进行冲突检测。wait_for_applier_module_recovery 函数返回的时机取决于 group_replication_recovery_complete_at。如果 group_replication_recovery_complete_at 设置的是 TRANSACTIONS_CERTIFIED，则会等待缓存事务完成冲突检测。如果设置的是 TRANSACTIONS_APPLIED，则会等待缓存事务应用完。

(6) 如果恢复的过程中没有报错，组会通过 GCS 模块向集群发送一个恢复结束消息 RECOVERY_END_MESSAGE。集群节点（包括节点自身）收到这个消息后，会将新节点的状态置为 ONLINE。

(7) 如果恢复的过程中报错，会执行退出操作。

- 删除组内其他成员的信息。
- 将节点的状态置为 ERROR。
- 将节点设置为只读。

如果设置了 group_replication_autorejoin_tries，则会自动重试，重新加入集群。

(8) 销毁恢复线程。

11.8.2 分布式恢复的相关参数

- **group_replication_recovery_retry_count、group_replication_recovery_reconnect_interval**

在全局恢复阶段，Joiner 会与 Donor 建立一个主从复制关系。如果复制建立失败（例如，复制账号权限不够），则会进行重试。重试的总次数由 group_replication_recovery_retry_count 决定，默认为 10。在重试的时候，会依次轮询（Round Robin）组内的所有 Donor，轮询的过程中不会等待。轮询结束后，如果还是失败，则会等待一段时间，然后进行下一轮的轮询。等待的具体时长由 group_replication_recovery_reconnect_interval 决定，默认为 60 秒。

- **group_replication_advertise_recovery_endpoints**

在 MySQL 8.0.21 之前，当进行分布式恢复操作时，在选择 Donor 后，默认的恢复地址取自下面的 hostname 和 port。

```
if (report_host)
  *hostname = report_host;
else
  *hostname = glob_hostname;

if (report_port)
  *port = report_port;
else
  *port = mysqld_port;
```

从 MySQL 8.0.21 开始，可通过 group_replication_advertise_recovery_endpoints 自定义恢复地址。例如：

```
group_replication_advertise_recovery_endpoints=
"127.0.0.1:3306,127.0.0.1:4567,[::1]:3306,localhost:3306"
```

注意，列表中指定的端口必须来自 port、report_port 或者 admin_port，而主机名只需是服务器上的有效地址，无须在 bind_address 或 admin_address 中指定。另外，如果是通过 admin_port 进行分布式恢复操作，还需要授予用户 SERVICE_CONNECTION_ADMIN 权限。注意，Donor 是随机选择的，与是否设置 group_replication_advertise_recovery_endpoints 无关，并不会因为某个节点设置了 group_replication_advertise_recovery_endpoints 而一定会选它作为 Donor。

- **group_replication_clone_threshold**

当差异事务的数量大于 group_replication_clone_threshold 时,会选择克隆操作恢复数据。这里的差异事务数等于集群中其他节点 gtid_executed 的超集减去 Joiner 的 gtid_executed。注意,在线上 group_replication_clone_threshold 不宜设置得过小,否则每次实例重启之后,都会选择克隆操作恢复数据,这样新节点就永远无法加入集群。

- **group_replication_recovery_complete_at**

将新节点标记为 ONLINE 的时间点,可取值如下。

❑ TRANSACTIONS_CERTIFIED:等待缓存事务完成冲突检测。
❑ TRANSACTIONS_APPLIED:等待缓存事务应用完。

由此来看,如果使用前者,应用有可能会读取到旧的数据。如果对旧数据进行写操作,这些写操作会因为冲突检测失败而被回滚。建议使用后者,默认值也是 TRANSACTIONS_APPLIED。

- **group_replication_view_change_uuid**

指定 View_change_log_event 对应的 GTID 的 UUID 部分。默认为 AUTOMATIC,此时,UUID 由 group_replication_group_name 决定。设置时注意,取值不能和 group_replication_group_name 一样。如果要部署 InnoDB ClusterSet,需显式指定 group_replication_view_change_uuid。

- **group_replication_recovery_compression_algorithms**

如果是使用 binlog 进行分布式恢复,在带宽有限的情况下,可指定压缩算法,可选值有 zlib、zstd 和 uncompressed,默认为 uncompressed(不压缩)。如果是使用克隆插件进行分布式恢复,则需通过 clone_enable_compression 开启压缩。

- **group_replication_recovery_zstd_compression_level**

指定 zstd 算法的压缩等级。等级越高,压缩的效果越明显。有效值为 1 ~ 22,默认为 3。

11.9 组复制的冲突检测

冲突检测是组复制的两大基石之一,它保证了所有节点会以相同的行为处理同一事务,要么全部认证通过,事务提交;要么全部认证失败,事务回滚。

组复制的冲突检测是基于 write_set 和冲突检测数据库实现的。接下来,我们看看什么是 write_set 及冲突检测数据。

11.9.1 write_set

以下面这条 INSERT 操作为例,我们看看它对应的 write_set 是什么。

```
create table slowtech.t1(id int primary key, c1 int, c2 int, unique key (c1, c2));
insert into slowtech.t1 values(1,2,3);
```

首先，通过 add_pke 函数生成 pke。

```cpp
bool add_pke(TABLE *table, THD *thd, const uchar *record) {
  DBUG_TRACE;
  assert(record == table->record[0] || record == table->record[1]);

  Rpl_transaction_write_set_ctx *ws_ctx =
      thd->get_transaction()->get_transaction_write_set_ctx();
  bool writeset_hashes_added = false;

  if (table->key_info && (table->s->primary_key < MAX_KEY)) {
    ptrdiff_t ptrdiff = record - table->record[0];
    // 定义一个字符串
    std::string pke_schema_table;
    pke_schema_table.reserve(NAME_LEN * 3);
    // HASH_STRING_SEPARATOR 是字符串之间的分割符，默认为½
    pke_schema_table.append(HASH_STRING_SEPARATOR);
    // 添加 db 名
    pke_schema_table.append(table->s->db.str, table->s->db.length);
    pke_schema_table.append(HASH_STRING_SEPARATOR);
    // 添加 db 的长度
    pke_schema_table.append(std::to_string(table->s->db.length));
    // 添加表名
    pke_schema_table.append(table->s->table_name.str,
                            table->s->table_name.length);
    pke_schema_table.append(HASH_STRING_SEPARATOR);
    // 添加表的长度
    pke_schema_table.append(std::to_string(table->s->table_name.length));
    // 对于 slowtech.t1,最后会生成 "½slowtech½8t1½2" 的字符串

    std::string pke;
    pke.reserve(NAME_LEN * 5);

    for (uint key_number = 0; key_number < table->s->keys; key_number++) {
      // 如果是非唯一索引，则跳过
      if (!((table->key_info[key_number].flags & (HA_NOSAME)) == HA_NOSAME))
        continue;

      pke.clear();
      // table->key_info[key_number].name 是索引名。如果是主键，则是 PRIMARY
      // 首先添加索引名，接着是上面生成的 pke_schema_table
      pke.append(table->key_info[key_number].name);
      pke.append(pke_schema_table);

      uint i = 0;
      Field *mv_field = nullptr;
      // 循环遍历索引中的所有列，毕竟有的索引是复合索引
      for (/*empty*/; i < table->key_info[key_number].user_defined_key_parts;
           i++) {
        int index = table->key_info[key_number].key_part[i].fieldnr;
        Field *field = table->field[index - 1];

        // 如果是 NULL，则直接跳过，因为 NULL 和任何值都不相等
        if (field->is_null(ptrdiff)) break;
        // 如果列是个数组（多值索引），则跳过，数组会在下面的 generate_mv_hash_pke 中单独处理
        if (field->is_array()) {
          assert(!mv_field);
          mv_field = field;
          continue;
        }
```

```
      field->move_field_offset(ptrdiff);
      const CHARSET_INFO *cs = field->charset();
      int max_length = cs->coll->strnxfrmlen(cs, field->pack_length());
      std::unique_ptr<uchar[]> pk_value(new uchar[max_length + 1]());

      size_t length = field->make_sort_key(pk_value.get(), max_length);
      pk_value[length] = 0;

      // 添加列值
      pke.append(pointer_cast<char *>(pk_value.get()), length);
      pke.append(HASH_STRING_SEPARATOR);
      // 添加列的长度
      pke.append(std::to_string(length));

      field->move_field_offset(-ptrdiff);
    }

    // 调用 generate_hash_pke, 这个函数会将生成的 pke 添加到 write_sets 中
    if (i == table->key_info[key_number].user_defined_key_parts) {
      // 如果是多值索引, 则调用 generate_mv_hash_pke 处理, 否则会调用 generate_hash_pke 处理
      if (mv_field) {
        mv_field->move_field_offset(ptrdiff);
        if (generate_mv_hash_pke(pke, thd, mv_field))
          return true;
        mv_field->move_field_offset(-ptrdiff);
      } else {
        if (generate_hash_pke(pke, thd))
          return true;
      }
      writeset_hashes_added = true;
    } else {
      assert(key_number != 0);
    }
  }
  ...
}

if (!writeset_hashes_added) ws_ctx->set_has_missing_keys();
return false;
}
```

基于代码的实现逻辑, 接下来, 我们看看 `insert into slowtech.t1 values(1,2,3)` 这条记录对应的 pke, 一共有两条。

- 主键: "PRIMARY½slowtech½t1½2\x80\0\0\U00000001¼4"。这里面的\x80\0\0\U00000001 用来表示 4 字节的 1, 最后的 4 是 INT 类型的长度。
- 唯一索引: "c1½slowtech½t1½2\x80\0\0\U00000002½4\x80\0\0\U00000003½4"。这里面的\x80\0\0\U00000002 和 \x80\0\0\U00000003 分别用来表示 2 和 3。

除了主键和唯一索引, 如果表上存在外键, 还会计算外键对应的 pke。

计算完 pke, 接着会通过 generate_hash_pke 生成 pke 对应的哈希值, 并将其添加到 write_set 中。write_set 是一个值为 uint64 的动态数组。

```
static bool generate_hash_pke(const std::string &pke, THD *thd) {
  DBUG_TRACE;
  assert(thd->variables.transaction_write_set_extraction != HASH_ALGORITHM_OFF);
  // 通过 transaction_write_set_extraction 指定的算法计算 pke 对应的哈希值
  uint64 hash = calc_hash<const char *>(
      thd->variables.transaction_write_set_extraction, pke.c_str(), pke.size());
  if (thd->get_transaction()->get_transaction_write_set_ctx()->add_write_set(
          hash))
    return true;

  DBUG_PRINT("info", ("pke: %s; hash: %" PRIu64, pke.c_str(), hash));
  return false;
}
```

write_set 中最后会有两条记录。

- 18368573604689368734：对应 "PRIMARY½slowtech½8t1½2\x80\0\0\U00000001¼4" 这个 pke。
- 5486073988594004197：对应 "c1½slowtech½8t1½2\x80\0\0\U00000002¼4\x80\0\0\U00000003¼4" 这个 pke。

实际上，在 group_replication_trans_before_commit 函数中，在通过 add_write_set(tcle, write_set) 将事务的 write_set 添加到 Transaction_context_log_event 的过程中，还会通过 Base64 算法将 write_set 中的哈希值进一步编码为字符串。上面两个哈希值编码后的结果分别是 "nlYgYl9I6v4=" 和 "5fCn5TlzIkw="。

所以，write_set 的生成一共涉及 3 个步骤。

(1) 生成 pke。
(2) 基于 transaction_write_set_extraction 指定的算法计算 pke 对应的哈希值。
(3) 通过 Base64 算法将哈希值进一步编码为字符串。

11.9.2 冲突检测数据库

冲突检测数据库是一个键为字符串、值为 Gtid_set_ref 的无序字典类型。

```
typedef std::unordered_map<std::string, Gtid_set_ref *> Certification_info;
```

其中，键是 write_set 中的一个元素，比如上面的 "nlYgYl9I6v4=" 或 "5fCn5TlzIkw="。

接下来看看 Gtid_set_ref 的定义。

```
class Gtid_set_ref : public Gtid_set {
 public:
  Gtid_set_ref(Sid_map *sid_map, int64 parallel_applier_sequence_number)
      : Gtid_set(sid_map),
        reference_counter(0),
        parallel_applier_sequence_number(parallel_applier_sequence_number) {}
```

Gtid_set_ref 类包括 3 个成员变量。

- Gtid_set：GTID 集。存储的是 snapshot_version。snapshot_version 是执行 SQL 时，实例当前的 gtid_executed。

- reference_counter：引用计数器。引用次数为 0 代表这条记录可以安全删除。
- parallel_applier_sequence_number：事务的 sequence_number。

冲突检测数据库的插入及更新逻辑是在 Certifier::add_item 中定义的。

```
bool Certifier::add_item(const char *item, Gtid_set_ref *snapshot_version,
                        int64 *item_previous_sequence_number) {
  DBUG_TRACE;
  mysql_mutex_assert_owner(&LOCK_certification_info);
  bool error = true;
  std::string key(item);
  // 查找键在 certification_info 中的值
  Certification_info::iterator it = certification_info.find(key);
  snapshot_version->link();
  // 如果 it 到了迭代器的末尾，则意味着键在 certification_info 中不存在。
  // 这个时候会将 key 和 snapshot_version 插入 certification_info
  if (it == certification_info.end()) {
    std::pair<Certification_info::iterator, bool> ret =
        certification_info.insert(
            std::pair<std::string, Gtid_set_ref *>(key, snapshot_version));
    error = !ret.second;
  } else {
    // 如果键在 certification_info 中存在，则替换之前的 snapshot_version
    *item_previous_sequence_number =
        it->second->get_parallel_applier_sequence_number();

    if (it->second->unlink() == 0) delete it->second;

    it->second = snapshot_version;
    error = false;
  }

  return error;
}
```

代码的逻辑比较简单。注意，在执行更新操作时，会将记录旧的 sequence_number 赋值给 item_previous_sequence_number。item_previous_sequence_number 会用来生成事务的 last_committed。这在后面会提到。

11.9.3 冲突检测的实现细节

冲突检测的检测流程是在 Certifier::certify 中定义的。我们分析一下该函数的实现逻辑。

```
rpl_gno Certifier::certify(Gtid_set *snapshot_version,
                          std::list<const char *> *write_set,
                          bool generate_group_id, const char *member_uuid,
                          Gtid_log_event *gle, bool local_transaction) {
  DBUG_TRACE;
  rpl_gno result = 0;
  const bool has_write_set = !write_set->empty();

  if (!is_initialized()) return -1; /* purecov: inspected */

  mysql_mutex_lock(&LOCK_certification_info);
  int64 transaction_last_committed = parallel_applier_last_committed_global;
```

```cpp
// 如果开启了冲突检测
if (conflict_detection_enable) {
  // 循环遍历 write_set 中的记录
  for (std::list<const char *>::iterator it = write_set->begin();
       it != write_set->end(); ++it) {
    // 获取记录在冲突检测数据库中的 snapshot_version
    Gtid_set *certified_write_set_snapshot_version =
        get_certified_write_set_snapshot_version(*it);

    // 如果冲突检测数据库中该记录对应的 snapshot_version 不是事务 snapshot_version 的子集，
    // 则意味着冲突检测数据库中该条记录的版本比事务中的新。
    // 换言之，冲突检测失败，直接执行 end 处的代码
    if (certified_write_set_snapshot_version != nullptr &&
        !certified_write_set_snapshot_version->is_subset(snapshot_version))
      goto end;
  }
}

if (certifying_already_applied_transactions &&
    !group_gtid_extracted->is_subset_not_equals(group_gtid_executed)) {
  certifying_already_applied_transactions = false;
}

// 如果当前事务没有指定 GTID，则会基于组 UUID 生成一个新的 GTID
if (generate_group_id) {
  if (snapshot_version->ensure_sidno(get_group_sidno()) != RETURN_STATUS_OK) {
    LogPluginErr(
        ERROR_LEVEL,
        ER_GRP_RPL_UPDATE_TRANS_SNAPSHOT_VER_ERROR); /* purecov: inspected */
    goto end;                                        /* purecov: inspected */
  }

  // 获取下一个可用的 GTID
  result = get_group_next_available_gtid(member_uuid);
  if (result < 0) goto end;


  // 将新生成的 GTID 添加到 snapshot_version 中
  snapshot_version->_add_gtid(get_group_sidno(), result);


  // 最后一个通过冲突检测的事务的 GTID
  // last_conflict_free_transaction 会显示在 performance_schema.replication_group_member_stats 中
  last_conflict_free_transaction.set(group_gtid_sid_map_group_sidno, result);

  DBUG_PRINT("info", ("Group replication Certifier: generated transaction "
                      "identifier: %" PRId64,
                      result));
} else {
  ...
}

if (has_write_set) {
  int64 transaction_sequence_number =
      local_transaction ? -1 : parallel_applier_sequence_number;
  // 初始化 Gtid_set_ref
  Gtid_set_ref *snapshot_version_value = new Gtid_set_ref(
      certification_info_sid_map, transaction_sequence_number);
```

```cpp
    if (snapshot_version_value->add_gtid_set(snapshot_version) !=
        RETURN_STATUS_OK) {
      result = 0;                      /* purecov: inspected */
      delete snapshot_version_value;   /* purecov: inspected */
      LogPluginErr(
          ERROR_LEVEL,
          ER_GRP_RPL_UPDATE_TRANS_SNAPSHOT_REF_VER_ERROR); /* purecov: inspected
                                                            */
      goto end; /* purecov: inspected */
    }
    // 将事务的 write_set 添加到冲突检测数据库中
    for (std::list<const char *>::iterator it = write_set->begin();
         it != write_set->end(); ++it) {
      int64 item_previous_sequence_number = -1;

      add_item(*it, snapshot_version_value, &item_previous_sequence_number);

      if (item_previous_sequence_number > transaction_last_committed &&
          item_previous_sequence_number != parallel_applier_sequence_number)
        transaction_last_committed = item_previous_sequence_number;
    }
  }

  // 更新远程事务的 last_committed 和 sequence_number
  if (!local_transaction) {
    bool update_parallel_applier_last_committed_global = false;
    // 对于 CREATE TABLE ... AS SELECT 操作,对应的 gle->last_committed 和 gle->sequence_number 等于 0
    if (0 == gle->last_committed && 0 == gle->sequence_number) {
      update_parallel_applier_last_committed_global = true;
    }
    // parallel_applier_sequence_number - 1其实是上一个事务的 sequence_number
    // 对于 DDL 操作,没有 write_set,不能跟其他 DML 操作并行,所以需要将上一个事务的
gle->sequence_number 赋值给 transaction_last_committed
    if (!has_write_set || update_parallel_applier_last_committed_global) {
      transaction_last_committed = parallel_applier_sequence_number - 1;
    }

    gle->last_committed = transaction_last_committed;
    gle->sequence_number = parallel_applier_sequence_number;
    assert(gle->last_committed >= 0);
    assert(gle->sequence_number > 0);
    assert(gle->last_committed < gle->sequence_number);
    // 将 parallel_applier_sequence_number 自增加 1
    // 如果是 DDL 或者 CREATE TABLE ... AS SELECT 操作,则会将 parallel_applier_sequence_number 赋值给
parallel_applier_last_committed_global
    increment_parallel_applier_sequence_number(
        !has_write_set || update_parallel_applier_last_committed_global);
  }
end:
  update_certified_transaction_count(result > 0, local_transaction);

  mysql_mutex_unlock(&LOCK_certification_info);
  DBUG_PRINT(
      "info",
      ("Group replication Certifier: certification result: %" PRId64, result));
  return result;
}
```

该函数的处理流程如下。

(1) 只有 conflict_detection_enable 为 true，才进行冲突检测。conflict_detection_enable 在两种情况下会为 true：多主模式；单主模式下，在故障切换后，新主在应用旧主 relay log 的过程中。
(2) 冲突检测逻辑很简单。循环遍历 write_set 中的记录，如果冲突检测数据库中存在这条记录，且记录对应的 snapshot_version 不是事务 snapshot_version 的子集，则意味着冲突检测数据库中的数据比事务中的要新，这个时候冲突检测失败。反之，如果冲突检测数据中没有对应的记录，或者虽然有但记录对应的 snapshot_version 是事务 snapshot_version 的子集，则意味着冲突检测成功。
(3) 如果冲突检测成功，且当前事务没有指定 GTID，则会基于组 UUID 生成一个新的 GTID，同时将新生成的 GTID 添加到事务的 snapshot_version 中。
(4) 如果事务存在 write_set，则将事务的 write_set 添加到冲突检测数据库中。
(5) 对于远程事务，会更新事务的 last_committed 和 sequence_number。
(6) 更新状态信息。

接下来重点说说远程事务 last_committed 和 sequence_number 的生成逻辑。

(1) 初始化 transaction_last_committed。将 parallel_applier_last_committed_global 赋值给 transaction_last_committed。
(2) 冲突检测通过后，会将事务的 write_set 添加到冲突检测数据库中。
(3) 在添加的过程中，如果发现记录在冲突检测数据库中存在，则会执行更新操作，用记录新的 snapshot_version 替代旧的 snapshot_version。在更新之前，会将记录旧的 sequence_number 赋值给 item_previous_sequence_number。
(4) 如果 item_previous_sequence_number 大于当前事务的 transaction_last_committed，则会将 item_previous_sequence_number 赋值给 transaction_last_committed。
(5) 如果当前操作是 DDL，则会将上一个事务的 sequence_number 赋值给 transaction_last_committed。
(6) 最后将 transaction_last_committed 赋值给 last_committed，将 parallel_applier_sequence_number 赋值给 sequence_number。
(7) 调用 increment_parallel_applier_sequence_number 让 parallel_applier_sequence_number 自增加 1。如果当前操作是 DDL，还会将 parallel_applier_sequence_number 赋值给 parallel_applier_last_committed_global。

看上去很复杂，概括起来实际上就两条。

- 对于 DML 操作，会将冲突检测数据库中对应记录的 sequence_number 作为新事务的 last_committed。这样就可以保证涉及同一行数据的不同事务，在从库重放时，不会并行执行。
- 对于 DDL 操作，没有 write_set，不能跟其他操作并行重放，所以会将上一个事务的 sequence_number 作为新事务的 last_committed。

11.9.4 冲突检测数据库的清理逻辑

冲突检测数据库的清除逻辑是在 Certifier::garbage_collect 中定义的。

```
void Certifier::garbage_collect() {
  ...
  mysql_mutex_lock(&LOCK_certification_info);

  Certification_info::iterator it = certification_info.begin();
  stable_gtid_set_lock->wrlock();
  while (it != certification_info.end()) {
    if (it->second->is_subset_not_equals(stable_gtid_set)) {
      if (it->second->unlink() == 0) delete it->second;
      certification_info.erase(it++);
    } else
      ++it;
  }
  stable_gtid_set_lock->unlock();

  mysql_mutex_unlock(&LOCK_certification_info);
  ...
}
```

清除逻辑很简单，就是循环遍历冲突检测数据库（certification_info）中的所有元素，判断这个元素对应的 snapshot_version 是否是 stable_gtid_set 的子集。如果是，则将这个元素从 certification_info 中删除。

接下来，我们看看 garbage_collect() 的调用时机。

garbage_collect() 是在 Certifier::stable_set_handle() 中调用的。

```
void Certifier_broadcast_thread::dispatcher() {
  ...
  while (!aborted) {
    ...
    if (broadcast_counter % broadcast_gtid_executed_period == 0) {
      broadcast_gtid_executed();
    }

    Certification_handler *cert = applier_module->get_certification_handler();
    Certifier_interface *cert_module = (cert ? cert->get_certifier() : nullptr);

    if (cert_module) {
      cert_module->stable_set_handle();
    }
    ...
    struct timespec abstime;
    set_timespec(&abstime, 1);
    mysql_cond_timedwait(&broadcast_dispatcher_cond, &broadcast_dispatcher_lock,
                        &abstime);
    mysql_mutex_unlock(&broadcast_dispatcher_lock);

    broadcast_counter++;
  }
  ...
}
```

虽然 stable_set_handle() 在 dispatcher() 中每秒调用一次，但在实际执行的时候，它会判断 Certifier 模块 incoming 队列的长度。如果队列的长度小于 ONLINE 状态的节点数，函数会直接返回 0，不做任何处理。如果队列的长度等于 ONLINE 状态的节点数，则会取队列中各个节点 gtid_executed 的交集作为 stable_gtid_set，然后调用 garbage_collect() 清理冲突检测数据库中 snapshot_version 是 stable_gtid_set 子集的元素。

因为 incoming 队列中的元素是通过 pop 方式弹出的，所以，在获取完各个节点的数据后，incoming 就是一个空队列了。重点来了，什么时候会向 incoming 队列中推送数据呢？在执行 broadcast_gtid_executed() 的时候。而 broadcast_gtid_executed() 在 dispatcher() 中的执行周期由 broadcast_gtid_executed_period 决定。broadcast_gtid_executed_period 取自 BROADCAST_GTID_EXECUTED_PERIOD，后者目前是硬编码在代码中的。

```
static const int BROADCAST_GTID_EXECUTED_PERIOD = 60;  // seconds
```

因此，冲突检测数据库会每 60 秒清理一次。

11.10 组复制的故障检测

故障检测（Failure Detection）是组复制的一个核心功能模块，通过它可以及时识别集群中的故障节点，并将故障节点从集群中剔除。如果不将故障节点及时剔除，一方面会影响集群的性能，另一方面会阻止集群拓扑的变更。

下面结合一个具体的示例，看看组复制的故障检测流程。

11.10.1 模拟网络分区

测试集群的节点信息如表 11-4 所示。

表 11-4 测试集群的节点信息

IP 地址	主 机 名	角 色
192.168.244.10	node1	Primary
192.168.244.20	node2	Primary
192.168.244.30	node3	Primary

首先模拟网络分区故障。在 node3 上执行以下命令。

```
# iptables -A INPUT  -p tcp -s 192.168.244.10 -j DROP
# iptables -A OUTPUT -p tcp -d 192.168.244.10 -j DROP

# iptables -A INPUT  -p tcp -s 192.168.244.20 -j DROP
# iptables -A OUTPUT -p tcp -d 192.168.244.20 -j DROP

# date "+%Y-%m-%d %H:%M:%S"
2022-07-31 13:03:01
```

其中，iptables 命令会断开 node3 与 node1、node2 之间的网络连接。date 记录了命令执行的时间。

命令执行完 5 秒（这个时间是固定的，在源码中通过 DETECTOR_LIVE_TIMEOUT 指定），各个节点开始响应（从节点的日志中可以观察到这一点）。

首先看看 node1 的日志及集群状态。

```
2022-07-31T13:03:07.582519-00:00 0 [Warning] [MY-011493] [Repl] Plugin group_replication reported:
'Member with address 192.168.244.30:3306 has become unreachable.'

mysql> select member_id,member_host,member_port,member_state,member_role from
performance_schema.replication_group_members;
+--------------------------------------+-----------------+-------------+--------------+-------------+
| member_id                            | member_host     | member_port | member_state | member_role |
+--------------------------------------+-----------------+-------------+--------------+-------------+
| 207db264-0192-11ed-92c9-02001700754e | 192.168.244.10  |        3306 | ONLINE       | PRIMARY     |
| 2cee229d-0192-11ed-8eff-02001700f110 | 192.168.244.20  |        3306 | ONLINE       | PRIMARY     |
| 4cbfdc79-0192-11ed-8b01-02001701bd0a | 192.168.244.30  |        3306 | UNREACHABLE  | PRIMARY     |
+--------------------------------------+-----------------+-------------+--------------+-------------+
3 rows in set (0.00 sec)
```

从 node1、node2 的角度来看，此时 node3 处于 UNREACHABLE 状态。

接下来看看 node3 的日志及集群状态。

```
2022-07-31T13:03:07.690416-00:00 0 [Warning] [MY-011493] [Repl] Plugin group_replication reported:
'Member with address 192.168.244.10:3306 has become unreachable.'
2022-07-31T13:03:07.690492-00:00 0 [Warning] [MY-011493] [Repl] Plugin group_replication reported:
'Member with address 192.168.244.20:3306 has become unreachable.'
2022-07-31T13:03:07.690504-00:00 0 [ERROR] [MY-011495] [Repl] Plugin group_replication reported: 'This
server is not able to reach a majority of members in the group. This server will now block all updates.
The server will remain blocked until contact with the majority is restored. It is possible to use
group_replication_force_members to force a new group membership.'

mysql> select member_id,member_host,member_port,member_state,member_role from
performance_schema.replication_group_members;
+--------------------------------------+-----------------+-------------+--------------+-------------+
| member_id                            | member_host     | member_port | member_state | member_role |
+--------------------------------------+-----------------+-------------+--------------+-------------+
| 207db264-0192-11ed-92c9-02001700754e | 192.168.244.10  |        3306 | UNREACHABLE  | PRIMARY     |
| 2cee229d-0192-11ed-8eff-02001700f110 | 192.168.244.20  |        3306 | UNREACHABLE  | PRIMARY     |
| 4cbfdc79-0192-11ed-8b01-02001701bd0a | 192.168.244.30  |        3306 | ONLINE       | PRIMARY     |
+--------------------------------------+-----------------+-------------+--------------+-------------+
3 rows in set (0.00 sec)
```

从 node3 的角度来看，此时 node1、node2 处于 UNREACHABLE 状态。

3 个节点的集群，只有一个节点处于 ONLINE 状态，因而不满足组复制的多数派原则。此时，node3 只能查询，写操作会被阻塞。

```
mysql> select * from slowtech.t1 where id=1;
+----+------+
| id | c1   |
+----+------+
|  1 | a    |
+----+------+
1 row in set (0.00 sec)

mysql> delete from slowtech.t1 where id=1;
阻塞中……
```

又过了 16 秒（这里的 16 秒实际上与 group_replication_member_expel_timeout 参数有关），node1、node2 会将 node3 驱逐出集群。此时，集群中只有两个节点。

看看 node1 的日志及集群状态。

```
2022-07-31T13:03:23.576960-00:00 0 [Warning] [MY-011499] [Repl] Plugin group_replication reported:
'Members removed from the group: 192.168.244.30:3306'
2022-07-31T13:03:23.577091-00:00 0 [System] [MY-011503] [Repl] Plugin group_replication reported:
'Group membership changed to 192.168.244.10:3306, 192.168.244.20:3306 on view 16592724636525403:3.'

mysql> select member_id,member_host,member_port,member_state,member_role from
performance_schema.replication_group_members;
+--------------------------------------+-----------------+-------------+--------------+-------------+
| member_id                            | member_host     | member_port | member_state | member_role |
+--------------------------------------+-----------------+-------------+--------------+-------------+
| 207db264-0192-11ed-92c9-02001700754e | 192.168.244.10  |        3306 | ONLINE       | PRIMARY     |
| 2cee229d-0192-11ed-8eff-02001700f110 | 192.168.244.20  |        3306 | ONLINE       | PRIMARY     |
+--------------------------------------+-----------------+-------------+--------------+-------------+
2 rows in set (0.00 sec)
```

再来看看 node3 的日志及集群状态。日志没有新的输出，节点状态也没变化。

```
mysql> select member_id,member_host,member_port,member_state,member_role from
performance_schema.replication_group_members;
+--------------------------------------+-----------------+-------------+--------------+-------------+
| member_id                            | member_host     | member_port | member_state | member_role |
+--------------------------------------+-----------------+-------------+--------------+-------------+
| 207db264-0192-11ed-92c9-02001700754e | 192.168.244.10  |        3306 | UNREACHABLE  | PRIMARY     |
| 2cee229d-0192-11ed-8eff-02001700f110 | 192.168.244.20  |        3306 | UNREACHABLE  | PRIMARY     |
| 4cbfdc79-0192-11ed-8b01-02001701bd0a | 192.168.244.30  |        3306 | ONLINE       | PRIMARY     |
+--------------------------------------+-----------------+-------------+--------------+-------------+
3 rows in set (0.00 sec)
```

接下来恢复 node3 与 node1、node2 之间的网络连接。

```
# iptables -F

# date "+%Y-%m-%d %H:%M:%S"
2022-07-31 13:07:30
```

首先看看 node3 的日志。

```
2022-07-31T13:07:30.464179-00:00 0 [Warning] [MY-011494] [Repl] Plugin group_replication reported:
'Member with address 192.168.244.10:3306 is reachable again.'
2022-07-31T13:07:30.464226-00:00 0 [Warning] [MY-011494] [Repl] Plugin group_replication reported:
'Member with address 192.168.244.20:3306 is reachable again.'
2022-07-31T13:07:30.464239-00:00 0 [Warning] [MY-011498] [Repl] Plugin group_replication reported:
'The member has resumed contact with a majority of the members in the group. Regular operation is restored
and transactions are unblocked.'
2022-07-31T13:07:37.458761-00:00 0 [ERROR] [MY-011505] [Repl] Plugin group_replication reported:
'Member was expelled from the group due to network failures, changing member status to ERROR.'
2022-07-31T13:07:37.459011-00:00 0 [Warning] [MY-011630] [Repl] Plugin group_replication reported:
'Due to a plugin error, some transactions were unable to be certified and will now rollback.'
2022-07-31T13:07:37.459037-00:00 0 [ERROR] [MY-011712] [Repl] Plugin group_replication reported: 'The
server was automatically set into read only mode after an error was detected.'
2022-07-31T13:07:37.459431-00:00 31 [ERROR] [MY-011615] [Repl] Plugin group_replication reported:
'Error while waiting for conflict detection procedure to finish on session 31'
2022-07-31T13:07:37.459478-00:00 31 [ERROR] [MY-010207] [Repl] Run function 'before_commit' in plugin
'group_replication' failed
2022-07-31T13:07:37.459811-00:00 33 [System] [MY-011565] [Repl] Plugin group_replication reported:
'Setting super_read_only=ON.'

2022-07-31T13:07:37.465738-00:00 34 [System] [MY-013373] [Repl] Plugin group_replication reported:
'Started auto-rejoin procedure attempt 1 of 3'
2022-07-31T13:07:37.496466-00:00 0 [System] [MY-011504] [Repl] Plugin group_replication reported:
```

```
'Group membership changed: This member has left the group.'
2022-07-31T13:07:37.498813-00:00 36 [System] [MY-010597] [Repl] 'CHANGE MASTER TO FOR CHANNEL
'group_replication_applier' executed'. Previous state master_host='<NULL>', master_port= 0,
master_log_file='', master_log_pos= 351, master_bind=''. New state master_host='<NULL>', master_port=
0, master_log_file='', master_log_pos= 4, master_bind=''.
2022-07-31T13:07:39.653028-00:00 34 [System] [MY-013375] [Repl] Plugin group_replication reported:
'Auto-rejoin procedure attempt 1 of 3 finished. Member was able to join the group.'
2022-07-31T13:07:40.653484-00:00 0 [System] [MY-013471] [Repl] Plugin group_replication reported:
'Distributed recovery will transfer data using: Incremental recovery from a group donor'
2022-07-31T13:07:40.653822-00:00 0 [System] [MY-011503] [Repl] Plugin group_replication reported:
'Group membership changed to 192.168.244.10:3306, 192.168.244.20:3306, 192.168.244.30:3306 on view
16592724636525403:4.'
2022-07-31T13:07:40.670530-00:00 46 [System] [MY-010597] [Repl] 'CHANGE MASTER TO FOR CHANNEL
'group_replication_recovery' executed'. Previous state master_host='<NULL>', master_port= 0,
master_log_file='', master_log_pos= 4, master_bind=''. New state master_host='192.168.244.20',
master_port= 3306, master_log_file='', master_log_pos= 4, master_bind=''.
2022-07-31T13:07:40.682990-00:00 47 [Warning] [MY-010897] [Repl] Storing MySQL user name or password
information in the master info repository is not secure and is therefore not recommended. Please consider
using the USER and PASSWORD connection options for START SLAVE; see the 'START SLAVE Syntax' in the
MySQL Manual for more information.
2022-07-31T13:07:40.687566-00:00 47 [System] [MY-010562] [Repl] Slave I/O thread for channel
'group_replication_recovery': connected to master 'repl@192.168.244.20:3306',replication started in
log 'FIRST' at position 4
2022-07-31T13:07:40.717851-00:00 46 [System] [MY-010597] [Repl] 'CHANGE MASTER TO FOR CHANNEL
'group_replication_recovery' executed'. Previous state master_host='192.168.244.20', master_port=
3306, master_log_file='', master_log_pos= 4, master_bind=''. New state master_host='<NULL>',
master_port= 0, master_log_file='', master_log_pos= 4, master_bind=''.
2022-07-31T13:07:40.732297-00:00 0 [System] [MY-011490] [Repl] Plugin group_replication reported:
'This server was declared online within the replication group.'
2022-07-31T13:07:40.732511-00:00 53 [System] [MY-011566] [Repl] Plugin group_replication reported:
'Setting super_read_only=OFF.'
```

日志的输出包括两部分，以空行为分界线。

☐ 当网络连接恢复后，node3 与 node1、node2 重新建立起了连接。node3 发现自己已经被集群驱逐，于是进入 ERROR 状态。

```
mysql> select member_id,member_host,member_port,member_state,member_role from
performance_schema.replication_group_members;
+--------------------------------------+----------------+-------------+--------------+-------------+
| member_id                            | member_host    | member_port | member_state | member_role |
+--------------------------------------+----------------+-------------+--------------+-------------+
| 4cbfdc79-0192-11ed-8b01-02001701bd0a | 192.168.244.30 |        3306 | ERROR        |             |
+--------------------------------------+----------------+-------------+--------------+-------------+
1 row in set (0.00 sec)
```

节点进入 ERROR 状态后，会被自动设置为只读，即日志中看到的 super_read_only=ON。注意，将 ERROR 状态的节点设置为只读是默认行为，与后面提到的 group_replication_exit_state_action 参数无关。

☐ 如果 group_replication_autorejoin_tries 不为 0，对于 ERROR 状态的节点，会自动重试，让其重新加入集群（auto-rejoin）。重试的次数由 group_replication_autorejoin_tries 决定，从 MySQL 8.0.21 开始，默认为 3。重试的时间间隔是 5 分钟。重试成功后，会进入分布式恢复阶段。

接下来看看 node1 的日志。

```
2022-07-31T13:07:39.555613-00:00 0 [System] [MY-011503] [Repl] Plugin group_replication reported:
'Group membership changed to 192.168.244.10:3306, 192.168.244.20:3306, 192.168.244.30:3306 on view
16592724636525403:4.'
2022-07-31T13:07:40.732568-00:00 0 [System] [MY-011492] [Repl] Plugin group_replication reported: 'The
member with address 192.168.244.30:3306 was declared online within the replication group.'
```

node3 又重新加入了集群中。

11.10.2 故障检测流程

结合上面的示例，我们来看看组复制的故障检测流程。

(1) 集群中每个节点都会定期（每秒 1 次）向其他节点发送心跳信息。如果在 5 秒（固定值，无参数调整）内没有收到其他节点的心跳信息，则会将该节点标记为可疑节点，同时会将该节点的状态设置为 `UNREACHABLE`。如果集群中有一半或一半以上的节点显示为 `UNREACHABLE`，则该集群不能对外提供写服务。

(2) 如果在 `group_replication_member_expel_timeout`（从 MySQL 8.0.21 开始，该参数的默认值为 5，单位是秒。最大可设置值为 3600，即 1 小时）时间内，可疑节点恢复正常，则会直接应用 XCom Cache 中的消息。XCom Cache 的大小由 `group_replication_message_cache_size` 决定，默认为 1GB。

(3) 如果在 `group_replication_member_expel_timeout` 时间内，可疑节点没有恢复正常，则会被驱逐出集群。

(4) 少数派节点不会自动离开集群，而是会一直维持当前的状态，直到满足下面两个条件之一。

- 网络恢复正常。
- 达到 `group_replication_unreachable_majority_timeout` 的限制。注意，该参数的起始计算时间是连接断开 5 秒之后，不是可疑节点被驱逐出集群的时间。该参数默认为 0。

无论哪种情况，都会触发：

- 节点状态从 `ONLINE` 切换到 `ERROR`；
- 回滚当前被阻塞的写操作。

```
mysql> delete from slowtech.t1 where id=1;
ERROR 3100 (HY000): Error on observer while running replication hook 'before_commit'.
```

(5) `ERROR` 状态的节点会自动设置为只读。

(6) 如果 `group_replication_autorejoin_tries` 不为 0，对于 `ERROR` 状态的节点，会自动重试，使其重新加入集群。

(7) 如果 `group_replication_autorejoin_tries` 为 0 或重试失败，则会执行 `group_replication_exit_state_action` 指定的操作。可选的操作如下。

- `READ_ONLY`：只读模式。在这种模式下，会将 `super_read_only` 设置为 `ON`。默认值。
- `OFFLINE_MODE`：离线模式。在这种模式下，会将 `offline_mode` 和 `super_read_only` 设置为 `ON`，此时，只有 CONNECTION_ADMIN（SUPER）权限的用户才能登录，普通用户不能登录。

```
# mysql -h 192.168.244.3. -P 3306 -ut1 -p123456
ERROR 3032 (HY000): The server is currently in offline mode
```

- ABORT_SERVER：关闭实例。

11.10.3　XCom Cache

XCom Cache 是 XCom 使用的消息缓存，用来缓存集群节点之间交换的消息。缓存的消息是共识协议的一部分。如果网络不稳定，可能会出现节点失联的情况。如果节点在一定时间（由 group_replication_member_expel_timeout 决定）内恢复正常，它会首先应用 XCom Cache 中的消息。如果 XCom Cache 没有它需要的所有消息，这个节点会被驱逐出集群。驱逐出集群后，如果 group_replication_autorejoin_tries 不为 0，它会重新加入集群。重新加入集群会使用分布式恢复补齐差异数据。相对于直接使用 XCom Cache 中的消息，通过分布式恢复加入集群需要的时间更长，过程也更复杂，并且集群的性能也会受到影响。所以，我们在设置 XCom Cache 的大小时，需预估 group_replication_member_expel_timeout 加 5 秒这段时间内的内存使用量。至于如何预估，后面会介绍相关的系统表。

下面模拟一下 XCom Cache 不足的场景。

(1) 将 group_replication_message_cache_size 调整为最小值（128MB），重启组复制，使其生效。

```
mysql> set global group_replication_message_cache_size=134217728;
Query OK, 0 rows affected (0.00 sec)

mysql> stop group_replication;
Query OK, 0 rows affected (4.15 sec)

mysql> start group_replication;
Query OK, 0 rows affected (3.71 sec)
```

(2) 将 group_replication_member_expel_timeout 调整为 3600。这样，我们才有充足的时间进行测试。

```
mysql> set global group_replication_member_expel_timeout=3600;
Query OK, 0 rows affected (0.01 sec)
```

(3) 断开 node3 与 node1、node2 之间的网络连接。

```
# iptables -A INPUT   -p tcp -s 192.168.244.10 -j DROP
# iptables -A OUTPUT  -p tcp -d 192.168.244.10 -j DROP

# iptables -A INPUT   -p tcp -s 192.168.244.20 -j DROP
# iptables -A OUTPUT  -p tcp -d 192.168.244.20 -j DROP
```

(4) 反复执行大事务。

```
mysql> insert into slowtech.t1(c1) select c1 from slowtech.t1 limit 1000000;
Query OK, 1000000 rows affected (10.03 sec)
Records: 1000000  Duplicates: 0  Warnings: 0
```

(5) 观察错误日志。

如果 node1 或 node2 的错误日志中提示以下信息，则意味着 node3 需要的消息已经被从 XCom Cache 中逐出了。

```
[Warning] [MY-011735] [Repl] Plugin group_replication reported: '[GCS] Messages that are needed to
recover node 192.168.244.30:33061 have been evicted from the message cache. Consider resizing the
maximum size of the cache by  setting group_replication_message_cache_size.'
```

(6) 查看系统表。

除了错误日志，我们还可以通过系统表来判断 XCom Cache 的使用情况。

```
mysql> select * from performance_schema.memory_summary_global_by_event_name where event_name like
"%GCS_XCom::xcom_cache%"\G
*************************** 1. row ***************************
                  EVENT_NAME: memory/group_rpl/GCS_XCom::xcom_cache
                 COUNT_ALLOC: 23678
                  COUNT_FREE: 22754
   SUM_NUMBER_OF_BYTES_ALLOC: 154713397
    SUM_NUMBER_OF_BYTES_FREE: 28441492
              LOW_COUNT_USED: 0
          CURRENT_COUNT_USED: 924
             HIGH_COUNT_USED: 20992
   LOW_NUMBER_OF_BYTES_USED: 0
CURRENT_NUMBER_OF_BYTES_USED: 126271905
   HIGH_NUMBER_OF_BYTES_USED: 146137294
1 row in set (0.00 sec)
```

其中各项解释如下。

- COUNT_ALLOC：缓存过的消息数量。
- COUNT_FREE：从缓存中删除的消息数量。
- CURRENT_COUNT_USED：当前正在缓存的消息数量，等于 COUNT_ALLOC - COUNT_FREE。
- SUM_NUMBER_OF_BYTES_ALLOC：分配的内存大小。
- SUM_NUMBER_OF_BYTES_FREE：被释放的内存大小。
- CURRENT_NUMBER_OF_BYTES_USED：当前正在使用的内存大小，等于 SUM_NUMBER_OF_BYTES_ALLOC - SUM_NUMBER_OF_BYTES_FREE。
- LOW_COUNT_USED 和 HIGH_COUNT_USED：CURRENT_COUNT_USED 的历史最小值和最大值。
- LOW_NUMBER_OF_BYTES_USED 和 HIGH_NUMBER_OF_BYTES_USED：CURRENT_NUMBER_OF_BYTES_USED 的历史最小值和最大值。

如果断开连接之后，在反复执行大事务的过程中，发现 COUNT_FREE 发生了变化，同样意味着 node3 需要的消息已经被从 XCom Cache 中驱逐了。

(7) 恢复 node3 与 node1、node2 之间的网络连接。

在 group_replication_member_expel_timeout 期间，如果网络恢复了，而 node3 需要的消息在 XCom Cache 中不存在了，则 node3 同样会被驱逐出集群。以下是这种场景下 node3 的错误日志。

```
[ERROR] [MY-011735] [Repl] Plugin group_replication reported: '[GCS] Node 0 is unable to get message
{4aec99ca 7562 0}, since the group is too far ahead. Node will now exit.'
[ERROR] [MY-011505] [Repl] Plugin group_replication reported: 'Member was expelled from the group due
to network failures, changing member status to ERROR.'
[ERROR] [MY-011712] [Repl] Plugin group_replication reported: 'The server was automatically set into
read only mode after an error was detected.'
[System] [MY-011565] [Repl] Plugin group_replication reported: 'Setting super_read_only=ON.'
[System] [MY-013373] [Repl] Plugin group_replication reported: 'Started auto-rejoin procedure attempt
1 of 3'
```

11.10.4 注意事项

如果集群中存在 UNREACHABLE 的节点，会有以下限制和不足。

- 不能调整集群的拓扑，包括添加和删除节点。
- 在单主模式下，如果 Primary 节点出现故障了，无法选择新主。
- 如果组复制的一致性级别等于 AFTER 或 BEFORE_AND_AFTER，则写操作会一直等待，直到 UNREACHABLE 节点 ONLINE 并应用该操作。
- 集群吞吐量会下降。如果是单主模式，可将 group_replication_paxos_single_leader（MySQL 8.0.27 引入的）设置为 ON 来解决这个问题。

所以，在线上 group_replication_member_expel_timeout 不宜设置得过大。

11.10.5 参考资料

- 博客文章 "Extending replication instrumentation: account for memory used in Xcom"，作者：André Negrão。
- 博客文章 "MySQL Group Replication - Default response to network partitions has changed"，作者：Pedro Ribeiro。
- 博客文章 "No Ping Will Tear Us Apart - Enabling member auto-rejoin in Group Replication"，作者：Ricardo Ferreira。

11.11 组复制的事务一致性

我们先基于图 11-10 回顾一下组复制的实现原理。图中的 M1、M2 和 M3 是 3 个 Primary 节点，T1 和 T2 是 2 个事务。

- 事务 T1 在 M1 上发起。
- 提交时，事务信息会通过 GCS 模块发送给集群的所有节点。
- 当集群的大多数节点确认接受后，每个成员开始独立进行冲突检测。
  - 如果存在冲突，则回滚事务。
  - 如果没有冲突，M1 会直接提交事务。M2 和 M3 则会先将事务写入 relay log，然后重放。

既然事务在 M2 和 M3 中是异步处理的，就有可能出现下面这种情况：事务 T2 在 M3 上发起，虽然是在 T1 之后执行的，但由于它执行时，T1 对应的操作还没在 M3 上重放，所以 T2 读取的数据就不一定是最新的。

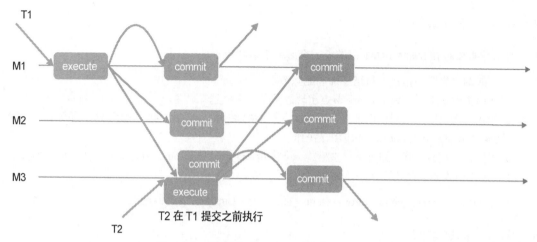

图 11-10　组复制（EVENTUAL 级别）的原理图

从这里我们也能看到，组复制只能保证事务的最终一致性，并不能保证在非写入节点读取的数据是最新的。如果要读取最新数据，可通过 MySQL 8.0.14 引入的 group_replication_consistency 参数实现。

11.1.1　group_replication_consistency

参数 group_replication_consistency 可设置为以下值。

- **EVENTUAL**

最终一致性。即图 11-10 描述的行为，也是 group_replication_consistency 参数的默认值。

- **BEFORE_ON_PRIMARY_FAILOVER**

当发生故障切换时，必须等待新主应用完积压队列中的事务，才能开始响应业务的读写请求，这样能避免在故障切换时业务读取到旧数据。

- **BEFORE**

一个事务会等待它之前的事务执行完再开始执行，这样能确保读取到的数据一定是最新的。图 11-11 是 BEFORE 级别的原理图。

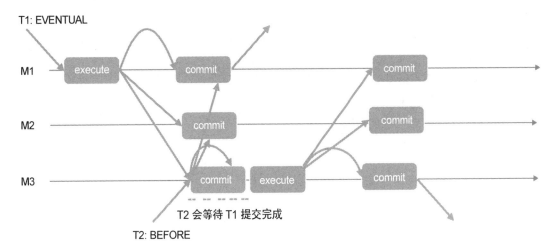

图 11-11 BEFORE 级别的原理图

接下来，我们看看 BEFORE 这个级别的实现细节。

- T2 在 M3 上发起。在执行之前，会通过 GCS 模块发送一条 Sync_before_execution_message 消息。为什么要发送这条消息呢？这实际上利用了 GCS 消息全局有序的特性。一旦收到对这条消息的反馈，就意味着这条消息之前的所有消息 M3 都已经被接受了，自然也包括 T1 这条事务。
- 通过 transactions_latch->waitTicket 暂停事务。
- 处理到 Sync_before_execution_message 对应的数据包时，会调用 transactions_latch->releaseTicket 让事务继续执行。
- M3 获取 group_replication_applier 通道的 retrieved_gtid_set。retrieved_gtid_set 代表 group_replication_applier 通道已经接受的 GTID 集，对应 performance_schema.replication_connection_status 中的 RECEIVED_TRANSACTION_SET。

  ```
  SELECT RECEIVED_TRANSACTION_SET FROM performance_schema.replication_connection_status WHERE
  channel_name = "group_replication_applier"
  ```

- 等待 M3 重放完 retrieved_gtid_set 中的所有事务。
- T2 开始执行。

由此可见，在 BEFORE 级别下，为了确保读取到的数据是最新的，只有 M3 上的事务需要等待，其他两个节点上的事务不受影响。

- AFTER

事务执行时，会等到它在所有的节点上都应用完才给客户端反馈。这样就能确保 AFTER 事务一旦执行完，后续其他节点的事务都能读取到它的最新值。图 11-12 是 AFTER 级别的原理图。

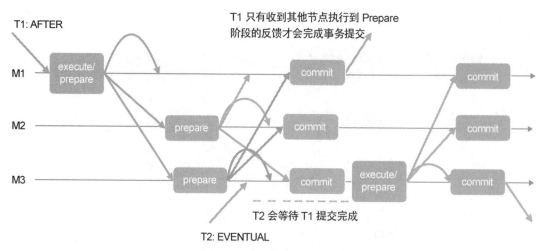

图 11-12 AFTER 级别的原理图

接下来，我们看看 AFTER 这个级别的实现细节。

- T1 在执行到 group_replication_trans_before_commit 时，会将事务消息封装成 Transaction_with_guarantee_message 发送给集群的其他节点。Transaction_with_guarantee_message 会带上 consistency_level，即 group_replication_consistency 的会话值。发送完事务消息，会暂停事务。
- 其他节点接收到这个消息后，首先会进行冲突检测。冲突检测没问题，则写 relay log。在重放 relay log 的过程中，其实也涉及事务的提交，同样会执行 group_replication_trans_before_commit。对于远程事务，如果它的 consistency_level 是 AFTER 或 BEFORE_AND_AFTER，则会向集群的其他节点发送一条 Transaction_prepared_message 消息，代表这个事务在当前节点已经执行到了 Prepare 阶段。
- M1 在收到 M2 和 M3 发送的 Transaction_prepared_message 消息后，会让暂停的事务继续执行。

所以，严格来说，AFTER 级别的事务需满足以下两个条件才会给客户端反馈。

- 事务在当前节点提交。
- 事务在其他节点已经执行到了 Prepare 阶段。

注意，其他节点的事务，如果是在 T1 Prepare 之后、Commit 之前发起的，则会一直等待，直到 T1 提交完成，即使这个事务是 EVENTUAL 级别，如 M3 上的 T2 事务。

所以，对于 AFTER 级别的事务，它会影响其他节点事务的提交。由此带来的好处是，AFTER 级别的事务一旦反馈给客户端，后续其他节点都能读取到它的最新值。

- **BEFORE_AND_AFTER**

BEFORE_AND_AFTER 实际上是 BEFORE 和 AFTER 两种模式的结合，即事务在执行时，

- 会等待它之前的事务执行完才开始执行；

- 会等到它在所有节点上都应用完才给客户端反馈。

图 11-13 是 BEFORE_AND_AFTER 级别的原理图。

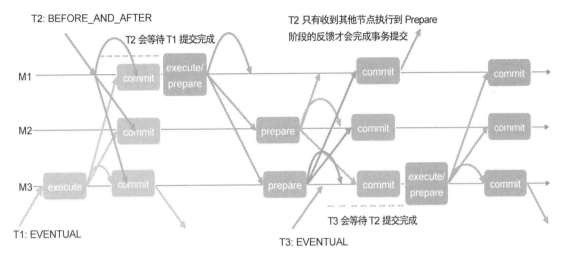

图 11-13　BEFORE_AND_AFTER 级别的原理图

11.11.2　总结

- group_replication_consistency 既可在全局级别设置，也可在会话级别设置。级别越高，对性能的影响也越大。
- 无论是 BEFORE 还是 AFTER，设置的初衷都是为了能读取到最新的数据。只不过两者的作用对象不同，BEFORE 作用在读操作上，只会影响当前会话的事务。而 AFTER 则作用在写操作上，一旦事务提交，后续其他节点都能读到该事务的最新数据，但它会影响其他节点事务的提交。

11.11.3　参考资料

- 官方文档 "Transaction Consistency Guarantees"。
- 博客文章 "Group Replication – Consistent Reads Deep Dive"，作者：Nuno Carvalho。
- 博客文章 "MySQL InnoDB Cluster – consistency levels"，作者：lefred。

11.12　组复制的流量控制机制

组复制是一种无共享（shared-nothing）架构，每个节点都会保存一份数据。虽然支持多点写入，但实际上系统的吞吐量是由处理能力最弱的那个节点决定的。如果各个节点的处理能力参差不齐，那么在写入量比较大的场景中，处理能力弱的节点就容易出现事务堆积。在事务堆积的时候，如果处理能力强的节点出现了故障，能否让处理能力弱（且存在事务堆积）的节点接受业务流量呢？

- 如果不等待堆积事务应用完，直接接受业务流量，一方面会读取到旧数据，另一方面也容易出现写冲突。毕竟基于旧数据进行的写操作，它的 snapshot_version 肯定小于冲突检测数据库中对应记录的 snapshot_version。
- 如果等待堆积事务应用完才接受业务流量，会影响数据库服务的可用性。

为了避免出现这样的两难场景，组复制引入了流量控制（以下简称流控）机制。在实现上，组复制的流控模块会定期检查各个节点的事务堆积情况，如果超过一定值，则会触发流控。流控会基于上一周期各个节点的事务认证情况和事务应用情况，决定当前节点（注意是当前节点，不是其他节点）下个周期的写入配额。超过写入配额的事务操作会被阻塞，等到下个周期才能执行。

接下来，我们通过源码分析一下流控的实现原理，主要包括 3 部分。

- 触发流控的条件。
- 配额的计算逻辑。
- 配额的作用时机。

11.12.1 触发流控的条件

默认情况下，节点的状态信息是每秒发送一次（节点的状态信息是在 flow_control_step 中发送的，发送周期由 group_replication_flow_control_period 决定）。当接收到其他节点的状态信息时，会调用 Flow_control_module::handle_stats_data 来处理。

```
int Flow_control_module::handle_stats_data(const uchar *data, size_t len,
                                            const std::string &member_id) {
  DBUG_TRACE;
  int error = 0;
  Pipeline_stats_member_message message(data, len);

  m_flow_control_module_info_lock->wrlock();
  // m_info 是个字典，定义是 std::map<std::string, Pipeline_member_stats>，
  // 其中，键是节点的地址，值是节点的状态信息
  Flow_control_module_info::iterator it = m_info.find(member_id);
  // 如果 member_id 对应节点的状态信息在 m_info 中不存在，则插入
  if (it == m_info.end()) {
    Pipeline_member_stats stats;

    std::pair<Flow_control_module_info::iterator, bool> ret = m_info.insert(
        std::pair<std::string, Pipeline_member_stats>(member_id, stats));
    error = !ret.second;
    it = ret.first;
  }
  // 更新节点的统计信息
  it->second.update_member_stats(message, m_stamp);

  // 检查是否需要流控
  if (it->second.is_flow_control_needed()) {
    ++m_holds_in_period;
#ifndef NDEBUG
    it->second.debug(it->first.c_str(), m_quota_size.load(),
                     m_quota_used.load());
#endif
  }
```

```
m_flow_control_module_info_lock->unlock();
return error;
}
```

首先判断节点的状态信息是否在 m_info 中存在。如果不存在,则插入。接着通过 update_member_stats 更新节点的统计信息。

更新后的统计信息包括以下两部分数据。

- 当前数据,如 m_transactions_waiting_certification(当前等待认证的事务数)和 m_transactions_waiting_apply(当前等待应用的事务数)。
- 上一周期的增量数据,如 m_delta_transactions_certified(上一周期进行认证的事务数)。

m_delta_transactions_certified 等于 m_transactions_certified(这一次的采集数据)− previous_transactions_certified(上一次的采集数据)。

最后会通过 is_flow_control_needed 判断是否需要流控。如果需要流控,则会将 m_holds_in_period 自增加 1。

如果是 debug 版本,将 log_error_verbosity 设置为 3。当需要流控时,会在错误日志中打印以下信息。

```
[Note] [MY-011726] [Repl] Plugin group_replication reported: 'Flow control - update member stats:
127.0.0.1:33071 stats certifier_queue 0, applier_queue 20 certified 387797 (308), applied 387786 (289),
local 0 (0), quota 400 (274) mode=1'
```

什么时候会触发流控呢?接下来看看 is_flow_control_needed 函数的处理逻辑。

```
bool Pipeline_member_stats::is_flow_control_needed() {
  return (m_flow_control_mode == FCM_QUOTA) &&
         (m_transactions_waiting_certification >
              get_flow_control_certifier_threshold_var() ||
          m_transactions_waiting_apply >
              get_flow_control_applier_threshold_var());
}
```

由此来看,触发流控需满足以下条件。

(1) group_replication_flow_control_mode 设置为 QUOTA。
(2) 当前等待认证的事务数大于 group_replication_flow_control_certifier_threshold。当前等待认证的事务数可通过 performance_schema.replication_group_member_stats 中的 COUNT_TRANSACTIONS_IN_QUEUE 查看。
(3) 当前等待应用的事务数大于 group_replication_flow_control_applier_threshold。当前等待应用的事务数可通过 performance_schema.replication_group_member_stats 中的 COUNT_TRANSACTIONS_REMOTE_IN_APPLIER_QUEUE 查看。

除了条件(1)必须满足,条件(2)和条件(3)满足一个即可。

当需要流控时,会将 m_holds_in_period 自增加 1。m_holds_in_period 这个变量会在 Flow_control_module::flow_control_step 中使用。而 Flow_control_module::flow_control_step 是在 Certifier_ broadcast_thread::dispatcher() 中调用的,每秒执行一次。

```
void Certifier_broadcast_thread::dispatcher() {
  ...
  while (!aborted) {
    ...
    applier_module->run_flow_control_step();
    ...
    struct timespec abstime;
    // 定义超时时长 1 秒
    set_timespec(&abstime, 1);
    mysql_cond_timedwait(&broadcast_dispatcher_cond, &broadcast_dispatcher_lock,
                         &abstime);
    mysql_mutex_unlock(&broadcast_dispatcher_lock);

    broadcast_counter++;
  }
}

void run_flow_control_step() override {
  flow_control_module.flow_control_step(&pipeline_stats_member_collector);
}
```

11.12.2 配额的计算逻辑

接下来重点分析 flow_control_step 函数的处理逻辑。这个函数非常关键,它是整个流控模块的核心,主要用来计算 m_quota_size 和 m_quota_used。m_quota_size 决定了下个周期允许提交的事务数,即我们所说的配额。m_quota_used 用来统计下个周期已经提交的事务数,在该函数中会重置为 0。

```
void Flow_control_module::flow_control_step(
    Pipeline_stats_member_collector *member) {
  // 这里的 seconds_to_skip 实际上就是 group_replication_flow_control_period, 后面会有定义。
  // 虽然 flow_control_step 每秒调用一次,但实际起作用的还是 group_replication_flow_control_period
  if (--seconds_to_skip > 0) return;

  // holds 即 m_holds_in_period
  int32 holds = m_holds_in_period.exchange(0);
  // get_flow_control_mode_var() 即 group_replication_flow_control_mode
  Flow_control_mode fcm =
      static_cast<Flow_control_mode>(get_flow_control_mode_var());
  // get_flow_control_period_var() 即 group_replication_flow_control_period
  seconds_to_skip = get_flow_control_period_var();
  // 计数器
  m_stamp++;
  // 发送当前节点的状态信息
  member->send_stats_member_message(fcm);

  switch (fcm) {
    case FCM_QUOTA: {
      // get_flow_control_hold_percent_var() 即 group_replication_flow_control_hold_percent, 默认为 10。
      // 所以 HOLD_FACTOR 默认为 0.9
      double HOLD_FACTOR =
          1.0 -
          static_cast<double>(get_flow_control_hold_percent_var()) / 100.0;
      // get_flow_control_release_percent_var() 即 group_replication_flow_control_release_percent,
      // 默认为 50。
```

11.12 组复制的流量控制机制

```cpp
      // 所以 RELEASE_FACTOR 默认为 1.5
      double RELEASE_FACTOR =
          1.0 +
          static_cast<double>(get_flow_control_release_percent_var()) / 100.0;
      // get_flow_control_member_quota_percent_var() 即
      // group_replication_flow_control_member_quota_percent, 默认为 0。
      // 所以 TARGET_FACTOR 默认为 0
      double TARGET_FACTOR =
          static_cast<double>(get_flow_control_member_quota_percent_var()) /
          100.0;
      // get_flow_control_max_quota_var() 即 group_replication_flow_control_max_quota, 默认为 0
      int64 max_quota = static_cast<int64>(get_flow_control_max_quota_var());

      // 将上一个周期的 m_quota_size, m_quota_used 赋值给 quota_size, quota_used, 同时自身重置为 0
      int64 quota_size = m_quota_size.exchange(0);
      int64 quota_used = m_quota_used.exchange(0);
      int64 extra_quota = (quota_size > 0 && quota_used > quota_size)
                              ? quota_used - quota_size
                              : 0;

      if (extra_quota > 0) {
        mysql_mutex_lock(&m_flow_control_lock);
        // 发送一个信号, 释放 do_wait() 处等待的事务
        mysql_cond_broadcast(&m_flow_control_cond);
        mysql_mutex_unlock(&m_flow_control_lock);
      }
      // m_holds_in_period 大于 0, 则意味着需要进行流控
      if (holds > 0) {
        uint num_writing_members = 0, num_non_recovering_members = 0;
        // MAXTPS 是 INT 的最大值, 即 2147483647
        int64 min_certifier_capacity = MAXTPS, min_applier_capacity = MAXTPS,
              safe_capacity = MAXTPS;

        m_flow_control_module_info_lock->rdlock();
        Flow_control_module_info::iterator it = m_info.begin();
        // 循环遍历所有节点的状态信息
        while (it != m_info.end()) {
          // 这一段源码中没有, 加到这里可以直观地看到触发流控时每个节点的状态信息
#ifndef NDEBUG
          it->second.debug(it->first.c_str(), quota_size,
                  quota_used);
#endif
          if (it->second.get_stamp() < (m_stamp - 10)) {
            // 如果节点的状态信息在最近 10 个周期内都没有更新, 则将其清理掉
            m_info.erase(it++);
          } else {
            if (it->second.get_flow_control_mode() == FCM_QUOTA) {
              // 如果 group_replication_flow_control_certifier_threshold 大于 0,
              // 且上一个周期进行认证的事务数大于 0,
              // 且当前等待认证的事务数大于 group_replication_flow_control_certifier_threshold,
              // 且上一个周期进行认证的事务数小于 min_certifier_capacity,
              // 则会将上一个周期进行认证的事务数赋予 min_certifier_capacity
              if (get_flow_control_certifier_threshold_var() > 0 &&
                  it->second.get_delta_transactions_certified() > 0 &&
                  it->second.get_transactions_waiting_certification() -
                          get_flow_control_certifier_threshold_var() >
                      0 &&
                  min_certifier_capacity >
                      it->second.get_delta_transactions_certified()) {
                min_certifier_capacity =
```

```cpp
            it->second.get_delta_transactions_certified();
      }

      if (it->second.get_delta_transactions_certified() > 0)
        // safe_capacity 取 safe_capacity 和 it->second.get_delta_transactions_certified()
        // 中的较小值
        safe_capacity =
            std::min(safe_capacity,
                     it->second.get_delta_transactions_certified());

      // 针对的是 applier，逻辑同 certifier 一样
      if (get_flow_control_applier_threshold_var() > 0 &&
          it->second.get_delta_transactions_applied() > 0 &&
          it->second.get_transactions_waiting_apply() -
                  get_flow_control_applier_threshold_var() >
              0) {
        if (min_applier_capacity >
            it->second.get_delta_transactions_applied())
          min_applier_capacity =
              it->second.get_delta_transactions_applied();

        if (it->second.get_delta_transactions_applied() > 0)
          // 如果上一个周期有事务应用，说明该节点不是 recovering 节点
          num_non_recovering_members++;
      }

      if (it->second.get_delta_transactions_applied() > 0)
        // safe_capacity 取 safe_capacity 和 it->second.get_delta_transactions_applied()
        // 中的较小值
        safe_capacity = std::min(
            safe_capacity, it->second.get_delta_transactions_applied());

      if (it->second.get_delta_transactions_local() > 0)
        // 如果上一个周期有本地事务，则意味着该节点存在写入
        num_writing_members++;
    }
    ++it;
  }
}
m_flow_control_module_info_lock->unlock();

num_writing_members = num_writing_members > 0 ? num_writing_members : 1;
// min_capacity 取 min_certifier_capacity 和 min_applier_capacity 中的较小值
int64 min_capacity = (min_certifier_capacity > 0 &&
                      min_certifier_capacity < min_applier_capacity)
                         ? min_certifier_capacity
                         : min_applier_capacity;

// lim_throttle 是最小配额
int64 lim_throttle = static_cast<int64>(
    0.05 * std::min(get_flow_control_certifier_threshold_var(),
                    get_flow_control_applier_threshold_var()));
// get_flow_control_min_recovery_quota_var() 即 group_replication_flow_control_min_recovery_quota
if (get_flow_control_min_recovery_quota_var() > 0 &&
    num_non_recovering_members == 0)
  lim_throttle = get_flow_control_min_recovery_quota_var();
// get_flow_control_min_quota_var() 即 group_replication_flow_control_min_quota
if (get_flow_control_min_quota_var() > 0)
  lim_throttle = get_flow_control_min_quota_var();
```

```cpp
      // min_capacity 不能太小，不能低于 lim_throttle
      min_capacity =
          std::max(std::min(min_capacity, safe_capacity), lim_throttle);

      // HOLD_FACTOR 默认为 0.9
      quota_size = static_cast<int64>(min_capacity * HOLD_FACTOR);

      // max_quota 是由 group_replication_flow_control_max_quota 定义的，即 quota_size 不能
      // 超过 max_quota
      if (max_quota > 0) quota_size = std::min(quota_size, max_quota);

      // num_writing_members 是有实际写操作的节点数
      if (num_writing_members > 1) {
        // 如果没有设置 group_replication_flow_control_member_quota_percent，则按照节点数
        // 平分 quota_size
        if (get_flow_control_member_quota_percent_var() == 0)
          quota_size /= num_writing_members;
        else
          // 如果设置了，则当前节点的 quota_size 等于
          // quota_size * group_replication_flow_control_member_quota_percent / 100
          quota_size = static_cast<int64>(static_cast<double>(quota_size) *
                                         TARGET_FACTOR);
      }
      // quota_size 还会减去上个周期超额使用的配额
      quota_size =
          (quota_size - extra_quota > 1) ? quota_size - extra_quota : 1;
#ifndef NDEBUG
      LogPluginErr(INFORMATION_LEVEL, ER_GRP_RPL_FLOW_CONTROL_STATS,
                   quota_size, get_flow_control_period_var(),
                   num_writing_members, num_non_recovering_members,
                   min_capacity, lim_throttle);
#endif
    } else {
      // 对应 m_holds_in_period = 0 的场景，RELEASE_FACTOR 默认为 1.5
      if (quota_size > 0 && get_flow_control_release_percent_var() > 0 &&
          (quota_size * RELEASE_FACTOR) < MAXTPS) {
        // 当流控结束后，quota_size = 上一个周期的 quota_size * 1.5
        int64 quota_size_next =
            static_cast<int64>(quota_size * RELEASE_FACTOR);
        quota_size =
            quota_size_next > quota_size ? quota_size_next : quota_size + 1;
      } else
        quota_size = 0;
    }

    if (max_quota > 0)
      // quota_size 会取 quota_size 和 max_quota 中的较小值
      quota_size =
          std::min(quota_size > 0 ? quota_size : max_quota, max_quota);
    // 最后，将 quota_size 赋值给 m_quota_size，m_quota_used 重置为 0
    m_quota_size.store(quota_size);
    m_quota_used.store(0);
    break;
  }

  // 如果 group_replication_flow_control_mode 为 DISABLED，
  // 则会将 m_quota_size 和 m_quota_used 置为 0，这个时候会禁用流控
  case FCM_DISABLED:
    m_quota_size.store(0);
```

```
        m_quota_used.store(0);
        break;

    default:
        assert(0);
    }

    if (local_member_info->get_recovery_status() ==
        Group_member_info::MEMBER_IN_RECOVERY) {
      applier_module->get_pipeline_stats_member_collector()
          ->compute_transactions_deltas_during_recovery();
    }
}
```

代码的逻辑看上去有点复杂。接下来，我们结合一个具体的示例看看 flow_control_step 函数的实现逻辑。

测试集群由 3 个节点组成：127.0.0.1:33061、127.0.0.1:33071 和 127.0.0.1:33081。测试集群运行在多主模式下。使用 sysbench 对 127.0.0.1:33061 进行插入测试（oltp_insert）。为了更容易触发流控，这里将 127.0.0.1:33061 节点的 group_replication_flow_control_applier_threshold 设置为了 10。

以下是触发流控时 127.0.0.1:33061 的日志信息。

```
[Note] [MY-011726] [Repl] Plugin group_replication reported: 'Flow control - update member stats:
127.0.0.1:33061 stats certifier_queue 0, applier_queue 0 certified 7841 (177), applied 0 (0), local
7851 (177), quota 146 (156) mode=1'
[Note] [MY-011726] [Repl] Plugin group_replication reported: 'Flow control - update member stats:
127.0.0.1:33071 stats certifier_queue 0, applier_queue 0 certified 7997 (186), applied 8000 (218), local
0 (0), quota 146 (156) mode=1'
[Note] [MY-011726] [Repl] Plugin group_replication reported: 'Flow control - update member stats:
127.0.0.1:33081 stats certifier_queue 0, applier_queue 15 certified 7911 (177), applied 7897 (195),
local 0 (0), quota 146 (156) mode=1'
[Note] [MY-011727] [Repl] Plugin group_replication reported: 'Flow control: throttling to 149 commits
per 1 sec, with 1 writing and 1 non-recovering members, min capacity 177, lim throttle 0'
```

以 127.0.0.1:33081 的状态数据为例，我们看看输出中各项的具体含义。

- certifier_queue 0：认证队列的长度。
- applier_queue 15：应用队列的长度。
- certified 7911 (177)：7911 是已经认证的总事务数，177 是上一周期进行认证的事务数（m_delta_transactions_certified）。
- applied 7897 (195)：7897 是已经应用的总事务数，195 是上一周期应用的事务数（m_delta_transactions_applied）。
- local 0 (0)：本地事务数。括号中的 0 是上一周期的本地事务数（m_delta_transactions_local）。
- quota 146 (156)：146 是上一周期的 quota_size，156 是上一周期的 quota_used。
- mode=1：mode 等于 1 意味着开启流控。

因为 127.0.0.1:33081 中 applier_queue 的长度（15）超过 127.0.0.1:33061 中的 group_replication_flow_control_applier_threshold（10），所以会触发流控。触发流控后，会调用 flow_control_step 计算下一周期的 m_quota_size。

(1) 循环遍历各节点的状态信息。集群的吞吐量（min_capacity）取各个节点 m_delta_transactions_certified 和 m_delta_transactions_applied 的最小值。具体在本例中，min_capacity 等于 min(177, 186, 218, 177, 195)，即 177。

(2) min_capacity 不能太小，不能低于 lim_throttle。lim_throttle 的取值逻辑如下。

- 初始值 = 0.05 * min (group_replication_flow_control_applier_threshold, group_replication_flow_control_certifier_threshold)。具体在本例中：min_capacity = 0.05 * min(10, 25000) = 0.5。
- 如果设置了 group_replication_flow_control_min_recovery_quota 且 num_non_recovering_members 为 0，则会将 group_replication_flow_control_min_recovery_quota 赋值给 min_capacity。num_non_recovering_members 什么时候会为 0 呢？在新节点加入时，且流控是因为认证队列中积压的事务超过阈值而触发的。
- 如果设置了 group_replication_flow_control_min_quota，则会将 group_replication_flow_control_min_quota 赋值给 min_capacity。

(3) quota_size = min_capacity * 0.9 = 177 × 0.9 ≈ 159。这里的 0.9 是通过 1 - group_replication_flow_control_hold_percent / 100 得到的。之所以要预留部分配额，主要是为了处理积压事务。

(4) quota_size 不能太大，不能超过 group_replication_flow_control_max_quota。

(5) 注意，这里计算的 quota_size 是集群的吞吐量，不是单个节点的吞吐量。如果要计算当前节点的吞吐量，最简单的办法是将 quota_size 除以有实际写操作的节点数（num_writing_members）。怎么判断一个节点是否进行了实际的写操作呢？很简单，上一个周期有本地事务提交，即 m_delta_transactions_local 大于 0。具体在本例中，只有一个写节点，所以当前节点的 quota_size 就等于集群的 quota_size，即 159。除了均分这个简单直接的方法，如果希望某些节点比其他节点承担更多的写操作，也可通过 group_replication_flow_control_member_quota_percent 设置权重。这个时候，当前节点吞吐量的计算公式为：quota_size * group_replication_flow_control_member_quota_percent / 100。

(6) 最后，当前节点的 quota_size 还会减去上一个周期超额使用的配额（extra_quota）。上一个周期的 extra_quota = 上一个周期的 quota_used - quota_size，即 156 - 146 = 10。所以，当前节点的 quota_size 就等于 159 - 10 = 149，和日志中的输出完全一致。为什么会出现配额超额使用的情况呢？这个问题后面会提到。

(7) 当 m_holds_in_period 又恢复为 0 时，就意味着流控结束。这个时候，MGR 不会完全放开配额的限制，否则写入量太大，容易出现突刺。MGR 采取的是一种渐进式的恢复策略：下一个周期的 quota_size = 上一个周期的 quota_size * (1 + group_replication_flow_control_release_percent / 100)。

(8) 如果 group_replication_flow_control_mode 是 DISABLED，则会将 m_quota_size 和 m_quota_used 置为 0。m_quota_size 置为 0，实际上会禁用流控。

11.12.3 配额的作用时机

既然我们已经计算出了下一个周期的 m_quota_size，什么时候使用它呢？答案是，在事务提交之后，GCS 广播事务消息之前。

```
int group_replication_trans_before_commit(Trans_param *param) {
  ...
  // 判断事务是否需要等待
  applier_module->get_flow_control_module()->do_wait();

  // 广播事务消息
  send_error = gcs_module->send_transaction_message(*transaction_msg);
  ...
}
```

接下来，我们看看 do_wait 函数的处理逻辑。

```
int32 Flow_control_module::do_wait() {
  DBUG_TRACE;
  // 首先加载 m_quota_size
  int64 quota_size = m_quota_size.load();
  // m_quota_used 自增加 1
  int64 quota_used = ++m_quota_used;

  if (quota_used > quota_size && quota_size != 0) {
    struct timespec delay;
    set_timespec(&delay, 1);

    mysql_mutex_lock(&m_flow_control_lock);
    mysql_cond_timedwait(&m_flow_control_cond, &m_flow_control_lock, &delay);
    mysql_mutex_unlock(&m_flow_control_lock);
  }

  return 0;
}
```

如果 quota_size 等于 0，do_wait 会直接返回，不会执行任何等待操作。这也就是为什么当 m_quota_size 等于 0 时，会禁用流控操作。

如果 quota_used 大于 quota_size 且 quota_size 不等于 0，则意味着当前周期的配额用完了。这个时候，会调用 mysql_cond_timedwait 触发等待。这里的 mysql_cond_timedwait 会在两种情况下退出：收到 m_flow_control_cond 信号（该信号会在 flow_control_step 函数中发出）；超时。这里的超时时间是 1 秒。

需要注意的是，m_quota_used 是自增在前，然后才进行判断，这也就是为什么配额会出现超额使用的情况。在等待的过程中，如果客户端是多线程并发写入，这里会等待多个事务，并且超额使用的事务数不会多于客户端的并发线程数。所以，在上面的示例中，quota_used（156）比 quota_size（146）多 10，这实际上是 sysbench 并发线程数的数量。

接下来，我们看看示例中的这 156 个事务在 do_wait 处的等待时间。

```
...
0.000020
0.000017
0.000023
0.000073
0.000023
0.000018
0.570180
0.567999
```

```
0.561916
0.561162
0.558930
0.557714
0.556683
0.550581
0.548102
0.547176
```

前 146 个事务的平均等待时间是 0.000 035 秒，后 10 个事务的平均等待时间是 0.558 044 秒。很显然，后 10 个事务是被流控了，最后被 flow_control_step（默认一秒执行一次）中发送的 m_flow_control_cond 信号释放的。

11.12.4 流控的相关参数

- **group_replication_flow_control_mode**

是否开启流控。默认为 QUOTA，基于配额进行流控。如果设置为 DISABLED，则会关闭流控。

- **group_replication_flow_control_period**

流控周期。有效值是 1~60，单位是秒。默认为 1。注意，各个节点的流控周期应保持一致，否则会将周期较短的节点配额作为集群配额。

看下面这个示例，127.0.0.1:33061 这个节点的 group_replication_flow_control_period 是 10，而其他两个节点的 group_replication_flow_control_period 是 1。

```
2022-08-27T19:01:50.699939+08:00 63 [Note] [MY-011726] [Repl] Plugin group_replication reported: 'Flow control - update member stats: 127.0.0.1:33061 stats certifier_queue 0, applier_queue 0 certified 217069 (1860), applied 1 (0), local 217070 (1861), quota 28566 (1857) mode=1'
2022-08-27T19:01:50.699955+08:00 63 [Note] [MY-011726] [Repl] Plugin group_replication reported: 'Flow control - update member stats: 127.0.0.1:33071 stats certifier_queue 0, applier_queue 2 certified 218744 (157), applied 218746 (165), local 0 (0), quota 28566 (1857) mode=1'
2022-08-27T19:01:50.699967+08:00 63 [Note] [MY-011726] [Repl] Plugin group_replication reported: 'Flow control - update member stats: 127.0.0.1:33081 stats certifier_queue 16383, applier_queue 0 certified 0 (0), applied 0 (0), local 0 (0), quota 28566 (1857) mode=1'
2022-08-27T19:01:50.699979+08:00 63 [Note] [MY-011726] [Repl] Plugin group_replication reported: 'Flow control: throttling to 141 commits per 10 sec, with 1 writing and 0 non-recovering members, min capacity 157, lim throttle 100'
```

最后，会将 127.0.0.1:33071 这个节点 1 秒的配额（157 × 0.9）当作 127.0.0.1:33061 这个节点 10 秒的配额。所以，我们会观察到下面这个现象。

```
执行时间      TPS
19:01:50      49
19:01:51      93
19:01:52       1
19:01:53       1
19:01:54       1
19:01:55       1
19:01:56       1
19:01:57       1
19:01:58       1
19:01:59       1
19:02:00       1
```

节点 127.0.0.1:33061 在前 2 秒内就使用完了所有配额，导致后面的事务需要等待 1 秒（mysql_cond_timedwait 的超时时长）之后才能被处理。因为模拟时指定的并发线程数是 1，所以这里的 TPS 是 1。

- **group_replication_flow_control_applier_threshold**

待应用的事务数如果超过 group_replication_flow_control_applier_threshold 的设置，则会触发流控，该参数默认为 25000。

- **group_replication_flow_control_certifier_threshold**

待认证的事务数如果超过 group_replication_flow_control_certifier_threshold 的设置，则会触发流控，该参数默认为 25000。

- **group_replication_flow_control_min_quota、group_replication_flow_control_min_recovery_quota**

两个参数都会决定当前节点下一个周期的最小配额，只不过 group_replication_flow_control_min_recovery_quota 适用于新节点加入时的分布式恢复阶段，group_replication_flow_control_min_quota 则适用于所有场景。如果两者同时设置了，则 group_replication_flow_control_min_quota 的优先级更高。两者默认都为 0，即不限制。

- **group_replication_flow_control_max_quota**

当前节点下一个周期的最大配额。默认为 0，即不限制。

- **group_replication_flow_control_member_quota_percent**

分配给当前成员的配额比例。有效值是 0 ~ 100。默认为 0，此时，节点配额等于集群配额除以上一个周期写节点的数量。注意，这里的写节点指的是有实际写操作的节点，不是仅指 Primary 节点。

- **group_replication_flow_control_hold_percent**

预留配额的比例。有效值是 0 ~ 100，默认为 10。预留的配额可用来处理落后节点积压的事务。

- **group_replication_flow_control_release_percent**

当流控结束后，会逐渐增加吞吐量以避免出现突刺，公式为：下一周期的 quota_size = 上一周期的 quota_size * (1 + group_replication_flow_control_release_percent / 100)。有效值是 0 ~ 1000，默认为 50。

11.12.5 总结

- 从可用性的角度出发，不建议在线上关闭流控。虽然 Primary 节点出现故障的概率很小，但按照墨菲定律的说法，任何有可能发生的事情最后一定会发生。在线上还是不能心存侥幸。
- 流控限制的是当前节点的流量，不是其他节点的。
- 流控参数在各节点应保持一致，尤其是 group_replication_flow_control_period。

11.12.6 参考资料

- 官方工作日志"WL#9838: Group Replication: Flow-control fine tuning"。
- 知乎专栏"MySQL Group Replication 流控实现分析",作者:温正湖。

11.13 组复制的重点参数

截止到 MySQL 8.0.27,组复制一共有 61 个参数。反观 MySQL 5.7,只有 36 个参数。在新增的参数里,很多是跟新特性有关的。这也是为什么如果要使用组复制,一定要使用 MySQL 8.0,而不是 MySQL 5.7。

因为大部分参数在前面解释过,所以这里不再赘述,只介绍其他需要注意的参数。

其他重点参数

- **group_replication_allow_local_lower_version_join**

是否允许低版本的节点加入集群,默认为 OFF,即不允许。这又分为两种情况。

- 如果集群的最低版本高于等于 MySQL 8.0.17,则会比较 patch_version(如 MySQL 8.0.27 中的 27),任何低于这个最低版本的节点都不能加入集群。例如,节点版本分别是 8.0.26、8.0.27、8.0.27 的集群,就不允许 8.0.26 之前版本的节点加入。
- 如果集群的最低版本低于 MySQL 8.0.17,则会比较 major_version(如 MySQL 8.0.27 中的 8),任何低于这个最低版本 major_version 的节点都不能加入集群。例如,节点版本分别是 8.0.15、8.0.16、8.0.16 的集群,就不允许 MySQL 5.7 的节点加入。

将 group_replication_allow_local_lower_version_join 设置为 ON,则允许低版本的节点加入集群。低版本的节点加入集群后,建议只允许低版本的节点接受写操作,否则容易出现不兼容的情况。

- **group_replication_communication_debug_options**

设置 GCS 和 XCom 的调试信息级别,调试信息会打印在数据目录下的 GCS_DEBUG_TRACE 文件中。支持的选项如下。

- GCS_DEBUG_NONE:不打印任何调试信息。默认值。
- GCS_DEBUG_BASIC:打印 GCS 中的基本调试信息。
- GCS_DEBUG_TRACE:打印 GCS 中的详细调试信息。
- XCOM_DEBUG_BASIC:打印 XCom 中的基本调试信息。
- XCOM_DEBUG_TRACE:打印 XCom 中的详细调试信息。
- GCS_DEBUG_ALL:打印 GCS 和 XCom 中的所有调试信息。

可将以上选项组合起来使用,例如:

```
set global group_replication_communication_debug_options='GCS_DEBUG_BASIC,XCOM_DEBUG_BASIC';
```

- **group_replication_communication_stack**

在 MySQL 8.0.27 之前，组成员之间的连接通信栈只有一种，即 XCom。在 XCom 通信栈中，XCom 自己实现了安全协议，包括 TLS/SSL 及 IP 白名单（group_replication_ip_allowlist）。但它不支持身份验证和网络命名空间（network namespace）。网络命名空间常用在容器和虚拟环境中，每个命名空间都有自己的 IP 地址、网络接口和路由表。

从 MySQL 8.0.27 开始，MySQL 通信栈得到了支持。MySQL 通信栈是 MySQL Server 层实现的，支持身份验证和网络命名空间。注意，在使用 MySQL 通信栈时，用来进行分布式恢复的复制用户需要被授予 GROUP_REPLICATION_STREAM 和 CONNECTION_ADMIN 权限。

- **group_replication_components_stop_timeout**

当关闭组复制时，会关闭组复制的所有模块。正常情况下，模块都能很快关闭，但在某些特殊场景，模块关闭操作可能会被阻塞。例如，节点上的全局读锁就会阻塞 Applier 线程关闭。通过 group_replication_components_stop_timeout 可以限制模块关闭时的超时时长。该参数从 MySQL 8.0.27 开始默认为 300 秒，在此之前是 31 536 000 秒（365 天）。

以下是因为全局读锁导致 STOP GROUP_REPLICATION 命令超时报错的一个示例。

```
mysql> stop group_replication;
ERROR 3095 (HY000): The STOP GROUP_REPLICATION command execution is incomplete: The applier thread got the stop signal while it was busy. The applier thread will stop once the current task is complete.
```

需要注意的是，在本例中，虽然 STOP GROUP_REPLICATION 执行报错了，但实际上这个节点在命令执行后不久就已经成功地离开了集群，这一点从错误日志中可以看出来。

```
2022-10-04T11:47:27.388621-00:00 0 [System] [MY-011504] [Repl] Plugin group_replication reported: 'Group membership changed: This member has left the group.'
2022-10-04T11:52:27.406373-00:00 342 [ERROR] [MY-011654] [Repl] Plugin group_replication reported: 'On shutdown there was a timeout on the Group Replication applier termination.'
2022-10-04T11:52:27.406639-00:00 342 [System] [MY-011651] [Repl] Plugin group_replication reported: 'Plugin 'group_replication' has been stopped.'
2022-10-04T11:52:29.406755-00:00 370 [ERROR] [MY-011535] [Repl] Plugin group_replication reported: 'Failed to stop the group replication applier thread.'
```

- **group_replication_compression_threshold**

如果组复制发送的消息大小超过 group_replication_compression_threshold，则会使用 LZ4 算法进行压缩。该参数默认为 1 000 000 字节（约 0.95MB）。如果设置为 0，则会禁用消息压缩。

- **group_replication_gtid_assignment_block_size**

为每个节点预留的 GTID 块的大小，默认为 1000000。这么做主要是为了避免高并发场景下 GTID 相关操作的争用问题。

下面是在多主模式下 3 个节点都有写入时，GTID 的生成情况。可以看到，每个节点的 GTID 初始值都不一样。

```
mysql> show master status;
+------------------+----------+--------------+------------------+-------------------------------------------------------------------+
| File             | Position | Binlog_Do_DB | Binlog_Ignore_DB | Executed_Gtid_Set                                                 |
+------------------+----------+--------------+------------------+-------------------------------------------------------------------+
| mysql-bin.000001 |     4196 |              |                  | d754479e-ffe4-11ec-a7ef-02001700754e:1-6:1000005-1000006:2000005-2000006 |
+------------------+----------+--------------+------------------+-------------------------------------------------------------------+
1 row in set (0.00 sec)
```

注意，如果某个节点生成的 GTID 超过了预留值，则会与其他节点的 GTID 进行合并，并不会冲突。

- **group_replication_paxos_single_leader**

默认情况下，组复制中的每个节点都有一个 Paxos Group，每个节点都是各自 Paxos Group 的 Leader。当节点出现故障时，就需要跳过故障节点相关的槽。而这会导致集群吞吐下降。

下面测试了两种场景。

☐ 使用 kill -9 关闭 Secondary 节点。

```
[ 20s ] thds: 30 tps: 562.81 qps: 11242.42 (r/w/o: 7875.38/2241.42/1125.61) lat (ms,95%): 108.68 err/s: 0.00 reconn/s: 0.00
[ 30s ] thds: 30 tps: 178.40 qps: 3584.43 (r/w/o: 2503.02/724.61/356.80) lat (ms,95%): 893.56 err/s: 0.00 reconn/s: 0.00
[ 40s ] thds: 30 tps: 43.10 qps: 862.00 (r/w/o: 603.40/172.40/86.20) lat (ms,95%): 1352.03 err/s: 0.00 reconn/s: 0.00
[ 50s ] thds: 30 tps: 47.50 qps: 893.00 (r/w/o: 623.00/178.00/92.00) lat (ms,95%): 4437.27 err/s: 0.00 reconn/s: 0.00
[ 60s ] thds: 30 tps: 564.79 qps: 11343.99 (r/w/o: 7947.02/2264.38/1132.59) lat (ms,95%): 121.08 err/s: 0.00 reconn/s: 0.00
```

可以看到，TPS 下降得很明显。TPS 下降的时间，其实就是节点被检测出故障到被踢出集群的这段时间。注意，如果通过 STOP GROUP_REPLICATION 命令让节点正常离开，则 TPS 不会下降。

☐ 节点重启后，重新加入集群。

这同样会导致 TPS 下降。

```
[ 110s ] thds: 30 tps: 576.82 qps: 11530.10 (r/w/o: 8071.21/2305.56/1153.33) lat (ms,95%): 101.13 err/s: 0.00 reconn/s: 0.00
[ 120s ] thds: 30 tps: 122.00 qps: 2455.80 (r/w/o: 1716.70/494.80/244.30) lat (ms,95%): 909.80 err/s: 0.00 reconn/s: 0.00
[ 130s ] thds: 30 tps: 85.60 qps: 1706.98 (r/w/o: 1197.39/338.40/171.20) lat (ms,95%): 909.80 err/s: 0.00 reconn/s: 0.00
[ 140s ] thds: 30 tps: 528.70 qps: 10574.77 (r/w/o: 7401.28/2116.19/1057.30) lat (ms,95%): 116.80 err/s: 0.00 reconn/s: 0.00
```

将 group_replication_paxos_single_leader 设置为 ON，可以解决上述问题。此时，组复制会将默认的 Multiple Leaders 模式切换为 Single Leader 模式。在 Single Leader 模式下，只有 Primary 节点才是 Leader。

还是上面的两个测试场景，我们看看把 group_replication_paxos_single_leader 设置为 ON 的效果。

- 使用 kill -9 关闭 Secondary 节点。

```
[ 20s ] thds: 30 tps: 527.61 qps: 10551.83 (r/w/o: 7386.16/2110.35/1055.32) lat (ms,95%): 112.67 err/s: 0.00 reconn/s: 0.00
[ 30s ] thds: 30 tps: 568.50 qps: 11378.75 (r/w/o: 7966.13/2275.61/1137.00) lat (ms,95%): 121.08 err/s: 0.00 reconn/s: 0.00
[ 40s ] thds: 30 tps: 580.20 qps: 11603.04 (r/w/o: 8121.76/2320.89/1160.39) lat (ms,95%): 112.67 err/s: 0.00 reconn/s: 0.00
```

- 节点重新加入集群。

```
[ 110s ] thds: 30 tps: 581.10 qps: 11616.28 (r/w/o: 8132.55/2321.62/1162.11) lat (ms,95%): 125.52 err/s: 0.00 reconn/s: 0.00
[ 120s ] thds: 30 tps: 583.41 qps: 11664.88 (r/w/o: 8165.10/2332.96/1166.83) lat (ms,95%): 114.72 err/s: 0.00 reconn/s: 0.00
[ 130s ] thds: 30 tps: 569.88 qps: 11409.34 (r/w/o: 7983.65/2286.03/1139.66) lat (ms,95%): 116.80 err/s: 0.00 reconn/s: 0.00
```

可以看到，无论是使用 kill -9 关闭实例还是节点加入集群，都不会导致集群吞吐量下降。

注意，`group_replication_paxos_single_leader` 是 MySQL 8.0.27 引入的，只适用于单主模式。

11.14 本章总结

本章系统地介绍了组复制的引入背景、部署、监控及常见管理操作。

组复制支持两种部署方式：单主模式和多主模式。注意，两者的本质区别是多主模式会开启冲突检测，而单主模式不会。组复制在部署和使用时虽然有一定的要求和限制，但细究下来，很多限制其实是针对多主模式的。这也就是为什么在线上更推荐使用单主模式。

组复制是基于数据库状态机实现的，数据库状态机的原理可简单理解为：只要初始数据一致，增量数据一致，就能保证数据的最终一致。具体到组复制中，初始数据一致是通过分布式恢复实现的，增量数据一致是通过 Xcom 和冲突检测模块实现的。Xcom 是组复制的消息通信层，它保证了消息的全局有序和安全传递。安全传递指的是消息必须发送给所有非故障节点，且只有在大多数节点确认接受后才能发送给上层应用。而冲突检测模块则保证了节点会以确定的行为处理同一事务，要么全部认证通过，要么全部认证失败。

在 MySQL 8.0.17 之前，分布式恢复只支持 binlog 这一种同步方式。从 MySQL 8.0.17 开始，得益于克隆插件的引入，开始支持物理备份恢复方式。

组复制的冲突检测依赖于 `write_set` 和冲突检测数据库。冲突检测数据库每 60 秒清理一次。

在通过冲突检测后，远程事务首先会写入 relay log，然后重放，这就意味着我们在非写入节点读取的数据并不一定是最新的。如果想读取到最新数据，可通过 `group_replication_consistency` 来实现。实现的方式有两种。

- 在读取时实现，将 `group_replication_consistency` 设置为 BEFORE。
- 在写入时实现，将 `group_replication_consistency` 设置为 AFTER。这样，其他节点想要读取这个事务的最新数据，就无须在读取时设置 `group_replication_consistency = BEFORE` 了。

组复制虽然支持多点写入，但无共享的架构决定了系统的吞吐量是由处理能力最弱的那个节点决定的。如果各个节点的处理能力不一样，事务堆积在所难免。当 Primary 节点出现故障时，事务堆积会影响数据库服务的可用性。从可用性的角度出发，不建议在线上关闭流控。

为了避免节点出现故障（或添加新的节点）时导致集群吞吐下降，在单主模式下，建议将 group_replication_paxos_single_leader 设置为 ON。

最后，虽然组复制是 MySQL 5.7 引入的，但很多参数、新特性只在 MySQL 8.0 中才有，所以建议使用 MySQL 8.0 而不是 MySQL 5.7 来部署组复制。

重点问题回顾

- 组复制的部署。
- 单主模式和多主模式的区别。
- 如何监控组复制。
- 组复制的限制。
- 组复制的新主选举算法。
- 大事务的危害。
- 如何降低大事务对组复制的影响。
- 事务在组复制中的处理流程。
- 分布式恢复的实现原理。
- 使用克隆插件执行恢复操作的前提条件。
- 什么是 write_set？
- 组复制是如何进行冲突检测的？
- 冲突检测数据库多久清理一次？
- 组复制的故障检测流程。
- 为什么 group_replication_member_expel_timeout 不宜设置得过大？
- 组复制的流控机制。
- 为什么在单主模式下建议将 group_replication_paxos_single_leader 设置为 ON？

第 12 章
InnoDB Cluster

InnoDB Cluster 是 MySQL 官方推出的一个全栈高可用的解决方案。

图 12-1 是 InnoDB Cluster 的原理图。

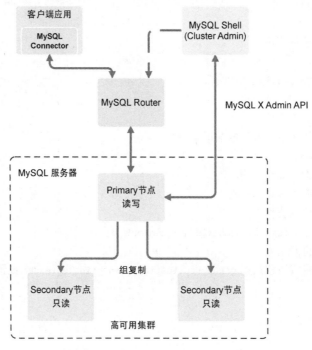

图 12-1　InnoDB Cluster 的原理图

InnoDB Cluster 中的核心组件有 3 个。

- 组复制

    后端通过组复制来提供高可用性。

- MySQL Shell

    MySQL Shell 是一个高级的 MySQL 客户端。在 InnoDB Cluster 中，主要用来部署和管理组复制。

❑ MySQL Router

　　MySQL Router 是一个轻量级的 MySQL 中间件。在 InnoDB Cluster 中，主要用来进行路由转发和负载均衡。这样即使组复制的拓扑发生了变化，对客户端也是透明的。

　　InnoDB Cluster 并不是将这 3 个组件简单地堆砌在一起，而是在易用性方面做了很多实实在在的工作。无论是部署还是日常管理，都可通过简单的一条或多条命令完成，极大地减少了 DBA 在管理组复制过程中的工作量。

　　本章主要包括以下内容。

❑ MySQL Shell。
❑ MySQL Router。
❑ InnoDB Cluster 的搭建。
❑ InnoDB Cluster 的管理操作。

12.1　MySQL Shell

　　MySQL Shell 的推出与 Document Store 有关。Document Store 是 MySQL 5.7.12 引入的一个新特性。基于这个特性，可将 MySQL 作为文档数据库来使用。

　　文档数据库是一种非关系数据库，可将数据以类 JSON 文档的方式进行存储。相对于关系数据库，文档数据库无须提前定义 schema，因而更加灵活，也更适合存储半结构化数据。文档数据库最典型的代表是 MongoDB。

　　图 12-2 是 Document Store 的架构图。

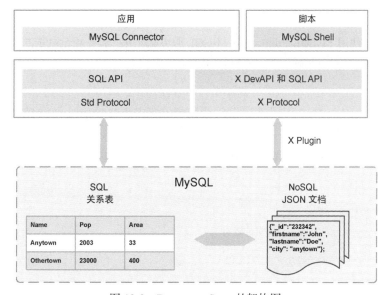

图 12-2　Document Store 的架构图

在这张架构图中，与 Document Store 相关的组件有 4 个：X Protocol、X Plugin、X DevAPI、MySQL Shell。这 4 个组件的关系如下。

- X Protocol：一个基于 Protobuf 库实现的新的客户端协议，可对文档和表进行 CRUD 操作。
- X Plugin：MySQL 服务端只有开启了 X Plugin，才能与实现了 X Protocol 的客户端进行通信。
- X DevAPI：基于 X Protocol 定义的 API。目前，支持的开发语言有 Java、Python、JavaScript、C++、C#、PHP。
- MySQL Shell：命令行工具，类似于 mysql 客户端，同时支持 Python、JavaScript、SQL 这 3 种语言。

Document Store 的核心是 X Protocol。

12.1.1 MySQL Shell 的安装

与 MySQL 一样，官网提供了 MySQL Shell 多个版本的下载。这里使用 Linux 二进制版本（Linux - Generic）。

```
# cd /usr/local/
# tar xvf mysql-shell-8.0.27-linux-glibc2.12-x86-64bit.tar.gz
# ln -s mysql-shell-8.0.27-linux-glibc2.12-x86-64bit mysql-shell
# export PATH=$PATH:/usr/local/mysql-shell/bin
```

12.1.2 MySQL Shell 的使用

在 MySQL Shell 中，一个很重要的概念是 Session。顾名思义，Session 就是会话。MySQL Shell 中的 Session 分为两类。

- Classic Session

  传统会话。使用 MySQL Protocol 与 MySQL 服务端交互，服务端无须安装 X Plugin。Classic Session 支持的操作可通过 \help ClassicSession 命令查看。

- X Session

  使用 X Protocol 与 MySQL 服务端交互，服务端必须安装 X Plugin。如果要对文档执行 CRUD 操作，必须使用 X Session。

MySQL Shell 支持 3 种语言模式：Python、JavaScript 和 SQL。不同模式之间可相互切换。切换命令如下所示。

```
mysql-js> \py
Switching to Python mode...
mysql-py> \sql
Switching to SQL mode... Commands end with ;
mysql-sql> \js
Switching to JavaScript mode...
```

默认是登录到 JavaScript 模式。

Session 的创建很简单,建立连接即可。看下面这两条命令。

```
# mysqlsh --mysql -h 127.0.0.1 -P 3306 -uroot -p
mysql-js> session
<ClassicSession:root@127.0.0.1:3306>

# mysqlsh --mysqlx -h 127.0.0.1 -P 33060 -uroot -p
mysql-js> session
<Session:root@127.0.0.1:33060>
```

指定 --mysql 会创建一个 Classic Session,指定 --mysqlx 会创建一个 X Session。

除了命令行指定,还可在登录后通过 shell.connect 建立连接。

```
# mysqlsh
mysql-js> \connect mysqlx://root@127.0.0.1:33060

mysql-js> shell.connect({host:'localhost', user:'root', socket:'/tmp/mysqlx.sock'})

mysql-js> shell.connect('root:123456@127.0.0.1:3306')
```

无论是命令行,还是 shell.connect,创建的 Session 都是一个全局 Session。即使模式发生了切换,这个 Session 依然存在。例如:

```
mysql-js> session
<ClassicSession:root@127.0.0.1:3306>
mysql-js> \py
Switching to Python mode...
mysql-py> session
<ClassicSession:root@127.0.0.1:3306>
```

除了全局 Session,还有另外一种 Session。这种 Session 只存在于创建时使用的模式中,一旦切换到其他模式就不存在了。下面看一个简单的示例。

```
mysql-js> var mySession = mysqlx.getSession('root:123456@127.0.0.1:33060');
mysql-js> mySession
<Session:root@127.0.0.1:33060>
mysql-js> \py
Switching to Python mode...
mysql-py> mySession
Traceback (most recent call last):
  File "<string>", line 1, in <module>
NameError: name 'mySession' is not defined
```

除了 Session,MySQL Shell 中另一个比较重要的概念是对象。几乎所有的 API 操作都是依附于对象的。建立连接后,MySQL Shell 会默认创建以下几个全局对象。

```
GLOBAL OBJECTS

The following modules and objects are ready for use when the shell starts:

 - dba      Used for InnoDB cluster administration.
 - mysql    Support for connecting to MySQL servers using the classic MySQL
            protocol.
 - mysqlx   Used to work with X Protocol sessions using the MySQL X DevAPI.
 - os       Gives access to functions which allow to interact with the operating
            system.
 - plugins  Plugin to manage MySQL Shell plugins
```

```
- session    Represents the currently open MySQL session.
- shell      Gives access to general purpose functions and properties.
- sys        Gives access to system specific parameters.
- util       Global object that groups miscellaneous tools like upgrade checker
             and JSON import.
```

下面看看这些对象的具体作用。

- dba：通过 AdminAPI 来管理 InnoDB Cluster、InnoDB ReplicaSet、InnoDB ClusterSet。
- mysql：创建 Classic Session。
- mysqlx：创建 X Session。
- os：调用可与操作系统交互的函数。
- plugins：管理插件。从 MySQL Shell 8.0.17 开始，可自定义插件。
- session：查看当前的会话。
- shell：通过 shell 调用 MySQL Shell 函数。
- sys：访问系统相关参数，支持 sys.argv 和 sys.path 这两个方法，其中前者用来获取脚本后面的参数，后者用来定义模块的搜索路径。
- util：MySQL Shell 提供的工具集。

如果要查看对象实现了哪些方法，可使用以下命令。

```
mysql-js> dba.help()
mysql-js> \help dba
```

如果要查看方法具体如何使用，可使用以下命令。

```
mysql-js> dba.help('createCluster');
mysql-js> \? dba.createCluster
```

最后看看 mysqlsh 支持的 shell 命令。

```
SHELL COMMANDS

The shell commands allow executing specific operations including updating the
shell configuration.

The following shell commands are available:

 - \                      Start multi-line input when in SQL mode.
 - \connect      (\c)     Connects the shell to a MySQL server and assigns the
                          global session.
 - \disconnect            Disconnects the global session.
 - \edit         (\e)     Launch a system editor to edit a command to be executed.
 - \exit                  Exits the MySQL Shell, same as \quit.
 - \help         (\?,\h)  Prints help information about a specific topic.
 - \history               View and edit command line history.
 - \js                    Switches to JavaScript processing mode.
 - \nopager               Disables the current pager.
 - \nowarnings   (\w)     Don't show warnings after every statement.
 - \option                Allows working with the available shell options.
 - \pager        (\P)     Sets the current pager.
 - \py                    Switches to Python processing mode.
 - \quit         (\q)     Exits the MySQL Shell.
 - \reconnect             Reconnects the global session.
```

```
- \rehash                Refresh the autocompletion cache.
- \show                  Executes the given report with provided options and
                         arguments.
- \source       (\.)     Loads and executes a script from a file.
- \sql                   Executes SQL statement or switches to SQL processing
                         mode when no statement is given.
- \status       (\s)     Print information about the current global session.
- \system       (\!)     Execute a system shell command.
- \use          (\u)     Sets the active schema.
- \warnings     (\W)     Show warnings after every statement.
- \watch                 Executes the given report with provided options and
                         arguments in a loop.
```

大部分命令与 mysql 客户端一样，只不过 mysqlsh 的命令前面加了反斜杠（\）。

下面重点看看 mysqlsh 中的新增命令。

- \：SQL 模式下，开启多行输入。

    ```
    mysql-sql> \
            > begin;
            > insert into slowtech.t1 values(1);
            > delete from slowtech.t1;
            > commit;
            >
    Query OK, 0 rows affected (0.0001 sec)
    Query OK, 1 row affected (0.0010 sec)
    Query OK, 1 row affected (0.0003 sec)
    Query OK, 0 rows affected (0.0072 sec)
    ```

- \sql：无须切换模式，即可执行 SQL 命令，但只支持单条命令。

    ```
    mysql-js> \sql select 1;
    +---+
    | 1 |
    +---+
    | 1 |
    +---+
    1 row in set (0.0002 sec)
    ```

- \show、\watch：与 mysqlsh 支持的报表功能有关。

12.1.3　X DevAPI 的关键特性

下面介绍 X DevAPI 的关键特性。

- CRUD 操作

    CRUD 是 CREATE、READ、UPDATE 和 DELETE 这 4 种操作的缩写。

    我们来具体看看在 mysqlsh 中如何通过 X DevAPI 对文档进行 CRUD 操作。

    ```
    mysql-js> var mysqlx = require('mysqlx');
    mysql-js> var session = mysqlx.getSession( {host: '127.0.0.1', port: 33060, user: 'root', password:
    '123456'}) // 获取 X Session

    mysql-js> var db = session.getSchema('slowtech') // 获取数据库对象
    ```

```
mysql-js> db.createCollection('collection1') // 创建一个集合，类似于关系数据库中的表
<Collection:collection1>

mysql-js> db.collection1.add({ name: 'john', age: 20, gender: 'M'}) // 添加一个文档
Query OK, 1 item affected (0.0150 sec)

mysql-js> db.collection1.find("name='john'") // 基于条件查询文档
{
    "_id": "000061d657870000000000000001",
    "age": 20,
    "name": "john",
    "gender": "M"
}
1 document in set (0.0006 sec)

mysql-js> db.collection1.modify("name='john'").set("age", 21) // 基于条件修改文档
Query OK, 1 item affected (0.0142 sec)

Rows matched: 1  Changed: 1  Warnings: 0

mysql-js> db.collection1.remove("name='john'") // 基于条件删除文档
Query OK, 1 item affected (0.0070 sec)

mysql-js> db.dropCollection("collection1") // 删除集合
```

除此之外，X DevAPI 还支持表的 DML 操作。

```
mysql-js> var mysqlx = require('mysqlx');
mysql-js> var session = mysqlx.getSession( {host: '127.0.0.1', port: 33060, user: 'root', password: '123456'})
mysql-js> var db = session.getSchema('slowtech')

mysql-js> db.getTables() // 获取当前 schema 中的所有表
[
    <Table:t1>
]

mysql-js> db.t1.select() // 全表扫描
+----+------+
| id | name |
+----+------+
|  1 | a    |
|  2 | b    |
+----+------+
2 rows in set (0.0004 sec)

mysql-js> db.t1.select(['name']).where('id = 1') // 基于条件查询记录
+------+
| name |
+------+
| a    |
+------+
1 row in set (0.0005 sec)

mysql-js> db.t1.update().set('name', 'c').where('id = 1') // 基于条件更新记录
Query OK, 1 item affected (0.0091 sec)

Rows matched: 1  Changed: 1  Warnings: 0

mysql-js> db.t1.delete().limit(1) // 随机删除一条记录
```

```
Query OK, 1 item affected (0.0052 sec)

mysql-js> db.t1.insert('id', 'name').values(3,'c'); // 插入记录
Query OK, 1 item affected (0.0059 sec)

mysql-js> db.dropCollection('t1') // dropCollection 方法也可用来删除表
```

- 支持事务

  原生支持事务，而 MongoDB 直到 4.0 版本才支持跨文档事务。

  ```
  var c2 = db.createCollection('collection2');
  session.startTransaction(); // 开启事务
  try {
    c2.add({id: 1, name: 'a'}).execute();
    c2.add({id: 2, name: 'b'}).execute();
    c2.add({id: 3, name: 'c'}).execute();
    session.commit(); // 事务提交
    print('Successed');
  }
  catch (err) {
    session.rollback(); // 事务回滚
    print('Failed:' + err.message);
  }
  ```

- 支持链式编程

  链式编程是将多个操作（方法调用）通过点号链接在一起。本质上，每次的方法调用都会返回一个对象，这样，该对象又可继续调用其他方法。链式编程具有如下优点。

  - 代码简洁，无须通过额外的变量来保存中间结果。
  - 可读性强，符合人类的认知习惯。

  链式编程是一种典型的语法糖。

  例如，上面对 t1 表的查询操作可简化为以下操作。

  ```
  mysql-js> session.getSchema('slowtech').getTable('t1').select();
  +----+------+
  | id | name |
  +----+------+
  |  1 | a    |
  |  2 | b    |
  +----+------+
  2 rows in set (0.0002 sec)
  ```

- 支持操作的异步执行

  这样，在与数据库交互时，无须等待结果返回，即可执行其他操作。常见的实现方式是定义一个回调函数，异步处理操作的返回结果。例如：

  ```
  var t1 = db.getTable('t1');
  t1.select('id', 'c1')
    .where('id=1')
    .execute(function (row) {
      // do something with a row
    })
  ```

```
      .catch(err) {
        // Handle error
      });
```

- 支持绑定变量

示例如下。

```
mysql-js> var query=session.getSchema('slowtech').getCollection('collection1').find('name
 = :param1 and age = :param2')
mysql-js> query.bind('param1','john').bind('param2',20).execute()
{
    "_id": "000061d6578700000000000000005",
    "age": 20,
    "name": "john",
    "gender": "M"
}
1 document in set (0.0006 sec)
```

12.1.4 MySQL Shell 工具集

除了支持 X DevAPI 和 AdminAPI，MySQL Shell 自身也提供了很多实用工具。

```
mysql-js> \help util
NAME
      util - Global object that groups miscellaneous tools like upgrade checker
             and JSON import.

DESCRIPTION
      Global object that groups miscellaneous tools like upgrade checker and
      JSON import.

FUNCTIONS
      checkForServerUpgrade([connectionData][, options])
            Performs series of tests on specified MySQL server to check if the
            upgrade process will succeed.

      dumpInstance(outputUrl[, options])
            Dumps the whole database to files in the output directory.

      dumpSchemas(schemas, outputUrl[, options])
            Dumps the specified schemas to the files in the output directory.

      dumpTables(schema, tables, outputUrl[, options])
            Dumps the specified tables or views from the given schema to the
            files in the target directory.

      exportTable(table, outputUrl[, options])
            Exports the specified table to the data dump file.

      help([member])
            Provides help about this object and it's members

      importJson(file[, options])
            Import JSON documents from file to collection or table in MySQL
            Server using X Protocol session.

      importTable(files[, options])
            Import table dump stored in files to target table using LOAD DATA
```

```
                LOCAL INFILE calls in parallel connections.

    loadDump(url[, options])
            Loads database dumps created by MySQL Shell.
```

dumpInstance、dumpSchemas、dumpTables、loadDump 在第 5 章提到过，我们重点看看其他几个工具。

- checkForServerUpgrade：检测目标实例能否升级到指定版本。
- exportTable：导出指定表。
- importTable：导入文本中的数据。
- importJson：导入 JSON 格式的数据。

这几个工具的具体用法如下。

1. util.checkForServerUpgrade([connectionData][, options])

检测目标实例是否满足升级到指定版本的条件。

connectionData 指定目标实例的连接信息，options 可指定的配置如下。

- configPath：目标实例配置文件的路径。
- outputFormat：检测结果的输出格式。可指定为 TEXT 或 JSON，默认为 TEXT。
- targetVersion：待升级的版本。若不指定，则为 MySQL Shell 的当前版本。
- password：指定目标实例的密码。

在使用该工具时，有两点需要注意。

- 不支持 MySQL 5.7 之前的版本。
- 只支持 GA 版本之间的升级检测。

看看下面这个示例。

```
mysql-js> util.checkForServerUpgrade('root@127.0.0.1:3307', {"password":"123456",
"targetVersion":"8.0.27", "outputFormat":"TEXT", "configPath":"/etc/my_3307.cnf"})
The MySQL server at 127.0.0.1:3307, version 5.7.36-log - MySQL Community Server
(GPL), will now be checked for compatibility issues for upgrade to MySQL
8.0.27...

1) Usage of old temporal type
  No issues found

2) Usage of db objects with names conflicting with new reserved keywords
  No issues found

3) Usage of utf8mb3 charset
  No issues found

4) Table names in the mysql schema conflicting with new tables in 8.0
  No issues found

5) Partitioned tables using engines with non native partitioning
  No issues found

6) Foreign key constraint names longer than 64 characters
  No issues found
```

```
7) Usage of obsolete MAXDB sql_mode flag
  No issues found

8) Usage of obsolete sql_mode flags
  Notice: The following DB objects have obsolete options persisted for
    sql_mode, which will be cleared during upgrade to 8.0.
  More information:
    https://dev.mysql.com/doc/refman/8.0/en/mysql-nutshell.html#mysql-nutshell-removals

  global system variable sql_mode - defined using obsolete NO_AUTO_CREATE_USER
    option

9) ENUM/SET column definitions containing elements longer than 255 characters
  No issues found

10) Usage of partitioned tables in shared tablespaces
  No issues found

11) Circular directory references in tablespace data file paths
  No issues found

12) Usage of removed functions
  No issues found

13) Usage of removed GROUP BY ASC/DESC syntax
  No issues found

14) Removed system variables for error logging to the system log configuration
  No issues found

15) Removed system variables
  Error: Following system variables that were detected as being used will be
    removed. Please update your system to not rely on them before the upgrade.
  More information:
    https://dev.mysql.com/doc/refman/8.0/en/added-deprecated-removed.html#optvars-removed

  innodb_large_prefix - is set and will be removed

16) System variables with new default values
  Warning: Following system variables that are not defined in your
    configuration file will have new default values. Please review if you rely on
    their current values and if so define them before performing upgrade.
  More information:
    https://mysqlserverteam.com/new-defaults-in-mysql-8-0/

  back_log - default value will change
  character_set_server - default value will change from latin1 to utf8mb4
  collation_server - default value will change from latin1_swedish_ci to
    utf8mb4_0900_ai_ci
  event_scheduler - default value will change from OFF to ON
  explicit_defaults_for_timestamp - default value will change from OFF to ON
  innodb_autoinc_lock_mode - default value will change from 1 (consecutive) to
    2 (interleaved)
  innodb_flush_method - default value will change from NULL to fsync (Unix),
    unbuffered (Windows)
  innodb_flush_neighbors - default value will change from 1 (enable) to 0
    (disable)
  innodb_max_dirty_pages_pct - default value will change from 75 (%)  90 (%)
  innodb_max_dirty_pages_pct_lwm - default value will change from 0 (%) to 10
    (%)
```

```
      innodb_undo_log_truncate - default value will change from OFF to ON
      innodb_undo_tablespaces - default value will change from 0 to 2
      log_error_verbosity - default value will change from 3 (Notes) to 2 (Warning)
      max_allowed_packet - default value will change from 4194304 (4MB) to 67108864
        (64MB)
      max_error_count - default value will change from 64 to 1024
      optimizer_trace_max_mem_size - default value will change from 16KB to 1MB
      performance_schema_consumer_events_transactions_current - default value will
        change from OFF to ON
      performance_schema_consumer_events_transactions_history - default value will
        change from OFF to ON
      slave_rows_search_algorithms - default value will change from 'INDEX_SCAN,
        TABLE_SCAN' to 'INDEX_SCAN, HASH_SCAN'
      table_open_cache - default value will change from 2000 to 4000
      transaction_write_set_extraction - default value will change from OFF to
        XXHASH64

  17) Zero Date, Datetime, and Timestamp values
      No issues found

  18) Schema inconsistencies resulting from file removal or corruption
      No issues found

  19) Tables recognized by InnoDB that belong to a different engine
      No issues found

  20) Issues reported by 'check table x for upgrade' command
      No issues found

  21) New default authentication plugin considerations
      Warning: The new default authentication plugin 'caching_sha2_password' offers
        more secure password hashing than previously used 'mysql_native_password'
        (and consequent improved client connection authentication). However, it also
        has compatibility implications that may affect existing MySQL installations.
        If your MySQL installation must serve pre-8.0 clients and you encounter
        compatibility issues after upgrading, the simplest way to address those
        issues is to reconfigure the server to revert to the previous default
        authentication plugin (mysql_native_password). For example, use these lines
        in the server option file:

      [mysqld]
      default_authentication_plugin=mysql_native_password

      However, the setting should be viewed as temporary, not as a long term or
      permanent solution, because it causes new accounts created with the setting
      in effect to forego the improved authentication security.
      If you are using replication please take time to understand how the
      authentication plugin changes may impact you.
      More information:
        https://dev.mysql.com/doc/refman/8.0/en/upgrading-from-previous-series.html#upgrade-caching-
      sha2-password-compatibility-issues
        https://dev.mysql.com/doc/refman/8.0/en/upgrading-from-previous-series.html#upgrade-caching-
      sha2-password-replication

Errors:   1
Warnings: 22
Notices:  1

1 errors were found. Please correct these issues before upgrading to avoid compatibility issues.
```

示例中一共进行了 21 项检查，主要是检查升级过程中需要注意的功能点。

我们看看最后的检测结果。

- `Errors`：会导致升级失败，需要提前修复的检查项。具体在本例中，是 `innodb_large_prefix` 这个参数在 MySQL 8.0 中已经被移除了。
- `Warnings` 和 `Notices`：需要注意，但不会导致升级失败的检查项。具体在本例中，涉及 3 大项：MySQL 8.0 中的 `sql_mode` 不包括 `NO_AUTO_CREATE_USER`；`back_log`、`character_set_server` 等参数的默认值发生了变化；默认的认证插件由 `mysql_native_password` 变更为了 `caching_sha2_password`。

2. `util.exportTable(table, outputUrl[, options])`

将表的数据导出到文本文件中，效果与 `SELECT INTO OUTFILE` 一样。例如：

```
util.exportTable('sbtest.sbtest1','/data/backup/sbtest1.txt')
```

它与 `dumpTables` 的不同之处有两个。

- 这里的 `outputUrl` 不是目录名，而是文件名。
- `exportTable` 是单线程备份。

3. `util.importTable(files[, options])`

将文本中的数据导入表中。例如：

```
mysql-js> util.importTable("/data/backup/sbtest1.txt", {schema: "slowtech", table: "t1", dialect: "default", showProgress: true})
Importing from file '/data/backup/sbtest1.txt' to table `slowtech`.`t1` in MySQL Server at 127.0.0.1:3306 using 4 threads
[Worker000] sbtest1.txt: Records: 226231  Deleted: 0  Skipped: 0  Warnings: 0
[Worker001] sbtest1.txt: Records: 257732  Deleted: 0  Skipped: 0  Warnings: 0
[Worker003] sbtest1.txt: Records: 258305  Deleted: 0  Skipped: 0  Warnings: 0
[Worker002] sbtest1.txt: Records: 257732  Deleted: 0  Skipped: 0  Warnings: 0
100% (193.89 MB / 193.89 MB), 1.94 MB/s
File '/data/backup/sbtest1.txt' (193.89 MB) was imported in 21.9001 sec at 8.85 MB/s
Total rows affected in slowtech.t1: Records: 1000000  Deleted: 0  Skipped: 0  Warnings: 0
```

在导入时，该工具会启动多个线程并行执行 `LOAD DATA LOCAL INFILE` 操作。

接下来看看 `options` 中的选项。

- `schema`：指定 schema 名。
- `table`：指定表名。
- `columns`：指定列名。
- `fieldsTerminatedBy`：指定字段之间的分隔符，默认为 `"\t"`。
- `fieldsEnclosedBy`：指定将字段括起来的字符，默认为空字符。
- `fieldsEscapedBy`：指定转义字符，默认为 `"\"`。
- `fieldsOptionallyEnclosed`：设置为 `false`（默认值），代表所有字段都由 `fieldsEnclosedBy` 指定的字符括起来。
- `linesTerminatedBy`：指定行之间的分隔符，默认为 `"\n"`。
- `replaceDuplicates`：是否覆盖重复数据，默认为 `false`。

- threads：指定并发线程数，默认为 8。每个线程负责一个块的导入。
- bytesPerChunk：指定每个 chunk 的大小，默认为 50MB。
- maxRate：限制单个线程的最大导入速率，默认为 0，即不限制。
- showProgress：是否打印进度信息。
- skipRows：忽略最初若干行数据，默认为 0，即不忽略。
- dialect：指定数据文件的解析规则，不同的解析规则对应不同的 fieldsTerminatedBy、fieldsEnclosedBy、fieldsOptionallyEnclosed、fieldsEscapedBy 和 linesTerminatedBy。支持的解析规则有 default、csv、tsv、json、csv-unix，默认为 default。

4. util.importJson(file[, options])

将 JSON 格式的数据导入 MySQL。

常见的使用场景是将 MongoDB 中通过 mongoexport 导出的数据导入 MySQL。

file 是文件路径，options 支持以下选项。

- schema：指定库名，库必须存在。
- collection：指定集合名，如果集合不存在，则会自动创建。
- table：指定表名。如果表不存在，同样会自动创建。
- tableColumn：与 table 一起使用，指定 JSON 要导入的列名，默认是 doc。
- convertBsonTypes：是否将 BSON 类型转换为 MySQL 数据类型。BSON 是 Binary JSON 的缩写，是 JSON 的一种二进制存储方案。
- convertBsonOid：是否转换 BSON 类型中的 ObjectId。
- extractOidTime：是否提取 ObjectId 中的时间戳信息，以单独一个列存储。列名由 extractOidTime 指定。

下面通过具体的示例看看各个配置项的作用。

```
# 在 MongoDB 中插入一条测试数据
rs:PRIMARY> db.t1.insert({"dateNow":new Date(),"longQuoted":"9223372036854775807","ts":new Timestamp()});

# 通过 mongoexport 导出这条数据
# mongoexport --host 127.0.0.1:27017 --authenticationDatabase admin -u root -p 123456 -d test -c t1 -o /tmp/test.json

# cat /tmp/test.json
{"_id":{"$oid":"61e804e8f5f0197925e1be6f"},"dateNow":{"$date":"2022-01-19T12:32:40.263Z"},"longQuoted":"9223372036854775807","ts":{"$timestamp":{"t":1642595560,"i":1}}}
```

这里的 oid（ObjectId）、date 和 timestamp 都是 BSON 自己实现的数据类型，在 JSON 中并不存在。

首先，不指定 BSON 相关的任何选项，看看导入的结果。

```
mysql-js> util.importJson("/tmp/test.json", { schema: "slowtech", collection: "c1" })
Importing from file "/tmp/test.json" to collection `slowtech`.`c1` in MySQL Server at 127.0.0.1:33060
```

```
Processed 169 bytes in 1 document in 0.0018 sec (1.00 document/s)
Total successfully imported documents 0 (0.00 documents/s)
Util.importJson: Data too long for column '_id' at row 1 (MYSQLSH 1406)
```

导入报错。虽然记录导入失败，但集合依然被创建了。我们看看 c1 的表结构。

```
mysql> show create table c1\G
*************************** 1. row ***************************
       Table: c1
Create Table: CREATE TABLE `c1` (
  `doc` json DEFAULT NULL,
  `_id` varbinary(32) GENERATED ALWAYS AS (json_unquote(json_extract(`doc`,_utf8mb4'$._id'))) STORED NOT NULL,
  `_json_schema` json GENERATED ALWAYS AS (_utf8mb4'{"type":"object"}') VIRTUAL,
  PRIMARY KEY (`_id`),
  CONSTRAINT `$val_strict_ADEF3FEF8D9EE26E61EEC9A3B5E4003CFC8E6FB1` CHECK (json_schema_valid(`_json_schema`,`doc`)) /*!80016 NOT ENFORCED */
) ENGINE=InnoDB DEFAULT CHARSET=utf8mb4 COLLATE=utf8mb4_0900_ai_ci
1 row in set (0.00 sec)
```

c1 虽然是一个集合，但从表的角度来看，它只有 3 个字段，其中 2 个字段如下。

- doc：JSON 类型，用于存储 JSON 数据。
- _id：虚拟列，用于存储 ObjectId。

接着，看看将 convertBsonOid 设置为 true 的效果。

```
mysql-js> util.importJson("/tmp/test.json", {schema: "slowtech", collection: "c1", convertBsonOid: true})

mysql> select * from c1\G
*************************** 1. row ***************************
         doc: {"ts": {"$timestamp": {"i": 1, "t": 1642595560}}, "_id": "61e804e8f5f0197925e1be6f", "dateNow": {"$date": "2022-01-19T12:32:40.263Z"}, "longQuoted": "9223372036854775807"}
         _id: 0x363165383034653866356630313937393235653162653666
_json_schema: {"type": "object"}
1 row in set (0.00 sec)
```

文档导入成功。_id 是文档中 ObjectId 的 ASCII 码。

接下来，看看 extractOidTime 的效果。

```
mysql-js> util.importJson("/tmp/test.json", {schema: "slowtech", collection: "c2", convertBsonOid: true, extractOidTime: "insert_time"})

mysql> select * from c2\G
*************************** 1. row ***************************
         doc: {"ts": {"$timestamp": {"i": 1, "t": 1642595560}}, "_id": "61e804e8f5f0197925e1be6f", "dateNow": {"$date": "2022-01-19T12:32:40.263Z"}, "longQuoted": "9223372036854775807", "insert_time": "2022-01-19 12:32:40"}
         _id: 0x363165383034653866356630313937393235653162653666
_json_schema: {"type": "object"}
1 row in set (0.00 sec)
```

相对于 c1，c2 新增了一列 insert_time，它的值是从 ObjectId 中提取出来的。

最后看看将 convertBsonTypes 设置为 true 的效果。

```
mysql-js> util.importJson("/tmp/test.json", {schema: "slowtech", collection: "c3", convertBsonOid: true, convertBsonTypes: true})
```

```
mysql> select * from c3\G
*************************** 1. row ***************************
        doc: {"ts": "2022-01-19 12:32:40", "_id": "61e804e8f5f0197925e1be6f", "dateNow":
"2022-01-19T12:32:40.263Z", "longQuoted": "9223372036854775807"}
       _id: 0x363165383034653866356630313937393235653162653666
_json_schema: {"type": "object"}
1 row in set (0.00 sec)
```

相对于 c1 和 c2，c3 并没有 oid、date 之类的类型标识，而是直接将其转化为了 MySQL 对应的数据类型。

12.1.5　MySQL Shell 的使用技巧

- MySQL Shell 日志的默认路径是 ~/.mysqlsh/mysqlsh.log，这里的 ~ 是当前用户的家目录。如果要修改为其他路径，需设置环境变量 `MYSQLSH_USER_CONFIG_HOME`，它会替代 ~/.mysqlsh/。
- 在使用 MySQL Shell 管理组复制时，如果想看 AdminAPI 执行了哪些 SQL 命令，可设置 `--dba-log-sql`，它默认为 `0`，即不记录。若设置为 `1`，会记录除了 `SELECT` 和 `SHOW` 之外的其他 SQL 命令。若设置为 `2`，则会记录所有 SQL 命令。
- 上面的示例都基于 MySQL Shell 经典模式的提示符 `"mysql-js> "`，默认的提示符是`" MySQL JS > "`。下面看看调整方法。

```
# cd /usr/local/mysql-shell
# mkdir -p ~/.mysqlsh
# cp share/mysqlsh/prompt/prompt_classic.json ~/.mysqlsh/prompt.json
```

这里创建的 ~/.mysqlsh/ 是 MySQL Shell 默认的家目录。

12.2　MySQL Router

MySQL Router 是 MySQL 官方推出的一个轻量级中间件。之所以说它是轻量级的，是因为它只做包的转发，不会解析包的内容。所以，相对于其他中间件，它的性能损耗很小。按照官方的说法，相对于直连，它只有 1% 的性能损耗。不过，不解析包的内容，就意味着 MySQL Router 目前并不支持分库分表、对应用透明的读写分离等特性。

MySQL Router 虽然支持读写分离，但在实现上，是通过不同端口来区分读写操作的。这种实现方式有两点不足：需要对业务端做改造，包括区分读写操作，配置多个 `DataSource`；如果后端节点的角色发生了变化，如主库降级为从库，MySQL Router 实际上是感知不到的。

从目前的趋势来看，MySQL Router 很难作为一个独立、功能丰富的中间件来使用。相反，作为 InnoDB Cluster 的 3 大核心组件之一，它与组复制的耦合度越来越高。对于上面提到的读写分离，如果后端节点是组复制，就无须担心在其节点角色发生变化时，MySQL Router 感知不到的问题。

下面，我们看看 MySQL Router 的工作流程。

- 应用向 MySQL Router 发起连接。
- MySQL Router 检查并选择一个可用的后端节点（MySQL Server）。

- MySQL Router 与这个后端节点建立连接。
- MySQL Router 来回转发应用与后端节点之间的数据包。
- 如果后端节点出现问题，MySQL Router 会断开应用端的连接。应用端重试后，MySQL Router 会选择另外一个可用节点重复上述流程。

12.2.1　MySQL Router 的安装

MySQL Router 同样提供了多个版本的下载，这里使用 Linux 二进制版本（Linux - Generic）。

```
# cd /usr/local/
# tar xvf mysql-router-8.0.27-linux-glibc2.12-x86_64.tar.xz
# ln -s mysql-router-8.0.27-linux-glibc2.12-x86_64 mysql-router
# export PATH=$PATH:/usr/local/mysql-router/bin
```

12.2.2　MySQL Router 的使用

首先，基于现有的组复制集群初始化 MySQL Router。

```
# mkdir -p /data/myrouter/6446
# mysqlrouter --bootstrap cluster_admin@192.168.244.10:3306 --user=mysql --directory
# /data/myrouter/6446 --conf-use-sockets --report-host='192.168.244.128'
Please enter MySQL password for cluster_admin:
# Bootstrapping MySQL Router instance at '/data/myrouter/6446'...

- Creating account(s) (only those that are needed, if any)
- Verifying account (using it to run SQL queries that would be run by Router)
- Storing account in keyring
- Adjusting permissions of generated files
- Creating configuration /data/myrouter/6446/mysqlrouter.conf

# MySQL Router configured for the InnoDB Cluster 'myCluster'

After this MySQL Router has been started with the generated configuration

    $ mysqlrouter -c /data/myrouter/6446/mysqlrouter.conf

InnoDB Cluster 'myCluster' can be reached by connecting to:

## MySQL Classic protocol

- Read/Write Connections: 192.168.244.128:6446, /data/myrouter/6446/mysql.sock
- Read/Only Connections:  192.168.244.128:6447, /data/myrouter/6446/mysqlro.sock

## MySQL X protocol

- Read/Write Connections: 192.168.244.128:6448, /data/myrouter/6446/mysqlx.sock
- Read/Only Connections:  192.168.244.128:6449, /data/myrouter/6446/mysqlxro.sock
```

命令行中各参数的具体含义如下。

- `--bootstrap`：基于指定的组复制集群初始化 MySQL Router。`cluster_admin@192.168.244.10:3306` 是集群中任意一个节点的连接信息，用户名可不指定，默认为 root。
- `--user`：MySQL Router 的运行用户。

- `--directory`：指定 MySQL Router 的工作目录，该目录相当于 MySQL 中的 basedir。注意，`--directory` 只能与 `--bootstrap` 一起使用。
- `--conf-use-sockets`：开启套接字文件通信。
- `--report-host`：显式指定主机 IP。

除了上述参数，还有一个参数需要注意。

- `--name`：给 Router 实例取个别名，默认为空。如果要在一台主机上部署多个指向同一集群的 Router 实例，那么实例与实例之间的别名不能重复，否则会有冲突。

  这一点，从 `mysql_innodb_cluster_metadata.routers` 的表结构也可以看出来，address 和 router_name 上存在唯一约束。

接着看看 `--directory` 目录中的内容。

```
# tree /data/myrouter/6446/
/data/myrouter/6446/
├── data
│   ├── ca-key.pem
│   ├── ca.pem
│   ├── keyring
│   ├── router-cert.pem
│   ├── router-key.pem
│   └── state.json
├── log
│   └── mysqlrouter.log
├── mysqlrouter.conf
├── mysqlrouter.key
├── run
├── start.sh
└── stop.sh

3 directories, 11 files
```

其中，data 是数据目录，log 是日志目录，mysqlrouter.conf 是配置文件，mysqlrouter.key 是密钥，run 是运行目录，start.sh 是启动脚本，stop.sh 是停止脚本。

MySQL Router 在初始化的过程中会执行以下操作。

- 在组复制中，维护 Router 的元数据信息。

```
mysql> select * from mysql_innodb_cluster_metadata.routers\G
*************************** 1. row ***************************
    router_id: 1
  router_name: 
 product_name: MySQL Router
      address: 192.168.244.128
      version: 8.0.27
 last_check_in: NULL
   attributes: {"ROEndpoint": "6447", "RWEndpoint": "6446", "ROXEndpoint": "6449", "RWXEndpoint": "6448", "MetadataUser": "mysql_router1_zitvheqzuiuh", "bootstrapTargetType": "cluster"}
   cluster_id: 7822af5a-79ab-11ec-b87f-000c29f66609
      options: NULL
clusterset_id: NULL
1 row in set (0.03 sec)
```

这里的 address 即命令行中的 --report-host，如果不显式指定，则默认为主机名。

- 创建 Router 访问用户。

  Router 后续会用这个用户获取集群的状态信息，更新 Router 的元数据信息。

  这个用户加密后的密码会存储在 /data/myrouter/6446/data/keyring 中。

- 将集群的元数据信息持久化到 /data/myrouter/6446/data/state.json 文件中，这样即使 MySQL Router 重启，也不会丢失后端的集群信息。

```
# cat /data/myrouter/6446/data/state.json
{
    "metadata-cache": {
        "group-replication-id": "720113f3-79ab-11ec-b87f-000c29f66609",
        "cluster-metadata-servers": [
            "mysql://192.168.244.10:3306",
            "mysql://192.168.244.20:3306",
            "mysql://192.168.244.30:3306"
        ]
    },
    "version": "1.0.0"
}
```

接下来，我们分析一下配置文件的内容。

```
# cat /data/myrouter/6446/mysqlrouter.conf
# File automatically generated during MySQL Router bootstrap
[DEFAULT]
user=mysql
logging_folder=/data/myrouter/6446/log
runtime_folder=/data/myrouter/6446/run
data_folder=/data/myrouter/6446/data
keyring_path=/data/myrouter/6446/data/keyring
master_key_path=/data/myrouter/6446/mysqlrouter.key
connect_timeout=15
read_timeout=30
dynamic_state=/data/myrouter/6446/data/state.json
client_ssl_cert=/data/myrouter/6446/data/router-cert.pem
client_ssl_key=/data/myrouter/6446/data/router-key.pem
client_ssl_mode=PREFERRED
server_ssl_mode=AS_CLIENT
server_ssl_verify=DISABLED

[logger]
level = INFO

[metadata_cache:myCluster]
cluster_type=gr
router_id=1
user=mysql_router1_zitvheqzuiuh
metadata_cluster=myCluster
ttl=0.5
auth_cache_ttl=-1
auth_cache_refresh_interval=2
use_gr_notifications=0
```

```
[routing:myCluster_rw]
bind_address=0.0.0.0
bind_port=6446
socket=/data/myrouter/6446/mysql.sock
destinations=metadata-cache://myCluster/?role=PRIMARY
routing_strategy=first-available
protocol=classic

[routing:myCluster_ro]
bind_address=0.0.0.0
bind_port=6447
socket=/data/myrouter/6446/mysqlro.sock
destinations=metadata-cache://myCluster/?role=SECONDARY
routing_strategy=round-robin-with-fallback
protocol=classic

[routing:myCluster_x_rw]
bind_address=0.0.0.0
bind_port=6448
socket=/data/myrouter/6446/mysqlx.sock
destinations=metadata-cache://myCluster/?role=PRIMARY
routing_strategy=first-available
protocol=x

[routing:myCluster_x_ro]
bind_address=0.0.0.0
bind_port=6449
socket=/data/myrouter/6446/mysqlxro.sock
destinations=metadata-cache://myCluster/?role=SECONDARY
routing_strategy=round-robin-with-fallback
protocol=x

[http_server]
port=8443
ssl=1
ssl_cert=/data/myrouter/6446/data/router-cert.pem
ssl_key=/data/myrouter/6446/data/router-key.pem

[http_auth_realm:default_auth_realm]
backend=default_auth_backend
method=basic
name=default_realm

[rest_router]
require_realm=default_auth_realm

[rest_api]

[http_auth_backend:default_auth_backend]
backend=metadata_cache

[rest_routing]
require_realm=default_auth_realm

[rest_metadata_cache]
require_realm=default_auth_realm
```

整个配置文件由多个 section 组成，section 的语法如下。

```
[section name:optional section key]
option = value
option = value
option = value
```

其中，`section name` 是必填项。

对于同一个 `section name`，可设置多个 section。不同的 section 通过 `optional section key` 来区分，如上面的 `routing`。

常用的 section 及其所属模块如下。

- `DEFAULT`：通用模块。
- `logger`：日志模块。
- `metadata_cache`：集群元数据模块。
- `routing`：路由模块。

下面，我们结合配置文件的内容，看看不同 section 中各参数的具体含义。

- **DEFAULT**

  - `user`：MySQL Router 的运行用户。
  - `dynamic_state`：指定文件用于存储集群的元数据信息，这样即使 MySQL Router 重启，也不会丢失后端的集群信息。该文件在 MySQL Router 初始化（`--bootstrap`）时自动生成。

- **logger**

  - `level`：设置日志级别，可设置为 `INFO`、`DEBUG`、`WARNING`、`ERROR`、`FATAL`。默认为 `INFO`。

- **metadata_cache**

  - `cluster_type`：集群的类型，可设置为 `gr` 和 `rs`，其中前者是组复制，后者是副本集（InnoDB ReplicaSet）。
  - `router_id`：Router ID，对应 `mysql_innodb_cluster_metadata.routers` 中的 `router_id`。
  - `user`：Router 用来访问 MySQL 的用户，是随机生成的。从 MySQL Router 8.0.19 开始，可通过 `--account` 显式指定。
  - `metadata_cluster`：集群名。
  - `ttl`：集群元数据的缓存时间，默认为 0.5 秒。每隔 0.5 秒，MySQL Router 就会向集群发起一次查询请求，获取集群的元数据信息。
  - `use_gr_notifications`：开启组复制的通知功能。开启后，MySQL Router 会异步监听集群的状态变化。当集群状态发生变化时，MySQL Router 会自动更新集群的元数据信息。如果开启了该特性，`ttl` 依然有效，只不过会作为一个辅助手段，此时可适当调大 `ttl` 的值。

- **routing**

  - `bind_address`：监听地址，由 `--conf-bind-address` 参数决定，默认为 0.0.0.0。
  - `bind_port`：监听端口。默认情况下，会监听 4 个端口：6446（对应 Classic 协议的读写操作）、

6447（对应 Classic 协议的只读操作）、6448（对应 X 协议的读写操作）、6449（对应 X 协议的只读操作）。可通过 --conf-base-port 显式设置 Router 的起始监听端口，如 6688，则开启的 4 个端口将依次为 6688、6689、6690、6691。
- socket：套接字文件地址。默认不生成，除非指定 --conf-use-sockets 参数。
- destinations：指定目标节点的地址。支持以下两种指定方式。
    - 显式指定目标节点的地址。
        ```
        destinations=192.168.244.10:3306,192.168.244.20:3306,192.168.244.30:3306
        ```
    - 指定 metadata-cache。
        ```
        destinations=metadata-cache://myCluster/default?role=PRIMARY
        ```
        metadata-cache 只能用在 InnoDB Cluster 中。

        指定的选项中，
        - myCluster 是集群名。
        - role 是节点的角色，可设置为 PRIMARY、SECONDARY、PRIMARY_AND_SECONDARY。

        除了 role，还可设置以下两个选项。
        - disconnect_on_promoted_to_primary：当 Secondary 提升为 Primary 时，是否断开 Secondary 当前的客户端连接。默认为 no。
        - disconnect_on_metadata_unavailable：当组复制负载过高时，是否断开当前的客户端连接。默认为 no。
- routing_strategy：路由策略。可设置的路由策略如下。
    - round-robin：轮询。如果节点不可用，会自动从轮询列表中剔除。节点恢复，又会重新加入到轮询列表中。
    - round-robin-with-fallback：同 round-robin 类似，只不过当所有的 Secondary 节点不可用时，会轮询 Primary 节点。
    - first-available：选择目标列表中的第一个可用节点。如果第一个节点不可用，则会依次选择目标列表中的其他节点。
    - next-available：同 first-available 类似，只不过当节点恢复后，不会重新加入目标列表。该策略在 InnoDB Cluster 中不可用，只是为了向后兼容。

    对于 Primary 节点，可设置的策略有 first-available 和 round-robin。如果不设置 routing_strategy，则默认为 round-robin。如果通过 --bootstrap 初始化，则它会被设置为 first-available。

    对于 Secondary 节点，可设置的策略有 first-available、round-robin 和 round-robin-with-fallback，默认为 round-robin。若通过 --bootstrap 初始化，则它会被设置为 round-robin-with-fallback。

    对于 Primary_and_secondary 节点，可设置的策略有 first-available 和 round-robin，默认为 round-robin。

- protocol：支持的协议。可设置为 classic、x。
- max_connections：最大连接数，默认为 512。MySQL Router 不支持连接池，所以应用发往 Router 的连接与 Router 发往 MySQL 的连接是一一对应的。

至于其他模块和选项，可参考官方文档 "Configuration File Options"。

12.2.3 启动 MySQL Router

要启动 MySQL Router，可直接使用 --directory 目录下的启动脚本。

```
# sh /data/myrouter/6446/start.sh
```

12.2.4 测试 MySQL Router

接下来，我们围绕 InnoDB Cluster 的功能点，进行以下两方面的测试。

- InnoDB Cluster 的高可用性。

  当 Primary 节点宕机时，集群能否继续对外提供写服务。

- 读写分离场景。

  主要考察读操作的负载均衡，以及当节点角色发生变化时，MySQL Router 能否自动感知。

集群的节点信息如表 12-1 所示。

表 12-1　集群的节点信息

IP 地址	主　机　名	角　　色
192.168.244.10	node1	Primary
192.168.244.20	node2	Secondary
192.168.244.30	node3	Secondary

首先，我们测试一下读操作的负载均衡。测试命令如下。

```
# mysql -h 192.168.244.128 -P 6447 -u root -p123456 -e " select concat(@@hostname,':',@@port)"
+-----------------------------+
| concat(@@hostname,':',@@port) |
+-----------------------------+
| node2:3306                  |
+-----------------------------+

# mysql -h 192.168.244.128 -P 6447 -u root -p123456 -e " select concat(@@hostname,':',@@port)"
+-----------------------------+
| concat(@@hostname,':',@@port) |
+-----------------------------+
| node3:3306                  |
+-----------------------------+
```

这里用了个小技巧，通过查看主机名来判断查询落在了哪个节点上。可以看到，同一个操作执行了两次，第一次落到了 node2 上，第二次落到了 node3 上。重复执行多次，依然如此。可见，MySQL

Router 确实能实现读操作的负载均衡。

下面测试一下 InnoDB Cluster 的高可用性。主要测试当 Primary 节点出现故障时，读写端口能否继续对外提供服务。

(1) 针对读写端口进行查询。

```
# mysql -h 192.168.244.128 -P 6446 -u root -p123456 -e " select concat(@@hostname,':',@@port)"
+-----------------------------+
| concat(@@hostname,':',@@port) |
+-----------------------------+
| node1:3306                  |
+-----------------------------+
```

返回的是 node1，即 Primary 节点的主机名。

(2) 模拟 Primary 节点故障。

```
# mysql -h 192.168.244.10 -P 3306 -u root -p123456 -e "shutdown"
```

(3) 针对读写端口再次查询。

```
# mysql -h 192.168.244.128 -P 6446 -u root -p123456 -e "select concat(@@hostname,':',@@port)"
+-----------------------------+
| concat(@@hostname,':',@@port) |
+-----------------------------+
| node3:3306                  |
+-----------------------------+
```

虽然 node1 上的实例关闭了，但查询依然可行，整个集群依然可用。只不过，此时的 Primary 节点切换到 node3 上了。

重启之前的 Primary 节点，再来看看之前的读操作。

```
# /usr/local/mysql/bin/mysqld_safe --defaults-file=/etc/my_3306.cnf &
# mysql -h 192.168.244.128 -P 6447 -u root -p123456 -e "select concat(@@hostname,':',@@port)"
+-----------------------------+
| concat(@@hostname,':',@@port) |
+-----------------------------+
| node2:3306                  |
+-----------------------------+

# mysql -h 192.168.244.128 -P 6447 -u root -p123456 -e " select concat(@@hostname,':',@@port)"
+-----------------------------+
| concat(@@hostname,':',@@port) |
+-----------------------------+
| node1:3306                  |
+-----------------------------+
```

读操作落到了 node1 和 node2 上，之前是 node2 和 node3。可见，MySQL Router 确实能自动感知组复制的节点角色变化。

12.2.5　MySQL Router 的注意事项

- 如果组复制的节点发生了变化，如新增或删除了一个节点，dynamic_state 文件不会同步更新，除非重启 MySQL Router。

- MySQL Router 本身是无状态的，所以建议部署多个，在前端通过 LVS、HAproxy、F5 等进行负载均衡。
- 官方建议将 MySQL Router 部署在应用节点上，通过套接字文件进行通信。这么做的效率无疑是最高的，但我们也要考虑到这种方式的可维护性。毕竟，应用节点一般是由业务运维人员来管理和维护的，而 MySQL Router 是数据库服务，一般是由 DBA 负责管理的。
- 读写分离，并不意味着只读端口只能接收只读请求。事实上，读端口也能接收 DML 和 DDL 请求，只不过 Secondary 节点设置了 --super-read-only。

```
# mysql -h 192.168.244.128 -P 6447 -u root -p123456 -e " CREATE DATABASE  slowtech"
ERROR 1290 (HY000) at line 1: The MySQL server is running with the --super-read-only option so it cannot execute this statement
```

12.3 InnoDB Cluster 的搭建

下面使用 MySQL Shell 搭建一个三节点的 InnoDB Cluster。

该集群的节点信息如表 12-2 所示。

表 12-2 集群的节点信息

IP 地址	主 机 名	角 色
192.168.244.10	node1	Primary
192.168.244.20	node2	Secondary
192.168.244.30	node3	Secondary

12.3.1 准备安装环境

为了方便测试，这里清空防火墙规则并关闭 SELinux。具体命令如下。

清空防火墙规则：

```
# iptables -F
```

关闭 SELinux：

```
# setenforce 0
```

12.3.2 初始化 MySQL 实例

初始化 MySQL 实例的具体步骤如下。在 3 个节点上均要执行这些步骤。

(1) 准备安装包。

```
# cd /usr/local/
# tar xvf mysql-8.0.27-linux-glibc2.12-x86_64.tar.xz
# ln -s mysql-8.0.27-linux-glibc2.12-x86_64 mysql
```

(2) 编辑 MySQL 配置文件。

node1 节点：

```
# vim /etc/my_3306.cnf
[client]
socket = /data/mysql/3306/data/mysql.sock

[mysqld]
#server configuration
user = mysql
datadir = /data/mysql/3306/data
basedir = /usr/local/mysql
port = 3306
socket = /data/mysql/3306/data/mysql.sock
log_timestamps = system
log_error = /data/mysql/3306/data/mysqld.err
skip_name_resolve
report_host = 192.168.244.10
disabled_storage_engines = MyISAM,BLACKHOLE,FEDERATED,ARCHIVE,MEMORY
report_port = 3306

#Replication Framework
server_id = 1
log_bin = mysql-bin
binlog_format = ROW
gtid_mode = ON
enforce_gtid_consistency = ON
log_slave_updates = ON
master_info_repository = TABLE
relay_log_info_repository = TABLE
binlog_transaction_dependency_tracking = WRITESET

#Multi-threaded Replication
slave-parallel-workers = 8
slave-preserve-commit-order = ON
slave-parallel-type = LOGICAL_CLOCK

#Group Replication Settings
transaction_write_set_extraction = XXHASH64
```

node2 节点:

参数基本相同,只需修改 report_host 和 server_id。

```
report_host = 192.168.244.20
server_id = 2
```

node3 节点:

```
report_host = 192.168.244.30
server_id=3
```

(3) 创建数据目录。

```
# mkdir -p /data/mysql/3306/data
```

(4) 初始化实例。

```
# /usr/local/mysql/bin/mysqld --defaults-file=/etc/my_3306.cnf --initialize-insecure
```

(5) 启动数据库。

```
# /usr/local/mysql/bin/mysqld_safe --defaults-file=/etc/my_3306.cnf &
```

12.3.3 创建超级管理员账号

方便起见，这里直接创建可远程登录的 root 账号。在 3 个节点上都要创建。

```
mysql> set session sql_log_bin=0;
mysql> create user root@'%' identified by '123456';
mysql> grant all on *.* to root@'%' with grant option;
mysql> set session sql_log_bin=1;
```

12.3.4 配置实例

然后配置实例，以满足 InnoDB Cluster 的要求。

3 个节点上均要执行此步骤，首先配置 node1。

```
# mysqlsh
mysql-js> dba.configureInstance('root:123456@192.168.244.10:3306', { clusterAdmin: 'cluster_admin', clusterAdminPassword: 'cluster_pass' })
Configuring MySQL instance at 192.168.244.10:3306 for use in an InnoDB cluster...

This instance reports its own address as 192.168.244.10:3306
Assuming full account name 'cluster_admin'@'%' for cluster_admin

applierWorkerThreads will be set to the default value of 4.

The instance '192.168.244.10:3306' is valid to be used in an InnoDB cluster.

Cluster admin user 'cluster_admin'@'%' created.
The instance '192.168.244.10:3306' is already ready to be used in an InnoDB cluster.

Successfully enabled parallel appliers.
```

命令中的 `clusterAdmin` 和 `clusterAdminPassword` 用来指定集群的管理账号及密码。

该命令看上去只是简单地创建了一个账号，但它实际上在背后做了大量的检测工作，包括检测数据库的版本、`performance_schema` 是否开启、参数是否满足 InnoDB Cluster 的要求等。

不仅如此，该命令还能修改配置，甚至重启实例。

下面看一个稍微复杂一点儿的案例。

```
mysql-js> dba.configureInstance('root:123456@192.168.244.10:3306', { mycnfPath:'/etc/my_3306.cnf', clusterAdmin:'cluster_admin', clusterAdminPassword:'cluster_pass' })
Configuring local MySQL instance listening at port 3306 for use in an InnoDB cluster...

This instance reports its own address as 192.168.244.10:3306
Assuming full account name 'cluster_admin'@'%' for cluster_admin

applierWorkerThreads will be set to the default value of 4.

NOTE: Some configuration options need to be fixed:
+--------------------------------------+----------------+----------------+------------------------------------------------------+
| Variable                             | Current Value  | Required Value | Note                                                 |
+--------------------------------------+----------------+----------------+------------------------------------------------------+
| binlog_transaction_dependency_tracking | COMMIT_ORDER | WRITESET       | Update the server variable and the config file       |
| enforce_gtid_consistency             | OFF            | ON             | Update the config file and restart the server        |
```

```
| gtid_mode                              | OFF          | ON         | Update the config file and restart the server |
| report_port                            | <not set>    | 3306       | Update the config file                        |
+----------------------------------------+--------------+------------+-----------------------------------------------+

Some variables need to be changed, but cannot be done dynamically on the server.
Do you want to perform the required configuration changes? [y/n]: y
Do you want to restart the instance after configuring it? [y/n]: y

Cluster admin user 'cluster_admin'@'%' created.
Configuring instance...
The instance '192.168.244.10:3306' was configured to be used in an InnoDB cluster.
Restarting MySQL...
NOTE: MySQL server at 192.168.244.10:3306 was restarted.
```

命令中的 `mycnfPath` 是实例配置文件的路径。

在上面这个案例中，该命令执行了以下操作。

(1) 创建集群管理账号。

(2) 检查参数是否满足 InnoDB Cluster 的要求。对于不满足要求的参数，会执行以下操作。

- 修改配置文件。
- 修改参数的内存值。

    如果是 MySQL 5.7，会通过 `SET GLOBAL` 命令进行修改。

    如果是 MySQL 8.0，则通过 `SET PERSIST` 命令进行修改。该命令不仅会修改参数，还会将配置持久化到文件（mysqld-auto.cnf）中。

    ```
    SET PERSIST `replica_parallel_workers` = 4
    SET PERSIST `binlog_transaction_dependency_tracking` = 'WRITESET'
    SET PERSIST_ONLY `enforce_gtid_consistency` = 'ON'
    SET PERSIST_ONLY `gtid_mode` = 'ON'
    ```

(3) 重启。

   如果是 MySQL 8.0，会直接执行 `RESTART` 操作。如果是 MySQL 5.7，则会提示手动执行重启操作。

   ```
   NOTE: MySQL server needs to be restarted for configuration changes to take effect.
   ```

在执行时，有以下几点需要注意。

- 如果要修改实例的配置文件，建议在实例本地执行 `dba.configureInstance`。
- 如果目标实例是 MySQL 8.0，其实无须指定 `mycnfPath` 选项，因为 MySQL 8.0 支持参数的持久化。
- 如果只想检查实例是否满足 InnoDB Cluster 的要求，而不进行任何修改，可使用 `dba.check-InstanceConfiguration` 命令。

除了命令中指定的选项，`configureInstance` 还支持以下选项。

- `outputMycnfPath`：将配置直接写入 `outputMycnfPath` 指定的文件，不直接修改实例的配置文件。
- `password`：指定实例的连接密码。

- clearReadOnly：是否允许将 super_read_only 设置为 OFF。会在后续版本中移除。
- interactive：是否开启交互模式。默认为 true。
- restart：是否允许自动重启实例。

接下来，配置 node2 和 node3。

```
mysql-js> dba.configureInstance('root:123456@192.168.244.20:3306', { clusterAdmin:'cluster_admin',
clusterAdminPassword:'cluster_pass' })

mysql-js> dba.configureInstance('root:123456@192.168.244.30:3306', { clusterAdmin:'cluster_admin',
clusterAdminPassword:'cluster_pass'})
```

12.3.5 创建 InnoDB Cluster

因为 node1 是 Primary 节点，所以这里登录 node1 执行创建操作。

```
mysql-js> shell.connect('cluster_admin:cluster_pass@192.168.244.10:3306')
<ClassicSession:cluster_admin@192.168.244.10:3306>

mysql-js> dba.createCluster('myCluster', { disableClone: false })
A new InnoDB cluster will be created on instance '192.168.244.10:3306'.

Validating instance configuration at 192.168.244.10:3306...

This instance reports its own address as 192.168.244.10:3306

Instance configuration is suitable.
NOTE: Group Replication will communicate with other members using '192.168.244.10:33061'. Use the
localAddress option to override.

Creating InnoDB cluster 'myCluster' on '192.168.244.10:3306'...

Adding Seed Instance...
Cluster successfully created. Use Cluster.addInstance() to add MySQL instances.
At least 3 instances are needed for the cluster to be able to withstand up to
one server failure.

<Cluster:myCluster>
```

该命令会执行以下操作。

(1) 检测实例是否满足 InnoDB Cluster 的要求，检测内容与 dba.checkInstanceConfiguration 命令相同。
(2) 开启组复制，包括加载组复制插件，修改并持久化组复制的相关参数。

```
INSTALL PLUGIN `group_replication` SONAME 'group_replication.so'
SET PERSIST `super_read_only` = 'ON'
SET PERSIST `group_replication_group_name` = '720113f3-79ab-11ec-b87f-000c29f66609'
SET PERSIST `group_replication_view_change_uuid` = '7201dc45-79ab-11ec-b87f-000c29f66609'
SET PERSIST `group_replication_enforce_update_everywhere_checks` = 'OFF'
SET PERSIST `group_replication_single_primary_mode` = 'ON'
SET PERSIST `group_replication_recovery_use_ssl` = 'ON'
SET PERSIST `group_replication_ssl_mode` = 'REQUIRED'
SET PERSIST `group_replication_local_address` = '192.168.244.10:33061'
SET PERSIST `group_replication_start_on_boot` = 'ON'
SET PERSIST `auto_increment_increment` = 1
```

```
SET PERSIST `auto_increment_offset` = 2
SET GLOBAL `group_replication_bootstrap_group` = 'ON'
START GROUP_REPLICATION
SET GLOBAL `group_replication_bootstrap_group` = 'OFF'
SET GLOBAL read_only= 0
```

(3) 创建 mysql_innodb_cluster_metadata 库，用于管理 InnoDB Cluster 的元数据。

(4) 加载克隆插件。

命令中的 disableClone: false 会开启克隆插件。除此之外，该命令还可指定以下选项。

- gtidSetIsComplete：初始化节点的 GTID_EXECUTED 是否对应所有已经执行过的事务，默认为 false。在添加节点时，这个选项会影响恢复方式的选择。如果初始化节点中存在以下场景下执行的事务，则该参数一定不能设置为 true：关闭 GTID；执行 RESET MASTER；将 sql_log_bin 设置为 OFF。
- multiPrimary：多主模式。默认为 false，即单主模式。
- force：默认情况下，如果设置了多主模式，会有一个确认动作。若要跳过这个动作，可将该参数设置为 true。
- interactive：是否开启交互模式，默认为 true。
- adoptFromGR：是否基于已有的组复制创建 InnoDB Cluster。
- memberSslMode：是否开启 SSL 通信，可设置为 REQUIRED（开启）、VERIFY_CA、VERIFY_IDENTITY、DISABLED（关闭）、AUTO（如果节点支持则开启，否则关闭）。默认为 AUTO。
- ipAllowlist：设置白名单，对应组复制中的 group_replication_ip_whitelist 参数。
- groupName：设置组名，即 group_replication_group_name。
- localAddress：设置当前节点的地址，即 group_replication_local_address。
- groupSeeds：设置种子节点的地址，即 group_replication_group_seeds。
- manualStartOnBoot：实例启动后，是否需要手动开启组复制。默认为 false，即会自动拉起组复制，对应组复制中的 group_replication_start_on_boot 参数。
- exitStateAction：节点被驱逐出集群时应该采取的动作，即 group_replication_exit_state_action，默认为 READ_ONLY。
- memberWeight：设置节点权重，即 group_replication_member_weight。
- consistency：设置事务的一致性级别，即 group_replication_consistency，默认为 EVENTUAL。
- failoverConsistency：类似于 consistency，会在后续版本中移除。
- expelTimeout：设置 group_replication_member_expel_timeout 的值，默认为 0。
- autoRejoinTries：设置节点被驱逐后的重试次数，即 group_replication_autorejoin_tries。
- clearReadOnly：是否允许将 super_read_only 设置为 OFF。会在后续版本中移除。
- multiMaster：类似于 multiPrimary，会在后续版本中移除。

12.3.6 添加节点

添加节点依赖于 Cluster 对象，所以在执行添加操作之前，必须通过 dba.getCluster 命令获取一个 Cluster 对象。

```
mysql-js> cluster=dba.getCluster('myCluster')
<Cluster:myCluster>
```

接下来添加 node2。

```
mysql-js> cluster.addInstance('cluster_admin:cluster_pass@192.168.244.20:3306')

NOTE: The target instance '192.168.244.20:3306' has not been pre-provisioned (GTID set is empty). The
Shell is unable to decide whether incremental state recovery can correctly provision it.
The safest and most convenient way to provision a new instance is through automatic clone provisioning,
which will completely overwrite the state of '192.168.244.20:3306' with a physical snapshot from an
existing cluster member. To use this method by default, set the 'recoveryMethod' option to 'clone'.

The incremental state recovery may be safely used if you are sure all updates ever executed in the cluster
were done with GTIDs enabled, there are no purged transactions and the new instance contains the same
GTID set as the cluster or a subset of it. To use this method by default, set the 'recoveryMethod' option
to 'incremental'.

Please select a recovery method [C]lone/[I]ncremental recovery/[A]bort (default Clone):
Validating instance configuration at 192.168.244.20:3306...

This instance reports its own address as 192.168.244.20:3306

Instance configuration is suitable.
NOTE: Group Replication will communicate with other members using '192.168.244.20:33061'. Use the
localAddress option to override.

A new instance will be added to the InnoDB cluster. Depending on the amount of
data on the cluster this might take from a few seconds to several hours.

Adding instance to the cluster...

Monitoring recovery process of the new cluster member. Press ^C to stop monitoring and let it continue
in background.
Clone based state recovery is now in progress.

NOTE: A server restart is expected to happen as part of the clone process. If the
server does not support the RESTART command or does not come back after a
while, you may need to manually start it back.

* Waiting for clone to finish...
NOTE: 192.168.244.20:3306 is being cloned from 192.168.244.10:3306
** Stage DROP DATA: Completed
** Clone Transfer
    FILE COPY  ############################################################  100%  Completed
    PAGE COPY  ############################################################  100%  Completed
    REDO COPY  ############################################################  100%  Completed

NOTE: 192.168.244.20:3306 is shutting down...

* Waiting for server restart... ready
* 192.168.244.20:3306 has restarted, waiting for clone to finish...
** Stage RESTART: Completed
* Clone process has finished: 72.61 MB transferred in 5 sec (14.52 MB/s)

Incremental state recovery is now in progress.

* Waiting for distributed recovery to finish...
NOTE: '192.168.244.20:3306' is being recovered from '192.168.244.10:3306'
* Distributed recovery has finished

The instance '192.168.244.20:3306' was successfully added to the cluster.
```

该命令在执行的过程中，会提示我们选择恢复的方式。

该命令会执行以下操作。

(1) 检查 cluster_admin 的权限。
(2) 检查实例的配置是否满足 InnoDB Cluster 的要求。
(3) 加载组复制插件，将 group_replication_clone_threshold 设置为 1，安装克隆插件。
(4) 修改并持久化组复制的相关参数，开启组复制。
(5) 基于克隆插件进行全量恢复。

```
CLONE INSTANCE FROM 'mysql_innodb_cluster_2'@'192.168.244.10':3306 IDENTIFIED BY <secret> REQUIRE SSL
```

(6) 基于 binlog 进行增量恢复。

```
CHANGE REPLICATION SOURCE TO SOURCE_USER = 'mysql_innodb_cluster_2', SOURCE_PASSWORD = <secret> FOR CHANNEL 'group_replication_recovery'
```

这里，该命令中只给出了实例的基本连接信息。除此之外，还可指定以下选项。

- label：给实例打上标签。
- recoveryMethod：指定恢复的方式，可设置以下值。
  - incremental：基于 binlog 的增量恢复。
  - clone：基于克隆插件的全量恢复。
  - auto：基于目标实例是否支持克隆插件以及 group_replication_clone_threshold 的大小来自动选择恢复方式。它是默认值。
- waitRecovery：终端是否等待恢复操作完成。可设置为 0、1、2、3，其中 0 代表不等待，让恢复操作在后台运行；1 代表等待，直到恢复操作完成；2 代表在 1 的基础上打印进度信息；3 代表在 2 的基础上打印进度条信息。默认是 3。
- password：指定实例的连接密码。
- memberSslMode：是否开启 SSL 通信，默认为 auto。
- ipAllowlist：设置白名单，即 group_replication_ip_whitelist。
- localAddress：设置当前节点的地址，即 group_replication_local_address。
- groupSeeds：设置种子节点的地址，即 group_replication_group_seeds。
- interactive：是否开启交互模式。
- exitStateAction：节点被驱逐出集群时应该采取的动作，即 group_replication_exit_state_action，默认为 READ_ONLY。
- memberWeight：设置节点权重，即 group_replication_member_weight。
- autoRejoinTries：设置节点被驱逐后的重试次数，即 group_replication_autorejoin_tries。

接下来添加 node3。

```
mysql-js> cluster.addInstance('cluster_admin:cluster_pass@192.168.244.30:3306')
```

12.3.7 查看集群的状态

通过 cluster.status() 命令查看集群的状态。

```
mysql-js> cluster.status()
{
    "clusterName": "myCluster",
    "defaultReplicaSet": {
        "name": "default",
        "primary": "192.168.244.10:3306",
        "ssl": "REQUIRED",
        "status": "OK",
        "statusText": "Cluster is ONLINE and can tolerate up to ONE failure.",
        "topology": {
            "192.168.244.10:3306": {
                "address": "192.168.244.10:3306",
                "memberRole": "PRIMARY",
                "mode": "R/W",
                "readReplicas": {},
                "replicationLag": null,
                "role": "HA",
                "status": "ONLINE",
                "version": "8.0.27"
            },
            "192.168.244.20:3306": {
                "address": "192.168.244.20:3306",
                "memberRole": "SECONDARY",
                "mode": "R/O",
                "readReplicas": {},
                "replicationLag": null,
                "role": "HA",
                "status": "ONLINE",
                "version": "8.0.27"
            },
            "192.168.244.30:3306": {
                "address": "192.168.244.30:3306",
                "memberRole": "SECONDARY",
                "mode": "R/O",
                "readReplicas": {},
                "replicationLag": null,
                "role": "HA",
                "status": "ONLINE",
                "version": "8.0.27"
            }
        },
        "topologyMode": "Single-Primary"
    },
    "groupInformationSourceMember": "192.168.244.10:3306"
}
```

下面看看输出中部分项的具体含义。

- address：节点地址。
- memberRole：节点角色。
- mode：节点的模式。R/W 代表读写，R/O 代表只读。
- replicationLag：复制延迟时间。

- role：角色。目前只有 HA，即高可用（high availability）。
- status：除了组复制的 5 种状态（ONLINE、OFFLINE、RECOVERING、UNREACHABLE、ERROR），还新增了一个 MISSING 状态。对于元数据中存在，但集群中不存在的节点，会标记为 MISSING。

cluster.status() 可指定以下选项。

- extended：设置输出的详细等级，可设置为 0、1、2、3，等级越高，输出越详尽。默认为 0。
- queryMembers：作用同 extended 类似，会在后续版本中移除。

12.3.8 部署 MySQL Router

MySQL Router 的下载和安装可参考 12.2 节。

这里，直接基于刚刚创建的集群初始化 MySQL Router。

```
# mkdir -p /data/myrouter/6446
# mysqlrouter --bootstrap cluster_admin@192.168.244.10:3306 --user=mysql --directory
/data/myrouter/6446 --conf-use-sockets --report-host='192.168.244.128'
```

接下来，启动 MySQL Router。

```
# sh /data/myrouter/6446/start.sh
```

除了使用脚本，也可通过 mysqlrouter 命令直接启动。

```
# mysqlrouter -c /data/myrouter/6446/mysqlrouter.conf --user=mysql &
```

默认会开启 5 个监听端口。

```
# netstat -ntlup | grep mysqlrouter
tcp        0      0 0.0.0.0:6446            0.0.0.0:*               LISTEN      3070/mysqlrouter
tcp        0      0 0.0.0.0:6447            0.0.0.0:*               LISTEN      3070/mysqlrouter
tcp        0      0 0.0.0.0:6448            0.0.0.0:*               LISTEN      3070/mysqlrouter
tcp        0      0 0.0.0.0:6449            0.0.0.0:*               LISTEN      3070/mysqlrouter
tcp        0      0 0.0.0.0:8443            0.0.0.0:*               LISTEN      3070/mysqlrouter
```

在这 5 个端口中，6446 对应 Classic 协议的读写操作，6447 对应 Classic 协议的只读操作，6448 对应 X 协议的读写操作，6449 对应 X 协议的只读操作，8443 提供 REST API 的 HTTP 端口。

在初始化过程中，MySQL Router 的元数据信息会被更新到 InnoDB Cluster 中。

与集群相关的 Router 信息可通过 cluster.listRouters() 命令查看。

```
mysql-js> cluster.listRouters()
{
    "clusterName": "myCluster",
    "routers": {
        "192.168.244.128::": {
            "hostname": "192.168.244.128",
            "lastCheckIn": "2022-01-20 15:03:29",
            "roPort": 6447,
            "roXPort": 6449,
            "rwPort": 6446,
            "rwXPort": 6448,
            "version": "8.0.27"
```

        }
    }
}

至此，InnoDB Cluster 搭建完毕，相关功能点的测试可参考 12.2 节。

12.4 InnoDB Cluster 的管理操作

在 InnoDB Cluster 技术栈中，所有的管理操作都是通过 MySQL Shell 来执行的，这些管理操作主要体现在两个对象上：dba 和 cluster。在这里，我们首先通过 \help 命令看看这两个对象支持的方法，然后基于这些方法看看 InnoDB Cluster 支持的管理操作及注意事项。

12.4.1 dba 对象支持的操作

首先通过 \help 命令看看 dba 对象支持的操作。

```
mysql-js> \help dba
NAME
      dba - InnoDB cluster and replicaset management functions.

DESCRIPTION
      Entry point for AdminAPI functions, including InnoDB clusters and replica
      sets.

      InnoDB clusters

      The dba.configureInstance() function can be used to configure a MySQL
      instance with the settings required to use it in an InnoDB cluster.

      InnoDB clusters can be created with the dba.createCluster() function.

      Once created, InnoDB cluster management objects can be obtained with the
      dba.getCluster() function.

      InnoDB ReplicaSets

      The dba.configureReplicaSetInstance() function can be used to configure a
      MySQL instance with the settings required to use it in a replicaset.

      ReplicaSets can be created with the dba.createReplicaSet() function.

      Once created, replicaset management objects can be obtained with the
      dba.getReplicaSet() function.

      Sandboxes

      Utility functions are provided to create sandbox MySQL instances, which
      can be used to create test clusters and replicasets.

PROPERTIES
      session
            The session the dba object will use by default.

      verbose
            Controls debug message verbosity for sandbox related dba
```

operations.

FUNCTIONS
    checkInstanceConfiguration(instance[, options])
        Validates an instance for MySQL InnoDB Cluster usage.

    configureInstance([instance][, options])
        Validates and configures an instance for MySQL InnoDB Cluster
        usage.

    configureLocalInstance(instance[, options])
        Validates and configures a local instance for MySQL InnoDB Cluster
        usage.

    configureReplicaSetInstance([instance][, options])
        Validates and configures an instance for use in an InnoDB
        ReplicaSet.

    createCluster(name[, options])
        Creates a MySQL InnoDB cluster.

    createReplicaSet(name[, options])
        Creates a MySQL InnoDB ReplicaSet.

    deleteSandboxInstance(port[, options])
        Deletes an existing MySQL Server instance on localhost.

    deploySandboxInstance(port[, options])
        Creates a new MySQL Server instance on localhost.

    dropMetadataSchema(options)
        Drops the Metadata Schema.

    getCluster([name][, options])
        Retrieves a cluster from the Metadata Store.

    getClusterSet()
        Returns an object representing a ClusterSet.

    getReplicaSet()
        Returns an object representing a ReplicaSet.

    help([member])
        Provides help about this object and it's members

    killSandboxInstance(port[, options])
        Kills a running MySQL Server instance on localhost.

    rebootClusterFromCompleteOutage([clusterName][, options])
        Brings a cluster back ONLINE when all members are OFFLINE.

    startSandboxInstance(port[, options])
        Starts an existing MySQL Server instance on localhost.

    stopSandboxInstance(port[, options])
        Stops a running MySQL Server instance on localhost.

    upgradeMetadata([options])
        Upgrades (or restores) the metadata to the version supported by the
        Shell.

```
SEE ALSO

- For general information about the AdminAPI use: \? AdminAPI
- For help on a specific function use: \? dba.<functionName>

e.g. \? dba.deploySandboxInstance
```

操作的对象分为 3 类。

- InnoDB Cluster。
- InnoDB ReplicaSet：InnoDB 副本集，同 InnoDB Cluster 类似，只不过它的后端存储节点是 MySQL 主从，不是组复制。
- Sandbox：沙盒。只要 mysqld 在系统路径（$PATH）下，只需提供端口，就能快速部署一个测试实例。可用来搭建 InnoDB Cluster 的测试环境。

下面重点说说 InnoDB Cluster 对象的相关操作。

1. dba.checkInstanceConfiguration(instance[, options])

检测实例是否满足 InnoDB Cluster 的要求，具体如下。

- 检查用户权限。
- 检查 @@report_host（@@hostname）是否有效。
- 检查目标实例是否开启了异步复制。
- 检查克隆插件的状态。
- 检查表的兼容性，包括表上是否存在主键，表的引擎是否为 InnoDB 或 MEMORY。
- 检查参数是否满足指定值。
  - innodb_page_size 需大于 4KB
  - performance_schema 需开启
  - server_id 需有效
  - log_bin 需开启
  - binlog_format = ROW
  - log_slave_updates = ON
  - enforce_gtid_consistency = ON
  - gtid_mode = ON
  - master_info_repository = TABLE
  - relay_log_info_repository = TABLE
  - binlog_checksum = NONE（针对 MySQL 8.0.21 之前的版本）
  - transaction_write_set_extraction = XXHASH64
  - report_port 必须设置
  - binlog_transaction_dependency_tracking = WRITESET
  - slave_parallel_type = LOGICAL_CLOCK
  - slave_preserve_commit_order = ON

具体的检测逻辑可参考源码：modules/adminapi/common/instance_validations.cc 和 modules/adminapi/common/provision.cc。

下面，我们看一个具体的示例。

```
mysql-js> dba.checkInstanceConfiguration('root:123456@192.168.244.10:3306',{ 'mycnfPath':'/etc/
my_3306.cnf' })
Validating local MySQL instance listening at port 3306 for use in an InnoDB cluster...

This instance reports its own address as node1:3306
Clients and other cluster members will communicate with it through this address by default. If this
is not correct, the report_host MySQL system variable should be changed.

Checking whether existing tables comply with Group Replication requirements...
ERROR: The following tables use a storage engine that are not supported by Group Replication:
slowtech.t1

ERROR: The following tables do not have a Primary Key or equivalent column:
slowtech.t2

Group Replication requires tables to use InnoDB and have a PRIMARY KEY or PRIMARY KEY Equivalent (non-null
unique key). Tables that do not follow these requirements will be readable but not updateable when used
with Group Replication. If your applications make updates (INSERT, UPDATE or DELETE) to these tables,
ensure they use the InnoDB storage engine and have a PRIMARY KEY or PRIMARY KEY Equivalent.
If you can't change the tables structure to include an extra visible key to be used as PRIMARY KEY,
you can make use of the INVISIBLE COLUMN feature available since 8.0.23:
https://dev.mysql.com/doc/refman/8.0/en/invisible-columns.html

Checking instance configuration...
Configuration file /etc/my_3306.cnf will also be checked.

NOTE: Some configuration options need to be fixed:
+--------------------------------------+----------------+----------------+--------------------------------------------+
| Variable                             | Current Value  | Required Value | Note                                       |
+--------------------------------------+----------------+----------------+--------------------------------------------+
| binlog_transaction_dependency_tracking | COMMIT_ORDER  | WRITESET       | Update the server variable and the config file |
| enforce_gtid_consistency             | OFF            | ON             | Update the config file and restart the server |
| gtid_mode                            | OFF            | ON             | Update the config file and restart the server |
| report_port                          | <not set>      | 3306           | Update the config file                     |
+--------------------------------------+----------------+----------------+--------------------------------------------+

Some variables need to be changed, but cannot be done dynamically on the server.
NOTE: Please use the dba.configureInstance() command to repair these issues.

{
    "config_errors": [
        {
            "action": "server_update+config_update",
            "current": "COMMIT_ORDER",
            "option": "binlog_transaction_dependency_tracking",
            "required": "WRITESET"
        },
        {
            "action": "config_update+restart",
            "current": "OFF",
            "option": "enforce_gtid_consistency",
            "required": "ON"
        },
        {
```

```
            "action": "config_update+restart",
            "current": "OFF",
            "option": "gtid_mode",
            "required": "ON"
        },
        {
            "action": "config_update",
            "current": "<not set>",
            "option": "report_port",
            "required": "3306"
        }
    ],
    "status": "error"
}
```

命令中的 `mycnfPath` 是实例配置文件的路径,可不指定。如果没有指定或指定的文件不存在,则只会检查参数的内存值。建议指定。

再来看看 `action` 的取值。

- `server_update`:需要修改参数的内存值。
- `config_update`:需要修改配置文件的值。
- `restart`:重启实例。对于只读参数,修改后需要重启实例才能生效。

最后看看 `status` 的取值。

- `error`:实例不满足 InnoDB Cluster 的要求,具体在本例中,slowtech.t1 不是 InnoDB 表,slowtech.t2 上没有主键,参数不满足要求。
- `ok`:实例满足 InnoDB Cluster 的要求。

2. `dba.configureInstance([instance][, options])`

配置实例,使其满足 InnoDB Cluster 的要求,具体可参考 12.3 节。

3. `dba.configureLocalInstance(instance[, options])`

同 `configureInstance` 类似,只不过待配置的实例在本地。

4. `dba.createCluster(name[, options])`

创建 InnoDB Cluster,具体可参考 12.3 节。

5. `dba.getCluster([name][, options])`

获取 Cluster 对象。

`options` 支持以下选项。

- `name`:集群名,可不指定,默认是当前节点所属的集群。
- `options`:目前只支持一个选项,即 `connectToPrimary`,默认为 `true`,代表 MySQL Shell 会自动连接到集群的 Primary 节点。

6. dba.upgradeMetadata([options])

升级 mysql_innodb_cluster_metadata 库的版本，该库用来存储 InnoDB Cluster 的元数据信息。在 MySQL Shell 8.0.19 之前，这个库中只有 6 张表。从 MySQL Shell 8.0.19 开始，metadata 库的版本升级到了 2.0.0，不仅表的数量增加了，表结构也发生了变化。

考虑到在升级 MySQL Shell 的同时，MySQL Router 也要同步升级，所以在执行具体的升级操作时，需遵循以下步骤。

(1) 升级 MySQL Shell。
(2) 执行 dba.upgradeMetadata()，一旦检测到集群中存在旧版本的 Router 实例，该命令就会终止。这个时候即可升级 MySQL Router。
(3) 升级 MySQL Router。
(4) 再次执行 dba.upgradeMetadata()。

options 支持以下选项。

- dryRun：试运行。
- interactive：是否开启交互模式。

```
mysql-js> dba.upgradeMetadata()
NOTE: Installed metadata at '192.168.244.30:3306' is up to date (version 2.1.0).
Metadata state is consistent and a restore is not necessary.
```

7. dba.dropMetadataSchema(options)

删除 mysql_innodb_cluster_metadata 库，常在重新搭建集群时执行。

```
mysql-js> dba.dropMetadataSchema()
Are you sure you want to remove the Metadata? [y/N]: y
ERROR: The MySQL instance at '192.168.244.30:3306' currently has the super_read_only system variable
set to protect it from inadvertent updates from applications.
You must first unset it to be able to perform any changes to this instance.
For more information see:
https://dev.mysql.com/doc/refman/en/server-system-variables.html#sysvar_super_read_only.
You must first unset it to be able to perform any changes to this instance.
For more information see:
https://dev.mysql.com/doc/refman/en/server-system-variables.html#sysvar_super_read_only.

Do you want to disable super_read_only and continue? [y/N]: y

Metadata Schema successfully removed.
```

options 支持以下选项。

- force：若设置为 true，则没有第一步的确认操作。
- clearReadOnly：是否允许将 super_read_only 设置为 OFF。

8. dba.rebootClusterFromCompleteOutage([clusterName][, options])

当集群的所有节点都处于 OFFLINE 状态，无法对外提供服务时，可用该命令恢复整个集群。

```
mysql-js> shell.connect('cluster_admin:cluster_pass@192.168.244.10:3306')
mysql-js> dba.rebootClusterFromCompleteOutage()
Restoring the cluster 'myCluster' from complete outage...

The instance '192.168.244.30:3306' was part of the cluster configuration.
Would you like to rejoin it to the cluster? [y/N]: y

The instance '192.168.244.20:3306' was part of the cluster configuration.
Would you like to rejoin it to the cluster? [y/N]: y

* Waiting for seed instance to become ONLINE...
192.168.244.10:3306 was restored.
Rejoining '192.168.244.30:3306' to the cluster.
Rejoining instance '192.168.244.30:3306' to cluster 'myCluster'...

The instance '192.168.244.30:3306' was successfully rejoined to the cluster.

Rejoining '192.168.244.20:3306' to the cluster.
Rejoining instance '192.168.244.20:3306' to cluster 'myCluster'...

The instance '192.168.244.20:3306' was successfully rejoined to the cluster.

The cluster was successfully rebooted.

<Cluster:myCluster>
```

该命令主要会执行以下操作。

首先，初始化组复制，在当前节点上执行。注意，这个节点必须包含最新的数据，建议在操作之前，通过对比各个节点的 `gtid_executed` 来选择最新节点。

接着，开启其他节点的组复制，让其他节点加入集群。

命令中的 `clusterName` 是集群名，可不指定，默认从当前节点获取。

`options` 支持以下选项。

- `user`：指定实例的连接用户。
- `password`：指定实例的连接密码。
- `removeInstances`：需要从集群中移除的实例列表。
- `rejoinInstances`：需要重新加入集群的实例列表。
- `clearReadOnly`：是否允许将 `super_read_only` 设置为 `OFF`。

该命令的常见使用场景如下。

- 集群通过 `cluster.dissolve()` 命令解散了。
- 集群所有节点同时宕机了。

12.4.2 cluster 对象支持的操作

首先通过 `\help` 命令看看 cluster 对象支持的操作。

```
mysql-js> \help cluster
NAME
```

```
            Cluster - Represents an InnoDB cluster.

DESCRIPTION
            The cluster object is the entry point to manage and monitor a MySQL
            InnoDB cluster.

            A cluster is a set of MySQLd Instances which holds the user's data.

            It provides high-availability and scalability for the user's data.

PROPERTIES
            name
                    Retrieves the name of the cluster.

FUNCTIONS
            addInstance(instance[, options])
                    Adds an Instance to the cluster.

            checkInstanceState(instance)
                    Verifies the instance gtid state in relation to the cluster.

            createClusterSet(domainName[, options])
                    Creates a MySQL InnoDB ClusterSet from an existing standalone
                    InnoDB Cluster.

            describe()
                    Describe the structure of the cluster.

            disconnect()
                    Disconnects all internal sessions used by the cluster object.

            dissolve([options])
                    Dissolves the cluster.

            forceQuorumUsingPartitionOf(instance[, password])
                    Restores the cluster from quorum loss.

            getClusterSet()
                    Returns an object representing a ClusterSet.

            getName()
                    Retrieves the name of the cluster.

            help([member])
                    Provides help about this class and it's members

            listRouters([options])
                    Lists the Router instances.

            options([options])
                    Lists the cluster configuration options.

            rejoinInstance(instance[, options])
                    Rejoins an Instance to the cluster.

            removeInstance(instance[, options])
                    Removes an Instance from the cluster.

            removeRouterMetadata(routerDef)
                    Removes metadata for a router instance.
```

```
rescan([options])
      Rescans the cluster.

resetRecoveryAccountsPassword(options)
      Reset the password of the recovery accounts of the cluster.

setInstanceOption(instance, option, value)
      Changes the value of an option in a Cluster member.

setOption(option, value)
      Changes the value of an option for the whole Cluster.

setPrimaryInstance(instance)
      Elects a specific cluster member as the new primary.

setupAdminAccount(user, options)
      Create or upgrade an InnoDB Cluster admin account.

setupRouterAccount(user, options)
      Create or upgrade a MySQL account to use with MySQL Router.

status([options])
      Describe the status of the cluster.

switchToMultiPrimaryMode()
      Switches the cluster to multi-primary mode.

switchToSinglePrimaryMode([instance])
      Switches the cluster to single-primary mode.

For more help on a specific function use: cluster.help('<functionName>')

e.g. cluster.help('addInstance')
```

下面分类阐述各个操作的具体作用。

1. Cluster 相关

- **<Cluster>.getName()**

获取 Cluster 的名字。

```
mysql-js> cluster.getName()
myCluster
```

- **<Cluster>.disconnect()**

断开 Cluster 对象与组复制节点之间的连接。如果要重新建立连接，必须执行 dba.getCluster()。

2. 节点相关

- **<Cluster>.checkInstanceState(instance)**

检查实例的状态。具体来说，是检查目标实例的 GTID 是否与集群兼容。

```
mysql-js> cluster.checkInstanceState('cluster_admin:cluster_pass@192.168.244.30:3306')
Analyzing the instance '192.168.244.30:3306' replication state...
```

The instance contains additional transactions in relation to the cluster. However, Clone is available
and if desired can be used to overwrite the data and add the instance to a cluster.

```
{
    "reason": "diverged",
    "state": "warning"
}
```

返回值有两个。

- state：实例的状态。
- reason：显示为这个状态的原因。常见的原因有如下几个。
  - new：目标实例是新实例，且集群的 gtid_purged 为空。
  - recoverable：目标实例的 gtid_executed 是集群 gtid_executed 的子集，且集群 gtid_purged 为空或集群 gtid_purged 是目标实例 gtid_executed 的子集。
  - diverged：目标实例存在集群没有的事务。
  - lost_transactions：目标实例的 gtid_executed 是集群 gtid_purged 的子集，代表目标实例需要的 binlog 已经被集群删除了。

若 reason 为 new 或 recoverable，则 state 显示为 ok，代表目标实例的 GTID 与集群的兼容。

若 reason 为 diverged 或 lost_transactions，代表目标实例的 GTID 与集群的 GTID 不兼容。这又可细分为两种情况：支持克隆插件，则 state 显示为 warning；不支持，则显示为 error。

- **<Cluster>.addInstance(instance[, options])**

添加实例，具体可参考 12.3 节。

- **<Cluster>.removeInstance(instance[, options])**

删除实例。

```
mysql-js> cluster.removeInstance('cluster_admin:cluster_pass@192.168.244.20:3306')
The instance will be removed from the InnoDB cluster. Depending on the instance
being the Seed or not, the Metadata session might become invalid. If so, please
start a new session to the Metadata Storage R/W instance.

* Waiting for instance to synchronize with the primary...
** Transactions replicated  ############################################################  100%
Instance '192.168.244.20:3306' is attempting to leave the cluster...

The instance '192.168.244.20:3306' was successfully removed from the cluster.
```

该命令主要会执行以下操作。

(1) 对于集群，会删除实例相关的元数据信息。

```
DROP USER IF EXISTS 'mysql_innodb_cluster_2'@'%'
UPDATE mysql_innodb_cluster_metadata.instances SET attributes = json_remove(attributes,
'$.recoveryAccountUser', '$.recoveryAccountHost') WHERE mysql_server_uuid =
'ffa2043d-79a9-11ec-9150-000c297b8a24'
DELETE FROM mysql_innodb_cluster_metadata.instances WHERE addresses->'$.mysqlClassic' =
'192.168.244.20:3306'
```

(2) 对于目标实例，会关闭组复制，重置组复制的相关通道，同时将组复制的相关参数恢复为默认值。

```
STOP GROUP_REPLICATION
RESET REPLICA ALL FOR CHANNEL 'group_replication_applier'
RESET REPLICA ALL FOR CHANNEL 'group_replication_recovery'
SET PERSIST `group_replication_start_on_boot` = 'OFF'
SET PERSIST `group_replication_enforce_update_everywhere_checks` = 'OFF'
SET PERSIST `group_replication_bootstrap_group` = DEFAULT
SET PERSIST `group_replication_group_seeds` = DEFAULT
SET PERSIST `group_replication_local_address` = DEFAULT
```

options 支持以下选项。

- password：实例的连接密码，它会覆盖 instance 中的密码。
- force：如果实例不能访问，在执行该命令时，会提示是否继续执行。将 force 设置为 true，会跳过提示，直接删除集群中的元数据。
- interactive：是否开启交互模式。

在删除实例时，如果该实例上存在待应用的事务，默认会等待 60 秒，让其应用完。倘若没有应用完且 force 未设置为 true，该命令会报错并终止当前的删除操作。这里的 60 秒由 MySQL Shell 中的配置项 dba.gtidWaitTimeout 决定，可通过以下方式修改。

```
mysql-js> shell.options["dba.gtidWaitTimeout"]=120
```

- **<Cluster>.rejoinInstance(instance[, options])**

让既非 ONLINE 又非 RECOVERING 状态的节点重新加入集群。

```
mysql-js> cluster.rejoinInstance('192.168.244.20:3306')
Rejoining instance '192.168.244.20:3306' to cluster 'myCluster'...

The instance '192.168.244.20:3306' was successfully rejoined to the cluster.
```

该命令会执行以下操作。

(1) 判断节点是否满足重新加入集群的条件。

- 节点属于这个集群。
- 节点的状态不能是 ONLINE 或 RECOVERING。

(2) 检查节点是否满足 InnoDB Cluster 的要求。

(3) 只有满足上述要求，才执行加入操作。具体来说，

- 如果发现节点在自动重连，则会关闭组复制（STOP GROUP_REPLICATION）。
- 修改组复制的相关参数。
- 启动组复制（START GROUP_REPLICATION）。

倘若 rejoinInstance 命令执行失败，可分析目标实例的错误日志。

options 支持以下选项。

- password：指定实例的连接密码。

- memberSslMode：是否开启 SSL 通信，默认为 AUTO。
- interactive：是否开启交互模式。
- ipWhitelist：设置白名单，即 group_replication_ip_whitelist。

该命令的常见使用场景有两个。

- 实例重启后，没有自动加入集群，常见的原因是组复制的相关参数没有被持久化到配置文件中。
- 网络中断过久，导致节点被驱逐出集群。

- **<Cluster>.rescan([options])**

基于集群的实际拓扑，更新集群的元数据信息。

```
mysql-js> cluster.rescan()
Rescanning the cluster...

Result of the rescanning operation for the 'myCluster' cluster:
{
    "name": "myCluster",
    "newTopologyMode": null,
    "newlyDiscoveredInstances": [],
    "unavailableInstances": [
        {
            "host": "192.168.244.30:3306",
            "label": "192.168.244.30:3306",
            "member_id": "0d1aecac-a43f-11eb-bd84-fa163eb6f223"
        }
    ],
    "updatedInstances": []
}
The instance '192.168.244.30:3306' is no longer part of the cluster.
The instance is either offline or left the HA group. You can try to add it to the cluster again with
the cluster.rejoinInstance('192.168.244.30:3306') command or you can remove it from the cluster
configuration.
Would you like to remove it from the cluster metadata? [Y/n]: y
Removing instance from the cluster metadata...
The instance '192.168.244.30:3306' was successfully removed from the cluster metadata.
```

options 支持以下选项。

- addInstances：待添加实例的连接信息。例如：

    ```
    mysql-js> cluster.rescan({addInstances:['192.168.244.30:3306']})
    ```

    也可设置为 auto，即自动添加。

- interactive：是否开启交互模式。
- removeInstances：待删除实例的连接信息，用法同 addInstances 一样。
- updateTopologyMode：是否自动更新集群的拓扑模式（单主模式或多主模式），默认为 false。

该命令的常见使用场景如下。

- 集群中存在手动添加（删除）的节点。
- 通过 SQL，而不是 AdminAPI 变更了集群的拓扑模式。

3. 集群状态相关

- `<Cluster>.status([options])`

打印集群的状态信息。具体可参考 12.3 节。

- `<Cluster>.describe()`

打印集群的拓扑结构。

```
mysql-js> cluster.describe()
{
    "clusterName": "myCluster",
    "defaultReplicaSet": {
        "name": "default",
        "topology": [
            {
                "address": "192.168.244.10:3306",
                "label": "192.168.244.10:3306",
                "role": "HA"
            },
            {
                "address": "192.168.244.20:3306",
                "label": "192.168.244.20:3306",
                "role": "HA"
            },
            {
                "address": "192.168.244.30:3306",
                "label": "192.168.244.30:3306",
                "role": "HA"
            }
        ],
        "topologyMode": "Single-Primary"
    }
}
```

4. MySQL Router 相关

- `<Cluster>.listRouters([options])`

输出集群中的 Router 实例信息。

```
mysql-js> cluster.listRouters()
{
    "clusterName": "myCluster",
    "routers": {
        "192.168.244.128::": {
            "hostname": "192.168.244.128",
            "lastCheckIn": "2022-01-20 15:19:58",
            "roPort": 6447,
            "roXPort": 6449,
            "rwPort": 6446,
            "rwXPort": 6448,
            "version": "8.0.27"
        }
    }
}
```

输出中的 "192.168.244.128::" 是 Router 实例的唯一标识，由 mysql_innodb_cluster_metadata. routers 表中的 address、router_name 两个字段组成。

目前只支持一个选项。

☐ onlyUpgradeRequired：设置为 true，则只输出需要升级的 router 实例。

- **<Cluster>.removeRouterMetadata(routerDef)**

删除指定 Router 实例的元数据信息。routerDef 是 Router 实例的唯一标识，可通过 cluster. listRouters() 查看。

```
mysql-js> cluster.removeRouterMetadata('192.168.244.128::')
```

对应 MySQL 的操作如下。

```
DELETE FROM mysql_innodb_cluster_metadata.routers WHERE router_id = 1
```

5. 配置相关

这里的配置指的是组复制相关的参数配置。

- **<Cluster>.options([options])**

打印集群的参数配置。

```
mysql-js> cluster.options()
{
    "clusterName": "myCluster",
    "defaultReplicaSet": {
        "globalOptions": [
            {
                "option": "groupName",
                "value": "720113f3-79ab-11ec-b87f-000c29f66609",
                "variable": "group_replication_group_name"
            },
            {
                "option": "memberSslMode",
                "value": "REQUIRED",
                "variable": "group_replication_ssl_mode"
            },
            {
                "option": "disableClone",
                "value": false
            }
        ],
        "tags": {
            "192.168.244.10:3306": [],
            "192.168.244.20:3306": [],
            "192.168.244.30:3306": [],
            "global": []
        },
        "topology": {
            "192.168.244.10:3306": [
                {
                    "option": "autoRejoinTries",
                    "value": "3",
                    "variable": "group_replication_autorejoin_tries"
```

```
            },
            {
                "option": "consistency",
                "value": "EVENTUAL",
                "variable": "group_replication_consistency"
            },
            ...
            {
                "value": "XXHASH64",
                "variable": "transaction_write_set_extraction"
            }
        ],
        "192.168.244.20:3306": [
            ...
        ],
        "192.168.244.30:3306": [
            ...
        ]
        }
    }
}
```

options 目前只支持下面这一个选项。

❑ all：若设置为 true，则会打印组复制相关的所有参数。

```
mysql-js> cluster.options( { all:true } )
```

● **<Cluster>.setInstanceOption(instance, option, value)**

设置单个实例的配置。

❑ instance：实例的配置信息。
❑ option：配置项。目前支持的配置有 tag、exitStateAction、memberWeight、autoRejoinTries、label。
❑ value：值。具体每个配置项支持哪些值可通过 \? Cluster.setInstanceOption 查看。

看下面这个示例。

```
mysql-js> cluster.setInstanceOption('192.168.244.10:3306', 'memberWeight', 40)
Setting the value of 'memberWeight' to '40' in the instance: '192.168.244.10:3306' ...

Successfully set the value of 'memberWeight' to '40' in the cluster member: '192.168.244.10:3306'.
```

tag 有两个内置属性。

❑ _hidden：将节点设置为隐藏节点。对于隐藏节点，Router 不会转发客户端请求。
❑ _disconnect_existing_sessions_when_hidden：当节点设置为隐藏节点时，是否断开已有的连接。

两者的用法如下。

```
mysql-js> cluster.setInstanceOption('192.168.244.30:3306',"tag:_hidden", true)
mysql-js>
cluster.setInstanceOption('192.168.244.30:3306',"tag:_disconnect_existing_sessions_when_hidden", true)
```

12.4 InnoDB Cluster 的管理操作

- `<Cluster>.setOption(option, value)`

同 setInstanceOption 类似，只不过 setOption 设置的是整个集群的配置。

目前，支持的配置有 tag、clusterName、exitStateAction、memberWeight、consistency、expelTimeout、autoRejoinTries、disableClone。

6. 账号相关

InnoDB Cluster 相关的账号有 3 个：集群管理员账号、Router 账号和恢复通道账号。基于合规和安全需求，我们需要定期修改这些账号的密码。这个时候，就需要用到下面这些命令。

- `<Cluster>.resetRecoveryAccountsPassword(options)`

重置恢复通道账号的密码。

```
mysql-js> cluster.resetRecoveryAccountsPassword()
The recovery account passwords of all the cluster instances' were successfully reset.
```

修改完密码后，会自动执行 CHANGE REPLICATION SOURCE TO 操作让其生效。

options 支持以下选项。

- force：如果集群中存在非 Online 状态的节点，在执行该命令时，会提示是否继续执行。若设置为 true，则会跳过这个提示，强制执行。
- interactive：是否开启交互模式。

- `<Cluster>.setupAdminAccount(user, options)`

新建一个集群管理员账号。如果账号存在，还可使用该命令更新账号的权限或密码。

更新账号的权限，常用在以下场景。

- 账号是手动创建的。
- 账号是之前版本的 MySQL Shell 创建的。

options 支持以下选项。

- password：新的密码。
- dryRun：试运行，若设置为 true，则只会打印账号需要的权限，不会实际创建账号。默认为 false。
- interactive：是否开启交互模式。
- update：账号存在时，更新权限或密码。

```
mysql-js> cluster.setupAdminAccount('cluster_admin_test', { password: '123456', update: true })
Updating user cluster_admin_test@%.
Updating user password.
Account cluster_admin_test@% was successfully updated.
```

- `<Cluster>.setupRouterAccount(user, options)`

用法与 setupAdminAccount 一样，针对的是 MySQL Router 用户。

7. 模式切换

- `<Cluster>.setPrimaryInstance(instance)`

单主模式下设置新的 Primary 节点。

```
mysql-js> cluster.setPrimaryInstance('192.168.244.20:3306')
Setting instance '192.168.244.20:3306' as the primary instance of cluster 'myCluster'...

Instance '192.168.244.30:3306' remains SECONDARY.
Instance '192.168.244.10:3306' was switched from PRIMARY to SECONDARY.
Instance '192.168.244.20:3306' was switched from SECONDARY to PRIMARY.

WARNING: The cluster internal session is not the primary member anymore. For cluster management
operations please obtain a fresh cluster handle using dba.getCluster().

The instance '192.168.244.20:3306' was successfully elected as primary.
```

该命令实际调用的是 SELECT group_replication_set_as_primary(member_uuid)。

- `<Cluster>.switchToMultiPrimaryMode()`

单主模式切换为多主模式。

```
mysql-js> cluster.switchToMultiPrimaryMode()
Switching cluster 'myCluster' to Multi-Primary mode...

Instance '192.168.244.30:3306' was switched from SECONDARY to PRIMARY.
Instance '192.168.244.10:3306' was switched from SECONDARY to PRIMARY.
Instance '192.168.244.20:3306' remains PRIMARY.

The cluster successfully switched to Multi-Primary mode.
```

该命令主要会执行以下操作。

(1) 调用函数切换到多主模式。

```
SELECT group_replication_switch_to_multi_primary_mode()
```

(2) 调整组复制的相关参数。

```
SET GLOBAL read_only= 0
SET PERSIST_ONLY group_replication_single_primary_mode = OFF
SET PERSIST_ONLY group_replication_enforce_update_everywhere_checks = ON
SET PERSIST `auto_increment_increment` = 7
SET PERSIST `auto_increment_offset` = 7
```

- `<Cluster>.switchToSinglePrimaryMode([instance])`

多主模式切换到单主模式。

可通过 instance 指定单主模式下的 Primary 节点，例如：

```
mysql-js> cluster.switchToSinglePrimaryMode('192.168.244.10:3306')
Switching cluster 'myCluster' to Single-Primary mode...

Instance '192.168.244.30:3306' was switched from PRIMARY to SECONDARY.
Instance '192.168.244.10:3306' remains PRIMARY.
Instance '192.168.244.20:3306' was switched from PRIMARY to SECONDARY.
```

```
WARNING: The cluster internal session is not the primary member anymore. For cluster management
operations please obtain a fresh cluster handle using dba.getCluster().

WARNING: Existing connections that expected a R/W connection must be disconnected, i.e. instances that
became SECONDARY.

The cluster successfully switched to Single-Primary mode.
```

如果不指定，则会按照默认的选举算法选择新主。

该命令主要会执行以下操作。

(1) 调用函数切换到单主模式。

```
SELECT group_replication_switch_to_single_primary_mode('e37b0e70-7e51-11ec-8b0f-525400d51a16')
```

(2) 调整组复制的相关参数。

```
SET GLOBAL super_read_only= 1
SET GLOBAL read_only= 0   # 只在新的 Primary 节点上执行
SET PERSIST_ONLY group_replication_enforce_update_everywhere_checks = OFF
SET PERSIST_ONLY group_replication_single_primary_mode = ON
SET PERSIST `auto_increment_increment` = 1
SET PERSIST `auto_increment_offset` = 2
```

8. 集群相关

- `<Cluster>.dissolve([options])`

解散集群。

```
mysql-js> cluster.dissolve()
The cluster still has the following registered instances:
{
    "clusterName": "myCluster",
    "defaultReplicaSet": {
        "name": "default",
        "topology": [
            {
                "address": "192.168.244.10:3306",
                "label": "192.168.244.10:3306",
                "role": "HA"
            },
            {
                "address": "192.168.244.30:3306",
                "label": "192.168.244.30:3306",
                "role": "HA"
            },
            {
                "address": "192.168.244.20:3306",
                "label": "192.168.244.20:3306",
                "role": "HA"
            }
        ],
        "topologyMode": "Single-Primary"
    }
}
WARNING: You are about to dissolve the whole cluster and lose the high availability features provided
```

```
by it. This operation cannot be reverted. All members will be removed from the cluster and replication
will be stopped, internal recovery user accounts and the cluster metadata will be dropped. User data
will be maintained intact in all instances.
Are you sure you want to dissolve the cluster? [y/N]: y

* Waiting for instance to synchronize with the primary...
** Transactions replicated  ############################################################  100% *
Waiting for instance to synchronize with the primary...
** Transactions replicated  ############################################################  100% *
Waiting for instance to synchronize with the primary...
** Transactions replicated  ############################################################  100% *
Waiting for instance to synchronize with the primary...
** Transactions replicated  ############################################################  100%
Instance '192.168.244.30:3306' is attempting to leave the cluster...
* Waiting for instance to synchronize with the primary...
** Transactions replicated  ############################################################  100%
Instance '192.168.244.20:3306' is attempting to leave the cluster...
Instance '192.168.244.10:3306' is attempting to leave the cluster...

The cluster was successfully dissolved.
Replication was disabled but user data was left intact.
```

该命令主要会执行以下操作。

(1) 登录 Primary 节点清除集群相关的元数据，包括删除恢复账号，删除 mysql_innodb_cluster_metadata 库中的记录。

(2) 从 Secondary 节点到 Primary 节点依次退出集群。退出动作包括关闭组复制，将 super_read_only 设置为 1，重置组复制的相关通道，同时将组复制的相关参数恢复为默认值。具体命令如下。

```
STOP GROUP_REPLICATION
SET GLOBAL super_read_only= 1
RESET REPLICA ALL FOR CHANNEL 'group_replication_applier'
RESET REPLICA ALL FOR CHANNEL 'group_replication_recovery'
SET PERSIST `group_replication_start_on_boot` = 'OFF'
SET PERSIST `group_replication_enforce_update_everywhere_checks` = 'OFF'
SET PERSIST `group_replication_bootstrap_group` = DEFAULT
SET PERSIST `group_replication_group_seeds` = DEFAULT
SET PERSIST `group_replication_local_address` = DEFAULT
```

注意，通过 cluster.dissolve() 解散集群，并不会删掉 mysql_innodb_cluster_metadata 库，只会删除库中的记录。

- **<Cluster>.forceQuorumUsingPartitionOf(instance[, password])**

第 11 章提到过，如果组内的大多数节点不可用，则剩下的少数节点是不会对外提供写服务的。此时，可设置 group_replication_force_members，让剩下的少数节点组建一个新的集群。实际上，forceQuorumUsingPartitionOf 也是基于该命令，只不过它做了很多的检测，以避免"脑裂"的发生。

在使用该命令时，只需给出某个 Online 实例的连接信息，它会自动查出其他 Online 节点。例如：

```
mysql-js> shell.connect('cluster_admin:cluster_pass@192.168.244.30:3306')

mysql-js> cluster=dba.getCluster()
WARNING: Cluster has no quorum and cannot process write transactions: Group has no quorum
WARNING: You are connected to an instance in state 'Read Only'
Write operations on the InnoDB cluster will not be allowed.
```

```
<Cluster:myCluster>
mysql-js> cluster.forceQuorumUsingPartitionOf('cluster_admin:cluster_pass@192.168.244.30:3306')
Restoring cluster 'myCluster' from loss of quorum, by using the partition composed of
[192.168.244.30:3306]

Restoring the InnoDB cluster ...

The InnoDB cluster was successfully restored using the partition from the instance
'cluster_admin@192.168.244.30:3306'.

WARNING: To avoid a split-brain scenario, ensure that all other members of the cluster are removed or
joined back to the group that was restored.
```

该命令的实现逻辑如下。

(1) 获取集群中所有 ONLINE 状态的实例列表，这个列表即输出中的 [192.168.244.30:3306]，在设置 group_replication_force_members 时会用到。

(2) 判断给定实例的地址、group_replication_group_name 是否与集群元数据一致。若一致，则说明实例属于该集群。

(3) 判定给定实例的状态是否为 ONLINE。只有 ONLINE 状态的节点才可用来恢复集群。

(4) 判定集群中 UNREACHABLE 状态的节点是否超过总节点数的一半。若超过，才代表集群内的大多数节点不可用。

(5) 获取集群所有节点的实例地址。与每个节点建立连接，如果该节点在给定实例中的状态既非 ONLINE，又非 RECOVERING，且能连上的，则会关闭该节点的组复制，以避免"脑裂"的发生。

(6) 设置 group_replication_force_members，这里的实例列表是第一步获取的。

```
SET GLOBAL `group_replication_force_members` = '192.168.244.30:33061'
```

(7) 最后，再清除 group_replication_force_members 的设置。

```
SET GLOBAL `group_replication_force_members` = ''
```

(8) 关闭只读。

```
SET GLOBAL read_only= 0
```

12.5　本章总结

本章主要介绍了 InnoDB Cluster 的 3 个核心组件及 InnoDB Cluster 常见的管理操作。

对 MySQL 不熟悉的人，经常会将 InnoDB Cluster 与 MySQL Cluster 混淆。实际上，它们是两个完全不同的产品。MySQL Cluster 的全称是 MySQL NDB Cluster，是一套基于内存、无共享的高可用方案，底层使用的是 NDB 存储引擎。

3 个组件都可单独使用，尤其是 MySQL Shell 内置了很多实用工具。

推荐使用 MySQL Shell 来管理组复制，原因有二。

- 易用性确实让人称赞，无论是对于部署还是日常管理来说。

- InnoDB Cluster 的管理操作内置了很多检查项，相对来说更安全。

注意，使用 MySQL Shell 来管理组复制，并不意味着就一定要使用 MySQL Router，也可使用其他中间件。只不过，其他的中间件不能通过 MySQL Shell 来管理。

重点问题回顾

- MySQL Shell 工具集的使用。
- MySQL Router 的读写分离是通过指定不同的端口来实现的。
- InnoDB Cluster 的搭建。
- `dba` 对象的常见操作及操作背后的实现原理。
- `cluster` 对象的常见操作及操作背后的实现原理。

附录 A
JSON

JSON 数据类型是从 MySQL 5.7.8 开始支持的。在此之前，只能通过字符类型（CHAR、VARCHAR 或 TEXT）来保存 JSON 文档。

相对于字符类型，原生的 JSON 类型具有以下优势。

- 在插入时能自动校验文档是否满足 JSON 格式的要求。
- 优化了存储格式。无须读取整个文档就能快速访问某个元素的值。

在 MySQL 引入 JSON 类型之前，如果我们想获取 JSON 文档中的某个元素，必须首先读取整个 JSON 文档，然后在客户端将其转换为 JSON 对象，最后通过对象获取指定元素的值。

下面是在 Python 中的获取方式。

```
import json

# JSON 字符串
x = '{ "name":"John", "age":30, "city":"New York"}'

# 将 JSON 字符串转换为 JSON 对象
y = json.loads(x)

# 读取 JSON 对象中指定元素的值
print(y["age"])
```

这种方式有两个弊端：一是消耗磁盘 I/O；二是消耗网络带宽，如果 JSON 文档比较大，在高并发场景下，有可能会"打爆"网卡。

如果使用的是 JSON 类型，直接使用 SQL 命令就可以满足相同的需求。这不仅能节省网络带宽，结合后面提到的函数索引，还能降低磁盘 I/O 消耗。

```
mysql> create table t(c1 json);
Query OK, 0 rows affected (0.09 sec)

mysql> insert into t values('{ "name":"John", "age":30, "city":"New York"}');
Query OK, 1 row affected (0.01 sec)

mysql> select c1->"$.age" from t;
+-------------+
| c1->"$.age" |
+-------------+
| 30          |
+-------------+
1 row in set (0.00 sec)
```

本附录主要包含以下内容。

- 什么是 JSON。
- JSON 字段的增删查改操作。
- 如何对 JSON 字段创建索引。
- 如何将存储 JSON 字符串的字符字段升级为 JSON 字段。
- 使用 JSON 的注意事项。
- Partial Update。
- 其他 JSON 函数。

A.1 什么是 JSON

JSON 是 JavaScript Object Notation（JavaScript 对象表示法）的缩写，是一种轻量级、基于文本、跨语言的数据交换格式，易于阅读和编写。

JSON 的基本数据类型如下。

- 数值：十进制数，不能有前导 0，可以为负数或小数，也可以为 e 或 E 表示的指数。
- 字符串：字符串必须用双引号引起来。
- 布尔值：true 和 false。
- 数组：一个由零或多个值组成的有序序列。每个值可以为任意类型。数组使用方括号（[]）括起来，元素之间用逗号（,）分隔。例如：

  [1, "abc", null, true, "10:27:06.000000", {"id": 1}]

- 对象：一个由零或者多个键值对组成的无序集合。键必须是字符串，值可以为任意类型。对象使用花括号（{}）括起来，键值对之间使用逗号（,）分隔，键与值之间用冒号（:）分隔。例如：

  {"db": ["mysql", "oracle"], "id": 123, "info": {"age": 20}}

- 空值：null。

A.2 JSON 字段的增删改查操作

下面我们看看 JSON 字段常见的增删改查操作。

A.2.1 插入操作

可直接插入 JSON 格式的字符串。

```
mysql> create table t(c1 json);
Query OK, 0 rows affected (0.03 sec)

mysql> insert into t values('[1, "abc", null, true, "08:45:06.000000"]');
Query OK, 1 row affected (0.01 sec)
```

```
mysql> insert into t values('{"id": 87, "name": "carrot"}');
Query OK, 1 row affected (0.01 sec)
```

也可使用函数，常用的有 `JSON_ARRAY()` 和 `JSON_OBJECT()`，前者用于构造 JSON 数组，后者用于构造 JSON 对象。例如：

```
mysql> select json_array(1, "abc", null, true,curtime());
+--------------------------------------------+
| json_array(1, "abc", null, true,curtime()) |
+--------------------------------------------+
| [1, "abc", null, true, "10:12:25.000000"]  |
+--------------------------------------------+
1 row in set (0.01 sec)

mysql> select json_object('id', 87, 'name', 'carrot');
+-----------------------------------------+
| json_object('id', 87, 'name', 'carrot') |
+-----------------------------------------+
| {"id": 87, "name": "carrot"}            |
+-----------------------------------------+
1 row in set (0.00 sec)
```

对于 JSON 文档，KEY 名不能重复。

如果插入的值中存在重复的 KEY，在 MySQL 8.0.3 之前，遵循 "first duplicate key wins" 原则，会保留第一个 KEY，后面的将被丢弃掉。

从 MySQL 8.0.3 开始，遵循的是 "last duplicate key wins" 原则，只会保留最后一个 KEY。

下面通过一个具体的示例来看看两者的区别。

MySQL 5.7.36：

```
mysql> select json_object('key1',10,'key2',20,'key1',30);
+--------------------------------------------+
| json_object('key1',10,'key2',20,'key1',30) |
+--------------------------------------------+
| {"key1": 10, "key2": 20}                   |
+--------------------------------------------+
1 row in set (0.02 sec)
```

MySQL 8.0.27：

```
mysql> select json_object('key1',10,'key2',20,'key1',30);
+--------------------------------------------+
| json_object('key1',10,'key2',20,'key1',30) |
+--------------------------------------------+
| {"key1": 30, "key2": 20}                   |
+--------------------------------------------+
1 row in set (0.00 sec)
```

A.2.2 查询操作

1. JSON_EXTRACT(json_doc, path[, path] ...)

json_doc 是 JSON 文档，path 是路径。该函数会从 JSON 文档提取指定路径（path）的元素。如果指定路径不存在,会返回 NULL。可指定多个路径,匹配到的多个值会以数组形式返回。

下面我们结合一些具体的示例来看看路径和 JSON_EXTRACT 的用法。

首先看看数组。

数组的路径是通过下标来表示的。第一个元素的下标是 0。

```
mysql> select json_extract('[10, 20, [30, 40]]', '$[0]');
+--------------------------------------------+
| json_extract('[10, 20, [30, 40]]', '$[0]') |
+--------------------------------------------+
| 10                                         |
+--------------------------------------------+
1 row in set (0.00 sec)

mysql> select json_extract('[10, 20, [30, 40]]', '$[0]', '$[1]','$[2][0]');
+--------------------------------------------------------------+
| json_extract('[10, 20, [30, 40]]', '$[0]', '$[1]','$[2][0]') |
+--------------------------------------------------------------+
| [10, 20, 30]                                                 |
+--------------------------------------------------------------+
1 row in set (0.00 sec)
```

除此之外,还可通过 [M to N] 获取数组的子集。

```
mysql> select json_extract('[10, 20, [30, 40]]', '$[0 to 1]');
+-------------------------------------------------+
| json_extract('[10, 20, [30, 40]]', '$[0 to 1]') |
+-------------------------------------------------+
| [10, 20]                                        |
+-------------------------------------------------+
1 row in set (0.00 sec)

# 这里的 last 代表最后一个元素的下标
mysql> select json_extract('[10, 20, [30, 40]]', '$[last-1 to last]');
+---------------------------------------------------------+
| json_extract('[10, 20, [30, 40]]', '$[last-1 to last]') |
+---------------------------------------------------------+
| [20, [30, 40]]                                          |
+---------------------------------------------------------+
1 row in set (0.00 sec)
```

也可通过 [*] 获取数组中的所有元素。

```
mysql> select json_extract('[10, 20, [30, 40]]', '$[*]');
+--------------------------------------------+
| json_extract('[10, 20, [30, 40]]', '$[*]') |
+--------------------------------------------+
| [10, 20, [30, 40]]                         |
+--------------------------------------------+
1 row in set (0.00 sec)
```

接下来,我们看看对象。

对象的路径是通过 KEY 来表示的。

```
mysql> set @j='{"a": 1, "b": [2, 3], "a c": 4}';
Query OK, 0 rows affected (0.00 sec)
```

```
# 如果KEY在路径表达式中不合法（如存在空格），则在引用这个KEY时，需用双引号括起来。
mysql> select json_extract(@j, '$.a'), json_extract(@j, '$."a c"'), json_extract(@j, '$.b[1]');
+-------------------------+-----------------------------+----------------------------+
| json_extract(@j, '$.a') | json_extract(@j, '$."a c"') | json_extract(@j, '$.b[1]') |
+-------------------------+-----------------------------+----------------------------+
| 1                       | 4                           | 3                          |
+-------------------------+-----------------------------+----------------------------+
1 row in set (0.00 sec)
```

除此之外，还可通过 .* 获取对象中的所有元素。

```
mysql> select json_extract('{"a": 1, "b": [2, 3], "a c": 4}', '$.*');
+--------------------------------------------------------+
| json_extract('{"a": 1, "b": [2, 3], "a c": 4}', '$.*') |
+--------------------------------------------------------+
| [1, [2, 3], 4]                                         |
+--------------------------------------------------------+
1 row in set (0.00 sec)

# 这里的 $**.b 匹配 $.a.b 和 $.c.b
mysql> select json_extract('{"a": {"b": 1}, "c": {"b": 2}}', '$**.b');
+---------------------------------------------------------+
| json_extract('{"a": {"b": 1}, "c": {"b": 2}}', '$**.b') |
+---------------------------------------------------------+
| [1, 2]                                                  |
+---------------------------------------------------------+
1 row in set (0.00 sec)
```

2. column->path

它和后面讲到的 column->>path 都是语法糖，会在实际使用的时候转化为 JSON_EXTRACT。

column->path 等同于 JSON_EXTRACT(column, path)，只能指定一条路径。

```
create table t(c2 json);

insert into t values('{"empno": 1001, "ename": "jack"}'), ('{"empno": 1002, "ename": "mark"}');

mysql> select c2, c2->"$.ename" from t;
+----------------------------------+---------------+
| c2                               | c2->"$.ename" |
+----------------------------------+---------------+
| {"empno": 1001, "ename": "jack"} | "jack"        |
| {"empno": 1002, "ename": "mark"} | "mark"        |
+----------------------------------+---------------+
2 rows in set (0.00 sec)

mysql> select * from t where c2->"$.empno" = 1001;
+------+----------------------------------+
| c1   | c2                               |
+------+----------------------------------+
|    1 | {"empno": 1001, "ename": "jack"} |
+------+----------------------------------+
1 row in set (0.00 sec)
```

3. column->>path

同 column->path 类似，只不过它返回的是字符串。以下三者是等价的。

- JSON_UNQUOTE(JSON_EXTRACT(column, path))
- JSON_UNQUOTE(column -> path)
- column->>path

```
mysql> select c2->'$.ename',json_extract(c2, "$.ename"),json_unquote(c2->'$.ename'),c2->>'$.ename' from t;
+----------------+-----------------------------+-----------------------------+----------------+
| c2->'$.ename'  | json_extract(c2, "$.ename") | json_unquote(c2->'$.ename') | c2->>'$.ename' |
+----------------+-----------------------------+-----------------------------+----------------+
| "jack"         | "jack"                      | jack                        | jack           |
| "mark"         | "mark"                      | mark                        | mark           |
+----------------+-----------------------------+-----------------------------+----------------+
2 rows in set (0.00 sec)
```

A.2.3 修改操作

1. JSON_INSERT(json_doc, path, val[, path, val] ...)

插入新值。

仅当指定位置或指定 KEY 的值不存在时，才执行插入操作。另外，如果指定的路径是数组下标，且 json_doc 不是数组，该函数会先将 json_doc 转化为数组，再插入新值。

下面来看几个示例。

```
mysql> select json_insert('1','$[0]',"10");
+------------------------------+
| json_insert('1','$[0]',"10") |
+------------------------------+
| 1                            |
+------------------------------+
1 row in set (0.00 sec)

mysql> select json_insert('1','$[1]',"10");
+------------------------------+
| json_insert('1','$[1]',"10") |
+------------------------------+
| [1, "10"]                    |
+------------------------------+
1 row in set (0.01 sec)

mysql> select json_insert('["1","2"]','$[2]',"10");
+--------------------------------------+
| json_insert('["1","2"]','$[2]',"10") |
+--------------------------------------+
| ["1", "2", "10"]                     |
+--------------------------------------+
1 row in set (0.00 sec)

mysql> set @j = '{ "a": 1, "b": [2, 3]}';
Query OK, 0 rows affected (0.00 sec)

mysql> select json_insert(@j, '$.a', 10, '$.c', '[true, false]');
+----------------------------------------------------+
| json_insert(@j, '$.a', 10, '$.c', '[true, false]') |
+----------------------------------------------------+
```

```
| {"a": 1, "b": [2, 3], "c": "[true, false]"}    |
+------------------------------------------------+
1 row in set (0.00 sec)
```

2. JSON_SET(json_doc, path, val[, path, val] ...)

插入新值,并替换已经存在的值。

换言之,如果指定位置或指定 KEY 的值不存在,会执行插入操作;如果存在,则执行更新操作。

```
mysql> set @j = '{ "a": 1, "b": [2, 3]}';
Query OK, 0 rows affected (0.00 sec)

mysql> select json_set(@j, '$.a', 10, '$.c', '[true, false]');
+-------------------------------------------------+
| json_set(@j, '$.a', 10, '$.c', '[true, false]') |
+-------------------------------------------------+
| {"a": 10, "b": [2, 3], "c": "[true, false]"}    |
+-------------------------------------------------+
1 row in set (0.00 sec)
```

3. JSON_REPLACE(json_doc, path, val[, path, val] ...)

替换已经存在的值。

```
mysql> set @j = '{ "a": 1, "b": [2, 3]}';
Query OK, 0 rows affected (0.00 sec)

mysql> select json_replace(@j, '$.a', 10, '$.c', '[true, false]');
+-----------------------------------------------------+
| json_replace(@j, '$.a', 10, '$.c', '[true, false]') |
+-----------------------------------------------------+
| {"a": 10, "b": [2, 3]}                              |
+-----------------------------------------------------+
1 row in set (0.00 sec)
```

A.2.4 删除操作

JSON_REMOVE(json_doc, path[, path] ...)

删除 JSON 文档指定位置上的元素。

```
mysql> set @j = '{ "a": 1, "b": [2, 3]}';
Query OK, 0 rows affected (0.00 sec)

mysql> select json_remove(@j, '$.a');
+------------------------+
| JSON_REMOVE(@j, '$.a') |
+------------------------+
| {"b": [2, 3]}          |
+------------------------+
1 row in set (0.00 sec)

mysql> set @j = '["a", ["b", "c"], "d", "e"]';
Query OK, 0 rows affected (0.00 sec)

mysql> select json_remove(@j, '$[1]');
```

```
+------------------------+
| JSON_REMOVE(@j, '$[1]') |
+------------------------+
| ["a", "d", "e"]         |
+------------------------+
1 row in set (0.00 sec)

mysql> select json_remove(@j, '$[1]','$[2]');
+--------------------------------+
| JSON_REMOVE(@j, '$[1]','$[2]') |
+--------------------------------+
| ["a", "d"]                     |
+--------------------------------+
1 row in set (0.00 sec)

mysql> select json_remove(@j, '$[1]','$[1]');
+--------------------------------+
| JSON_REMOVE(@j, '$[1]','$[1]') |
+--------------------------------+
| ["a", "e"]                     |
+--------------------------------+
1 row in set (0.00 sec)
```

在最后一个查询中，虽然两个 path 都是 $[1]，但其作用对象不一样：第一个 path 的作用对象是 ["a", "b", "c"], "d", "e"]；第二个 path 的作用对象是删除 $[1] 后的数组，即 ["a", "d", "e"]。

A.3 如何对 JSON 字段创建索引

同 TEXT、BLOB 字段一样，JSON 字段不允许直接创建索引。

```
mysql> create table t(c1 json, index (c1));
ERROR 3152 (42000): JSON column 'c1' supports indexing only via generated columns on a specified JSON path.
```

即使支持，实际意义也不大，因为我们一般会基于文档中的元素进行查询，很少会基于整个 JSON 文档进行查询。

对文档中的元素进行查询，就需要用到 MySQL 5.7 引入的虚拟列了。

我们来看一个具体的示例。

```
# c2 即虚拟列
# index (c2) 对虚拟列添加索引
create table t ( c1 json, c2 varchar(10) as (JSON_UNQUOTE(c1 -> "$.name")), index (c2) );

insert into t (c1) values ('{"id": 1, "name": "a"}'), ('{"id": 2, "name": "b"}'), ('{"id": 3, "name": "c"}'), ('{"id": 4, "name": "d"}');

mysql> explain select * from t where c2 = 'a';
+----+-------------+-------+------------+------+---------------+------+---------+-------+------+----------+-------+
| id | select_type | table | partitions | type | possible_keys | key  | key_len | ref   | rows | filtered | Extra |
+----+-------------+-------+------------+------+---------------+------+---------+-------+------+----------+-------+
|  1 | SIMPLE      | t     | NULL       | ref  | c2            | c2   | 43      | const |    1 |   100.00 | NULL  |
+----+-------------+-------+------------+------+---------------+------+---------+-------+------+----------+-------+
1 row in set, 1 warning (0.00 sec)
```

```
mysql> explain select * from t where c1->'$.name' = 'a';
+----+-------------+-------+------------+------+---------------+------+---------+-------+------+----------+-------+
| id | select_type | table | partitions | type | possible_keys | key  | key_len | ref   | rows | filtered | Extra |
+----+-------------+-------+------------+------+---------------+------+---------+-------+------+----------+-------+
|  1 | SIMPLE      | t     | NULL       | ref  | c2            | c2   | 43      | const |    1 |   100.00 | NULL  |
+----+-------------+-------+------------+------+---------------+------+---------+-------+------+----------+-------+
1 row in set, 1 warning (0.00 sec)
```

可以看到，无论是使用虚拟列，还是使用文档中的元素来查询，都可以利用上索引。

注意，在创建虚拟列时需指定 JSON_UNQUOTE，将 c1 -> "$.name" 的返回值转换为字符串。从 MySQL 8.0.13 开始，得益于函数索引的引入，可直接对 JSON 字段创建函数索引。

```
create table t ( c1 json, index index_name((cast(c1->>"$.name" as char(10)) collate utf8mb4_bin)));
```

A.4 如何将存储 JSON 字符串的字符字段升级为 JSON 字段

在 MySQL 支持 JSON 类型之前，JSON 文档一般是以字符串的形式存储在字符类型（VARCHAR 或 TEXT）中的。

在 JSON 类型被引入之后，如何将这些字符字段升级为 JSON 字段呢？

为方便演示，这里首先构建测试数据。

```
create table t (id int auto_increment primary key, c1 text);
```

```
insert into t (c1) values ('{"id": "1", "name": "a"}'), ('{"id": "2", "name": "b"}'), ('{"id": "3", "name": "c"}'), ('{"id", "name", "d"}');
```

注意，最后一个文档有问题，不是合格的 JSON 文档。

如果使用 DDL 直接修改字段的数据类型，会报错。

```
mysql> alter table t modify c1 json;
ERROR 3140 (22032): Invalid JSON text: "Missing a colon after a name of object member." at position 5 in value for column '#sql-7e1c_1f6.c1'.
```

下面我们看看具体的升级步骤。

(1) 使用 json_valid 函数找出不满足 JSON 格式要求的文档。

```
mysql> select * from t where json_valid(c1) = 0;
+----+---------------------+
| id | c1                  |
+----+---------------------+
|  4 | {"id", "name": "d"} |
+----+---------------------+
1 row in set (0.00 sec)
```

(2) 处理不满足 JSON 格式要求的文档。

```
mysql> update t set c1='{"id": "4", "name": "d"}' where id=4;
Query OK, 1 row affected (0.01 sec)
Rows matched: 1  Changed: 1  Warnings: 0
```

(3) 将 TEXT 字段修改为 JSON 字段。

```
mysql> select * from t where json_valid(c1) = 0;
Empty set (0.00 sec)

mysql> alter table t modify c1 json;
Query OK, 4 rows affected (0.13 sec)
Records: 4  Duplicates: 0  Warnings: 0
```

A.5 使用 JSON 的注意事项

对于 JSON 类型，有以下几点需要注意。

- 在 MySQL 8.0.13 之前，不允许对 BLOB、TEXT、GEOMETRY、JSON 字段设置默认值。从 MySQL 8.0.13 开始，取消了这个限制。

  在设置时，注意默认值需通过圆括号（()）括起来，否则还是会提示 JSON 字段不允许设置默认值。

  ```
  mysql> create table t(c1 json not null default (''));
  Query OK, 0 rows affected (0.03 sec)

  mysql> create table t(c1 json not null default '');
  ERROR 1101 (42000): BLOB, TEXT, GEOMETRY or JSON column 'c1' can't have a default value
  ```

- 不允许直接创建索引，可创建函数索引。
- JSON 列的最大大小和 LONGBLOB（或 LONGTEXT）一样，都是 4GB。
- 在插入时，单个文档的大小受到 max_allowed_packet 的限制，该参数最大是 1GB。

A.6 Partial Update

在 MySQL 5.7 中，对 JSON 文档进行更新的处理策略是，先删除旧的文档，再插入新的文档。即使这个修改很微小，只涉及几字节，也会替换掉整个文档。很显然，这种处理方式的效率较低。

MySQL 8.0 针对 JSON 文档，引入了一项新特性——Partial Update（部分更新），支持 JSON 文档的原地更新。得益于这个特性，JSON 文档的处理性能得到了极大的提升。

下面我们具体来看看。

A.6.1 使用 Partial Update 的条件

为方便阐述，这里先构造测试数据。

```
create table t (id int auto_increment primary key, c1 json);

insert into t (c1) values ('{"id": 1, "name": "a"}'), ('{"id": 2, "name": "b"}'), ('{"id": 3, "name": "c"}'), ('{"id": 4, "name": "d"}');

mysql> select * from t;
+----+------------------------+
| id | c1                     |
+----+------------------------+
|  1 | {"id": 1, "name": "a"} |
```

```
| 2 | {"id": 2, "name": "b"} |
| 3 | {"id": 3, "name": "c"} |
| 4 | {"id": 4, "name": "d"} |
+----+------------------------+
4 rows in set (0.00 sec)
```

使用 Partial Update 需满足以下 4 个条件。

- 被更新的列是 JSON 类型。
- 使用 JSON_SET、JSON_REPLACE 和 JSON_REMOVE 进行 UPDATE 操作。例如：

   update t set c1=json_remove(c1,'$.id') where id=1;

   如果不使用这 3 个函数，而显式赋值，就不会进行部分更新。例如：

   update t set c1='{"id": 1, "name": "a"}' where id=1;

- 输入列和目标列必须是同一列。例如：

   update t set c1=json_replace(c1,'$.id',10) where id=1;

   否则不会进行部分更新。例如：

   update t set c1=json_replace(c2,'$.id',10) where id=1;

- 变更前后，JSON 文档的空间使用不会增加。

关于最后一个条件，我们看看下面这个示例。

```
mysql> select *,json_storage_size(c1),json_storage_free(c1) from t where id=1;
+----+--------------------+-----------------------+-----------------------+
| id | c1                 | json_storage_size(c1) | json_storage_free(c1) |
+----+--------------------+-----------------------+-----------------------+
|  1 | {"id": 1, "name": "a"} |                27 |                     0 |
+----+--------------------+-----------------------+-----------------------+
1 row in set (0.00 sec)

mysql> update t set c1=json_remove(c1,'$.id') where id=1;
Query OK, 1 row affected (0.01 sec)
Rows matched: 1  Changed: 1  Warnings: 0

mysql> select *,json_storage_size(c1),json_storage_free(c1) from t where id=1;
+----+--------------+-----------------------+-----------------------+
| id | c1           | json_storage_size(c1) | json_storage_free(c1) |
+----+--------------+-----------------------+-----------------------+
|  1 | {"name": "a"} |                   27 |                     9 |
+----+--------------+-----------------------+-----------------------+
1 row in set (0.00 sec)

mysql> update t set c1=json_set(c1,'$.id',3306) where id=1;
Query OK, 1 row affected (0.01 sec)
Rows matched: 1  Changed: 1  Warnings: 0

mysql> select *,json_storage_size(c1),json_storage_free(c1) from t where id=1;
+----+------------------------+-----------------------+-----------------------+
| id | c1                     | json_storage_size(c1) | json_storage_free(c1) |
+----+------------------------+-----------------------+-----------------------+
|  1 | {"id": 3306, "name": "a"} |                 27 |                     0 |
+----+------------------------+-----------------------+-----------------------+
```

```
1 row in set (0.00 sec)

mysql> update t set c1=json_set(c1,'$.id','mysql') where id=1;
Query OK, 1 row affected (0.01 sec)
Rows matched: 1  Changed: 1  Warnings: 0

mysql> select *,json_storage_size(c1),json_storage_free(c1) from t where id=1;
+----+----------------------------+-----------------------+-----------------------+
| id | c1                         | json_storage_size(c1) | json_storage_free(c1) |
+----+----------------------------+-----------------------+-----------------------+
|  1 | {"id": "mysql", "name": "a"} |                  33 |                    0 |
+----+----------------------------+-----------------------+-----------------------+
1 row in set (0.00 sec)
```

示例中用到了两个函数：JSON_STORAGE_SIZE 和 JSON_STORAGE_FREE，前者用来获取 JSON 文档的空间使用情况，后者用来获取 JSON 文档在执行原地更新后的空间释放情况。

这里一共执行了 3 次 UPDATE 操作，前两次是原地更新，第三次不是。同样是 JSON_SET 操作，为什么第一次是原地更新，而第二次不是呢？

因为第一次的 JSON_SET 复用了 JSON_REMOVE 释放的空间；而第二次的 JSON_SET 执行的是更新操作，且 "mysql" 比 3306 需要更多的存储空间。

A.6.2 如何在 binlog 中开启 Partial Update

Partial Update 不仅适用于存储引擎层，还可用于主从复制场景。

在主从复制场景下开启 Partial Update，只需将参数 binlog_row_value_options（默认为空）设置为 PARTIAL_JSON。

下面具体来看看，对于下面的这个 UPDATE 操作，开启和不开启 Partial Update 在 binlog 中的记录有何区别。

```
update t set c1=json_replace(c1,'$.id',10) where id=1;
```

不开启：

```
### UPDATE `slowtech`.`t`
### WHERE
###   @1=1
###   @2='{"id": "1", "name": "a"}'
### SET
###   @1=1
###   @2='{"id": 10, "name": "a"}'
```

开启：

```
### UPDATE `slowtech`.`t`
### WHERE
###   @1=1
###   @2='{"id": 1, "name": "a"}'
### SET
###   @1=1
###   @2=JSON_REPLACE(@2, '$.id', 10)
```

对比 binlog 的内容，可以看到：不开启，无论是修改前的镜像（`before_image`）还是修改后的镜像（`after_image`），记录的都是完整的文档；而开启后，修改后的镜像记录的是命令，而不是完整的文档，这样可节省近一半的空间。

在将 `binlog_row_value_options` 设置为 `PARTIAL_JSON` 后，对于可使用 Partial Update 的操作，在 binlog 中不再通过 `ROWS_EVENT` 来记录，而是新增了一个名为 `PARTIAL_UPDATE_ROWS_EVENT` 的事件类型。

需要注意的是，在 binlog 中使用 Partial Update 时，只需满足存储引擎层使用 Partial Update 的前 3 个条件，无须考虑变更前后 JSON 文档的空间使用是否会增加。

A.6.3　关于 Partial Update 的性能测试

首先构造测试数据，t 表一共有 16 个文档，每个文档近 10 MB。

```
create table t(id int auto_increment primary key,
               json_col json,
               name varchar(100) as (json_col->>'$.name'),
               age int as (json_col->'$.age'));

insert into t(json_col) values
(json_object('name', 'Joe', 'age', 24,
             'data', repeat('x', 10 * 1000 * 1000))),
(json_object('name', 'Sue', 'age', 32,
             'data', repeat('y', 10 * 1000 * 1000))),
(json_object('name', 'Pete', 'age', 40,
             'data', repeat('z', 10 * 1000 * 1000))),
(json_object('name', 'Jenny', 'age', 27,
             'data', repeat('w', 10 * 1000 * 1000)));

insert into t(json_col) select json_col from t;
insert into t(json_col) select json_col from t;
```

接下来，测试 SQL 语句

```
update t set json_col = json_set(json_col, '$.age', age + 1);
```

在以下 4 种场景下的执行时间：

- MySQL 5.7.36
- MySQL 8.0.27
- MySQL 8.0.27，binlog_row_value_options=PARTIAL_JSON
- MySQL 8.0.27，binlog_row_value_options=PARTIAL_JSON 且 binlog_row_image=MINIMAL

分别执行 10 次，去掉最大值和最小值后求平均值。

最后的测试结果如图 A-1 所示。

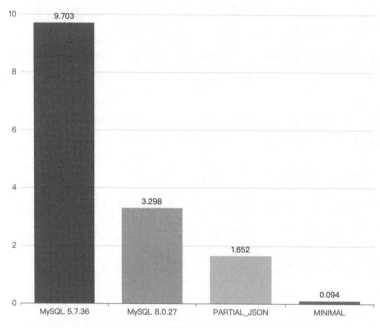

图 A-1　不同场景下的执行时间

以 MySQL 5.7.36 的查询时间作为基准。

(1) MySQL 8.0 只开启存储引擎层的 Partial Update，查询时间比 MySQL 5.7 快 1.94 倍。

(2) MySQL 8.0 同时开启存储引擎层和 binlog 中的 Partial Update，查询时间比 MySQL 5.7 快 4.87 倍。

(3) 如果在(2)的基础上，同时将 `binlog_row_image` 设置为 `MINIMAL`，查询时间比 MySQL 5.7 快 102.22 倍。

当然，在生产环境中，我们一般很少将 `binlog_row_image` 设置为 `MINIMAL`。

即使如此，只开启存储引擎层和 binlog 中的 Partial Update，查询时间也比 MySQL 5.7 快 4.87 倍，性能提升还是比较明显的。

A.7　其他 JSON 函数

A.7.1　查询相关

1. JSON_CONTAINS(target, candidate[, path])

判断 target 文档是否包含 candidate 文档。如果包含，返回 1，否则返回 0。

```
mysql> set @j = '{"a": [1, 2], "b": 3, "c": {"d": 4}}';
Query OK, 0 rows affected (0.00 sec)
```

```
mysql> select json_contains(@j, '1', '$.a'),json_contains(@j, '1', '$.b');
+-------------------------------+-------------------------------+
| json_contains(@j, '1', '$.a') | json_contains(@j, '1', '$.b') |
+-------------------------------+-------------------------------+
|                             1 |                             0 |
+-------------------------------+-------------------------------+
1 row in set (0.00 sec)

mysql> select json_contains(@j,'{"d": 4}','$.a'),json_contains(@j,'{"d": 4}','$.c');
+------------------------------------+------------------------------------+
| json_contains(@j,'{"d": 4}','$.a') | json_contains(@j,'{"d": 4}','$.c') |
+------------------------------------+------------------------------------+
|                                  0 |                                  1 |
+------------------------------------+------------------------------------+
1 row in set (0.00 sec)
```

2. JSON_CONTAINS_PATH(json_doc, one_or_all, path[, path] ...)

判断指定的路径是否存在。如果存在，返回 1，否则返回 0。

函数中的 one_or_all 可指定为 one 或 all，其中 one 是指任意一个路径存在就返回 1，all 则是指所有路径都存在才返回 1。

```
mysql> set @j = '{"a": [1, 2], "b": 3, "c": {"d": 4}}';
Query OK, 0 rows affected (0.00 sec)

mysql> select json_contains_path(@j, 'one', '$.a', '$.e'),json_contains_path(@j, 'all', '$.a', '$.e');
+---------------------------------------------+---------------------------------------------+
| json_contains_path(@j, 'one', '$.a', '$.e') | json_contains_path(@j, 'all', '$.a', '$.e') |
+---------------------------------------------+---------------------------------------------+
|                                           1 |                                           0 |
+---------------------------------------------+---------------------------------------------+
1 row in set (0.00 sec)

mysql> select json_contains_path(@j, 'one', '$.c.d'),json_contains_path(@j, 'one', '$.a.d');
+----------------------------------------+----------------------------------------+
| json_contains_path(@j, 'one', '$.c.d') | json_contains_path(@j, 'one', '$.a.d') |
+----------------------------------------+----------------------------------------+
|                                      1 |                                      0 |
+----------------------------------------+----------------------------------------+
1 row in set (0.00 sec)
```

3. JSON_SEARCH(json_doc, one_or_all, search_str[, escape_char[, path] ...])

返回某个字符串（search_str）在 JSON 文档中的位置，各个参数的含义如下。

- one_or_all：匹配的次数，其中 one 是指只匹配一次，all 是指匹配所有。如果匹配到多个，结果会以数组的形式返回。
- search_str：子串，支持使用%和_进行模糊匹配。
- escape_char：转义符，如果该参数不填或为 NULL，则取默认转义符（\）。
- path：查找路径。

```
mysql> set @j = '["abc", [{"k": "10"}, "def"], {"x":"abc"}, {"y":"bcd"}]';
Query OK, 0 rows affected (0.00 sec)
```

```
mysql> select json_search(@j, 'one', 'abc'),json_search(@j, 'all', 'abc'),json_search(@j, 'all', 'ghi');
+-------------------------------+-------------------------------+-------------------------------+
| json_search(@j, 'one', 'abc') | json_search(@j, 'all', 'abc') | json_search(@j, 'all', 'ghi') |
+-------------------------------+-------------------------------+-------------------------------+
| "$[0]"                        | ["$[0]", "$[2].x"]            | NULL                          |
+-------------------------------+-------------------------------+-------------------------------+
1 row in set (0.00 sec)

mysql> select json_search(@j, 'all', '%b%', NULL, '$[1]'), json_search(@j, 'all', '%b%', NULL, '$[3]');
+---------------------------------------------+---------------------------------------------+
| json_search(@j, 'all', '%b%', NULL, '$[1]') | json_search(@j, 'all', '%b%', NULL, '$[3]') |
+---------------------------------------------+---------------------------------------------+
| NULL                                        | "$[3].y"                                    |
+---------------------------------------------+---------------------------------------------+
1 row in set (0.00 sec)
```

4. JSON_KEYS(json_doc[, path])

返回 JSON 文档最外层的 KEY，如果指定了路径，则返回该路径对应元素最外层的 KEY。

```
mysql> select json_keys('{"a": 1, "b": {"c": 30}}');
+---------------------------------------+
| json_keys('{"a": 1, "b": {"c": 30}}') |
+---------------------------------------+
| ["a", "b"]                            |
+---------------------------------------+
1 row in set (0.00 sec)

mysql> select json_keys('{"a": 1, "b": {"c": 30}}', '$.b');
+----------------------------------------------+
| json_keys('{"a": 1, "b": {"c": 30}}', '$.b') |
+----------------------------------------------+
| ["c"]                                        |
+----------------------------------------------+
1 row in set (0.00 sec)
```

5. JSON_VALUE(json_doc, path)

MySQL 8.0.21 引入，从 JSON 文档提取指定路径的元素，。

该函数的完整语法如下：

```
JSON_VALUE(json_doc, path [RETURNING type] [on_empty] [on_error])

on_empty:
    {NULL | ERROR | DEFAULT value} ON EMPTY

on_error:
    {NULL | ERROR | DEFAULT value} ON ERROR
```

各个参数解释如下。

- RETURNING type：返回值的类型，如果不指定，则默认是 VARCHAR(512)。如果不指定字符集，则默认是 utf8mb4，且区分大小写。
- on_empty：如果指定路径没有值，会触发 on_empty 子句，默认返回 NULL，也可指定 ERROR 抛出错误，或者通过 DEFAULT value 返回默认值。

❑ on_error：在 3 种情况下会触发 on_error 子句。一是从数组或对象中提取元素时，解析到多个值；二是类型转换错误，如将 "abc" 转换为 unsigned 类型；三是值被截断了。默认返回 NULL。

```
mysql> select json_value('{"item": "shoes", "price": "49.95"}', '$.item');
+-----------------------------------------------------------+
| json_value('{"item": "shoes", "price": "49.95"}', '$.item') |
+-----------------------------------------------------------+
| shoes                                                     |
+-----------------------------------------------------------+
1 row in set (0.00 sec)

mysql> select json_value('{"item": "shoes", "price": "49.95"}', '$.price' returning decimal(4,2)) as price;
+-------+
| price |
+-------+
| 49.95 |
+-------+
1 row in set (0.00 sec)

mysql> select json_value('{"item": "shoes", "price": "49.95"}', '$.price1' error on empty);
ERROR 3966 (22035): No value was found by 'json_value' on the specified path.

mysql> select json_value('[1, 2, 3]', '$[1 to 2]' error on error);
ERROR 3967 (22034): More than one value was found by 'json_value' on the specified path.

mysql> select json_value('{"item": "shoes", "price": "49.95"}', '$.item' returning unsigned error on error) as price;
ERROR 1690 (22003): UNSIGNED value is out of range in 'json_value'
```

6. value MEMBER OF(json_array)

判断 value 是否是 JSON 数组的一个元素。如果是，返回 1，否则返回 0。

```
mysql> select 17 member of('[23, "abc", 17, "ab", 10]');
+-------------------------------------------+
| 17 member of('[23, "abc", 17, "ab", 10]') |
+-------------------------------------------+
|                                         1 |
+-------------------------------------------+
1 row in set (0.00 sec)

mysql> select cast('[4,5]' as json) member of('[[3,4],[4,5]]');
+--------------------------------------------------+
| cast('[4,5]' as json) member of('[[3,4],[4,5]]') |
+--------------------------------------------------+
|                                                1 |
+--------------------------------------------------+
1 row in set (0.00 sec)
```

7. JSON_OVERLAPS(json_doc1, json_doc2)

MySQL 8.0.17 引入，比较两个 JSON 文档是否有相同的键值对或数组元素。如果有，返回 1，否则返回 0。如果两个参数都是标量，则判断这两个标量是否相等。

```
mysql> select json_overlaps('[1,3,5,7]', '[2,5,7]'),json_overlaps('[1,3,5,7]', '[2,6,8]');
+---------------------------------------+---------------------------------------+
| json_overlaps('[1,3,5,7]', '[2,5,7]') | json_overlaps('[1,3,5,7]', '[2,6,8]') |
+---------------------------------------+---------------------------------------+
|                                     1 |                                     0 |
+---------------------------------------+---------------------------------------+
1 row in set (0.00 sec)

mysql> select json_overlaps('{"a":1,"b":2}', '{"c":3,"d":4,"b":2}');
+-------------------------------------------------------+
| json_overlaps('{"a":1,"b":2}', '{"c":3,"d":4,"b":2}') |
+-------------------------------------------------------+
|                                                     1 |
+-------------------------------------------------------+
1 row in set (0.00 sec)

mysql> select json_overlaps('{"a":1,"b":2}', '{"c":3,"d":4,"b":10}');
+--------------------------------------------------------+
| json_overlaps('{"a":1,"b":2}', '{"c":3,"d":4,"b":10}') |
+--------------------------------------------------------+
|                                                      0 |
+--------------------------------------------------------+
1 row in set (0.00 sec)

mysql> select json_overlaps('5', '5'),json_overlaps('5', '6');
+-------------------------+-------------------------+
| json_overlaps('5', '5') | json_overlaps('5', '6') |
+-------------------------+-------------------------+
|                       1 |                       0 |
+-------------------------+-------------------------+
1 row in set (0.00 sec)
```

从 MySQL 8.0.17 开始，InnoDB 支持多值索引，可用在 JSON 数组中。当我们使用 JSON_CONTAINS、MEMBER OF 或 JSON_OVERLAPS 进行数组的相关操作时，可使用多值索引来加快查询。

A.7.2 修改相关

1. JSON_ARRAY_APPEND(json_doc, path, val[, path, val] ...)

向数组的指定位置追加元素。如果指定路径不存在，则不添加。

```
mysql> set @j = '["a", ["b", "c"], "d"]';
Query OK, 0 rows affected (0.00 sec)

mysql> select json_array_append(@j, '$[0]', 1, '$[1][0]', 2, '$[3]', 3);
+-----------------------------------------------------------+
| json_array_append(@j, '$[0]', 1, '$[1][0]', 2, '$[3]', 3) |
+-----------------------------------------------------------+
| [["a", 1], [["b", 2], "c"], "d"]                          |
+-----------------------------------------------------------+
1 row in set (0.00 sec)

mysql> set @j = '{"a": 1, "b": [2, 3], "c": 4}';
Query OK, 0 rows affected (0.00 sec)

mysql> select json_array_append(@j, '$.b', 'x', '$', 'z');
```

```
+----------------------------------------------+
| json_array_append(@j, '$.b', 'x', '$', 'z') |
+----------------------------------------------+
| [{"a": 1, "b": [2, 3, "x"], "c": 4}, "z"]   |
+----------------------------------------------+
1 row in set (0.00 sec)
```

2. JSON_ARRAY_INSERT(json_doc, path, val[, path, val] ...)

向数组的指定位置插入元素。

```
mysql> set @j = '["a", ["b", "c"],{"d":"e"}]';
Query OK, 0 rows affected (0.00 sec)

mysql> select json_array_insert(@j, '$[0]', 1);
+----------------------------------+
| json_array_insert(@j, '$[0]', 1) |
+----------------------------------+
| [1, "a", ["b", "c"], {"d": "e"}] |
+----------------------------------+
1 row in set (0.00 sec)

mysql> select json_array_insert(@j, '$[1]', cast('[1,2]' as json));
+------------------------------------------------------+
| json_array_insert(@j, '$[1]', cast('[1,2]' as json)) |
+------------------------------------------------------+
| ["a", [1, 2], ["b", "c"], {"d": "e"}]                |
+------------------------------------------------------+
1 row in set (0.00 sec)

mysql> select json_array_insert(@j, '$[5]', 2);
+----------------------------------+
| json_array_insert(@j, '$[5]', 2) |
+----------------------------------+
| ["a", ["b", "c"], {"d": "e"}, 2] |
+----------------------------------+
1 row in set (0.00 sec)
```

3. JSON_MERGE_PATCH(json_doc, json_doc[, json_doc] ...)

MySQL 8.0.3 引入，合并多个 JSON 文档，其合并规则如下。

- 如果两个文档不都是 JSON 对象，则合并后的结果是第二个文档。
- 如果两个文档都是 JSON 对象，且不存在同名 KEY，则合并后的文档包括两个文档的所有元素；如果存在同名 KEY，则第二个文档的值会覆盖第一个的。

```
mysql> select json_merge_patch('[1, 2]', '[3, 4]'), json_merge_patch('[1, 2]', '{"a": 123}');
+--------------------------------------+------------------------------------------+
| json_merge_patch('[1, 2]', '[3, 4]') | json_merge_patch('[1, 2]', '{"a": 123}') |
+--------------------------------------+------------------------------------------+
| [3, 4]                               | {"a": 123}                               |
+--------------------------------------+------------------------------------------+
1 row in set (0.00 sec)

mysql> select json_merge_patch('{"a": 1}', '{"b": 2}'),json_merge_patch('{ "a": 1, "b":2 }','{ "a": 3, "c":4 }');
+------------------------------------------+-----------------------------------------------------------+
| json_merge_patch('{"a": 1}', '{"b": 2}') | json_merge_patch('{ "a": 1, "b":2 }','{ "a": 3, "c":4 }') |
```

```
+------------------------------------------+------------------------------------------+
| {"a": 1, "b": 2}                         | {"a": 3, "b": 2, "c": 4}                 |
+------------------------------------------+------------------------------------------+
1 row in set (0.00 sec)

# 如果第二个文档存在 NULL 值，文档合并后不会输出对应的 KEY
mysql> select json_merge_patch('{"a":1, "b":2}', '{"a":3, "b":null}');
+---------------------------------------------------------+
| json_merge_patch('{"a":1, "b":2}', '{"a":3, "b":null}') |
+---------------------------------------------------------+
| {"a": 3}                                                |
+---------------------------------------------------------+
1 row in set (0.00 sec)
```

4. JSON_MERGE_PRESERVE(json_doc, json_doc[, json_doc] ...)

MySQL 8.0.3 引入，用来代替 `JSON_MERGE`。它虽然也是用于合并文档的，但合并规则与 `JSON_MERGE_PATCH` 的有所不同。

- 在两个文档中，只要有一个是数组，则另一个会被合并到该数组中。
- 如果两个文档都是 JSON 对象，且存在同名 KEY，则第二个文档并不会覆盖第一个，而是会将值附加（append）到第一个文档中。

```
mysql> select json_merge_preserve('1','2'),json_merge_preserve('[1, 2]', '[3, 4]');
+------------------------------+-----------------------------------------+
| json_merge_preserve('1','2') | json_merge_preserve('[1, 2]', '[3, 4]') |
+------------------------------+-----------------------------------------+
| [1, 2]                       | [1, 2, 3, 4]                            |
+------------------------------+-----------------------------------------+
1 row in set (0.00 sec)

mysql> select json_merge_preserve('[1,2]', '{"a": 123}'),json_merge_preserve('{"a": 123}', '[3,4]');
+--------------------------------------------+--------------------------------------------+
| json_merge_preserve('[1,2]', '{"a": 123}') | json_merge_preserve('{"a": 123}', '[3,4]') |
+--------------------------------------------+--------------------------------------------+
| [1, 2, {"a": 123}]                         | [{"a": 123}, 3, 4]                         |
+--------------------------------------------+--------------------------------------------+
1 row in set (0.00 sec)

mysql> select json_merge_preserve('{"a": 1}','{"b": 2}'),json_merge_preserve('{ "a": 1,"b":2 }','{ "a": 3, "c":4 }');
+--------------------------------------------+--------------------------------------------------------------+
| json_merge_preserve('{"a": 1}', '{"b": 2}')| json_merge_preserve('{ "a": 1,"b":2 }','{ "a": 3, "c":4 }')  |
+--------------------------------------------+--------------------------------------------------------------+
| {"a": 1, "b": 2}                           | {"a": [1, 3], "b": 2, "c": 4}                                |
+--------------------------------------------+--------------------------------------------------------------+
1 row in set (0.00 sec)
```

5. JSON_MERGE(json_doc, json_doc[, json_doc] ...)

与 `JSON_MERGE_PRESERVE` 的作用一样，从 MySQL 8.0.3 开始不建议使用，后续会移除。

A.7.3 其他辅助函数

1. JSON_QUOTE(string)

生成有效的 JSON 字符串，主要是对一些特殊字符（如双引号）进行转义。

```
mysql> select json_quote('null'), json_quote('"null"'), json_quote('[1, 2, 3]');
+--------------------+----------------------+-------------------------+
| json_quote('null') | json_quote('"null"') | json_quote('[1, 2, 3]') |
+--------------------+----------------------+-------------------------+
| "null"             | "\"null\""           | "[1, 2, 3]"             |
+--------------------+----------------------+-------------------------+
1 row in set (0.00 sec)
```

除此之外，也可通过 CAST(value AS JSON) 进行类型转换。

2. JSON_UNQUOTE(json_val)

将 JSON 转义成字符串输出。

```
mysql> select c2->'$.ename',json_unquote(c2->'$.ename'),
    -> json_valid(c2->'$.ename'),json_valid(json_unquote(c2->'$.ename')) from t;
+---------------+-----------------------------+---------------------------+-----------------------------------------+
| c2->'$.ename' | json_unquote(c2->'$.ename') | json_valid(c2->'$.ename') | json_valid(json_unquote(c2->'$.ename')) |
+---------------+-----------------------------+---------------------------+-----------------------------------------+
| "jack"        | jack                        |                         1 |                                       0 |
| "mark"        | mark                        |                         1 |                                       0 |
+---------------+-----------------------------+---------------------------+-----------------------------------------+
2 rows in set (0.00 sec)
```

直观地看，没加 JSON_UNQUOTE 的字符串会被用双引号引起来，加了 JSON_UNQUOTE 的则不会。但在本质上，前者是 JSON 中的字符串（STRING）类型，后者是 MySQL 中的字符类型，这一点可通过 JSON_VALID 来判断。

3. JSON_OBJECTAGG(key, value)

取表中的两列作为参数，其中第一列是键，第二列是值，它会返回 JSON 对象。

```
mysql> select * from emp;
+--------+----------+--------+
| deptno | ename    | sal    |
+--------+----------+--------+
|     10 | emp_1001 | 100.00 |
|     10 | emp_1002 | 200.00 |
|     20 | emp_1003 | 300.00 |
|     20 | emp_1004 | 400.00 |
+--------+----------+--------+
4 rows in set (0.00 sec)

mysql> select json_objectagg(ename,sal) from emp;
+----------------------------------------------------------------------------+
| json_objectagg(ename,sal)                                                  |
+----------------------------------------------------------------------------+
| {"emp_1001": 100.00, "emp_1002": 200.00, "emp_1003": 300.00, "emp_1004": 400.00} |
+----------------------------------------------------------------------------+
1 row in set (0.00 sec)

mysql> select deptno,json_objectagg(ename,sal) from emp group by deptno;
+--------+------------------------------------------+
| deptno | json_objectagg(ename,sal)                |
+--------+------------------------------------------+
|     10 | {"emp_1001": 100.00, "emp_1002": 200.00} |
|     20 | {"emp_1003": 300.00, "emp_1004": 400.00} |
+--------+------------------------------------------+
2 rows in set (0.00 sec)
```

4. JSON_ARRAYAGG(col_or_expr)

将列的值聚合成 JSON 数组，注意，JSON 数组中元素的顺序是随机的。

```
mysql> select json_arrayagg(ename) from emp;
+-------------------------------------------------+
| json_arrayagg(ename)                            |
+-------------------------------------------------+
| ["emp_1001", "emp_1002", "emp_1003", "emp_1004"] |
+-------------------------------------------------+
1 row in set (0.00 sec)

mysql> select deptno,json_arrayagg(ename) from emp group by deptno;
+--------+--------------------------+
| deptno | json_arrayagg(ename)     |
+--------+--------------------------+
|     10 | ["emp_1001", "emp_1002"] |
|     20 | ["emp_1003", "emp_1004"] |
+--------+--------------------------+
2 rows in set (0.00 sec)
```

5. JSON_PRETTY(json_val)

将 JSON 格式化输出。

```
mysql> select json_pretty("[1,3,5]");
+------------------------+
| json_pretty("[1,3,5]") |
+------------------------+
| [
  1,
  3,
  5
]                        |
+------------------------+
1 row in set (0.00 sec)
mysql> select json_pretty('{"a":"10","b":"15","x":"25"}');
+---------------------------------------------+
| json_pretty('{"a":"10","b":"15","x":"25"}') |
+---------------------------------------------+
| {
  "a": "10",
  "b": "15",
  "x": "25"
}                                             |
+---------------------------------------------+
1 row in set (0.00 sec)
```

6. JSON_STORAGE_FREE(json_val)

它是 MySQL 8.0 新增的，与 Partial Update 有关，用于计算 JSON 文档在部分更新后的剩余空间。

7. JSON_STORAGE_SIZE(json_val)

MySQL 5.7.22 引入，计算 JSON 文档的空间使用情况。

8. JSON_DEPTH(json_doc)

返回 JSON 文档的最大深度。对于空数组、空对象、标量值，深度为 1。

```
mysql> select json_depth('{}'),json_depth('[10, 20]'),json_depth('[10, {"a": 20}]');
+------------------+------------------------+-------------------------------+
| json_depth('{}') | json_depth('[10, 20]') | json_depth('[10, {"a": 20}]') |
+------------------+------------------------+-------------------------------+
|                1 |                      2 |                             3 |
+------------------+------------------------+-------------------------------+
1 row in set (0.00 sec)
```

9. JSON_LENGTH(json_doc[, path])

返回 JSON 文档的长度,其计算规则如下。

- 如果是标量值,长度为 1。
- 如果是数组,长度为数组元素的个数。
- 如果是对象,长度为对象元素的个数。
- 不包括嵌套数据和嵌套对象的长度。

```
mysql> select json_length('"abc"');
+----------------------+
| json_length('"abc"') |
+----------------------+
|                    1 |
+----------------------+
1 row in set (0.00 sec)

mysql> select json_length('[1, 2, {"a": 3}]');
+---------------------------------+
| json_length('[1, 2, {"a": 3}]') |
+---------------------------------+
|                               3 |
+---------------------------------+
1 row in set (0.00 sec)

mysql> select json_length('{"a": 1, "b": {"c": 30}}');
+-----------------------------------------+
| json_length('{"a": 1, "b": {"c": 30}}') |
+-----------------------------------------+
|                                       2 |
+-----------------------------------------+
1 row in set (0.00 sec)

mysql> select json_length('{"a": 1, "b": {"c": 30}}', '$.a');
+------------------------------------------------+
| json_length('{"a": 1, "b": {"c": 30}}', '$.a') |
+------------------------------------------------+
|                                              1 |
+------------------------------------------------+
1 row in set (0.00 sec)
```

10. JSON_TYPE(json_val)

返回 JSON 值的类型。

```
mysql> select json_type('123');
+------------------+
| json_type('123') |
+------------------+
| INTEGER          |
+------------------+
```

```
1 row in set (0.00 sec)

mysql> select json_type('"abc"');
+--------------------+
| json_type('"abc"') |
+--------------------+
| STRING             |
+--------------------+
1 row in set (0.00 sec)

mysql> select json_type(cast(now() as json));
+--------------------------------+
| json_type(cast(now() as json)) |
+--------------------------------+
| DATETIME                       |
+--------------------------------+
1 row in set (0.00 sec)

mysql> select json_type(json_extract('{"a": [10, true]}', '$.a'));
+-----------------------------------------------------+
| json_type(json_extract('{"a": [10, true]}', '$.a')) |
+-----------------------------------------------------+
| ARRAY                                               |
+-----------------------------------------------------+
1 row in set (0.00 sec)
```

11. JSON_VALID(val)

判断给定的值是否是有效的 JSON 文档。

```
mysql> select json_valid('hello'), json_valid('"hello"');
+---------------------+-----------------------+
| json_valid('hello') | json_valid('"hello"') |
+---------------------+-----------------------+
|                   0 |                     1 |
+---------------------+-----------------------+
1 row in set (0.00 sec)
```

12. JSON_TABLE(expr, path COLUMNS (column_list) [AS] alias)

从 JSON 文档中提取数据并以表格的形式返回。

该函数的完整语法如下：

```
JSON_TABLE(
    expr,
    path COLUMNS (column_list)
) [AS] alias

column_list:
    column[, column][, ...]

column:
    name FOR ORDINALITY
    | name type PATH string_path [on_empty] [on_error]
    | name type EXISTS PATH string_path
    | NESTED [PATH] path COLUMNS (column_list)
```

```
on_empty:
    {NULL | DEFAULT json_string | ERROR} ON EMPTY
on_error:
    {NULL | DEFAULT json_string | ERROR} ON ERROR
```

各个参数解释如下。

- expr：返回 JSON 文档的表达式。可以是一个标量（JSON 文档）、列名或者函数调用（JSON_EXTRACT(t1.json_data,'$.post.comments')）。
- path：JSON 的路径表达式。
- column：列的类型，支持以下 4 种类型。
    - name FOR ORDINALITY：序号，其中的 name 是列名。
    - name type PATH string_path [on_empty] [on_error]：提取指定路径（string_path）的元素。name 是列名，type 是 MySQL 中的数据类型。
    - name type EXISTS PATH string_path：指定路径（string_path）的元素是否存在。
    - NESTED [PATH] path COLUMNS (column_list)：将嵌套对象或数组与来自父对象或父数组的 JSON 值扁平化为一行输出。

```
select *
 from
   json_table(
    '[{"x":2, "y":"8", "z":9, "b":[1,2,3]}, {"x":"3", "y":"7"}, {"x":"4", "y":6, "z":10}]',
    "$[*]" columns(
       id for ordinality,
       xval varchar(100) path "$.x",
       yval varchar(100) path "$.y",
       z_exist int exists path "$.z",
       nested path '$.b[*]' columns (b INT PATH '$')
    )
  ) as t;
+------+------+------+---------+------+
| id   | xval | yval | z_exist | b    |
+------+------+------+---------+------+
|   1  |  2   |  8   |    1    |  1   |
|   1  |  2   |  8   |    1    |  2   |
|   1  |  2   |  8   |    1    |  3   |
|   2  |  3   |  7   |    0    | NULL |
|   3  |  4   |  6   |    1    | NULL |
+------+------+------+---------+------+
5 rows in set (0.00 sec)
```

13. JSON_SCHEMA_VALID(schema,document)

判断 document（JSON 文档）是否满足 schema（JSON 对象）定义的规范要求。完整的规范要求可参考"Draft 4 of the JSON Schema specification"。如果不满足，可通过 JSON_SCHEMA_VALIDATION_REPORT() 获取具体的原因。

以下面这个 schema 为例。

```
set @schema = '{
    "type": "object",
```

```
    "properties": {
      "latitude": {
        "type": "number",
        "minimum": -90,
        "maximum": 90
      },
      "longitude": {
        "type": "number",
        "minimum": -180,
        "maximum": 180
      }
    },
    "required": ["latitude", "longitude"]
}';
```

它的要求如下。

- document 必须是 JSON 对象。
- JSON 对象必需的两个属性是 latitude 和 longitude。
- latitude 和 longitude 必须是数值类型，且两者的大小范围分别为-90 ~ 90 和-180 ~ 180。

下面通过具体的 document 测试一下。

```
mysql> set @document = '{"latitude": 63.444697,"longitude": 10.445118}';
Query OK, 0 rows affected (0.00 sec)

mysql> select json_schema_valid(@schema, @document);
+---------------------------------------+
| json_schema_valid(@schema, @document) |
+---------------------------------------+
|                                     1 |
+---------------------------------------+
1 row in set (0.00 sec)

mysql> set @document = '{"latitude": 63.444697}';
Query OK, 0 rows affected (0.00 sec)

mysql> select json_schema_valid(@schema, @document);
+---------------------------------------+
| json_schema_valid(@schema, @document) |
+---------------------------------------+
|                                     0 |
+---------------------------------------+
1 row in set (0.00 sec)

mysql> select json_pretty(json_schema_validation_report(@schema, @document))\G
*************************** 1. row ***************************
json_pretty(json_schema_validation_report(@schema, @document)): {
  "valid": false,
  "reason": "The JSON document location '#' failed requirement 'required' at JSON Schema location '#'",
  "schema-location": "#",
  "document-location": "#",
  "schema-failed-keyword": "required"
}
1 row in set (0.00 sec)

mysql> set @document = '{"latitude": 91,"longitude": 0}';
Query OK, 0 rows affected (0.00 sec)
```

```
mysql> select json_schema_valid(@schema, @document);
+---------------------------------------+
| json_schema_valid(@schema, @document) |
+---------------------------------------+
|                                     0 |
+---------------------------------------+
1 row in set (0.00 sec)

mysql> select json_pretty(json_schema_validation_report(@schema, @document))\G
*************************** 1. row ***************************
json_pretty(json_schema_validation_report(@schema, @document)): {
  "valid": false,
  "reason": "The JSON document location '#/latitude' failed requirement 'maximum' at JSON Schema location '#/properties/latitude'",
  "schema-location": "#/properties/latitude",
  "document-location": "#/latitude",
  "schema-failed-keyword": "maximum"
}
1 row in set (0.00 sec)
```

A.8 总结

如果要使用 JSON 类型，推荐使用 MySQL 8.0。相比于 MySQL 5.7，Partial Update 带来的性能提升还是十分明显的。

Partial Update 在存储引擎层是默认开启的，在 binlog 中是否开启取决于 binlog_row_value_options。该参数默认为空，即不开启 Partial Update，建议设置为 PARTIAL_JSON。

注意使用 Partial Update 的前提条件。

当我们使用 JSON_CONTAINS、MEMBER OF、JSON_OVERLAPS 进行数组相关的操作时，可使用 MySQL 8.0.17 引入的多值索引来加快查询。

A.9 参考资料

- 维基百科"JSON"词条。
- 官方文档"The JSON Data Type"。
- 官方文档"JSON Functions"。
- 博客文章"Upgrading JSON data stored in TEXT columns"，作者：Matt Lord。
- 博客文章"Indexing JSON documents via Virtual Columns"，作者：Matt Lord。
- 博客文章"Partial update of JSON values"，作者：Knut Anders Hatlen。
- 博客文章"MySQL 8.0: InnoDB Introduces LOB Index For Faster Updates"，作者：Annamalai Gurusami。

附录 B
MySQL 8.0 的新特性

MySQL 8.0 提供了 300 多个新特性，本附录从中选取了 112 个 MySQL 用户平时较为关注、使用频率相对较高的特性。这些新特性涉及的范围比较广，包括备份、升级、DDL、慢日志、迁移、日常维护、字符集、语法、索引、从 MySQL 5.7 升级到 MySQL 8.0 需要注意的不兼容项、参数、hint、优化、复制、组复制、安全、账号、密码等。

掌握这些新特性有助于我们更好地使用 MySQL 8.0。

本章将按照管理、开发、优化器、InnoDB、数据字典、复制、组复制、安全和账号、密码管理、performance_schema 等 10 个类别对这些新特性进行分类阐述。

除了新特性，本章最后还介绍了 MySQL 8.0 中新增和移除的函数，及默认值相对于 MySQL 5.7 发生变化的参数列表。

B.1 管理

- 可持久化全局变量。

  持久化后的变量会存储在数据目录下的 mysqld-auto.cnf 文件中。以下是持久化变量相关的命令。持久化后的变量既可在 mysqld-auto.cnf 中查看，也可通过 performance_schema.persisted_variables 查看。

  ```
  # 持久化变量，同时修改变量的内存值
  SET PERSIST max_connections = 2000;
  # 只持久化变量，不修改变量的内存值，适用于只读参数的调整
  SET PERSIST_ONLY back_log = 2000;
  # 从 mysqld-auto.cnf 中删除所有持久化变量
  RESET PERSIST;
  # 从 mysqld-auto.cnf 中删除指定的变量。如果变量不存在，会报错
  RESET PERSIST system_var_name;
  # 从 mysqld-auto.cnf 中删除指定的变量。如果变量不存在，会提示 warning，不报错
  RESET PERSIST IF EXISTS system_var_name;
  ```

- 可设置管理 IP 和端口。

  管理 IP 通过 admin_address 参数设置，管理端口通过 admin_port 参数设置。管理连接的数量没有限制，但仅允许具有 SERVICE_CONNECTION_ADMIN 权限的用户连接。默认情况下，管理接口没有自己的独立线程，可将 create_admin_listener_thread 设置为 ON 开启。

建议设置管理 IP 和端口，这样即使连接数满了，也不用担心因登录不上实例而无法调整 max_connections 的大小。

- 安装包。

从 MySQL 8.0.16 开始，MySQL 针对通用二进制包（Linux - Generic）提供了一个最小化版本。最小化版本移除了 debug 相关的二进制文件。

MySQL 8.0.31 普通版本（mysql-8.0.31-linux-glibc2.12-x86_64.tar.xz）包的大小是 576.8MB，而最小化版本（mysql-8.0.31-linux-glibc2.17-x86_64-minimal.tar.xz）只有 57.4MB。后者的大小只是前者的约 9.95%。

从 MySQL 8.0.31 开始，通用二进制包还提供了 Linux - Generic (glibc 2.17) (ARM, 64-bit) 版本的下载。

- 资源组（Resource Group）。

资源组可用来控制组内线程的优先级及其能使用的资源。目前，能被管理的资源只有 CPU。

```
# 创建资源组
CREATE RESOURCE GROUP Batch
TYPE = USER # 资源组的类型，可设置为 USER（用户资源组）或 SYSTEM（系统资源组）
VCPU = 0-1 # 设置 CPU 亲和性，让线程运行在指定的 CPU 上。不设置，则默认会使用所有的 CPU
THREAD_PRIORITY = 10; # 设置线程优先级，有效值是-20（最高优先级）到 19（最低优先级）。不设置，则默认为 0
                     # 对于系统资源组，可设置的优先级范围是-20 到 0，对于用户资源组，可设置的
优先级范围是 0 到 19
```

以下是资源组的几种常用方式。

```
SET RESOURCE GROUP Batch FOR 702,703; # 将指定线程分配给资源组。702 是线程 ID，对应
performance_schema.threads 中的 THREAD_ID
SET RESOURCE GROUP Batch; # 将当前会话的线程分配给资源组
SELECT /*+ RESOURCE_GROUP(Batch) */ COUNT(*) FROM sbtest.sbtest1;
```

- ALTER DATABASE 支持 READ ONLY 选项。

设置为只读模式的库将禁止任何更新操作。适用于数据库迁移场景。

```
# 将 mydb 设置为只读模式
ALTER DATABASE mydb READ ONLY = 1;
# 关闭只读模式
ALTER DATABASE mydb READ ONLY = 0;
```

- 设置 SHOW PROCESSLIST 的实现方式。

SHOW PROCESSLIST 默认是从线程管理器中获取线程信息的。这种实现方式会持有全局互斥锁，对数据库的性能会有一定的影响。所以一般推荐使用 performance_schema.processlist，这种方式不会持有全局锁。

在 MySQL 8.0.22 中，引入了 performance_schema_show_processlist 参数，用来设置 SHOW PROCESSLIST 的实现方式。设置为 ON，则会使用 performance_schema.processlist 这种实现方式，默认为 OFF。

- 在 MySQL 8.0.27 中，引入了 innodb_ddl_threads 和 innodb_ddl_buffer_size 提升索引的创建速度。
- 操作系统查看 MySQL 的线程名。

  从 MySQL 8.0.27 开始，通过 ps 命令可以直接查看 MySQL 的线程名。

  ```
  # ps -p 22307 H -o "pid tid cmd comm"
  PID    TID   CMD                             COMMAND
  22307  22307 /usr/local/mysql/bin/mysqld     mysqld
  22307  22316 /usr/local/mysql/bin/mysqld     ib_io_ibuf
  22307  22318 /usr/local/mysql/bin/mysqld     ib_io_log
  22307  22319 /usr/local/mysql/bin/mysqld     ib_io_rd-1
  22307  22331 /usr/local/mysql/bin/mysqld     ib_io_rd-2
  ...
  ```

- 控制连接的内存使用量。

  从 MySQL 8.0.28 开始，引入了 connection_memory_limit 参数限制单个用户连接可以使用的最大内存量，引入了 global_connection_memory_limit 参数限制所有用户连接可以使用的内存总量。注意，这里说的内存不包括 InnoDB 缓冲池。

  ```
  mysql> SELECT LENGTH(GROUP_CONCAT(f1 ORDER BY f2)) FROM t1;
  ERROR 4082 (HY000): Connection closed. Connection memory limit 2097152 bytes exceeded. Consumed 2456976 bytes.
  ```

- 慢日志。

  在 MySQL 8.0.14 中，引入了 log_slow_extra 参数，可以将更详细的信息记录到慢日志中。看下面这个示例，对比下参数开启前后的输出。

  ```
  # Time: 2022-12-11T08:19:52.135515Z
  # User@Host: root[root] @ localhost []  Id:   660
  # Query_time: 10.000188  Lock_time: 0.000000 Rows_sent: 1  Rows_examined: 1
  SET timestamp=1670746782;
  select sleep(10);

  # Time: 2022-12-11T08:20:54.397597Z
  # User@Host: root[root] @ localhost []  Id:   662
  # Query_time: 10.000194  Lock_time: 0.000000 Rows_sent: 1  Rows_examined: 1 Thread_id: 662 Errno:
  0 Killed: 0 Bytes_received: 23 Bytes_sent: 57 Read_first: 0 Read_last: 0 Read_key: 0 Read_next: 0
  Read_prev: 0 Read_rnd: 0 Read_rnd_next: 0 Sort_merge_passes: 0 Sort_range_count: 0 Sort_rows: 0
  Sort_scan_count: 0 Created_tmp_disk_tables: 0 Created_tmp_tables: 0 Start:
  2022-12-11T08:20:44.397403Z End: 2022-12-11T08:20:54.397597Z
  SET timestamp=1670746844;
  select sleep(10);
  ```

  除此之外，SET timestamp 现在记录的是语句的开始时间，不再是语句的结束时间。

- 克隆插件。

  克隆插件是 MySQL 8.0.17 引入的一个重大特性。

  有了克隆插件，只需一条命令就能很方便地添加一个新的节点，无论是在组复制还是普通的主从环境中都是如此。

- 备份锁。

  注意，引入备份锁是为了阻塞备份过程中的 DDL，不是为了替代全局读锁。XtraBackup 8.0 和 MySQL Enterprise Backup 之所以在备份的过程中不再加全局读锁，主要是因为 performance_schema.log_status 的引入。

- 数据库升级。

  数据库升级无须再执行 mysql_upgrade 脚本。升级逻辑已内置到 mysqld 的启动流程中。

  升级之前，可通过 MySQL Shell 中的 util.checkForServerUpgrade() 检查实例是否满足升级条件。

- mysql 客户端。

  mysql 客户端默认会开启 --binary-as-hex。

  开启后，mysql 客户端会使用十六进制表示法显示二进制数据。例如：

  ```
  mysql8.0> SELECT UNHEX(41);
  +-----------------------+
  | UNHEX(41)             |
  +-----------------------+
  | 0x41                  |
  +-----------------------+
  1 row in set (0.00 sec)

  mysql5.7> SELECT UNHEX(41);
  +-----------+
  | UNHEX(41) |
  +-----------+
  | A         |
  +-----------+
  1 row in set (0.00 sec)
  ```

  如果要禁用十六进制表示法，需设置 --skip-binary-as-hex。

- 可通过 RESTART 命令重启 MySQL 实例。

  能使用 RESTART 命令的前提是，mysqld 是通过 mysqld_safe 或 systemctl 等守护进程启动的。

  ```
  mysql> restart;
  ERROR 3707 (HY000): Restart server failed (mysqld is not managed by supervisor process).
  ```

- 在 MySQL 8.0.30 中，mysqldump 新增了 --mysqld-long-query-time 选项，允许自定义 long_query_time 的会话值。

  这样可避免将备份相关的查询语句记录在慢日志中。

B.2 开发

- 字符集。

  默认字符集由 latin1 调整为 utf8mb4。

在 MySQL 8.0 中，utf8mb4 默认的校对集是 utf8mb4_0900_ai_ci，在 MySQL 5.7 中则是 utf8mb4_general_ci。

utf8mb4_0900_ai_ci 中的 0900 指的是 Unicode 9.0 规范；ai 是 accent insensitivity 的缩写，指的是不区分音调；ci 是 case insensitivity 的缩写，指的是不区分大小写。

❑ 公用表表达式（common table expression，CTE）。

公用表表达式，简单来说，就是一个命名的临时结果集。只需定义一次，即可多次使用。使用公用表表达式不仅让 SQL 语句变得简洁，同时也提升了 SQL 语句的可读性。

```
# 普通的公用表表达式
WITH
cte1 AS (SELECT a, b FROM table1),
cte2 AS (SELECT c, d FROM table2)
SELECT b, d FROM cte1 JOIN cte2
WHERE cte1.a = cte2.c;

# 递归公用表表达式
WITH RECURSIVE cte (n) AS
(
SELECT 1
UNION ALL
SELECT n + 1 FROM cte WHERE n < 5
)
SELECT * FROM cte;
```

❑ 窗口函数。

窗口函数，也称为分析函数，可针对一组行进行计算，并为每行返回一个结果。这一点与聚合函数不同。聚合函数只能为每个分组返回一个结果。窗口函数中的 OVER 子句定义了要计算行的行窗口。

看下面这个示例，它们都是为了实现行号，只不过在 MySQL 8.0 之前，需借助于自定义变量，而在 MySQL 8.0 中，可以直接使用 ROW_NUMBER()。

```
# MySQL 5.7
SET @row_number = 0;
SELECT dept_no, dept_name,
  (@row_number:=@row_number + 1) AS row_num
FROM departments ORDER BY dept_no;

# MySQL 8.0
SELECT dept_no, dept_name,
  ROW_NUMBER() OVER (ORDER BY dept_no) AS row_num
FROM departments;
```

❑ 支持将表达式作为默认值。

下面是官方文档中的一个示例。

```
CREATE TABLE t1 (
# 文本默认值
i INT          DEFAULT 0,
c VARCHAR(10) DEFAULT '',
```

```
# 表达式默认值
f FLOAT         DEFAULT (RAND() * RAND()),
b BINARY(16)    DEFAULT (UUID_TO_BIN(UUID())),
d DATE          DEFAULT (CURRENT_DATE + INTERVAL 1 YEAR),
p POINT         DEFAULT (Point(0,0)),
j JSON          DEFAULT (JSON_ARRAY())
);
```

注意，表达式默认值必须放到括号内。

从 MySQL 8.0.13 开始，BLOB、TEXT、GEOMETRY 和 JSON 字段允许设置表达式默认值。例如：

```
CREATE TABLE t2 (b BLOB DEFAULT ('abc'));
```

- 支持 CHECK 约束。

  看下面这个示例。

```
CREATE TABLE t1
(
c1 INT CHECK (c1 > 10),
c2 INT CONSTRAINT c2_positive CHECK (c2 > 0),
c3 INT CHECK (c3 < 100),
CONSTRAINT c1_nonzero CHECK (c1 <> 0),
CHECK (c1 > c3)
);

mysql> SHOW CREATE TABLE t1\G
*************************** 1. row ***************************
       Table: t1
Create Table: CREATE TABLE `t1` (
  `c1` int DEFAULT NULL,
  `c2` int DEFAULT NULL,
  `c3` int DEFAULT NULL,
  CONSTRAINT `c1_nonzero` CHECK ((`c1` <> 0)),
  CONSTRAINT `c2_positive` CHECK ((`c2` > 0)),
  CONSTRAINT `t1_chk_1` CHECK ((`c1` > 10)),
  CONSTRAINT `t1_chk_2` CHECK ((`c3` < 100)),
  CONSTRAINT `t1_chk_3` CHECK ((`c1` > `c3`))
) ENGINE=InnoDB DEFAULT CHARSET=utf8mb4 COLLATE=utf8mb4_0900_ai_ci
1 row in set (0.00 sec)
```

- 隐藏列。

  隐藏列是 MySQL 8.0.23 引入的新特性。对于隐藏列，只有显式指定才能访问，无论是在查询还是 DML 语句中都是如此。

  如果是通过 SELECT * 查询，则不会返回隐藏列的内容。

```
mysql> CREATE TABLE t1 (c1 INT, c2 INT INVISIBLE);
Query OK, 0 rows affected (0.03 sec)

mysql> INSERT INTO t1 (c1, c2) VALUES(1, 2);
Query OK, 1 row affected (0.01 sec)

mysql> INSERT INTO t1 VALUES(3);
Query OK, 1 row affected (0.00 sec)

mysql> SELECT * FROM t1;
```

```
+------+
| c1   |
+------+
|    1 |
|    3 |
+------+
2 rows in set (0.00 sec)

mysql> SELECT c1, c2 FROM t1;
+------+------+
| c1   | c2   |
+------+------+
|    1 |    2 |
|    3 | NULL |
+------+------+
2 rows in set (0.00 sec)
```

- 不可见索引。

  对于冗余索引，在执行删除操作之前，可以先将其设置为不可见，然后观察一段时间，确定对业务没有影响后再执行删除操作。

  ```
  # 创建不可见索引
  ALTER TABLE t1 ADD INDEX k_idx (k) INVISIBLE;
  # 将索引设置为不可见
  ALTER TABLE t1 ALTER INDEX i_idx INVISIBLE;
  # 将索引设置为可见
  ALTER TABLE t1 ALTER INDEX i_idx VISIBLE;
  ```

  优化器默认不会使用不可见索引。如果要使用，可设置 optimizer_switch。

  ```
  SET SESSION optimizer_switch='use_invisible_indexes=on';
  SELECT /*+ SET_VAR(optimizer_switch = 'use_invisible_indexes=on') */ * FROM t1 WHERE k = 1;
  ```

- 降序索引。

  对于涉及多列但排序顺序又不一致的排序操作，可以通过降序索引来优化。如 ORDER BY c1 ASC, c2 DESC 这个排序操作就可以通过下面这个索引来优化。

  ```
  ALTER TABLE t add INDEX idx_1 (c1 ASC, c2 DESC);
  ```

- 函数索引。

  函数索引允许对表达式创建索引，在引入它之前，只能对列或列的前缀创建索引。

  ```
  CREATE TABLE tbl (
  col1 LONGTEXT,
  INDEX idx1 ((SUBSTRING(col1, 1, 10)))
  );
  # 对查询列使用相同的函数，可以使用索引
  mysql> EXPLAIN SELECT * FROM tbl WHERE SUBSTRING(col1, 1, 10) = '1234567890';
  +----+-------------+-------+------------+------+---------------+------+---------+-------+------+----------+-------+
  | id | select_type | table | partitions | type | possible_keys | key  | key_len | ref   | rows | filtered | Extra |
  +----+-------------+-------+------------+------+---------------+------+---------+-------+------+----------+-------+
  |  1 | SIMPLE      | tbl   | NULL       | ref  | idx1          | idx1 | 33      | const |    1 |   100.00 | NULL  |
  +----+-------------+-------+------------+------+---------------+------+---------+-------+------+----------+-------+
  1 row in set, 1 warning (0.00 sec)
  ```

- VALUES。

  VALUES 是 MySQL 8.0.19 开始支持的语法，它会以表的形式返回一行或多行数据。行数据通过 ROW()函数来构造。函数中的元素既可以是标量值，也可以是表达式。列名是 column_x，其中 x 是序号，从 0 开始递增。

  ```
  mysql> VALUES ROW(1,now()), ROW(2,now());
  +----------+---------------------+
  | column_0 | column_1            |
  +----------+---------------------+
  |        1 | 2022-12-13 11:01:40 |
  |        2 | 2022-12-13 11:01:40 |
  +----------+---------------------+
  2 rows in set (0.00 sec)
  ```

- INTERSECT 和 EXCEPT。MySQL 8.0.31 开始支持 INTERSECT 和 EXCEPT，分别用来取两个集合的交集和差集。

  ```
  mysql> VALUES ROW(1,2), ROW(3,4) INTERSECT VALUES ROW(1,2);
  +----------+----------+
  | column_0 | column_1 |
  +----------+----------+
  |        1 |        2 |
  +----------+----------+
  1 row in set (0.00 sec)

  mysql> VALUES ROW(1,2), ROW(3,4) EXCEPT VALUES ROW(1,2);
  +----------+----------+
  | column_0 | column_1 |
  +----------+----------+
  |        3 |        4 |
  +----------+----------+
  1 row in set (0.00 sec)
  ```

- 唯一键冲突的报错信息会输出表名，在 MySQL 5.7 中，只会输出唯一键名。

  ```
  # MySQL 8.0
  ERROR 1062 (23000): Duplicate entry '1' for key 't1.PRIMARY'
  # MySQL 5.7
  ERROR 1062 (23000): Duplicate entry '1' for key 'PRIMARY'
  ```

- 查询改写插件在 MySQL 8.0.12 之前只支持 SELECT 语句，从 MySQL 8.0.12 开始支持 INSERT、REPLACE、UPDATE 和 DELETE 语句。
- JSON 字段支持部分更新，极大提升了 JSON 字段的处理性能。
- NOWAIT 和 SKIP LOCKED。

  SELECT ... FOR SHARE 和 SELECT ... FOR UPDATE 语句中引入 NOWAIT 和 SKIP LOCKED 选项，用来解决电商场景中的热点行问题。

  ```
  session1> CREATE TABLE t(id INT PRIMARY KEY);
  Query OK, 0 rows affected (0.06 sec)

  session1> INSERT INTO t VALUES(1),(2),(3);
  Query OK, 3 rows affected (0.01 sec)
  Records: 3  Duplicates: 0  Warnings: 0
  ```

```
session1> BEGIN;
Query OK, 0 rows affected (0.00 sec)

session1> SELECT * FROM t WHERE id=2 FOR UPDATE;
+----+
| id |
+----+
|  2 |
+----+
1 row in set (0.00 sec)

# 如果需要加锁的行被其他事务锁定，指定 NOWAIT 会立即报错，不会等到锁超时
session2> SELECT * FROM t WHERE id = 2 FOR UPDATE NOWAIT;
ERROR 3572 (HY000): Statement aborted because lock(s) could not be acquired immediately and NOWAIT is set.

# 指定 SKIP LOCKED 则会跳过锁定行
session2> SELECT * FROM t FOR UPDATE SKIP LOCKED;
+----+
| id |
+----+
|  1 |
|  3 |
+----+
2 rows in set (0.00 sec)
```

- 不再支持 GROUP BY ASC/DESC 语法。例如：

  ```
  GROUP BY dept_no ASC;
  ```

  如果要对分组列进行排序，需显式指定排序列，例如：

  ```
  GROUP BY dept_no ORDER BY dept_no;
  ```

- 通用的分区接口（Handler）已从代码层移除。在 MySQL 8.0 中，如果要使用分区表，只能使用 InnoDB 存储引擎。

- 在 MySQL 8.0 中，正则表达式底层库由 Henry Spencer 调整为了 International Components for Unicode（ICU）。Spencer 库的部分语法不再支持。

- 引入了 sql_require_primary_key 参数，可要求表上必须存在主键。默认为 OFF。

  ```
  mysql> CREATE TABLE slowtech.t1(id INT);
  ERROR 3750 (HY000): Unable to create or change a table without a primary key, when the system variable 'sql_require_primary_key' is set. Add a primary key to the table or unset this variable to avoid this message. Note that tables without a primary key can cause performance problems in row-based replication, so please consult your DBA before changing this setting.
  ```

- 在 MySQL 8.0.30 中，引入了 sql_generate_invisible_primary_key 参数，可为没有显式设置主键的表创建一个隐式主键。默认为 OFF。

  看下面这个示例。

  ```
  mysql> CREATE TABLE t1(c1 INT);
  Query OK, 0 rows affected (0.04 sec)

  mysql> SHOW CREATE TABLE t1\G
  *************************** 1. row ***************************
         Table: t1
  ```

```
Create Table: CREATE TABLE `t1` (
  `my_row_id` bigint unsigned NOT NULL AUTO_INCREMENT /*!80023 INVISIBLE */,
  `c1` int DEFAULT NULL,
  PRIMARY KEY (`my_row_id`)
) ENGINE=InnoDB DEFAULT CHARSET=utf8mb4 COLLATE=utf8mb4_0900_ai_ci
1 row in set (0.01 sec)
```

- hint（提示）。

  引入了多个 hint。在 MySQL 5.7 中，hint 只有 13 个。在 MySQL 8.0 中，则新增到了 37 个。

  新增的 hint 中，有一个是 SET_VAR，可在语句级别调整参数的会话值。例如：

  ```
  SELECT /*+ SET_VAR(max_execution_time = 1000) */ * FROM employees.employees;

  mysql> CREATE TABLE t_parent (id INT PRIMARY KEY);
  Query OK, 0 rows affected (0.04 sec)

  mysql> CREATE TABLE t_child (id INT PRIMARY KEY, parent_id INT, FOREIGN KEY (parent_id) REFERENCES t_parent(ID));
  Query OK, 0 rows affected (0.04 sec)

  mysql> INSERT INTO t_child VALUES(1,1);
  ERROR 1452 (23000): Cannot add or update a child row: a foreign key constraint fails (`slowtech`.`t_child`, CONSTRAINT `t_child_ibfk_1` FOREIGN KEY (`parent_id`) REFERENCES `t_parent` (`id`))

  mysql> INSERT /*+ SET_VAR(foreign_key_checks = OFF) */ INTO t_child VALUES(1,1);
  Query OK, 1 row affected (0.01 sec)
  ```

- 秒级加列。

  从 MySQL 8.0.12 开始，Online DDL 开始支持 INSTANT 算法。使用这个算法进行加列操作，只需修改表的元数据信息，操作瞬间就能完成。不过在 MySQL 8.0.29 之前，列只能被添加到表的最后位置。从 MySQL 8.0.29 开始，则移除了这一限制，新增列可以添加到表的任意位置。不仅如此，从 MySQL 8.0.29 开始，删列操作也可以使用 INSTANT 算法。

- 提升了 DROP TABLE、TRUNCATE TABLE、DROP TABLESPACE 操作的性能。

  这个优化是 MySQL 8.0.23 引入的。在之前的版本中，这些操作会遍历整个缓冲池，删除对应表（或表空间）的数据页。在遍历的过程中，会加锁（latch）。加锁期间，会阻塞所有的 DML 操作。注意，阻塞时间与缓冲池的大小有关，与表的大小无关。缓冲池越大，遍历时间会越长，相应地，阻塞时间也会越长。优化后，待删除的数据页会做异步处理。

B.3 优化器

- 直方图。

  直方图可用来统计列中的数据分布情况，从而帮助优化器选择最优的执行计划。

  类似的效果通过索引也能实现，那在什么情况下应考虑直方图呢？

  - 维护索引是有成本的，而直方图只在创建或更新时才有开销。
  - 估算给定范围内的记录数时，使用直方图比索引的 index dive 操作成本更低。

直方图相对于索引有以下不足。

- 在进行 DML 操作时，直方图信息不会自动更新。如果在数据变化比较频繁的列上创建直方图，直方图信息可能会不准确。
- 索引可用来减少查询需要扫描的数据量，而直方图不能。

满足以下条件的列适合创建直方图。

- 用在 WHERE 子句或连接条件中。
- 数据分布不均匀。
- 选择性差。
- 列上没有索引。

直方图的相关操作如下。

```
# 创建或更新直方图
ANALYZE TABLE tbl_name UPDATE HISTOGRAM ON col_name [, col_name] WITH N BUCKETS;
# 删除直方图
ANALYZE TABLE tbl_name DROP HISTOGRAM ON col_name [, col_name];
```

❑ 哈希连接（hash join）。

经典的哈希连接算法由两个阶段组成：构建阶段和探测阶段。

- 构建阶段：选择一张小表（build table）构造一个内存哈希表。这个哈希表的 KEY 是使用哈希函数对该表关联列进行计算后的结果。
- 探测阶段：逐行遍历另外一张表（probe table），使用相同的哈希函数对该表的关联列进行计算，判断计算后的结果是否在哈希表中。如果在，则输出记录。

哈希连接的适用场景如下。

- probe table 的关联列上没有合适的索引，包括没有索引或者索引的区分度不高。
- 返回的结果集比较大。

在 MySQL 8.0.20 之前，哈希连接只适用于等值连接和笛卡儿积。

```
SELECT * FROM t1 JOIN t2 ON (t1.c1 = t2.c1); # 等值连接
SELECT * FROM t1 JOIN t2; # 笛卡儿积
```

从 MySQL 8.0.20 开始，哈希连接支持非等值内连接（inner non-equi-join）、半连接（semijoin）、反连接（antijoin）、左外连接（left outer join）、右外连接（right outer join）。

```
SELECT * FROM t1 JOIN t2 ON t1.c1 < t2.c1; # 非等值内连接
SELECT * FROM t1 WHERE t1.c1 IN (SELECT t2.c2 FROM t2); # 半连接
SELECT * FROM t2 WHERE NOT EXISTS (SELECT * FROM t1 WHERE t1.c1 = t2.c1); # 反连接
SELECT * FROM t1 LEFT JOIN t2 ON t1.c1 = t2.c1; # 左外连接
SELECT * FROM t1 RIGHT JOIN t2 ON t1.c1 = t2.c1; # 右外连接
```

❑ EXPLAIN ANALYZE。

与 EXPLAIN 不一样的是，EXPLAIN ANALYZE 会实际执行 SQL，并输出各个迭代器的实际成本。

```
mysql> EXPLAIN ANALYZE SELECT COUNT(*) FROM employees WHERE emp_no IN ( SELECT emp_no FROM dept_emp
WHERE dept_no='d001' )\G
*************************** 1. row ***************************
EXPLAIN: -> Aggregate: count(0)  (cost=21212.78 rows=1) (actual time=37.711..37.711 rows=1 loops=1)
    -> Nested loop inner join   (cost=17385.18 rows=38276) (actual time=0.233..36.683 rows=20211
loops=1)
        -> Filter: (dept_emp.dept_no = 'd001')  (cost=3988.58 rows=38276) (actual time=0.034..9.150
rows=20211 loops=1)
            -> Covering index lookup on dept_emp using dept_no (dept_no='d001')  (cost=3988.58
rows=38276) (actual time=0.032..5.878 rows=20211 loops=1)
        -> Single-row covering index lookup on employees using PRIMARY (emp_no=dept_emp.emp_no)
(cost=0.25 rows=1) (actual time=0.001..0.001 rows=1 loops=20211)

1 row in set (0.04 sec)
```

- 引入并行查询特性，提升了 SELECT COUNT (*) FROM TABLE 操作的执行效率。
- 在执行完 EXPLAIN 之后，通过 SHOW WARNINGS 可以查看 SQL 语句改写后的情况。在 MySQL 8.0.12 之前，只支持 SELECT 操作。从 MySQL 8.0.12 开始支持 DELETE、INSERT、REPLACE 和 UPDATE 操作。
- UPDATE/DELETE 子查询支持半连接优化。

  简单来说，以下两类操作无须再手动改写为两表关联操作。

  ```
  UPDATE t1 SET x=y WHERE z IN (SELECT * FROM t2);
  DELETE FROM t1 WHERE z IN (SELECT * FROM t2);
  ```

- 优化器成本模型数据库设置了默认值。

  优化器成本模式数据库由 mysql.server_cost 和 mysql.engine_cost 两张表组成，其中，mysql.server_cost 用来配置服务器相关操作的优化器成本，mysql.engine_cost 用来配置存储引擎相关操作的优化器成本。

  在 MySQL 8.0 中，mysql.engine_cost 为 io_block_read_cost（从磁盘读取数据块的成本）和 memory_block_read_cost（从缓冲池读取数据块的成本）设置了不同的默认值。而在 MySQL 5.7 中，这两个配置的默认值都是 NULL。在这种情况下，数据无论是在磁盘还是在缓冲池中，对优化器来说，成本都是一样的。

B.4 InnoDB

- redo log 的优化。

  相关的优化包括：对 redo log 进行了无锁设计；允许多个用户线程并发写入 redo log buffer；可动态修改 innodb_log_buffer_size 的大小。

- 使用了基于 CATS（Contention-Aware Transaction Scheduling，竞争感知事务调度）的事务调度算法。

  当有多个事务竞争同一把锁时，在 MySQL 8.0 之前，使用的是 FIFO（First In First Out，先进先出）算法，该算法会将锁优先分配给最先请求的事务。在 MySQL 8.0 中，引入了 CATS 算法，该算法会计算每个事务阻塞的事务数，最后将锁优先分配给阻塞事务最多的事务。

- 自增主键的持久化。
- 支持通过 SQL 语句来管理回滚表空间。

  ```
  # 添加回滚表空间，文件扩展名必须是.ibu
  CREATE UNDO TABLESPACE tablespace_name ADD DATAFILE 'file_name.ibu';
  # 删除回滚表空间
  DROP UNDO TABLESPACE tablespace_name;
  # 在删除回滚表空间之前，必须先将它设置为 INACTIVE 状态
  ALTER UNDO TABLESPACE tablespace_name SET INACTIVE;
  # 将回滚表空间设置为 ACTIVE 状态
  ALTER UNDO TABLESPACE tablespace_name SET ACTIVE;
  ```

- 默认开启回滚表空间。在 MySQL 5.7 中，默认没有开启。若要开启，只能在初始化时设置。
- 默认的内存临时表由 MEMORY 引擎更改为 TempTable 引擎，相比于前者，后者支持以变长方式存储 VARCHAR、VARBINARY 等变长字段。从 MySQL 8.0.13 开始，TempTable 引擎支持 BLOB 字段。
- 引入了 innodb_dedicated_server 参数，可基于服务器的内存动态设置 innodb_buffer_pool_size、innodb_log_file_size 和 innodb_flush_method。
- 引入了 innodb_deadlock_detect 参数关闭死锁检测。
- 支持在线回收临时表空间的磁盘空间。

  在 MySQL 5.7 中，用户创建的临时表和磁盘临时表会存储在全局临时表空间（ibtmp1）中。ibtmp1 一旦增长，就不会收缩。如果要回收 ibtmp1 的空间，只能重启实例。在 MySQL 8.0 中，用户创建的临时表和磁盘临时表会存储在会话临时表空间中，会话临时表空间默认位于 #innodb_temp 目录下。一个会话最多分配两个临时表空间，分别用来存储用户临时表和磁盘临时表。当会话的连接断开时，会截断这两个临时表空间。

- 设置独立的双写缓冲区（doublewrite buffer）。

  在 MySQL 8.0.20 之前，双写缓冲区默认存储在 InnoDB 系统表空间（ibdata1）中。从 MySQL 8.0.20 开始，双写缓冲区会存储在单独的#ib_16384_0.dblwr 和#ib_16384_1.dblwr 文件中。从 MySQL 8.0.30 开始，innodb_doublewrite 在 ON（开启双写缓冲区）和 OFF（关闭双写缓冲区）的基础上新增了两个选项：DETECT_AND_RECOVER 和 DETECT_ONLY。DETECT_AND_RECOVER 的作用与 ON 一样，都是开启双写缓冲区：在进行脏页刷新时，会首先将数据页写入双写缓冲区；在进行故障恢复时，会通过双写缓冲区来修复数据文件中不完整的数据页。DETECT_ONLY 也会开启双写缓冲区，但在进行脏页刷新时，只会记录数据页的元数据信息，不会存储数据页的内容，所以在进行故障恢复时，即使数据文件中存在不完整的数据页，也不能通过双写缓冲区来修复。该选项主要用来检测实例启动时是否有不完整的数据页。

- 禁用重做日志记录（redo logging）。

  从 MySQL 8.0.21 开始，可以使用 ALTER INSTANCE DISABLE INNODB REDO_LOG 命令禁用重做日志记录。该特性适用于数据加载场景，在线上生产环境中切记不要禁用。禁用后，可通过 ALTER INSTANCE ENABLE INNODB REDO_LOG 命令开启。

- 查看表缓存在缓冲池中数据页的数量。

  可通过 information_schema.innodb_cached_indexes 查看表缓存在缓冲池中数据页的数量。

```
SELECT
tables.name, indexes.name, cached.n_cached_pages
FROM
information_schema.innodb_cached_indexes AS cached,
information_schema.innodb_indexes AS indexes,
information_schema.innodb_tables AS tables
WHERE
cached.index_id = indexes.index_id AND indexes.table_id = tables.table_id;
+------------+------------+----------------+
| table_name | index_name | n_cached_pages |
+------------+------------+----------------+
| db1/t1     | idx_c1     |            279 |
| db1/t1     | PRIMARY    |           4756 |
+------------+------------+----------------+
2 rows in set (0.00 sec)
```

- 支持动态调整 redo log 的容量。

  从 MySQL 8.0.30 开始，可通过 innodb_redo_log_capacity 参数来动态调整 redo log 的容量。在 MySQL 8.0.30 之前，InnoDB 默认会在数据目录下创建两个 redo log。redo log 的数量和大小分别由 innodb_log_files_in_group 和 innodb_log_file_size 决定。随着 innodb_redo_log_capacity 的引入，InnoDB 会在数据目录的#innodb_redo 目录下创建 32 个 redo log。redo log 的总大小由 innodb_redo_log_capacity 决定。innodb_redo_log_capacity 如果没有显式设置，则默认等于 innodb_log_file_size * innodb_log_files_in_group。

B.5 数据字典

- 引入了原生的、基于 InnoDB 的数据字典。数据字典表位于 mysql 库中，对用户不可见。它同 mysql 库的其他系统表一样，保存在数据目录下的 mysql.ibd 文件中。
- 移除了之前版本的 frm、par、TRN、TRG、isl、db.opt、ddl_log.log 等文件。这些文件之前会存储部分元数据信息。
- 原子 DDL。
- information_schema 中的部分表已重构为基于数据字典的视图，在此之前，这是通过临时表实现的。
- information_schema.statistics 和 information_schema.tables 两张表中表相关的统计信息会被缓存起来以提升查询性能。缓存的时间由 information_schema_stats_expiry 参数决定，默认是 86 400 秒（24 小时）。如果要更新某张表的统计信息，可执行 ANALYZE TABLE。
- 因为数据字典的引入，MySQL 不再识别手动创建的数据库目录。
- 数据库对象的元数据信息除了存储在数据字典中，也会再存储一份 SDI（序列化的字典信息）。当数据字典不可用时，我们可以基于 SDI 提取对象的元数据信息。对于 InnoDB 表，SDI 会存储在 .ibd 文件中。对于其他存储引擎的表，SDI 会存储在 .sdi 文件中。
- 对于存储在 .ibd 文件中的 SDI，需通过 ibd2sdi 命令查看。对于存储在 .sdi 文件中的 SDI，可直接使用 cat 命令查看。

B.6 复制

- 支持在 channel 级别设置过滤规则。在 MySQL 8.0 之前，过滤规则只能在全局级别设置。
  ```
  CHANGE REPLICATION FILTER REPLICATE_DO_DB=(db1) FOR CHANNEL channel_1;
  ```
- 在设置 GTID_PURGED 时，无须 GTID_EXECUTED 为空。
- GTID 复制支持在事务、存储过程、函数、触发器中执行 CREATE TEMPORARY TABLE 或 DROP TEMPORARY TABLE 语句。
- GTID 复制支持 CREATE TABLE ... SELECT ... 语句。
- binlog 中记录表的列名信息和主键信息。

  默认情况下，binlog 中只会记录表有限的元数据信息，如果要记录完整的元数据信息，需将 binlog_row_metadata 设置为 FULL。在通过 mysqlbinlog 解析 binlog 时，如果要查看完整的元数据信息，需指定 --print-table-metadata。

  ```
  BEGIN
  /*!*/;
  # at 347
  #221207  9:22:09 server id 1  end_log_pos 433 CRC32 0x101d2cd6     Table_map: `sbtest`.`sbtest1` mapped to number 132
  # Columns(`id` INT NOT NULL,
  #         `k` INT NOT NULL,
  #         `c` CHAR(120) NOT NULL CHARSET utf8mb4 COLLATE utf8mb4_0900_ai_ci,
  #         `pad` CHAR(60) NOT NULL CHARSET utf8mb4 COLLATE utf8mb4_0900_ai_ci)
  # Primary Key(id)
  # at 433
  #221207  9:22:09 server id 1  end_log_pos 658 CRC32 0x41a0ae3d     Delete_rows: table id 132 flags: STMT_END_F
  ### DELETE FROM `sbtest`.`sbtest1`
  ### WHERE
  ###   @1=6
  ###   @2=500585
  ###   @3='37216201353-39109531021-11197415756-87798784755-02463049870-83329763120-57551308766-61100580113-80090253566-30971527105'
  ###   @4='05161542529-00085727016-35134775864-52531204064-98744439797'
  # at 658
  #221207  9:22:09 server id 1  end_log_pos 689 CRC32 0x2b6cc7fc     Xid = 1820
  COMMIT/*!*/;
  ```

- binlog 中记录事务的长度信息。例如，transaction_length=492。
- binlog 中会记录事务在源主库（执行写操作的节点）提交的时间戳（original_commit_timestamp）和当前节点提交的时间戳（immediate_commit_timestamp），这两个时间戳可用来计算主从延迟。
- binlog 可设置秒级别的过期时间。

  在 MySQL 8.0 之前，binlog 的过期时间由 expire_logs_days 参数控制，而 expire_logs_days 的单位是天。在 MySQL 8.0 中，引入了 binlog_expire_logs_seconds，可设置秒级别的过期时间。binlog_expire_logs_seconds 默认是 2 592 000 秒，即 30 天。如果两个参数同时设置了非零值，binlog_expire_logs_seconds 的优先级更高。

  从 MySQL 8.0.29 开始，引入了 binlog_expire_logs_auto_purge，可用来禁用 binlog 的自动清

理功能。在此之前，该需求只能通过将 binlog_expire_logs_seconds 和 expire_logs_days 设置为 0 来实现。

- 在搭建复制时，新增了以下选项。
  - PRIVILEGE_CHECKS_USER：指定重放用户，这样就可以通过限制重放用户的权限，来达到限制重放操作的目的。
  - REQUIRE_ROW_FORMAT：检查接收的二进制日志事件是否为 ROW 格式。若不是，SQL 线程会中断。
  - REQUIRE_TABLE_PRIMARY_KEY_CHECK：检查表上是否存在主键。
  - ASSIGN_GTIDS_TO_ANONYMOUS_TRANSACTIONS：是否为匿名事务分配 GTID。
  - SOURCE_CONNECTION_AUTO_FAILOVER：当主库出现故障时，是否自动切换到已定义的其他节点上。
  - GTID_ONLY：设置为 1，则只会使用 GTID 来作为事务是否已经重放的依据，不再更新 mysql.slave_relay_log_info 和 mysql.slave_worker_info 表的位置点信息。将 GTID_ONLY 设置为 1 需 SOURCE_AUTO_POSITION = 1 且 REQUIRE_ROW_FORMAT = 1。
- InnoDB ReplicaSet：基于 MySQL Shell、MySQL Router 和主从复制的解决方案。架构同 InnoDB Cluster 类似，只不过被管理对象是主从复制。

B.7 组复制

- 更细粒度的流控机制。
- group_replication_exit_state_action 引入了一个新的选项：OFFLINE_MODE。
- 引入了 group_replication_communication_debug_options，可以将 GCS 和 XCom 详细的调试信息打印在数据目录下的 GCS_DEBUG_TRACE 文件中。
- 引入了 group_replication_set_as_primary() 函数，用来在线切换单主模式下的 Primary 节点。
- 引入了 group_replication_switch_to_single_primary_mode() 和 group_replication_switch_to_multi_primary_mode() 函数，用来在线调整集群模式。
- 引入了 group_replication_get_write_concurrency() 和 group_replication_set_write_concurrency() 函数，用来查看和设置一个组可以并行执行的最大共识实例数。
- 引入了 group_replication_get_communication_protocol() 和 group_replication_set_communication_protocol() 函数，用来查看和设置组使用的组复制通信协议版本。
- 引入了 group_replication_member_expel_timeout 参数，可自定义可疑节点被驱逐出集群所需的时间。
- 引入了 group_replication_consistency 参数，用来设置事务的一致性级别。
- 从 MySQL 8.0.14 开始，支持 IPv6。
- 引入了 group_replication_autorejoin_tries 参数，用来设置节点被驱逐出集群后自动重试的次数。
- 引入了 group_replication_message_cache_size 参数，用来调整 XCom Cache 的大小，默认为 1GB，最小可设置到 128MB。

- 在 MySQL 8.0.16 中，组复制引入了消息分片特性。基于这个特性，任何大于 group_replication_communication_max_message_size（默认是 10MB）的消息将被自动拆分为多个分片进行传输，这样可以在很大程度上规避大事务导致节点被驱逐出集群的问题。
- 从 MySQL 8.0.17 开始，支持通过克隆插件来进行分布式恢复，在此之前，只能使用 binlog。
- 在 MySQL 8.0.21 之前，binlog_checksum 只能设置为 NONE。从 MySQL 8.0.21 开始，可将 binlog_checksum 设置为 CRC32。
- 从 MySQL 8.0.21 开始，可以在执行 START GROUP_REPLICATION 命令时指定分布式恢复使用的账号（USER）和密码（PASSWORD）。

  在此之前，分布式恢复使用的账号和密码是通过 CHANGE MASTER TO 命令指定的，指定的账号和密码会明文存储在 mysql.slave_master_info 表中，存在一定的安全风险。如果是在执行 START GROUP_REPLICATION 时指定的，账号和密码就只会存储在内存中。
- 在 MySQL 8.0.21 中，引入了 group_replication_advertise_recovery_endpoints 用来暴露当前节点可供其他节点进行分布式恢复操作的地址。
- 在 MySQL 8.0.26 中，引入了 group_replication_disable_member_action 函数，可以让单主模式下的集群在发生切换后，不再将新的主节点的 super_read_only 设置为 OFF。适用于灾备集群。

```
# 集群在发生切换（AFTER_PRIMARY_ELECTION）后，执行 mysql_disable_super_read_only_if_primary 操作。
# mysql_disable_super_read_only_if_primary 会将主节点的 super_read_only 设置为 OFF
SELECT group_replication_enable_member_action("mysql_disable_super_read_only_if_primary",
"AFTER_PRIMARY_ELECTION");

# 集群在发生切换后，不执行 mysql_disable_super_read_only_if_primary 操作
SELECT group_replication_disable_member_action("mysql_disable_super_read_only_if_primary",
"AFTER_PRIMARY_ELECTION");
```

- 从 MySQL 8.0.27 开始，冲突检测数据库的清理工作由组通信系统（GCS）线程迁移至后台线程，这样在清理冲突检测数据库时就不会阻塞消息的发送和接收。
- 在 MySQL 8.0.27 中引入了 group_replication_paxos_single_leader 参数，可以将单主模式下默认的 Multiple Leaders 模式切换为 Single Leader 模式，提升了单主模式下，在集群节点发生变化时，集群性能的稳定性。
- MySQL InnoDB ClusterSet：基于多个 InnoDB Cluster 的跨数据中心的灾备解决方案。
- 从 MySQL 8.0.30 开始，可以查看组复制的各个常见组件的内存使用情况。在此之前，只能查看 XCom Cache 的内存使用情况。

B.8 安全和账号

- role（角色）。

  role 是一组权限的组合。如果一个实例中的账号比较多，通过 role 可以简化 DBA 的日常管理操作。

```
CREATE ROLE 'app_read';
GRANT SELECT ON app_db.* TO 'app_read';
CREATE USER 'read_user1'@'localhost' IDENTIFIED BY 'read_user1pass';
GRANT 'app_read' TO 'read_user1'@'localhost';
```

- 在通过 GRANT 进行授权操作时，如果账号不存在，不再隐式创建账号。

  ```
  # MySQL 5.7
  mysql> GRANT SELECT ON db1.* TO 'u1'@'%' IDENTIFIED BY '123456';
  Query OK, 0 rows affected, 1 warning (0.00 sec)

  # 在 MySQL 8.0，相同的授权语句会直接报错
  mysql> GRANT SELECT ON db1.* TO 'u1'@'%' IDENTIFIED BY '123456';
  ERROR 1064 (42000): You have an error in your SQL syntax; check the manual that corresponds to your
  MySQL server version for the right syntax to use near 'IDENTIFIED BY '123456'' at line 1

  # 规范的授权方式应是先创建账号，再进行授权操作
  mysql> CREATE USER 'u1'@'%' IDENTIFIED BY '123456';
  Query OK, 0 rows affected (0.01 sec)

  mysql> GRANT SELECT ON db1.* TO 'u1'@'%';
  Query OK, 0 rows affected (0.01 sec)
  ```

- 默认的认证插件由 mysql_native_password 更改为 caching_sha2_password。
- 移除了 PASSWORD() 函数。现在不能通过 SET PASSWORD ... = PASSWORD('auth_string') 命令修改账号的密码。
- 可剔除某些 Schema 的授权。

  例如，我有 100 个库，想创建一个账号，只对其中的 99 个库进行授权。在 MySQL 8.0 之前，如果要实现这个目的，需执行 GRANT 操作 99 次。在 MySQL 8.0 中，可先对所有库授权，再回收其中一个库的权限。看下面这个示例。

  ```
  mysql> CREATE USER u1;
  Query OK, 0 rows affected (0.01 sec)

  mysql> GRANT SELECT, INSERT ON *.* TO u1;
  Query OK, 0 rows affected (0.01 sec)

  mysql> REVOKE INSERT ON db1.* FROM u1;
  ERROR 1141 (42000): There is no such grant defined for user 'u1' on host '%'

  mysql> SET GLOBAL partial_revokes=ON;
  Query OK, 0 rows affected (0.00 sec)

  mysql> REVOKE INSERT ON db1.* FROM u1;
  Query OK, 0 rows affected (0.01 sec)
  ```

- 给账号添加备注信息。

  创建用户时，可以通过 COMMENT/ATTRIBUTE 子句对账号添加备注信息。设置的 COMMENT 或 ATTRIBUTE 可在 information_schema.user_attributes 中查看。

  ```
  CREATE USER 'u1'@'localhost' COMMENT 'This is u1';
  CREATE USER 'u2'@'localhost' ATTRIBUTE '{"ename":"u1", "empno":"10001", "job":"DBA"}';
  ```

- 多因子认证。

  多因子认证是 MySQL 8.0.27 引入的，在此之前，MySQL 只支持密码这一种认证方式。

看下面这个示例，了解如何开启密码加上套接字插件的双因子认证。

```
# 通过authentication_policy设置认证因子的数量及认证方式，'*,*'是双因子认证(最多支持三因子认证)，
因为设置的是'*'，所以认证方式没有限制
mysql> SET GLOBAL authentication_policy='*,*';
Query OK, 0 rows affected (0.00 sec)

# 安装套接字插件，该插件会验证使用套接字文件的操作系统用户名是否与MySQL用户名相同
mysql> INSTALL PLUGIN auth_socket SONAME 'auth_socket.so';
Query OK, 0 rows affected (0.00 sec)

# 创建用户时指定密码加上套接字插件的双因子认证
mysql> CREATE USER 'u1'@'localhost' IDENTIFIED WITH caching_sha2_password BY '123456' AND IDENTIFIED
WITH auth_socket;
Query OK, 0 rows affected (0.01 sec)

# 只能在u1这个操作系统用户下通过u1用户来登录
[root@slowtech ~]# su - u1
[u1@slowtech ~]$ mysql -uu1 -p'123456' -S /data/mysql/3306/data/mysql.sock
```

- MySQL 5.7引入的表空间加密特性可对redo log、undo log、binlog、通用表空间、系统表空间进行加密。
- 引入了更多细粒度的权限来替代SUPER权限，现在授予SUPER权限会提示warning。
- 授权相关的系统表，现在是InnoDB存储引擎。在MySQL 8.0之前，是MyISAM引擎。
- 密码复杂度验证插件（validate_password）在MySQL 8.0中是通过Component（组件）实现的。
- 企业版支持LDAP认证。
- 企业版支持Kerberos认证。
- 引入了print_identified_with_as_hex，可将账号在caching_sha2_password下生成的密码以十六进制的形式打印出来。这样可实现复制账号的目的。

看下面这个示例。

```
mysql> CREATE USER 'u1'@'localhost' IDENTIFIED BY '123456';
Query OK, 0 rows affected (0.02 sec)

mysql> SHOW CREATE USER 'u1'@'localhost'\G
*************************** 1. row ***************************
CREATE USER for u1@localhost: CREATE USER `u1`@`localhost` IDENTIFIED WITH 'caching_sha2_password'
AS '$A$005$H7\\~2-H^<rC8l1~RRAtY9f7c5TJ2qZ7ETb7/iIpelUicKT3QbOfNh/Iuw5' REQUIRE NONE PASSWORD
EXPIRE DEFAULT ACCOUNT UNLOCK PASSWORD HISTORY DEFAULT PASSWORD REUSE INTERVAL DEFAULT PASSWORD
REQUIRE CURRENT DEFAULT
1 row in set (0.00 sec)

mysql> SET SESSION print_identified_with_as_hex=ON;
Query OK, 0 rows affected (0.00 sec)

mysql> SHOW CREATE USER 'u1'@'localhost'\G
*************************** 1. row ***************************
CREATE USER for u1@localhost: CREATE USER `u1`@`localhost` IDENTIFIED WITH 'caching_sha2_password'
AS
0x2441243030352448101C375C7E322D485E013C724317386C317E015252417459396637633554 4A32715A3745546237
2F694970656C5569634B543351624F664E682F49757735 REQUIRE NONE PASSWORD EXPIRE DEFAULT ACCOUNT UNLOCK
PASSWORD HISTORY DEFAULT PASSWORD REUSE INTERVAL DEFAULT PASSWORD REQUIRE CURRENT DEFAULT
1 row in set (0.00 sec)
```

```
mysql> CREATE USER 'u2'@'localhost' IDENTIFIED WITH 'caching_sha2_password' AS
0x24412430303052448101C375C7E322D485E013C724317386C317E0152524174593966376335544A32715A3745546237
2F694970656C5569634B543351624F664E682F49757735;
Query OK, 0 rows affected (0.01 sec)
```

B.9 密码管理

- 可设置密码的复用策略。

    密码的复用策略既可通过以下参数进行全局配置，也可通过 CREATE USER 或 ALTER USER 命令针对具体用户进行个性化设置。

    ```
    # 相关参数
    password_history=6 # 禁止重用最近 6 个密码
    password_reuse_interval=365 # 禁止重用 365 天之内的密码

    CREATE USER 'u1'@'localhost' PASSWORD HISTORY 5 PASSWORD REUSE INTERVAL 365 DAY;
    ALTER USER 'u1'@'localhost' PASSWORD HISTORY 5 PASSWORD REUSE INTERVAL 365 DAY;
    # 恢复为默认值，这个时候起作用的就是 password_history 和 password_reuse_interval
    ALTER USER 'u1'@'localhost' PASSWORD HISTORY DEFAULT PASSWORD REUSE INTERVAL DEFAULT;
    ```

- 修改当前用户密码时需指定之前的密码。

    该策略既可通过以下参数进行全局配置，也可通过 CREATE USER 或 ALTER USER 命令针对具体用户进行个性化设置。

    ```
    # 相关参数
    password_require_current=ON

    CREATE USER 'u1'@'localhost' PASSWORD REQUIRE CURRENT; # 要求指定之前的密码
    ALTER USER 'u1'@'localhost' PASSWORD REQUIRE CURRENT; # 要求指定之前的密码
    ALTER USER 'u1'@'localhost' PASSWORD REQUIRE CURRENT OPTIONAL; # 无须指定之前的密码
    ALTER USER 'u1'@'localhost' PASSWORD REQUIRE CURRENT DEFAULT; # 恢复为默认值，这个时候起作用的就是 password_require_current
    # 通过 REPLACE 子句指定之前的密码
    ALTER USER 'u1'@'localhost' IDENTIFIED BY 'new_password' REPLACE 'old_password';
    ```

- 一个账号可同时设置两个密码。

    ```
    # 在设置新密码的同时保留之前的密码
    ALTER USER 'u1'@'localhost' IDENTIFIED BY 'new_password' RETAIN CURRENT PASSWORD;
    # 删除之前的密码
    ALTER USER 'u1'@'localhost' DISCARD OLD PASSWORD;
    ```

- 生成随机密码。

    具体用法如下。

    ```
    mysql> CREATE USER 'u1'@'localhost' IDENTIFIED BY RANDOM PASSWORD;
    +------+-----------+--------------------+-------------+
    | user | host      | generated password | auth_factor |
    +------+-----------+--------------------+-------------+
    | u1   | localhost | h!7E[R%tStL1M3c(ClkG |           1 |
    +------+-----------+--------------------+-------------+
    1 row in set (0.01 sec)

    mysql> ALTER USER 'u1'@'localhost' IDENTIFIED BY RANDOM PASSWORD;
    ```

```
+------+-----------+---------------------+-------------+
| user | host      | generated password  | auth_factor |
+------+-----------+---------------------+-------------+
| u1   | localhost | M,xkZVUjcnVxwxUNzIZe |          1 |
+------+-----------+---------------------+-------------+
1 row in set (0.01 sec)

mysql> SET PASSWORD FOR 'u1'@'localhost' TO RANDOM;
+------+-----------+---------------------+-------------+
| user | host      | generated password  | auth_factor |
+------+-----------+---------------------+-------------+
| u1   | localhost | O!n4{B:Si4z81%OHPE7B |          1 |
+------+-----------+---------------------+-------------+
1 row in set (0.00 sec)
```

其中，随机密码的长度由参数 generated_random_password_length 决定，默认为 20。

- 多次登录失败锁定账号。注意，这里说的登录失败指的是因密码输入错误而导致的失败。

```
CREATE USER 'u1'@'localhost' IDENTIFIED BY 'password' FAILED_LOGIN_ATTEMPTS 5 PASSWORD_LOCK_TIME 3;
ALTER USER 'u2'@'localhost' FAILED_LOGIN_ATTEMPTS 5 PASSWORD_LOCK_TIME UNBOUNDED;
```

命令中的 FAILED_LOGIN_ATTEMPTS 是连续失败次数，PASSWORD_LOCK_TIME 是锁定时间，单位是天，可设置为 0 ~ 32767 或 UNBOUNDED（永久锁定）。如果账号被锁定了，再登录会提示以下错误信息。

```
# mysql -uu1 -ppassword1
mysql: [Warning] Using a password on the command line interface can be insecure.
ERROR 3955 (HY000): Access denied for user 'u1'@'localhost'. Account is blocked for 3 day(s) (3 day(s) remaining) due to 2 consecutive failed logins.
```

对于锁定的账号，在以下情况下会解锁。

- 实例重启。
- 执行 FLUSH PRIVILEGES。
- 锁定时间结束。
- 通过 ALTER USER 命令重新设置了 FAILED_LOGIN_ATTEMPTS 或 PASSWORD_LOCK_TIME。
- 通过 ALTER USER ... UNLOCK 命令解锁账号。

B.10 performance_schema

- performance_schema 查询性能提升，在很多常用表上创建了索引。
- 引入了 events_errors_summary_xxx 表统计客户端的报错信息。

```
mysql> SELECT * FROM performance_schema.events_errors_summary_global_by_error WHERE
SUM_ERROR_RAISED <> 0 limit 2,1\G
*************************** 1. row ***************************
    ERROR_NUMBER: 1049
      ERROR_NAME: ER_BAD_DB_ERROR
       SQL_STATE: 42000
SUM_ERROR_RAISED: 5
SUM_ERROR_HANDLED: 0
      FIRST_SEEN: 2022-12-09 22:02:27
       LAST_SEEN: 2022-12-10 18:06:47
1 row in set (0.00 sec)
```

- 引入了 performance_schema.events_statements_histogram_by_digest 表，用来统计查询的响应时间分布情况。
- 移除了 information_schema 中的 innodb_locks 和 innodb_lock_waits 表，取而代之的分别是 performance_schema 中的 data_locks 和 data_lock_waits 表。

  这里重点说说 data_lock_waits 表，它会直观地呈现 SQL 语句的加锁类型。这一点对于我们分析锁相关的问题相当有用。看下面这个示例。

```
session1> BEGIN;
Query OK, 0 rows affected (0.00 sec)

session1> DELETE FROM slowtech.t1 WHERE id=1;
Query OK, 1 row affected (0.00 sec)

session2> DELETE FROM slowtech.t1 WHERE id=1;
...阻塞中

session3> SELECT
thread_id,object_schema,object_name,index_name,lock_type,lock_mode,lock_status,lock_data FROM
performance_schema.data_locks;
+-----------+---------------+-------------+------------+-----------+---------------+-------------+-----------+
| thread_id | object_schema | object_name | index_name | lock_type | lock_mode     | lock_status | lock_data |
+-----------+---------------+-------------+------------+-----------+---------------+-------------+-----------+
|      5356 | slowtech      | t1          | NULL       | TABLE     | IX            | GRANTED     | NULL      |
|      5356 | slowtech      | t1          | PRIMARY    | RECORD    | X,REC_NOT_GAP | WAITING     | 1         |
|      5359 | slowtech      | t1          | NULL       | TABLE     | IX            | GRANTED     | NULL      |
|      5359 | slowtech      | t1          | PRIMARY    | RECORD    | X,REC_NOT_GAP | GRANTED     | 1         |
+-----------+---------------+-------------+------------+-----------+---------------+-------------+-----------+
4 rows in set (0.00 sec)
```

- performance_schema.events_statements_summary_by_digest 新增了 QUERY_SAMPLE_TEXT 列，用来记录具体的 SQL 语句，在此之前只会记录 DIGEST_TEXT（SQL 语句规范化后的文本）。
- 从 MySQL 8.0.22 开始，可通过 performance_schema.error_log 查看错误日志。
- 引入了 performance_schema.variables_info 表，记录了参数的来源及修改情况。

B.11　函数

新增的函数可分为以下几类。

- 窗口函数：CUME_DIST、DENSE_RANK、FIRST_VALUE、LAG、LAST_VALUE、LEAD、NTH_VALUE、NTILE、PERCENT_RANK、RANK、ROW_NUMBER。
- 正则表达式相关函数：REGEXP_REPLACE、REGEXP_INSTR、REGEXP_LIKE、REGEXP_SUBSTR。
- JSON 相关函数：JSON_OVERLAPS、JSON_SCHEMA_VALID、JSON_SCHEMA_VALIDATION_REPORT、JSON_STORAGE_FREE。
- performance_schema 相关函数：PS_CURRENT_THREAD_ID、PS_THREAD_ID、FORMAT_BYTES、FORMAT_PICO_TIME。
- 加密相关函数：STATEMENT_DIGEST_TEXT、STATEMENT_DIGEST。

- 其他函数：IS_UUID、BIN_TO_UUID、SOURCE_POS_WAIT、UUID_TO_BIN、GROUPING、ROLES_GRAPHML、CURRENT_ROLE、ICU_VERSION。

移除的函数有 ENCRYPT、DES_ENCRYPT、DES_DECRYPT、ENCODE、DECODE、PASSWORD。

B.12 参数

在 MySQL 8.0 中，很多参数的默认值发生了变化。表 B-1 中列举了部分常用参数。

表 B-1 MySQL 8.0 中默认值发生变化的参数列表

参数	旧的默认值	新的默认值
Server 相关		
character_set_server	latin1	utf8mb4
collation_server	latin1_swedish_ci	utf8mb4_0900_ai_ci
explicit_defaults_for_timestamp	OFF	ON
max_allowed_packet	4194304（4MB）	67108864（64MB）
event_scheduler	OFF	ON
table_open_cache	2000	4000
log_error_verbosity	3（notes）	2（warning）
local_infile	ON	OFF
InnoDB 相关		
innodb_undo_tablespaces	0	2
innodb_undo_log_truncate	OFF	ON
innodb_flush_method	NULL	fsync（Unix）、unbuffered（Windows）
innodb_autoinc_lock_mode	1	2
innodb_flush_neighbors	1	0
innodb_max_dirty_pages_pct_lwm	0	10
innodb_max_dirty_pages_pct	75	90
performance_schema 相关		
performance-schema-instrument='wait/lock/metadata/sql/%=ON'	OFF	ON
performance-schema-instrument='memory/%=COUNTED'	OFF	COUNTED
performance-schema-consumer-events-transactions-current=ON	OFF	ON
performance-schema-consumer-events-transactions-history=ON	OFF	ON
performance-schema-instrument='transaction%=ON'	OFF	ON
复制相关		
log_bin	OFF	ON
server_id	0	1

（续）

参　　数	旧的默认值	新的默认值
log-slave-updates	OFF	ON
expire_logs_days	0	30
master-info-repository	FILE	TABLE
relay-log-info-repository	FILE	TABLE
transaction-write-set-extraction	OFF	XXHASH64
slave_parallel_type	DATABASE	LOGICAL_CLOCK
slave_parallel_workers	0	4
slave_preserve_commit_order	OFF	ON
slave_rows_search_algorithms	INDEX_SCAN、TABLE_SCAN	INDEX_SCAN、HASH_SCAN

注意，lave_parallel_type、slave_parallel_workers、slave_preserve_commit_order 这 3 个参数的默认值是从 MySQL 8.0.27 开始调整的。

除此之外，tx_read_only 和 tx_isolation 这 2 个参数在 MySQL 8.0 中已重命名为 transaction_read_only 和 transaction_isolation。